History OF THE
Theory OF Numbers

Volume I

History OF THE Theory OF Numbers

Volume I
Divisibility and Primality

Leonard Eugene Dickson

Dover Publications, Inc.
Mineola, New York

Bibliographical Note

This Dover edition, first published in 2005, is an unabridged republication of the work originally published as Publication Number 256 in Washington, D.C. by the Carnegie Institute of Washington in 1919.

International Standard Book Number: 0-486-44232-2

Manufactured in the United States of America
Dover Publications, Inc., 31 East 2nd Street, Mineola, N.Y. 11501

PREFACE.

The efforts of Cantor and his collaborators show that a chronological history of mathematics down to the nineteenth century can be written in four large volumes. To cover the last century with the same elaborateness, it has been estimated that about fifteen volumes would be required, so extensive is the mathematical literature of that period. But to retain the chronological order and hence devote a large volume to a period of at most seven years would defeat some of the chief purposes of a history, besides making it very inconvenient to find all of the material on a particular topic. In any event there is certainly need of histories which treat of particular branches of mathematics up to the present time.

The theory of numbers is especially entitled to a separate history on account of the great interest which has been taken in it continuously through the centuries from the time of Pythagoras, an interest shared on the one extreme by nearly every noted mathematician and on the other extreme by numerous amateurs attracted by no other part of mathematics. This history aims to give an adequate account of the entire literature of the theory of numbers. The first volume presents in twenty chapters the material relating to divisibility and primality. The concepts, results, and authors cited are so numerous that it seems appropriate to present here an introduction which gives for certain chapters an account in untechnical language of the main results in their historical setting, and for the remaining chapters the few remarks sufficient to clearly characterize the nature of their contents.

Perfect numbers have engaged the attention of arithmeticians of every century of the Christian era. It was while investigating them that Fermat discovered the theorem which bears his name and which forms the basis of a large part of the theory of numbers. A perfect number is one, like $6 = 1+2+3$, which equals the sum of its divisors other than itself. Euclid proved that $2^{p-1}(2^p-1)$ is a perfect number if 2^p-1 is a prime. For $p = 2$, 3, 5, 7, the values 3, 7, 31, 127 of 2^p-1 are primes, so that 6, 28, 496, 8128 are perfect numbers, as noted by Nicomachus (about A. D. 100). A manuscript dated 1456 correctly gave 33550336 as the fifth perfect number; it corresponds to the value 13 of p. Very many early writers believed that 2^p-1 is a prime for every odd value of p. But in 1536 Regius noted that

$$2^9 - 1 = 511 = 7 \cdot 73, \qquad 2^{11} - 1 = 2047 = 23 \cdot 89$$

are not primes and gave the above fifth perfect number. Cataldi, who founded at Bologna the most ancient known academy of mathematics, noted in 1603 that $2^p - 1$ is composite if p is composite and verified that it is a prime for $p = 13$, 17, and 19; but he erred in stating that it is also a prime for $p = 23$, 29, and 37. In fact, Fermat noted in 1640 that $2^{23} - 1$ has the factor 47, and $2^{37} - 1$ the factor 223, while Euler observed in 1732 that $2^{29} - 1$ has the factor 1103. Of historical importance is the statement made by Mersenne in 1644 that the first eleven perfect numbers are given by $2^{p-1}(2^p - 1)$ for $p = 2$, 3, 5, 7, 13, 17, 19, 31, 67, 127, 257; but he erred at least in including 67 and excluding 61, 89, and 107. That $2^{67} - 1$ is composite was proved by Lucas in 1876, while its actual factors were found by Cole in 1903. The primality of $2^{61} - 1$, a number of 19 digits, was established by Pervušin in 1883, Seelhoff in 1886, and Hudelot in 1887. Both Powers and Fauquembergue proved in 1911–14 that $2^{89} - 1$ and $2^{107} - 1$ are primes. The primality of $2^{31} - 1$ and $2^{127} - 1$ had been established by Euler and Lucas respectively. Thus $2^p - 1$ is known to be a prime, and hence lead to a perfect number, for the twelve values 2, 3, 5, 7, 13, 17, 19, 31, 61, 89, 107 and 127 of p. Since $2^p - 1$ is known (pp. 15–31) to be composite for 32 primes $p \leq 257$, only the eleven values $p = 137$, 139, 149, 157, 167, 193, 199, 227, 229, 241, 257 now remain in doubt.

Descartes stated in 1638 that he could prove that every even perfect number is of Euclid's type and that every odd perfect number must be of the form ps^2, where p is a prime. Euler's proofs (p. 19) were published after his death. An immediate proof of the former fact was given by Dickson (p. 30). According to Sylvester (pp. 26–27), there exists no odd perfect number with fewer than six distinct prime factors, and none with fewer than eight if not divisible by 3. But the question of the existence of odd perfect numbers remains unanswered.

A multiply perfect number, like 120 and 672, is one the sum of whose divisors equals a multiple of the number. They were actively investigated during the years 1631–1647 by Mersenne, Fermat, St. Croix, Frenicle, and Descartes. Many new examples have been found recently by American writers.

Two numbers are called amicable if each equals the sum of the aliquot divisors of the other, where an aliquot divisor of a number means a divisor other than the number itself. The pair 220 and 284 was known to the Pythagoreans. In the ninth century, the Arab Thâbit ben Korrah noted that $2^n ht$ and $2^n s$ are amicable numbers if $h = 3 \cdot 2^n - 1$, $t = 3 \cdot 2^{n-1} - 1$ and $s = 9 \cdot 2^{2n-1} - 1$ are all primes, and $n > 1$. This result leads to amicable numbers for $n = 2$ (giving the above pair), $n = 4$ and $n = 7$, but for no further value ≤ 200 of n. The chief investigation of amicable numbers is that by Euler who listed (pp. 45, 46) 62 pairs. At the age of 16, Paganini announced in 1866 the remarkable new pair 1184 and 1210. A few new pairs of very large numbers have been found by Legendre, Seelhoff, and Dickson.

Interesting amicable triples and amicable numbers of higher order have been recently found by Dickson and Poulet (p. 50).

Although it had been employed in the study of perfect and amicable numbers, the explicit expression for the sum $\sigma(n)$ of all the divisors of n is reserved for Chapter II, in which is presented the history of Fermat's two problems to solve $\sigma(x^3) = y^2$ and $\sigma(x^2) = y^3$ and John Wallis's problem to find solutions other than $x = 4$ and $y = 5$ of $\sigma(x^2) = \sigma(y^2)$.

Fermat stated in 1640 that he had a proof of the fact, now known as Fermat's theorem, that, if p is any prime and x is any integer not divisible by p, then $x^{p-1} - 1$ is divisible by p. This is one of the fundamental theorems of the theory of numbers. The case $x = 2$ was known to the Chinese as early as 500 B. C. The first published proof was given by Euler in 1736. Of first importance is the generalization from the case of a prime p to any integer n, published by Euler in 1760: if $\phi(n)$ denotes the number of positive integers not exceeding n and relatively prime to n, then $x^{\phi(n)} - 1$ is divisible by n for every integer x relatively prime to n. Another elegant theorem states that, if p is a prime, $1 + \{1 \cdot 2 \cdot 3 \ldots (p-1)\}$ is divisible by p; it was first published by Waring in 1770, who ascribed it to Sir John Wilson. This theorem was stated at an earlier date in a manuscript by Leibniz, who with Newton discovered the calculus. But Lagrange was the first one to publish (in 1773) a proof of Wilson's theorem and to observe that its converse is true. In 1801 Gauss stated and suggested methods to prove the generalization of Wilson's theorem: if P denotes the product of the positive integers less than A and prime to A, then $P + 1$ is divisible by A if $A = 4$, p^m or $2p^m$, where p is an odd prime, while $P - 1$ is divisible by A if A is not of one of these three forms. A very large number of proofs of the preceding theorems are given in the first part of Chapter III. Various generalizations are then presented (pp. 84–91). For instance, if $N = p_1^{e_1} \ldots p_s^{e_s}$, where $p_1, \ldots, p_s$ are distinct primes,

$$a^N - (a^{N/p_1} + \ldots + a^{N/p_s}) + (a^{N/p_1 p_2} + \ldots) - \ldots + (-1)^s a^{N/p_1 \ldots p_s}$$

is divisible by N, a fact due to Gauss for the case in which a is a prime.

Many cases have been found in which $a^{n-1} - 1$ is divisible by n for a composite number n. But Lucas proved the following converse of Fermat's theorem: if $a^x - 1$ is divisible by n when $x = n - 1$, but not when x is a divisor $< n - 1$ of $n - 1$, then n is a prime.

Any integral symmetric function of degree d of $1, 2, \ldots, p-1$ with integral coefficients is divisible by the prime p if d is not a multiple of $p-1$. A generalization to the case of a divisor p^n is due to Meyer (p. 101). Nielsen proved in 1893 that, if p is an odd prime and if k is odd and $1 < k < p-1$, the sum of the products of $1, 2, \ldots, p-1$ taken k at a time is divisible by p^2. Taking $k = p - 2$, we see that if p is a prime > 3 the numerator of the fraction equal to $1+1/2+1/3+\ldots+1/(p-1)$ is divisible by p^2, a result first proved by Wolstenholme in 1862. Sylvester stated in 1866 that the sum of all products of n distinct numbers chosen from $1, 2, \ldots, m$ is divisible by each prime $> n+1$ which is contained in any term of the set $m-n+1, \ldots, m, m+1$. There are various theorems analogous to these.

In Chapter IV are given properties of the quotient $(u^{p-1}-1)/p$, which plays an important rôle in recent investigations on Fermat's last theorem (the impossibility of $x^p+y^p=z^p$ if $p>2$), the history of which will be treated in the final chapter of Volume II. Some of the present papers relate to $(u^{\phi(n)}-1)/n$, where n is not necessarily a prime.

While Euler's ϕ-function was defined above in order to state his generalization of Fermat's theorem, its numerous properties and generalizations are reserved for the long Chapter V. In 1801 Gauss gave the result that $\phi(d_1)+\ldots+\phi(d_k)=n$, if $d_1, \ldots, d_k$ are the divisors of n; this was generalized by Laguerre in 1872, H. G. Cantor in 1880, Busche in 1888, Zsigmondy in 1893, Vahlen in 1895, Elliott in 1901, and Hammond in 1916. In 1808 Legendre proved a simple formula for the number of integers $\leq n$ which are divisible by no one of any given set of primes. The asymptotic value of $\phi(1)+\ldots+\phi(G)$ for G large was discussed by Dirichlet in 1849, Mertens in 1874, Perott in 1881, Sylvester in 1883 and 1897, Cesàro in 1883 and 1886–8, Berger in 1891, and Kronecker in 1901. The solution of $\phi(x)=g$ was treated by Cayley in 1857, Minin in 1897, Pichler in 1900, Carmichael in 1907–9, Ranum in 1908, and Cunningham in 1915. H. J. S. Smith proved in 1875 that the m-rowed determinant, having as the element in the ith row and jth column any function $f(\delta)$ of the greatest common divisor δ of i and j, equals the product of $F(1), F(2), \ldots, F(m)$, where

$$F(m)=f(m)-\Sigma f\left(\frac{m}{p}\right)+\Sigma f\left(\frac{m}{pq}\right)-\ldots, \qquad m=p^\alpha q^\beta\ldots.$$

In particular, $F(m)=\phi(m)$ if $f(\delta)=\delta$. In several papers (pp. 128–130) Cesàro considered analogous determinants. The fact that 30 is the largest number such that all smaller numbers relatively prime to it are primes was first proved by Schatunowsky in 1893.

A. Thacker in 1850 evaluated the sum $\phi_k(n)$ of the kth powers of the integers $\leq n$ which are prime to n. His formula has been expressed in various symbolic forms by Cesàro and generalized by Glaisher and Nielsen. Crelle had noted in 1845 that $\phi_1(n)=\frac{1}{2}n\phi(n)$. In 1869 Schemmel considered the number of sets of n consecutive integers each $<m$ and prime to m. In connection with linear congruence groups, Jordan evaluated the number of different sets of k positive integers $\leq n$ whose greatest common divisor is prime to n. This generalization of Euler's ϕ-function has properties as simple as the latter function and occurs in many papers under a variety of notations. It in turn has been generalized (pp. 151–4).

The properties of the set of all irreducible fractions, arranged in order of magnitude, whose numerators are $\leq m$ and denominators are $\leq n$ (called a Farey series if $m = n$), have been discussed by many writers and applied to the approximation of numbers, to binary quadratic forms, to the composition of linear fractional substitutions, and to geometry (pp. 155–8).

Some of the properties of periodic decimal fractions are already familiar to the reader in view of his study of arithmetic and the chapter of algebra dealing with the sum to infinity of a geometric progression. For the generalization to periodic fractions to any base b, not necessarily 10, the length of the period of the periodic fraction for $1/d$, where d is prime to b, is the least positive exponent e such that $b^e - 1$ is divisible by d. Hence this Chapter VI, which reports upon more than 160 papers, is closely related to the following chapter and furnishes a concrete introduction to it.

The subject of exponents and primitive roots is one of the important topics of the theory of numbers. To present the definitions in the customary, compact language, we shall need the notion of congruence. If the difference of two integers a and b is divisible by m, they are called congruent modulo m and we write $a \equiv b \pmod{m}$. For example, $8 \equiv 2 \pmod 6$. If $n^e \equiv 1 \pmod m$, but $n^s \not\equiv 1 \pmod m$ for $0 < s < e$, we say that n belongs to the exponent e modulo m. For example, 2 and 3 belong to the exponent 4 modulo 5, while 4 belongs to the exponent 2. In view of Euler's generalization of Fermat's theorem, stated above, e never exceeds $\phi(m)$. If n belongs to this maximum exponent $\phi(n)$ modulo m, n is called a primitive root of m. For example, 2 and 3 are primitive roots of 5, while 1 and 4 are not. Lambert stated in 1769 that there exists a primitive root of any prime p, and Euler gave a defective proof in 1773. In 1785 Legendre proved that there are exactly $\phi(e)$ numbers belonging modulo p to any exponent e which divides $p - 1$. In 1801 Gauss proved that there exist primitive roots of m if and only if $m = 2, 4, p^k$ or $2p^k$, where p is an odd prime. In particular, for a primitive root a of a prime modulus p and any integer N not divisible by p, there is an exponent ind N, called the index of N by Gauss, such that $N \equiv a^{\mathrm{ind}\,N} \pmod p$. Indices play a rôle similar to logarithms, but we require two companion tables for each modulus p. The extension to a power of prime modulus is immediate. For a general modulus, systems of indices were employed by Dirichlet in 1837 and 1863 and by Kronecker in 1870. Jacobi's Canon Arithmeticus of 1839 gives companion tables of indices for each prime and power of a prime < 1000. Cunningham's Binary Canon of 1900 gives the residues of the successive powers of 2 when divided by each prime or power of a prime < 1000 and companion tables showing the powers of 2 whose residues are $1, 2, 3, \ldots$. In 1846 Arndt proved that, if g is a primitive root of the odd prime p, g belongs to the exponent $p^{n-1}(p-1)$ modulo p^n if and only if $G = g^{p-1} - 1$ is divisible by p^λ, but not by $p^{\lambda+1}$, where

$\lambda < n$; taking $\lambda = 1$, we see that, if G is not divisible by p^2, g is a primitive root of p^2 and of all higher powers of p. This Chapter VII presents many more theorems on exponents, primitive roots, and binomial congruences, and cites various lists of primitive roots of primes < 10000. Lagrange proved easily that a congruence of degree n has at most n roots if the modulus is a prime. Lebesgue found the number of sets of solutions of $a_1 x_1^m + \ldots + a_k x_k^m \equiv a \pmod{p}$, when p is a prime such that $p-1$ is divisible by m. König (p. 226) employed a cyclic determinant and its minors to find the exact number of real roots of any congruence in one unknown; Gegenbauer (p. 228) and Rados (p. 233) gave generalizations to congruences in several unknowns.

Galois's introduction of imaginary roots of congruences has not only led to an important extension of the theory of numbers, but has given rise to wide generalizations of theorems which had been obtained in subjects like linear congruence groups by applying the ordinary theory of numbers. Instead of the residues of integers modulo p, let us consider the residues of polynomials in a variable x with integral coefficients with respect to two moduli, one being a prime p and the other a polynomial $f(x)$ of degree n which is irreducible modulo p. The residues are the p^n polynomials in x of degree $n-1$ whose coefficients are chosen from the set $0, 1, \ldots, p-1$. These residues form a Galois field within which can be performed addition, subtraction, multiplication, and division (except by zero). As a generalization of Fermat's theorem, Galois proved that the power $p^n - 1$ of any residue except zero is congruent to unity with respect to our pair of moduli p and $f(x)$. He avoided our second modulus $f(x)$ by introducing an undefined imaginary root i of $f(x) \equiv 0 \pmod{p}$ and considering the residues modulo p of polynomials in i; but the above use of the two moduli affords the only logical basis of the theory. In view of the fullness of the reports in the text (pp. 233–252) of the papers on this subject, further comments here are unnecessary. The final topics of this long Chapter VIII are cubic congruences and miscellaneous results on congruences and possess little general interest.

In Chapter IX are given Legendre's expression for the exponent of the highest power of a prime p which divides the factorial $1\cdot2\ldots m$, and the generalization to the product of any integers in arithmetical progression; many theorems on the divisibility of one product of factorials by another product and on the residues of multinomial coefficients; various determinations of the sign in $1\cdot2\ldots (p-1)/2 \equiv \pm 1 \pmod{p}$; and miscellaneous congruences involving factorials.

In the extensive Chapter X are given many theorems and formulas concerning the sum of the kth powers of all the divisors of n, or of its even or odd divisors, or of its divisors which are exact sth powers, or of those divisors whose complementary divisors are even or odd or are exact sth powers, and the excess of the sum of the kth powers of the divisors of the form $4m+1$ of a number over the sum of the kth powers of the divisors of the form $4m+3$, as well as more technical sums of divisors defined on pages 297, 301–2, 305, 307–8, 314–5 and 318. For the important case $k=0$, such a sum becomes the number of the divisors in question. There are theorems on the number of sets of positive integral solutions of $u_1u_2\ldots u_k=n$ or of $x^ay^b=n$. Also Glaisher's cancellation theorems on the actual divisors of numbers (pp. 310–11, 320–21). Scattered through the chapter are approximation and asymptotic formulas involving some of the above functions.

In Chapter XI occur Dirichlet's theorem on the number of cases in the division of n by $1, 2, \ldots, p$ in turn in which the ratio of the remainder to the divisor is less than a given proper fraction, and the generalizations on pp. 330–1; theorems on the number of integers $\leq n$ which are divisible by no exact sth power >1; theorems on the greatest divisor which is odd or has specified properties; many theorems on greatest common divisor and least common multiple; and various theorems on mean values and probability.

The casting out of nines or of multiples of 11 or 7 to check arithmetical computations is of early origin. This topic and the related one of testing the divisibility of one number by another have given rise to the numerous elementary papers cited in Chapter XII.

The frequent need of the factors of numbers and the excessive labor required for their direct determination have combined to inspire the construction of factor tables of continually increasing limit. The usual method is essentially that given by Eratosthenes in the third century B. C. A special method is used by Lebon (pp. 355–6). Attention is called to Lehmer's Factor Table for the First Ten Millions and his List of Prime Numbers from 1 to 10,006,721, published in 1909 and 1914 by the Carnegie Institution of Washington. Since these tables were constructed anew with the greatest care and all variations from the chief former tables were taken account of, they are certainly the most accurate tables extant. Absolute accuracy is here more essential than in ordinary tables of continuous functions. Besides giving the history of factor tables and lists of primes, this Chapter XIII cites papers which enumerate the primes in various intervals, prime pairs (as 11, 13), primes of the form $4n+1$, and papers listing primes written to be base 2 or large primes.

Chapter XIV cites the papers on factoring a number by expressing it as a difference of two squares, or as a sum of two squares in two ways, or by use of binary quadratic forms, the final digits, continued fractions, Pell equations, various small moduli, or miscellaneous methods.

Fermat expressed his belief that $F_n=2^{2^n}+1$ is a prime for every value of n. While this is true if $n=1, 2, 3, 4$, it fails for $n=5$ as noted by Euler. Later,

Gauss proved that a regular polygon of m sides can be constructed by ruler and compasses if m is a product of a power of 2 and distinct odd primes each of the form F_n, and stated correctly that the construction is impossible if m is not such a product. In view of the papers cited in Chapter XV, F_n is composite if $n = 5$, 6, 7, 8, 9, 11, 12, 18, 23, 36, 38 and 73, while nothing is known for other values > 4 of n. No comment will be made on the next chapter which treats of the factors of numbers of the form $a^n \pm b^n$ and of certain trinomials.

In Chapter XVII are treated questions on the divisors of terms of a recurring series and in particular of Lucas' functions

$$ u_n = \frac{a^n - b^n}{a - b}, \qquad v_n = a^n + b^n, $$

where a and b are roots of $x^2 - Px + Q = 0$, P and Q being relatively prime integers. By use of these functions, Lucas obtained an extension of Euler's generalization of Fermat's theorem, which requires the correction noted by Carmichael (p. 406), as well as various tests for primality, some of which have been employed in investigations on perfect numbers. Many papers on the algebraic theory of recurring series are cited at the end of the chapter.

Euclid gave a simple and elegant proof that the number of primes is infinite. For the generalization that every arithmetical progression n, $n+m$, $n+2m, \ldots$, in which n and m are relatively prime, contains an infinitude of primes, Legendre offered an insufficient proof, while Dirichlet gave his classic proof by means of infinite series and the classes of binary quadratic forms, and extended the theorem to complex integers. Mertens and others obtained simpler proofs. For various special arithmetical progressions, the theorem has been proved in elementary ways by many writers. Dirichlet also obtained the theorems that, if a, $2b$, and c have no common factor, $ax^2 + 2bxy + cy^2$ represents an infinitude of primes, while an infinitude of these primes are representable by any given linear form $Mx + N$ with M and N relatively prime, provided a, b, c, M, N are such that the quadratic and linear forms can represent the same number.

No complete proof has been found for Goldbach's conjecture in 1742 that every even integer is a sum of two primes. One of various analogous unproved conjectures is that every even integer is the difference of two consecutive primes in an infinitude of ways (in particular, there exists an infinitude of pairs of primes differing by 2). No comment will be made on the further topics of this Chapter XVIII: polynomials representing numerous primes, primes in arithmetical progression, tests for primality, number of primes between assigned limits, Bertrand's postulate of the existence of at least one prime between x and $2x - 2$ for $x > 3$, miscellaneous results on primes, diatomic series, and asymptotic distribution of primes.

If $F(m) = \Sigma f(d)$, summed for all the divisors d of m, we can express $f(m)$ in terms of F by an inversion formula given in Chapter XIX along with generalizations and related formulas. Bougaief called $F(m)$ the numerical integral of $f(m)$.

The final Chapter XX gives many elementary results involving the digits of numbers mainly when written to the base 10.

Since the history of each main topic is given separately, it has been possible without causing confusion to include reports on minor papers and isolated problems for the sake of completeness. In the cases of books and journals not usually accessible, the reports are quite full with some indication of the proofs. In other cases, proofs are given only when necessary to differentiate the paper from others deriving the same result.

The references were selected mainly from the Subject Index of the Royal Society of London Catalogue of Scientific Papers, volume I, 1908 (with which also the proof-sheets were checked), and the supplementary annual volumes forming the International Catalogue of Scientific Literature, Jahrbuch über die Fortschritte der Mathematik, Revue semestrielle des publications mathématiques, Poggendorff's Handwörterbuch, Klügel's Mathematische Wörterbuch, Wölffing's Mathematischer Bücherschatz (a list of mathematical books and pamphlets of the nineteenth century), historical journals, such as Bulletino di bibliografia e di storia delle scienze matematiche e fisiche, Bolletino...., Bibliotheca Mathematica, Abhandlungen zur Geschichte der mathematischen Wissenschaften, various histories and encyclopedias, including the Enclyclopédie des sciences mathématiques. Further, the author made a direct examination at the stacks of books and old journals in the libraries of Chicago, California, and Cambridge Universities, and Trinity College, Cambridge, and the excellent John Crerar Library at Chicago. He made use of G. A. Plimpton's remarkable collection, in New York, of rare books and manuscripts. In 1912 the author made an extended investigation in the libraries of the British Museum, Kensington Museum, Royal Society, Cambridge Philosophical Society, Bibliothèque Nationale, Université de Paris, St. Geneviève, l'Institut de France, University of Göttingen, and the Königliche Bibliothek of Berlin (where there is a separate index of the material on the theory of numbers). Many books have since been borrowed from various libraries; the Ladies' and other Diaries were loaned by R. C. Archibald.

At the end of the volume is a separate index of authors for each of the twenty chapters, which will facilitate the tracing of the relation of a paper to kindred papers and hence will be of special service in the case of papers inaccessible to the reader. The concluding volume will have a combined index of authors from which will be omitted minor citations found in the chapter indices.

The subject index contains a list of symbols; while [x] usually denotes the greatest integer $\leq x$, occasionally such square brackets are used to inclose an addition to a quotation. The symbol * before an author's name signifies that his paper was not available for report. The symbol † before a date signifies date of death. Initials are given only in the first of several immediately successive citations of an author.

Although those volumes of Euler's Opera Omnia which contain his Commentationes Arithmeticæ Collectæ have been printed, they are not yet available; a table showing the pages of the Opera and the corresponding pages in the present volume of this history will be given in the concluding volume.

The author is under great obligations to the following experts in the theory of numbers for numerous improvements resulting from their reading the initial page proofs of this volume: R. D. Carmichael, L. Chanzy, A. Cunningham, E. B. Escott, A. Gérardin, A. J. Kempner, D. N. Lehmer, E. Maillet, L. S. Shively, and H. J. Woodall; also the benefit of D. E. Smith's accurate and extensive acquaintance with early books and writers was fortunately secured; and the author's special thanks are due to Carmichael and Kempner, who read the final page proofs with the same critical attention as the initial page proofs and pointed out various errors and obscurities. To these eleven men who gave so generously of their time to perfect this volume, and especially to the last two, is due the gratitude of every devotee of number theory who may derive benefit or pleasure from this history. In return, such readers are requested to further increase the usefulness of this work by sending corrections, notices of omissions, and abstracts of papers marked not available for report, for insertion in the concluding volume.

Finally, this laborious project would doubtless have been abandoned soon after its inception seven years ago had not President Woodward approved it so spontaneously, urged its completion with the greatest thoroughness, and given continued encouragement.

November, 1918.

L. E. DICKSON.
Professor of Mathematics
University of Chicago

History OF THE
Theory OF Numbers

Volume I

TABLE OF CONTENTS.

1

CHAPTER I.

PERFECT, MULTIPLY PERFECT, AND AMICABLE NUMBERS.

PERFECT, ABUNDANT, AND DEFICIENT NUMBERS.

By the aliquot parts or divisors of a number are meant the divisors, including unity, which are less than the number. A number, like $6-1+2+3$, which equals the sum of its aliquot divisors is called perfect (vollkommen, vollständig). If the sum of the aliquot divisors is less than the number, as is the case with 8, the number is called deficient (diminute, defective, unvollkommen, unvollständig, mangelhaft). If the sum of the aliquot divisors exceeds the number, as is the case with 12, the number is called abundant (superfluos, plus quam-perfectus, redundantem, excédant, übervollständig, überflussig, überschiessende).

Euclid[1] proved that, if $p=1+2+2^2+\ldots+2^n$ is a prime, $2^n p$ is a perfect number. He showed that $2^n p$ is divisible by $1, 2, \ldots, 2^n, p, 2p, \ldots, 2^{n-1}p$, but by no further number less than itself. By the usual theorem on geometrical progressions, he showed that the sum of these divisors is $2^n p$.

The early Hebrews[1a] considered 6 to be a perfect number.

Philo Judeus[1b] (first century A. D.) regarded 6 as the most productive of all numbers, being the first perfect number.

Nicomachus[2] (about A. D. 100) separated the even numbers (book I, chaps. 14, 15) into abundant (citing 12, 24), deficient (citing 8, 14), and perfect, and dwelled on the ethical import of the three types. The perfect (I, 16) are between excess and deficiency, as consonant sound between acuter and graver sounds. Perfect numbers will be found few and arranged with fitting order; 6, 28, 496, 8128 are the only perfect numbers in the respective intervals between 1, 10, 100, 1000, 10000, and they have the property of ending alternately in 6 and 8. He stated that Euclid's rule gives all the perfect numbers without exception.

Theon of Smyrna[3] (about A. D. 130) distinguished between perfect (citing 6, 28), abundant (citing 12) and deficient (citing 8) numbers.

---

[1] Elementa, liber IX, prop. 36. Opera, 2, Leipzig, 1884, 408.

[1a] S. Rubin, "Sod Hasfiroth" (secrets of numbers), Wien, 1873, 59; citation of the Bible, Kings, II, 13, 19.

[1b] Treatise on the account of the creation of the world as given by Moses, C. D. Young's transl. of Philo's works, London, 1854, vol. 1, p. 3.

[2] Nicomachi Gerasini arithmeticæ libri duo. Nunc primùm typis excusi, in lucem eduntur. Parisiis, 1538. In officina Christiani Wecheli. (Greek.)

Theologumena arithmeticæ. Accedit Nicomachi Gerasini institutio arithmetica ad fidem codicum Monacensium emendata. Ed., Fridericus Astius. Lipsiae, 1817. (Greek.)

Nicomachi Geraseni Pythagorei introductionis arithmeticæ libri ii. Recensvit Ricardus Hoche. Lipsiae, 1866. (Greek.)

[3] Theonis Smyrnaei philosophi Platonici expositio rerum mathematicarum ad legendum Platonem utilium. Ed., Ed. Hiller, Leipzig, 1878, p. 45.

Theonis Smyrnaei Platonici, Latin by Ismaele Bullialdo. Paris, 1644, chap. 32, pp. 70–72.

Iamblichus[4] (about 283–330) repeated in effect the remarks by Nico-machus on perfect, abundant, and deficient numbers, but made erroneous additions. He stated that there is one and but one perfect number in the successive intervals between 1, 10, 100, ..., 100000, etc., to infinity. "Examples of a perfect number are 6, and 28, and 496, and 8128, and the like numbers, alternately ending in 6 and 8." He remarked that the Pythag-oreans called the perfect number 6 marriage, and also health and beauty (on account of the integrity of its parts and the agreement existing in it).

Aurelius Augustinus[5] (354–430) remarked that, 6 being the first perfect number, God effected the creation in 6 days rather than at once, since the perfection of the work is signified by the number 6. The sum of the aliquot parts of 9 falls short of it; likewise for 10. But the sum of the aliquot parts of 12 exceeds it.

Anicius Manlius Severinus Boethius[6] (about 481–524), in a Latin exposi-tion of the arithmetic of Nicomachus, stated that perfect numbers are rare, easily counted, and generated in a very regular order, while abundant (superfluos) and deficient (diminutos) numbers are found to an unlimited extent and not in regular order. The perfect numbers below 10000 are 6, 28, 496, 8128. And these numbers always end alternately in 6 and 8.

Munyos[7] stated that Boethius added to Euclid's idea of perfect number that of deficient (diminute) and abundant (redundantem) numbers.

Isidorus of Seville[8] (570–636) distinguished even and odd numbers, perfect and abundant numbers, linear, flächen and Körper Zahlen (primes, products of two, products of three factors).

Alcuin[9] (735–804), of York and Tours, explained the occurrence of the number 6 in the creation of the universe on the ground that 6 is a perfect number. The second origin of the human race arose from the deficient number 8; indeed, in Noah's ark there were 8 souls from which sprung the entire human race, showing that the second origin was more imperfect than the first, which was made according to the number 6.

[4]Iamblichus Chalcidensis ex Coele-Syria in Nicomachi Geraseni arithmeticam introduc-tionem, et de Fato. Accedit Joachimi Camerarii explicatio in duos libros Nicomachi. Ed., Samuel Tennulius. Arnhemiae, 1668, pp. 43–47. (Greek text and Latin translation in parallel columns.)
Iamblichi in Nicomachi arithmeticam introductionem liber ad fidem codicis Florentini. Ed., H. Pistelli. Lipsiae, 1894. (Greek.)
[5]De Civitate Dei, liber XI, cap. XXX, ed., B. Dombart, Lipsiae, 1877, I, p. 504. The reference by Frizzo[29] in to lib. II, cap. 39.
[6]Arithmetica boetij, Augsburg, 1488; Cologne, 1489; Leipzig, 1490; Venice, 1491–2, 1499; Paris, [1496, 1501], 1503, etc.; lib. 1, cap. 20, "De generatione numeri perfecti."
Opera Boetii, Venice, 1491–2, etc.; ed., Friedlein, Leipzig, 1867.
[7]Institvtiones arithmeticae ad percipiendam astrologiam et mathematicas facultates neces-sariae. Auctore Hieronymo Munyos, Valentiae, 1566, f. 5, verso.
[8]Incipit epistola Isidori iunioris hispalensis . . . Finit liber etymologiarum . . . [Augsburg, 1472]; Venice, 1483, etc. In this book of etymologies, arithmetic is treated very briefly in Book 3, beginning f. 15.
[9]Bibliotheca Rerum Germanicarum, tomus sextus: Monumenta Alcuiniana, Berlin, 1873, epistolae 259, pp. 818–821. Cf. Migne, Patrologiae, vol. 100, 1851, p. 665; Hankel, Geschichte Math., p. 311.

Thâbit ben Korrah,[10] in a manuscript composed the last half of the ninth century, attributed to Pythagoras and his school the employment of perfect and amicable numbers in illustration of their philosophy. Let $s=1+2+\ldots+2^n$. Then (prop. 5), $2^n s$ is a perfect number if s is a prime; $2^n p$ is abundant if p is a prime $<s$, deficient if p is a prime $>s$, and the excess or deficiency of the sum of all the divisors over the number equals the difference of s and p. Let (prop. 6) p' and p'' be distinct primes >2; the sum of the divisors $<N$ of $N=p'p''2^n$ is

$$a=(2^{n+1}-1)(1+p'+p'')+(2^n-1)p'p''.$$

Hence N is abundant or deficient according as

$$a-N=(2^{n+1}-1)(1+p'+p'')-p'p''>0 \text{ or } <0.$$

Hrotsvitha,[11] a nun in Saxony, in the second half of the tenth century, mentioned the perfect numbers 6, 28, 496, 8128.

Abraham Ibn Ezra[11a] (†1167), in his commentary to the Pentateuch, Ex. 3, 15, stated that there is only one perfect number between any two successive powers of 10.

Rabbi Josef b. Jehuda Ankin[11b], at the end of the twelfth century, recommended the study of perfect numbers in the program of education laid out in his book "Healing of Souls."

Jordanus Nemorarius[12] (†1236) stated (in Book VII, props. 55, 56) that every multiple of a perfect or abundant number is abundant, and every divisor of a perfect number is deficient. He attempted to prove (VII, 57) the erroneous statement that all abundant numbers are even.

Leonardo Pisano, or Fibonacci, cited in his Liber Abbaci[13] of 1202, revised about 1228, the perfect numbers

$$\tfrac{1}{2}\,2^2(2^2-1)=6, \qquad \tfrac{1}{2}\,2^3(2^3-1)=28, \qquad \tfrac{1}{2}\,2^5(2^5-1)=496,$$

excluding the exponent 4 since 2^4-1 is not prime. He stated that by proceeding so, you can find an infinitude of perfect numbers.

---

[10]Manuscript 952, 2, Suppl. Arabe, Bibliothèque impériale, Paris. Textual transl., except of the proofs which are given in modern algebraic notation as foot-notes [as numbers were represented by line. in the manuscript], by Franz Woepcke, Journal Asiatique, (4), 20, 1852, 420–9.

[11]See Ch. Magnin, Théatre de Hrotsvitha, Paris, 1845.

[11a]Mikrooth Gedoloth, Warsaw, 1874 ("Large Bible" in Hebrew). Samuel Ben Sáadias Ibn Motot, a Spaniard, wrote in 1370 a commentary on Ibn Ezra's commentary, Perush ai Perush Ibn Ezra, Venice, 1554, p. 19, noting the perfect numbers 6, 28, 496, 8128, and citing Euclid's rule. Steinschneider, in his book on Ibn Ezra, Abh. Geschichte Math. Wiss., 1880, p. 92, stated that Ibn Ezra gave a rule for finding all perfect numbers. As this rule is not given in the Mikrooth Gedoloth of 1874, Mr. Ginsburg of Columbia University infers the existence of a fuller version of Ibn Ezra's commentary.

[11b]Quoted by Güdeman, Das Jüdische Unterrichtswesen während der Spanish Arabischen Periode, Wien, 1873.

[12]In hoc opere contenta. Arithmetica decem libris demonstrata . . . . Epitome i libros arithmeticos diui Seuerini Boetij . . . , Paris, 1496, 1503, etc. It contains Jordanus' "Elementa arithmetica decem libris, demonstrationibus Jacobi Fabri Stapulensis," and "Jacobi Fabri Stapulensis epitome in duos libros arithmeticos diui Seuerini Boetij."

[13]Il Liber Abbaci di Leonardo Pisano. Roma, 1857, p. 283 (Scritti, vol. 1).

In the manuscript[14] Codex lat. Monac. 14908, a part dated 1456 and a part 1461, the first four perfect numbers are given (f. 33') as usual and the fifth perfect number is stated correctly to be 33550336.

Nicolas Chuquet[15] defined perfect, deficient, and abundant numbers, indicated a proof of Euclid's rule and stated incorrectly that perfect numbers end alternately in 6 and 8.

Luca Paciuolo, de Borgo San Sepolcro,[16] gave (f. 6) Euclid's rule, saying one must find by experiment whether or not the factor $1+2+4+\ldots$ is prime, stated (f. 7) that the perfect numbers end alternately in 6 and 8, as 6, 28, 496, etc., to infinity. In the fifth article (ff. 7, 8), he illustrated the finding of the aliquot divisors of a perfect number by taking the case of the fourteenth perfect number 9007199187632128. He gave its half, then the half of the quotient, etc., until after 26 divisions by 2, the odd number 134217727, marked "Indivisibilis" [prime]. Dividing the initial number by these quotients, he obtained further factors $[1, 2, \ldots, 2^{26}$, but written at length]. The proposed number is said to be evidently perfect, since it is the sum of these factors [but he has not employed all the factors, since the above odd number equals $2^{27}-1$ and has the factor $2^3-1=7$]. Although Paciuolo did not list the perfect numbers between 8128 and 90...8, the fact that he called the latter the fourteenth perfect number implies the error expressly committed by Bovillus.[20]

Thomas Bradwardin[17] (1290–1349) stated that there is only one perfect number (6) between 1 and 10, one (28) up to 100, 496 up to 1000, 8128 up to 10000, from which these numbers, taken in order, end alternately in 6 and 8. He then gave Euclid's rule.

Faber Stapulensis[18] or Jacques Lefèvre (born at Étaples 1455, †1537) stated that all perfect numbers end alternately in 6 and 8, and that Euclid's rule gives all perfect numbers.

Georgius Valla[19] gave the first four perfect numbers and observed that

---

[14]The manuscript is briefly described by Gerhardt, Monatsber. Berlin Ak., 1870, 141–2. See Catalogus codicum latinorum bibliothecae regiae Monacensis, Tomi II, pars II, codices num. 11001–15028 complectus, Munich, 1876, p. 250. An extract of ff. 32'–34 on perfect numbers was published by Maximilian Curtze, Bibliotheca Mathematica, (2), 9, 1895, 39–42.

[15]Triparty en la science des nombres, manuscript No. 1436, Fonds Français, Bibliothèque Nationale de Paris, written at Lyons, 1484. Published by Aristide Marre, Bull. Bibl. Storia Sc. Mat. et Fis.. 13 (1880), 593–659, 693–814; 14 (1881), 417–460. See Part 1, Ch. III, 3, 619–621, manuscript, ff. 20–21.

[16]Summa de Arithmetica geometria proportioni et proportionalita. [Suma . . . , Venice, 1494.] Toscolano, 1523 (two editions substantially the same).

[17]Arithmetica thome brauardini. Tractatus perutilis. In arithmetica speculativa a magistro thoma Brauardini ex libris euclidis boecij & aliorum qua optimne excerptus. Parisiis, 1495, 7th unnumbered page.
Arithmetica Speculativa Thome Brauardini nuper mendis Plusculis tersa et diligenter Impressa, Parisiis [1502], 6th and 7th unnumbered pages. Also undated edition [1510], 3d page.

[18]Epitome (iii) of the arithmetic of Boethius in Faber's edition of Jordanus,[12] 1496, etc. Also in Introductio Jacobi fabri Stapulēsis in Arithmecam diui Seuerini Boetij pariter Jordani, Paris, 1503, 1507. Also in Stapulensis, Jacobi Fabri, Arithmetica Boëthi epitome, Basileae, 1553, 40.

[19]De expetendis et fvgiendis rebvs opvs, Aldus, 1501. Liber I (=Arithmeticae I), Cap. 12.

"these happen to end in 6 or 8...and these terminal numbers will always be found alternately."

Carolus Bovillus[20] or Charles de Bouvelles (1470–1553) stated that every perfect number is even, but his proof applies only to those of Euclid's type. He corrected the statement of Jordanus[12] that every abundant number is even, by citing 45045 [$=5 \cdot 9 \cdot 7 \cdot 11 \cdot 13$] and its multiples. He stated that $2^n - 1$ is a prime if n is odd, explicitly citing 511 [$=7 \cdot 73$] as a prime. He listed as perfect numbers $2^{n-1}(2^n - 1)$, n ranging over all the odd numbers $\leqq 39$ [Cataldi[44] later indicated that 8 of these are not perfect]. He repeated the error that all perfect numbers end alternately in 6 and 8. He stated (f. 175, No. 25) that if the sum of the digits of a perfect number > 6 be divided by 9, the remainder is unity [proved for perfect numbers of Euclid's type by Cataldi,[44] p. 43]. He noted (f. 178) that any divisor of a perfect number is deficient, any multiple abundant. He stated (No. 29) that one or both of $6n \pm 1$ are primes and (No. 30) conversely any prime is of the form $6n \pm 1$ [Cataldi,[44] p. 45, corrects the first statement and proves the second]. He stated (f. 174) that every perfect number is triangular, being $2^n(2^n - 1)/2$.

Martinus[21] gave the first four perfect numbers and remarked that they end alternately in 6 and 8.

Gasper Lax[22] stated that the perfect numbers end alternately in 6 and 8.

V. Rodulphus Spoletanus[23] was cited by Cataldi,[44] with the implication of errors on perfect numbers. [Copy not seen.]

Girardus Ruffus[24] stated that every perfect number is even, that most odd numbers are deficient, that, contrary to Jordanus,[12] the odd number 45045 is abundant, and that for n odd $2^n - 1$ always leads to a perfect number, citing 7, 31, 127, 511, 2047, 8191 as primes [the fourth and fifth are composite].

Feliciano[25] stated that all perfect numbers end alternately in 6 and 8.

Regius[26] defined a perfect number to be an even number equal to the sum of its aliquot divisors, indicated that 511 and 2047 are composite, gave correctly 33550336 as the fifth perfect number, but said the perfect numbers

---

[20]Caroli Bouilli Samarobrini Liber De Perfectis Numeris (dated 1509 at end), one (ff. 172–180) of 13 tracts in his work, Que hoc volumine continētur: Liber de intellectu, . . . De Numeris Perfectis, . . . , dated on last page, 1510, Paris, ex officina Henrici Stephani. Biography in G. Maupin, Opinions et Curiosités touchant la Math., Paris, 1, 1901, 186–94.

[21]Ars Arithmetica Ioannis Martini, Silicei: in theoricen & praxim. 1513, 1514. Arithmetica Ioannis Martini, Scilicei, Paris, 1519.

[22]Arithmetica speculatiua magistri Gasparis Lax. Paris, 1515, Liber VII, No. 87 (end).

[23]De proportione proportionvm dispvtatio, Rome, 1515.

[24]Divi Severini Boetii Arithmetica, dvobvs discreta libris, Paris, 1521; ff. 40–44 of the commentary by G. Ruffus.

[25]Libro di Arithmetica & Geometria speculatiua & praticale: Composto per maestro Francesco Feliciano da Lazisio Veronese Intitulato Scala Grimaldelli: Nouamente stampato. Venice, 1526 (p. 3), 1527, 1536 (p. 4), 1545, 1550, 1560, 1570, 1669, Padoua, 1629, Verona, 1563, 1602.

[26]Vtrivsqve Arithmetices, epitome ex uariis authoribus concinnata per Hvdalrichum Regium. Strasburg, 1536. Lib. I, Cap. VI: De Perfecto. Hvdalrichvs Regius, Vtrivsque. . . ex variis . . . , Friburgi, 1550 [and 1543], Cap. VI, fol. 17–18.

end alternately in 6 and 8. A multiple of an abundant or perfect number is abundant, a divisor of a perfect number is deficient.

Cardan[27] (1501–1576) stated that perfect numbers were to be formed by Euclid's rule and always end with 6 or 8; and that there is one between any two successive powers of ten.

De la Roche[28] stated in effect that $2^{n-1}(2^n-1)$ is perfect for every odd n, citing in particular 130816 and 2096128, given by $n=9$, $n=11$. This erroneous law led him to believe that the successive perfect numbers end alternately in 6 and 8.

Noviomagus[29] or Neomagus or Jan Bronckhorst (1494–1570) gave Euclid's rule correctly and stated that among the first 10 numbers, 6 alone is perfect,..., among the first 10000 numbers, 6, 28, 496, 8128 alone are perfect, etc., etc. [implying falsely that there is one and but one perfect number with any prescribed number of digits]. In Lib. II, Cap. IV, is given the sieve (or crib) of Eratosthenes, with a separate column for the multiples of 3, a separate one for the multiples of 5, etc.

Willichius[30] (†1552) listed the first four perfect numbers and stated that to these are to be added a very few others, whose nature is that they end either in 6 or 8.

Michael Stifel[31] (1487–1567) stated that all perfect numbers except 6 are multiples of 4, while $4(8-1)$, $16(32-1)$, $64(128-1)$, $256(512-1)$, etc., to infinity, are perfect [error, Kraft[85]]. He later[32] repeated the latter error, listing as perfect

$$2\times3,\ 4\times7,\ 16\times31,\ 64\times127,\ 256\times511,\ 1024\times2047,$$

"& so fort an ohn end." Every perfect number is triangular.

Peletier[33] (1517–1582) stated (1549, V left; 1554, p. 20) that the perfect numbers end in 6 or 8, that there is a single perfect number between any two successive powers of 10, and (1549, C III left; 1554, pp. 270–1) that $4(8-1)$, $16(32-1)$, $64(128-1)$, $256(511)$,...are perfect. The first two statements were also given later by Peletier.[34]

---

[27]Hieronimi C. Cardani Medici Mediolanensis, Practica Arithmetice, & Mensurandi singularis. Milan, 1537, 1539; Nürnberg, 1541, 1542, Cap. 42, de proprietatibus numerorum mirificis. Opera IV, Lyon, 1663.

[28]Larismetique & Geometrie de maistre Estienne de la Roche dict Ville Franche, Nouuellement Imprimee & des fautes corrigee, Lyon, 1538, fol. 2, verso. Ed. 1, 1520.

[29]De Nvmeris libri dvo . . . . authore Ioanne Nouiomago, Paris, 1539, Lib. II, Cap. III. Reprinted, Cologne, 1544; Deventer, 1551. Edition by G. Frizzo, Verona, 1901, p. 132.

[30]Iodoci Vvillichii Reselliani, Arithmeticae libri tres, Argentorati, 1540, p. 37.

[31]Arithmetica Integra, Norimbergae, 1544, ff. 10, 11.

[32]Die Coss Christoffs Rudolffs Die schönen Exempeln der Coss Durch Michael Stifel Gebessert vnd sehr gemehrt, Königsperg in Preussen, 1553, Anhang Cap. I, f. 10 verso, f. 11 (f. 27 v.), and 1571.

[33]L'Arithmetiqve de Iacqves Peletier dv Mans, departie en quatre Liures, Poitiers, 1549, 1550, 1553. . . . , ff. 77 v, 78 r. Revüe e augmentee par l' Auteur, Lion, 1554. . . . . Troisieme edition, reucuë et augmentee, par Iean de Tovrnes, 1607.

[34]Arithmeticae Practicae methodvs facilis, per Gemmam Frisivm, Medicvm, ac Mathematicum conscripta . . . . In eandem Ioannis Steinii & Iacobi Peletarii Annotationes. Antverpiae, 1581, p. 10.

Postello[35] stated erroneously that 130816 [= 256·511] is perfect. Lodoico Baëza[36] stated that Euclid's rule gives all perfect numbers. Pierre Forcadel[37] (†1574) gave 130816 as the fifth perfect number, implying incorrectly that 511 is a prime.

Tartaglia[38] (1506–1559) gave an erroneous [Kraft[85]] list of the first twenty perfect numbers, viz., the expanded forms of $2^{n-1}(2^n-1)$, for $n=2$ and the successive odd numbers as far as $n=39$. He stated that the sums $1+2+4$, $1+2+4+8,\ldots$ are alternately prime and composite; and that the perfect numbers end alternately in 6 and 8. The third "notable property" mentioned is that any perfect number except 6 yields the remainder 1 when divided by 9.

Robert Recorde[39] (about 1510–1558) stated that all the perfect numbers under $6 \cdot 10^9$ are 6, 28, 496, 8128, 130816, 2096128, 33550336, 536854528 [the fifth, sixth, eighth of these are not perfect].

Petrus Ramus[40] (1515–1572) stated that in no interval between successive powers of 10 can you find more than one perfect number, while in many intervals you will find none. At the end of Book I (p. 29) of his Arithmeticae libri tres, Paris, 1555, Ramus had stated that 6, 28, 496, 8128 are the only perfect numbers less than 100000.

Franciscus Maurolycus[41] (1494–1575) gave an argument to show that every perfect number is hexagonal and hence triangular.

Peter Bungus[42] (†1601) gave (1584, pars altera, p. 68) a table of 20 numbers stated erroneously to be the perfect numbers with 24 or fewer digits [the same numbers had been given by Tartaglia[38]]. In the editions of 1591, etc., p. 468, the table is extended to include a perfect number of 25 digits, one of 26, one of 27, and one of 28. He stated (1584, pp. 70–71; 1591, pp. 471–2) that all perfect numbers end alternately in 6 and 28; employing Euclid's formula, he observed that the product of a power of 2 ending in 4 by a number ending in 7 itself ends in 28, while the product of one ending in 6 by one ending in 1 ends in 6. He verified (1585, pars

---

[35]Theoricae Arithmetices Compendium à Guilielmo Postello, Lutetiae, 1552, a syllabus on one large sheet of arithmetic definitions.

[36]Nvmerandi Doctrina, Lvtetiae, 1555, fol. 27–28.

[37]L'Arithmeticqve de P. Forcadel de Beziers, Paris, 1556–7. Livre I (1556), fol. 12 verso.

[38]La seconda Parte del General Trattato di Nvmeri, et Misvre di Nicolo Tartaglia, Vinegia, 1556, f. 146 verso.

L' Arithmetiqve de Nicolas Tartaglia Brescian . . . . Recueillie, & traduite d'Italien en François, par Gvillavme Gosselin de Caen, ... Paris, 1578, f. 98 verso, f. 99.

[39]The Whetstone of witte, whiche is the seconde parte of Arithmetike, London, 1557, eighth unnumbered page.

[40]Petri Rami Scholarum Mathematicarum, Libri unus et triginta, à Lazaro Schonero recogniti & emendati, Francofvrti, 1599, Libr. IV (Arith.), p. 127, and Basel, 1578.

[41]Arithmeticorvm libri dvo, Venetiis, 1575, p. 10; 1580. Published with separate paging, at end of Opuscula mathematica.

[42]Mysticae nvmerorvm significationis liber in dvas divisvs partes, R. D. Petro Bongo Canonico Bergomate avctore. Bergomi. Pars prior, 1583, 1585. Pars altera, 1584.

Petri Bungi Bergomatis Numerorum mysteria, Bergomi, 1591, 1599, 1614, Lutetiae Parisiorum, 1618, all four with the same text and paging. Classical and biblical citations on numbers (400 pages on 1, 2, . . , 12). On the 1618 edition, see Fontés, Mém. Acad. Sc. Toulouse, (9), 5, 1893, 371–380.

prior, p. 238; 1591, p. 343) for the first seven numbers of his table [two being imperfect, however] that the sum of the digits of a perfect number exceeds by unity a multiple of 9. Every perfect number is triangular (1591, p. 270). Every multiple of a perfect number is abundant, every divisor deficient (1591, p. 464).

Unicornus[43] (1523–1610) cited Bungus and repeated his error that $2^{n-1}(2^n-1)$ is always perfect for n odd and that all perfect numbers end alternately in 6 and 8.

Cataldi[44] (1548–1626) noted in his Preface that Paciuolo's[16] fourteenth perfect number 90...8 is in fact abundant since it arose from $1+2+4+...$ $+2^{26} = 134217727$, which is divisible by 7, whereas Paciuolo said it was prime. Citing the error of the latter, Bovillus,[20] and others, that all perfect numbers end alternately in 6 and 8, Cataldi observed (p. 42) that the fifth perfect number is 33550336 and the sixth is 8589869056, from $8191 = 2^{13}-1$ and $131071 = 2^{17}-1$, respectively, proved to be primes (pp. 12–17) by actually trying as possible divisor every prime less than their respective square roots. He gave (pp. 17–22) the corresponding work showing $2^{19}-1$ to be prime. He stated (p. 11) that 2^n-1 is a prime for $n=2, 3, 5, 7, 13, 17, 19, 23, 29, 31, 37$, remarking that the prime $n=11$ does not yield a perfect number since (p. 5) $2^{11}-1 = 2047 = 23·89$, while it is composite if n is composite. He proved (p. 8) that the perfect numbers given by Euclid's rule end in 6 or 8. He gave (pp. 28–40, 48) a table of all divisors of all even and odd numbers ≤ 800, and a table of primes < 750.

Georgius Henischiib[45] (1549–1618) stated that the perfect numbers end alternately in 6 and 8, and that one occurs between any two successive powers of 10. He applied Euclid's formula without restricting the factor 2^n-1 to primes.

Johan Rudolff von Graffenried[46] stated that all perfect numbers are given by Euclid's rule, which he applied without restricting 2^n-1 to primes, expressly citing 256×511 as the fifth perfect number. Every perfect number is triangular.

Bachet de Mézirac[47] (1581–1638) gave (f. 102) a lengthy proof of Euclid's theorem that $2^n p$ is perfect if $p = 1+2+...+2^n$ is a prime, but

---

[43]De l'arithmetica vniversale del Sig. Ioseppo Vnicorno, Venetia, 1598, f. 57.

[44]Trattato de nvmeri perfetti di Pierto Antonio Cataldo, Bologna, 1603. According to the Preface, this work was composed in 1588. Cataldi founded at Bologna the Academia Erigende, the most ancient known academy of mathematics; his interest in perfect numbers from early youth is shown by the end of the first of his "due lettioni fatte nell' Academia di Perugia" (G. Libri, Hist. Sc. Math. en Italie, 2d ed., vol. 4, Halle, 1865, p. 91). G. Wertheim, Bibliotheca Math., (3), 3, 1902, 76–83, gave a summary of the Trattato.

[45]Arithmetica Perfecta et Demonstrata, Georgii Henischiib, Augsburg [1605], 1609, pp. 63–64.

[46]Arithmeticae Logistica Popularis Librii IIII. In welchen der Algorithmus in gantzen Zahlen u. Fracturen . . . . , Bern, 1618, 1619, pp. 236–7.

[47]Elementorum arithmeticorum libri XIII auctori D . . . , a Latin manuscript in the Bibliothèque de l'Institut de France. On the inside of the front cover is a comment on the sale of the manuscript by the son of Bachet to Dalibert, treasurer of France. A general account of the contents of the manuscript was given by Henry, Bull. Bibl. Storia Sc. Mat. e Fis., 12, 1879, pp. 619–641. The present detailed account of Book 4, on perfect numbers, was taken from the manuscript.

(f. 103, verso) is abundant if p is composite. Every multiple of a perfect or abundant number is abundant, every divisor of a perfect number is deficient (ff. 104 verso, 105). The product of two primes, other than 2×3, is deficient (f. 105 verso). The odd number 945 is abundant, the sum of its aliquot divisors being 975 (f. 107). Commenting (f. 111 verso, f. 112) on the statement of Boethius[6] and Cardan[27] that the perfect numbers end alternately in 6 and 8, he stated that the fourth is 8128 and the fifth is 2096128 [an error], the fifth not being $130816 = 256 \times 511$, since $511 = 7 \times 73$.

Jean Leurechon[48] (about 1591–1670) stated that there are only ten perfect numbers between 1 and 10^{12}, listed them (noting the admirable property that they end alternately in 6 and 8) and gave the twentieth perfect number. [They are the same as in Tartaglia's[38] list.]

Lantz[49] stated that the perfect numbers are $2(4-1)$, $4(8-1)$, $16(32-1)$, $64(128-1)$, $256(512-1)$, $1024(2048-1)$, etc.

Hugo Sempilius[50] or Semple (Scotland, 1594–Madrid, 1654) stated that there are only seven perfect numbers up to 40,000,000; they end alternately in 6 and 8.

Casper Ens[51] stated that there are only seven perfect numbers $< 4 \cdot 10^7$, viz., 6, 28, 496, 8128, 130816, 1996128 [for 2096128], 33550336, and that they end alternately in 6 and 8.

Daniel Schwenter[52] (1585–1636) made the same error as Casper Ens.[51]

Erycius Puteanus[53] quoted from Martiano Capella, lib. VII, De Nuptiis Philologiae, to the effect that the perfect number 6 is attributed to Venus; for it is made by the union of the two sexes, that is, from triad, which is male since it is odd, and from diad, which is feminine since it is even. Puteanus said that the perfect numbers in order are 6, 28, 496, 8128, 130816, 2096128, 33550336, and gave all their divisors [implying that 511, 2047, 8191 are primes], and stated that these seven and all the remaining end alternately in 6 and 8. Between any two successive powers of 10 is one perfect number. That they are all triangular adds perfection to the perfect.

Joannes Broscius[54] or Brocki remarked that there is no perfect number between 10000 and 10000000, contrary to Stifel,[31] Bungus,[42] Sempilius,[50] Puteanus,[53] and the author of Selectarum Propositionum Mathematicarum, quas propugnavit, Mussiponti, Anno 1622, Maximilianus Willibaldus, Baro

---

[48]Récréations mathématiques, Pont-à-Mousson, 1624; London, 1633, 1653, 1674 (these three English editions by Wm. Oughtred), p. 92. The authorship is often attributed to Leurechon's pupil Henry Van Etten, whose name is signed to the dedicatory epistle. Cf. Poggendorff, Handwörterbuch, 1863, 2, p. 250 (under C. Mydorge); Bibliotheque des écrivains de la compagnie de Jésus, par A. de Backer, 2, 1872, 731; Biographie Générale, 31, 1872, 10.

[49]Institutionum Arithmeticarum libri quatuor à Ioanne Lantz, Coloniae Agrippinae, 1630, p. 54.

[50]De Mathematicis Disciplinis libri Duodecim, Antverpiae, 1635, Lib. 2, Cap. 3, N. 10, p. 46. There is (pp. 263-5) an index of writers on geometry and one for arithmetic.

[51]Thaumaturgus Math., Munich, 1636, p. 101; Coloniae, 1636, 1651; Venice, 1706.

[52]Deliciae Physico-Mathematicae oder Mathemat: vnd Philosophische Erquickstunden, part I (574 pp.), Nürnberg, 1636, p. 108.

[53]De Bissexto Liber: nova temporis facula qua intercalandi arcana . . . . Lovanii, 1637; 1640, pp. 103-7. Reproduced by J. G. Graevius, Thesaurus Antiquitatum Romanarum (12 vols., 1694-9), Lugduni Batavorum, vol. 8.

in Waldpurg. While they considered 511×256 and 2047×1024 as perfect, 511 has the factor 7, and (as pointed out to him by Stanislaus Pudlowski) 2047 has the factor 23. Broscius stated that

$2^n - 1$ has the factor 3 5 7 11 13 17 19 23 29 31
if n is a multiple of 2 4 3 10 12   8 18 11 28   5.

The contents of the second dissertation are given below under the date 1652.

René Descartes,[55] in a letter to Mersenne, November 15, 1638, thought he could prove that every even perfect number is of Euclid's type, and that every odd perfect number must have the form ps^2, where p is a prime. He saw no reason why an odd perfect number may not exist. For $p = 22021$, $s = 3 \cdot 7 \cdot 11 \cdot 13$, ps^2 would be perfect if p were prime [but $p = 61 \cdot 19^2$]. In a letter to Frenicle, January 9, 1639, Oeuvres, 2, p. 476, he expressed his belief that an odd perfect number could be found by replacing 7, 11, 13 in s by other values.

Fermat[56] stated that he possessed a method of solving all questions relating to aliquot parts. Citing this remark, Frenicle[57] challenged Fermat to find a perfect number of 20 or 21 digits. Fermat[58] replied that there is none with 20 or 21 digits, contrary to the opinion of those who believe that there is a perfect number between any two consecutive powers of 10.

Fermat,[59] in a letter to Mersenne, June (?), 1640, stated three propositions which he had proved not without considerable trouble and which he called the basis of the discovery of perfect numbers: if n is composite, $2^n - 1$ is composite; if n is a prime, $2^n - 2$ is divisible by $2n$, and $2^n - 1$ is divisible by no prime other than those of the form $2kn + 1$ [cf. Euler[87]]. For example, $2^{11} - 1 = 23 \cdot 89$, $2^{37} - 1$ has the factor 223. Also $2^{23} - 1$ has the factor 47, Oeuvres, 2, p. 210, letter to Frenicle, October 18, 1640.

Mersenne[60] (1588–1648) stated that, of the 28 numbers* exhibited by

[54]De numeris perfectis disceptatio qua ostenditur a decem millibus ad centies centena millia, nullum esse perfectum numerum atque ideo ab unitate usque ad centies centena millia quatuor tantum perfectos numerari, Amsterdam, 1638. Reproduced as the first (pp. 115–120) of two dissertations on perfect numbers, they forming pp. 111–174 of Apologia pro Aristotele & Evclide, contra Petrvm Ramvm, & alios. Addititiae sunt Dvae Disceptationes de Nvmeris Perfectis. Authore Ioanne Broscio, Dantisci, 1652 (with a somewhat different title, Amsterdam, 1699).

[55]Oeuvres de Descartes, II, Paris, 1898, p. 429.

[56]Oeuvres de Fermat, 2, Paris, 1894, p. 176; letter to Mersenne, Dec. 26, 1638.

[57]Oeuvres de Fermat, 2, p. 185; letter to Mersenne, March, 1640.

[58]Oeuvres, 2, p. 194; letter to Mersenne, May (?), 1640.

[59]Oeuvres de Fermat, 2, pp. 198–9; Varia Opera Math. d. Petri de Fermat, Tolosae, 1679, p. 177; Précis des Oeuvres math. de P. Fermat et de l' Arithmétique de Diophante, par E. Brassinne, Mém. Ac. Imp. Sc. Toulouse, (4), 3, 1853, 149–150.

[60]F. Marini Mersenni minimi Cogitata Physico Mathematica, Parisiis, 1644. Praefatio Generalis, No. 19. C. Henry (Bull. Bibl. Storia Sc. Mat. e Fis., 12, 1879, 524–6) believed that these remarks were taken from letters from Fermat and Frenicle, and that Mersenne had no proof. A similar opinion was expressed by W. W. Rouse Ball, Messenger Math., 21, 1892, 39 (121). On documents relating to Mersenne see l'intermédiaire des math., 2, 1895, 6; 8, 1901, 105; 9, 1902, 101, 297; 10, 1903, 184. Cf. Lucas.[118]

*Only 24 were given by Bungus. While his table has 28 lines, one for each number of digits, there are no entry of numbers of 5, 11, 17, 23 digits.

Bungus,[42] chap. 28, as perfect numbers, 20 are imperfect and only 8 are perfect:

6,   28,   496,   8128,   23550336 [for 33. . .],   8589869056,
137438691328,   2305843008139952128,

which occur at the lines marked 1, 2, 3, 4, 8, 10, 12 and 29 [for 19] of Bungus' table [indicating the number of digits]. Perfect numbers are so rare that only eleven are known, that is, three different from those of Bungus; nor† is there any perfect number other than those eight, unless you should surpass the exponent 62 in $1+2+2^2+\ldots$ The ninth perfect number is the power with the exponent 68 less 1; the tenth, the power 128 less 1; the eleventh, the power 258 less 1, $i.\ e.$, the power 257, decreased by unity, multiplied by the power 256. [The first 11 perfect numbers are thus said to be $2^{n-1}(2^n-1)$ for $n=2, 3, 5, 7, 13, 17, 19, 31, 67, 127, 257$, in error as to $n=61, 67, 89, 107$ at least.] He who would find 11 others will know that all analysis up to the present will have been exceeded, and will remember in the meantime that there is no perfect number from the power 17000 to 32000, and no interval of powers can be assigned so great but that it can be given without perfect numbers. For example, if the exponent be 1050000, there is no larger exponent n up to 2090000 for which 2^n-1 is a prime. One of the greatest difficulties in mathematics is to exhibit a prescribed number of perfect numbers; and to tell if a given number of 15 or 20 digits is prime or not, all time would not suffice for the test, whatever use is made of what is already known.

Mersenne[61] stated that 2^p-1 is a prime if p is a prime which exceeds by 3, or by a smaller number, a power of 2 with an even exponent. Thus 2^7-1 is a prime since $7=2^2+3$; again, since $67=3+2^6$, $2^{67}+1=1\ldots7$ [for $2^{67}-1$] is a prime and leads to a perfect number [error corrected by Cole[173]]. Understand this only of primes 2^p-1. Wherefore this property does not belong to the prime 5, but to 3, 7, 31, 127, 8191, 131071, 524287, 2147483647, and all such. Numbers expressible as the sum or difference of two squares in several ways are composite, as $65=1+64=16+49$. As he speaks of Frenicle's knowledge of numbers, at least part of his results are doubtless due to the latter.

In 1652, J. Broscius (Apologia,[54] p. 121) observed that while perfect numbers were deduced by Euclid from geometrical progressions, they may be derived from arithmetical progressions:

$$6=1+2+3, \qquad 28=1+2+3+4+5+6+7, \qquad 496=1+2+3+\ldots+31.$$

---

†Neque enim vllus est alius perfectus ab illis octo, nisi superes exponentem numerum 62, progressionis duplae ab 1 incipientis. Nonus enim perfectus est potestas exponentis 68, minus 1. Decimus, potestas exponentis 128, minus 1. Vndecimus denique, potestas 258, minus 1, hoc est potestas 257, unitate decurtata, multiplicata per potestatem 256.

[61]F. Marini Mersenni Novarvm Observationvm Physico-Mathematicarum, Tomvs III, Parisiis, 1647, Cap. 21, p. 182. The Reflectiones Physico-Math. begin with p. 63; Cap. 21 is quoted in Oeuvres de Fermat, 4, 1912, pp. 67–8.

He stated that while perfect numbers end with 6 or 28, the proof by Bungus[42] does not show that they end alternately with 6 and 28, since Bungus included imperfect as well as perfect numbers. The numbers 130816 and 2096128, cited as perfect by Puteanus,[53] are abundant. After giving a table of the expanded form of 2^n for $n=0, 1, \ldots, 100$, Broscius (p. 130, seq.) gave a table of the prime divisors of 2^n-1 $(n=1, \ldots, 100)$, but showing no prime factor when n is any one of the primes, other than 11 and 23, less than 100. For $n=11$, the factors are 23, 89; for $n=23$, the factor 47 is given. Thus omitting unity, there remain only 23 numbers out of the first hundred which can possibly generate perfect numbers. Contrary to Cardan,[27] but in accord with Bungus,[42] there is (p. 135) no perfect number between 10^4 and 10^5. Of Bungus' 24 numbers, only 10 are perfect (pp. 135-140): those with 1, 2, 3, 4, 8, 10, 12, 18, 19, 22 digits, and given by $2^{n-1}(2^n-1)$ for $n=2, 3, 5, 7, 13, 17, 19, 29, 31, 37$, respectively. The primality of the last three was taken on the authority of unnamed predecessors.

There are only 21 abundant numbers between 10 and 100, and all of them are even; the only odd abundant number <1000 is 945, the sum of whose aliquot divisors is 975 (p. 146). The statement by Lucas, Théorie des nombres, 1, Paris, 1891, p. 380, Ex. 5, that $3^3 \cdot 5 \cdot 79$ [deficient] is the smallest abundant number is probably a misprint for $945=3^3 \cdot 5 \cdot 7$. This error is repeated in Encyclopédie Sc. Math., I, 3, Fas. 1, p. 56.

Johann Jacob Heinlin[62] (1588-1660) stated that the only perfect numbers $<4 \cdot 10^7$ are 6, 28, 496, 8128, 130816, 2096128, 33550336, and that all perfect numbers end alternately in 6 and 8.

Andrea Tacquet[63] (Antwerp, 1612-1660) stated (p. 86) that Euclid's rule gives all perfect numbers. Referring to the 11 numbers given as perfect by Mersenne,[60] Tacquet said that the reason why not more have been found so far is the greatness of the numbers 2^n-1 and the vast labor of testing their primality.

Frenicle[64] stated in 1657 that Euclid's formula gives all the even perfect numbers, and that the odd perfect numbers, if such exist, are of the form pk^2, where p is a prime of the form $4n+1$ [cf. Euler[98]].

Frans van Schooten[65] (the younger, 1615-1660) proposed to Fermat that he prove or disprove the existence of perfect numbers not of Euclid's type.

Joh. A. Leuneschlos[66] remarked that the infinite multitude of numbers contains only ten perfect numbers; he who will find ten others will know

[62]Joh. Jacobi Heinlini, Synopsis Math. praecipuas totius math. . . . Tubingae, 1653. Synopsis Math. Universalis, ed. III, Tubingae, 1679, p. 6. English translation of last by Venterus Mandey, London, 1709, p. 5.

[63]Arithmeticae Theoria et Praxis, Lovanii, 1656 and 1682 (same paging), [1664, 1704]. His opera math., Antwerpiae, 1669, does not contain the Arithmetic.

[64]Correspondence of Chr. Huygens, No. 389; Oeuvres de Fermat, 3, Paris, 1896, p. 567.

[65]Oeuvres de Huygens, II, Correspondence, No. 378, letter from Schooten to J. Wallis, Mar. 18, 1658. Oeuvres de Fermat, 3, Paris, 1896, p. 558.

[66]Mille de Quantitate Paradoxa Sive Admiranda, Heildelbergae, 1658, p. 11, XLVI, XLVII.

that he has surpassed all analysis up to the present. Goldbach[67] called Euler's attention to these remarks and stated that they were probably taken from Mersenne, the true sense not being followed.

Wm. Leybourn[68] listed as the first ten perfect numbers and the twentieth those which occur in the table of Bungus.[42] "The number 6 hath an eminent Property, for his parts are equal to himself."

Samuel Tennulius, in his notes (pp. 130–1) on Iamblicus,[4] 1668, stated that the perfect numbers end alternately in 6 and 8, and included $130816 = 256 \cdot 511$ and $2096128 = 1024 \cdot 2047$ among the perfect numbers.

Tassius[69] stated that all perfect numbers end in 6 or 8. Any multiple of a perfect or abundant number is abundant, any divisor of a perfect number is deficient. He gave as the first eight known perfect numbers the first eight listed by Mersenne.[60]

Joh. Wilh. Pauli[70] (Philiatrus) noted that if $2^n - 1$ is a prime, n is, but not conversely. For $n = 2, 3, 5, 7, 13, 17, 19, 2^n - 1$ is a prime; but $2^{11} - 1$ is divisible by 23, $2^{23} - 1$ by 47, and $2^{41} - 1$ by 83, the three divisors being $2n + 1$.

G. W. Leibniz[71] quoted in 1679 the facts stated by Pauli and set himself the problem to find the basis of these facts. Returning about five years later to the subject of perfect numbers, Leibniz implied incorrectly that $2^p - 1$ is a prime if and only if p is.

Jean Prestet[72] (†1690) stated that the fifth, . . . , ninth perfect numbers are

23550336 [for 33 . . .], 8589869056, 137438691328, 238584300813952128 [for 2305 . . . 39952128], $2^{513} - 2^{256}$.

[Hence $2^{n-1}(2^n - 1)$ for $n = 13, 17, 19, 31, 257$. The numerical errors were noted by E. Lucas,[124] p. 784.]

Jacques Ozanam[73] (1640–1717) stated that there is an infinitude of perfect numbers and that all are given by Euclid's rule, which is to be applied only when the odd factor is a prime.

Charles de Neuveglise[74] proved that the products $3 \cdot 4, . . . , 8 \cdot 9$ of two consecutive numbers are abundant. All multiples of 6 or an abundant number are abundant.

---

[67]Correspondence Math. Phys., ed., Fuss, I, 1843; letters to Euler, Oct. 7, 1752 (p. 584), Nov. 18 (p. 593).

[68]Arithmetical Recreations; or Enchriridion of Arithmetical Questions both Delightful and Profitable, London, 1667, p. 143.

[69]Arithmeticae Empiricae Compendium, Johannis Adolfi Tassii. Ex recensione Henrici Siveri, Hamburgi, 1673, pp. 13, 14.

[70]De numero perfecto, Leipzig, 1678, Magister-disputation.

[71]Manuscript in the Hannover Library. Cf. D. Mahnke, Bibliotheca Math., (3), 13, 1912–3, 53–4, 260.

[72]Nouveaux elemens des Mathematiques, ou Principes generaux de toutes les sciences, Paris, 1689, I, 154–5.

[73]Recreations mathematiques et physiques, Paris and Amsterdam, 2 vols., 1696, I, 14, 15.

[74]Traité methodique et abregé de toutes les mathématiques, Trevoux, 1700, tome 2 (L'arithmétique ou Science des nombres), 241–8.

John Harris,[75] D. D., F. R. S., stated that there are but ten perfect numbers between unity and one million of millions.

John Hill[76] stated that there are only nine perfect numbers up to a hundred thousand million. He gave (pp. 147-9) a table of values of 2^n for $n=1,\ldots,144$.

Christian Wolf[77] (1679-1754) discussed perfect numbers of the form $y^n x$ [where x, y are primes]. The sum of its aliquot parts is

$$1+y+\ldots+y^n+x+yx+\ldots+y^{n-1}x,$$

which must equal $y^n x$. Thus

$$x=(1+y+\ldots+y^n)/d, \qquad d=y^n-1-y-\ldots-y^{n-1}.$$

He stated* that x is an integer only when $d=1$, and that this requires $y=2$, $x=1+2+\ldots+2^n$. Then if this x is a prime, $2^n x$ is a perfect number. This is said to be the case for $n=8$ and $n=10$, since $2^9-1=511$ and $2^{11}-1=2047$ are primes, errors pointed out by Euler.[83] A. G. Kästner[78] was not satisfied with the argument leading to the conclusion $y=2$.

Jacques Ozanam[79] listed as perfect numbers

$$2(4-1),\ 4(8-1),\ 16(32-1),\ 64(128-1),\ 256(512-1),\ 1024(2048-1),\ldots$$

without explicit mention of the condition that the final factor shall be prime, and stated that perfect numbers are rare, only ten being known, and all end in 6 and 8 alternately. [Criticisms by Montucla,[99] Grüson.[100]]

Johann Georg Liebnecht[80] said there were scarcely 5 or 6 perfect numbers up to 4.10^7; they always end alternately in 6 and 8.

Alexander Malcolm[81] observed that it is not yet proved that there is no perfect number not in Euclid's set. He stated that, if pA is a perfect number, where p is a prime, and if $M<p$ and M is not a factor of A, then MM is an abundant number [probably a misprint for MA, as the conditions are satisfied when $p=7$, $A=4$, $M=5$, and $MA=20$ is abundant, while $M^2=25$ is deficient].

Christian Wolf[82] made the same error as Casper Ens.[51]

[75]Lexicon Technicum, or an Universal English Dictionary of Arts and Sciences, vol. I, London, 1704; ed. 5, vol. 2, London, 1736.

[76]Arithmetik, London, ed. 2, 1716, p. 3.

[77]Elementa Matheseos Universae, Halae Magdeburgicae, vol. I, 1730 and 1742, pp. 383-4, of the five volume editions [first printed 1713-41]; vol. I, 1717, 315-6, of the two volume edition. Quoted, with other errors, Ladies' Diary, 1733, Q. 166; Leybourn's ed., 1, 1817, 218; Hutton's ed., 2, 1775, 10; Diarian Repository, by Soc. Math., 1774, 289.

*"Jam ut x sit numerus integer, nec in casu speciali, si y per numerum explicetur, numerus partium aliquotarum diversus sit a numero earundem in formula generali; necesse est ut $d=1$."

[78]Math. Anfangsgründe, I, 2 (Fortsetzung der Rechnenkunst, ed. 2, 1801, 546-8).

[79]Recreations math., new ed. of 4 vols., 1723, 1724, 1735, etc., I, 29-30.

[80]Grund-Sätze der gesammten Math. Wiss. u. Lehren, Giessen u. Franckfurt, 1724, p. 21.

[81]A new system of arithmetik, theoretical & practical, London, 1730, p. 394.

[82]Mathematisches Lexicon, I, 1734 (under Vollkommen Zahl).

Leonard Euler[83] (1707–1783) noted that $2^n - 1$ may be composite for n a prime; for instance, $2^{11} - 1 = 23 \cdot 89$, contrary to Wolf.[77] If $n = 4m - 1$ and $8m - 1$ are primes, $2^n - 1$ has the factor $8m - 1$, so that $2^n - 1$ is composite for $n = 11$, 23, 83, 131, 179, 191, 239, etc. [Proof by Lucas.[123]] Furthermore, $2^{37} - 1$ has the factor 223, $2^{43} - 1$ the factor 431, $2^{29} - 1$ the factor 1103, $2^{73} - 1$ the factor 439, etc. "However, I venture to assert that aside from the cases noted, every prime less than 50, and indeed than 100, makes $2^{n-1}(2^n - 1)$ a perfect number, whence the eleven values 1, 2, 3, 5, 7, 13, 17, 19, 31, 41, 47 of n yield perfect numbers. I derived these results from the elegant theorem, of whose truth I am certain, although I have no proof: $a^n - b^n$ is divisible by the prime $n+1$, if neither a nor b is." [For later proofs by Euler, see Chapter III on Fermat's theorem.] Euler's errors as to $n = 41$ and 47 were corrected by Winsheim,[90] Euler[93] himself, and Plana.[110]

Michael Gottlieb Hansch[84] stated that $2^n - 1$ is a prime if n is any of the twenty-two primes ≤ 79 [error, Winsheim,[90] Kraft[93]].

George Wolfgang Kraft[85] corrected Stifel's[31] error that $511 \cdot 256$ is perfect and the error of Ozanam (Elementis algebrae, p. 290) that the sum of all the divisors of 2^{4n} is a prime, by noting that the sum for $n = 2$ is $511 = 7 \cdot 73$; and noted that false perfect numbers were listed by Ozanam.[79] Kraft presented (pp. 9–11) an incomplete proof, communicated to him by Tobias Maier [cf. Fontana[101]], that every perfect number is of Euclid's type. Let 1, m, n,...,p, A,...be the aliquot parts of any perfect number pA, where p and A are the middle factors [as 4 and 7 in 28]. Then

$$1 + m + n + r + q + p + A + \frac{pA}{q} + \frac{pA}{r} + \frac{pA}{n} + \frac{pA}{m} = pA.$$

Solving for A, he stated that the denominator must be unity, whence $p = 2q/D$, $D = q - 1 - q/r - q/n - q/m$. Again, $D = 1$, whence $q = 2r/D'$, $D' = r - 1 - r/n - r/m$. From $D' = 1$, $r = 2n/D''$, $D'' = n - 1 - n/m$. From $D'' = 1$, $n = 2m/(m - 1)$, $m - 1 = 1$, $m = 2$, $n = 4$, $r = 8$, etc. Thus the aliquot parts up to the middle must be the successive powers of 2, and A must be a prime, since otherwise there would be new divisors. For $p = 2^{n-1}$, we get $A = 2^n - 1$. Kraft observed that if we drop from Tartaglia's[38] list of 20 numbers those shown to be imperfect by Euler's[83] results, we have left only eight perfect numbers $2^{n-1}(2^n - 1)$ for $n \leq 39$, viz., those for $n = 2, 3, 5, 7, 13, 17, 19, 31$. For these, other than the first, as well as for the false ones of Tartaglia, if we add the digits, then add the digits of that sum, etc., we finally get unity (p. 14) [proof by Wantzel[106]]. All perfect numbers end in 6 or 28.

[83]Comm. Acad. Petropol., 6, 1738, ad annos 1732–3, p. 103. Commentationes Arithmeticae Collectae, I, Petropoli, 1849, p. 2.

[84]Epistola ad mathematicos de theoria arithmetices nouis a se inuentis aucta, Vindobonae [Vienna], 1739.

[85]De numeris perfectis, Comm. Acad. Petrop., 7, 1740, ad annos 1734–5, 7–14.

Johann Christoph Heilbronner[86] stated that the perfect numbers up to .4·10⁷ are 6, 28, 496, 8128, 130816, 2096128. "The fathers of the early church and many writers always held this number 6 in high esteem. God completed the creation in 6 days and since all things created by Him came out perfect, he wished the work of creation completed according to the number 6 as being a perfect number."

L. Euler[87] deduced from Fermat's theorem, which he here proved by use of the binomial theorem, the result* that, if m is a prime, $2^m - 1$, when composite, has no prime factors other than those of the form $mn + 1$.

J. Landen[88] noted that 196 is the least number $4x^n$, where x is prime, the sum of whose aliquot parts exceeds the number by 7.

L. Euler[89] gave a table of the prime factors of $2^n - 1$ for $n \leqq 37$.

C. N. de Winsheim[90] noted that $2^{47} - 1$ has the factor 2351, and stated that $2^n - 1$ is a prime for $n = 2$, 3, 5, 7, 13, 17, 19, 31, composite for the remaining $n < 48$, but was doubtful as to $n = 41$, thus reducing the list of perfect numbers given by Euler[83] by one or perhaps two. He suspected that $n = 41$ leads to an imperfect number since it was excluded by the acute Mersenne,[60] who gave instead $2^{66}(2^{67} - 1)$ as the ninth perfect number. He remarked that the basis of Mersenne's assertion is doubtless to be found in the stupendous genius of Mersenne which perhaps recognized more truths than he could demonstrate. He discussed the error of Hansch[84] that $2^n - 1$ is a prime if n is a prime $\leqq 79$.

G. W. Kraft[91] considered perfect numbers AP, where P is a prime [not dividing A]. Thus $a(P+1) = 2AP$, where a is the sum of all the divisors of A. Hence $a/(2A - a)$ equals the prime P. Let $2A - a = 1$, a property holding for $A = 2^m$. Then $P = 2^{m+1} - 1$ and the resulting numbers are of Euclid's type.

L. Euler,[92] in a letter to Goldbach, October 28, 1752, stated that he knew only seven perfect numbers, viz., $2^{p-1}(2^p - 1)$ for $p = 2$, 3, 5, 7, 13, 17, 19, and was uncertain whether $2^{31} - 1$ is prime or not (a factor is necessarily of the form $64n + 1$ and none are < 2000).

---

[86]Historia matheseos universae. Accedit recensio elementorum compendiorum et operum math. atque historia arithmetices ad nostra tempora, Lipsiae, 1742, 755–6. There is a 63-page list of arithmetics of the 16th century.

[87]Novi Comm. Ac. Petrop., 1, 1747–8, 20; Comm. Arith., I, 56, §39.

*We may simplify the proof by using the fact that 2 belongs to an exponent e modulo p (p a prime) such that e divides $p - 1$. For, if p is a factor of $2^m - 1$, m is a multiple of e, whence e equals the prime m. Thus $p - 1 = nm$. If we take $m > 2$, we see that n is even since p is odd and conclude with Fermat[59] that, if m is an odd prime, $2^m - 1$ is divisible by no primes other than those of the form $2km + 1$.

[88]Ladies' Diary, 1748, Question 305. The Diarian Repository, Collection of all the mathematical questions from the Ladies' Diary, 1704–1760, by a society of mathematicians, London, 1774, 509. Hutton's The Diarian Miscellany (from Ladies' Diary, 1704–1773), London, 1775, vol. 2, 271. Leybourn's Math. Quest. proposed in Ladies' D., 2, 1817, 9–10.

[89]Opuscula varii argumenti, Berlin, 2, 1750, 25; Comm. Arith., 1, 1849, 104.

[90]Novi Comm. Ac. Petrop., 2, 1751, ad annum 1749, mem., 68–99.

[91]Ibid., mem., 112–3.

[92]Corresp. Math. Phys. (ed., Fuss), I, 1843, 590, 597–8.

G. W. Kraft[93] stated (p. 114) that Euler had communicated to him privately in 1741 the fact that $2^{47}-1$ is divisible by 2351. He stated (p. 121) that if 2^p-1 is composite (p being prime), it has a factor of the form $2q^m p+1$, where q is a prime [including unity], using as illustrations the factorizations noted by Euler.[83] Of the numbers 2^n-1, n a prime ≤ 71, stated to be prime by Hansch,[84] six are composite, while the cases $53, \ldots, 71$ are in doubt (p. 115).

A. Saverien[94] repeated the remarks by Ens[51] without reference.

L. Euler[95] stated in a letter to Bernoulli that he had verified that $2^{31}-1$ is a prime by examining the primes up to 46339 which are contained in the possible forms $248n+1$ and $248n+63$ of divisors.

L. Euler[96] gave a prime factor of $2^n\pm1$ for various values of n, but no new cases 2^n-1 with n a prime.

L. Euler,[97] in a posthumous paper, proved that every even perfect number is of Euclid's type. Let $a=2^n b$ be perfect, where b is odd. Let B denote the sum of the divisors of b. The sum $(2^{n+1}-1)B$ of the divisors of a must equal $2a$. Thus $b/B=(2^{n+1}-1)/2^{n+1}$, a fraction in its lowest terms. Hence $b=(2^{n+1}-1)c$. If $c=1$, $b=2^{n+1}-1$ must be a prime since the sum of its divisors is $B=2^{n+1}$, whence Euclid's formula. If $c>1$, the sum B of the divisors of b is not less than $b+2^{n+1}-1+c+1$; hence

$$\frac{B}{b} \geq \frac{2^{n+1}(c+1)}{b} > \frac{2^{n+1}}{2^{n+1}-1},$$

contrary to the earlier equation. The proof given in another posthumous paper by Euler[98] is not complete.

L. Euler[98] proved that any odd perfect number must be of the form $r^{4\lambda+1}P^2$, where r is a prime of the form $4n+1$ [Frenicle[64]]. Express it as a product $ABC\ldots$ of powers of distinct primes. Denote by a, b, $c,\ldots$ the sums of the divisors of $A, B, C,\ldots$, respectively. Then $abc\ldots=2ABC\ldots$. Thus one of the numbers a, b, $\ldots$, say a, is the double of an odd number, and the remaining ones are odd. Thus $B, C,\ldots$ are even powers of primes, while $A=r^{4\lambda+1}$. In particular, no odd perfect number has the form $4n+3$. Amplifications of this proof have been given by Lionnet,[128] Stern,[137] Sylvester,[149] Lucas.[157] See also Liouville[30] in Chapter X.

Montucla[99] remarked that Euclid's rule does not give as many perfect numbers as believed by various writers; the one often cited [Paciuolo[16]] as the fourteenth perfect number is imperfect; the rule by Ozanam[79] is false since 511 and 2047 are not primes.

---

[93]Novi Comm. Ac. Petrop., 3, 1753, ad annos 1750–1.

[94]Dictionnaire universel de math. et physique, two vols., Paris, 1753, vol. 2, p. 216.

[95]Nouv. Mém. Acad. Berlin, année 1772, hist., 1774, p. 35; Euler, Comm. Arith., 1, 1849, 584.

[96]Opusc. anal., 1, 1773, 242; Comm. Arith., 2, p. 8.

[97]De numeris amicabilibus, Comm. Arith., 2, 1849, 630; Opera postuma, 1, 1862, 88.

[98]Tractatus de numerorum doctrina, Comm. Arith., 2, 514; Opera postuma, 1, 14–15.

[99]Récréations math. et physiques par Ozanam, nouvelle éd. par M., Paris, 1, 1778, 1790, p. 33. Engl. transl. by C. Hutton, London, 1803, p. 35.

Johann Philipp Grüson[100] made the same criticism of Ozanam[79] and noted that, if $2^n x$ is perfect and x is an odd prime,

$$1+2+\ldots+2^n = 2^n x - x - 2x - \ldots - 2^{n-1}x = x.$$

M. Fontana[101] noted that the theorem that all perfect numbers are triangular is due to Maurolycus[41] and not to T. Maier (cf. Kraft[85]). Thomas Taylor[102] stated that only éight perfect numbers have been found so far [the 8 listed are those of Mersenne[60]].

J. Struve[103] considered abundant numbers which are products abc of three distinct primes in ascending order; thus

$$\frac{ab+a+b+1}{ab-a-b-1}>c, \qquad \frac{2}{1-\dfrac{1}{a}-\dfrac{1}{b}-\dfrac{1}{ab}}>c+1.$$

The case $a\geq 3$ is easily excluded, also $a=2$, $b\geq 5$ [except 2·5·7]. For $a=2$, $b=3$, c any prime >3, $6c$ is abundant. Next, $abcd$ is abundant if

$$\frac{2abc}{abc-(ab+ac+bc+a+b+c+1)}>d+1.$$

For $a=2$, $b=3$, $c=5$ or 7, and for $a=2$, $b=5$, $c=7$, $abcd$ is abundant for any prime d $[>c]$. Of the numbers ≤ 1000, 52 are abundant.

J. Westerberg[104] gave the factors of $2^n \pm 1$ for $n=1,\ldots,32$, and of $10^n \pm 1$, $n=1,\ldots,15$.

O. Terquem[105] listed $2^{41}-1$ and $2^{47}-1$ as primes.

L. Wantzel[106] proved the remark of Kraft[85] that if N_1 be the sum of the digits of a perfect number $N>6$ [of Euclid's type], and N_2 the sum of the digits of N_1, etc., a certain N_i is unity. Since $N\equiv 1\pmod 9$, each $N_i\equiv 1 \pmod 9$, while the N_i's decrease.

V. A. Lebesgue[107] stated that he had a proof that there is no odd perfect number with fewer than four distinct prime factors. For an even perfect number $2^\alpha y^\beta z^\gamma \ldots$,

$$y^\beta z^\gamma \ldots + \frac{y^\beta z^\gamma \ldots}{2^{\alpha+1}-1}=(1+y+\ldots+y^\beta)(1+z+\ldots+z^\gamma)\ldots,$$

---

[100]Enthüllte Zaubereyen und Geheimnisse der Arithmetik, erster Theil, Berlin, 1796, p. 85, and Zusatz (end of Theil I).

[101]Memorie dell' Istituto Nazionale Ital., mat., 2, pt. 1, 1808, 285–6.

[102]The elements of a new arithmetical notation and of a new arithmetic of infinites, with an appendix....of perfect, amicable and other numbers no less remarkable than novel, London, 1823, 131.

[103]Ueber die so gennannten numeri abundantes oder die Ueberfluss mit sich führenden Zahlen, besonders im ersten Tausend unsrer Zahlen, Altona, 1827, 20 pp.

[104]De factoribus numerorum compositorum dignoscendis, Disquisitio Acad. Carolina, Lundae, 1838. In the volume, Meditationum Math.....publice defendent C. J. D. Hill, Pt. II, 1831.

[105]Nouv. Ann. Math., 3, 1844, 219 (cf. 553).

[106]Ibid., p. 337.

[107]Ibid., 552–3.

the impossibility of which is evident when the exponents $\beta, \gamma, \ldots$ are other than 1, 0, 0, ..., a case giving Euclid's solution [cf. Desboves[127]].

C. G. Reuschle[108] gave in his table C the exponent to which 2 belongs modulo p, for each prime $p < 5000$. Thus $2^n - 1$ has the factor 1399 for $n = 233$, the factor 2687 for $n = 79$, and 3391 for $n = 113$ [as stated explicitly by Le Lasseur[119,132]]; also 2351·4513 for $n = 47$, 1433 for $n = 179$, and 1913 for $n = 239$. In the addition (p. 22) to Table A, he gave the prime factors of $2^n - 1$ for various n's to 156, 37 being the least n for which the decomposition is not given completely, while 41 is the least n for which no factor is known. For 34 errata in Table C, see Cunningham[110] of Ch. VII.

F. Landry[109] gave a new proof that $2^{31} - 1$ is a prime.

Jean Plana[110] gave (p. 130) the factorization into two primes:

$$2^{41} - 1 = 13367 \times 164511353.$$

His statement (p. 141) that $2^{53} - 1$ has no factor < 50033 was corrected by Landry[113] (quoted by Lucas,[119] p. 280) and Gérardin.[177]

Giov. Nocco[111] showed that an odd perfect number has at least three distinct prime factors. For, if $a^m b^n$ is perfect,

$$2a^m = \frac{b^{n+1} - 1}{b - 1}, \qquad b^n = \frac{a^{m+1} - 1}{a - 1},$$

whence

$$\frac{a}{2(b-1)} = \frac{a^{m+1}}{2(b-1)a^m} = \frac{(a-1)b^n + 1}{b^{n+1} - 1},$$

$$a + b(ab^n + 2b^{n-1} + 2) = 2 + b(2b^n + 2ab^{n-1}).$$

But the minimum values of a, b are 3, 5. Thus $b(a-2) > 2a - 2$,

$$ab^n - 2b^n = b^{n-1} \cdot b(a-2) > b^{n-1}(2a-2), \qquad ab^n + 2b^{n-1} > 2b^n + 2ab^{n-1},$$

contrary to the earlier equation. In attempting to prove that every even perfect number $2^m b^n c^r d^s \ldots$ is of Euclid's type, he stated without proof that

$$2^{m+1} b^n c^r \ldots = (2^{m+1} - 1)BC \ldots, \qquad B = \frac{b^{n+1} - 1}{b - 1}, \ C = \frac{c^{r+1} - 1}{c - 1}, \ldots$$

require that $2^{m+1} = B$, $b^n = 2^{m+1} - 1$, $d^s = C, \ldots$ (the first two of which results yield Euclid's formula).

F. Landry[112] stated (p. 8) that he possessed the complete decomposition of $2^n \pm 1 (n \le 64)$ except for $2^{61} \pm 1$, $2^{64} + 1$, and gave (pp. 10–11) the factors of $2^{75} - 1$ and of $2^n + 1$ for $n = 65, 66, 69, 75, 90, 105$.

[108]Mathematische Abhandlung, enthaltend neue Zahlentheoretische Tabellen sammt einer dieselben betreffenden Correspondenz mit dem verewigten C. G. J. Jacobi. Prog., Stuttgart, 1856, 61 pp. Described by Kummer, Jour. für Math., 53, 1857, 379.

[109]Procédés nouveaux pour demontrer que le nombre 2147483647 est premier. Paris, 1859. Reprinted in Sphinx-Oedipe, Nancy, 1909, 6–9.

[110]Mem. Reale Ac. Sc. Torino, (2), 20, 1863, dated Nov. 20, 1859.

[111]Alcune teorie su'numeri pari, impari, e perfetti, Lecce, 1863.

[112]Aux mathématiciens de toutes les parties du monde: communication sur la décomposition des nombres en leurs facteurs simples, Paris, 1867, 12 pp.

F. Landry[113] soon published his table.   It includes the entries (quoted by Lucas[120, 122]):

$$2^{43}-1 = 431 \cdot 9719 \cdot 2099863, \ 2^{47}-1 = 2351 \cdot 4513 \cdot 13264529,$$
$$2^{53}-1 = 6361 \cdot 69431 \cdot 20394401, \ 2^{59}-1 = 179951 \cdot 3203431780337,$$

the least factors of the first two of which had been given by Euler.[83, 93] This table was republished by Lucas[123] (p. 239), who stated that only three entries remain in. doubt: $2^{61}-1$, $(2^{61}+1)/3$, $2^{64}+1$, each being conjectured a prime by Landry.   The second was believed to be prime by Kraitchik.[113a] Landry's factors of 2^n+1, for $28 \leq n \leq 64$ were quoted elsewhere.[113b]

Jules Carvallo[114] announced that he had a proof that there exists no odd perfect number.   Without indication of proof, he stated that an odd perfect number must be a square and that the ratio of the sum of the divisors of an odd square to itself cannot be 2.   The first statement was abandoned in his published erroneous proof,[133] while the second follows at once from the fact that, when p is an odd prime, the sum of the $2n+1$ divisors, each odd, of p^{2n} is odd.

E. Lucas[115] stated that long calculations of his indicated that $2^{67}-1$ and $2^{89}-1$ are composite [cf. Cole,[173] Powers[185]].   See Lucas[20] of Ch. XVII.

E. Lucas[116] stated that $2^{31}-1$ and $2^{127}-1$ are primes.

E. Catalan[116] remarked that, if we admit the last statement, and note that 2^2-1, 2^3-1, 2^7-1 are primes, we may state empirically that, up to a certain limit, if 2^n-1 is a prime p, then 2^p-1 is a prime q, 2^q-1 is a prime, etc. [cf. Catalan[135]].

G. de Longchamps[117] suggested that the composition of $2^n \pm 1$ might be obtained by continued multiplications, made by simple displacements from right to left, of the primes written to the base 2.

E. Lucas[118] verified once only that $2^{127}-1$, a number of 39 digits, is a prime.   The method will be given in Ch. XVII, where are given various results relating indirectly to perfect numbers.   He stated (p. 162) that he had the plan of a mechanism which will permit one to decide almost instantaneously whether the assertions of Mersenne and Plana that 2^n-1 is a prime for $n=53$, 67, 127, 257 are correct.   The inclusion of $n=53$ is an error of citation.   He tabulated prime factors of 2^n-1 for $n \leq 40$.

E. Lucas[119] gave a table of primes with 12 to 16 digits occurring as a factor in 2^n-1 for $n=49$, 59, 65, 69, 87, and in 2^n+1 for $n=43$, 47, 49, 53, 69, 72, 75, 86, 94, 98, 99, 135, and several even values of $n>100$.   The

[113]Décomposition des nombres $2^n \pm 1$ en leurs facteurs premiers de $n=1$ à $n=64$, moins quatre, Paris, 1869, 8 pp.
[113a]Sphinx-Oedipe, 1911, 70, 95.
[113b]L'intermédiaire des math., 9, 1902, 186.
[114]Comptes Rendus Paris, 81, 1875, 73-75.
[115]Sur la théorie des nombres premiers, Turin, 1876, p. 11; Théorie des nombres, 1891, 376.
[116]Nouv. Corresp. Math., 2, 1876, 96.
[117]Comptes Rendus Paris, 85, 1877, 950-2.
[118]Bull. Bibl. Storia Sc. Mat. e Fis., 10, 1877, 152 (278-287).   Lucas[18, 23] of Ch. XVII.
[119]Atti R. Ac. Sc. Torino, 13, 1877-8, 279.

verification of the primality was made by H. Le Lasseur. To the latter is attributed (p. 283) the factorization of 2^n-1 for $n=73$, 79, 113. These had been given without reference by Lucas.[120]

E. Lucas[121] proposed as a problem the proof that if $8q+7$ is a prime, $2^{4q+3}-1$ is not.

E. Lucas[122] stated as new the assertion of Euler[83] that if $4m-1$ and $8m-1$ are primes, the latter divides $A=2^{4m-1}-1$.

E. Lucas[123] proved the related fact that if $8m-1$ is a prime, it divides A. For, by Fermat's theorem, it divides $2^{8m-2}-1$ and hence divides A or $2^{4m-1}+1$. That the prime $8m-1$ divides A and not the latter, follows from Euler's criterion that $2^{(p-1)/2}-1$ is divisible by the prime p if 2 is a quadratic residue of p, which is the case if $p=8m\pm1$. No reference was made to Euler, who gave the first seven primes $4m-1$ for which $8m-1$ is a prime. Lucas gave the new cases 251, 359, 419, 431, 443, 491. Lucas[124] elsewhere stated that the theorem results from the law of reciprocity for quadratic residues, again without citing Euler. Later, Lucas[125] again expressly claimed the theorem as his own discovery.

T. Pepin[126] noted that if p is a prime and $q=2^p-1$ is a quadratic non-residue of a prime $4n+1=a^2+b^2$, then q is a prime if and only if $(a-bi)/(a+bi)$ is a quadratic non-residue of q.

A. Desboves[127] amplified the proof by Lebesgue[107] that every even perfect number is of Euclid's type by noting that the fractional expression in Lebesgue's equation must be an integer which divides $y^\beta z^\gamma\ldots$ and hence is a term of the expansion of the second member. Hence this expansion produces only the two terms in the left member, so that $(\beta+1)(\gamma+1)\ldots=2$. Thus one of the exponents, say β, is unity and the others are zero. The same proof has been given by Lucas[123] (pp. 234–5) and Théorie des Nombres, 1891, p. 375. Desboves (p. 490, exs. 31–33) stated that no odd perfect number is divisible by only 2 or 3 distinct primes, and that in an odd perfect number which is divisible by just n distinct primes the least prime is less than 2^n.

F. J. E. Lionnet[128] amplified Euler's[98] proof about odd perfect numbers.

F. Landry[129] stated that $2^{61}\pm1$ are the only cases in doubt in his table.[113]

Moret-Blanc[130] gave another proof that $2^{31}-1$ is a prime.

[120]Assoc. franç. avanc. sc., 6, 1877, 165.

[121]Nouv. Corresp. Math., 3, 1877, 433.

[122]Mess. Math., 7, 1877–8, 186. Also, Lucas.[119]

[123]Amer. Jour. Math., 1, 1878, 236.

[124]Bull. Bibl. Storia Sc. Mat. e Fis., 11, 1878, 792. The results of this paper will be cited in Ch. XVI.

[125]Récréations math., ed. 2, 1891, 1, p. 236.

[126]Comptes Rendus Paris, 86, 1878, 307–310.

[127]Questions d'algèbre élémentaire, ed. 2, Paris, 1878, 487–8.

[128]Nouv. Ann. Math., (2), 18, 1879, 306.

[129]Bull. Bibl. Storia Sc. Mat., 13, 1880, 470, letter to C. Henry.

[130]Nouv. Ann. Math., (2), 20, 1881, 263. Quoted, with Lucas' proof, Sphinx-Oedipe, 4, 1909, 9–12.

H. LeLasseur found after[131] 1878 and apparently just before[132] 1882 that 2^n-1 has the prime factor 11447 if $n=97$, 15193 if $n=211$, 18121 if $n=151$, 18287 if $n=223$, and that there is no divisor <30000 of 2^n-1 for the 24 prime values of n, $n\leqq 257$, which remain in doubt, viz. [cf. Lucas[136]],

$$61,\ 67,\ 71,\ 89, 101, 103, 107, 109, 127, 137, 139, 149,$$
$$157, 163, 167, 173, 181, 193, 197, 199, 227, 229, 241, 257.$$

J. Carvallo[133] attempted again[114] to prove the non-existence of odd perfect numbers $y^n z^p \ldots u^r$, where $y, \ldots u$ are distinct odd primes. He began by noting that one and only one of the exponents $n, \ldots, r$ is odd [Euler[98]]. Let $y<z<\ldots<u$, and call their number μ. From the definition of a perfect number,

$$\frac{y-1/y^n}{y-1}\cdots\frac{u-1/u^r}{u-1}=2, \qquad \frac{y}{y-1}\cdots\frac{u}{u-1}>2.$$

The fractions in this inequality form a decreasing series. Hence

$$\left(\frac{y}{y-1}\right)^\mu>2, \qquad y<\frac{2^{1/\mu}}{2^{1/\mu}-1}, \qquad k\cdot\frac{u}{u-1}>2, \qquad k\equiv\left(\frac{y}{y-1}\right)^{\mu-1}.$$

Thus $u(2-k)<2$. By a petitio principii (the division by $2-k$, not known to be positive), it was concluded (p. 10) that

$$u<\frac{2}{2-k}, \qquad k<2, \qquad y>\frac{2^{1/(\mu-1)}}{2^{1/(\mu-1)}-1}.$$

[This error, repeated on p. 15, was noted by P. Mansion.[134]] For a given μ, there is at most one prime between the two limits (of difference <2) for y. A superior limit is found for z as a function of y. An incomplete computation is made to show that, if $\mu>8$, $z<y+1$.

It is shown (p. 7) that an odd perfect number has a prime factor greater than the prime factor w entering to an odd power, since $w+1$ divides the sum of the divisors. In a table (p. 30) of the first ten perfect numbers, $2^{29}-1$ and $2^{41}-1$ are entered as primes [contrary to Euler[83] and Plana[110]].

E. Catalan[135] stated that 2^p-1 is a prime if p is a prime of the form $2^\lambda-1$. If correct this would imply that $2^{127}-1$ is a prime [cf. Catalan[116]].

E. Lucas[136] repeated the remark of LeLasseur[132] on the 24 prime values of $n\leqq 257$ for which the composition of 2^n-1 is in doubt. According to a

[131]Since these four values of n are included in the list by Lucas[124] of the 28 values of $n\leqq 257$ for which the composition of 2^n-1 is unknown. Cf. Lucas[123], p. 236.

[132]Lucas, Récréations math., 1, 1882, 241; 2, 1883, 230. Later, Lucas[125] credited LeLasseur with these four cases as well as $n=73$ [Euler[83]] and $n=79$, 113, 233 [cf. Reuschle[108]]. The last four cases were given by Lucas[124], while the last three do not occur in the table (Lucas[124], pp. 788–9) by LeLasseur of the proper divisors of 2^n-1 for each odd n, $n<79$, and for a few larger composite n's. The last three were given also by Lucas[123] (p. 236) without reference.

[133]Théorie des nombres parfaits, par M. Jules Carvallo, Paris, 1883, 32 pp.

[134]Mathesis, 6, 1886, 147.

[135]Mélanges Math., Bruxelles, 1, 1885, 376.

[136]Mathesis, 6, 1886, 146.

communication from Pellet, $2^n - 1$ is divisible by $6n + 1$ if n and $6n + 1$ are primes such that $6n + 1 = 4L^2 + 27M^2$ [provided* $n \equiv 1 \pmod 4$, $i.\,e.$, L is odd].

M. A. Stern[137] amplified Euler's[98] proof concerning odd perfect numbers.

E. Lucas[138] repeated the statement [Desboves[127]] that an odd perfect number must contain at least four distinct primes.

G. Valentin[139] gave a table, computed in 1872, showing factors of $2^n - 1$ for $n = 79$, 113, 233, etc., but not the new cases of LeLasseur.[132]

The primality of $N = 2^{61} - 1$, a number of 19 digits, considered composite by Mersenne and prime by Landry, was established by J. Pervušin[140] and P. Seelhoff[141] independently. The latter claimed to verify that there is no factor $< N^{1/3}$ of the form $8n + 7$, abbreviating the work by use of various numbers of which N is a quadratic residue; thus N is a prime or the product of two primes. Since $N = 2(2^{30})^2 - 1$, 2 is a quadratic residue of any prime factor of N, so that the factor is $8n \pm 1$. It was verified that $3^\beta \equiv 1 \pmod N$, where $\beta = (N - 1)/9$. If $N = fF$, where F is the prime factor $8n + 1$, then $3^\beta \equiv 1 \pmod F$ and, by Fermat's theorem, $3^{F-1} \equiv 1 \pmod F$. It is stated without proof that one of the exponents β and $F - 1$ divides the other. Cole[173] regarded the proof as unsatisfactory.

Seelhoff proved that a perfect number of the form $p^\pi r^\rho$ is of Euclid's type if p and r are primes and $p < r$. The condition is

$$p^\pi = \frac{r^{\rho+1} - 1}{r^{\rho+1}(2 - p) - 2r^\rho(1 - p) - p}.$$

If $p > 2$, the denominator is negative. Hence $p = 2$ and

$$2^\pi = \frac{r^{\rho+1} - 1}{2r^\rho - 2}, \qquad 2^{\pi+1} = r + \frac{r - 1}{r^\rho - 1}, \qquad \rho = 1, \qquad r = 2^{\pi+1} - 1.$$

His statements (p. 177) about the factors of $2^n - 1$, $n = 37$, 47, 53, 59, were corrected by him (ibid., p. 320) to accord with Landry.[113]

P. Seelhoff[142] obtained the known factors of these $2^n - 1$ and proved that $2^{31} - 1$ is a prime, by use of his method of quadratic residues.

H. Novarese[143] proved that every perfect number of Euclid's type ends in 6 or 28, and that each one > 6 is of the form $9k + 1$.

Jules Hudelot[144] verified in 54 hours that $2^{61} - 1$ is a prime by use of the test by Lucas, Récréations math., 2, 1883, 233.

---

*Correction by Kraitchik, Sphinx-Oedipe, 6, 1911, 73; Pellet, 7, 1912, 15.

[137]Mathesis, 6, 1886, p. 248.

[138]Ibid., p. 250.

[139]Archiv Math. Phys., (2), 4, 1886, 100-3.

[140]Bull. Acad. Sc. St. Pétersb., (3), 31, 1887, p. 532; Mélanges math. astr. ac. St. Pétersb., 6, 1881-8, 553; communicated Nov. 1883.

[141]Zeitschr. Math. Phys., 31, 1886, 174-8.

[142]Archiv Math. Phys., (2), 2, 1885, 327; 5, 1887, 221-3 (misprint for $n = 41$).

[143]Jornal de sciencias math. e astr., 8, 1887, 11-14. [Servais[145].]

[144]Mathesis, 7, 1887, 46. Sphinx-Oedipe, 1909, 16.

Cl. Servais[145] republished the proofs by Novarese[143] and proved that $a^m b^n$ is not perfect if a and b are odd primes. For, by the equations [Nocco[111]]

$$a^{m+1} - 1 = b^n(a-1), \qquad b^{n+1} - 1 = 2a^m(b-1),$$

we obtain, by subtraction,

$$(2a^m - b^n)(a+b-1) = a^{m+1}.$$

Thus $2a^m > b^n$. Since $a \geq 3$, $a^{m+1} \geq 3a^m > a^m + b^n > a + b - 1$. He next proved that, if an odd perfect number is divisible by only three distinct primes a, b, c, two of them are 3 and 5, since [as by Carvallo[133]]

$$\left(1 - \frac{1}{a}\right)\left(1 - \frac{1}{b}\right)\left(1 - \frac{1}{c}\right) < \frac{1}{2}.$$

Taking $a = 3$, $b = 5$, we have $c < 16$, whence $c = 7$, 11, or 13. He quoted from a letter from Catalan that the sum of the reciprocals of the divisors of a perfect number equals 2.

E. Cesàro[146] proved that in an odd perfect number containing n distinct prime factors, the least prime factor is $\leq n\sqrt{2}$.

Cl. Servais[147] showed that it does not exceed n since, if $a < b < c < \ldots$,

$$\frac{b}{b-1} < \frac{a+1}{a}, \qquad \frac{c}{c-1} < \frac{a+2}{a+1}, \ldots$$

$$2 < \frac{a}{a-1} \cdot \frac{b}{b-1} \cdots < \frac{a}{a-1} \cdot \frac{a+1}{a} \cdot \frac{a+2}{a+1} \cdots \frac{a+n-1}{a+n-2},$$

whence $2(a-1) < a+n-1$, $a < n+1$. If l is the $(m-1)$th prime factor and s is the mth, and if

$$\frac{a}{a-1} \cdot \frac{b}{b-1} \cdots \frac{l}{l-1} \leq L < 2,$$

then

$$L \cdot \frac{s}{s-1} \cdot \frac{s+1}{s} \cdots \frac{s+n-m}{s+n-m+1} > 2, \qquad s < \frac{L(n-m)+2}{2-L}.$$

J. J. Sylvester[148] reproduced Euler's[97] proof that every even perfect number is of Euclid's type. From the fact that $\frac{3}{2} \cdot \frac{5}{4} < 2$, he concluded that there is no odd perfect number $a^m b^n$. For the case of three prime factors he obtained the result of Servais[145] in the same manner. He proved that no odd perfect number is divisible by 105 and stated that there is none with fewer than six distinct prime factors.

Sylvester[149] and Servais[150] gave complete proofs that there exists no odd perfect number with only three distinct prime factors.

[145]Mathesis, 7, 1887, 228–230.
[146]Ibid., 245–6.
[147]Mathesis, 8, 1888, 92–3.
[148]Nature, 37, Dec. 15, 1887, 152 (minor correction, p. 179); Coll. Math. Papers, 4, 1912, 588.
[149]Comptes Rendus Paris, 106, 1888, 403–5 (correction, p. 641); reproduced with notes by P. Mansion, Mathesis, 8, 1888, 57–61. Sylvester's Coll. Math. Papers, 4, 1912, 604, 615.
[150]Mathesis, 8, 1888, 135.

Sylvester[151] proved there is no odd perfect number not divisible by 3 with fewer than eight distinct prime factors.

Sylvester[152] proved there is no odd perfect number with four distinct prime factors.

Sylvester[153] spoke of the question of the non-existence of odd perfect numbers as a "problem of the ages comparable in difficulty to that which previously to the labors of Hermite and Lindemann environed the subject of the quadrature of the circle." He gave a theorem useful for the investigation of this question: For r an integer other than 1 or -1, the sum $1+r+r^2+\ldots+r^{p-1}$ contains at least as many distinct prime factors as p contains divisors >1, with a possible reduction by one in the number of prime factors when $r=-2$, p even, and when $r=2$, p divisible by 6.

E. Catalan[154] proved that if an odd perfect number is not divisible by 3, 5, or 7, it has at least 26 distinct prime factors and thus has at least 45 digits. In fact, the usual inequality gives

$$\frac{10}{11}\frac{12}{13}\cdots\frac{l-1}{l}<\frac{1}{2}, \qquad P(l)\equiv\frac{2}{3}\cdot\frac{4}{5}\cdot\frac{6}{7}\cdot\frac{10}{11}\cdots\frac{l-1}{l}<\frac{1}{2}\cdot\frac{2}{3}\cdot\frac{4}{5}\cdot\frac{6}{7}<0.2285.$$

By Legendre's table IX, Théorie des nombres, ed. 2, 1808; ed. 3, 1830, of the values of $P(w)$ up to $w=1229$, we see that $l\geqq127$. But 127 is the 27th prime >7.

R. W. D. Christie[155] erroneously considered $2^{41}-1$ and $2^{47}-1$ as primes.

E. Lucas[156] proved that every even perfect number, aside from 6 and 496, ends with 16, 28, 36, 56, or 76; any one except 28 is of the form $7k\pm1$; any one except 6 has the remainder 1, 2, 3, or 8 when divided by 13, etc.

E. Lucas[157] reproduced his[156] proofs and the proof by Euler,[98] and gave (p. 375) a list of known factorizations of 2^n-1.

Genaille[158] stated that his machine "piano arithmétique" gives a practical means of applying in a few hours the test by Lucas (ibid., 5, 1876, 61) for the primality of 2^n-1.

J. Fitz-Patrick and G. Chevrel[159] stated that $2^{28}(2^{29}-1)$ is perfect.

E. Fauquembergue[160] found that $2^{67}-1$ is composite by a process not yielding its factors [cf. Mersenne,[60] Lucas,[115] Cole[173]].

A. Cunningham[161] called 2^p-1 a Lucassian if p is a prime of the form $4k+3$ such that also $2p+1$ is a prime, stating that Lucas[123] had proved that 2^p-1 has the factor $2p+1$. Cunningham listed all such primes $p<2500$

[151]Comptes Rendus Paris, 106, 1888, 448–450; Coll. M. Papers, IV, 609–610.
[152]Ibid., 522–6; Coll. M. Papers, IV, 611–4.
[153]Nature, 37, 1888, 417–8; Coll. M. Papers, IV, 625–9.
[154]Mathesis, 8, 1888, 112–3. Mém. soc. sc. Liège, (2), 15, 1888, 205–7 (Mélanges math., III).
[155]Math. Quest. Educat. Times, 48, 1888, p. xxxvi, 183; 49, p. 85.
[156]Mathesis, 10, 1890, 74–76.
[157]Théorie des nombres, 1891, 424–5.
[158]Assoc. franç. avanc. sc., 20, I, 1891, 159.
[159]Exercices d'Arith., Paris, 1893, 363.
[160]L'intermédiaire des math., 1, 1894, 148; 1915, 105, for representations by u^2+67v^2.
[161]British Assoc. Reports, 1894, 563.

and considered it probable that primes of the forms $2^x \pm 1$, $2^x \pm 3$ (if not yielding Lucassians) generally yield prime values of $2^p - 1$, and that no other primes will. All known and conjectured primes $2^p - 1$, with p prime, fall under this rule.

In a letter to Tannery,[162] Lucas stated that Mersenne[60,61] implied that a necessary and sufficient condition that $2^p - 1$ be a prime is that p be a prime of one of the forms $2^{2n} + 1$, $2^{2n} \pm 3$, $2^{2n+1} - 1$. Tannery expressed his belief that the theorem was empirical and due to Frenicle, rather than to Fermat, and noted that the sufficient condition would be false if $2^{67} - 1$ is composite [as is the case, Fauquembergue[160]].

Goulard and Tannery[163] made minor remarks on the subject of the last two papers.

A. Cunningham[164] found that $2^{197} - 1$ has the factor 7487. This contradicts LeLasseur's[132] statement on divisors < 30000 of Mersenne's numbers.

A. Cunningham[165] found 13 new cases (317, 337, 547, 937,...) in which $2^q - 1$ is composite, and stated that for the 22 outstanding primes $q \leqq 257$ [above list[132] except 61, 197] $2^q - 1$ has no divisor $< 50,000$ (error as to $q = 181$, see Woodall[184]). The factors obtained in the mentioned 13 cases were found after much labor by the indirect method of Bickmore,[166] who gave the factors 1913 and 5737 of $2^{239} - 1$.

A. Cunningham[167] gave a factor of $2^q - 1$ for $q = 397$, 1801, 1367, 5011 and for five larger primes q.

C. Bourlet[168] proved that the sum of the reciprocals of all the divisors d_i of a perfect number n equals 2 [Catalan[145]], by noting that n/d_i ranges with d_i over the divisors of n, so that $2n = \Sigma n/d_i$. The same proof occurs in Il Pitagora, Palermo, 16, 1909–10, 6–7.

M. Stuyvaert[169] remarked that an odd perfect number, if it exists, is a sum of two squares since it is of the form pk^2, where p is a prime $4n + 1$ [Frenicle,[64] Euler[98]].

T. Pepin[170] proved that an odd perfect number relatively prime to $3 \cdot 7$, $3 \cdot 5$ or $3 \cdot 5 \cdot 7$ contains at least 11, 14 or 19 distinct prime factors, respectively, and can not have the form $6k + 5$.

F. J. Studnička[171] called $E_p = 2^{p-1}(2^p - 1)$ an Euclidean number if $2^p - 1$ is a prime. The product of all the divisors $< E_p$ of E_p is E_p^{p-1}. When E_p is written in the diadic system (base 2), it has $2p - 1$ digits, the first p of which are unity and the last $p - 1$ are zero.

[162]L'intermédiaire des math., 2, 1895, 317.
[163]Ibid., 3, 1896, 115, 188, 281.
[164]Nature, 51, 1894–5, 533; Proc. Lond. Math. Soc., 26, 1895, 261; Math. Quest. Educat. Times, 5, 1904, 108, last footnote.
[165]British Assoc. Reports, 1895, 614.
[166]On the numerical factors of $a^n - 1$, Messenger Math., 25, 1895–6, 1–44; 26, 1896–7, 1–38. French transl. by Fitz-Patrick, Sphinx-Oedipe, 1912, 129–144, 155–160.
[167]Proc. London Math. Soc., 27, 1895–6, 111.
[168]Nouv. Ann. Math., (3), 15, 1896, 299.
[169]Mathesis, (2), 6, 1896, 132.
[170]Memoire Accad. Pont. Nuovi Lincei, 13, 1897, 345–420.
[171]Sitzungsber. Böhm. Gesell., Prag, 1899, math. nat., No. 30.

Mario Lazzarini[172] attempted to prove that there is no odd perfect number $a^{\alpha}b^{\beta}c^{\gamma}$, but made the error of thinking that a is relatively prime to $b^{\beta}+\ldots+b+1$. He attempted to show that $p=2^{\alpha}-1$ is a prime if and only if p divides $N=3^{k}+1$, where $k=2^{\alpha-1}-1$ [false for $\alpha=2$, since $p=3$, $N=4$]. He restricted his argument to the case a odd, whence $p\equiv1\pmod 3$. Then, if p is a prime, -3 is a quadratic residue of p, so that $(-3)^{(p-1)/2}\equiv1$ $\pmod p$, whence p divides N. Conversely, when this congruence holds, he concluded falsely that $z^{2}\equiv-3\pmod p$ has two and only two roots, so that p is expressible in a single way as a sum of a square and the triple of a square and hence is prime. To show the error, let $p=ab$, where $a=23$, $b=3851$ are primes; then

$$(-3)^{11}+1=-2ab,\ (-3)^{\frac{b-1}{2}}\equiv-1\ (\mathrm{mod}\ b),\ (-3)^{\frac{b-1}{2}}\equiv(-3)^{11\cdot175}\equiv-1\ (\mathrm{mod}\ a),$$

whence $(-3)^{(p-1)/2}\equiv1\pmod p$. Cipolla remarked (p. 288) that we may deduce from a result of Lucas[120] that p is a prime if it divides N without dividing $3^{\delta}+1$ for any divisor δ of $p=2^{\alpha-1}-1$.

F. N. Cole[173] found that $2^{67}-1$ is the product of the two primes 193707721, 761838257287. In the footnote to p. 136, he criticized the proof by Seelhoff[141] of the primality of $N=2^{61}-1$ and stated he had verified that N is prime by an actual computation of a series of primes of which N is a quadratic residue.

R. D. Carmichael[174] proved that any even perfect number $2^{a}p_{2}^{a_{1}}\ldots p_{n}^{a_{n}}$ is of Euclid's type. Write d for $2^{a+1}-1$. Then, as usual,

$$\frac{2^{a+1}}{d}=\Pi\frac{(p_{i}^{a_{i}}+\ldots+p_{i}+1)}{p_{i}^{a_{i}}},\qquad 1+\frac{1}{d}\geq\Pi\left(1+\frac{1}{p_{i}}\right).$$

If $n>2$, p_{i} is less than d, being an aliquot divisor of it, so that $1+1/p_{i}$ exceeds the left member of the inequality. Hence $n=2$, $p_{2}=d$.

A. Cunningham[175] gave the residues of $k=2^{2^{n}}$, 2^{k}, etc., modulo $2^{q}-1$ for primes $q\leq101$.

A. Turčaninov[176] (Turtschaninov) proved that an odd perfect number has at least four distinct prime factors and exceeds 2000000.

A. Gérardin[177] noted the error by Plana.[110]

A. Gérardin[178] stated the empirical laws: If n is a prime of the form $24x+11$ and if $2^{n}-1$ is composite, the least factor is of the form $24y+23$

[172]Periodico di mat. insegn. sec., 18, 1903, 203; criticized by C. Ciamberlini, p. 283, and by M. Cipolla, p. 285.
[173]Bull. Amer. Math. Soc., 10, 1903–4, 134–7. French transl., Sphinx-Oedipe, 1910, 122–4. Cf. Fauquembergue.[160]
[174]Annals of Math., (2), 8, 1906–7, 149.
[175]Proc. London Math. Soc., (2), 5, 1907, 259 [250].
[176] Věst. opytn. fiziki (Spaczinskis Bote), Odessa, 1908, No. 461 (pp. 106–113), No. 463 (162–3), No. 465–6 (213–9), No. 470 (314–8). In Russian. Cf. Bourlet.[168]
[177]L'intermédiaire des math., 15, 1908, 230–1.
[178]Sphinx-Oedipe, Nancy, 3, 1908–9, 113–123; Assoc. franç. avanc. sc., 1909, 145–156. In Wiskundig Tijdschrift, 10, 1913, 61, he added that in the remaining three cases <257, n=107, 167, 227, the least divisor (necessarily >1 million) is respectively 5136 y+2783, 8016 y+335, 10896 y+5903.

($e.$ $g.$, $n=11$, 59, 83, 131, 179, 251). If n is a prime $24x+23$ and 2^n-1 is composite, the least factor is of the form $48y+47$ ($e.$ $g.$, $n=47$, $y=48$, factor 2351; $n=23$, 71, 191, 239). Gérardin[179] gave tables of the possible, but (unverified, factors of 2^n-1, $n<257$.

A. Cunningham[180] gave the factor 150287 of $2^{163}-1$.

A. Cunningham[181] found the factor 228479 of $2^{71}-1$.

T. M. Putnam[182] proved that not all of the r distinct prime factors of a perfect number exceed $1+r/\log_e 2$ and hence do not all equal or exceed $1+3r/2$.

L. E. Dickson[183] gave an immediate proof that every even perfect number is of Euclid's type. Let $2^n q$ be perfect, where q is odd and $n>0$. Then $(2^{n+1}-1)s = 2^{n+1}q$, where s is the sum of all the divisors of q. Thus $s=q+d$, where $d=q/(2^{n+1}-1)$. Hence d is an integral divisor of q, so that q and d are the only divisors of q. Hence $d=1$ and q is a prime.

H. J. Woodall[184] obtained the factor 43441 of $2^{181}-1$.

R. E. Powers[185] verified that $2^{89}-1$ is a prime by use of Lucas' test on the series 4, 14, 194,.... H. Tarry[186] made an incomplete examination. E. Fauquembergue[187] proved that $2^{89}-1$ is a prime by writing the residues of that series to base 2.

A. Cunningham[188] noted that 2^q-1 is composite for three primes of 8 digits. On the proof-sheets of this history, he noted that the first two should be

$$q=67108493, \quad p=134216987; \quad q=67108913, \quad p=134217827.$$

A Gérardin[188a] observed that $2^{2n+1}-1 = F^2-2G^2$, $F = 2^{n+1}\pm1 = 2m+1$, $G = 2^n \pm 1$, $G^2 = m^2+(m+1)^2-(2^n)^2$.

H. Tarry[188b] verified for the known composite numbers 2^p-1, where p is a prime, that, if a is the least factor, 2^a-1 is composite.

A. Gérardin added empirically that, if p is any number and a any divisor of 2^p-1, $a=8m\pm1$ not being of the form 2^u-1 then 2^a-1 is composite.

A. Cunningham[189] noted that, if q is a prime,

$$M_q = 2^q-1 = T^2-2(qu)^2 = (qt)^2-2U^2.$$

If M_q is a prime it can be expressed in the forms $A^2+3B^2 = G^2+6H^2$, and in one or the other of the pairs of forms $t^2\pm au^2$ ($a=7$, 14, 21, 42). He discussed M_q to the base 2.

[179]Sphinx-Oedipe, 3, 1908-9, 118-120, 161-5, 177-182; 4, 1909, 1-5, 158, 168; 1910, 149, 166.
[180]Proc. London Math. Soc., (2), 6, 1908, p. xvii.
[181]L'intermédiaire des math., 16, 1909, 252; Sphinx-Oedipe, 4, 1909, 4e Trimestre, 36-7.
[182]Amer. Math. Monthly, 17, 1910, 167.   [183]Ibid., 18, 1911, 109.
[184]Bull. Amer. Math. Soc., 16, 1910-11, 540 (July, 1911). Proc. London Math. Soc., (2), 9, 1911, p. xvi. Mem. and Proc. Manchester Literary and Phil. Soc., 56, 1911-12, No. 1, 5 pp. Sphinx-Oedipe, 1911, 92. Verification by J. Hammond, Math. Quest. Solutions, 2, 1916, 30-2.
[185]Bull. Amer. Math. Soc., 18, 1911-12, 162 (report of meeting Oct., 1911). Amer. Math. Monthly, 18, 1911, 195. Sphinx-Oedipe, Feb., 1912, 17-20.
[186]Sphinx-Oedipe, Dec., 1911, p. 192; 1912, 15. (Proc. London Math. Soc., (2), 10, 1912, Records of Meetings, 1911-12, p. ii.)
[187]Ibid., 1912, 20-22.   [188]Messenger Math., 41, 1911, 4.
[188a]Bull. Soc. Philomatiques de Paris, (10), 3, 1911, 221.   [188b]Sphinx-Oedipe, 6, 1911, 174 186, 192.
[189]Math. Quest. Educ. Times, (2), 19, 1911, 81-2; 20, 1911, 90-1, 105-6; 21, 1912, 58-9, 73.

A. Cunningham[190] found the factor 730753 of $2^{173}-1$.

V. Ramesam[191] verified that the quotient of $2^{71}-1$ by the factor 228479 [Cunningham[181]] is the product of the primes 48544121 and 212885833.

A. Aubry[192] stated erroneously that "Mersenne affirmed that 2^n-1 is a prime, for $n \leq 257$, only for $n=1$, 2, 3, 4, 8, 10, 12, 29, 61, 67, 127, 257 (which has now been almost proved); this proposition seems to be due to Frenicle.[57]" What Mersenne[60] actually stated was that the first 8 perfect numbers occur at the lines marked 1, 2, 3, 4, 8, etc., in the table by Bungus.

A. Cunningham[192a] noted that M_{113}, M_{151}, M_{251} have the further factors 23279·65993, 55871, 54217, respectively. Cf. Reuschle[108], Lucas[123].

A. Gérardin[192b] noted that there is no divisor < 1000000 of the composite Mersenne numbers not already factored. Let d denote the least divisor of 2^q-1, q a prime ≤ 257. If $q=60u+43$, then $d \equiv 47 \pmod{96}$, except for the cases given by Euler's[83] theorem (verified for 43, 163, 223). If $q=40u+33$, $d \equiv 7 \pmod{24}$, verified for 73, 113, 233. If $q=30m+1$, $d \equiv 1 \pmod{24}$, verified for 31, 61, 151, 181, 211.

E. Fauquembergue[192c] proved that $2^{101}-1$ is composite by means of Lucas' test with 4, 14, 194,..., written to base 2 (Ch. XVII).

L. E. Dickson[193] called a non-deficient number *primitive* if it is not a multiple of a smaller non-deficient number, and proved that there is only a finite number of primitive non-deficient numbers having a given number of distinct odd prime factors and a given number of factors 2. As a corollary, there is not an infinitude of odd perfect numbers with any given number of distinct prime factors. There is no odd abundant number with fewer than three distinct prime factors; the primitive ones with three are

$$3^3 5 \cdot 7, \quad 3^2 5^2 7, \quad 3^2 5 \cdot 7^2, \quad 3^3 5^2 11, \quad 3^5 5^2 13, \quad 3^4 5^3 13, \quad 3^4 5^2 13^2, \quad 3^3 5^3 13^2.$$

There is given a list of the numerous primitive odd abundant numbers with four distinct prime factors and lists of even non-deficient numbers of certain types. In particular, all primitive non-deficient numbers <15000 are determined (23 odd and 78 even). In view of these lists, there is no odd perfect number with four or fewer distinct prime factors (cf. Sylvester[148-153]).

A. Cunningham[194] gave a summary of the known results on the composition of the 56 Mersenne numbers $M_q = 2^q - 1$, q a prime ≤ 257. Of these, 12 have been proved prime: M_q, $q=1$, 2, 3, 5, 7, 13, 17, 19, 31, 61, 89, 127; while 29 of them have been proved composite. Thus only 15 remain in

[190]British Assoc. Reports, 1912, 406–7.   Sphinx-Oedipe, 7, 1912, 38 (1910, 170, that 730753 is a possible factor).   Cf. Cunningham[194].

[191]Nature, 89, 1912, p. 87; Sphinx-Oedipe, 1912, 38.   Jour. of Indian Math. Soc., Madras, 4 1912, 56.

[192]Oeuvres de Fermat, 4, 1912, 250, note to p. 67.

[192a] Mem. and Proc. Manchester Lit. and Phil. Soc., 56, 1911–2, No. 1.

[192b] Sphinx-Oedipe, 7, 1912, numéro spécial, 15–16.

[192c] Ibid., Nov., 1913, 176.

[193]Amer. Jour. Math., 35, 1913, 413–26.

[194]Proc. Fifth International Congress, I, Cambridge, 1913, 384–6.   Proc. London Math. Soc., (2), 11, 1913, Record of Meeting, Apr. 11, 1912, xxiv.   British Assoc. Reports, 1911, 321.   Math. Quest. Educat. Times, (2), 23, 1913, 76.

doubt: M_q, $q = 101$, 103, 107, 109, 137, 139, 149, 157, 167, 193, 199, 227, 229, 241, 257. The last has no factor under one million, as verified by R. E. Powers.[194a] No one of the other 14 has a factor under one million, as verified twice with the collaboration of A. Gérardin. Up to the present three errors have been found in Mersenne's assertion; M_{67} has been proved composite (Lucas,[115] Cole[173]), while M_{61} and M_{89} have been proved prime (Pervušin,[140] Seelhoff,[141] Cole,[173] Powers[185]). It is here announced that M_{173} has the factor 730753, found with the collaboration of A. Gérardin.

J. McDonnell[195] commented on a test by Lucas in 1878 for the primality of $2^n - 1$.

L. E. Dickson[196] gave a table of the even abundant numbers < 6232.

R. Niewiadomski[197] noted that $2^{761} - 1$ has the factor 4567 and gave known factors of $2^n - 1$. He gave the formula
$$2^{6m+1} - 1 = (2^{2m} + 2^m - 1)^3 + (2^{2m} - 2^m - 1)^3 + 1.$$

G. Ricalde[198] gave relations between the primes p, q and least solutions of
$$2^{2n+1} - 1 = pq, \qquad a^2 - 2b^2 = p, \qquad c^2 - 2d^2 = q.$$

R. E. Powers[199] proved that $2^{107} - 1$ is a prime by means of Lucas'[31] test in Ch. XVII.

E. Fauquembergue[200] proved that $2^p - 1$ is prime for $p = 107$ and 127, composite for $p = 101$, 103, 109.

T. E. Mason[201] described a mechanical device for applying Lucas'[118] method for testing the primality of $2^{4q+3} - 1$.

R. E. Powers[202] proved that $2^{103} - 1$ and $2^{109} - 1$ are composite by means of Lucas' tests with 3, 7, 47,... and 4, 14, 194... (Ch. XVII), respectively.

A. Gérardin[203] gave a history of perfect numbers and noted that $2^p - 1$ can be factored if we find t such that $m = 2pt + 1$ is a prime not dividing $s = 1 + 2^p + 2^{2p} + \ldots + 2^{(2t-1)p}$, since $2^{2pt} - 1 \equiv (2^p - 1)s \pmod{m}$. Or we may seek to express $2^p - 1$ in two ways in the form $x^2 - 2y^2$.

On tables of exponents to which 2 belongs, see Ch. VII, Cunningham and Woodall[128], Kraitchik.[125]

ADDITIONAL PAPERS OF A MERELY EXPOSITORY CHARACTER.

E. Catalan, Mathesis, (1), 6, 1886, 100–1, 178.
W. W. Rouse Ball, Messenger Math., 21, 1891–2, 34–40, 121.
Fontés (on Bovillus[20]), Mém. Ac. Sc. Toulouse, (9), 6, 1894, 155–67.
J. Bezdiček, Casopis Mat. a Fys., Prag, 25, 1896, 221–9.
Hultsch (on Iamblichus), Nachr. Kgl. Sächs. Gesell., 1895–6.
H. Schubert, Math. Mussestunden, I, Leipzig, 1900, 100–5.
M. Nassò, Revue de math. (Peano), 7, 1900–1, 52–53.

[194a]Sphinx-Oedipe, 1913, 49–50.
[195]London Math. Soc., Records of Meeting, Dec., 1912, v-vi.
[196]Quart. Jour. Math., 44, 1913, 274–7.
[197]L'intermédiaire des math., 20, 1913, 78, 167.
[198]Ibid., 7–8, 149–150; cf. 140–1.
[199]Proc. London Math. Soc., (2), 13, 1914, Records of meetings, xxxix. Bull. Amer. Math. Soc., 20, 1913–4, 531. Sphinx-Oedipe, 1914, 103–8.
[200]Sphinx-Oedipe, June, 1914, 85; l'intermédiaire des math., 24, 1917, 33.
[201]Proc. Indiana Acad. Science, 1914, 429–431.
[202]Proc. London Math. Soc., (2), 15, 1916, Records of meetings, Feb. 10, 1916, xxii.
[203]Sphinx-Oedipe, 1909, 1–26.

G. Wertheim, Anfangsgründe der Zahlentheorie, 1902.
G. Giraud, Periodico di Mat., 21, 1906, 124–9.
F. Ferrari, Suppl. al Periodico di Mat., 11, 1908, 36–8, 53, 75–6 (Cipolla).
P. Bachmann, Niedere Zahlentheorie, II, 1910, 97–101.
A. Aubry, Assoc. franç. avanc. sc., 40, 1911, 53–4; 42, 1913; l'enseignement math., 1911, 399; 1913, 215–6, 223.
*M. Kiseljak, Beiträge zur Theorie der vollkommenen Zahlen, Progr. Agram, 1911.
*J. Vaës, Wiskundig Tijdschrift, 8, 1911, 31, 173; 9, 1912, 120, 187.
J. Fitz-Patrick, Exercices Math., ed. 3, 1914, 55–7.

MULTIPLY PERFECT NUMBERS.

A multiply perfect or pluperfect number n is one the sum of whose divisors, including n and 1, is a multiple of n. If the sum is mn, m is called the multiplicity of n. For brevity, a multiply perfect number of multiplicity m shall be designated by P_m. Thus an ordinary perfect number is a P_2. Although Robert Recorde[39] in 1557 cited 120 as an abundant number, since the sum of its parts is 240, such numbers were first given names and investigated by French writers in the seventeenth century. As a P_3 equals one-half of the sum of its aliquot divisors or parts (divisors $< P_3$), it was called a sous-double; a P_4 equals one-third of the sum of its aliquot parts and was called a sous-triple; a P_5 a sous-quadruple; etc.

F. Marin Mersenne proposed to R. Descartes[300] the problem to find a sous-double other than $P_3^{(1)} = 120 = 2^3 3 \cdot 5$. The latter did not react on the question until seven years later.

Mersenne[301] mentioned (in the Epistre) the problem to find a P_4, a P_5 or a P_m, a P_3 besides 120, and a rule to find as many as one pleases. He remarked (p. 211) that the P_3 120, the P_4 240 [for 30240?] and all other abundant numbers can signify the most fruitful natures.

Pierre de Fermat[302] referred in 1636 to his former [lost] letter in which he gave "the proposition concerning aliquot parts and the construction to find an infinitude of numbers of the same nature." He[303] found the second P_3, viz., $P_3^{(2)} = 672 = 2^5 3 \cdot 7$.

Mersenne[304] stated that Fermat found the P_3 672 and knew infallible rules and analysis to find an infinitude of such numbers. He[305] later gave [Fermat's] method of finding such P_3: Begin with the geometric

1	3	7	15...
2	4	8	16...
3	5	9	17...

[300]Oeuvres de Descartes, 1, Paris, 1897, p. 229, line 28, letter from Descartes to Mersenne, Oct or Nov., 1631.
[301]Les Preludes de l'Harmonie Universelle ou Questions Curiouses, Utiles aux Predicateurs, aux Theologiens, Astrologues, Medecins, & Philosophes, Paris, 1634.
[302]Oeuvres de Fermat, 2, Paris, 1894, p. 20, No. 3, letter to Mersenne, June 24, 1636.
[303]Oeuvres de Fermat, 2, p. 66 (French transl. 3, p. 288), 2, p. 72, letters to Mersenne and Roberval, Sept., 1636.
[304]Harmonie Universelle, Paris, 1636, Premiere Preface Generale (preceded by a preface of two pages), unnumbered page 9, remark 10. Extract in Oeuvres de Fermat, 2, 1894, 20–21.
[305]Mersenne, Seconde Partie de l'Harmonie Universelle, Paris, 1637. Final subdivision: Nouvelles Observations Physiques et Mathématiques, p. 26, Observation 13. Extract in Oeuvres de Fermat, 2, 1894, p. 21.

progression 2, 4, 8, ... Subtract unity and place the remainders above the
former. Add unity and place the sums below. Then if the quotient of
the $(n+3)$th number of the top line by the nth number of the bottom
line is a prime, its triple multiplied by the $(n+2)$th number of the middle
line is a P_3. Thus if $n=1$, 15/3 is a prime and $3 \cdot 5 \cdot 8 = 120$ is a P_3. For
$n=3$, 63/9 is a prime and $3 \cdot 7 \cdot 32 = 672$ is a P_3. [This rule thus states in
effect that $3 \cdot 2^{n+2} p$ is a P_3 if $p = (2^{n+3}-1)/(2^n+1)$ is a prime.]

The third P_3, discovered by André Jumeau, Prior of Sainte-Croix, is

$$P_3^{(3)} = 523776 = 2^9 3 \cdot 11 \cdot 31.$$

In April, 1638, he communicated it to Descartes[306] and asked for the fourth
P_3 (the fifth and last of St. Croix's challenge problems).

Descartes[307] stated that the rule [305] of Fermat furnishes no P_3 other than
120 and 672 and judged that Fermat did not find these numbers by the
formula, but accommodated the formula to them, after finding them by trial.

Descartes[308] answered the challenge of St. Croix with the fourth P_3,

$$P_3^{(4)} = 1476304896 = 2^{13} 3 \cdot 11 \cdot 43 \cdot 127.$$

Soon afterwards Descartes[309] announced the following six P_4:

$$P_4^{(1)} = 30240 = 2^5 3^3 5 \cdot 7,$$
$$P_4^{(2)} = 32760 = 2^3 3^2 5 \cdot 7 \cdot 13,$$
$$P_4^{(3)} = 23569920 = 2^9 3^3 5 \cdot 11 \cdot 31,$$
$$P_4^{(4)} = 142990848 = 2^9 3^2 7 \cdot 11 \cdot 13 \cdot 31,$$
$$P_4^{(5)} = 66433720320 = 2^{13} 3^3 5 \cdot 11 \cdot 43 \cdot 127,$$
$$P_4^{(6)} = 403031236608 = 2^{13} 3^2 7 \cdot 11 \cdot 13 \cdot 43 \cdot 127,$$

and the sous-quadruple

$$P_5^{(1)} = 14182439040 = 2^7 3^4 5 \cdot 7 \cdot 11^2 17 \cdot 19.$$

He stated that his analysis had led him to a method which would require
time to explain in the form of a rule, but that he could find, for example,
a sous-centuple, necessarily very large.

Fermat apparently responded to the fifth challenge problem of St. Croix
on the fourth P_3. Without warrant, Descartes[310] suspected that Fermat
had not found independently the fourth P_3, but had learned from some one
in Paris of its earlier discovery by Descartes. Fermat[311] indicated that he
possessed an analytic method by which he could solve all questions con-

[306]Oeuvres de Descartes, 2, Paris, 1898, p. 428, p. 167 (latter without name of St. Croix); cf.
    Oeuvres de Fermat, 2, 1894, pp. 63–64.
[307]Oeuvres de Descartes, 2, 1898, p. 148, letter to Mersenne, May 27, 1638.
[308]Oeuvres de Descartes, 2, 1898, 167, letter to Mersenne, June 3, 1638.
[309]Oeuvres de Descartes, 2, 1898, 250–1, letter to Mersenne, July 13, 1638. In June, 1645,
    Descartes, 4, 1901, p. 229, again mentioned the first two of these P_4.
[310]Oeuvres de Descartes, 2, 1898, 273, letter to Mersenne, July 27, 1638.
[311]Oeuvres de Fermat, 2, 1894, p. 165, No. 4; p. 176, No. 1; letters to Mersenne, Aug. 10 and
    Dec. 26, 1638.

cerning aliquot parts, apart from the testing of the primality of a number n, knowing no method except the trial of each number $< \sqrt{n}$ as a divisor. Descartes[312] gave the following rules for multiply perfect numbers:

I. If n is a P_3 not divisible by 3, then $3n$ is a P_4.

II. If a P_3 is divisible by 3, but by neither 5 nor 9, then $45P_3$ is a P_4.

III. If a P_3 is divisible by 3, but not by 7, 9 or 13, then $3 \cdot 7 \cdot 13 \, P_3$ is a P_4.

IV. If n is divisible by 2^9, but by no one of the numbers 2^{10}, 31, 43, 127, then $31n$ and $16 \cdot 43 \cdot 127n$ are proportional to the sums of their aliquot parts.

V. If n is not divisible by 3 and if $3n$ is a P_{4k}, then n is a P_{3k}.

By applying rule II to $P_3{}^{(2)}$, $P_3{}^{(3)}$, $P_3{}^{(4)}$, Descartes obtained his $P_4{}^{(1)}$, $P_4{}^{(3)}$, $P_4{}^{(5)}$. By applying rule III to $P_3{}^{(1)}$, $P_3{}^{(3)}$, $P_3{}^{(4)}$, he obtained his $P_4{}^{(2)}$, $P_4{}^{(4)}$, $P_4{}^{(6)}$.

In the same letter, Descartes expressed to Mersenne a desire to know what Frenicle de Bessy had found on this subject. Frenicle wrote direct to Descartes, who in his reply[313] expressed his astonishment that Frenicle should regard as sterile the above rules for finding P_4, since Descartes had deduced by them six P_4 from four P_3, at a time when Mersenne had stated to Descartes that it was thought to be impossible to find any at all. Descartes stated that, since one can find an infinity of such rules, one has the means of finding an infinitude of P_m. From one of Frenicle's P_5 (communicated to Descartes by Mersenne),

$$P_5{}^{(2)} = 30823866178560 = 2^{10} 3^5 5 \cdot 7^2 13 \cdot 19 \cdot 23 \cdot 89,$$

Descartes (p. 475) derived the smaller P_5:

$$P_5{}^{(3)} = 31998395520 = 2^7 3^5 5 \cdot 7^2 \cdot 13 \cdot 17 \cdot 19.$$

Mersenne[314] listed various P_m due to his correspondents, without citation of names. He listed the above $P_3{}^{(i)}$ ($i = 1, 2, 3, 4$) and remarked that "un excellent esprit"[315] found that when

$$P_3{}^{(5)} = 459818240 = 2^8 5 \cdot 7 \cdot 19 \cdot 37 \cdot 73$$

is multiplied by 3, the product is a P_4:

$$P_4{}^{(7)} = 2^8 3 \cdot 5 \cdot 7 \cdot 19 \cdot 37 \cdot 73,$$

attributed to Lucas[321] by Carmichael.[334]

---

[312]Oeuvres, 2, 1898, 427–9, letter to Mersenne, Nov. 15, 1638.

[313]Oeuvres de Descartes, 2, 1898, 471, letter to Frenicle, Jan. 9, 1639.

[314]Les Nouvelles Pensees de Galilei, traduit d'Italien en François, Paris, 1639, Preface, pp. 6–7. Quoted in Oeuvres de Descartes, 10, Paris, 1908, pp. 564–6, and in Oeuvres de Fermat, 4, 1912, pp. 65–66.

[315]Frenicle de Bessy, according to the editors of the Oeuvres de Fermat, 2, 1894, p. 255, note 2; 4, 1912, p. 65, note 2 (citing Oeuvres de Descartes, 2, letter Descartes to Mersenne, Nov. 15, 1638, pp. 419–448 [p. 429]). It is clear that the discoverers Fermat, St. Croix, and Descartes of the $P_4{}^{(i)}$ ($i = 2, 3, 4$) are not meant. It is attributed to Legendre[319] by Carmichael.[334]

There are listed Descartes' six P_4 and $P_5^{(1)}$, Frenicle's $P_5^{(2)}$, and also

$$P_4^{(8)} = 45532800 = 2^7 3^3 5^2 17\cdot 31,$$
$$P_4^{(9)} = 43861478400 = 2^{10} 3^3 5^2 23\cdot 31\cdot 89,$$

and the erroneous P_5 508666803200 (not divisible by 5^2+5+1), probably a misprint for the correct P_5 (in the list by Lehmer[323]):

$$P_5^{(4)} = 518666803200 = 2^{11} 3^3 5^2 7^2 13\cdot 19\cdot 31.$$

A part of these P_m, but no new ones, were mentioned by Mersenne[60] in 1644; the least P_3 is stated to be 120. (Oeuvres de Fermat, 4, 66–7.)

In 1643 Fermat[316] cited a few of the P_m he had found:

$$P_3^{(6)} = 51001180160 = 2^{14} 5\cdot 7\cdot 19\cdot 31\cdot 151,$$
$$P_4^{(10)} = 3P_3^{(6)},$$
$$P_4^{(11)} = 14942123276641920 = 2^7 3^6 5\cdot 17\cdot 23\cdot 137\cdot 547\cdot 1093,$$
$$P_5^{(5)} = 1802582780370364661760 = 2^{20} 3^3 5\cdot 7^2 13^2 19\cdot 31\cdot 61\cdot 127\cdot 337,$$
$$P_5^{(6)} = 87934476737668055040 = 2^{17} 3^5 5\cdot 7^3 13\cdot 19^2 37\cdot 73\cdot 127,$$
$$P_6^{(1)} = 2^{23} 3^7 5^3 7^4 11^3 13^3 17^2 31\cdot 41\cdot 61\cdot 241\cdot 307\cdot 467\cdot 2801,$$
$$P_6^{(2)} = 2^{27} 3^5 5^3 7\cdot 11\cdot 13^2 19\cdot 29\cdot 31\cdot 43\cdot 61\cdot 113\cdot 127.$$

He stated that he possessed a general method of finding all P_m.

Replying to Mersenne's query as to the ratio of

$$P_6^{(3)} = 2^{36} 3^8 5^5 11\cdot 13^2 19\cdot 31^2 43\cdot 61\cdot 83\cdot 223\cdot 331\cdot 379\cdot 601\cdot 757$$
$$\times 1201\cdot 7019\cdot 823543\cdot 616318177\cdot 100895598169$$

to the sum of its aliquot parts, Fermat[317] stated that it is a P_6, the prime factors of the final factor being 112303 and 898423 [on the finding of these factors, see Ch. XIV, references 23, 92, 94, 103]. Note that $823543 = 7^7$.

Descartes[318] constructed $P_3^{(2)} = 672 = 21.32$ by starting with 21 and noting that $\sigma(21) = 32$, $\sigma(32) = 63 = 3\cdot 21$, for σ defined as on p. 53.

Mersenne[61] noted that if a P_3 is not divisible by 3, then $3P_3$ is a P_4 [rule I of Descartes[312]]; if a P_5 is not divisible by 5, then $5P_5$ is a P_6, etc. He stated that there had been found 34 P_4, 18 P_5, 10 P_6, 7 P_7, but no P_8 so far.

In 1652, J. Broscius (Apologia,[54] p. 162) cited the $P_4^{(1)}$ [of Descartes[309]]. The P_3 120 and 672 are mentioned in the 1770 edition of Ozanam's[79] Récréations, I, p. 35, and in Hutton's translation of Montucla's[99] edition, I, p. 39.

A. M. Legendre[319] determined the P_m of the form $2^n a\beta\gamma\ldots$, where a, β, γ, $\ldots$ are distinct odd primes, for $m=3$, $n\leq 8$; $m=4$, $n=3$, 5; $m=5$, $n=7$. No new P_n were found.

[316]Oeuvres, 2, 1894, p. 247 (261), letter to Carcavi; Varia opera, p. 178; Précis des oeuvres math. de Fermat, par E. Brassinne, Toulouse, 1853, p. 150.

[317]Oeuvers de Fermat, 2, 1894, 255, letter to Mersenne, April 7, 1643. The editors (p. 256, note) explained the method of factoring probably used by Fermat. The sum of the aliquot parts of 2^{36} is $223N$, where $N = 616318177$, and the sum of the aliquot parts of N is $2\cdot 7^7 M$ $M = 898423$. As M does not occur elsewhere in P_6, it is to be expected as a factor of the final factor of P_6.

[318]Manuscript published by C. Henry, Bull. Bibl. Storia Sc. Mat. e Fis., 12, 1879. 714.

[319]Thoréie des nombres, 3d ed., vol. 2, Paris, 1830, 146–7; German transl. by H. Maser, Leipzig, 2, 1893, 141–3. The work for $m=3$ was reproduced by Lucas[320] without reference.

E. Lucas[320] gave a table of P_m of the form $2^{n-1}(2^n-1)N$ which includes only 15 of the 26 P_m given above and no additional P_m, $m>2$, except two erroneous P_5:

$$2^{10}3^45\cdot7^2\cdot11^2\cdot19\cdot23\cdot89, \qquad 2^{17}5\cdot7^213\cdot19^237\cdot73\cdot127,$$

attributed elsewhere[321] by him to Fermat. If we replace 7^2 by 7 in the former, we obtain a correct P_5 listed by Carmichael:[332]

$$P_5^{(7)}=2^{10}3^45\cdot7\cdot11^219\cdot23\cdot89.$$

If in the second, we replace $5\cdot7^2$ by $3^5\cdot5\cdot7^3$ we obtain Fermat's $P_5^{(6)}$.

A. Desboves[322] noted that 120 and 672 are the only P_3 of the form $2^n\cdot3\cdot p$, where p is a prime.

D. N. Lehmer[323] gave the additional P_m:

$$P_4^{(12)}=2^23^25\cdot7^213\cdot19,$$
$$P_4^{(13)}=2^83^27^213\cdot19^237\cdot73\cdot127,$$
$$P_5^{(8)}=2^{21}3^65^27\cdot19\cdot23^231\cdot79\cdot89\cdot137\cdot547\cdot683\cdot1093,$$
$$P_6^{(4)}=2^{19}3^65^37^211\cdot13\cdot19\cdot23\cdot31\cdot41\cdot137\cdot547\cdot1093,$$
$$P_6^{(5)}=2^{24}3^85\cdot7^211\cdot13\cdot17\cdot19^231\cdot43\cdot53\cdot127\cdot379\cdot601\cdot757\cdot1801.$$

He readily proved that a P_3 contains at least 3 distinct prime factors, a P_4 at least 4, a P_5 at least 6, a P_6 at least 9, a P_7 at least 14.

J. Westlund[324] proved that $2^33\cdot5$ and $2^53\cdot7$ are the only P_3 of the form $p_1{}^ap_2{}^bp_3$, where the p's are primes and $p_1<p_2<p_3$. He[325] proved that the only $P_3=p_1{}^ap_2p_3p_4$, $p_1<p_2<p_3<p_4$, is $P_3^{(3)}=2^93.11.31$.

A. Cunningham[326] considered P_m of the form $2^{q-1}(2^q-1)F$, where F is to be suitably determined. There exists at least one such P_m for every q up to 39, except 33, 35, 36, and one for $q=45$, 51, 62. Of the 85 P_m found, the only one published is the largest one, viz., for $q=62$, giving $P_6^{(6)}$ with

$$F=3^75^47^211\cdot13\cdot19^223\cdot59\cdot71\cdot79\cdot127\cdot157\cdot379\cdot757\cdot43331\cdot3033169;$$

while none have $m>6$, and for $m=3$ at most one has a given q. He found in 1902 (but did not publish) the two $P_7=2^{46}(2^{47}-1)F$, where

$$F=C\cdot19^2127 \text{ or } C\cdot19^4151\cdot911,$$
$$C=3^{15}\cdot5^3\cdot7^5\cdot11\cdot13\cdot17\cdot23\cdot31\cdot37\cdot41\cdot43\cdot61\cdot89\cdot97\cdot193\cdot442151.$$

R. D. Carmichael[328] has shown that there exists no odd P_m with only three distinct prime factors; that $2^33\cdot5$ and $2^53\cdot7$ are the only P_m with only

[320]Bull. Bibl. e Storia Mat. e Fis., 10, 1877, 286. In $2^53\cdot5\cdot7$, listed as a P_4, 3 is a misprint for 3^3.

[321]Lucas, Théorie des Nombres, 1, Paris, 1891, 380. Here the factor 11^3 13^3 of Fermat's $P_6^{(1)}$ is given erroneously as $11\cdot13^2$, while the $P_6^{(1)}$ of Descartes is attributed to Fermat.

[322]Questions d'Algèbre, 2d ed., 1878, p. 490, Ex. 24.

[323]Annals of Math., (2), 2, 1900–1, 103–4.

[324]Annals of Math., (2), 2, 1900–1, 172–4.

[325]Annals of Math., (2), 3, 1901–2, 161–3.

[326]British Association Reports, 1902, 528–9.

[328]American Math. Monthly, 13, Feb., 1906, 35–36.

three distinct prime factors;[329] that those with only four distinct prime factors are[330] the $P_3{}^{(3)}$ of St. Croix[306] and the $P_4{}^{(1)}$ of Descartes;[309] and that the even P_m with five[331] distinct prime factors are $P_3{}^{(4)}$, $P_4{}^{(2)}$, $P_4{}^{(3)}$ of Descartes[308, 309] and $P_4{}^{(8)}$ of Mersenne.[314] Carmichael[331a] stated and J. Westlund proved that if $n>4$, no P_n has only n distinct prime factors.

Carmichael's[332] table of multiply perfect numbers contains the misprint 1 for the final digit 0 of Descartes' $P_4{}^{(3)}$, and the erroneous entry 919636480 in place of its half, viz., $P_3{}^{(5)}$ of Mersenne.[314] The only new P_m is

$$P_6{}^{(7)} = 2^{15}3^5 5^2 7^2 11 \cdot 13 \cdot 17 \cdot 19 \cdot 31 \cdot 43 \cdot 257.$$

All $P_m < 10^9$ were determined; only known ones were found.

Carmichael[333] gave an erroneous P_5 and the new P_4:

$$P_4{}^{(14)} = 2^{14}3^2 7^2 13 \cdot 19^2 31 \cdot 127 \cdot 151,$$
$$P_4{}^{(15)} = 2^{25}3^3 5^2 19 \cdot 31 \cdot 683 \cdot 2731 \cdot 8191,$$
$$P_4{}^{(16)} = 2^{25}3^6 5 \cdot 19 \cdot 23 \cdot 137 \cdot 547 \cdot 683 \cdot 1093 \cdot 2731 \cdot 8191.$$

Carmichael and T. E. Mason[334] gave a table which includes the above listed 10 P_2, 6 P_3, 16 P_4, 8 P_5, 7 P_6, together with 204 new multiply perfect numbers P_i $(i=3,\ldots, 7)$. Of the latter, 29 are of multiplicity 7, each having a very large number of prime factors. No P_7 had been previously published.

[As a generalization, consider numbers n the sum of the kth powers of whose divisors $<n$ is a multiple of n. For example, $n=2p$, where p is a prime $8h \pm 3$ and k is such that 2^k+1 is divisible by p; cases are $p=3$, $k=1$; $p=5$, $k=2$; $p=11$, $k=5$; $p=13$, $k=6$.]

AMICABLE NUMBERS.

Two numbers are called amicable* if each equals the sum of the aliquot divisors of the other.

According to Iamblichus[4] (pp. 47–48), "certain men steeped in mistaken opinion thought that the perfect number was called love by the Pythagoreans on account of the union of different elements and affinity which exists in it; for they call certain other numbers, on the contrary, amicable numbers, adopting virtues and social qualities to numbers, as 284 and 220, for the parts of each have the power to generate the other, according to the rule of friendship, as Pythagoras affirmed. When asked what is a friend, he replied, 'another I,' which is shown in these numbers. Aristotle so defined a friend in his Ethics."

[329]Annals of Math., (2), 7, 1905–6, 153; 8, 1906–7, 49–56; 9, 1907–8, 180, for a simpler proof that there is no $P_3 = p_1{}^a p_2{}^b p_4{}^c$, $c>1$.
[330]Annals of Math., (2), 8, 1906–7, 149–158.
[331]Bull. Amer. Math. Soc., 15, 1908–9, pp. 7–8. Fr. transl., Sphinx-Oedipe, Nancy, 5, 1910, 164–5.
[331a]Amer. Math. Monthly, 13, 1906, 165.
[332]Bull. Amer. Math. Soc., 13, 1906–7, 383–6. Fr. transl., Sphinx-Oedipe, Nancy, 5, 1910, 161–4.
[333]Sphinx-Oedipe, Nancy, 5, 1910, 166.
[334]Proc. Indiana Acad. Sc., 1911, 257–270.
*Amiable, agreeable, befreundete, verwandte.

In the ninth century the Arab Thâbit ben Korrah[10] (prop. 10) noted that $2^n ht$ and $2^n s$ are amicable numbers if

(1) $\qquad h = 3 \cdot 2^n - 1, \qquad t = 3 \cdot 2^{n-1} - 1, \qquad s = 9 \cdot 2^{2n-1} - 1$

are primes > 2, literally, if

$h = z + 2^n, \quad t = z - 2^{n-1}, \quad z = 1 + 2 + \ldots + 2^n, \quad s = (2^{n+1} + 2^{n-2})2^{n+1} - 1.$

The term used for amicable numbers was *se invicem amantes*. In the article in which F. Woepcke[10] translated this Arabic manuscript into French, he noted that a definition of these numbers, called *congeneres*, occurs in the 51st treatise (on arithmetic) of Ikhovân Alçafâ, manuscript 1105, anciens fonds arabes, p. 15, of the National Library of Paris.

Among Jacob's presents to Esau were 200 she-goats and 20 he-goats, 200 ewes and 20 rams (Genesis, XXXII, 14). Abraham Azulai[349] (1570–1643), in commenting on this passage from the Bible, remarked that he had found written in the name of Rau Nachshon (ninth century A. D.): Our ancestor Jacob prepared his present in a wise way. This number 220 (of goats) is a hidden secret, being one of a pair of numbers such that the parts of it are equal to the other one 284, and conversely. And Jacob had this in mind; this has been tried by the ancients in securing the love of kings and dignatories.

Ibn Khaldoun[350] related "that persons who have concerned themselves with talismans affirm that the amicable numbers 220 and 284 have an influence to establish a union or close friendship between two individuals. To this end a theme is prepared for each individual, one during the ascendency of Venus, when that planet is in its exaltation and presents to the moon an aspect of love or benevolence; for the second theme the ascendency should be in the seventh. On each of these themes is written one of the specified numbers, the greater (or that with the greater sum of its aliquot parts?) being attributed to the person whose friendship is sought."

The Arab El Madschrîtî,[351] or el-Maǧrîtî, (†1007) of Madrid related that he had himself put to the test the erotic effect of "giving any one the smaller number 220 to eat, and himself eating the larger number 284."

Ibn el-Hasan[351a] (†1320) wrote several works, including the "Memory of Friends," on the explanation of amicable numbers.

Ben Kalonymos[351b] discussed amicable numbers in 1320 in a work written for Robert of Anjou, a fragment of which is in Munich (Hebr. MS. 290, f. 60). A knowledge of amicable numbers was considered necessary by Jochanan Allemanno (fifteenth century) to determine whether an aspect of the planets was friendly or not.

---

[349]Baale Brith Abraham [Commentary on the Bible], Wilna, 1873, 22. Quotation supplied by Mr. Ginsburg.

[350]Prolégoménes hist. d'Ibn Khaldoun, French transl. by De Slane, Notices et Extraits des Manuscrits de la Bibl. Impériale, Paris, 21, I, 1868, 178–9.

[351]Manuscript Maǧriti; Steinschneider, Zur pseudoepigraphischen Literatur inbesondere der geheimen Wissenschaften des Mittelalters, Berlin, 1862, p. 37 (cf. p. 41).

[351a]H. Suter, Abh. Gesch. Math. Wiss., 10, 1900, 159, § 389.

[351b]Hebr. Bibl., VII, 91. Steinschneider, Zeitschrift der Morgenländischen Ges., 24, 1870, 369.

Alkalacadi,[352] a Spanish Arab (†1486), showed the method of finding the least amicable numbers 220, 284.

Nicolas Chuquet[15] in 1484 and de la Roche[28] in 1538 cited the amicable numbers 220, 284, "de merueilleuse familiarite lung auec laultre." In 1553, Michael Stifel[32] (folios 26v–27v) mentioned only this pair of amicable numbers. The same is true of Cardan,[27] of Peter Bungus[42] (Mysticae numerorum signif., 1585, 105), and of Tartaglia.[353] Reference may be made also to Schwenter.[52]

In 1634 Mersenne[301] (p. 212) remarked that "220 and 284 can signify the perfect friendship of two persons since the sum of the aliquot parts of 220 is 284 and conversely, as if these two numbers were only the same thing."

According to Mersenne's[304] statement in 1636, Fermat[354] found the second pair of amicable numbers

$$17296 = 2^4 \cdot 23 \cdot 47, \qquad 18416 = 2^4 \cdot 1151,$$

and communicated to Mersenne[305] the general rule: Begin with the geometric progression 2, 4, 8, ..., write the products by 3 in the line below; subtract 1 from the products and enter in the top row. The bottom row is $6 \cdot 12 - 1$, $12 \cdot 24 - 1$, ... When a number of the last row is a prime (as 71) and the one (11) above it in the top row is a prime,

5	11	23	47
2	4	8	16
6	12	24	48
	71	287	1151

and the one (5) preceding that is also a prime, then $71 \cdot 4 = 284$, $5 \cdot 11 \cdot 4 = 220$ are amicable. Similarly for

$$1151 \cdot 16 = 18416, \qquad 23 \cdot 47 \cdot 16 = 17296,$$

and so to infinity. [The rule leads to the pair $2^n ht$, $2^n s$, where h, t, s are given by (1).]

Descartes[355] gave the rule: Take (2 or) any power of 2 such that its triple less 1, its sextuple less 1, and the 18-fold of its square less 1 are all primes;* the product of the last prime by the double of the assumed power of 2 is one of a pair of amicable numbers. Starting with the powers 2, 8, 64, we get 284, 18416, 9437056, whose aliquot parts make 220, etc. Thus the third pair is

$$9363584 = 2^7 \cdot 191 \cdot 383, \qquad 9437056 = 2^7 \cdot 73727.$$

Descartes[356] stated that Fermat's rule agrees exactly with his own.

Although we saw that Mersenne quoted in 1637 the rule in Fermat's form and expressly attributed it to Fermat, curiously enough Mersenne[314] gave in 1639 the rule in Descartes' form, attributing it to "un excellent Géomètre" (meaning without doubt Descartes, according to C. Henry[357]),

---

[352]Manuscript in Bibliothèque Nationale Paris, a commentary on the arithmetic Talkhys of Ibn Albanna (13th cent.). Cf. E. Lucas, L'arithmétique amusante, Paris, 1895, p. 64.

[353]Quesiti et Inventione, 1554, fol. 98 v.

[354]Oeuvres de Fermat, 2, 1894, p. 72, letter to Roberval, Sept. 22, 1636; p. 208, letter to Frenicle, Oct. 18, 1640.

[355]Oeuvres de Descartes, 2, 1898, 93–94, letter to Mersenne, Mar. 31, 1638.

*Evidently the numbers (1) if the initial power of 2 be 2^{n-1}.

[356]Oeuvres de Descartes, 2, 1898, 148, letter to Mersenne, May 27, 1638.

[357]Bull. Bibl. Storia Sc. Mat. e Fis., 12, 1879, 523.

and derived as did Descartes the first three pairs of amicable numbers from 2, 8, 64. We shall see that various later writers attributed the rule to Descartes.

Mersenne[60] again in 1644 gave the above three pairs of amicable numbers, the misprints in both[358] of the numbers of the third pair being noticed at the end of his book, and stated there are others innumerable.

Mersenne[61] in 1647 gave without citation of his source the rule in the form $2 \cdot 2^n s$, $2 \cdot 2^n h t$, where $t = 3 \cdot 2^n - 1$, $h = 2t + 1$, $s = ht + h + t$ are primes [as in (1)].

Frans van Schooten,[359] the younger, showed how to find amicable numbers by indeterminate analysis. Consider the pair $4x$, $4yz$ [x, y, z odd primes]; then

$$7 + 3x = 4yz, \qquad 7 + 7y + 7z + 3yz = 4x.$$

Eliminating x, we get $z = 3 + 16/(y - 3)$. The case $y = 5$ gives $z = 11$, $x = 71$, yielding 284, 220. He proved that there are none of the type $2x$, $2yz$, or $8x$, $8yz$, and argued that no pair is smaller than 284, 220. For $16x$, $16yz$, he found $z = 15 + 256/(y - 15)$, which for $y = 47$ yields the second known pair. There are none of the type $32x$, $32yz$, or type $64x$, $64yz$. For $128x$, $128yz$, he got $z = 127 + 16384/(y - 127)$, which for $y = 191$ yields the third known pair. Finally, he quoted the rule of Descartes.

W. Leybourn[68] stated in 1667 that "there is a fine harmony between these two numbers 220 and 284, that the aliquot parts of the one do make up the other . . . and this harmony is not to be found in many other numbers."

In 1696, Ozanam[73] gave in great detail the derivation of the three known pairs of "amiable" numbers by the rule as stated by Descartes, whose name was not cited. Nothing was added in the later editions.[79, 99]

Paul Halcke[360] gave Stifel's[32] rule, as expressed by Descartes.[355]

E. Stone[361] quoted Descartes' rule in the incorrect form that $2^{2n}pq$ and $3 \cdot 2^n p$ are amicable if $p = 3 \cdot 2^n - 1$ and $q = 6 \cdot 2^n - 1$ are primes.

Leonard Euler[362] remarked that Descartes and van Schooten found only three pairs of amicable numbers, and gave, without details, a list of 30 pairs, all included in the later paper by Euler.[364]

G. W. Kraft[363] considered amicable numbers of the type APQ, AR, where P, Q, R are primes not dividing A. Let a be the sum of all the divisors of A. Then

$$R + 1 = (P + 1)(Q + 1), \qquad (R + 1)a = APQ + AR.$$

Assuming prime values of P and Q such that the resulting R is prime, he sought a number A for which A/a has the derived value. For $P = 3$, $Q = 11$,

---

[358]Not noticed in the correction (left in doubt) in Oeuvres de Fermat, 4, 1912, p. 250 (on pp. 66–7). One error is noted in Broscius[54], Apologia, 1652, p. 154.

[359]Exercitationum mathematicarum libri quinque, Ludg. Batav., 1657, liber V: sectiones triginta miscellaneas, sect. 9, 419–425. Quoted by J. Landen.[88]

[360]Deliciae Mathematicae, oder Math. Sinnen-Confect, Hamburg, 1719, 197–9.

[361]New Mathematical Dictionary, 1743 (under amicable).

[362]De numeris amicabilibus, Nova Acta Eruditorum, Lipsiae, 1747, 267–9; Comm. Arith. Coll., II, 1849, 637–8.

[364]Novi Comm. Ac. Petrop., 2, 1751, ad annum 1749, Mem., 100–18.

then $A:a=3:5$; he took $A=3B$, 3^2B, 3^3B, but found no solution. For $P=5$, $Q=41$, we have $R=251$, $38A=21a$; set $A=49B$, whence $3\cdot57b=38\cdot7B$, where b is the sum of the divisors of B; set $B=9C$, whence $C:c=13:14$, $C=13$, yielding the amicable numbers $5\cdot41A$, $251A$, where $A=3^2\cdot7^213=5733$ [the pair VII in Euler's[362] list and (7) in the table below]. Again, to make $A/a=3/8$, set $A=3B$, whence $a=4b$ and the condition is $b=2B$, whence B is a perfect number prime to 3. Using $B=28$, we get $A=84$. For use in such questions, Kraft gave a table of the sum of the divisors of each number ≤150. He quoted the rule of Descartes.

L. Euler[364] obtained, in addition to two special pairs, 62 pairs [including two false pairs] of amicable numbers of the type am, an, in which the common factor a is relatively prime to both m and n. He wrote $\int m$ for the sum of all the divisors of m. The conditions are therefore

$$\int m = \int n, \qquad \int a \cdot \int m = a(m+n).$$

If m and n are both primes, then $m=n$ and we have a repeated perfect number. Euler treated five problems.

(1) Euler's problem 1 is to find amicable numbers apq, ar, where p, q, r, are distinct primes not dividing the given number a. From the first condition we have $r=xy-1$, where $x=p+1$, $y=q+1$. From the second,

$$xy \int a = a(2xy-x-y).$$

Let $a/(2a-\int a)$ equal b/c, a fraction in its lowest terms. Then

$$y=bx/(cx-b), \qquad (cx-b)(cy-b)=b^2.$$

Thus x and y are to be found by expressing b^2 as a product of two factors, increasing each by b, and dividing the results by c.

(1_1) First, take $a=2^n$. Then $b=2^n$, $c=1$, x, $y=2^{n\pm k}+2^n$. Let $n-k=m$. Then

$$p=2^m(2^{2k}+2^k)-1, \qquad q=2^m(1+2^k)-1, \qquad r=2^{2m}(2^{2k+1}+2^{3k}+2^k)-1.$$

When these three are primes, $2^{m+k}pq$ and $2^{m+k}r$ are amicable. Euler noted that the rule communicated by Descartes to van Schooten is obtained by taking $k=1$, and stated that 1, 3, 6 are the only values ≤8 of m which yield amicable numbers (above[355]). For $k=2$ or 4, Euler remarked that r is divisible by 3; for $k=3$, $m<6$, and for $k=5$, $m\leq2$, p, q, or r is composite.

(1_2) Take $a=2^nf$, where $f=2^{n+1}+e$ is a prime. Then $2a-\int a=e+1$. If $e+1$ divides a, we have $c=1$. Set $e+1=2^k$, $k\leq m$, $n=m+k$. Then

$$f=2^k(2^{m+1}+1)-1, \qquad a=2^{m+k}f, \qquad b=2^mf, \qquad b^2=(x-b)(y-b).$$

For $k=1$, $f=2^{m+2}+1$ is to be a prime, whence $m+2$ is a power of 2. If $m=0$, $b=f=5$, and either $x=y$, $p=q$; or x, $y=6$, 30; p, $q=5$, 29, whereas p and q are to be distinct and prime to 10. If $m=2$, $f=17$, 68^2 is to be resolved into distinct even factors; in the four resulting cases, p, q, r are

---

[364]De numeris amicabilibus, Opuscula varii argumenti, 2, 1750, 23–107, Berlin; Comm. Arith., 1, 1849, 102–145. French transl. in Sphinx-Oedipe, Nancy, 1, 1906-7, Supplément I–LXXVI.

not all prime. In the next case $m=6$, $f=257$, Euler examined only the case* $x-b=2^5 \cdot 257$, finding q composite.

For $k=2$, Euler excluded $m=1, 3$ [$m=4$ is easily excluded].

(1_3) For $k \geqq n$ in (1_2), $c=2^m$, where $m=k-n$. Then

$$b = 2^{n+1} + 2^{m+n} - 1 = f$$

must be a prime. Thus we must take as the factors of b^2

$$2^m x - b = 1, \qquad 2^m y - b = b^2,$$

whence $x = 2^n + 2^{n+1-m}$, $y = bx$. If $m=1$, one of

$$f = 2^{n+2} - 1, \qquad p = 2^{n+1} - 1$$

has the factor 3 and yet must be a prime; hence $n=1$, $q=27$. If $m=2$, Euler treated the cases $n \leqq 5$ and found (for $n=2$) the pair (4) of the table. [For $6 \leqq n \leqq 17$, f or p is composite.] For m odd and >1, f or p has the factor 3. For $m=4$, $n \leqq 17$, no solution results.

(1_4) For $a = 2^n(g-1)(h-1)$, where the last two factors are prime, set $d = 2a - \int a$. Then

$$(g - 2^{n+1})(h - 2^{n+1}) = d - 2^{n+1} + 2^{2n+2}.$$

Euler treated the cases $n \leqq 3$, $d = 4, 8, 16$, finding only the pair (9).

(1_5) Special odd values of a led (§§56–65) to seven pairs (5)–(8), (11)–(13). The cases $a = 3^3 \cdot 5$, $3^2 \cdot 7^2 \cdot 13 \cdot 19$ were unfruitful.

(2) Euler's problem 2 is to find amicable numbers apq, ars, where p, q, r, s are distinct primes not dividing the given number a. Since $\int p \cdot \int q = \int r \cdot \int s$, we may set

$$p = ax - 1, \qquad q = \beta y - 1, \qquad r = \beta x - 1, \qquad s = ay - 1.$$

We set $\int a : a = 2b - c : b$, where b and c are relatively prime. The second condition $\int a \cdot \int pq = a(pq + rs)$ gives

$$ca\beta xy = b(a+\beta)(x+y) - 2b.$$

Multiply it by $ca\beta$. Then

$$[ca\beta x - b(a+\beta)][ca\beta y - b(a+\beta)] = b^2(a+\beta)^2 - 2bca\beta.$$

Given a, β, a and hence b, c, we are to express the second member as a product of two factors and then find x, y.

For $a=1$, $\beta=3$, $a=2^n$, Euler obtained the pairs (a), (28). For $a=2$, $\beta=3$, $a=3^2 \cdot 5 \cdot 13$, he got (32); for $a=1$, $\beta=4$, $a=3^3 \cdot 5$, (30). The ratio $a:\beta$ may be more complex, as $5:21$ or $1:102$, in (γ). As noted by K. Hunrath,[364a] the numbers (γ) are not amicable. Nor are the ratios as given, although these ratios result if we replace 8563 by $8567 = 13 \cdot 659$. This false pair occurs as XIII in Euler's[362] list.

(3) Problem 3 is derived from problem 2 by replacing s by a number f not necessarily prime. Let h be the greatest common divisor of $\int f = hg$ and $p+1 = hx$. Then $r+1 = xy$, $q+1 = gy$. Also

$$ghxy \int a \equiv \int (afr) = a(pq + fr) \equiv a\{(hx-1)(gy-1) + f(xy-1)\}.$$

---

*All the remaining cases are readily excluded.
[364a]Bibliotheca Math., (3), 10, 1909–10, 80–81.

Multiply by b/a and replace $b\int a$ by $2ab-ac$ [see case (1)]. Thus

$$exy-bhx-bgy=b(f-1), \qquad e\equiv bf-bgh+cgh.$$

Thus $L\equiv b^2gh+be(f-1)$ is to be expressed as the product PQ of two factors and they are to be equated to $ex-bg$, $ey-bh$. The case $a=2$ is unfruitful.

(3_1) Let $a=4$. Then $b=4$, $c=1$, $e=4f-3gh$. The case $f=3$ is excluded since it gives $e=0$. For $f=5$, $g=2$, $h=3$, we again get (a) and also (β). For $f=5$, $g=1$, $h=6$, we get only the same two pairs. For a prime $f\geqq 7$, no new solutions are found. For $f=5\cdot13$, (51) results.

(3_2) Let $a=8$, whence $b=8$, $c=1$. The cases $f=11$, 13 are fruitless, while $f=17$ yields (16). The least composite f yielding solutions is $11\cdot23$, giving (44), (45), (46). This fruitful case led Euler to the more convenient notations (§88) $M=hP$, $N=gQ$, $L=PQ$. The problem is now to resolve $L\int f$ into two factors, M, N, such that

$$p=\frac{M+b\int f}{e}-1, \qquad q=\frac{N+b\int f}{e}-1$$

are integers and primes, while in $r+1=(p+1)(q+1)/\int f$, r is a prime.

(3_3) Let $a=16$. For $f=17$, we obtain the pairs (21), (22); for $f=19$, (23); for $f=23$, (17), (19), (20); for $f=47$, (18); for $f=17\cdot167$, (49). Cases $f=31$, $17\cdot151$ are fruitless [the last since 129503 has the factor 11, not noticed by Euler].

(3_4) For $a=3^3\cdot5$ or $3^2\cdot7\cdot13$, $b=9$, $c=2$; the first a with $f=7$ yields (30).

(4) Problem 4 relates to amicable numbers $agpq$, ahr, where p, q, r are primes. Eventually he took also g and h as primes. We may then set $g+1=km$, $h+1=kn$. For $m=1$, $n=3$, $a=4$ or 8, no amicables are found. For $m=3$, $n=1$, the cases $a=10$, $k=8$ and $a=3^3\cdot5$, $k=8$, yield (38), (55).

(5) Euler's final problem 5 is of a new type. He discussed amicable numbers zap, zbq, where a and b are given numbers, p and q are unknown primes, while z is unknown but relatively prime to a, b, p, q. Set $\int a:\int b=m:n$, where m and n are relatively prime. Since $(p+1)\int a=(q+1)\int b$, we may set $p+1=nx$, $q+1=mx$. The usual second condition gives

$$nx\int a\cdot\int z=za(nx-1)+zb(mx-1), \qquad \frac{z}{\int z}=\frac{nx\int a}{(na+mb)x-a-b}.$$

Let the latter fraction in its lowest terms be r/s. Then $z=kr$, $\int z=ks$. Since $\int(kr)\geqq k\int r$, we have $s\geqq\int r$. Hence we have the useful theorem: if $z:\int z=r':s'$, $s'<\int r'$, then r' and s' have a common factor >1.

(5_1) The unfruitful case $a=3$, $b=1$, was treated like the next.

(5_2) Let $a=5$, $b=1$, whence $m=6$, $n=1$, $z:\int z=6x:11x-6$. By the theorem in (5), x must be divisible by 2 or 3. Euler treated the cases $x=3(3t+1)$, $x=2(2t+1)$. But this classification is both incomplete and overlapping. Since $p=x-1$ is to be prime, x is even (since $x=3$ makes z divisible by $p=2$). Hence $x=2P, z: \int z=6P:11P-3$. By the theorem in (5), $6P$ and $11P-3$ have a common factor 2 or 3, so that P is either odd or divisible by 6. For $P=6l$, the ratio is that of $12b$ to $22l-1$, which as before must have the common factor 3, whence $l=3t+1$. Then $z: \int z=4(3t+1):22t+7$, a ratio of relatively prime numbers, whence $22t+7\geqq \int 4(3t+1)$, and hence $t=2k$, $k=0$ or $k>3$. For $k=0$, we obtain the pair 220, 284. The next value >3 of k for which $p=x-1$ and $q=6x-1$ are primes is $k=6$, giving $p=443$, $q=2663$, numbers much larger than those in the (unnecessary) cases treated by Euler. Then $z: \int z=4\cdot37:271$; set $z=37^e d$, d not divisible by 37; the cases $e=1, 2, 3$ are excluded by the theorem in (5). For the remaining case P odd, $P=2Q+1$, Euler treated those values $\leqq 100$ of Q, and also $Q=244$, for which p and q are primes and obtained the pair in (1_3), two pairs in (1_5), and (14), (15).

(5_3) Euler treated in §§112–7 various sets a, b, and obtained (a) and nine new pairs given in the table.

In the following table of the 64 pairs of amicable numbers obtained by Euler, the numbering of any pair is the same as in Euler's list, but the pairs have been rearranged so that it becomes easy to decide if any proposed pair is one of Euler's. As noted by F. Rudio,[364b] (37) contained the misprint 3^3 for 3^2, while (γ) and (34) are erroneous, 220499 being composite (311·709); he checked that all other entries are correct.

(39) $2\cdot5\begin{Bmatrix}7\cdot19\cdot107\\47\cdot359\end{Bmatrix}$	(38) $2\cdot5\begin{Bmatrix}7\cdot60659\\23\cdot29\cdot673\end{Bmatrix}$	(1) $2^2\begin{Bmatrix}5\cdot11\\71\end{Bmatrix}$	(51) $2^2\begin{Bmatrix}5\cdot13\cdot1187\\43\cdot2267\end{Bmatrix}$
(4) $2^2\cdot23\begin{Bmatrix}5\cdot137\\827\end{Bmatrix}$	(a) $2^2\begin{Bmatrix}5\cdot131\\17\cdot43\end{Bmatrix}$	(β) $2^3\begin{Bmatrix}5\cdot251\\13\cdot107\end{Bmatrix}$	(29) $2^2\cdot11\begin{Bmatrix}17\cdot263\\43\cdot107\end{Bmatrix}$
(9) $2^2\cdot13\cdot17\begin{Bmatrix}389\cdot509\\198899\end{Bmatrix}$	(46) $2^3\begin{Bmatrix}11\cdot23\cdot1619\\647\cdot719\end{Bmatrix}$	(45) $2^3\begin{Bmatrix}11\cdot23\cdot1871\\467\cdot1151\end{Bmatrix}$	(44) $2^3\begin{Bmatrix}11\cdot23\cdot2543\\383\cdot1907\end{Bmatrix}$
(47) $2^3\begin{Bmatrix}11\cdot29\cdot239\\191\cdot449\end{Bmatrix}$	(43) $2^3\begin{Bmatrix}11\cdot59\cdot173\\47\cdot2609\end{Bmatrix}$	(40) $2^3\begin{Bmatrix}11\cdot163\cdot191\\31\cdot11807\end{Bmatrix}$	(16) $2^3\begin{Bmatrix}17\cdot79\\23\cdot59\end{Bmatrix}$
(48) $2^3\begin{Bmatrix}17\cdot4799\\29\cdot47\cdot59\end{Bmatrix}$	(60) $\begin{Bmatrix}2^3\cdot19\cdot41\\2^5\cdot199\end{Bmatrix}$	(61) $\begin{Bmatrix}2^3\cdot41\cdot467\\2^5\cdot19\cdot233\end{Bmatrix}$	(49) $2^4\begin{Bmatrix}17\cdot167\cdot13679\\809\cdot51071\end{Bmatrix}$
(21) $2^4\begin{Bmatrix}17\cdot5119\\239\cdot383\end{Bmatrix}$	(22) $2^4\begin{Bmatrix}17\cdot10303\\167\cdot1103\end{Bmatrix}$	(23) $2^4\begin{Bmatrix}19\cdot1439\\149\cdot191\end{Bmatrix}$	
(γ) $2^4\begin{Bmatrix}19\cdot8563\\83\cdot2039\end{Bmatrix}$ (false)	(2) $2^4\begin{Bmatrix}23\cdot47\\1151\end{Bmatrix}$	(50) $2^4\begin{Bmatrix}23\cdot47\cdot9767\\1583\cdot7103\end{Bmatrix}$	
(20) $2^4\begin{Bmatrix}23\cdot467\\103\cdot107\end{Bmatrix}$	(19) $2^4\begin{Bmatrix}23\cdot479\\89\cdot127\end{Bmatrix}$	(17) $2^4\begin{Bmatrix}23\cdot1367\\53\cdot607\end{Bmatrix}$	(36) $2^4\cdot67\begin{Bmatrix}37\cdot2411\\227\cdot401\end{Bmatrix}$
(18) $2^4\begin{Bmatrix}47\cdot89\\53\cdot79\end{Bmatrix}$	(25) $2^5\begin{Bmatrix}37\cdot12671\\227\cdot2111\end{Bmatrix}$	(26) $2^5\begin{Bmatrix}53\cdot10559\\79\cdot7127\end{Bmatrix}$	(24) $2^5\begin{Bmatrix}59\cdot1103\\79\cdot827\end{Bmatrix}$
(27) $2^6\begin{Bmatrix}79\cdot11087\\383\cdot2309\end{Bmatrix}$	(3) $2^7\begin{Bmatrix}191\cdot383\\73727\end{Bmatrix}$	(28) $2^8\begin{Bmatrix}383\cdot9203\\1151\cdot3067\end{Bmatrix}$	
(37) $3^2\cdot5\begin{Bmatrix}7\cdot11\cdot29\\31\cdot89\end{Bmatrix}$	(5) $3^2\cdot7\cdot13\begin{Bmatrix}5\cdot17\\107\end{Bmatrix}$	(7) $3^2\cdot7^2\cdot13\begin{Bmatrix}5\cdot41\\251\end{Bmatrix}$	(52) $3^2\cdot7\cdot13\begin{Bmatrix}5\cdot17\cdot1187\\131\cdot971\end{Bmatrix}$
(15) $3^2\cdot7\cdot13\cdot41\cdot163\begin{Bmatrix}5\cdot977\\5867\end{Bmatrix}$	(14) $3^2\cdot7^2\cdot13\cdot97\begin{Bmatrix}5\cdot193\\1163\end{Bmatrix}$	(10) $3^2\cdot5\cdot19\cdot37\begin{Bmatrix}7\cdot897\\7103\end{Bmatrix}$	
(35) $3^2\cdot5\cdot19\begin{Bmatrix}7\cdot227\\37\cdot47\end{Bmatrix}$	(8) $3^2\cdot5\cdot7\begin{Bmatrix}53\cdot1889\\102059\end{Bmatrix}$	(6) $3^2\cdot5\cdot13\begin{Bmatrix}11\cdot19\\239\end{Bmatrix}$	

(31) $3^2 \cdot 5 \cdot 13 \begin{Bmatrix} 11 \cdot 199 \\ 29 \cdot 79 \end{Bmatrix}$ (54) $3^2 \cdot 5^2 \begin{Bmatrix} 11 \cdot 59 \cdot 179 \\ 17 \cdot 19 \cdot 359 \end{Bmatrix}$ (13) $3^2 \cdot 5 \cdot 13 \cdot 19 \begin{Bmatrix} 29 \cdot 569 \\ 17099 \end{Bmatrix}$

(33) $3^2 \cdot 5 \cdot 13 \cdot 19 \begin{Bmatrix} 37 \cdot 1583 \\ 227 \cdot 263 \end{Bmatrix}$ (32) $3^2 \cdot 5 \cdot 13 \begin{Bmatrix} 19 \cdot 47 \\ 29 \cdot 31 \end{Bmatrix}$ (12) $3^2 \cdot 7^2 \cdot 11 \cdot 13 \begin{Bmatrix} 41 \cdot 461 \\ 19403 \end{Bmatrix}$

(41) $3^2 \cdot 7 \cdot 13 \cdot 23 \begin{Bmatrix} 11 \cdot 19 \cdot 367 \\ 79 \cdot 1103 \end{Bmatrix}$ (34) $3^2 \cdot 7^2 \cdot 13 \cdot 19 \begin{Bmatrix} 11 \cdot 220499 \\ 89 \cdot 29399 \end{Bmatrix}$ (false)

(30) $3^2 \cdot 5 \begin{Bmatrix} 7 \cdot 71 \\ 17 \cdot 31 \end{Bmatrix}$ (55) $3^3 \cdot 5 \begin{Bmatrix} 7 \cdot 21491 \\ 17 \cdot 23 \cdot 397 \end{Bmatrix}$ (42) $3^3 \cdot 5 \cdot 23 \begin{Bmatrix} 11 \cdot 19 \cdot 367 \\ 79 \cdot 1103 \end{Bmatrix}$

(11) $3^4 \cdot 5 \cdot 11 \begin{Bmatrix} 29 \cdot 89 \\ 2699 \end{Bmatrix}$ (56) $3^4 \cdot 7 \cdot 11^2 \cdot 19 \begin{Bmatrix} 47 \cdot 7019 \\ 389 \cdot 863 \end{Bmatrix}$ (57) $3^4 \cdot 7 \cdot 11^2 \cdot 19 \begin{Bmatrix} 53 \cdot 6959 \\ 179 \cdot 2087 \end{Bmatrix}$

(53) $3^5 \cdot 7^2 \cdot 13 \cdot 53 \begin{Bmatrix} 11 \cdot 211 \\ 2543 \end{Bmatrix}$ (58) $3^5 \cdot 7^2 \cdot 13 \cdot 19 \begin{Bmatrix} 47 \cdot 7019 \\ 389 \cdot 863 \end{Bmatrix}$ (59) $3^5 \cdot 7^2 \cdot 13 \cdot 19 \begin{Bmatrix} 53 \cdot 6959 \\ 179 \cdot 2087 \end{Bmatrix}$

Euler's final list of 61 pairs did not include the pairs a, β, γ, although he had obtained a four times in the body of his paper, viz., in (2), (3₁), (5₃); β twice in (3₁); γ in (2). Moreover, these three unlisted pairs occur as VIII, IX, and XIII among the 30 pairs in Euler's[362] earlier list, a fact noted on p. XXVI and p. LVIII of the Preface by P. H. Fuss and N. Fuss to Euler's Comm. Arith. Coll., who failed to observe that these three pairs occur in the text of Euler's present paper. Nor did these editors note that the fourth mentioned case of divergence between the two lists is due merely to the misprint[364c] of 57 for 47 in (43) of the present list, so that the correctly printed pair XXVIII of the list of 30 is really this (43) and not a new pair, as supposed by them.

From the fact that Euler obtained in his posthumous tract[97] on amicable numbers the pairs a, β (once on p. 631 and again on p. 633 and finally on p. 635), the editors inferred, p. LXXXI of the Preface, that the tract differs in analysis from the long paper just discussed. But no new pairs are found, while the cases treated on pp. 631–2 are merely problems 1 and 2 of Euler's preceding paper. It is different with p. 634, where Euler started with two numbers like 71 and $5 \cdot 11$ which, by his table, have the same sum, 72, of divisors, and required a number a relatively prime to them such that $71a$ and $55a$ are amicable. The single condition is $72 \int a = (71+55)a$, whence $\int a : a = 7 : 4$. Thus a has the factor 4. If $a = 4b$, where b is odd, then $\int b = b = 1$, and the pair 284, 220 results. The case $a = 8b$ is impossible. This method was used in a special way by Kraft[363] who limited the numbers from which one starts to a prime and a product of two primes.

In the Encyclopédie Sc. Math., I, 3₁, p. 59, note 320, it is stated that this posthumous tract contains four pairs not in Euler's list of 61, two pairs being those of Fermat[354] and Descartes.[355] But these were listed as (2) and (3) by Euler and were obtained by him in case (1₁) and attributed to Descartes.

E. Waring[365] noted that $2^n x$, $2^n yz$ are amicable if

$$ x = \frac{2^n yz - 2^{n+1} + 1}{2^n - 1}, \qquad z = 2^n - 1 + \frac{2^{2n}}{y - 2^n + 1}, $$

where x, y, z are primes and $y - 2^n + 1$ divides 2^{2n}. He cited the first two such pairs of amicable numbers.

[364c]G. Eneström, Bibliotheca Math., (3), 9, 1909, 263.
[365]Meditationes algebraicae, 1770, 201; ed. 3, 1782, 342–3.

The first three pairs were given in an anonymous work.[366]
In 1796, J. P. Grüson[100] (p. 87) gave the usual rule (1) leading to the
three first known amicable pairs (verwandte Zahlen).

A. M. Legendre[367] attributed the rule (1) to Descartes.

G. S. Klügel[368] gave a process leading to the choice of P and Q, left
arbitrary by Kraft.[363] We have $A:a=R+1:PQ+R=2R-P-Q$. Thus
$P+Q=\{R(2A-a)-a\}/A$, while PQ is given by Kraft's second equation.
Hence P and Q are the roots of a quadratic equation. For example, if
$A=4$, then

$$8P, 8Q=R-7\pm\sqrt{R^2-62R-63}.$$

The positive root of $x^2-62x-63=0$ lies between 60 and 61. Thus we
try primes $\geqq 61$ for R, such that $R-7$ is divisible by 8. The first available
R is 71, giving $P=11$, $Q=5$ and the amicable pair 220, 284. In general,
the quantity $a^2R^2+2\beta R+\gamma$ under the radical sign can be made equal to the
square of $aR+p$ (p arbitrary) by choice of R.

John Gough[369] considered amicable numbers ax, ayz, where x, y, z are
distinct primes not dividing a. Let q be the sum of the aliquot divisors
of a. Then

$$a+q+qx=ayz, \qquad x+1=(y+1)(z+1).$$

If $q\leqq a/4$, the first gives $ayz<(1+x)a/4$, while $2y\cdot 2z>x+1$ by the second,
Thus $q>a/4$. Let $a=r^n$, where r is a prime >1. Then $q=(a-1)/(r-1)$,
which with $q>a/4$ implies $a(5-r)>4$, $r=2$ or 3. He proved that $r\neq 3$.
whence $r=2$, the case treated by van Schooten.[359]

J. Struve[103] cited his Osterprogramm, 1815, on amicable numbers.

A. M. Legendre[370] discussed the amicable numbers of the type (1_1) of
Euler[364] (with Euler's m, k replaced by $m-\mu$, μ). Legendre noted that
$r=2^{2m+k}(2^k+1)^2-1$ is of the form s^2-1 and hence composite, if k is even;
also that, if $k=3$, $p=9\cdot 2^{m+3}-1$, $q=9\cdot 2^m-1$, one of which is of the form
s^2-1. He considered the new case $k=7$ and found for $m=1$ that $p=33023$,
$q=257$, $r=8520191$, stating that if r be a prime we have the amicable numbers 2^8pq, 2^8r. This is in fact the case.[371] For $k=1$, we have the ancient
rule (1); he proved that for $n\leqq 15$ it gives only the known three pairs of
amicable numbers.

Paganini[372], at age 16, announced the amicable numbers $1184=2^5.37$,
$1210=2.5.11^2$, not in the list by Euler[364], but gave no indication of the
method of discovery.

[366]Encyclopédie méthodique...Amusemens des Sciences Math. et Phys., nouv. éd., Padoue,
    1793, I, 116. Cf. Les amusemens math., Lille, 1749, 315.
[367]Théorie des nombres, 1798, 463.
[368]Math. Wörterbuch, 1, 1803, 246–252 [5, 1831, 55].
[369]New Series of the Math. Repository (ed., Th. Leybourn), vol. 2, pt. 2, 1807, 34–39. He cited
    Hutton's Math. Dict., article Amicable Numbers, taken from van Schooten[359].
[370]Théorie des nombres, ed. 3, 1830, II, §472; p. 150. German transl. by H. Maser, Leipzig,
    1893, II, p. 145.
[371]Tchebychef, Jour. de Math., 16, 1851, 275; Werke, 1, 90. T. Pepin, Atti Acc. Pont. Nuovi
    Lincei, 48, 1889, 152–6. Kraitchik, Sphinx-Oedipe, 6, 1911, 92. Also by Lehmer's Fac-
    tor Table or Table of Primes.
[372]B. Nicolò I. Paganini, Atti della R. Accad. Sc. Torino, 2, 1866–7, 362. Cf. Cremona's Ital.
    transl. of Baltzer's Mathematik, pt. III.

P. Seelhoff[373] treated Euler's[364] problems 1 and 2 by Euler's methods (though the contrary is implied), and gave about 20 pairs of amicable numbers due to Euler, with due credit for only three pairs. The only new pairs (pp. 79, 84, 89) are

$$3^2 7^2 13 \cdot 19 \cdot 23 \begin{cases} 83 \cdot 1931 \\ 162287 \end{cases} \qquad 2^6 \begin{cases} 139 \cdot 863 \\ 167 \cdot 719. \end{cases}$$

E. Catalan[374] stated empirically that if n_1 is the sum of the divisors $< n$ of n, and n_2 is the sum of the divisors $< n_1$ of n_1, etc., then $n, n_1, n_2, \ldots$ have a limit λ, where λ is unity or a perfect number.

J. Perrott[375] [Perott] noted that there is no limit for $n = 220$, since

$$n_1 = n_3 = \ldots = 284, \qquad n_2 = n_4 = \ldots = 220.$$

H. LeLasseur[376] found that for $n < 35$ the numbers (1) are all odd primes, and hence give amicable numbers, only when $n = 2, 4, 7$.

Josef Bezdíček[377] gave a translation into Bohemian of Euler,[364] without credit to Euler, and a table of 65 pairs of amicable numbers.

Aug. Haas[378] proved that, if M and N are amicable numbers,

$$1/\Sigma \frac{1}{m} + 1/\Sigma \frac{1}{n} = 1,$$

where m and n range over all divisors of M and N, respectively. For, $\Sigma m = \Sigma n = M + N$, so that

$$\Sigma \frac{1}{m} = \frac{\Sigma m}{M} = \frac{M+N}{M}, \qquad \Sigma \frac{1}{n} = \frac{\Sigma n}{N} = \frac{M+N}{N}.$$

If $M = N$, N is perfect and the result becomes that of Catalan.[145]

A. Cunningham[379] considered the sum $s(n)$ of the divisors $< n$ of n and wrote $s^2(n)$ for $s\{s(n)\}$, etc. For most numbers, $s^k(n) = 1$ when k is sufficiently large. There is a small class of perfect and amicable numbers, and a small class of numbers n (even when $n < 1000$) for which $s^k(n)$ increases beyond the practical power of calculation [cf. Catalan[374]].

A. Gérardin[380] proved that the only pairs $2^2 \cdot 5x$, $2^2 yz$ of amicable numbers, where x, y, z are odd primes, are Euler's (α), (β); the only pairs $2^4 \cdot 23x$, $2^4 yz$ are Euler's (17), (19), (20). He cited the Exercices d'arithmétique of Fitz-Patrick and Chevrel; also Dupuis' Table de logarithmes, which gives 24 pairs of amicable numbers.

Gérardin[381] proved that the only pair $8xy$, $32z$ is Euler's (60). He made an incomplete examination of $16 \cdot 53x$, $16yz$, but found no new pairs.

[373]Archiv Math. Phys., 70, 1884, 75–89.
[374]Bull. Soc. Math. France, 16, 1887–8, 129. Mathesis, 8, 1888, 130.
[375]Ibid., 17, 1888–9, 155–6.
[376]Lucas, Théorie des nombres, 1, 1891, 381.
[377]Casopis mat. a fys., Praze (Prag), 25, 1896, 129–142, 209–221.
[378]Ibid., 349–350.
[379]Proc. London Math. Soc., 35, 1902–3, 40.
[380]Mathesis, 6, 1906, 41–44.
[381]Sphinx-Oedipe, Nancy, 1906–7, 14–15, 53.

Gérardin[382] proved that the three numbers (1) with $n = m+2$ are not all primes if $34 < m \leq 60$, the cases $m = 38$ and 53 not being decided. Replacing m by $m+1$ and k by $2g+1$ in case (1_1) of Euler[364], we get the pair $2^n pq$, $2^n r$, where $n = m+2g+2$,

$$p = 2^{m+2g+2}P - 1, \qquad q = 2^{m+1}P - 1, \qquad r = 2^{2m+2g+3}P^2 - 1,$$

with $P = 2^{2g+1} + 1$. For $g = 0$, we have the case (1) just mentioned; all values $m \leq 200$ are excluded except $m = 38, 74, 98, 146, 149, 182, 185, 197$. The case $g = 1$ is excluded since y or z is a difference of two squares. For $g = 2$, all values $m \leq 60$ are excluded except $m = 29, 34, 37, 49$. For $g = 3$, all values < 100 are excluded except $m = 8, 15, 23, 92$.

O. Meissner,[383] using the notation of Cunningham,[379] noted that n and $s(n)$ are amicable if $s^2(n) = n$ and raised the question of the existence of numbers n for which $s^k(n) = n$ for $k \geq 3$, so that $n, s(n), \ldots, s^{k-1}(n)$ would give amicable numbers of higher order. He asked if the repetition of the operation s, a finite number (k) of times always leads to a prime, a perfect or amicable number; also if k increases with n to infinity. On these questions, see Dickson[386] and Poulet.[387]

A. Gérardin[384] stated that the only values $n < 200$ for which the numbers (1) are all primes are the three known to Descartes.

L. E. Dickson[385] obtained the two new pairs of amicable numbers

$$2^4 \cdot 12959 \cdot 50231, \quad 2^4 \cdot 17 \cdot 137 \cdot 262079; \quad 2^4 \cdot 10103 \cdot 735263, \quad 2^4 \cdot 17 \cdot 137 \cdot 2990783,$$

by treating the type $16pq$, $16 \cdot 17 \cdot 137r$, where p, q, r are distinct odd primes. These are amicable if and only if

$$p = m + 9935, \qquad q = n + 9935, \qquad r = 4(m+n) + 88799, \qquad mn = 2^7 3^4 7 \cdot 23 \cdot 73.$$

Although Euler[364] mentioned this type (3_3) in §95, he made no discussion of it since r always exceeds the limit 100000 of the table of primes accessible to him. An examination of the 120 distinct cases led only to the above two amicable pairs.

Dickson[386] proved that there exist only five pairs of amicable numbers in which the smaller number is < 6233, viz., (1), (a), (β), (60) in Euler's[364] table, and Paganini's[372] pair. In the notation of Cunningham,[379] the chain $n, s(n), s^2(n), \ldots$ is said to be of period k if $s^k(n) = n$. The empirical theorem of Catalan[374] is stated in the corrected form that every non-periodic chain contains a prime and verified for a wide range of values of n. In particular, if $n < 6233$, there is no chain of period 3, 4, 5, or 6. For k odd and > 1, there is no chain $an_1, an_2, \ldots, an_k$ of period k in which $n_1, \ldots, n_k$ have no common factor and each n_j is prime to $a > 1$.

[382]Sphinx-Oedipe, 1907–8, 49–56, 65–71; some details are inaccurate, but the results correct.
[383]Archiv Math. Phys., (3), 12, 1907, 199; Math.-Naturw. Blätter, 4, 1907, 86 (for $k=3$).
[384]Assoc. franç. avanc. sc., 37, 1908, 36–48; l'intermédiaire des math., 1909, 104.
[385]Amer. Math. Monthly, 18, 1911, 109.
[386]Quart. Jour. Math., 44, 1913, 264–296.

P. Poulet[387] discovered the chain of period five,

$$n = 12496 = 2^4 \cdot 11 \cdot 71, \quad s(n) = 2^4 \cdot 19 \cdot 47, \quad s^2(n) = 2^4 \cdot 967, \quad s^3(n) = 2^3 \cdot 23 \cdot 79,$$
$$s^4(n) = 2^3 \cdot 1783,$$

with $s^5(n) = n$; and noted that 14316 leads a chain of 28 terms.

GENERALIZATIONS OF AMICABLE NUMBERS.

Daniel Schwenter[52] noted in 1636 that 27 and 35 have the same sum of aliquot parts. Kraft[363] noted in 1749 that this is true of the pairs 45, 3·29; 39, 55; 93, 145; and 45, 13·19. In 1823, Thomas Taylor[102] called two such numbers imperfectly amicable, citing the pairs 27, 35; 39, 55; 65, 77; 51, 91; 95, 119; 69, 133; 115, 187; 87, 247. George Peacock[400] used the same term.

E. B. Escott[401] asked if there exist three or more numbers such that each equals the sum of the [aliquot] divisors of the others.

A. Gérardin[402] called numbers with the same sum of aliquot parts nombres associés, citing 6 and 25; 5·19, 7·17, and 11·13, and many more sets. An equivalent definition is that the n numbers be such that the product of $n-1$ by the sum of the aliquot divisors of any one of them shall equal the sum of the aliquot divisors of the remaining $n-1$ numbers.

L. E. Dickson[403] defined an amicable triple to be three numbers such that the sum of the aliquot divisors of each equals the sum of the remaining two numbers. After developing a theory analogous to that by Euler[364] for amicable numbers, Dickson obtained eight sets of amicable triples in which two of the numbers are equal, and two triples of distinct numbers:

293·337a,	5·16561a,	99371a	$(a = 2^5 \cdot 3 \cdot 13)$,
3·89b,	11·29b,	359b	$(b = 2^{14} \cdot 5 \cdot 19 \cdot 31 \cdot 151)$.

[387]L'intermédiaire des math., 25, 1918, 100–1.
[400]Encyclopaedia Metropolitana, London, I, 1845, 422.
[401]L'intermédiaire des math., 6, 1899, 152.
[402]Sphinx-Oedipe, 1907–8, 81–83.
[403]Amer. Math. Monthly, 20, 1913, 84–92.

CHAPTER II.

FORMULAS FOR THE NUMBER AND SUM OF DIVISORS, PROBLEMS OF FERMAT AND WALLIS.

FORMULA FOR THE NUMBER OF THE DIVISORS OF A NUMBER.

Cardan[1] stated that a product P of k distinct primes has $1+2+2^2+\ldots+2^{k-1}$ aliquot parts (divisors $<P$).

Michael Stifel[2] proved this rule and found[3] the number of divisors of $2^4 3^3 5^2 P$, where $P=7\cdot11\cdot13\cdot17\cdot19\cdot23\cdot29$, by first noting that there are $1+2+\ldots+64$ divisors $<P$ of P according to Cardan's rule and hence 128 divisors of P. The factor 5^2 gives rise to $128+128$ more divisors, so that we now have 384 divisors. The factor 3^3 gives 3.384 more, so that we have 1536. Then the factor 2^4 gives 4.1536 more.

Mersenne[4] asked what number has 60 divisors; since $60=2\cdot2\cdot3\cdot5$, subtract unity from each prime factor and use the remainders 1, 1, 2, 4 as exponents; thus $3^2\cdot2^4\cdot7\cdot5=5040$ (so much lauded by Plato) has 60 divisors. It is no more difficult if a large number of aliquot parts is desired.

I. Newton[5] found all the divisors of 60 by dividing it by 2, the quotient 30 by 2, and the new quotient 15 by 3. Thus the prime divisors are 1, 2, 2, 3, 5. Their products by twos give 4, 6, 10, 15. The products by threes give 12, 20, 30. The product of all is 60. The commentator J. Castillionei, of the 1761 edition, noted that the process proves that the number of all divisors of $a^m b^n \ldots$ is $(m+1)(n+1)\ldots$ if $a, b,\ldots$ are distinct primes.

Frans van Schooten[6] devoted pp. 373–6 to proving that a product of k distinct primes has 2^k-1 aliquot parts and made a long problem (p. 379) of that to find the number of divisors of a given number. To find (pp. 380–4) the numbers having 15 aliquot parts, he factored $15+1$ in all ways and subtracted unity from each factor, obtaining $abcd$, a^3bc, a^3b^3, a^7b, a^{15}. By comparing the arithmetically least numbers of these various types, he found (pp. 387–9) the least number having 15 aliquot parts.

John Kersey[7] cited the long rule of van Schooten to find the number of aliquot parts of a number and then gave the simple rule that $a_1{}^{e_1}\ldots a_n{}^{e_n}$ has $(e_1+1)\ldots(e_n+1)$ divisors in all if $a_1,\ldots, a_n$ are distinct primes.

John Wallis[8] gave the last rule. To find a number with a prescribed number of divisors, factor the latter number in all possible ways; if the

[1]Practica Arith. & Mensurandi, Milan, 1537; Opera, IV, 1663.
[2]Arithmetica Integra, Norimbergae, 1544, lib. 1, fol. 101.
[3]Stifel's posthumous manuscript, fol. 12, preceding the printed text of Arith. Integra; cf. E. Hoppe, Mitt. Math. Gesell. Hamburg, 3, 1900, 413.
[4]Cogitata Physico Math., II, Hydravlica Pnevmatica, Preface, No. 14, Paris, 1644. (Quoted by Winsheim, Novi Comm. Ac. Petrop., II, ad annum 1749, Mem., 68–99). Also letter from Mersenne to Torricello, June 24, 1644, Bull. Bibl. Storia Sc. Mat., 8, 1875, 414–5.
[5]Arithmetica Universalis, ed. 1732, p. 37; ed. 1761, I, p. 61. De Inventione Divisorum.
[6]Exercitationum Math., Lugd. Batav., 1657.
[7]The Elements of Algebra, London, vol. 1, 1673, p. 199.
[8]A Treatise of Algebra, London, 1685, additional treatise, Ch. III.

factors are r, s,..., the required number is $p^{r-1}q^{s-1}$..., where p, q,...are any distinct primes. When the number of divisors is odd, the number itself is a square, and conversely. The number of ways $N = a^\alpha b^\beta$...can be expressed as a product of two factors is $k = \frac{1}{2}(\alpha+1)(\beta+1)$...or $\frac{1}{2}+k$, according as N is not or is a square.

Jean Prestet[9] noted that a product of k distinct primes has 2^k divisors, while the nth power of a prime has $n+1$ divisors. The divisors of $a^2b^3c^2$ are the 12 divisors of a^2b^3, their products by c and by c^2, the general rule not being stated explicitly.

Pierre Rémond de Montmort[10] stated in words that the number of divisors of $a_1^{e_1}...a_n^{e_n}$ is $(e_1+1)...(e_n+1)$ if the a's are distinct primes.

Abbé Deidier[11] noted that a product of k distinct primes has

$$1+k+\binom{k}{2}+\binom{k}{3}+\cdots$$

divisors, treating the problem as one on combinations (but did not sum the series and find 2^k). To find the number of divisors of $2^43^35^2$ he noted that five are powers of 2 (including unity). Since there are three divisors of 3^3, multiply 5 by 3 and add 5, obtaining 20. In view of the two divisors of 5^2, multiply 20 by 2 and add 20. The answer is 60.

E. Waring[12] proved that the number of divisors of $a^m b^n$...is $(m+1)$ $(n+1)$...if a, b,...are distinct primes, and that the number is a square if the number of its divisors is odd.

E. Lionnet[13] proved that if a, b, c,...are relatively prime in pairs, the number of divisors of abc...equals the product of the number of divisors of a by the number for b, etc. According as a number is a square or not, the number of its divisors is odd or even.

T. L. Pujo[14] noted the property last mentioned.

Emil Hain[15] derived the last theorem from $a^m = (t_1...t_m)^2$, where $t_1,..., t_m$ denote the divisors of a.

A. P. Minin[16] determined the smallest integer with a given number of divisors.

G. Fontené[17] noted that, if $2^\alpha 3^\beta...m^\mu n^\nu$ $(\alpha \geqq \beta \geqq ... \geqq \mu \geqq \nu)$ is the least number with a given number of divisors, then $\nu+1$ is a prime, and $\mu+1$ is a prime except for the least number 2^33 having eight divisors.

FORMULA FOR THE SUM OF THE DIVISORS OF A NUMBER.

R. Descartes,[21] in a manuscript, doubtless of date 1638, noted that, if p is a prime, the sum of the aliquot parts of p^n is $(p^n-1)/(p-1)$. If b is the

[9]Nouv. Elemens des Math., Paris, 1689, vol. 1, p. 149.
[10]Essay d'analyse sur les jeux de hazard, ed. 2, Paris, 1713, p. 55. Not in ed. 1, 1708.
[11]Suite de l'arithmétique des géométres, Paris, 1739, p. 311.
[12]Medit. Algebr., 1770, 200; ed. 3, 1782, 341.
[12]Nouv. Ann. Math., (2), 7, 1868, 68-72.
[14]Les Mondes, 27, 1872, 653-4.
[15]Archiv Math. Phys., 55, 1873, 290-3.
[16]Math. Soc. Moscow (in Russian), 11, 1883-4, 632.
[17]Nouv. Ann. Math., (4), 2, 1902, 288; proof by Chalde, 3, 1903, 471-3.
[21]"De partibus aliquotis numerorum," Opuscula Posthuma Phys. et Math., Amstelodami, 1701, p. 5; Oeuvres de Descartes (ed. Tannery and Adams, 1897-1909), vol. 10, pp. 300-2.

sum of the aliquot parts of a, the sum of the aliquot parts of ap is $bp+a+b$.
If b is the sum of the aliquot parts of a and if x is prime to a, the sum of the aliquot parts of ax^n is

$$\frac{bx^{n+1}+ax^n-a-b}{x-1} \qquad \left[=(b+a)\left(\frac{x^{n+1}-1}{x-1}\right)-ax^n\right].$$

Descartes[22] stated a result which may be expressed by the formula

(1) $\qquad \sigma(nm)=\sigma(n)\sigma(m)$ $\qquad$ (n, m relatively prime),

where $\sigma(n)$ is the sum of the divisors (including 1 and n) of n. Here he solved $n:\sigma(n)=5:13$. Thus n must be divisible by 5. Enter 5 in column A and $\sigma(5)=6$ in column B. Then enter the factor 2 in column A and $\sigma(2)=3$ in column B. Having two threes in column B, we enter 9 in column A and $\sigma(9)=13$ in B. Every number except 13 in column B is in column A. Hence the product $5\cdot2\cdot9=90$ is a solution n. Next, to solve $n:\sigma(n)=5:14$, we enter also 13 in column A and 14 in B, and obtain the solution $90\cdot13$. If n is a perfect number, $5n:\sigma(5n)=5:12$ and, if $n\neq6$, $15n:$ $\sigma(15n)=5:16$.

A	B
5	2·3
2	3
9	13

Descartes[23] stated that he possessed a general rule [illustrated above] for finding numbers having any given ratio to the sum of their aliquot parts. Fermat[24] had treated the same problem. Replying to Mersenne's remark that the sum of the aliquot parts of 360 bears to 360 the ratio 9 to 4, Fermat[25] noted that 2016 has the same property.

John Wallis[26] noted that Frenicle knew formula (1).

Wallis[27] knew the formula

(2) $\qquad \sigma(a^\alpha b^\beta \dots)=\dfrac{a^{\alpha+1}-1}{a-1}\cdot\dfrac{b^{\beta+1}-1}{b-1}\dots$

Thus these formulæ were known before 1685, the date set by Peano,[28] who attributed them to Wallis.[52]

G. W. Kraft[29] noted that the method of Newton[5] shows that the sum of the divisors of a product of distinct primes $P,\dots$, S is $(P+1)\dots(S+1)$. He gave formula (1) and also (2), a formula which Cantor[30] stated had probably not earlier been in print. To find a number the sum of whose divisors is a square, Kraft took PA, where P is a prime not dividing A. If $\sigma(A)=a$, then $\sigma(PA)=(P+1)a$ will be the square of $(P+1)B$ if $P=$

[22]"De la façon de trouver le nombres de parties aliquotes in ratione data," manuscript Fonds-français, nouv. acquisitions, No. 3280, ff. 156–7, Bibliothèque Nationale, Paris. Published by C. Henry, Bull. Bibl. Storia Sc. Mat. e Fis., 12, 1879, 713–5.

[23]Oeuvres, 2, p. 149, letter to Mersenne, May 27, 1638.

[24]Oeuvres de Fermat, 2, top of p. 73, letter to Roberval, Sept. 22, 1636.

[25]Oeuvres, 2, 179, letter to Mersenne, Feb. 20, 1639.

[26]Commercium Epistolicum, letter 32, April 13, 1658; French transl. in Oeuvres de Fermat, 3, 553.

[27]Commercium Epist., letter 23, March, 1658; Oeuvres de Fermat, 3, 515–7.

[28]Formulaire Math., 3, Turin, 1901, 100–1.

[29]Novi Comm. Ac. Petrop., 2, 1751, ad annum 1749, 100–109.

[30]Geschichte Math., 3, 595; ed. 2, 616.

a/B^2-1; for $A=14$, take $B=2$, whence $P=5$. Again, the sum of the aliquot parts of $3P^2$ is $(2+P)^2$. The numbers AP and BPQ have the same sum of divisors if $a(P+1)=b(P+1)(Q+1)$, i. e., if $Q=a/b-1$; taking $a=24$, $b=6$, we have $Q=3$, a prime, $A=14$, $B=5$ (by his table of the sum of the divisors of $1,\ldots,150$); this problem had been solved otherwise by Wolff.[31]

L. Euler[32] gave a table of the prime factors of $\sigma(p)$, $\sigma(p^2)$, and $\sigma(p^3)$ for each prime $p<1000$; also those of $\sigma(p^a)$ for various a's for $p\leqq23$ (for instance, $a\leqq36$ when $p=2$). He proved formulas (1) and (2) here and in his[33] posthumous tract, where he noted (p. 514) all the cases in which $\sigma(n)=\sigma(m)\leqq60$.

E. Waring[12] proved formula (2). He[34] noted that if $P=a^m b^n\ldots$ and $Q=a^\alpha b^\beta\ldots$, where $m-\alpha$, $n-\beta,\ldots$ are large, then $\sigma(PQ)/\sigma(P)$ is just greater than Q. If $A=(l-1)!$, $\sigma(lA)/\sigma(A)\leqq l+1$. If $a^x b^y\ldots=A$ and $(x+1)(y+1)\ldots$ is a maximum, then $a^{x+1}=b^{y+1}=\ldots$ For $a, b,\ldots$ distinct primes, $\sigma(A)$ is not a maximum. He cited numbers with equal sums of divisors: 6 and 11, 10 and 17, 14 and 15 and 23.

L. Kronecker[35] derived the formulas for the number and sum of the divisors of an integer by use of infinite series and products.

E. B. Escott[36] listed integers whose sum of divisors is a square.

PROBLEMS OF FERMAT AND WALLIS ON SUMS OF DIVISORS.

Fermat[40] proposed January 3, 1657, the two problems: (i) Find a cube which when increased by the sum of its aliquot parts becomes a square;* for example, $7^3+(1+7+7^2)=20^2$. (ii) Find a square which when increased by the sum of its aliquot parts becomes a cube.

John Wallis[41] replied that unity is a solution of both problems and proposed the new problem: (iii) Find two squares, other than 16 and 25, such that if each is increased by the sum of its aliquot parts the resulting sums are equal.

Brouncker[42] gave $1/n^6$ and $343/n^6$ as solutions (!) of problem (i).

[31]Elementa Analyseos, Cap. 2, prob. 87.
[32]Opuscula varii argumenti, 2, Berlin, 1750, p. 23; Comm. Arith., 1, 102 (p. 147 for table to 100). Opera postuma, I, 1862, 95–100. F. Rudio, Bibl. Math., (3), 14, 1915, 351, stated that there are fully 15 errors.
[33]Comm. Arith., 2, 512, 629. Opera postuma, I, 12–13.
[34]Meditationes Algebr., ed. 3, 1782, 343. (Not in ed. of 1770.)
[35]Vorlesungen über Zahlentheorie, I, 1901, 265–6.
[36]Amer. Math. Monthly, 23, 1916, 394.
*Erroneously given as "cube" in the French tr., Oeuvres de Fermat, 3, 311.
[40]Oeuvres, 2, 332, "premier défi aux mathématiciens;" also, pp. 341–2, Fermat to Digby, June 6, 1657, where 7^3 is said to be not the only solution. These two problems by Fermat were quoted in a letter by the Astronomer Jean Hévélius, Nov. 1, 1657, published by C. Henry, Bull. Bibl. Storia Sc. Mat. e Fis., 12, 1879, 683–5, along with extracts from the Commercium Epistolicum. Cf. G. Wertheim, Abh. Geschichte Math., 9, 1899, 558–561, 570–2 (=Zeitschr. Math. Phys., 44, Suppl. 14).
[41]Commercium Epistolicum de Wallis, Oxford, 1658; Wallis, Opera, 2, 1693. Letter II, from Wallis to Brouncker, Mar. 17, 1657; letter XVI, Wallis to Digby, Dec. 1, 1657. Oeuvres de Fermat, 3, 404, 414, 427, 482–3, 503–4, 513–5.
[42]Commercium, letter IX, Wallis to Digby; Fermat's Oeuvres, 3, 419.

Frenicle[43] expressed his astonishment that experienced mathematicians should not hesitate to present, for the third time, unity as a solution.

Wallis[44] tabulated $\sigma(x^3)$ for each prime $x<100$ and for low powers of 2, 3, 5, and then excluded those primes x for which $\sigma(x^3)$ has a prime factor not occurring elsewhere in the table. By similar eliminations and successive trials, he was led to the solutions[45] of (i):

$$a = 3^3 5 \cdot 11 \cdot 13 \cdot 41 \cdot 47, \quad b = 2 \cdot 3 \cdot 5 \cdot 13 \cdot 41 \cdot 47; \quad 7a, \ 7b,$$

adding that they are identical with the four numbers given by Frenicle.[46] Note that $\sigma(a)$ is the square of $2^8 3^2 5 \cdot 7 \cdot 11 \cdot 13 \cdot 17 \cdot 29 \cdot 61$, while $\sigma(b)$ is the square of $2^7 3^2 5^2 7 \cdot 13 \cdot 17 \cdot 29$. Wallis[47] gave the further solutions of $\sigma(x^3) = y^2$:

$$x = 17 \cdot 31 \cdot 47 \cdot 191, \qquad y = 2^{10} 3^2 5 \cdot 13 \cdot 17 \cdot 29 \cdot 37,$$
$$2 \cdot 3 \cdot 5 \cdot 13 \cdot 17 \cdot 31 \cdot 41 \cdot 191, \qquad 2^{12} 3^3 5^2 7 \cdot 13 \cdot 17 \cdot 29^2 37,$$
$$3^3 5 \cdot 11 \cdot 13 \cdot 17 \cdot 31 \cdot 41 \cdot 191, \qquad 2^{13} 3^3 5 \cdot 7 \cdot 11 \cdot 13 \cdot 17 \cdot 29^2 37 \cdot 61,$$

and the products of each x by 7.

Wallis[48] gave solutions of his problem (iii):

$$2^3 3 \cdot 37, \ 2 \cdot 19 \cdot 29; \qquad 2^2 3 \cdot 11 \cdot 19 \cdot 37, \ 2^3 7 \cdot 29 \cdot 67;$$
$$29 \cdot 67, \ 2 \cdot 3 \cdot 5 \cdot 37; \qquad 2^3 7 \cdot 29 \cdot 67, \ 3 \cdot 5 \cdot 11 \cdot 19 \cdot 37.$$

Frenicle[49] gave 48 solutions of Wallis' problem (iii), including $2 \cdot 163$ $11 \cdot 37$; $3 \cdot 11 \cdot 19$, $7 \cdot 107$; $2 \cdot 5 \cdot 151$, $3^3 \cdot 67$; also 83 sets of three squares having the same sum of divisors, for example, the squares of

$$2^2 11 \cdot 37 \cdot 151, \qquad 3^3 67 \cdot 163, \qquad 5 \cdot 11 \cdot 37 \cdot 151, \qquad \sigma = 3^2 7^3 19 \cdot 31 \cdot 67 \cdot 1093;$$

also various such sets of n squares (with prime factors <500) for $n \leqq 19$, for example, the squares of ac, ad, $4bd$, $4bc$, $5bd$, and $5bc$, where

$$a = 2 \cdot 5 \cdot 29 \cdot 47 \cdot 67 \cdot 139, \qquad b = 13 \cdot 37 \cdot 191 \cdot 359, \qquad c = 7 \cdot 107, \qquad d = 3 \cdot 11 \cdot 19.$$

Frans van Schooten[50] made ineffective attempts to solve problems (i), (ii).

Frenicle[51] gave the solution

$$x = 2^2 5 \cdot 7 \cdot 11 \cdot 37 \cdot 67 \cdot 163 \cdot 191 \cdot 263 \cdot 439 \cdot 499, \qquad y = 3^2 7^3 13 \cdot 19 \cdot 31^2 67 \cdot 109$$

of problem (ii), $\sigma(x^2) = y^3$; also a new solution of $\sigma(x^3) = y^2$:

$$x = 2^5 5 \cdot 7 \cdot 31 \cdot 73 \cdot 241 \cdot 243 \cdot 467, \qquad y = 2^{12} 3^2 5^3 11 \cdot 13^2 17 \cdot 37 \cdot 41 \cdot 113 \cdot 193 \cdot 257.$$

[43]Letter XXII, to Digby, Feb. 3, 1658. Cf. Leibnitii et Bernoullii Commercium philos. et math., I, 1795, 263, letter from Johann Bernoulli to Leibniz, Apr. 3, 1697.

[44]Letter XXIII, to Digby, Mar. 14, 1658.

[45]The same tentative process for finding this solution a was given by E.Waring, Meditationes Algebraicae, 1770, pp. 216–7; ed. 3, 1782, 377–8. The solution $b = 751530$ was quoted by Lucas, Théorie des nombres, 1891, 380, ex. 3.

[46]Solutio duorum problematum circa numeros cubos...1657, dedicated to Digby [lost work]. See Oeuvres de Fermat, p. 2. 434, Note; Wallis.[52]

[47]Letter XXVIII, March 25, 1658; Wallis, Opera, 2, 814; Wallis[52].

[48]Letter XXIX, Mar. 29, 1658; Wallis[52].

[49]Letter XXXI, Apr. 11, 1658.

[50]Letter XXXIII, Feb. 17, 1657 and Mar. 18, 1658.

[51]Letter XLIII, May 2, 1658.

Wallis[52] for use in problem (ii) gave a table showing the sum of the divisors of the square of each number < 500. Excluding numbers in whose divisor sum occurs a prime entering the table only once or twice, there are left the squares of 2, 4, 8, 3, 5, 7, 11, 19, 29, 37, 67, 107, 163, 191, 263, 439, 499. By a very long process of exclusion he found only two solutions within the limits of the table, viz., Frenicle's[51] and

$$\sigma\{(7\cdot11\cdot29\cdot163\cdot191\cdot439)^2\} = \{3\cdot7\cdot13\cdot19\cdot31\cdot67\}^3.$$

Jacques Ozanam[53] stated that Fermat had proposed the problem to find a square which with its aliquot parts makes a square (giving 81 as the answer) and the problem to find a square whose aliquot parts make a square. For the latter, Ozanam found 9 and 2401, whose aliquot parts make 4 and 400, and remarked that he did not believe that Fermat ever solved these questions, although he proposed them as if he knew how.

Ozanam[54] noted that the sum of $961 = 31^2$ and its aliquot parts 1 and 31 is 993, which equals the sum of the aliquot parts of $1156 = 34^2$. As examples of two squares with equal total sums of divisors [Wallis' problem (iii)], he cited 16 and 25, 326^2 and 407^2, while others may be derived by multiplying these by an odd square not divisible by 5. The sum of all the divisors of 9^2 is 11^2, that of 20^2 is 31^2. The numbers 99 and 63 have the property that the sum 57 of the aliquot parts of 99 exceeds the sum 41 of the aliquot parts of 63 by the square 16; similarly for 325 and 175.

E. Lucas[55] noted that the problem to find all integral solutions of

(1) $$1+x+x^2+x^3 = y^2$$

is equivalent to the solution of the system

(2) $$1+x = 2u^2, \qquad 1+x^2 = 2v^2, \qquad y = 2uv,$$

and stated that the complete solution is given by that of $2v^2 - x^2 = 1$.

E. Gerono[56] proved that the only solutions of (1) are

$$(x, y) = (-1, 0), \qquad (0, \pm1), \qquad (1, \pm2), \qquad (7, \pm20).$$

E. Lucas[57] stated that there is an infinitude of solutions of Fermat's problem (i); the least composite solution is the cube of $2\cdot3\cdot5\cdot13\cdot41\cdot47$, the sum of whose divisors is the square of $2^7 3^2 5^2 7\cdot13\cdot17\cdot29$. [This solution was given by Frenicle.[46]] For the case of a prime, the problem becomes (1).

A. S. Bang[58] gave for problem (i) the first of the three answers by Wallis;[47] for (ii), $\sigma(43098^2) = 1729^3$; for (iii), $29\cdot67$, $2\cdot3\cdot5\cdot37$ of Wallis[48] and the first two by Frenicle;[49] all without references.

[52]A Treatise of Algebra, 1685, additional treatises, Ch. IV.
[53]Letter to De Billy, Nov. 1, 1677, published by C. Henry, Bull. Bibl. Storia Sc. Mat. e Fis., 12, 1879, 519. Reprinted in Oeuvres de Fermat, 4, 1912, p. 140.
[54]Récréations Mathématiques et Phys., new ed., 1723, 1724, 1735, etc., Paris, I, 41–43.
[55]Nouv. Corresp. Math., 2, 1876, 87–8.
[56]Nouv. Ann. Math., (2), 16, 1877, 230–4.
[57]Bull. Bibl. Storia Sc. Mat. e Fis., 10, 1877, 287.
[58]Nyt Tidsskrift for Mat., 1878, 107–8; on problems in 1877, 180.

E. Fauquembergue,[59] after remarking that (1) is equivalent to the system (2), cited Fermat's[60] assertion that the first two equations (2) hold only for $x=7$ [aside from the evident solutions $x=\pm 1$, 0], which has been proved by Genocchi.[61]

H. Brocard[62] thought that Fermat's assertion that 7^3 is not the only solution of problem (i) implied a contradiction with Genocchi.[61] G. Vacca (*ibid.*, p. 384) noted the absence of contradiction as (i) leads to equation (1) only if x be a prime.

C. Moreau[63] treated the equation, of type (1),

$$x^4+x^3+x^2+x+1=y^2.$$

While he used the language of extracting the square root of $X=x^4+\ldots$ written to the base x, he in effect put $X=(x^2+a)^2$, $0<a<x$. Then $a^2=x+1$, $2ax^2=x^3+x^2$, whence $2a=a^2$, $a=2$, $x=3$, $y=11$.

E. Lucas[64] stated that $(x^5+y^5)/(x+y)=z^2$ has the solutions

$$(3,-1,11),\ (8,11,101),\ (123,35,13361),\ldots$$

Moret-Blanc[65] gave also the solutions (0, 1, 1), (1, 1, 1).

E. Landau[66] proved that the equation

$$\frac{x^n-1}{x-1}=y^2$$

is impossible in integers (aside from $x=0$, $y=\pm 1$) for an infinitude of values of n, viz., for all n's divisible by 3 such that the odd prime factors of $n/3$, if any, are all of the form $6v-1$ (the least such n being 6). For, setting $n=3m$, we see that y^2 is the product of x^2+x+1 and $F=x^{3m-3}+\ldots+x^3+1$. These two factors are relatively prime since $x^3\equiv 1$ gives $F\equiv m$ (mod x^2+x+1). Hence x^2+x+1 is a square, which is impossible for $x\neq 0$ since it lies between x^2 and $(x+1)^2$.

Brocard[62] had noted the solution $x=1$, $y=m$. if $n=m^2$.

A. Gérardin[67] obtained six new solutions of problem (i):

$x=2.47.193.239.701$, $y=2^7.3^3.5^3.13^3.17.97.149$.
$x=2.5.23.41.83.239$, $y=2^8.3^3.5^2.7.13^3.29.53$.
$x=3.13.23.47.83.239$, $y=2^{11}.3^2.5^3.7.13^3.17.53$.
$x=2.3.13.23.83.193.701$, $y=2^8.3^3.5^4.7.13.17.53.97.149$.
$x=3.5.13.41.193.239.701$, $y=2^9.3^3.5^3.7.13^3.17.29.97.149$.
$x=2.5.13.43.191.239.307$, $y=2^{13}.3^2.5^3.11^2.17.29.37.53.113.197.241.257$.

Also $\sigma(N^2)=S^2$ for $N=3.7.11.29.37$, $S=3.7.13.19.67$.

---

[59]Nouv. Ann. Math., (3), 3, 1884, 538–9.
[60]Oeuvres, 2, 434, letter to Carcavi, Aug., 1659.
[61]Nouv. Ann. Math., (3), 2, 1883, 306–10. Cf. Chapter on Diophantine Equations of order 2.
[62]L'intermédiaire des math., 7, 1900, 31, 84.
[63]Nouv. Ann. Math., (2), 14, 1875, 335.
[64]Ibid., 509.
[65]Ibid., (2), 20, 1881, 150.
[66]L'intermédiaire des math., 8, 1901, 149–150.
[67]Ibid., 22, 1915, 111–4, 127.

Gérardin[68] gave five new solutions of (i):

$x = 3.11.31.443.499,$ $y = 2^9.3.5^4.13.37.61.157.$

$x = 2.3^3 31.443.449,$ $y = 2^7 3.5^4 11.13.37.61.157.$

$x = 11.17.41.43.239.307.443.499,$

$\qquad y = 2^{12}.3^3.5^7.7.11.13^3.29^2.37.61.157.$

$x = 2.11.17.23.41.211.467.577.853,$

$\qquad y = 2^{10}.3^4.5^3.7.13^2.17.29^2.53.61.113.193.197.$

$x = 3^3 11.13.23.83.193.701,$

$\qquad y = 2^9 3^3 5^3 7.11.13.17.53.61.97.149,$

the last following from his[67] fourth pair in view of

$$\sigma(3^9 11^3): \ \sigma(2^3 3^3) = 2^5 3.11^2 61^2: 2^3 3.5^2 = 2^2 11^2 61^2: 5^2.$$

A. Cunningham and J. Blaikie[69] found solutions of the form $x = 2^r p$ of $s(x) = g^2$, where $s(n)$ is the sum of the divisors $< n$ of n.

PRODUCT OF ALIQUOT PARTS.

Paul Halcke[75] noted that the product of the aliquot parts of 12, 20, or 45 is the square of the number; the product for 24 or 40 is the cube; the product for 48, 80 or 405 is the biquadrate.

E. Lionnet[76] defined a perfect number of the second kind to be a number equal to the product of its aliquot parts. The only ones are p^3 and pq, where p and q are distinct primes.

[68]L'intermédiaire des math., 24, 1917, 132–3.
[69]Math. Quest. Educ. Times, (2), 7, 1905, 68–9.
[75]Deliciae Math. oder Math. Sinnen-Confect, Hamburg, 1719, 197, Exs. 150–2.
[76]Nouv. Ann. Math., (2), 18, 1879, 306–8. Lucas, Théorie des nombres, 1891, 373, Ex. 6

CHAPTER III.

FERMAT'S AND WILSON'S THEOREMS, GENERALIZATIONS AND CONVERSES; SYMMETRIC FUNCTIONS OF 1, 2, ..., P−1 MODULO P.

FERMAT'S AND WILSON'S THEOREMS; IMMEDIATE GENERALIZATIONS.

The Chinese[1] seem to have known as early as 500 B. C. that 2^p-2 is divisible by the prime p. This fact was rediscovered by P. de Fermat[2] while investigating perfect numbers. Shortly afterwards, Fermat[3] stated that he had a proof of the more general fact now known as Fermat's theorem: If p is any prime and x is any integer not divisible by p, then $x^{p-1}-1$ is divisible by p.

G. W. Leibniz[4] (1646–1716) left a manuscript giving a proof of Fermat's theorem. Let p be a prime and set $x=a+b+c+\ldots$ Then each multinominal coefficient appearing in the expansion of $x^p-\Sigma a^p$ is divisible by p. Take $a=b=c=\ldots=1$. Thus x^p-x is divisible by p for every integer x.

G. Vacca[5] called attention to this proof by Leibniz.

Vacca[6] cited manuscripts of Leibniz in the Hannover Library showing that he proved Fermat's theorem before 1683 and that he knew the theorem now known as Wilson's[17] theorem: If p is a prime, $1+(p-1)!$ is divisible by p. But Vacca did not explain an apparent obscurity in Leibniz's statement [cf. Mahnke[7]].

D. Mahnke[7] gave an extensive account of those results in the manuscripts of Leibniz in the Hannover Library which relate to Fermat's and Wilson's theorems. As early as January 1676 (p. 41) Leibniz concluded, from the expressions for the yth triangular and yth pyramidal numbers, that

$$(y+1)y\equiv y^2-y\equiv0 \ (\text{mod } 2), \qquad (y+2)(y+1)y\equiv y^3-y\equiv0 \ (\text{mod } 3),$$

and similarly for moduli 5 and 7, whereas the corresponding formula for modulus 9 fails for $y=2$,—thus forestalling the general formula by Lagrange.[18] On September 12, 1680 (p. 49), Leibniz gave the formula now known as Newton's formula for the sum of like powers and noted (by incomplete induction) that all the coefficients except the first are divisible by the exponent p, when p is a prime, so that

$$a^p+b^p+c^p+\ldots\equiv(a+b+c+\ldots)^p \qquad (\text{mod } p).$$

Taking $a=b=\ldots=1$, we obtain Fermat's theorem as above.[4] That the binomial coefficients in $(1+1)^p-1-1$ are divisible by the prime p was

[1]G. Peano, Formulaire math., 3, Turin, 1901, p. 96. Jeans.[220]
[2]Oeuvres de Fermat, Paris, 2, 1894, p. 198, 2°, letter to Mersenne, June (?), 1640; also p. 203, 2; p. 209.
[3]Oeuvres, 2, 209, letter to Frenicle de Bessy, Oct. 18, 1640; Opera Math., Tolosae, 1679, 163.
[4]Leibnizens Math. Schriften, herausgegeben von G. J. Gerhardt, VII, 1863, 180–1, "nova algebrae promotio."
[5]Bibliotheca math., (2), 8, 1894, 46–8.
[6]Bolletino di Bibliografia Storia Sc. Mat., 2, 1899, 113–6.
[7]Bibliotheca math., (3), 13, 1912–3, 29–61.

proved in 1681 (p. 50). Mahnke gave reasons (pp. 54–7) for believing that Leibniz rediscovered independently Fermat's theorem before he became acquainted, about 1681–2, with Fermat's Varia opera math. of 1679. In 1682 (p. 42), Leibniz stated that $(p-2)!\equiv 1 \pmod{p}$ if p is a prime [equivalent to Wilson's theorem], but that $(p-2)!\equiv m \pmod{p}$, if p is composite, m having a factor >1 in common with p.

De la Hire[8] stated that if k^{2r+1} is divided by $2(2r+1)$ we get k as a remainder, perhaps after adding a multiple of the divisor. For example, if k^5 is divided by 10 we get the remainder k. He remarked that Carré had observed that the cube of any number $k<6$ has the remainder k when divided by 6.

L. Euler[9] stated Fermat's theorem in the form: If $n+1$ is a prime dividing neither a nor b, then a^n-b^n is divisible by $n+1$. He was not able to give a proof at that time. He stated the generalization: If $e=p^{m-1}(p-1)$ and if p is a prime, the remainder obtained on dividing a^e by p^m is 0 or 1 [a special case of Euler[14]]. He stated also that if $m, n, p, \ldots$ are distinct primes not dividing a and if A is the l. c. m. of $m-1, n-1, p-1, \ldots$, then a^A-1 is divisible by $mnp\ldots$[and a^k-1 by $m^r n^s\ldots$if $k=A\ m^{r-1}n^{s-1}\ldots$].

Euler[10] first published a proof of Fermat's theorem. For a prime p,

$$2^p=(1+1)^p=1+p+\binom{p}{2}+\ldots+p+1=2+mp,$$

$$3^p=(1+2)^p=1+kp+2^p, \qquad 3^p-3-(2^p-2)=kp,$$

$$(1+a)^p=\ 1+np+a^p, \qquad (1+a)^p-(1+a)-(a^p-a)=np.$$

Hence if a^p-a is divisible by p, also $(1+a)^p-(1+a)$ is, and hence also $(a+2)^p-(a+2),\ldots,(a+b)^p-(a+b)$. For $a=2$, 2^p-2 was proved divisible by p. Hence, writing x for $2+b$, we conclude that x^p-x is divisible by p for any integer x.

G. W. Kraft[11] proved similarly that $2^p-2=mp$.

L. Euler's[12] second proof is based, like his first, on the binomial theorem. If a, b are integers and p is a prime, $(a+b)^p-a^p-b^p$ is divisible by p. Then, if a^p-a and b^p-b are divisible by p, also $(a+b)^p-a-b$ is divisible by p. Take $b=1$. Thus $(a+1)^p-a-1$ is divisible by p if a^p-a is. Taking $a=1, 2, 3, \ldots$ in turn, we conclude that $2^p-2, 3^p-3, \ldots, c^p-c$ are divisible by p.

L. Euler[13] preferred his third proof to his earlier proofs since it avoids the use of the binomial theorem. If p is a prime and a is any integer not

[8]Hist. Acad. Sc. Paris, année 1704, pp. 42–4; mém., 358–362.
[9]Comm. Ac. Petrop., 6, 1732–3, 106; Comm. Arith., 1, 1849, p. 2. [Opera postuma, I, 1862, 167–8 (about 1778)].
[10]Comm. Ac. Petrop., 8, ad annum 1736, p. 141; Comm. Arith., 1, p. 21.
[11]Novi Comm. Ac. Petrop., 3, ad annos 1750–1, 121–2.
[12]Novi Comm. Ac. Petrop., 1, 1747–8, 20; Comm. Arith., 1, 50. Also, letter to Goldbach, Mar. 6, 1742, Corresp. Math. Phys. (ed. Fuss), I, 1843, 117. An extract of the letter is given in Nouv. Ann. Math., 12, 1853, 47.
[13]Novi Comm. Ac. Petrop., 7, 1758–9, p. 70 (ed. 1761, p. 49); 18, 1773, p. 85; Comm. Arith., 1, 260–9, 518–9. Reproduced by Gauss, Disq. Arith., art. 49; Werke, 1, 1863, p. 40.

divisible by p, at most $p-1$ of the positive residues $<p$, obtained by dividing $1, a, a^2, \ldots$ by p, are distinct. Let, therefore, a^μ and a^v, where $\mu > v$, have the same residue. Then $a^{\mu-v}-1$ is divisible by p. Let λ be the least positive integer for which $a^\lambda-1$ is divisible by p. Then $1, a, a^2, \ldots, a^{\lambda-1}$ have distinct residues when divided by p, so that $\lambda \leq p-1$. If $\lambda = p-1$, Fermat's theorem is proved. If $\lambda < p-1$, there exists a positive integer k $(k<p)$ which is not the residue of a power of a. Then $k, ak, a^2k, \ldots, a^{\lambda-1}k$ have distinct residues, no one the residue of a power of a. Since the two sets give 2λ distinct residues, we have $2\lambda \leq p-1$. If $\lambda < (p-1)/2$, we start with a new residue s and see that $s, as, a^2s, \ldots, a^{\lambda-1}s$ have distinct residues, no one the residue of a power of a or of $a^\mu k$. Hence $\lambda \leq (p-1)/3$. Proceeding in this manner, we see that λ divides $p-1$. Thus $a^{p-1}-1$ is divisible by $a^\lambda-1$ and hence by p.

L. Euler[14] soon gave his fundamental generalization of Fermat's theorem from the case of a prime to any integer N:

Euler's theorem: If $n = \phi(N)$ is the number of positive integers not exceeding N and relatively prime to N, then x^n-1 is divisible by N for every integer x relatively prime to N.

Let v be the least positive integer for which x^v has the residue 1 when divided by N. Then the residues of $1, x, x^2, \ldots, x^{v-1}$ are distinct and prime to N. Thus $v \leq n$. If $v < n$, there is an additional positive integer a less than N and prime to N. Then, when $a, ax, ax^2, \ldots, ax^{v-1}$ are divided by N, the residues are distinct from each other and from those of the powers of x. Thus, $2v \leq n$. Similarly, if $2v < n$, then $3v \leq n$. It follows in this manner that v divides n.

J. H. Lambert[15] gave a proof of Fermat's theorem differing slightly from the first proof by Euler.[10] If b is not divisible by the prime p, $b^{p-1}-1$ is divisible by p. For, set $b = c+1$. Then

$$b^{p-1}-1 = -1 + c^{p-1} + (p-1)c^{p-2} + \ldots + 1$$
$$= -1 + c^{p-1} - c^{p-2} + c^{p-3} - \ldots + 1 + Ap,$$

where A is an integer. The intermediate terms equal

$$\frac{c^p+1}{c+1} = c^{p-1} - \frac{c^{p-1}-1}{c+1}.$$

Hence

$$\frac{b^{p-1}-1}{p} = \frac{c^{p-1}-1}{p} + A - f, \qquad f = \frac{c^{p-1}-1}{p(c+1)}.$$

The theorem will thus follow by induction if f is shown to be integral. [Take $p > 2$, so that $p-1$ is even.] Then $c^{p-1}-1$ is divisible by $c+1$, and by the hypothesis for the induction, by p. Since $c+1 = b$ is relatively prime to p, f is an integer.

[14]Novi Comm. Ac. Petrop., 8, 1760-1, p. 74; Comm. Arith., 1, 274-286; 2, 524-6.
[15]Nova Acta Eruditorum, Lipsiae, 1769, 109.

E. Waring[16] first published the theorem that [Leibniz[6]] $1+(p-1)!$ is divisible by the prime p, ascribing it to Sir John Wilson[17] (1741–1793). Waring (p. 207; ed. 3, p. 356) proved that if a^p-a is divisible by p, then $(a+1)^p-a-1$ is, since $(a+1)^p=a^p+pA+1$, "a property first invented by Dom. Beaufort and first proved by Euler."

J. L. Lagrange[18] was the first to publish a proof of Wilson's theorem. Let

$$(x+1)(x+2)\dots(x+p-1)=x^{p-1}+A_1x^{p-2}+\dots+A_{p-1}.$$

Replace x by $x+1$ and multiply the resulting equation by $x+1$. Comparing with the original equation multiplied by $x+p$, we get

$$(x+p)(x^{p-1}+\dots+A_{p-1})=(x+1)^p+A_1(x+1)^{p-1}+\dots+A_{p-1}(x+1).$$

Apply the binomial theorem and equate coefficients of like powers of x. Thus

$$A_1=\binom{p}{2},\ 2A_2=\binom{p}{3}+\binom{p-1}{2}A_1,\ 3A_3=\binom{p}{4}+\binom{p-1}{3}A_1+\binom{p-2}{2}A_2,\dots.$$

Let p be a prime. Then, for $0<k<p$, $\binom{p}{k}$ is an integer divisible by p. Hence $A_1, 2A_2,\dots, (p-2)A_{p-2}$ are divisible by p. Also,

$$(p-1)A_{p-1}=\binom{p}{p}+\binom{p-1}{p-1}A_1+\binom{p-2}{p-2}A_2+\dots=1+A_1+A_2+\dots+A_{p-2}.$$

Thus $1+A_{p-1}$ is divisible by p. By the original equation, $A_{p-1}=(p-1)!$, so that Wilson's theorem follows.

Moreover, if x is any integer, the proof shows that

$$x^{p-1}-1-(x+1)(x+2)\dots(x+p-1)$$

is divisible by the prime p. If x is not divisible by p, some one of the integers $x+1,\dots,x+p-1$ is divisible by p. Hence $x^{p-1}-1$ is divisible by p, giving Fermat's theorem.

Lagrange deduced Wilson's theorem from Fermat's. By the formula[19] for the differences of order $p-1$ of $1^{p-1},\dots, n^{p-1}$,

(1) $$(p-1)!=p^{p-1}-(p-1)(p-1)^{p-1}+\binom{p-1}{2}(p-2)^{p-1}$$
$$-\binom{p-1}{3}(p-3)^{p-1}+\dots+(-1)^{p-1}.$$

Dividing the second member by p, and applying Fermat's theorem, we obtain the residue

$$-(p-1)+\binom{p-1}{2}-\binom{p-1}{3}+\dots+(-1)^{p-1}=(1-1)^{p-1}-1=-1.$$

---

[16]Meditationes algebraicae, Cambridge, 1770, 218; ed. 3, 1782, 380.

[17]On his biography see Nouv. Corresp. Math., 2, 1876, 110–114; M. Cantor, Bibliotheca math., (3), 3, 1902, 412; 4, 1903, 91.

[18]Nouv. Mém. Acad. Roy. Berlin, 2, 1773, année 1771, p. 125; Oeuvres, 3, 1869, 425. Cf. N. Nielsen, Danske Vidensk. Selsk. Forh., 1915, 520.

[19]Euler, Novi Comm. Ac. Petrop., 5, 1754–5, p. 6; Comm. Arith., 1, p. 213; 2, p. 532; Opera postuma, Petropoli, 1, 1862, p. 32.

Finally, Lagrange proved the converse of Wilson's theorem: If n divides $1+(n-1)!$, then n is a prime. For $n=4m+1$, n is a prime if $(2\cdot3\ldots2m)^2$ has the remainder -1 when divided by n. For $n=4m-1$, if $(2m-1)!$ has the remainder ±1.

L. Euler[20] also proved by induction from $x=n$ to $n+1$ that

$$(2) \qquad x!=a^x-x(a-1)^x+\binom{x}{2}(a-2)^x-\binom{x}{3}(a-3)^x+\ldots,$$

which reduces to (1) for $x=p-1$, $a=p$; and more generally,

$$(3)\ a^x-n(a-1)^x+\binom{n}{2}(a-2)^x-\ldots+(-1)^k\binom{n}{k}(a-k)^x+\ldots=\begin{cases}0 & \text{if } x<n\\ n! & \text{if } x=n.\end{cases}$$

D'Alembert[21] stated that the theorem that the difference of order m of a^m is $m!$ had been long known and gave a proof.

L. Euler[22] made use of a primitive root a of the prime p to prove Wilson's theorem (though his proof of the existence of a was defective). When $1, a, a^2, \ldots, a^{p-2}$ are divided by p, the remainders are $1, 2, 3, \ldots, p-1$ in some order. Hence $a^{(p-1)(p-2)/2}$ has the same remainder as $(p-1)!$. Taking $p>2$, we may set $p=2n+1$. Since a^n has the remainder -1, then $a^n a^{2n(n-1)}$, and hence also $(p-1)!$, has the remainder -1.

P. S. Laplace[23] proved Fermat's theorem essentially by the first method of Euler[10] without citing him: If a is an integer $<p$ not divisible by the prime p,

$$\frac{a^p}{a}=\frac{1}{a}(a-1+1)^p=\frac{1}{a}\{(a-1)^p+p(a-1)^{p-1}+\ldots+1\},$$

$$a^{p-1}-1=\frac{1}{a}\{(a-1)^p+1-a+hp(a-1)\}=\frac{a-1}{a}\{(a-1)^{p-1}-1+hp\}.$$

Hence by induction $a^{p-1}-1$ is divisible by p. For $a>p$, set $a=np+q$ and use the theorem for q.

He gave a proof of Euler's[14] generalization by the method of powering: if $n=p^\mu p_1^{\mu_1}\ldots$, where $p, p_1, \ldots$ are distinct primes, and if a is prime to n, then a^v-1 is divisible by n, where

$$v=n\left(\frac{p-1}{p}\right)\left(\frac{p_1-1}{p_1}\right)\ldots=qr,$$

$$q=p^{\mu-1}(p-1), \qquad r=p_1^{\mu_1-1}(p_1-1)p_2^{\mu_1-1}(p_2-1)\ldots.$$

Set $a^q=x$. Then $a^v-1=x^r-1$ is divisible by $x-1$. Using the binomial theorem and $a^{p-1}-1=hp$, we find that $x-1$ is divisible by p^μ.

[20]Novi Comm. Ac. Petrop., 13, 1768, 28–30.

[21]Letter to Turgot, Nov. 11, 1772, in unedited papers in the Bibliothèque de l'Institut de France. Cf. Bull. Bibl. Storia Sc. Mat. e Fis., 18, 1885, 531.

[22]Opuscula analytica, St. Petersburg, 1, 1783 [Nov. 15, 1773], p. 329; Comm. Arith., 2, p. 44; letter to Lagrange (Oeuvres, 14, p. 235), Sept. 24, 1773; Euler's Opera postuma, I, 583.

[23]De la Place, Théorie abrégée des nombres premiers, 1776, 16–23. His proofs of Fermat's and Wilson's theorems were inserted at the end of Bossut's Algèbre, ed. 1776, and reproduced by S. F. Lacroix, Traité du Calcul Diff. Int., Paris, ed. 2, vol. 3, 1818, 722–4, on p. 10 of which is a proof of (2) for $a=x$ by the calculus of differences.

From the $(p-1)$th order of differences for $x^{p-1}-1$,

$$(x+p-1)^{p-1}-1-(p-1)\{(x+p-2)^{p-1}-1\}+\binom{p-1}{2}\{(x+p-3)^{p-1}-1\}$$
$$-\ldots+x^{p-1}-1=(p-1)!.$$

Set $x=1$ and use Fermat's theorem. Hence $1+(p-1)!$ is divisible by p. E. Waring,[16] 1782, 380–2, made use of

$$x^r=x(x-1)\ldots(x-r+1)+Px(x-1)\ldots(x-r+2)$$
$$+Qx(x-1)\ldots(x-r+3)+\ldots+Hx(x-1)+Ix,$$

where $P=1+2+\ldots+(r-1)$, $Q=PA^1-B$, etc., B denoting the sum of the products of $1, 2, \ldots, r-1$ two at a time, and $A^1=1+2+\ldots+(r-2)$. Then

$$1^r+2^r+\ldots+x^r=\frac{1}{r+1}(x+1)x(x-1)\ldots(x-r+1)+\frac{P}{r}(x+1)x\ldots(x-r+2)$$

$$+\frac{Q}{r-1}(x+1)x\ldots(x-r+3)+\ldots+\frac{H}{3}(x+1)x(x-1)+\frac{I}{2}(x+1)x.$$

Take $r=x$ and let $x+1$ be a prime. By Fermat's theorem, $1^x, 2^x, \ldots, x^x$ each has the remainder unity when divided by $x+1$, so that their sum has the remainder x. Thus $1+x!$ is divisible by $x+1$.

Genty[24] proved the converse of Wilson's theorem and noted that an equivalent test for the primality of p is that p divide $(p-n)!(n-1)!-(-1)^n$. For $n=(p+1)/2$, the latter expression is $\{(\frac{p-1}{2})!\}^2\pm 1$ [Lagrange[18]].

Franz von Schaffgotsch[25] was led by induction to the fact (of which he gave no proof) that, if n is a prime, the numbers $2, 3, \ldots, n-2$ can be paired so that the product of the two in any pair is of the form $xn+1$ and the two of a pair are distinct. Hence, by multiplication, $2\cdot 3\ldots(n-2)$ has the remainder unity when divided by n, so that $(n-1)!$ has the remainder $n-1$. For example, if $n=19$, the pairs are $2\cdot 10$, $4\cdot 5$, $3\cdot 13$, $7\cdot 11$, $6\cdot 16$, $8\cdot 12$, $9\cdot 17$, $14\cdot 15$. Similarly, for n any power of a prime p, we can so pair the integers $<n-1$ which are not divisible by p. But for $n=15$, 4 and 4 are paired, also 11 and 11. Euler[26] had already used these associated residues (residua sociata).

F. T. Schubert[26a] proved by induction that the nth order of differences of $1^n, 2^n, \ldots$ is $n!$.

A. M. Legendre[27] reproduced the second proof by Euler[12] of Fermat's theorem and used the theory of differences to prove (2) for $a=x$. Taking $x=p-1$ and using Fermat's theorem, we get $(p-1)!\equiv(1-1)^p-1 \pmod{p}$.

[24]Histoire et mém. de l'acad. roy. sc. insc. de Toulouse, 3, 1788 (read Dec. 4, 1783), p. 91.
[25]Abhandlungen d. Böhmischen Gesell. Wiss., Prag, 2, 1786, 134.
[26]Opusc. anal., 1, 1783 (1772), 64, 121; Novi Comm. Ac. Petrop., 18, 1773, 85, §26; Comm. Arith. 1, 480, 494, 519.
[26a]Nova Acta Acad. Petrop., 11, ad annum 1793, 1798, mem., 174–7.
[27]Théorie des nombres, 1798, 181–2; ed. 2, 1808, 166–7.

C. F. Gauss[28] proved that, if n is a prime, $2, 3, \ldots, n-2$ can be associated in pairs such that the product of the two of a pair is of the form $xn+1$. This step completes Schaffgotsch's[25] proof of Wilson's theorem.

Gauss[29] proved Fermat's theorem by the method now known to be that used by Leibniz[4] and mentioned the fact that the reputed proof by Leibniz had not then been published.

Gauss[30] proved that if a belongs to the exponent t modulo p, a prime, then $a \cdot a^2 \cdot a^3 \ldots a^t \equiv (-1)^{t+1}$ (mod p). In fact, a primitive root ρ of p may be chosen so that $a \equiv \rho^{(p-1)/t}$. Thus the above product is congruent to ρ^k, where

$$k = (1+2+\ldots+t)\left(\frac{p-1}{t}\right) = \frac{(t+1)(p-1)}{2}.$$

Thus $\rho^k = \left(\rho^{\frac{p-1}{2}}\right)^{t+1} \equiv (-1)^{t+1}$ (mod p). When a is a primitive root, a, $a^2, \ldots, a^{p-1}$ are congruent to $1, 2, \ldots, p-1$ in some order. Hence $(p-1)! \equiv (-1)^p$. This method of proving Wilson's theorem is essentially that of Euler.[22]

Gauss[31] stated the generalization of Wilson's theorem: The product of the positive integers $<A$ and prime to A is congruent modulo A to -1 if $A = 4$, p^m or $2p^m$, where p is an odd prime, but to $+1$ if A is not of one of these three forms. He remarked that a proof could be made by use of associated numbers[28] with the difference that $x^2 \equiv 1$ (mod A) may now have roots other than ± 1; also by use of indices and primitive roots[30] of a composite modulus.

S. F. Lacroix[32] reproduced Euler's[13] third proof of Fermat's theorem without giving a reference.

James Ivory[33] obtained Fermat's theorem by a proof later rediscovered by Dirichlet.[40] Let N be any integer not divisible by the prime p. When the multiples $N, 2N, 3N, \ldots, (p-1)N$ are divided by p, there result p distinct positive remainders $<p$, so that these remainders are $1, 2, \ldots, p-1$ in some order.[34] By multiplication, $N^{p-1}Q = Q + mp$, where $Q = (p-1)!$. Hence p divides $N^{p-1}-1$ since it does not divide Q.

Gauss[35] used the last method in his proof of the lemma (employed in his third proof of the quadratic reciprocity law): If k is not divisible by the odd prime p, and if exactly μ of the least positive residues of $k, 2k, \ldots,$ $\frac{1}{2}(p-1)k$ modulo p exceed $p/2$, then $k^{(p-1)/2} \equiv (-1)^\mu$ (mod p). [Cf. Grunert.[45]]

[28]Disquisitiones Arith., 1801, arts. 24, 77; Werke, 1, 1863, 19, 61.
[29]Disq. Arith., art. 51, footnote to art. 50.
[30]Disq. Arith., art. 75.
[31]Disq. Arith., art. 78.
[32]Complément des élémens d'algèbre, Paris, ed. 3, 1804, 298–303; ed. 4, 1817, 313–7.
[33]New Series of the Math. Repository (ed. Th. Leybourn), vol. 1, pt. 2, 1806, 6–8.
[34]A fact known to Euler, Novi Comm. Acad. Petrop., 8, 1760–1, 75; Comm. Arith., 1, 275; and to Gauss, Disq. Arith., art. 23. Cf. G. Tarry, Nouv. Ann. Math., 18, 1899, 149, 292.
[35]Comm. soc. reg. sc. Gottingensis, 16, 1808; Werke, 2, 1–8. Gauss' Höhere Arith., German transl. by H. Maser, Berlin, 1889, p. 458.

J. A. Grunert[36] considered the series

$$[m, n] = n^m - \binom{n}{1}(n-1)^m + \binom{n}{2}(n-2)^m - \ldots,$$

to which Euler's (3) reduces for $a = n$, $x = m$, and proved that

$$[m, n] = n\{[m-1, n-1] + [m-1, n]\}.$$

This recursion formula gives

$$[m, n] = 0 \ (m = 0, 1, \ldots, n-1); \qquad [n, n] = n! \ [\text{cf. } (2)],$$

$$[n+1, n] = n! \binom{n+1}{2}, \qquad [n+2, n] = n! \binom{n+2}{3} \cdot \frac{3n+1}{4},$$

$$[n+3, n] = n! \binom{n+1}{2}\binom{n+3}{4}.$$

Any $[m, n]$ is divisible by $n!$. As by the proof of Lagrange,[18] $[m, n] + (-1)^n$ is divisible by $m+1$ if the latter is a prime $> n$. Again,

$$m!h^m = (x+mh)^m - m\{x+(m-1)h\}^m + \binom{m}{2}\{x+(m-2)h\}^m + \ldots + (-1)^m x^m,$$

which for $x = 0$, $h = 1$, gives $[m, m] = m!$.

W. G. Horner[37] proved Euler's theorem by generalizing Ivory's[33] method. If $r_1, \ldots, r_\varphi$ are the integers $< m$ and prime to m, then $r_1 N, \ldots, r_\varphi N$ have the r's as their residues modulo m.

P. F. Verhulst[38] gave Euler's proof[22] in a slightly different form.

F. T. Poselger[39] gave essentially Euler's[10] first proof.

G. L. Dirichlet[40] derived Fermat's and Wilson's theorems from a common source. Call m and n corresponding numbers if each is less than the prime p and if $mn \equiv a \pmod{p}$, where a is a fixed integer not divisible by p (thus generalizing Euler's[26] associated numbers). Each number $1, 2, \ldots$, $p-1$ has one and but one corresponding number. If $x^2 \equiv a \pmod{p}$ has no integral solution, corresponding numbers are distinct and

$$(p-1)! \equiv a^{(p-1)/2} \pmod{p}.$$

But if k is a positive integer $< p$ such that $k^2 \equiv a \pmod{p}$, the second root is $p-k$, and the product of the numbers $1, \ldots, p-1$, other than k and $p-k$, has the same residue as $a^{(p-3)/2}$, whence

$$(p-1)! \equiv -a^{(p-1)/2} \pmod{p}.$$

The case $a = 1$ leads to Wilson's theorem. By the latter, we have

$$a^{(p-1)/2} \equiv \pm 1 \pmod{p},$$

[36]Math. Abhandlungen, Erste Sammlung, Altona, 1822, 67–93. Some of the results were quoted by Grunert, Archiv Math. Phys., 32, 1859, 115–8. For an interpretation in factoring of [m, n], see Minetola[166] of Ch. X.

[37]Annals of Phil. (Mag. Chem. . . .), new series, 11, 1826, 81.

[38]Corresp. Math. Phys. (ed. Quételet), 3, 1827, 71.

[39]Abhand. Ak. Wiss. Berlin (Math.), 1827, 21.

[40]Jour. für Math., 3, 1828, 390; Werke, 1, 1889, 105. Dirichlet,[85] §34.

the sign being $+$ or $-$ according as $k^2 \equiv a \pmod{p}$ has or has not integral solutions (Euler's criterion). Squaring, we obtain Fermat's theorem. Finally, Dirichlet rediscovered the proof by Ivory.[33] [Cf. Moreau.[123]] J. Binet[41] also rediscovered the proof by Ivory.[33] A. Cauchy[42] gave a proof analogous to that by Euler.[10]

An anonymous writer[43] proved that if n is a prime the binomial coefficient $(n-1)_k$ has the residue $(-1)^k$ modulo n, so that

$$(1+x)^{n-1}-1 \equiv -x+x^2-\ldots+x^{n-1}, \quad (1+x)\{(1+x)^{n-1}-1\} \equiv x(x^{n-1}-1),$$

modulo n. Thus Fermat's theorem follows by induction on x as in the proof by Euler.[12]

V. Bouniakowsky[44] gave a proof of Euler's theorem similar to that by Laplace.[23] If $a \pm b$ is divisible by a prime p, $a^{p^{n-1}} \pm b^{p^{n-1}}$ is divisible by p^n, provided $p > 2$ when the sign is plus. Hence if $p, p', \ldots$ are distinct primes, $a^t \pm b^t$ is divisible by $N = p^n p'^{n'} \ldots$, where $t = p^{n-1} p'^{n'-1} \ldots$, if $a \pm b$ is divisible by $pp' \ldots$, provided the p's are > 2 if the sign is plus. Replace a by its $(p-1)$th power and b by 1 and use Fermat's theorem; we see that $a^e - 1$ is divisible by N if $e = \phi(N)$. The same result gives a generalization of Wilson's theorem[6]

$$\{(p-1)!\}^{p^{n-1}} + 1 \equiv 0 \pmod{p^n}.$$

He gave (ibid., 563–4) Gauss'[30] proof of Wilson's theorem.

J. A. Grunert[45] used the known fact that, if $0 < k < p$, then $k, 2k, \ldots$, $(p-1)k$ are congruent to $1, 2, \ldots, p-1$ in some order modulo p, a prime, to show that $kx \equiv 1 \pmod{p}$ has a unique root x. Wilson's theorem then follows as by Gauss. If (ibid., p. 1095) we square Gauss' formula,[35] we get Fermat's theorem.

Giovanni de Paoli[46] proved Fermat's and Euler's theorems. In

$$(x+1)^p = x^p + 1 + pS_x,$$

where p is a prime, S_x is an integer. Change x to $x-1, \ldots, 2, 1$ and add the resulting equations. Thus

$$x^p - x = p \sum_{z=1}^{x-1} S_z.$$

Replace x by x^m, divide by x^m and set $y = x^{p-1}$. Thus

$$y^m - 1 = pX_m, \qquad X_m = \Sigma S_z^m / x^m = \text{integer}.$$

Replace m by $2m, \ldots, (p-1)m$, add the resulting equations, and set $Y_m = 1 + X_m + X_{2m} + \ldots + X_{(p-1)m}$. Thus

$$y^{mp} - 1 = p(y^m - 1)Y_m = p^2 X_m Y_m.$$

[41] Jour. de l'école polytechnique, 20, 1831, 291 (read 1827). Cauchy, Comptes Rendus Paris, 12, 1841, 813, ascribed the proof to Binet.
[42] Exer. de math., 4, 1829, 221; Oeuvres, (2), 9, 263. Résumé analyt., Turin, 1, 1833, 10.
[43] Jour. für Math., 6, 1830, 100–6.
[44] Mém. Ac. Sc. St. Pétersbourg, Sc. Math. Phys. et Nat., (6), 1, 1831, 139 (read Apr. 1, 1829).
[45] Klügel's Math. Wörterbuch, 5, 1831, 1076–9.
[46] Opuscoli Matematici e Fisici di Diversi Autori, Milano, 1, 1832, 262–272.

Change m to $mp, \ldots, mp^{n-2}$. Thus

$$y^{mp^2} - 1 = p(y^{mp} - 1)Y_{mp} = p^3 X_m Y_m Y_{mp},$$
$$y^{mp^3} - 1 = p^4 X_m Y_m Y_{mp} Y_{mp^2}, \ldots, \qquad y^{mp^{n-1}} - 1 = p^n(\ ).$$

Hence $x^{\varphi(N)} - 1$ is divisible by N for $N = p^n$ and so for any N.
For x odd, $x^2 - 1$ is divisible by 8, and $x^{4m} - 1$ by $2(x^{2m} - 1)$. As above, he found that $x^t - 1$ is divisible by 2^i for $t = m \cdot 2^{i-2}$, $i > 2$. Thus, if $N = 2^i n$, n odd, $x^k - 1$ is divisible by N for $k = 2^{i-2}\phi(n)$.

A. L. Crelle[47] employed a fixed quadratic non-residue v of the prime p, and set $j^2 \equiv r_j$, $vj^2 \equiv v_j \pmod{p}$. By multiplication of

$$(p-j)^2 \equiv r_j, \qquad vj^2 \equiv v_j \pmod{p} \qquad \left(j = 1, \ldots, \frac{p-1}{2}\right)$$

and use of $v^{(p-1)/2} \equiv -1$, we get

$$-\{(p-1)!\}^2 \equiv \Pi r_j v_j \equiv (p-1)! \pmod{p}.$$

F. Minding[48] proved the generalized Wilson theorem. Let P be the product of the π integers $a, \beta, \ldots, < A$ and relatively prime to A. Let $A = 2^\mu p^m q^n r^k \ldots$, where $p, q, r, \ldots$ are distinct odd primes, and $m > 0$. Take a quadratic non-residue t of p and determine a so that $a \equiv t \pmod{p}$, $a \equiv 1 \pmod{2qr \ldots}$. Then a is an odd quadratic non-residue of A. Let $ax \equiv a \pmod{A}$. For $\beta \neq x$, a, let $\beta y \equiv a \pmod{A}$. Then $y \neq a$, x, β. In this way the π numbers $a, \beta, \ldots$ can be paired so that the product of the two in any pair is $\equiv a \pmod{A}$, whence $P \equiv a^{\pi/2} \pmod{A}$.
First, let $A = 2^\mu p^m$. Then $a^s \equiv -1 \pmod{p^m}$, $s = p^{m-1}(p-1)/2$, whence $P \equiv -1 \pmod{A}$ if $\mu = 0$ or 1. But, if $\mu > 1$,

$$a^{\frac{\pi}{2}} \equiv (-1)^{2^{\mu-1}} \equiv 1 \pmod{p^m}, \qquad a^{\frac{\pi}{2}} \equiv a^{2^{\mu-1}} \equiv 1 \pmod{2^\mu}, \qquad P \equiv +1 \pmod{A}.$$

Next, let $m > 1$, $n > 1$, in A. Raising the above $a^s \equiv -1$ to the power $2^{\mu-1}q^{n-1}(q-1) \ldots$, we get $a^{\pi/2} \equiv +1 \pmod{p^n}$. A like congruence holds moduli $q^n, r^k, \ldots$, and 2^μ, whence $P \equiv +1 \pmod{A}$.
Finally, let $A = 2^\mu$, $\mu > 1$. Then $a = -1$ is a quadratic non-residue of 2^μ and, as above, $P \equiv (-1)^l \pmod{A}$, $l = 2^{\mu-2}$. The proof of Fermat's theorem due to Ivory[33] is given by Minding on p. 32.

J. A. Grunert[49] gave Horner's[37] proof of Euler's theorem, attributing the case of a prime to Dirichlet instead of Ivory.[33] A part of the generalized Wilson theorem was proved as follows: Let $r_1, \ldots, r_q$ denote the positive integers $< p$ and prime to p. Let a be prime to p. In the table

$$r_1 a^2 r_1, r_2 a^2 r_1, \ldots, r_q a^2 r_1$$
$$\cdots \cdots \cdots \cdots \cdots \cdots \cdots$$
$$r_1 a^2 r_q, r_2 a^2 r_q, \ldots, r_q a^2 r_q$$

[47]Abh. Ak. Wiss. Berlin (Math.), 1832, 66. Reprinted.[65]
[48]Anfangsgründe der Höheren Arith., 1832, 75–78.
[49]Math. Wörterbuch, 1831, pp. 1072–3; Jour. für Math. 8, 1832, 187.

a single term of a row is $\equiv 1 \pmod{p}$. If this term be $r_k a^2 r_k$, replace it by $(p - r_k) a^2 r_k \equiv -1$. Next, if $r_n a^2 r_k \equiv \mp 1$, $r_n a^2 r_l \equiv \pm 1$, then $r_k + r_l = p$ and one of the r_n is replaced by $p - r_n$. Hence we may separate $r_1 a, \ldots, r_q a$ into $q/2$ pairs such that the product of the two of a pair is $\equiv \pm 1 \pmod{p}$. Taking $a = 1$, we get $r_1 \ldots r_q \equiv \pm 1 \pmod{p}$. The sign was determined only for the case p a prime (by Gauss' method).

A. Cauchy[50] derived Wilson's theorem from (1), page 62 above.

*Caraffa[51] gave a proof of Fermat's theorem.

E. Midy[52] gave Ivory's[33] proof of Fermat's theorem.

W. G. Horner[53] gave Euler's[14] proof of his theorem.

G. Libri[54] reproduced Euler's proof[12] without a reference.

Sylvester[55] gave the generalized Wilson theorem in the incomplete form that the residue is ± 1.

Th. Schönemann[56] proved by use of symmetric functions of the roots that if $z^n + b_1 z^{n-1} + \ldots = 0$ is the equation for the pth powers of the roots of $x^n + a_1 x^{n-1} + \ldots = 0$, where the a's are integers and p is a prime, then $b_i \equiv a_i^p \pmod{p}$. If the latter equation is $(x-1)^n = 0$, the former is $z^n - (n^p + pQ) z^{n-1} + \ldots = 0$, and yet is evidently $(z-1)^n = 0$. Hence $n^p \equiv n \pmod{p}$.

W. Brennecke[57] elaborated one of Gauss'[31] suggestions for a proof of the generalized Wilson theorem. For $a > 2$, $x^2 \equiv 1 \pmod{2^a}$ has exactly four incongruent roots, ± 1, $\pm (1 + 2^{a-1})$, since one of the factors $x \pm 1$, of difference 2, must be divisible by 2 and the other by 2^{a-1}. For p an odd prime, let $r_1, \ldots, r_\mu$ be the positive integers $< p^a$ and prime to p^a, taking $r_1 = 1$, $r_\mu = p^a - 1$. For $2 \leqq s \leqq \mu - 1$, the root x of $r_s x \equiv 1 \pmod{p^a}$ is distinct from r_1, r_μ, r_s. Thus $r_2 \ldots, r_{\mu-1}$ may be paired so that the product of the two of a pair is $\equiv 1 \pmod{p^a}$. Hence $r_1 \ldots r_\mu \equiv -1 \pmod{p^a}$. This holds also for modulus $2p^a$. For $a > 2$,

$$(2^{a-1} - 1)(2^{a-1} + 1) \equiv -1, \qquad r_1 \ldots r_\mu \equiv +1 \pmod{2^a}.$$

Finally, let $N = p^a M$, where M is divisible by an odd prime, but not by p. Then $m = \phi(M)$ is even. The integers $< N$ and prime to p are

$$r_j, r_j + p^a, r_j + 2p^a, \ldots, r_j + (M-1)p^a \qquad (j = 1, \ldots, \mu).$$

For a fixed j, we obtain m integers $< N$ and prime to N. Hence if $\{N\}$ denotes the product of all the integers $< N$ and prime to N,

$$\{N\} \equiv (r_1 \ldots r_\mu)^m \equiv 1 \pmod{p^a}.$$

For $N = p^a q^\beta \ldots$, $\{N\} \equiv 1 \pmod{q^\beta}, \ldots$, whence $\{N\} \equiv 1 \pmod{N}$.

---

[50]Résumé analyt., Turin, 1, 1833, 35.
[51]Elem. di mat. commentati da Volpicelli, Rome, 1836, I, 89.
[52]De quelques propriétés des nombres, Nantes, 1836.
[53]London and Edinb. Phil. Mag., 11, 1837, 456.
[54]Mém. divers savants ac. sc. Institut de France (math.), 5, 1838, 19.
[55]Phil. Mag., 13, 1838, 454 (14, 1839, 47); Coll. Math. Papers, 1, 1904, 39.
[56]Jour. für Math., 19, 1839, 290; 31, 1846, 288. Cf. J. J. Sylvester, Phil. Mag., (4), 18, 1859, 281.
[57]Jour. für Math., 19, 1839, 319.

A. L. Crelle[58] proved the generalized Wilson theorem. By pairing each root σ of $x^2 \equiv 1$ (mod s) with the root $s-\sigma$, and each integer $a < s$, prime to s and not a root, with its associated number a', where $aa' \equiv 1$ (mod s), we see that the product of all the integers $< s$ and prime to s is $\equiv +1$ or -1 (mod s) according as the number n of pairs of roots σ, $s-\sigma$ is even or odd. To find n, express s in every way as a product of two factors u, v, whose g. c. d. is 1 or 2; in the respective cases, each factor pair gives a single root σ or two roots. Treating four subcases at length it is shown that the number of factor pairs is 2^k in each case, where k is the number of distinct odd primes dividing s; and then that n is odd if $s = 4$, p^m or $2p^n$, but even if n is not of one of these three forms.

A. Cauchy[58a] proved Fermat's theorem as had Leibniz.[4]

V[59] (S. Earnshaw?) proved Wilson's theorem by Lagrange's method and noted that, if S_r is the sum of the products of the roots of $A_0 x^m + A_1 x^{m-1} + \ldots$ $\equiv 0$ (mod p) taken r at a time, then $A_0 S_i - (-1)^i A_i$ is divisible by p.

Paolo Gorini[60] proved Euler's theorem $b^t \equiv 1$ (mod Δ), where $t = \phi(\Delta)$, by arranging in order of magnitude the integers (A) p', p'', $\ldots$, $p^{(t)}$ which are less than Δ and prime to Δ. After omitting the numbers in (A) which are divisible by b, we obtain a set (B) q', $\ldots$, $q^{(l)}$. Let $q^{(\omega)}$ be the least of the latter which when increased by Δ gives a multiple of b:

(C) $$q^{(\omega)} + \Delta = p^{(\alpha)} b.$$

The numbers* (A) coincide with those in sets (B) and (D):

(D) $$p'b, \; p''b, \ldots, \; p^{(\alpha-1)}b.$$

Hence by multiplication and cancellation of p', $\ldots$, $p^{(\alpha-1)}$,

(F) $$q' \ldots q^{(l)} b^{\alpha-1} = p^{(\alpha)} \ldots p^{(l)}.$$

To each number (B) add the least multiple of Δ which gives a sum divisible by b, say (G) $q' + g'\Delta, \ldots, q^{(l)} + g^{(l)}\Delta$. The least of these is $q^{(\omega)} + \Delta = p^{(\alpha)}b$, by (C). Each number (G) is $< b\Delta$ and all are distinct. The quotients obtained by dividing the numbers (G) by b are prime to Δ and hence included among the $p^{(\alpha)}, \ldots, p^{(t)}$, whose number is $t - a + 1 = l$, so that each arises as a quotient. Hence

(H) $$\prod_{i=1}^{l}(q^{(i)} + g^{(i)}\Delta) = P\Delta + q' \ldots q^{(l)} = p^{(\alpha)} p^{(\alpha+1)} \ldots p^{(l)} b^{t-\alpha+1}.$$

Combine this with (F) to eliminate the p's. We get

$$q' \ldots q^{(l)} b^{\alpha-1} b^{t-\alpha+1} = P\Delta + q' \ldots q^{(l)}, \qquad q' \ldots q^{(l)}(b^t - 1) = P\Delta, \qquad b^t - 1 = Q\Delta.$$

[58]Jour. für Math., 20, 1840, 29–56. Abstract in Bericht Akad. Wiss. Berlin, 1839, 133–5.
[58a]Mém. Ac. Sc. Paris, 17, 1840, 436; Oeuvres, (1), 3, 163–4.
[59]Cambr. Math. Jour., 2, 1841, 79–81.
[60]Annali di Fisica, Chimica e Mat. (ed., G. A. Majocchi), Milano, 1, 1841, 255–7.
*To follow the author's steps, take $\Delta = 15$, $b = 2$, whence $t = 8$, $l = 4$, (A) 1, 2, 4, 7, 8, 11, 13, 14; (B) 1, 7, 11, 13; (C) $1 + 15 = 8 \cdot 2$, $p^{(\alpha)} = 8$, $a = 5$; (D) 2, 4, 8, 14; (F) $1 \cdot 7 \cdot 11 \cdot 13 \cdot 2^4 = 8 \cdot 11 \cdot 13 \cdot 14$; (G) $1 + 15$, $7 + 15$, $11 + 15$; $13 + 15$, each $g = 1$; the quotients of the latter by 2 are 8, 11, 13, 14, viz., last four in (A); (H) $P.15 + 1.7.11.13 = 8.11.13.14.2^4$; the second member is $1 \cdot 7 \cdot 11 \cdot 13 \cdot 2^8$ by (F). Hence $1 \cdot 7 \cdot 11 \cdot 13 \; (2^8 - 1) = 15P$.

E. Lionnet[61] proved that, if p is an odd prime, the sum of the mth powers of $1, \ldots, p-1$ is divisible by p for $0 < m < p-1$. Hence the sum P_m of the products of $1, \ldots, p-1$ taken m at a time is divisible by p [Lagrange[18]]. Since

$$(1+1)(1+2) \ldots (1+p-1) = 1 + P_1 + P_2 + \ldots + P_{p-2} + (p-1)!,$$

$1 + (p-1)!$ is divisible by p.

E. Catalan[62] gave the proofs by Ivory[33] and Horner.[37] C. F. Arndt[63] gave Horner's proof; and proved the generalized Wilson theorem by associated numbers. O. Terquem[64] gave the proofs by Gauss[28] and Dirichlet.[40]

A. L. Crelle[65] republished his proof[47] of Wilson's theorem, as well as that by Gauss[30] and Dirichlet.[40] Crelle[66] gave two proofs of the generalized Wilson theorem, essentially that by Minding[48] and that given by himself.[58] If μ is the number of distinct odd prime factors of z, and 2^m is the highest power of 2 dividing z, and r is a quadratic residue of z, then (p. 150) the number n of pairs of roots $\pm x$ of $x^2 \equiv r \pmod{z}$ is $2^{\mu-1}$ if $m = 0$ or 1, 2^μ if $m = 2$, $2^{\mu+1}$ if $m > 2$. Using the fact (p. 122) that the quadratic residues of z are the $e = \phi(z)/(2n)$ roots of $r^e \equiv 1 \pmod{z}$, it is shown (p. 173) that, if v is any integer prime to z, $v^{\phi(z)/n} \equiv 1 \pmod{z}$, "a perfection of the Euler-Fermat theorem."

L. Poinsot[67] failed in his attempt to prove the generalized Wilson theorem. He began as had Crelle.[58] But he stated incorrectly that the number n of pairs of roots $\pm x$ of $x^2 \equiv 1 \pmod{s}$ equals the number v of ways of expressing s as a product of two factors P, Q whose g. c. d. is 1 or 2. For each pair $\pm x$, it is implied that $x-1$ and $x+1$ uniquely determine P, Q. For $s = 24$, $n = v = 4$; but for the root $x = 7$ (or for $x = 17$), $x \pm 1$ yield $P, Q = 3, 8$, and $6, 4$. To correct another error by Poinsot, let μ be the number of distinct odd prime factors of s and let 2^m be the highest power of 2 dividing s; then $v = 2^{\mu-1}$, 2^μ, $3 \cdot 2^{\mu-1}$ or $2^{\mu+1}$, according as $m = 0$, 1, 2, or ≥ 3, whereas [Crelle[66]] $n = 2^{\mu-1}$, $2^{\mu-1}$, 2^μ, $2^{\mu+1}$. No difficulty is met (pp. 53–5) in case the modulus is a power of a prime. He noted (p. 33) that if $r_1, r_2, \ldots$ are the integers $< N$ and prime to N, and π is their product, they are congruent modulo N to $\pi/r_1, \pi/r_2, \ldots$, whence $\pi \equiv \pi^{r-1} \pmod{N}$, where $\nu = \phi(N)$. Thus, by Euler's theorem, $\pi^2 \equiv 1$. This does not imply that $\pi \equiv \pm 1$ as cited by Aubry,[137] pp. 300–1.

Poinsot (p. 51) proved Euler's theorem by considering a regular polygon of N sides. Let x be prime to N and $< N$. Join any vertex with the xth vertex following it, the new vertex with the xth vertex following it, etc., thus defining a regular (star) polygon of N sides. With the same x, derive

[61]Nouv. Ann. Math., 1, 1842, 175–6.
[62]Ibid., 462–4.
[63]Archiv Math. Phys., 2, 1842, 7, 22, 23.
[64]Nouv. Ann. Math., 2, 1843, 193; 4, 1845, 379.
[65]Jour. für Math., 28, 1844, 176–8.
[66]Ibid., 29, 1845, 103–176.
[67]Jour. de Math., 10, 1845, 25–30.   German exposition by J. A. Grunert, Archiv Math. Phys., 7, 1846, 168, 367.

similarly a new N-gon, etc., until the initial polygon is reached.[68] The
number μ of distinct polygons thus obtained is seen to be a divisor of $\phi(N)$,
the number of polygons corresponding to the various x's. If in the initial
polygon we take the x^μth vertex following any one, etc., we obtain the
initial polygon. Hence x^μ and thus also $x^{\varphi(N)}$ has the remainder unity when
divided by N. [When completed this proof differs only slightly from that
by Euler.[14]]

E. Prouhet[69] modified Poinsot's method and obtained a correct proof
of the generalized Wilson theorem. Let r be the number of roots of $x^2 \equiv 1$
(mod N), and w the number of ways of expressing N as a product of two
relatively prime factors. If $N = 2^m p_1^{r_1} \ldots p_\mu^{r_\mu}$, where the p's are distinct
odd primes, evidently $w = 2^\mu$ if $m > 0$, $w = 2^{\mu-1}$ if $m = 0$. By considering
divisors of $x \pm 1$, it is proved that $r = 2w$ if $m = 0$ or 2, $r = w$ if $m = 1$, $r = 4w$ if
$m > 2$. Hence $r = 2^\mu$ if $m = 0$ or 1, $2^{\mu+1}$ if $m = 2$, $2^{\mu+2}$ if $m > 2$. By Crelle,[58]
the product P of the integers $< N$ and prime to N is $\equiv (-1)^{r/2}$ (mod N).
Thus for $\mu > 0$, $P \equiv +1$ unless $m = 0$ or 1, $\mu = 1$, viz., $N = p^r$ or $2p^r$; while, for
$\mu = 0$, $N = 2^m$, $m > 2$, we have $r = 4$, $P \equiv +1$.

Friderico Arndt[70] elaborated Gauss'[31] second suggestion for a proof of
the generalized Wilson theorem. Let g be a primitive root of the modulus
p^n or $2p^n$, where p is an odd prime. Set $v = \phi(p^n)$. Then $g, g^2, \ldots, g^v$ are
congruent to the numbers less than the modulus and prime to it. If P is
the product of the latter, $P \equiv g^{v(v+1)/2}$. But $g^{v/2} \equiv -1$. Hence $P \equiv -1$.
Next, if $n > 2$, the product of the incongruent numbers belonging to an
exponent 2^{n-m} is $\equiv 1$ (mod 2^n). Next, consider the modulus $M = AB$,
where A and B are relatively prime. The positive integers $< M$ and prime
to M are congruent modulo M to $Ay_i + Bx_j$, where the x_i are $< A$ and prime
to A, the y_i are $< B$ and prime to B. But, if $a = \phi(A)$,

$$\pi_1 = \prod_{j=1}^{a}(Ay_1 + Bx_j) \equiv B^a x_1 \ldots x_a \equiv x_1 \ldots x_a \pmod{A},$$
$$P = \pi_1 \pi_2 \ldots \equiv (x_1 \ldots x_a)^{\varphi(B)} \pmod{A}.$$

By resolving M into a product of powers of primes and applying the above
results, we determine the sign in $P \equiv \pm 1$ (mod M).

J. A. Grunert[71] proved that if a prime $n+1 > 2$ divides no one of the
integers $a_1, \ldots, a_n$, nor any of their differences, it divides $a_1 a_2 \ldots a_n + 1$, and
stated that this result is much more general than Wilson's theorem (the
case $a_j = j$). But the generalization is only superficial since $a_1, \ldots, a_n$ are
congruent modulo $n+1$ to $1, \ldots, n$ in some order. His proof employed
Fermat's theorem and certain complex equations involving products of
differences of n numbers and sums of products of n numbers taken m at
a time.

J. F. Heather[72] gave without reference the first results of Grunert.[36]

[68]Cf. P. Bachmann, Die Elemente der Zahlentheorie, 1892, 19–23.
[69]Nouv. Ann. Math., 4, 1845, 273–8.
[70]Jour. für Math., 31, 1846, 329–332.
[71]Archiv Math. Phys., 10, 1847, 312.
[72]The Mathematician, London, 2, 1847, 296.

A. Lista[73] gave Lagrange's proof of Wilson's theorem.

V. Bouniakowsky[74] gave Euler's[22] proof.

P. L. Tchebychef[75] concluded from Fermat's theorem that

$$(x-1)(x-2)\ldots(x-p+1)-x^{p-1}+1\equiv 0 \pmod{p}$$

is an identity if p is a prime.   Hence if s_j is the sum of the products of $1,\ldots,$ $p-1$ taken j at a time, $\varepsilon_j\equiv 0$ $(j<p-1)$, $s_{p-1}\equiv -1 \pmod{p}$, the last being Wilson's theorem.

Sir F. Pollock[76] gave an incomplete statement and proof of the generalized Wilson theorem by use of associated numbers.   Likewise futile was his attempt to extend Dirichlet's[40] method [not cited] of association into pairs with the product$\equiv a \pmod{m}$ to the case of a composite m.

E. Desmarest[77] gave Euler's[13] proof of Fermat's theorem.

O. Schlömilch[77a] considered the quotient

$$\{n^p - \binom{n}{1}(n-1)^p + \binom{n}{2}(n-2)^p - \ldots\}/n!.$$

J. J. Sylvester[78] took $x=1, 2,\ldots, p-1$ in turn in

$$(x-1)(x-2)\ldots(x-p+1)=x^{p-1}+A_1x^{p-2}+\ldots+A_{p-1},$$

where p is a prime.   Since $x^{p-1}\equiv 1 \pmod{p}$, there result $p-1$ congruences linear and homogeneous in $A_1,\ldots, A_{p-2}, A_{p-1}+1$, the determinant of whose coefficients is the product of the differences of $1, 2,\ldots, p-1$ and hence not divisible by p.   Thus $A_1\equiv 0,\ldots, A_{p-1}+1\equiv 0$, the last giving Wilson's theorem.

W. Brennecke[79] proved Euler's theorem by the methods of Horner[37] and Laplace,[23] noting that

$$(a^{p-1})^p\equiv 1 \pmod{p^2}, \qquad (a^{p-1})^{p^2}\equiv 1 \pmod{p^3},\ldots$$

He gave the proof by Tchebychef[75] and his own proof.[57]

J. T. Graves[80] employed $nx\equiv n+1 \pmod{p}$, where p is a prime, and stated that, for $n=1,\ldots, p-1$, then $x\equiv 2,\ldots, p$ in some order.   Also $x\equiv p$ for $n=p-1$.   Hence $2\cdot 3\ldots(p-1)\equiv p-1 \pmod{p}$.

H. Dürège[81] obtained (2) for $a=x$ and Grunert's[36] results on the series $[m, n]$ by use of partial fractions for the reciprocal of $x(x-1)\ldots(x-n)$.

E. Lottner[82] employed for the same purpose infinite trigonometric and algebraic series, obtaining recursion formulæ for the coefficients.

[73]Periodico Mensual Ciencias Mat. y Fis., Cadiz, 1, 1848, 63.

[74]Bull. Ac. Sc. St. Pétersbourg, 6, 1848, 205.

[75]Theorie der Congruenzen, 1849 (Russian); in German, 1889, §19.   Same proof by J. A. Serret, Cours d'algèbre supérieure, ed. 2, 1854, 324.

[76]Proc. Roy. Soc. London, 5, 1851, 664.

[77]Théorie des nombres, Paris, 1852, 223-5.

[77a]Jour. für Math., 44, 1852, 348.

[78]Cambridge and Dublin Math. Jour., 9, 1854, 84; Coll. Math. Papers, 2, 1908, 10.

[79]Einige Sätze aus den Anfangsgründen der Zahlenlehre, Progr. Realschule Posen, 1855.

[80]British Assoc. Report, 1856, 1-3.

[81]Archiv Math. Phys., 30, 1858, 163-6.

[82]Ibid., 32, 1859, 111-5.

J. Toeplitz[83] gave Lagrange's proof of Wilson's theorem.

M. A. Stern[84] made use of the series for log $(1-x)$ to show that

$$1+x+x^2+\ldots=\frac{1}{1-x}=e^{x+x^2/2+x^3/3+\cdots}.$$

Multiply together the series for e^x, $e^{x^2/2}$, etc. By the coefficient of x^p,

$$1=\frac{1}{p!}+s+\frac{1}{p}, \qquad s=\frac{1/2}{(p-2)!}+\ldots$$

Take p a prime. No term of s has a factor p in the denominator. Hence

$$(1-s)\cdot(p-1)!=\frac{1+(p-1)!}{p}=\text{integer}.$$

V. A. Lebesgue[85] obtained Wilson's theorem by taking $x=p-1$ in

$$p\sum_{k=1}^{x} k(k+1)\ldots(k+p-2)=x(x+1)\ldots(x+p-1).$$

If P is a composite number $\neq 4$, $(P-1)!$ is divisible by P. He (p. 74) attributed Ivory's[33] proof of Fermat's theorem to Gauss, without reference.

G. L. Dirichlet[86] gave Horner's[37] and Euler's[14] proof of Euler's theorem and derived it from Fermat's by the method of powering. His proof (§38) of the generalized Wilson theorem is by associated numbers, but is somewhat simpler than the analogous proofs.

Jean Plana[87] used the method of powering. Let $N=p^k p_1^{k_1}\ldots$ For M prime to N, $M^{p-1}=1+pQ$. Hence

$$M^{\varphi(p^k)}=(1+pQ)^{p^{k-1}}=1+p^k U, \qquad M^{\varphi(p_1^{k_1})}=1+p_1^{k_1}U_1,\ldots$$

Thus for $e=\varphi(p^k p_1^{k_1})$, M^e-1 is divisible by p^k and $p_1^{k_1}$ and hence by their product, etc. Plana gave also a modification of Lagrange's proof of Wilson's theorem by use of (2); take $x=a=p-1$, subtract the expansion of $(1-1)^{p-1}$ and write the resulting series in reverse order:

$$(p-1)!+1=\binom{p-1}{2}(2^{p-1}-1)-\binom{p-1}{3}(3^{p-1}-1)+\ldots$$
$$-\binom{p-1}{p-2}\{(p-2)^{p-1}-1\}+\{(p-1)^{p-1}-1\}.$$

H. F. Talbot[88] gave Euler's[12] proof of Fermat's theorem.

J. Blissard[88a] proved the last statement of Euler.[9]

C. Sardi[89] gave Lagrange's proof of Wilson's theorem.

P. A. Fontebasso[90] proved (2) for $x=a$ by finding the first term of the ath order of differences of y^a, $(y+h)^a$, $(y+2h)^a$, ... and then setting $y=0$, $h=1$.

[83]Archiv Math. Phys., 32, 1859, 104.
[84]Lehrbuch der Algebraischen Analysis, Leipzig, 1860, 391.
[85]Introd. théorie des nombres, Paris, 1862, 80, 17.
[86]Zahlentheorie (ed. Dedekind), §§19, 20, 127, 1863; ed. 2, 1871; ed. 3, 1879, ed. 4, 1894.
[87]Mem. Acad. Turin, (2), 20, 1863, 148-150.
[88]Trans. Roy. Soc. Edinburgh, 23, 1864, 45-52.
[88a]Math. Quest. Educ. Times, 6, 1866, 26-7.
[89]Giornale di Mat., 5, 1867, 371-6.
[90]Saggio di una introd. arit. trascendente, Treviso, 1867, 77-81.

C. A. Laisant and E. Beaujeux[91] used the period $a_1 \ldots a_n$ of the periodic fraction to base B for the irreducible fraction p_1/q, where q is prime to B. If $p_2, \ldots, p_n$ are the successive remainders,

$$Bp_1 = a_1q + p_2, \qquad Bp_2 = a_2q + p_3, \ldots, \qquad Bp_n = a_nq + p_1.$$

Starting with the second equation, we obtain the period $a_2 \ldots a_n a_1$ for p_2/q. Similarly for $p_3/q, \ldots, p_n/q$. Thus the $f = \varphi(q)$ irreducible fractions with denominator q separate into sets of n each. Hence $f = kn$. Since $B^n \equiv 1$, $B^f \equiv 1 \pmod{q}$.

L. Ottinger[92] employed differential calculus to show that, in

$$P = (a+d)(a+2d)\ldots\{a+(p-1)d\} = a^{p-1} + C_1^{p-1}a^{p-2}d + C_2^{p-1}a^{p-3}d^2 + \ldots,$$

$$rC_r^{p-1} = \sum_{q=1}^{r} \frac{qp(p-1)\ldots(p-q)}{q+1} C_{r-q}^{p-q-2} \qquad (r \leqq p-2),$$

C_r^k being the sum of the products of $1, 2, \ldots, k$ taken r at a time. Hence, if p is a prime, C_r^{p-1} $(r = 1, \ldots, p-2)$ is divisible by p, and

$$P \equiv a^{p-1} + d \cdot 2d \ldots (p-1)d \pmod{p}.$$

For $a = d = 1$, this gives $0 \equiv 1 + (p-1)! \pmod{p}$.

H. Anton[93] gave Gauss'[28] proof of Wilson's theorem.

J. Petersen[94] proved Wilson's theorem by dividing the circumference of a circle into p equal parts, where p is a prime, and marking the points $1, \ldots, p$. Designate by $12 \ldots p$ the polygon obtained by joining 1 with 2, 2 with 3, $\ldots$, p with 1. Rearranging these numbers we obtain new polygons, not all convex. While there are $p!$ rearrangements, each polygon can be designated in $2p$ ways [beginning with any one of the p numbers as first and reading forward or backward], so that we get $(p-1)!/2$ figures. Of these $\frac{1}{2}(p-1)$ are regular. The others are congruent in sets of p, since by rotation any one of them assumes p positions. Hence p divides $(p-1)!/2 - (p-1)/2$ and hence $(p-2)! - 1$. Cf. Cayley[101].

To prove Fermat's theorem, take p elements from q with repetitions in all ways, that is, in q^p ways. The q sets with elements all alike are not changed by a cyclic permutation of the elements, while the remaining $q^p - q$ sets are permuted in sets of p. Hence p divides $q^p - q$. [Cf. Perott,[126] Bricard.[131]]

F. Unferdinger[95] proved by use of series of exponentials that

$$z^n - \binom{m}{1}(z-1)^n + \binom{m}{2}(z-2)^n - \ldots + (-1)^m\binom{m}{m}(z-m)^n$$

[91]Nouv. Ann. Math., (2), 7, 1868, 292–3.
[92]Archiv Math. Phys., 48, 1868, 159–185.
[93]Ibid., 49, 1869, 297–8.
[94]Tidsskrift for Mathematik, (3), 2, 1872, 64–65 (Danish).
[95]Sitzungsberichte Ak. Wiss. Wien, 67, 1873, II, 363.

is zero if $n < m$, but, if $n \geq m$, equals

$$E_m + \binom{z-m}{1} E_{m+1} + \binom{z-m}{2} E_{m+2} + \ldots + \binom{z-m}{n-m} E_n,$$

where

$$E_k = k^n - \binom{k}{1}(k-1)^n + \binom{k}{2}(k-2)^n - \ldots + (-1)^{k-1}\binom{k}{k-1}1^n.$$

For $n = m$, the initial sum equals $E_m = m!$.

P. Mansion[96] noted that Euler's theorem may be identified with a property of periodic fractions [cf. Laisant[91]]. Let N be prime to R. Taking R as the base of a scale of notation, divide $100 \ldots$ by N and let $q_1 \ldots q_n$ be the repetend. Then $(R^n - 1)/N = q_1 \ldots q_n$. Unless the n remainders r_i exhaust the integers $< N$ and prime to N, we divide $r_1' 00 \ldots$ by N, where r_1' is one of the integers distinct from the r_i, and obtain n new remainders r_i'. In this way it is seen that n divides $\varphi(N)$, so that N divides $R^{\varphi(N)} - 1$. [At bottom this is Euler's[14] proof.]

P. Mansion[97] reproduced this proof, made historical remarks on the theorem and indicated an error by Poinsot.[67]

Franz Jorcke[98] reproduced Euler's[22] proof of Wilson's theorem.

G. L. P. v. Schaewen[99] proved (2) with a changed to $-p$, by expanding the binomials.

Chr. Zeller[100] proved that, for $n \neq 4$,

$$n^x - (n-1)(n-1)^x + \binom{n-1}{2}(n-2)^x - \binom{n-1}{3}(n-3)^x + \ldots$$

is divisible by n unless n is a prime such that $n-1$ divides x, in which case the expression is $\equiv -1 \pmod{n}$.

A. Cayley[101] proved Wilson's theorem as had Petersen.[94]

E. Schering[102] took a prime to $m = 2^\pi p_1^{\pi_1} \ldots p_\mu^{\pi_\mu}$, where the p's are distinct odd primes and proved that $x^2 \equiv a \pmod{m}$ has roots if and only if a is a quadratic residue of each p_i and if $a \equiv 1 \pmod 4$ when $\pi = 2$, $a \equiv 1 \pmod 8$ when $\pi > 2$, and then has $\psi(m)$ roots, where $\psi(m) = 2^\mu$, $2^{\mu+1}$ or $2^{\mu+2}$, according as $\pi < 2$, $\pi = 2$, or $\pi > 2$. Let a be a fixed quadratic residue of m and denote the roots by $\pm a_j$ $(j = 1, \ldots, \psi/2)$. Set $a_j' = m - a_j$. The $\phi(m) - \psi(m)$ integers $< m$ and prime to m, other than the a_j, a_j', may be denoted by a_j, a_j' $(j = \frac{1}{2}\psi + 1, \ldots, \frac{1}{2}\phi)$, where $a_j a_j' \equiv a \pmod m$. From the latter and $-a_j a_j' \equiv a$ $(j = 1, \ldots, \psi/2)$, we obtain, by multiplication,

$$a^{\frac{1}{2}\varphi(m)} \equiv (-1)^{\frac{1}{2}\psi(m)} r_1 \ldots r_\varphi \pmod m,$$

[96]Messenger Math., 5, 1876, 33 (140); Nouv. Corresp. Math., 4, 1878, 72–6.

[97]Théorie des nombres, 1878, Gand (tract).

[98]Über Zahlenkongruenzen, Progr. Fraustadt, 1878, p. 31.

[99]Die Binomial Coefficienten, Progr. Saarbrücken, 1881, p. 20.

[100]Bull. des sc. math. astr., (2), 5, 1881, 211–4.

[101]Messenger of Math., 12, 1882–3, 41; Coll. Math. Papers, 12, p. 45.

[102]Acta Math., 1, 1882, 153–170; Werke, 2, 1909, 69–86.

where the r_j are the integers $<m$ and prime to m. Taking $a=1$, we have the generalized Wilson theorem. Applying a like argument when a is a quadratic non-residue of m [Minding[48]], we get

$$a^{\frac{1}{2}\varphi(m)} \equiv r_1 \ldots r_\varphi \equiv (-1)^{\frac{1}{2}\psi(m)} \pmod{m}.$$

This investigation is a generalization of that by Dirichlet.[40]

E. Lucas[103] wrote X_p for $x(x+1)\ldots(x+p-1)$, and Γ_p^q for the sum of the products of $1,\ldots,p$ taken q at a time. Thus

$$x^p + \Gamma_{p-1}^1 x^{p-1} + \ldots + \Gamma_{p-1}^{p-1} x = X_p.$$

Replacing p by $1,\ldots,n$ in turn and solving, we get

$$x^n = X_n + \Delta_1 X_{n-1} + \ldots + \Delta_{n-1} X_1,$$

where

$$(-1)^{n-p+1}\Delta_{n-p+1} = \begin{vmatrix} \Gamma^1 \Gamma^2 \ldots & \Gamma^{n-p+1} \\ 1 \ \Gamma^1 \ldots & \Gamma^{n-p} \\ \cdot \ \cdot \ \cdot \ \cdot & \cdot \\ 0 \ 0 \ldots 1 \ \Gamma^1 \end{vmatrix},$$

the subscript $p-1$ on the Γ's being dropped. After repeating the argument by Tchebychef[75], Lucas noted that, if p is an odd prime, $\Delta_{n-p+1} \equiv 1$ or 0 (mod p), according as $p-1$ is or is not a divisor of n.

G. Wertheim[104] gave Dirichlet's[86] proof of the generalized Wilson theorem; also the first step in the proof by Arndt.[70]

W. E. Heal[105] gave without reference Euler's[14] proof.

E. Catalan[106] noted that if $2n+1$ is composite, but not the square of a prime, $n!$ is divisible by $2n+1$; if $2n+1$ is the square of a prime, $(n!)^2$ is divisible by $2n+1$.

C. Garibaldi[107] proved Fermat's theorem by considering the number N of combinations of ap elements p at a time, a single element being selected from each row of the table

$$e_{11} \ e_{12} \ldots e_{1a}$$
$$\ldots\ldots\ldots$$
$$e_{p1} \ e_{p2} \ldots e_{pa}.$$

From all possible combinations are to be omitted those containing elements from exactly n rows, for $n=1,\ldots,p-1$. Let A_n denote the number of combinations p at a time of an elements forming n rows, such that in each combination occur elements from each row. Then

$$N = \binom{ap}{p} - \sum_{n=1}^{p-1} \binom{p}{n} A_n.$$

[103]Bull. Soc. Math. France, 11, 1882–3, 69–71; Mathesis, 3, 1883, 25–8.
[104]Elemente der Zahlentheorie, 1887, 186–7; Anfangsgründe der Zahlenlehre, 1902, 343–5 (331–2).
[105]Annals of Math., 3, 1887, 97–98.
[106]Mém. soc. roy. sc. Liège, (2), 15, 1888 (Mélanges Math., III, 1887, 139).
[107]Giornale di Mat., 26, 1888, 197.

Take each $e_{ij}=1$; then $N=a^p$ since the number of the specified combinations becomes the sum of all products of p factors unity, one from each row of the table. Thus

$$a^p \equiv \binom{ap}{a} \equiv a \pmod{p}.$$

R. W. Genese[108] proved Euler's theorem essentially as did Laisant.[91] M. F. Daniëls[109] proved the generalized Wilson theorem. If $\psi(n)$ denotes the product of the integers $<n$ and prime to n, he proved by induction that $\psi(p^\pi) \equiv -1 \pmod{p^\pi}$ for p an odd prime. For, if $\rho_1, \ldots, \rho_n$ are the integers $<p^\pi$ and prime to it, then $\rho_1+jp^\pi, \ldots, \rho_n+jp^\pi$ $(j=0, 1, \ldots, p-1)$ are the integers $<p^{\pi+1}$ and prime to it. He proved similarly by induction that $\psi(2^\pi) \equiv +1 \pmod{2^\pi}$ if $\pi>2$. Evidently $\psi(2) \equiv 1 \pmod 2$, $\psi(4) \equiv -1 \pmod 4$. If $m=a^\alpha b^\beta \ldots$ and $n=l^\lambda$, where l is a new prime, then $\psi(m) \equiv \epsilon \pmod m$, $\psi(n) \equiv \eta \pmod n$ lead by the preceding method to $\psi(mn) \equiv \epsilon^{\varphi(n)} \pmod m$, viz., 1, unless $n=2$. The theorem now follows easily.

E. Lucas[110] noted that, if x is prime to $n=AB\ldots$, where $A, B, \ldots$ are powers of distinct primes, and if ϕ is the l. c. m. of $\phi(A), \phi(B), \ldots$, then $x^\phi \equiv 1 \pmod n$. In case $A=2^k, k>2$, we may replace $\phi(A)$ by its half. To get a congruence holding whether or not x is prime to n, multiply the former congruence by x^σ, where σ is the greatest exponent of the prime factors of n. Note that $\phi+\sigma<n$ [Bachmann[129, 143]]. Carmichael[139] wrote $\lambda(n)$ for ϕ.

E. Lucas[111] found $\Delta^{p-1}x^{p-1}$ in two ways by the theory of differences. Equating the two results, we have

$$(p-1)! = (p-1)^{p-1} - \binom{p-1}{1}(p-2)^{p-1} + \ldots - \binom{p-1}{p-2}1^{p-1}.$$

Each power on the right is $\equiv 1 \pmod p$. Thus

$$(p-1)! \equiv (1-1)^{p-1} - 1 \equiv -1 \pmod p.$$

P. A. MacMahon[112] proved Fermat's theorem by showing that the number of circular permutations of p distinct things n at a time, repetitions allowed, is

$$\frac{1}{n}\Sigma\phi(d)p^{n/d},$$

where d ranges over the divisors of n. For n a prime, this gives

$$p^n+(n-1)p \equiv 0, \qquad p^n \equiv p \pmod n.$$

Another specialization led to Euler's generalization.

E. Maillet[113] applied Sylow's theorem on subgroups whose order is the highest power p^h of a prime p dividing the order m of a group, viz.,

[108]British Association Report, 1888, 580–1.
[109]Lineaire Congruenties, Diss. Amsterdam, 1890, 104–114.
[110]Bull. Ac. Sc. St. Pétersbourg, 33, 1890, 496.
[111]Mathesis, (2), 1, 1891, 11; Théorie des nombres, 1891, 432.
[112]Proc. London Math. Soc., 23, 1891–2, 305–313.
[113]Recherches sur les substitutions, Thèse, Paris, 1892, 115.

$m=pN(1+np)$, when $h=1$. For the symmetric group on p letters, $m=p!$ and $N=p-1$, so that $(p-1)!\equiv-1$ (mod p). There is exhibited a special group for which $m=pa^p$, $N=a$, whence $a^p\equiv a$ (mod p).

G. Levi[114] failed in his attempt to prove Wilson's theorem. Let b and $a=(p-1)b$ have the least positive residues r_1 and r when divided by p. Then $r+r_1=p$. Multiply $b/p=q+r_1/p$ by $p-1$. Thus $r_1(p-1)$ has the same residue as a, so that

$$r_1(p-1)=r+mp, \qquad \frac{a}{p}=q(p-1)+m+\frac{r}{p}.$$

He concluded that $r_1(p-1)=r$, falsely, as the example $p=5$, $b=7$, shows. He added the last equation to $r+r_1=p$ and concluded that $r_1=1$, $r=p-1$, so that $(a+1)/p$ is an integer. The fact that this argument is independent of Levi's initial choice that $b=(p-2)!$ and his assumption that p is a prime shows that the proof is fallacious.

Axel Thue[115] obtained Fermat's theorem by adding

$$a^p-(a-1)^p=1+kp, \quad (a-1)^p-(a-2)^p=1+hp, \quad \ldots, \quad 1^p-0^p=1$$

[Paoli[46]]. Then the differences $\Delta^1 F(j)$ of the first order of $F(x)=x^{p-1}$ are divisible by p for $j=1,\ldots,p-2$; likewise $\Delta^2 F(1),\ldots,\Delta^{p-2}F(1)$. By adding

$$\Delta^{j+1}F(0)=\Delta^j F(1)-\Delta^j F(0) \quad (j=1,\ldots,p-2),$$

we get

$$-\Delta^{p-1}F(0)\doteq1+\Delta^1 F(1)-\Delta^2 F(1)+\ldots+\Delta^{p-2}F(1), \quad (p-1)!+1\equiv0 \text{ (mod }p).$$

N. M. Ferrers[116] repeated Sylvester's[78] proof of Wilson's theorem.

M. d'Ocagne[117] proved the identity in r:

$$(r+1)^{k+1}+\frac{(k+1)}{q!}\sum_{i=1}^{q} P_{k-1}^{(i-1)}P_q^{(q-i)}(r+1)^{k+1-2i}(-r)^i\equiv r^{k+1}+1,$$

where $q=[(k+1)/2]$ and $P_m^{(n)}$ is the product of n consecutive integers of which m is the largest, while $P_n^0=1$. Hence if $k+1$ is a prime, it divides $(r+1)^{k+1}-r^{k+1}-1$, and Fermat's theorem follows. The case $k=p-1$ shows that if p is a prime, $q=(p-1)/2$, and r is any integer,

$$\sum_{i=1}^{q} P_{p-2}^{(i-1)} P_q^{(q-i)}(r+1)^{p-2i}(-r)^i\equiv0 \text{ (mod }q!).$$

T. del Beccaro[118] used products of linear functions to obtain a very complicated proof of the generalized Wilson theorem.

A. Schmidt[119] regarded two permutations of $1, 2, \ldots, p$ as identical if one is derived from the other by a cyclic substitution of its elements. From one of the $(p-1)!$ distinct permutations he derived a second by adding

[114]Atti del R. Istituto Veneto di Sc., (7), 4, 1892–3, pp. 1816–42.
[115]Archiv Math. og Natur., Kristiania, 16, 1893, 255–265.
[116]Messenger Math., 23, 1893–4, 56.
[117]Jour. de l'école.polyt., 64, 1894, 200–1.
[118]Atti R. Ac. Lincei (Fis. Mat.), 1, 1894, 344–371.
[119]Zeitschrift Math. Phys., 40, 1895, 124.

unity to each element and replacing $p+1$ by 1. Let m be the least number of repetitions of this process which will yield the initial permutation. For p a prime, $m=1$ or p. There are $p-1$ cases in which $m=1$. Hence $(p-1)!-(p-1)$ is divisible by p. Cf. Petersen.[94] Many proofs of (3), p. 63, have been given.[120]

D. von Sterneck[121] gave Legendre's proof of Wilson's theorem.

L. E. Dickson[122] noted that, if p is a prime, $p(p-1)$ of the $p!$ substitutions on p letters have a linear representation $x' \equiv ax+b$, $a \not\equiv 0 \pmod{p}$, while the remaining ones are represented analytically by functions of degree >1 which fall into sets of $p^2(p-1)$ each, viz., $af(x+b)+c$, where a is prime to p. Hence $p!-p(p-1)$ is a multiple of $p^2(p-1)$, and therefore $(p-1)!+1$ is a multiple of p.

C. Moreau[123] gave without references Schering's[102] extension to any modulus of Dirichlet's[40] proof of the theorems of Fermat and Wilson.

H. Weber[124] deduced Euler's theorem from the fact that the integers $<m$ and prime to m form a group under multiplication, whence every integer belongs to an exponent dividing the order $\phi(m)$ of the group.

E. Cahen[125] proved that the elementary symmetric functions of $1, \ldots, p-1$ of order $<p-1$ are divisible by the prime p. Hence

$$(x-1)(x-2)\ldots(x-p+1) \equiv x^{p-1}+(p-1)! \pmod{p},$$

identically in x. The case $x=1$ gives Wilson's theorem, so that also Fermat's theorem follows.

J. Perott[126] gave Petersen's[94] proof of Fermat's theorem, using q^p "configurations" obtained by placing the numbers $1, 2, \ldots, q$ into p cases, arranged in a line. It is noted that the proof is not valid for p composite; for example, if $p=4$, $q=2$, the set of configurations derived from 1212 by cyclic permutations contains but one additional configuration 2121.

L. Kronecker[127] proved the generalized Wilson theorem essentially as had Brennecke.[57]

G. Candido[128] made use of the identity

$$a^p+b^p = (a+b)^p - pab(a+b)^{p-2} + \ldots$$
$$+(-1)^r \frac{p(p-2r+1)\ldots(p-r-1)}{1 \cdot 2 \ldots r} a^r b^r (a+b)^{p-2r} + \ldots$$

Take p a prime and $b=-1$. Thus $a^p - a \equiv (a-1)^p - (a-1) \pmod{p}$.

[120]L'intermédiaire des math., 3, 1896, 26–28, 229–231; 7, 1900, 22–30; 8, 1901, 164.   A. Capelli Giornale di Mat., 31, 1893, 310. S. Pincherle, *ibid.*, 40, 1902, 180–3.
[121]Monatshefte Math. Phys., 7, 1896, 145.
[122]Annals of Math., (1), 11, 1896–7, 120.
[123]Nouv. Ann. Math., (3), 17, 1898, 296–302.
[124]Lehrbuch der Algebra, II, 1896, 55; ed. 2, 1899, 61.
[125]Éléments de la théorie des nombres, 1900, 111–2.
[126]Bull. des Sc. Math., 24, I, 1900. 175.
[127]Vorlesungen über Zahlentheorie, 1901, I, 127–130.
[128]Giornale di Mat., 40, 1902, 223.

P. Bachmann[129] proved the first statement of Lucas.[110] He gave as a "new" proof of Euler's theorem (p. 320) the proof by Euler,[14] and of the generalized Wilson theorem (p. 336) essentially the proof by Arndt.[70] J. W. Nicholson[130] proved the last formula of Grunert.[36] Bricard[131] changed the wording of Petersen's[94] proof of Fermat's theorem. Of the q^p numbers with p digits written to the base q, omit the q numbers with a single repeated digit. The remaining $q^p - q$ numbers fall into sets each of p distinct numbers which are derived from one another by cyclic permutations of the digits.

G. A. Miller[132] proved the generalized Wilson theorem by group theory. The integers relatively prime to g taken modulo g form under multiplication an abelian group of order $\phi(g)$ which is the group of isomorphisms of a cyclic group of order g. But in an abelian group the product of all the elements is the identity if and only if there is a single element of period 2. It is shown that a cyclic group is of order p^a, $2p^a$ or 4 if its group of isomorphisms contains a single element of period 2.

V. d'Escamard[133] reproduced Sylvester's[78] proof of Wilson's theorem.

K. Petr[134] gave Petersen's[94] proof of Wilson's theorem.

Prompt[135] gave an obscure proof that $2^{p-1} - 1$ is divisible by the prime p.

G. Arnoux[136] proved Euler's theorem. Let λ be any one of the $v = \phi(m)$ integers $\alpha, \beta, \gamma, \ldots$, prime to m and $< m$. We can solve the congruences

$$\alpha\alpha' \equiv \beta\beta' \equiv \gamma\gamma' = \ldots \equiv \lambda \pmod{m}.$$

Here $\alpha', \beta', \ldots$ form a permutation of $\alpha, \beta, \ldots$. Thus

$$\alpha\alpha'\beta\beta' \ldots \equiv (\alpha\beta \ldots)^2 \equiv \lambda^v.$$

In particular, for $\lambda = 1$, we get $(\alpha\beta \ldots)^2 \equiv 1$. Hence for any λ prime to m, $\lambda^v \equiv 1 \pmod{m}$. [Cf. Dirichlet,[40] Schering,[102] C. Moreau.[123]]

R. A. Harris[136a] proved that $(\alpha\beta \ldots)^2 \equiv 1$ as did Arnoux[136], but inferred falsely that $\alpha.\beta \ldots \equiv \pm 1$.

A. Aubry[137] started, as had Waring in 1782, with

$$x^n = Y_n + AY_{n-1} + \ldots + MY_2 + Y_1,$$

where $Y_p = x(x-1) \ldots (x-p+1)$. Then

$$x^{n+1} - x^n = Y_{n+1} + AY_n + \ldots + MY_3 + Y_2.$$

Summing for $x = 1, \ldots, p-1$ and setting $s_k = 1^k + 2^k + \ldots + (p-1)^k$, we get

$$s_{n+1} - s_n = \frac{\{n+1\}}{n+2} + A\frac{\{n\}}{n+1} + \ldots + \frac{M\{3\}}{4} + \frac{\{2\}}{3},$$

[129]Niedere Zahlentheorie, I, 1902, 157–8.　　[130]Amer. Math. Monthly, 9, 1902, 187, 211.
[131]Nouv. Ann. Math., (4), 3, 1903, 340–2.
[132]Annals of Math., (2), 4, 1903, 188–190. Cf. V. d'Escamard, Giornale di Mat., 41, 1903, 203–4; U. Scarpis, ibid., 43, 1905, 323–8.
[133]Giornale di Mat., 43, 1905, 379–380.　　[134]Casopis, Prag, 34, 1905, 164.
[135]Remarques sur le théorème de Fermat, Grenoble, 1905, 32 pp.
[136]Arithmétique Graphique; Fonctions Arith., 1906, 24.
[136a]Math. Magazine, 2, 1904, 272.　　[137]L'enseignement math., 9, 1907, 434–5, 440.

where $\{k\} = p(p-1)\ldots(p-k)$. Hence, if p is a prime and $n < p-1$, $s_{n+1} - s_n \equiv 0$. But $s_1 \equiv 0$. Hence $s_n \equiv 0 (n < p-1)$, $s_{p-1} \equiv -(p-1)!$. Thus Wilson's theorem follows from Fermat's.

Without giving references, Aubry (p. 298) attributed Horner's[37] proof of Euler's theorem to Gauss; the proof (pp. 439–440) by Paoli[46] (and Thue[115]) of Fermat's theorem to Euler[12]; the proof (p. 458) by Laplace[23] of Euler's theorem by powering to Euler.

R. D. Carmichael[138] noted that, if L is the l. c. m. of all the roots z of $\phi(z) = a$, and if x is prime to L, then $x^a \equiv 1 \pmod{L}$. Hence except when n and $n/2$ are the only numbers whose ϕ-function is the same as that of n, $x^{\varphi(n)} \equiv 1$ holds for a modulus M which is some multiple of n. A practical method of finding M is given.

R. D. Carmichael[139] proved the first result by Lucas.[110]

J. A. Donaldson[140] deduced Fermat's theorem from the theory of periodic fractions.

W. A. Lindsay[141] proved Fermat's theorem by use of the binomial theorem.

J. I. Tschistjakov[142] extended Euler's theorem as had Lucas.[110]

P. Bachmann[143] proved the remarks by Lucas,[110] but replaced $\phi + \sigma < n$ by $n \geq \phi + \sigma$, stating that the sign is $>$ if n is divisible by at least two distinct primes.

A. Thue[144] noted that a different kinds of objects can be placed into n given places in a^n ways. Of these let U_a^n be the number of placings such that each is converted into itself by not fewer than n applications of the operation which replaces each by the next and the last by the first. Then U_a^n is divisible by n. If n is a prime, $U_a^n = a^n - a$ and we have Fermat's theorem. Next, $a^n = \Sigma U_a^d$, where d ranges over the divisors of n. Finally, if $p, q, \ldots, r$ are the distinct prime factors of n,

$$U_a^n = \Sigma(-1)^\theta a^{n/D} \equiv 0 \pmod{n},$$

where D ranges over the distinct divisors of $pq\ldots r$, while θ is the number of prime factors of D. Euler's theorem is deduced from this.

H. C. Pocklington[145] repeated Bricard's[131] proof.

U. Scarpis[146] proved the generalized Wilson theorem by a method similar to Arndt's.[70] The case of modulus 2^λ ($\lambda > 2$) is treated by induction. Assume that $\Pi r \equiv 1 \pmod{2^\lambda}$, where $r_1, \ldots, r_v$ are the $v = \phi(2^\lambda)$ odd integers $< 2^\lambda$. Then $r_1, \ldots, r_v, r_1 + 2^\lambda, \ldots, r_v + 2^\lambda$ are the residues modulo $2^{\lambda+1}$ and their product is seen to be $\equiv 1 \pmod{2^{\lambda+1}}$. Next, let the modulus be

[138]Bull. Amer. Math. Soc., 15, 1908–9, 221–2.
[139]Ibid., 16, 1909–10, 232–3.
[140]Edinburgh Math. Soc. Notes, 1909–11, 79–84.
[141]Ibid., 78–79.
[142]Tagbl. XII Vers. Russ. Nat., 124, 1910 (Russian).
[143]Niedere Zahlentheorie, II, 1910, 43–44.
[144]Skrifter Videnskabs–Selskabet, Christiania, 1910, No. 3, 7 pp.
[145]Nature, 84, 1910, 531.
[146]Periodico di Mat., 27, 1912, 231–3.

$n = p_1^{\alpha_1} \ldots p_h^{\alpha_h}$ $(h > 2)$, $n \neq 2p^\lambda$. Then a system of residues modulo n, each prime to n, is given by $\sum\limits_{i=1}^{h} A_i r_i$, with

$$A_i = \left(\frac{n}{p_i^{\alpha_i}}\right)^{\phi(p_i^{\alpha_i})},$$

where r_i ranges over a system of residues modulo $p_i^{\alpha_i}$, each prime to p_i. Let P be the product of these $\Sigma A_i r_i$. Since $A_i A_j$ is divisible by n if $i \neq j$,

$$P \equiv \sum_{i=1}^{h} A_i^{\varphi(n)} (\Pi r_i)^{\varphi\left(n/p_i^{\alpha_i}\right)} \pmod{n}.$$

Thus $P - 1$ is divisible by each $p_k^{\alpha_k}$ and hence by n.

*Illgner[147] proved Fermat's theorem.

A. Bottari[148] proved Wilson's theorem by use of a primitive root [Gauss[30]].

J. Schumacher[149] reproduced Cayley's[101] proof of Wilson's theorem.

A. Arévalo[150] employed the sum S_n of the products taken n at a time of $1, 2, \ldots, p-1$. By the known formula

$$S_n = \frac{1}{n}\left\{\binom{p}{n+1} + \binom{p-1}{n}S_1 + \binom{p-2}{n-1}S_2 + \ldots + \binom{p-n+1}{2}S_{n-1}\right\},$$

it follows by induction that S_n is divisible by the prime p if $n < p-1$. In the notation of Wronski, write $a^{p/r}$ for

$$a(a+r)\ldots\{a+(p-1)r\} = a^p + S_1 a^{p-1} r + \ldots + S_{p-1} a r^{p-1}.$$

For $a = r = 1$, we have $p! = 1 + S_1 + \ldots + S_{p-1}$, whence $S_{p-1} \equiv -1 \pmod{p}$, giving Wilson's theorem. Also, $a^{p/r} \equiv a^p - a \cdot r^{p-1}$. Dividing by a and taking $r = 1$, we have

$$(a+1)^{(p-1)/1} \equiv a^{p-1} - 1 \pmod{p}.$$

The left member is divisible by p if a is not. Hence we have Fermat's theorem. Another proof follows from Vandermonde's formula

$$(x+a)^{p/r} = \sum_{h=0}^{p} \binom{p}{h} x^{(p-h)/r} a^{h/r} \equiv x^{p/r} + a^{p/r} \pmod{p},$$

$$(x_1 + \ldots + x_a)^{p/r} \equiv x_1^{p/r} + \ldots + x_a^{p/r}, \qquad a^{p/r} \equiv a \cdot 1^{p/r}.$$

Remove the factor a and set $r = 0$; we obtain Fermat's theorem.

Prompt[151] gave Euler's[14] proof of his theorem and two proofs of the type sketched by Gauss of his generalization of Wilson's theorem; but obscured the proofs by lengthy numerical computations and the use of unconventional notations.

F. Schuh[152] proved Euler's theorem, the generalized Wilson theorem, and discussed the symmetric functions of the roots of a congruence for a prime modulus.

[147]Lehrsatz über $x^n - x$, Unterrichts Blätter für Math. u. Naturwiss., Berlin, 18, 1912, 15.
[148]Il Boll. Matematica Gior. Sc.-Didat., 11, 1912, 289.
[149]Zeitschrift Math.-naturwiss. Unterricht, 44, 1913, 263–4.
[150]Revista de la Sociedad Mat. Española, 2, 1913, 123–131.
[151]Démonstrations nouvelles des théorèmes de Fermat et de Wilson, Paris, Gauthier-Villars, 1913, 18 pp. Reprinted in l'intermédiaire des math., 20, 1913, end.
[152]Suppl. de Vriend der Wiskunde, 25, 1913, 33–59, 143–159, 228–259.

G. Frattini[153] noted that, if $F(a, \beta, \ldots)$ is a homogeneous symmetric polynomial, of degree g with integral coefficients, in the integers $a, \beta, \ldots$ less than m and prime to m, and if F is prime to m, then $k^g \equiv 1 \pmod{m}$ for every integer k prime to m. In fact,

$$F(a, \beta, \ldots) \equiv F(ka, k\beta, \ldots) \equiv k^g F(a, \beta, \ldots) \qquad \pmod{m}.$$

Taking F to be the product $a\beta\ldots$, we have Euler's theorem. Another corollary is

$$\prod_{j=1}^{p-1} (1+j) \equiv 1 + (p-1)! \qquad \pmod{p},$$

for p a prime, which implies Wilson's theorem.

*J. L. Wildschütz-Jessen[154] gave an historical account of Fermat's and Wilson's theorems.

E. Piccioli[155] repeated the work of Dirichlet.[40]

THE GENERALIZATION $F(a, N) \equiv 0 \pmod{N}$ OF FERMAT'S THEOREM.

C. F. Gauss[160] noted that, if $N = p_1^{e_1} \ldots p_s^{e_s}$ (p's distinct primes),

$$F(a, N) = a^N - \sum_{i=1}^{s} a^{N/p_i} + \sum_{i<j} a^{N/p_i p_j} - \sum_{i<j<k} a^{N/p_i p_j p_k} + \ldots + (-1)^s a^{N/p_1 \ldots p_s}$$

is divisible by N when a is a prime, the quotient being the number of irreducible congruences modulo a of degree N and highest coefficient unity. He proved that

(1) $$a^N = \Sigma F(a, d), \qquad F(a, 1) = a,$$

where d ranges over all the divisors of N, and stated that this relation readily leads to the above expression for $F(a, N)$. [See Ch. XIX on inversion.]

Th. Schönemann[161] gave the generalization that if a is a power p^n of a prime, the number of congruences of degree N irreducible in the Galois field of order a is $N^{-1}F(a, N)$.

An account of the last two papers and later ones on irreducible congruences will be given in Ch. VIII.

J. A. Serret[162] stated that, for any integers a and N, $F(a, N)$ is divisible by N. For $N = p^e$, p a prime, this implies that

$$a^{\phi(p^e)} \equiv 1 \pmod{p^e},$$

when a is prime to p, a case of Euler's theorem.

S. Kantor[163] showed that the number of cyclic groups of order N in any birational transformation of order a in the plane is $N^{-1}F(a, N)$. He obtained (1) and then the expression for $F(a, N)$ by a lengthy method completed for special cases.

---

[153]Periodico di Mat., 29, 1913, 49–53.
[154]Nyt Tidskrift for Mat., 25, A, 1914, 1–24, 49–68 (Danish).
[155]Periodico di Mat., 32, 1917, 132–4.
[160]Posthumous paper, Werke, 2, 1863, 222; Gauss-Maser, 611.
[161]Jour. für Math., 31, 1846, 269–325. Progr. Brandenburg, 1844.
[162]Nouv. Ann. Math., 14, 1855, 261–2.
[163]Annali di Mat., (2), 10, 1880, 64–73. Comptes Rendus Paris, 96, 1883, 1423.

Ed. Weyr[164], E. Lucas[165], and Pellet[165] gave direct proofs that $F(a, N)$ is divisible by N for any integers a, N.

H. Picquet[166] noted the divisibility of $F(3m-1, N)$ by N in an enumeration of certain curvilinear polygons of N sides, at the same time inscribed and circumscribed in a given cubic curve. He gave a proof of the divisibility of $F(a, N)$ by N, requiring various subcases. He stated that the function $F(a, N)$ is characterized by the two relations

(2) $\qquad F(a, np^s) = F(a^{p^s}, n) - F(a^{p^{s-1}}, n), \qquad F(a, p^s) = a^{p^s} - a^{p^{s-1}},$

where a is any integer, n an integer not divisible by the prime p.

A. Grandi[167] proved that $F(a, N)$ is divisible by N by writing it as

$$a^N - a^{N/p_1} - \{(a^{N/p_1} - a^{N/p_1p_2}) + (a^{N/p_3} - a^{N/p_2p_3}) + \ldots\}$$
$$+ \{(a^{N/p_2p_3} - a^{N/p_1p_2p_3}) + \ldots\} + \ldots$$

Each of these binomials is divisible by $p_1^{e_1}$ since

$$a^{(p-1)p^{e-1}} \equiv 1, \qquad a^{p^e} \equiv a^{p^{e-1}} \pmod{p^e}.$$

G. Koenigs[168] considered a uniform substitution $z' = \phi(z)$ and its nth power $z'' = \phi_n(z)$. Those roots of $z - \phi_n(z) = 0$ which satisfy no like equation of lower index are said to belong to the index n. If x belongs to the index n, so do also $\phi_i(x)$ for $i = 1, \ldots, n-1$. Thus the roots belonging to the index n are distributed into sets of n. If a is the degree of the polynomials in the numerator and denominator of $\phi(z)$, the number of roots belonging to the index n is $F(a, n)$, which is therefore divisible by n.

MacMahon's[112] paper contains in a disguised form the fact that $F(a, N)$ is divisible by N. Proofs were given by E. Maillet[113] by substitution groups, and by G. Cordone.[169]

Borel and Drach[170] made use of Gauss' result that $F(p, N)$ is divisible by N for every prime p and integer N, and Dirichlet's theorem that there exist an infinitude of primes p congruent modulo N to any given integer a prime to N, to conclude that $F(a, N)$ is divisible by N.

L. E. Dickson[171] proved by induction (from k to $k+1$ primes) that $F(a, N)$ is characterized by properties (2) and concluded by induction that $F(a, N)$ is divisible by N. A like conclusion was drawn from

$$\{F(a, N)\}^q - F(a, N) \equiv F(a, qN) \pmod{q},$$

where q is a prime. He gave the relations

$$F(a, nN) = F(a^N, n) - \sum_{i=1}^{s} F(a^{N/p_i}, n) + \sum_{i<j} F(a^{N/p_ip_j}, n) - \ldots$$
$$+ (-1)^s F(a^{N/p_1 \cdots p_s}, n),$$
$$F(a, N) = \Sigma \phi(d),$$

[164]Casopis, Prag, 11, 1882, 39.
[165]Comptes Rendus Paris, 96, 1883, 1300–2.
[166]Ibid., p. 1136, 1424. Jour. de l'école polyt., cah. 54, 1884, 61, 85–91.
[167]Atti R. Istituto Veneto di Sc., (6), 1, 1882–3, 809.
[168]Bull. des sciences math., (2), 8, 1884, 286.
[169]Rivista di Mat., Torino, 5, 1895, 25.
[170]Introd. théorie des nombres, 1895, 50.
[171]Annals of Math., (2), 1, 1899, 35. Abstr. in Comptes Rendus Paris, 128, 1899, 1083–5.

where d ranges over those divisors of a^N-1 which do not divide a^v-1 for $0 < v < N$; while, in the former, $p_1, \ldots, p_s$ are the distinct prime factors of N, and n is prime to N.

L. Gegenbauer[172] wrote $F(a, n)$ in the form $\Sigma\mu(d)a^{n/d}$, where d ranges over the divisors of n, and $\mu(d)$ is the function discussed in Chapter XIX on Inversion. As there shown, $\Sigma\mu(d)=0$ if $n>1$. This case $f(x)=\mu(x)$ is used to prove the generalization: If the function $f(x)$ has the property that $\Sigma f(d)$ is divisible by n, then for every integer a the function $\Sigma f(d)a^{n/d}$ is divisible by n, where in each sum d ranges over the divisors of n. Another special case, $f(x)=\phi(x)$, was noted by MacMahon.[112]

J. Westlund[173] considered any ideal A in a given algebraic number field, the distinct prime factors $P_1, \ldots, P_i$ of A, the norm $n(A)$ of A, and proved that if a is any algebraic integer,

$$a^{n(A)} - \Sigma a^{n(A)/n(P_1)} + \Sigma a^{n(A)/n(P_1P_2)} - \ldots + (-1)^i a^{n(A)/n(P_1\ldots P_i)}$$

is always divisible by A.

J. Vályi[174] noted that the number of triangles similar to their nth pedal but not to the dth pedal $(d<n)$ is

$$\chi(n) = \psi(n) - \Sigma\psi\left(\frac{n}{p_1}\right) + \Sigma\psi\left(\frac{n}{p_1p_2}\right) - \ldots,$$

if $p_1, p_2, \ldots$ are the distinct prime factors of n, and $\psi(k)=2^k(2^k-1)$. He proved that $\chi(n)$ is divisible by n, since if the nth pedal to ABC is the first one similar to ABC, a like property is true of the first pedal, $\ldots$, $(n-1)$th pedal, so that the $\chi(n)$ triangles fall into sets of n each of period n. [Note that $\chi(n) = F(4, n) - F(2, n)$.]

A. Axer[175] proved the following generalization of Gegenbauer's[172] theorem: If $G(r_1, \ldots, r_h)$ is any polynomial with integral coefficients, and if, when d ranges over all the divisors of n,

$$\Sigma f(d)G(r_1^{n/d}, \ldots, r_h^{n/d}) \equiv 0 \pmod{n}$$

for a particular function $G=G_0$ and a particular set of values $r_{10}, \ldots, r_{h0}$, not a set of solutions of G_0, and for which G_0 is prime to n, then it holds for every G and every set $r_1, \ldots, r_h$.

FURTHER GENERALIZATIONS OF FERMAT'S THEOREM.

For the generalization to Galois imaginaries, see Ch. VIII.

For the generalization by Lucas, see Ch. XVII, Lucas,[39] Carmichael.[89]

On $x^l \equiv 1 \pmod{n}$ for x prime to n, see Cauchy,[26] Moreau,[93] Epstein,[112] of Ch. VII.

O. H. Mitchell[178] considered the 2^i products s of distinct primes dividing $k = p_1^{e_1} \ldots p_i^{e_i}$ and denoted by $\tau_s(k)$ the number of positive integers $X_s < k$ which are divisible by s but by no prime factor of k not dividing s.

[172]Monatshefte Math. Phys., 11, 1900, 287–8.
[173]Proc. Indiana Ac. Sc., 1902, 78–79.
[174]Monatshefte Math. Phys., 14, 1903, 243–253.
[175]Monatshefte Math. Phys., 22, 1911, 187–194.
[178]Amer. Jour. Math., 3, 1880, 300; Johns Hopkins Univ. Circular, 1, 1880–1, 67, 97.

The products of the various X_s by any one of them are congruent modulo k to the X_s in some order. Hence

$$X_s{}^{\tau_s(k)} \equiv R_s \pmod{k},$$

where R_s is the corresponding one of the 2^i roots of $x^2 \equiv x \pmod{k}$. The analogous extension of Wilson's theorem is $\Pi X_s \equiv \pm R_s \pmod{k}$, the sign being minus only when $k/\sigma = p^\tau$, $2p^\tau$ or 4 and at the same time σ/s is odd. Here $\sigma = \Pi p_j{}^{e_j}$ if $s = \Pi p_j$. Cf. Mitchell,[50] Ch. V.

F. Rogel[179] proved that, if p is a prime not dividing n,

$$n^{p-1} = 1 + \binom{p}{1}(n-1) + \binom{p}{2}(n-1)^2 + \ldots + \binom{p}{k}(n-1)^k + \rho, \qquad k = \frac{p-1}{2},$$

where ρ is divisible by every prime lying between k and $p+1$.

Borel and Drach[180] investigated the most general polynominal in x divisible by m for all integral values of x, but not having all its coefficients divisible by m. If $m = p^\alpha q^\beta, \ldots$, where $p, q, \ldots$ are distinct primes, and if $P(x), Q(x), \ldots$ are the most general polynomials divisible by $p^\alpha, q^\beta, \ldots$, respectively, that for m is evidently

$$\{P(x) + p^\alpha f(x)\} \{Q(x) + q^\beta g(x)\} \ldots.$$

For $\alpha < p+1$, the most general $P(x)$ is proved to be

$$\sum_{k=1}^{\alpha} f_k(x)\phi_k(x), \qquad \phi_k(x) = p^{\alpha-k}(x^p - x)^k,$$

where the f's are arbitrary polynomials. For $\alpha < 2(p+1)$, the most general $P(x)$ is

$$\sum_{k=1}^{\alpha} f_k \phi_k + \sum_{k=1}^{\alpha-p} \psi_k g_k, \qquad \psi_k = \phi(x)(x^p - x)^{k-1} p^{\alpha-p-k},$$

where $\phi(x) = (x^p - x)^p - p^{p-1}(x^p - x)$, and the f's, g's are arbitrary polynomials. Note that $\phi^p(x) - p^{p^2-1}\phi(x)$ is divisible by p^{p^2+p+1}. Cf. Nielsen.[194]

E. H. Moore[181] proved the generalization of Fermat's theorem:

$$\begin{vmatrix} x_1^{p^{m-1}} & \ldots & x_m^{p^{m-1}} \\ \ldots\ldots\ldots\ldots \\ x_1^p & \ldots & x_m^p \\ x_1 & \ldots & x_m \end{vmatrix} \equiv \prod_{k=1}^{m} \prod_{c_{k+1}=0}^{p-1} \ldots \prod_{c_m=0}^{p-1} (x_k + c_{k+1}x_{k+1} + \ldots + c_m x_m) \pmod{p}.$$

F. Gruber[182] showed that, if n is composite and $a_1, \ldots, a_t$ are the $t = \phi(n)$ integers $< n$ and prime to n, the congruence

(1) $$x^t - 1 \equiv (x - a_1) \ldots (x - a_t) \pmod{n}$$

is an identity in x if and only if $n = 4$ or $2p$, where p is a prime $2^i + 1$.

[179]Archiv Math. Phys., (2), 10, 1891, 84–94 (210).
[180]Introduction théorie des nombres, 1895, 339–342.
[181]Bull. Amer. Math. Soc., 2, 1896, 189; cf. 13, 1906–7, 280.
[182]Math. Nat. Berichte aus Ungarn, 13, 1896, 413–7; Math. termés ertesito, 14, 1896, 22–25.

E. Malo[183] employed integers A_i' and set $u=x^\mu z$,

$$z=\Sigma A_i'x^i, \quad z^k=\Sigma A_i^{(k)}x^i, \qquad \theta=\frac{u^{n-1}du}{1-u^m}=\Sigma\omega_p x^{p-1}dx.$$

Since $\int_0^u \theta=\Sigma u^k/k$ $(k=n, m+n, 2m+n,\ldots)$,

$$\Sigma\frac{\omega_p}{p}x^p=\Sigma\frac{x^{\mu k}z^k}{k}, \qquad \frac{\omega_p}{p}=\Sigma\frac{1}{k}A_{p-\mu k}^{(k)},$$

where k takes the values $n, m+n,\ldots$ which are $\leq p/\mu$. If no prime factor of such a k occurs in the denominator of the expansion of ω_p/p, the latter is an integer; this is the case if p is a prime and $\mu\geq 2$. For $m=n=1$, $\mu=2$,

$$z(1-x)^a=\binom{a}{2}-\binom{a}{3}x+\ldots \mp ax^{a-3}\pm x^{a-2},$$

we get $\omega_p=a^p-a$ and hence Fermat's theorem.

L. Kronecker[184] generalized Fermat's and Wilson's theorems to modular systems.

R. Le Vavasseur[185] obtained a result evidently equivalent to that by Moore[181] for the non-homogeneous case $x_m=1$.

M. Bauer[186] proved that if $n=p^\pi m$, where m is not divisible by the odd prime p, and $a_1,\ldots, a_t$ are the $t=\phi(n)$ integers $<n$ and prime to n,

$$(x-a_1)\ldots(x-a_t)\equiv(x^{p-1}-1)^{t/(p-1)}(\bmod\ p^\pi),$$

identically in x. If $p=2$ and $\pi>1$, the product is identically congruent to $(x^2-1)^{t/2}$. Hence he found the values of d, n for which (1) holds modulo d, when d is a divisor of n. If p denotes an odd prime and q a prime 2^i+1, the values are

d	$2q$	4	p	2
n	$2q$	4	$p^a, 2p^a$	$2^a, 2^a q_1 q_2\ldots$

M. Bauer[187] determined how n and N must be chosen so that x^n-1 shall be congruent modulo N to a product of linear functions. We may restrict N to the case of a power of a prime. If p is an odd prime, x^n-1 is congruent modulo p^a to a product of linear functions only when $p\equiv 1$ (mod n), a arbitrary, or when $n=p^\pi m$, $a=1$, $p\equiv 1$ (mod m). For $p=2$, only when $n=2^\beta$, $a=1$, or $n=2$, a arbitrary. For the case n a prime, the problem was treated otherwise by Perott.[188]

M. Bauer[189] noted that, if $n=p^\pi m$, where m is not divisible by the odd prime p,

$$\prod_{i=1}^n(x-i)\equiv(x^p-x)^{n/p}\ (\bmod\ p^\pi).$$

[183]L'intermédiaire des math., 7, 1900, 281, 312.
[184]Vorlesungen über Zahlentheorie, I, 1901, 167, 192, 220-2.
[185]Comptes Rendus Paris, 135, 1902, 949; Mém. Ac. Sc. Toulouse, (10), 3, 1903, 39-48.
[186]Nouv. Ann. Math., (4), 2, 1902, 256-264.
[187]Math. Nat. Berichte aus Ungarn, 20, 1902, 34-38; Math. és Phys. Lapok, 10, 1901, 274-8
        (pp. 145-152 relate to the "theory of Fermat's congruence"; no report is available).
[188]Amer. Jour. Math., 11, 1888; 13, 1891.
[189]Math. és Phys. Lapok, 12, 1903, 159-160.

Richard Sauer[190] proved that, if a, b, $a-b$ are prime to k,

$$a^{\varphi}+a^{\varphi-1}b+a^{\varphi-2}b^2+\ldots+b^{\varphi}\equiv 1 \pmod{k}, \qquad \varphi=\varphi(k),$$

since $a^{\varphi+1}-b^{\varphi+1}\equiv a-b$. Changing alternate signs to minus, we have a congruence valid if a, b are prime to k, and if $a+b$ is not divisible by k. If p is an odd prime dividing $a \mp b$,

$$a^{p-1}\pm a^{p-2}b+\ldots+b^{p-1}$$

is divisible by p, but not by p^2.

A. Capelli[191] showed that, if a, b are relatively prime,

$$\frac{a^{\varphi(b)}+b^{\varphi(a)}-1}{ab}=\left[\frac{a^{\varphi(b)-1}}{b}\right]+\left[\frac{b^{\varphi(a)-1}}{a}\right]+1,$$

where $[x]$ is the greatest integer $\leq x$.

M. Bauer[192] proved that, if p is an odd prime and $m=p^a$ or $2p^a$, every integer x relatively prime to m satisfies the congruence

$$(x^{p-1}-1)^{p^{a-1}}\equiv(x+k_1)\ldots(x+k_l) \pmod{m},$$

where $k_1,\ldots,k_l$ denote the $l=\phi(m)$ integers $<m$ and prime to $m>2$. If m is not 4, p^a or $2p^a$, every integer x prime to m satisfies the congruence

$$(x^{\varphi(m)/2}-1)^2\equiv(x+k_1)\ldots(x+k_l) \pmod{m}.$$

L. E. Dickson[193] proved Moore's[181] theorem by invariantive theory.

N. Nielsen[194] proved that, if $\Phi(x)$ is a polynomial with integral coefficients not having a common factor >1, and if for every integral value of x the value of $\Phi(x)$ is divisible by the positive integer m, then

$$\Phi(x)=\phi(x)\,\omega_p(x)+\sum_{s=1}^{p-1} m_{p-s}\,A_s\,\omega_s(x), \qquad \omega_n(x)\equiv x(x+1)\ldots(x+n-1),$$

where $\phi(x)$ is a polynomial with integral coefficients, the A_s are integers, p is the least positive integer for which $p!$ is divisible by m, and m_{p-s} is the least positive integer l for which $s!l$ is divisible by m. Cf. Borel and Drach.[180]

H. S. Vandiver[195] proved that, if V ranges over a complete set of incongruent residues modulo $m=p_1{}^a\ldots p_k{}^k$, while U ranges over those V's which are prime to m,

$$\Pi(x-V)\equiv\sum_{s=1}^{k}t_s(x^{p_s}-x)^{m/p_s}, \qquad \Pi(x-U)\equiv\Sigma t_s(x^{p_s-1}-1)^{\varphi(m)/(p_s-1)},$$

modulo m, where $t_s=(m/p_s{}^{a_s})^e$, $e=\phi(p_s{}^{a_s})$. For $m=p^a$, the second congruence is due to Bauer.[186, 192]

[190]Eine polynomische Verallgemeinerung des Fermatschen Satzes, Diss., Giessen, 1905.
[191]Dritter Internat. Math. Kongress, Leipzig, 1905, 148–150.
[192]Archiv Math. Phys., (3), 17, 1910, 252–3. Cf. Bouniakowsky[36] of Ch. XI.
[193]Trans. Amer. Math. Soc., 12, 1911, 76; Madison Colloquium of the Amer. Math. Soc., 1914, 39–40.
[194]Nieuw Archief voor Wiskunde, (2), 10, 1913, 100–6.
[195]Annals of Math., (2), 18, 1917, 119.

FURTHER GENERALIZATIONS OF WILSON'S THEOREM; RELATED PROBLEMS.

J. Steiner[200] proved that, if A_k is the sum of all products of powers of a_1, a_2, ..., a_{p-k} of degree k, and the a's have incongruent residues $\neq 0$ modulo p, a prime, then A_1, ..., A_{p-2} are divisible by p.

He first showed by induction that

$$x^{p-1} = X_{p-1} + A_1 X_{p-2} + \ldots + A_{p-2}X_1 + A_{p-1},$$
$$X_k \equiv (x-a_1)\ldots(x-a_k), \quad A_1 = a_1 + \ldots + a_{p-1},$$
$$A_2 = a_1{}^2 + a_1 a_2 + \ldots + a_1 a_{p-2} + a_2{}^2 + a_2 a_3 + \ldots + a_{p-2}^2, \ldots$$

For example, to obtain x^3 he multiplied the respective terms of

$$x^2 = (x-a_1)(x-a_2) + (a_1+a_2)(x-a_1) + a_1{}^2$$

by x, $(x-a_3)+a_3$, $(x-a_2)+a_2$, $(x-a_1)+a_1$. Let $a_1, \ldots, a_{p-1}$ have the residues $1, \ldots, p-1$ in some order, modulo p. For $x-a_2$ divisible by p, $x^{p-1} \equiv A_{p-1} = a_1^{p-1}$ (mod p), so that $A_{p-2}X_1$ and hence also A_{p-2} is divisible by p. Then for $x \equiv a_3$, $A_{p-3}X_2$ and A_{p-3} are divisible by p. For $x=0$, $a_1=1$, the initial equation yields Wilson's theorem.

C. G. J. Jacobi[201] proved the generalization: If $a_1, \ldots, a_n$ have distinct residues $\neq 0$, modulo p, a prime, and P_{nm} is the sum of their multiplicative combinations with repetitions m at a time, P_{nm} is divisible by p for $m = p-n$, $p-n+1, \ldots, p-2$.

Note that Steiner's A_k is $P_{p-k,k}$. We have

$$(1) \quad \frac{1}{(x-a_1)\ldots(x-a_n)} = \frac{1}{x^n} + \frac{P_{n1}}{x^{n+1}} + \frac{P_{n2}}{x^{n+2}} + \ldots, \qquad P_{nm} = \sum_{j=1}^{n} a_j^{n+m-1}/D_j,$$

$$D_j = (a_j - a_1)\ldots(a_j - a_{j-1})(a_j - a_{j+1})\ldots(a_j - a_n), \qquad 0 = \sum_{j=1}^{n} a_j^k/D_j \quad (k < n-1).$$

Let $n+m-1 = k+\beta(p-1)$. Then $a_j^{n+m-1} \equiv a_j^k$ (mod p). Hence if $k < n-1$,

$$D_1 \ldots D_n P_{nm} \equiv D_1 \ldots D_n \Sigma a_j^k / D_j, \qquad P_{nm} \equiv 0 \ (\text{mod } p).$$

The theorem follows by taking $\beta=1$ and $k=0, 1, \ldots, n-2$ in turn.

H. F. Scherk[202] gave two generalizations of Wilson's theorem. Let p be a prime. By use of Wilson's theorem it is easily proved that

$$(p-n-1)! \equiv (-1)^n \frac{px-1}{n!} \ (\text{mod } p),$$

where x is an integer such that $px \equiv 1$ (mod $n!$). Next, let C_k^r denote the sum of the products of $1, 2, \ldots, k$ taken r at a time with repetitions. By use of partial fractions it is proved that

$$(p-r-1)! \, C_{p-r-1}^r + (-1)^r \equiv 0 \ (\text{mod } p) \qquad (r < p-1).$$

It is stated that

[200]Jour. für Math., 13, 1834, 356; Werke 2, p. 9.
[201]Ibid., 14, 1835, 64–5; Werke 6, 252–3.
[202]Bericht über die 24. Versammlung Deutscher Naturforscher und Aerzte in 1846, Kiel, 1847, 204–208.

$$C^r_{p-r-1}C^{p-r-1}_r + (-1)^r \equiv 0, \qquad C^m_m - m! \equiv 0 \pmod{p}, \qquad m = \frac{p-1}{2}.$$

H. F. Scherk[203] proved Jacobi's theorem and the following: Form the sum P_{nh} of the multiplicative combinations with repetitions of the hth class of any n numbers less than the prime p, and the sum of the combinations without repetitions out of the remaining $p-n-1$ numbers $<p$; then the sum or the difference of the two is divisible by p according as h is odd or even.

Let C^h_k denote the sum of the combinations with repetitions of the hth class of $1, 2, \ldots, k$; A^h_k the sum without repetitions. If $0 < h < p-1$,

$$C^j_k \equiv 0 \pmod{p}, \qquad j = p-k, \ldots, p-2; \qquad C^h_{np+k} \equiv C^h_k.$$

For $h = p-1$, $C^{p-1}_{np+k} \equiv n+1$ for $k = 1, \ldots, p$. For $h = m(p-1)+t$, $C^h_k \equiv C^t_k$ when $k < p+1$. For $1 < h < k$, the sum of C^h_k and A^h_k is divisible by $k^2(k+1)^2$; likewise, each C and A if h is odd. For $h < 2k$, $C^h_k - A^h_k$ is divisible by $2k+1$. The sum of the $2n$th powers of $1, \ldots, k$ is divisible by $2k+1$.

K. Hensel[204] has given the further generalization: If $a_1, \ldots, a_n, b_1, \ldots, b_v$ are $n+v = p-1$ integers congruent modulo p to $1, 2, \ldots, p-1$ in some order, and

$$\psi(x) = (x-b_1) \ldots (x-b_v) = x^v - B_1 x^{v-1} + \ldots \pm B_v,$$

then, for any j, $P_{nj} \equiv (-1)^{j_0} B_{j_0} \pmod{p}$, where j_0 is the least residue of j mod $p-1$ and $B_k = 0$ $(k > v)$.

For Steiner's X_n, $X_n \psi(x) \equiv x^{p-1} - 1 \pmod{p}$. Multiply (1) by $x^n(x^{p-1}-1)$. Thus

$$x^n \psi(x) \equiv x^{p-1} + P_{n1} x^{p-2} + \ldots + P_{n\,p-2} x + P_{n\,p-1} - 1 + \frac{P_{np} - P_{n1}}{x}$$
$$+ \frac{P_{n\,p+1} - P_{n2}}{x^2} + \ldots \pmod{p}.$$

Replace $\psi(x)$ by its initial expression and compare coefficients. Hence

$$P_{n\,i+p-1} \equiv P_{ni}, \qquad P_{n\,v+1} \equiv P_{n\,v+2} \equiv \ldots \equiv P_{n\,p-2} \equiv 0, \qquad P_{n\,p-1} \equiv 1,$$
$$P_{nj} \equiv (-1)^j B_j \, (j = 1, \ldots, v).$$

Taking $v = j = p-2$ and choosing $2, \ldots, p-1$ for $b_1, \ldots, b_v$, we get $1 \equiv -(p-1)! \pmod{p}$.

CONVERSE OF FERMAT'S THEOREM.

In a Chinese manuscript dating from the time of Confucius it is stated erroneously that $2^{n-1} - 1$ is not divisible by n if n is not prime (Jeans[220]).

Leibniz in September 1680 and December 1681 (Mahnke,[7] 49–51) stated incorrectly that $2^n - 2$ is not divisible by n if n is not a prime. If $n = rs$, where r is the least prime factor of n, the binomial coefficient $\binom{n}{r}$ was shown to be not divisible by n, since $n-1, \ldots, n-r+1$ are not divisible by r, whence not all the separate terms in the expansion of $(1+1)^n - 2$ are

[203]Ueber die Theilbarkeit der Combinationssummen aus den natürlichen Zahlen durch Primzahlen, Progr., Bremen, 1864, 20 pp.

[204]Archiv Math. Phys., (3), 1, 1901, 319; Kronecker's Zahlentheorie 1, 1901, 503.

divisible by n. From this fact Leibniz concluded erroneously that the expression itself is not divisible by n.

Chr. Goldbach[210] stated that $(a+b)^p - a^p - b^p$ is divisible by p also when p is any composite number. Euler (p. 124) points out the error by noting that $2^{35} - 2$ is divisible by neither 5 nor 7.

In 1769 J. H. Lambert[15] (p. 112) proved that, if $d^m - 1$ is divisible by a, and $d^n - 1$ by b, where a, b are relatively prime, then $d^c - 1$ is divisible by ab if c is the l. c. m. of m, n (since divisible by $d^m - 1$ and hence by a). This was used to prove that if g is odd [and prime to 5] and if the decimal fraction for $1/g$ has a period of $g-1$ terms, then g is a prime. For, if $g = ab$ [where a, b are relatively prime integers > 1], $1/a$ has a period of m terms, $m \leq a-1$, and $1/b$ a period of n terms, $n \leq b-1$, so that the number of terms in the period for $1/g$ is $\leq (a-1)(b-1)/2 < g-1$. Thus Lambert knew at least the case $k = 10$ of the converse of Fermat's theorem (Lucas[214, 217]).

An anonymous writer[211] stated that $2n+1$ is or is not a prime according as one of the numbers $2^n \pm 1$ is or is not divisible by n. F. Sarrus[212] noted the falsity of this assertion since $2^{170} - 1$ is divisible by the composite number 341.

In 1830 an anonymous writer[43] noted that $a^{n-1} - 1$ may be divisible by n when n is composite. In $a^{p-1} = kp+1$, where p is a prime, set $k = \lambda q$. Then $a^{(p-1)q} \equiv 1 \pmod{pq}$. Thus $a^{pq-1} \equiv 1$ if $a^{q-1} \equiv 1 \pmod{pq}$, and the last will hold if $q-1$ is a multiple of $p-1$; for example, if $p = 11$, $q = 31$, $a = 2$, whence $2^{340} \equiv 1 \pmod{341}$.

V. Bouniakowsky[213] proved that if N is a product of two primes and if $N-1$ is divisible by the least positive integer a for which $2^a \equiv 1$, whence $2^{N-1} \equiv 1 \pmod{N}$, then each of the two primes decreased by unity is divisible by a. He noted that $3^6 \equiv 1 \pmod{91 = 7 \cdot 13}$.

E. Lucas[214] noted that $2^{n-1} \equiv 1 \pmod{n}$ for $n = 37 \cdot 73$ and stated the true converse to Fermat's theorem: If $a^x - 1$ is divisible by p for $x = p-1$, but not for $x < p-1$, then p is a prime.

F. Proth[215] stated that, when a is prime to n, n is a prime if $a^x \equiv 1 \pmod{n}$ for $x = (n-1)/2$, but for no other divisor of $(n-1)/2$; also, if $a^x \equiv 1 \pmod{n}$ for $x = n-1$, but for no divisor $< \sqrt{n}$ of $n-1$. If $n = m \cdot 2^k + 1$, where m is odd and $< 2^k$, and if a is a quadratic non-residue of n, then n is a prime if and only if $a^{(n-1)/2} \equiv -1 \pmod{n}$. If p is a prime $> \frac{1}{2}\sqrt{n}$, $n = mp+1$ is a prime if $a^{n-1} - 1$ is divisible by n, but $a^m \pm 1$ is not.

*F. Thaarup[216] showed how to use $a^{n-1} \equiv 1 \pmod{n}$ to tell if n is prime.

E. Lucas[217] proved the converse of Fermat's theorem: If $a^x \equiv 1 \pmod{n}$ for $x = n-1$, but not for x a proper divisor of $n-1$, then n is a prime.

[210]Corresp. Math. Phys. (ed. Fuss), I, 1843, 122, letter to Euler, Apr. 12, 1742.
[211]Annales de Math. (ed. Gergonne), 9, 1818–9, 320.
[212]Ibid., 10, 1819–20, 184–7.
[213]Mém. Ac. Sc. St. Pétersbourg (math.), (6), 2, 1841 (1839), 447–69; extract in Bulletin, 6, 97–8.
[214]Assoc. franç. avanc. sc., 5, 1876, 61; 6, 1877, 161–2; Amer. Jour. Math., 1, 1878, 302.
[215]Comptes Rendus Paris, 87, 1878, 926.
[216]Nyt Tidsskr. for Mat., 2A, 1891, 49–52.
[217]Théorie des nombres, 1891, 423, 441.

G. Levi[114] was of the erroneous opinion that P is prime or composite according as it is or is not a divisor of $10^{P-1}-1$ [criticized by Cipolla,[229] p. 142].

K. Zsigmondy[218] noted that, if q is a prime $\equiv 1$ or 3 (mod 4), then $2q+1$ is a prime if and only if it divides $(2^q+1)/3$ or 2^q-1, respectively; $4q+1$ is a prime if and only if it divides $(2^{2q}+1)/5$.

E. B. Escott[219] noted that Lucas'[214] condition is sufficient but not necessary.

J. H. Jeans[220] noted that if p, q are distinct primes such that $2^p\equiv 2$ (mod q), $2^q\equiv 2$ (mod p), then $2^{pq}\equiv 2$ (mod pq), and found this to be the case for $pq = 11\cdot31$, $19\cdot73$, $17\cdot257$, $31\cdot151$, $31\cdot331$. He ascribed to Kossett the result $2^{n-1}\equiv 1$ (mod n) for $n = 645$.

A. Korselt[221] noted this case 645 and stated that $a^p\equiv a$ (mod p) if and only if p has no square factor and $p-1$ is divisible by the l. c. m. of $p_1-1,\ldots,$ p_n-1, where $p_1,\ldots,p_n$ are the prime factors of p.

J. Franel[222] noted that $2^{pq}\equiv 2$ (mod pq), where p, q are distinct primes, requires that $p-1$ and $q-1$ be divisible by the least integer a for which $2^a\equiv 1$ (mod pq). [Cf. Bouniakowsky.[213]]

L. Gegenbauer[222a] noted that $2^{pq-1}\equiv 1$ (mod pq) if $p = 2^r-1 = \kappa\rho\tau+1$ and $q = \kappa\tau+1$ are primes, as for $p = 31$, $q = 11$.

T. Hayashi[223] noted that 2^n-2 is divisible by $n = 11\cdot31$. If odd primes p and q can be found such that $2^p\equiv 2$, $2^q\equiv 2$ (mod pq), then $2^{pq}-2$ is divisible by pq. This is the case if $p-1$ and $q-1$ have a common factor p' for which $2^{p'}\equiv 1$ (mod pq), as for $p = 23$, $q = 89$, $p' = 11$.

Ph. Jolivald[224] asked whether $2^{N-1}\equiv 1$ (mod N) if $N = 2^p-1$ and p is a prime, noting that this is true if $p = 11$, whence $N = 2047$, not a prime. E. Malo[225] proved this as follows:

$$N-1 = 2(2^{p-1}-1) = 2pm, \qquad 2^{N-1} = (2^p)^{2m} = (N+1)^{2m}\equiv 1 \text{ (mod } N).$$

G. Ricalde[226] noted that a similar proof gives $a^{N-a+1}\equiv 1$ (mod N) if $N = a^p-1$, and a is not divisible by the prime p.

H. S. Vandiver[227] proved the conditions of J. Franel[222] and noted that they are not satisfied if $a < 10$. Solutions for $a = 10$ and $a = 11$ are $pq = 11\cdot31$ and $23\cdot89$, respectively.

H. Schapira[228] noted that the test for the primality of N that $a^q\equiv 1$

---

[218]Monatshefte Math. Phys., 4, 1893, 79.

[219]L'intermédiaire des math., 4, 1897, 270.

[220]Messenger Math., 27, 1897–8, 174.

[221]L'intermédiaire des math., 6, 1899, 143.

[222]Ibid., p. 142.

[222a]Monatshefte Math. Phys., 10, 1899, 373.

[223]Jour. of the Physics School in Tokio, 9, 1900, 143–4. Reprinted in Abhand. Geschichte Math. Wiss., 28, 1910, 25–26.

[224]L'intermédiaire des math., 9, 1902, 258.

[225]Ibid., 10, 1903, 88.

[226]Ibid., p. 186.

[227]Amer. Math. Monthly, 9, 1902, 34–36.

[228]Tchebychef's Theorie der Congruenzen, ed. 2, 1902, 306.

(mod N) for $q = N - 1$, but for no smaller q, is practical only if it be known that a small number a is a primitive root of N.

G. Arnoux[228a] gave numerical instances of the converse of Fermat's theorem.

M. Cipolla[229] stated that the theorem of Lucas[217] implies that, if p is a prime and $k = 2$, 4, 6, or 10, then $kp + 1$ is a prime if and only if $2^{kp} \equiv 1$ (mod $kp + 1$). He treated at length the problem to find a for which $a^{P-1} \equiv 1$ (mod P), given a composite P; and the problem to find P, given a. In particular, we may take P to be any odd factor of $(a^{2p} - 1)/(a^2 - 1)$ if p is an odd prime not dividing $a^2 - 1$. Again, $2^{P-1} \equiv 1$ (mod P) for $P = F_m F_n \cdots$ F_s, $m > n > \ldots > s$, if and only if $2^s > m$, where $F_s = 2^{2^s} + 1$ is a prime. If p and $q = 2p - 1$ are primes and a is any quadratic residue of q, then $a^{pq-1} \equiv 1$ (mod pq); we may take $a = 3$ if $p = 4n + 3$; $a = 2$ if $p = 4n + 1$; both $a = 2$ and $a = 3$ if $p = 12k + 1$; etc.

E. B. Escott[230] noted that $e^{n-1} \equiv 1$ (mod n) if $e^a - 1$ contains two or more primes whose product n is $\equiv 1$ (mod a), and gave a list of 54 such n's.

A. Cunningham[231] noted the solutions $n = F_3 F_4 F_5 F_6 F_7$, $n = F_4 \ldots F_{15}$, etc. [cf. Cipolla], and stated that there exist solutions in which n has more than 12 prime factors. One with 12 factors is here given by Escott.

T. Banachiewicz[232] verified that $2^N - 2$ is divisible by N for N composite and < 2000 only when N is

$$341 = 11 \cdot 31, \quad 561 = 3 \cdot 11 \cdot 17, \quad 1387 = 19 \cdot 73, \quad 1729 = 7 \cdot 13 \cdot 19, \quad 1905 = 3 \cdot 5 \cdot 127.$$

Since $2^N - 2$ is evidently divisible by N for every $N = F_k = 2^{2^k} + 1$, perhaps Fermat was thus led to his false conjecture that every F_k is a prime.

R. D. Carmichael[233] proved that there are composite values of n (a product of three or more distinct odd primes) for which $e^{n-1} \equiv 1$ (mod n) holds for every e prime to n.

J. C. Morehead[234] and A. E. Western proved the converse of Fermat's theorem.

D. Mahnke[7] (pp. 51–2) discussed Leibniz' converse of Fermat's theorem in the form that n is a prime if $x^{n-1} \equiv 1$ (mod n) for all integers x prime to n and noted that this is false when n is the square or higher power of a prime or the product of two distinct primes, but is true for certain products of three or more primes, as $3 \cdot 11 \cdot 17$, $5 \cdot 13 \cdot 17$, $5 \cdot 17 \cdot 29$, $5 \cdot 29 \cdot 73$, $7 \cdot 13 \cdot 19$.

R. D. Carmichael[235] used the result of Lucas[110] to prove that $a^{P-1} \equiv 1$ (mod P) holds for every a prime to P if and only if $P - 1$ is divisible by $\lambda(P)$. The latter condition requires that, if P is composite, it be a product of three or more distinct odd primes. There are found 14 products P of

[228a]Assoc. franç., 32, 1903, II, 113–4.
[229]Annali di Mat., (3), 9, 1903–4, 138–160.
[230]Messenger Math., 36, 1907, 175–6; French transl., Sphinx-Oedipe, 1907–8, 146–8.
[231]Math. Quest. Educat. Times, (2), 14, 1908, 22–23; 6, 1904, 26–7, 55–6.
[232]Spraw. Tow. Nauk, Warsaw, 2, 1909, 7–10.
[233]Bull. Amer. Math. Soc., 16, 1909–10, 237–8.
[234]Ibid., p. 2.
[235]Amer. Math. Monthly, 19, 1912, 22–7.

three primes, as well as $P = 13 \cdot 37 \cdot 73 \cdot 457$, for each of which the congruence holds for every a prime to P.

Welsch[236] stated that if $k = 4n+1$ is composite and <1000, $2^{k-1} \equiv 1 \pmod{k}$ only for $k = 561$ and 645; hence $n^n \equiv 1 \pmod{k}$ for these two k's.

P. Bachmann[237] proved that $x^{pq-1} \equiv 1 \pmod{pq}$ is never satisfied by all integers prime to pq if p and q are distinct odd primes [Carmichael[235]].

SYMMETRIC FUNCTIONS OF $1, 2, \ldots p-1$ MODULO p.

Report has been made above of the work on this topic by Lagrange,[18] Lionnet,[61] Tchebychef,[75] Sylvester,[78] Ottinger,[92] Lucas,[103] Cahen,[125] Aubry,[137] Arévalo,[150] Schuh,[152] Frattini,[153] Steiner,[200] Jacobi,[201] Hensel.[204]

We shall denote $1^n + 2^n + \ldots + (p-1)^n$ by s_n, and take p to be a prime.

E. Waring[250] wrote $a, \beta, \ldots$ for $1, 2, \ldots, x$, and considered

$$s = a^a \beta^b \gamma^c \ldots + a^b \beta^a \gamma^c \ldots + a^a \beta^b \gamma^d \ldots.$$

If $t = a+b+c+\ldots$ is odd and $<x$, and $x+1$ is prime, s is divisible by $(x+1)^2$. If $t < 2x$ and $a, b, \ldots$ are all even and prime to $2x+1$, s is divisible by $2x+1$.

V. Bouniakowsky[251] noted that s_m is divisible by p^2, if $p>2$ and m is odd and not $\equiv 1 \pmod{p-1}$; also if both $m \equiv 1 \pmod{p-1}$ and $m \equiv 0 \pmod{p}$.

C. von Staudt[252] proved that, if $S_n(x) = 1 + 2^n + \ldots + x^n$,

$$S_n(ab) \equiv bS_n(a) + naS_{n-1}(a)S_1(b-1) \pmod{a^2},$$
$$2S_{2n+1}(a) \equiv (2n+1)aS_{2n}(a) \pmod{a^2}.$$

If $a, b, \ldots, l$ are relatively prime in pairs,

$$\frac{S_n(ab\ldots l)}{ab\ldots l} - \frac{S_n(a)}{a} - \ldots - \frac{S_n(l)}{l} = \text{integer}.$$

A. Cauchy[253] proved that $1 + 1/2 + \ldots + 1/(p-1) \equiv 0 \pmod{p}$.

G. Eisenstein[254] noted that $s_m \equiv -1$ or $0 \pmod{p}$ according as m is or is not divisible by $p-1$. If m, n are positive integers $<p-1$,

$$\sum_{\sigma=1}^{p-2} \sigma^m (\sigma+1)^n \equiv 0 \text{ or } -\binom{n}{p-1-m} \pmod{p},$$

according as $m+n < \text{or} \geq p-1$.

L. Poinsot[255] noted that, when a takes the values $1, \ldots, p-1$, then $(ax)^n$ has the same residues modulo p as a^n, order apart. By addition, $s_n x^n \equiv s_n \pmod{p}$. Take x to be one of the numbers not a root of $x^n \equiv 1$. Hence $s_n \equiv 0 \pmod{p}$ if n is not divisible by $p-1$.

[236] L'intermédiaire des math., 20, 1913, 94.
[237] Archiv Math. Phys., (3), 21, 1913, 185–7.
[250] Meditationes algebraicae, ed. 3, 1782, 382.
[251] Bull. Ac. Sc. St. Pétersbourg, 4, 1838, 65–9.
[252] Jour. für Math., 21, 1840, 372–4.
[253] Mém. Ac. Sc. de l'Institut de France, 17, 1840, 340–1, footnote; Oeuvres, (1), 3, 81–2.
[254] Jour. für Math., 27, 1844, 292–3; 28, 1844, 232.
[255] Jour. de Math., 10, 1845, 33–4.

J. A. Serret[256] concluded by applying Newton's identities to $(x-1)\ldots$
$(x-p+1) \equiv 0$ that $s_n \equiv 0$ (mod p) unless n is divisible by $p-1$.

J. Wolstenholme[257] proved that the numerators of

$$1+\frac{1}{2}+\frac{1}{3}+\ldots+\frac{1}{p-1}, \qquad 1+\frac{1}{2^2}+\ldots+\frac{1}{(p-1)^2}$$

are divisible by p^2 and p respectively, if p is a prime >3. Proofs have also
been given by C. Leudesdorf[258], A. Rieke,[259] E. Allardice,[260] G. Osborn,[261]
L. Birkenmajer,[262] P. Niewenglowski,[263] N. Nielsen,[264] H. Valentiner,[265]
and others.[266]

V. A. Lebesgue[267] proved that s_m is divisible by p if m is not divisible
by $p-1$ by use of the identities

$$(n+1)\sum_{k=1}^{x} k(k+1)\ldots(k+n-1)=x(x+1)\ldots(x+n) \qquad (n=1,\ldots,p-1).$$

P. Frost[268] proved that, if p is a prime not dividing $2^{2r}-1$, the numera-
tors of σ_{2r}, σ_{2r-1}, $p(2r-1)\sigma_{2r}+2\sigma_{2r-1}$ are divisible by p, p^2, p^3, respectively,
where

$$\sigma_k=1+\frac{1}{2^k}+\ldots+\frac{1}{(p-1)^k}.$$

The numerator of the sum of the first half of the terms of σ_{2r} is divisible by
p; likewise that of the sum of the odd terms.

J. J. Sylvester[269] stated that the sum $S_{n, m}$ of all products of n distinct
numbers chosen from $1,\ldots, m$ is the coefficient of t^n in the expansion of
$(1+t)(1+2t)\ldots(1+mt)$ and is divisible by each prime $>n+1$ contained in
any term of the set $m-n+1,\ldots, m, m+1$.

E. Fergola[270] stated that, if $(a, b,\ldots, l)^n$ represents the expression
obtained from the expansion of $(a+b+\ldots+l)^n$ by replacing each numerical
coefficient by unity, then

$$(x, x+1,\ldots, x+r)^n=\sum_{j=0}^{n}\binom{r+n}{j}(1, 2,\ldots, r)^{n-j}x^j.$$

[256]Cours d'algèbre supérieure, ed. 2, 1854, 324.
[257]Quar. Jour. Math., 5, 1862, 35–39.
[258]Proc. London Math. Soc., 20, 1889, 207.
[259]Zeitschrift Math. Phys., 34, 1889, 190–1.
[260]Proc. Edinburgh Math. Soc., 8, 1890, 16–19.
[261]Messenger Math., 22, 1892–3, 51–2; 23, 1893–4, 58.
[262]Prace Mat. Fiz., Warsaw, 7, 1896, 12–14 (Polish).
[263]Nouv. Ann. Math., (4), 5, 1905, 103.
[264]Nyt Tidsskrift for Mat., 21, B, 1909–10, 8–10.
[265]Ibid., p. 36–7.
[266]Math. Quest, Educat. Times, 48, 1888, 115; (2), 22, 1912, 99; Amer. Math. Monthly, 22, 1915,
    103, 138, 170.
[267]Introd. à la théorie des nombres, 1862, 79–80, 17.
[268]Quar. Jour. Math., 7, 1866, 370–2.
[269]Giornale di Mat., 4, 1866, 344. Proof by Sharp, Math. Ques. Educ. Times, 47, 1887, 145–6;
    63, 1895, 38.
[270]Ibid., 318–9. Cf. Wronski[151] of Ch. VIII.

The number $(1, 2, \ldots, r)^n$ is divisible by every prime $> r$ which occurs in the series $n+2$, $n+3$, $\ldots$, $n+r$.

G. Torelli[271] proved that

$$(a_1, \ldots, a_n)^r = (a_1, \ldots, a_{n-1})^r + a_n(a_1, \ldots, a_n)^{r-1},$$
$$(a_1, \ldots, a_n, b)^r - (a_1, \ldots, a_n, c)^r = (b-c)(a_1, \ldots, a_n, b, c)^{r-1},$$
$$(x+a_0, x+a_1, \ldots, x+a_n)^r = \Sigma \binom{n+r}{j}(a_0, \ldots, a_n)^{r-j}x^j,$$

which becomes Fergola's for $a_i = i$ ($i = 0, \ldots, n$). Proof is given of Sylvester's[269] theorem and the generalization that $S_{j,i}$ is divisible by $\binom{i+1}{j+1}$.

Torelli[272] proved that the sum $\sigma_{n,m}$ of all products of n equal or distinct numbers chosen from $1, 2, \ldots, m$ is divisible by $\binom{n+m}{n+1}$, and gave recursion formulas for $\sigma_{n,m}$.

C. Sardi[273] deduced Sylvester's theorem from the equations $A_1 = \binom{p}{2}, \ldots$ used by Lagrange.[18] Solving them for $A_p = S_{p,n}$, we get

$$p!(-1)^{p+1}S_{p,n} = \begin{vmatrix} -1 & 0 & 0 & \ldots & 0 & \binom{n+1}{2} \\ \binom{n}{2} & -2 & 0 & \ldots & 0 & \binom{n+1}{3} \\ \binom{n}{3} & \binom{n-1}{2} & -3 & \ldots & 0 & \binom{n+1}{4} \\ \ldots & \ldots & \ldots & \ldots & \ldots & \ldots \\ \binom{n}{p} & \binom{n-1}{p-1} & \binom{n-2}{p-2} & \ldots & \binom{n-p+2}{2} & \binom{n+1}{p+1} \end{vmatrix}.$$

If $n+1$ is a prime we see by the last column that $S_{n-1,n}$ is divisible by $n+1$. When $p = n-1$, denote the determinant by D. Then if $n+1$ is a prime, D is evidently divisible by $n+1$. Conversely, if D is divisible by $n+1$ and the quotient by $(n-1)!$, then $n+1$ is a prime. It is shown that

$$mS_{m,n} = \sum_{p=1}^{m} (-1)^{p+1}r_p S_{m-p,n}, \qquad r_p = 1^p + \ldots + n^p.$$

Using this for $m = 1, \ldots, n$, we see that r_p is divisible by any integer prime to $2, 3, \ldots, p+1$ which occurs in $n+1$ or n. Hence if $n+1$ is a prime, it divides $r_1, \ldots, r_{n-1}$, while $r_n \equiv n$ (mod $n+1$). If $n+1$ divides r_{n-1} it is a prime.

Sardi[274] proved Sylvester's theorem and the formula

$$\sum_{r=0}^{k} (-1)^r S_{r, n+r-1}\sigma_{k-r, n+r} = 0,$$

stated by Fergola.[275]

[271]Giornale di Mat., 5, 1867, 110–120.
[272]Ibid., 250–3.
[273]Ibid., 371–6.
[274]Ibid., 169–174.
[275]Ibid., 4, 1866, 380.

Sylvester[276] stated that, if $p_1, p_2, \ldots$ are the successive primes $2, 3, 5, \ldots$,

$$S_{j;\,n} = \frac{(n+1)n(n-1)\ldots(n-j+1)}{p_1^{e_1}p_2^{e_2}\ldots}F_{j-1}(n),$$

where $F_k(n)$ is a polynomial of degree k with integral coefficients, and the exponent e of the prime p is given by

$$e = \sum_{k=0}^{\infty}\left[\frac{j}{(p-1)p^k}\right].$$

E. Cesàro[277] stated Sylvester's[269] theorem and remarked that $S_{n,\,m}-n!$ is divisible by $m-n$ if $m-n$ is a prime.

E. Cesàro[278] stated that the prime p divides $S_{m,\,p-2}-1$, $S_{p-1,\,p}+1$, and, except when $m=p-1$, $S_{m,\,p-1}$. Also (p. 401), each prime $p>(n+1)/2$ divides $S_{p-1,\,n}+1$, while a prime $p=(n+1)/2$ or $n/2$ divides $S_{p-1,\,n}+2$.

O. H. Mitchell[279] discussed the residues modulo k (any integer) of the symmetric functions of $0, 1, \ldots, k-1$. To this end he evaluated the residue of $(x-a)(x-\beta)\ldots$, where $a, \beta, \ldots$ are the s-totitives of k (numbers $<k$ which contain s but no prime factor of k not found in s). The results are extended to the case of moduli p, $f(x)$, where p is a prime [see Ch. VIII].

F. J. E. Lionnet[280] stated and Moret-Blanc proved that, if $p=2n+1$ is a prime >3, the sum of the powers with exponent $2a$ (between zero and $2n$) of $1, 2, \ldots, n$, and the like sum for $n+1, n+2, \ldots, 2n$, are divisible by p.

M. d'Ocagne[281] proved the first relation of Torelli.[271]

E. Catalan[282] stated and later proved[283] that s_k is divisible by the prime $p>k+1$. If p is an odd prime and $p-1$ does not divide k, s_k is divisible by p; while if $p-1$ divides k, $s_k \equiv -1 \pmod{p}$. Let $p=a^\alpha b^\beta \ldots$; if no one of $a-1, b-1, \ldots$ divides k, s_k is divisible by p; in the contrary case, not divisible. If p is a prime >2, and $p-1$ is not a divisor of $k+l$, then

$$S = 1^k(p-1)^l + 2^k(p-2)^l + \ldots + (p-1)^k 1^l$$

is divisible by p; but, if $p-1$ divides $k+l$, $S \equiv -(-1)^l \pmod{p}$. If k and l are of contrary parity, p divides S.

M. d'Ocagne[284] proved for Fergola's[270] symbol the relation

$$(a\ldots fg\ldots l\ldots v\ldots z)^n = \Sigma(a\ldots f)^\lambda(g\ldots l)^\mu \ldots (v\ldots z)^\rho,$$

summed for all combinations such that $\lambda+\mu+\ldots+\rho=n$. Denoting by $a^{(p)}$ the letter a taken p times, we have

$$(a^{(p)}ab\ldots l)^n = \sum_{i=0}^{n} a^i(1^{(p)})^i(ab\ldots l)^{n-i}.$$

[276]Nouv. Ann. Math., (2), 6, 1867, 48.
[277]Nouv. Corresp. Math., 4, 1878, 401; Nouv. Ann. Math., (3), 2, 1883, 240.
[278]Nouv. Corresp. Math., 4, 1878, 368.
[279]Amer. Jour. Math., 4, 1881, 25–38.
[280]Nouv. Ann. Math., (3), 2, 1883, 384; 3, 1884, 395–6.
[281]Ibid., (3), 2, 1883, 220–6. Cf. Cesàro, (3), 4, 1885, 67–9.
[282]Bull. Ac. Sc. Belgique, (3), 7, 1884, 448–9.
[283]Mém. Ac. R. Sc. Belgique, 46, 1886, No. 1, 16 pp.
[284]Nouv. Ann. Math., (3), 5, 1886, 257–272.

It is shown that $(1^{(p)})^n$ equals the number of combinations of $n+p-1$ things $p-1$ at a time. Various algebraic relations between binomial coefficients are derived.

L. Gegenbauer[285] considered the polynomial

$$f(x) = \sum_{i=0}^{p-2+k} b_i x^i \qquad (1-p<k\leqq p-1)$$

and proved that

$$\sum_{\lambda=1}^{p-1} f(\lambda)/\lambda^{p-2} \equiv -b_{p-2} \text{ (mod } p), \qquad k<p-1,$$

$$\sum_{\lambda=1}^{p-1} f(\lambda)/\lambda^{p-1} \equiv -b_{p-2}-b_{2p-3} \text{ (mod } p), \qquad k=p-1,$$

and deduced the theorem on the divisibility of s_n by p.

E. Lucas[286] proved the theorem on the divisibility of s_n by p by use of the symbolic expression $(s+1)^n-s^n$ for x^n-1.

N. Nielsen[286a] proved that if p is an odd prime and if k is odd and $1<k<p-1$, the sum of the products of $1,\ldots,p-1$ taken k at a time is divisible by p^2. For $k=p-2$ this result is due to Wolstenholme.[257]

N. M. Ferrers[287] proved that, if $2n+1$ is a prime, the sum of the products of $1, 2,\ldots, 2n$ taken r at a time is divisible by $2n+1$ if $r<2n$ [Lagrange[18]], while the sum of the products of the squares of $1,\ldots, n$ taken r at a time is divisible by $2n+1$ if $r<n$. [Other proofs by Glaisher.[294]]

J. Perott[288] gave a new proof that s_n is divisible by p if $n<p-1$.

R. Rawson[289] proved the second theorem of Ferrers.

G. Osborn[290] proved for $r<p-1$ that s_r is divisible by p if r is even, by p^2 if r is odd; while the sum of the products of $1,\ldots, p-1$ taken r at a time is divisible by p^2 if r is odd and $1<r<p$.

J. W. L. Glaisher[291] stated theorems on the sum $S_r(a_1,\ldots, a_i)$ of the products of $a_1,\ldots, a_i$ taken r at a time. If r is odd, $S_r(1,\ldots, n)$ is divisible by $n+1$ (special case $n+1$ a prime proved by Lagrange and Ferrers). If r is odd and >1, and if $n+1$ is a prime>3, $S_r(1,\ldots, n)$ is divisible by $(n+1)^2$ [Nielsen[286a]]. If r is odd and >1, and if n is a prime >2, $S_r(1,\ldots, n)$ is divisible by n^2. If $n+1$ is a prime, $S_r(1^2,\ldots, n^2)$ is divisible by $n+1$ for $r=1,\ldots, n-1$, except for $r=n/2$, when it is congruent to $(-1)^{1+n/2}$ modulo $n+1$. If p is a prime $\leqq n$, and k is the quotient obtained on dividing $n+1$ by p, then $S_{p-1}(1,\ldots, n) \equiv -k \text{ (mod } p)$; the case $n=p-1$ is Wilson's theorem.

[285]Sitzungsber. Ak. Wiss. Wien (Math.), 95 II, 1887, 616–7.
[286]Théorie des nombres, 1891, 437.
[286a]Nyt Tidsskrift for Mat., 4, B, 1893, 1–10.
[287]Messenger Math., 23, 1893–4, 56–58.
[288]Bull. des sc. math., 18, I, 1894, 64. Other proofs, Math. Quest. Educ. Times, 58, 1893, 109; 4, 1903, 42.
[289]Messenger Math., 24, 1894–5, 68–69.
[290]Ibid., 25, 1895–6, 68–69.
[291]Ibid., 28, 1898–9, 184–6. Proofs[294].

S. Monteiro[292] noted that $2n+1$ divides $(2n)! \Sigma_1^{2n} 1/r$.

J. Westlund[293] reproduced the discussion by Serret[256] and Tchebychef.[75] Glaisher[294] proved his[291] earlier theorems. Also, if $p = 2m+1$ is prime,

$$(m-t)pS_{2t}(1, \ldots, 2m) \equiv S_{2t+1}(1, \ldots, 2m) \pmod{p^3}$$

and, if $t > 1$, modulo p^4. According as n is odd or even,

$$S_{2t}(1, \ldots, n) \equiv S_{2t}(1, \ldots, n-1) \pmod{n^2 \text{ or } \tfrac{1}{2}n^2}.$$

For m odd and > 3, $S_{2m-3}(1, \ldots, 2m-1)$ is divisible by m^2, and

$$S_{m-2}(1^2, \ldots, \{m-1\}^2), \qquad S_{2m-4}(1, \ldots, 2m-1)$$

are divisible by m. He gave the values of $S_r(1, \ldots, n)$ and $A_r = S_r(1, \ldots, n-1)$ in terms of n for $r = 1, \ldots, 7$; the numerical values of $S_r(1, \ldots, n)$ for $n \leq 22$, and a list of known theorems on the divisors of A_r and S_r. For r odd, $3 \leq r \leq m-2$, $S_r(1, \ldots, 2m-1)$ is divisible by m and, if m is a prime > 3, by m^2. He proved (*ibid.*, p. 321) that, if $1 \leq r \leq (p-3)/2$, and B_r is a Bernoulli number,

$$\frac{2S_{2r+1}(1, \ldots, p-1)}{p^2} \equiv \frac{-(2r+1)S_{2r}(1, \ldots, p-1)}{p},$$

$$\frac{S_{2r}(1, \ldots, p-1)}{p} \equiv \frac{(-1)^r B_r}{2r} \pmod{p}.$$

Glaisher[295] gave the residues of σ_k [Frost[268]] modulo p^2 and p^3 and proved that $\sigma_2, \sigma_4, \ldots, \sigma_{p-3}$ are divisible by p, and $\sigma_3, \sigma_5, \ldots, \sigma_{p-2}$ by p^2, if p is a prime.

Glaisher[296] proved that, if p is an odd prime,

$$1 + \frac{1}{3^{2n}} + \frac{1}{5^{2n}} + \ldots + \frac{1}{(p-2)^{2n}} \equiv 0 \text{ or } -\tfrac{1}{2} \pmod{p},$$

according as $2n$ is not or is a multiple of $p-1$. He obtained (pp. 154–162) the residue of the sum of the inverses of like powers of numbers in arithmetical progression.

F. Sibirani[296a] proved for the $S_{n,m}$ of Sylvester[269] (designated $s_{n,m+1}$) that

$$S_{i,j} = jS_{i-1,j-1} + S_{i,j-1},$$

$$\begin{vmatrix} S_{n,n} & S_{n-1,n} \cdots & S_{n-k+1,n} \\ \cdots & \cdots & \cdots \\ S_{n+k-1,n+k-1} & S_{n+k-2,n+k-1} \cdots & S_{n,n+k-1} \end{vmatrix} = (n!)^k.$$

[292]Jornal Sc. Mat. Phys. e Nat., Lisbon, 5, 1898, 224.
[293]Proc. Indiana Ac. Sc., 1900, 103–4.
[294]Quar. Jour. Math., 31, 1900, 1–35.
[295]*Ibid.*, 329–39; 32, 1901, 271–305.
[296]Messenger Math., 30, 1900–1, 26–31.
[296a]Periodico di Mat., 16, 1900–1, 279–284.

K. Hensel[297] proved by the method of Poinsot[255] that any integral symmetric function of degree v of $1, \ldots, p-1$ with integral coefficients is divisible by the prime p if v is not a multiple of $p-1$.

W. F. Meyer[298] gave the generalization that, if $a_1, \ldots, a_{p-1}$ are incongruent modulo p^n, and each $a_i^{p-1}-1$ is divisible by p^n, any integral symmetric function of degree v of $a_1, \ldots, a_{p-1}$ is divisible by p^n if v is not a multiple of $p-1$. Of the $\phi(p^n)$ residues modulo p^n, prime to p, there are $p^k(p-1)^2$ for which $a^{p-1}-1$ is divisible by p^{n-1-k}, but by no higher power of p, where $k-1, \ldots, n-1$; the remaining $p-1$ residues give the above $a_1, \ldots, a_{p-1}$.

J. W. Nicholson[299] noted that, if p is a prime, the sum of the nth powers of p numbers in arithmetical progression is divisible by p if $n < p-1$, and $\equiv -1 \pmod{p}$ if $n = p-1$.

G. Wertheim[300] proved the same result by use of a primitive root.

A. Aubry[301] took $x = 1, 2, \ldots, p-1$ in

$$(x+1)^n - x^n = nx^{n-1} + Ax^{n-2} + \ldots + Lx + 1$$

and added the results. Thus

$$p^n = ns_{n-1} + As_{n-2} + \ldots + Ls_1 + p.$$

Hence by induction s_{n-1} is divisible by the prime p if $n < p$. He attributed this theorem to Gauss and Libri without references.

U. Concina[302] proved that s_n is divisible by the prime $p > 2$ if n is not divisible by $p-1$. Let δ be the g. c. d. of n, $p-1$, and set $\mu\delta = p-1$. The μ distinct residues r_i of nth powers modulo p are the roots of $x^\mu \equiv 1 \pmod{p}$, whence $\Sigma r_i \equiv 0 \pmod{p}$ for n not divisible by $p-1$. For each r_i, $x^n \equiv r_i$ has δ incongruent roots. Hence $s_n \equiv \delta\Sigma r_i \equiv 0$. He proved also that, if $p+1$ is a prime >3, and n is even and not divisible by p, $1^n + 2^n + \ldots + (p/2)^n$ is divisible by $p+1$.

W. H. L. Janssen van Raay[303] considered, for a prime $p > 3$,

$$A_h = \frac{(p-1)!}{h}, \qquad B_h = \frac{(p-1)!}{h(p-h)}$$

and proved that $B_1 + B_2 + \ldots + B_{(p-1)/2}$ is divisible by p, and

$$A_1 + \ldots + A_{p-1}, \qquad 1 + \frac{1}{2} + \frac{1}{3} + \ldots + \frac{1}{p-1}$$

are divisible by p^2.

U. Concina[304] proved that $S = 1 + 2^n + \ldots + k^n$ is divisible by the odd number k if n is not divisible by $p-1$ for any prime divisor of p of k. Next, let k be even. For n odd >1, S is divisible by k or only by $k/2$ according

[297]Archiv Math. Phys., (3), 1, 1901, 319. Inserted by Hensel in Kronecker's Vorlesungen über Zahlentheorie I, 1901, 104–5, 504.
[298]Archiv Math. Phys., (3), 2, 1902, 141. Cf. Meissner[39] of Ch. IV.
[299]Amer. Math. Monthly, 9, 1902, 212–3. Stated, 1, 1894, 188.
[300]Anfangsgründe der Zahlentheorie, 1902, 265–6.
[301]L'enseignement math., 9, 1907, 296.
[302]Periodico di Mat., 27, 1912, 79–83.
[303]Nieuw Archief voor Wiskunde, (2), 10, 1912, 172–7.
[304]Periodico di Mat., 28, 1913, 164–177, 267–270.

as k is or is not divisible by 4. For n even, S is divisible only by $k/2$ provided n is not divisible by any prime factor, diminished by unity, of k.

N. Nielsen[305] wrote C_p^r for the sum of the products r at a time of $1,\ldots,$ $p-1$, and

$$s_n(p) = \sum_{s=1}^{p} s^n, \qquad \sigma_n(p) = \sum_{s=1}^{p}(-1)^{p-s}s^n.$$

If p is a prime $> 2n+1$,

$$\sigma_{2n}(p-1) \equiv s_{2n}(p-1) \equiv 0 \pmod{p}, \quad s_{2n+1}(p-1) \equiv 0 \pmod{p^2}.$$

If $p = 2n+1$ is a prime > 3, and $1 \leq r \leq n-1$, C_p^{2r+1} is divisible by p^2.

Nielsen[306] proved that $2D_n^{2p+1}$ is divisible by $2n$ for $2p+1 \leq n$, where D_n^s is the sum of the products of $1, 3, 5, \ldots, 2n-1$ taken s at a time; also,

$$2^{2q+1}s_{2q}(n-1) \equiv 2^{2q}s_{2q}(2n-1) \pmod{4n^2},$$

and analogous congruences between sums of powers of successive even or successive odd integers, also when alternate terms are negative. He proved (pp. 258–260) relations between the C's, including the final formulas by Glaisher.[294]

Nielsen[307] proved the results last cited. Let p be an odd prime. If $2n$ is not divisible by $p-1$, $s_{2n}(p-1) \equiv 0 \pmod{p}$, $s_{2n+1}(p-1) \equiv 0 \pmod{p^2}$. But if $2n$ is divisible by $p-1$,

$$s_{2n}(p-1) \equiv -1, \quad s_{2n+1}(p-1) \equiv 0 \pmod{p}, \quad s_p(p-1) \equiv 0 \pmod{p^2}.$$

T. E. Mason[308] proved that, if p is an odd prime and i an odd integer > 1, the sum A_i of the products i at a time of $1, \ldots, p-1$ is divisible by p^2. If p is a prime > 3, s_k is divisible by p^2 when k is odd and not of the form $m(p-1)+1$, by p when k is even and not of the form $m(p-1)$, and not by p if k is of the latter form. If $k = m(p-1)+1$, s_k is divisible by p^2 or p according as k is or is not divisible by p. Let p be composite and r its least prime factor; then $r-1$ is the least integer t for which A_t is not divisible by p and conversely. Hence p is a prime if and only if $p-1$ is the least t for which A_t is not divisible by p. The last two theorems hold also if we replace A's by s's.

T. M. Putnam[309] proved Glaisher's[295] theorem that s_{-n} is divisible by p if n is not a multiple of $p-1$, and

$$\sum_{j=1}^{(p-1)/2} j^{p-2} \equiv \frac{2-2^p}{p} \pmod{p}.$$

W. Meissner[310] arranged the residues modulo p, a prime, of the successive

[305]K. Danske Vidensk. Selsk. Skrifter, (7), 10, 1913, 353.
[306]Annali di Mat., (3), 22, 1914, 81–94.
[307]Ann. sc. l'école norm. sup., (3), 31, 1914, 165, 196–7.
[308]Tôhoku Math. Jour., 5, 1914, 136–141.
[309]Amer. Math. Monthly, 21, 1914, 220–2.
[310]Mitt. Math. Gesell. Hamburg, 5, 1915, 159–182.

powers of a primitive root h of p in a rectangular table of t rows and τ columns, where $t\tau = p - 1$. For $p = 13$, $h = 2$, $t = 4$, the table
is shown here. Let R range over the numbers in any
column. Then ΣR and $\Sigma 1/R$ are divisible by p. If t is
even, $\Sigma 1/R$ is divisible by p^2, as $1/1 + 1/8 + 1/12 + 1/5 =$
$13^2/120$. For $t = p - 1$, the theorem becomes the first one
due to Wolstenholme.[257] Generalizations are given at the end of the
paper.

$$
\begin{array}{ccc}
1 & 2 & 4 \\
8 & 3 & 6 \\
12 & 11 & 9 \\
5 & 10 & 7
\end{array}
$$

N. Nielsen[311] proved his[286a] theorem and the final results of Glaisher.[294]
Nielsen[312] proceeded as had Aubry[301] and then proved

$$s_{2n+1} \equiv 0 \ (\mathrm{mod}\ p^2), \qquad \sum_{j=1}^{(p-1)/2} j^{2n} \equiv 0 \ (\mathrm{mod}\ p), \quad 1 \leqq n \leqq \frac{p-3}{2}.$$

Then by Newton's identities we get Wilson's theorem and Nielsen's[305] last
result.

E. Cahen[313] stated Nielsen's[286a] theorem.

F. Irwin stated and E. B. Escott[314] proved that if S_j is the sum of the
products j at a time of $1, 1/2, 1/3, .., 1/t$, where $t = (p-1)/2$, then $2S_2 - S_1^2$,
etc., are divisible by the odd prime p.

[311]Oversigt Danske Vidensk. Selsk. Forhandlinger, 1915, 171–180, 521.
[312]Ibid., 1916, 194–5.
[313]Comptes Rendus Séances Soc. Math. France, 1916, 29.
[314]Amer. Math. Monthly, 24, 1917, 471–2.

CHAPTER IV.

RESIDUE OF $(U^{p-1}-1)/P$ MODULO P.

N. H. Abel[1] asked if there are primes p and integers a for which

(1) $$a^{p-1} \equiv 1 \pmod{p^2}, \qquad 1 < a < p.$$

C. G. J. Jacobi[2] noted that, for $p \leq 37$, (1) holds only when $p=11$, $a=3$ or 9; $p=29$, $a=14$; $p=37$, $a=18$. Cf. Thibault[31] of Ch. VI.

G. Eisenstein[3] noted that, for p a prime, the function

$$q_u = (u^{p-1}-1)/p$$

has the properties

(2) $$q_{uv} \equiv q_u + q_v, \qquad q_{u+pv} \equiv q_u - \frac{v}{u} \pmod{p},$$

$$2q_2 \equiv 1 - \tfrac{1}{2} + \tfrac{1}{3} - \tfrac{1}{4} + \ldots - \frac{1}{p-1} \equiv \Sigma \frac{1}{s} \pmod{p},$$

where $s = (p+1)/2, \ldots, p-1$. All solutions of (1) are included in $a \equiv u + puq_u$, $0 < u < p$.

E. Desmarest[4] noted that (1) holds for $p=487$, $a=10$, and stated that $p=3$ and $p=487$ are the only primes <1000 for which 10 is a solution.

J. J. Sylvester[5] stated that, if p, r are distinct primes, $p>2$, then q_r is congruent modulo p to a sum of fractions with the successive denominators $p-1, \ldots, 2, 1$ and (as corrected) with numerators the repeated cycle of the positive integers $\leq r$ congruent modulo r to $1/p, 2/p, \ldots, r/p$. Thus, for $r=5$,

$$q_5 \equiv \frac{1}{p-1} + \frac{2}{p-2} + \frac{3}{p-3} + \frac{4}{p-4} + \frac{5}{p-5} + \frac{1}{p-6} + \ldots \qquad (p = 10k+1),$$

$$q_5 \equiv \frac{3}{p-1} + \frac{1}{p-2} + \frac{4}{p-3} + \frac{2}{p-4} + \frac{5}{p-5} + \frac{3}{p-6} + \ldots \qquad (p = 10k+7).$$

According as $p = 4k+1$ or $4k-1$, q_2 is congruent to

$$\frac{2}{p-3} + \frac{2}{p-4} + \frac{2}{p-7} + \frac{2}{p-8} + \frac{2}{p-11} + \ldots,$$

$$-\frac{2}{p-2} - \frac{2}{p-3} - \frac{2}{p-6} - \frac{2}{p-7} - \frac{2}{p-10} \cdots$$

[the signs were given $+$ erroneously]. For any p,

$$q_2 \equiv -\frac{1}{p-1} + \frac{1}{p-2} - \frac{1}{p-3} + \ldots \pmod{p}.$$

[1] Jour. für Math., 3, 1828, 212; Oeuvres, 1, 1881, 619.
[2] *Ibid.*, 301-2; Werke, 6, 238-9; Canon Arithmeticus, Berlin, 1839, Introd., xxxiv.
[3] Berlin Berichte, 1850, 41.
[4] Théorie des nombres, 1852, 295.
[5] Comptes Rendus Paris, 52, 1861, 161, 212, 307, 817; Phil. Mag., 21, 1861, 136; Coll. Math. Papers, II, 229-235, 241, 262-3.

Jean Plana[6] developed $\{(M-1)+1\}^p$ and obtained

$$M^p - M - \{(M-1)^p - (M-1)\} = pf(M),$$

$$f(M) = M - 1 + \frac{p-1}{2}(M-1)^2 + \frac{(p-1)(p-2)}{2 \cdot 3}(M-1)^3 + \ldots + (M-1)^{p-1}.$$

Take $M = m, m-1, \ldots, 1$ in the first equation and add. Thus

$$\frac{m^p - m}{p} = f(1) + \ldots + f(m) = s_1 + \frac{p-1}{2}s_2 + \ldots + s_{p-1},$$

where $s_i = 1^i + 2^i + \ldots + (m-1)^i$. For $j > 1$, we may replace p by j and get

$$m^j - m = js_{j-1} + \binom{j}{2}s_{j-2} + \binom{j}{3}s_{j-3} + \ldots + js_1,$$

a result obtained by Plana by a long discussion [Euler[41]]. He concluded erroneously that each s_i is divisible by m (for $m = 3$, $s_2 = 5$).

F. Proth[7] stated that, if p is a prime, $2^p - 2$ is not divisible by p^2 [error, see Meissner[33]].

M. A. Stern[8] proved that, if p is an odd prime,

$$\frac{m^p - m}{p} \equiv s_1 - \tfrac{1}{2}s_2 + \tfrac{1}{3}s_3 - \ldots - \frac{1}{p-1}s_{p-1} \equiv \sigma_{p-1} + \tfrac{1}{2}\sigma_{p-2} + \ldots + \frac{1}{p-1}\sigma_1$$

$$\equiv \tfrac{1}{2}\left(\frac{m}{p-1} + \frac{m^2}{p-2} + \ldots + m^{p-1}\right) + \frac{1}{p-2}s_2 + \frac{1}{p-4}s_4 + \ldots + s_{p-1} \pmod{p},$$

for s_i as by Plana and $\sigma_i = 1^i + 2^i + \ldots + m^i$. Proof is given of the formula below (2) of Eisenstein[3] and Sylvester's formulæ for q_2 (corrected), as well as several related formulæ.

L. Gegenbauer[9] used Stern's congruences to prove that the coefficient of the highest power of x in a polynomial $f(x)$ of degree $p-2$ is congruent to $(m^p - m)/p$ modulo p if $f(x)$ satisfies one of the systems of equations

$$f(\lambda) = (-1)^{\lambda+1}\lambda^{p-3}s_\lambda(m-1), \qquad f(\lambda) = \lambda^{p-3}s_{p-\lambda}(m) \qquad (\lambda = 1, \ldots, p-1).$$

E. Lucas[10] proved that q_2 is a square only for $p = 2, 3, 7$, and stated the result by Desmarest.[4]

F. Panizza[11] enumerated the combinations p at a time of ap distinct things separated into p sets of a each, by counting for each r the combinations of the things belonging to r of the p sets:

$$\binom{ap}{p} = \sum_{r=1}^{p} \binom{p}{r} \Sigma \binom{a}{i_1}\binom{a}{i_2} \ldots \binom{a}{i_r},$$

[6]Mem. Acad. Turin, (2), 20, 1863, 120.
[7]Comptes Rendus Paris, 83, 1876, 1288.
[8]Jour. fur Math., 100, 1887, 182–8.
[9]Sitzungsber. Ak. Wiss. Wien (Math.), 95, 1887, II, 616–7.
[10]Théorie des nombres, 1891, 423.
[11]Periodico di Mat., 10, 1895, 14–16, 54–58.

where $i_1 + \ldots + i_r = p$, $i_j > 0$. The term given by $r = p$ is a^p. For p a prime, the left member is $\equiv a \pmod p$ and we have Fermat's theorem. By induction on r,

$$\Sigma \binom{a}{i_1} \ldots \binom{a}{i_r} = \sum_{i=0}^{r-1} (-1)^i \binom{r}{i} \binom{(r-i)a}{p}.$$

Taking $r = p$, we have

$$\frac{a^p - a}{p} = \frac{1}{p}\left\{ \binom{ap}{p} - a \right\} + \sum_{i=1}^{p-1} (-1)^i \frac{1}{i} \binom{p-1}{i-1} \binom{(p-i)a}{p}.$$

D. Mirimanoff[12] wrote a_0 for the least positive integer making $a_0 p + 1$ divisible by the prime $r < p$, and denoted the quotient by $r^{e_i} b_1$, where b_1 is prime to r. Similarly, let a_i be the least positive integer such that $a_i p + b_i = r^{e_i} b_{i+1}$. We ultimately find an n for which $b_n = 1$. Then $b_{n+i} = b_i$. By (2),

$$q_{b_i} - \frac{a_i}{b_i} \equiv e_i q_r + q_{b_{i+1}}, \qquad -\sum_{i=0}^{n-1} \frac{a_i}{b_i} \equiv q_r \Sigma e_i \pmod p.$$

Let r belong to the exponent ω modulo p and set $e\omega = p - 1$. Then $\Sigma e_i = \omega$, while $1, b_1, \ldots, b_{n-1}$ are the distinct residues of the eth powers of the integers $< r$ and prime to r. Thus

$$q_r \equiv e \sum_{i=0}^{n-1} \frac{a_i}{b_i} \pmod p.$$

The formula obtained by taking r a primitive root of p is included in the following, which holds also for any prime r:

$$q_r \equiv \sum_{\beta_i=1}^{p-1} \frac{a_i}{\beta_i} \pmod p,$$

a_i being the least positive integer for which $a_i p + \beta_i \equiv 0 \pmod r$. Set $\beta_i = p - \delta$, $p' p \equiv 1 \pmod r$, $0 < p' < r$. Then $a_i \equiv p'\delta - 1 \pmod r$,

$$q_r \equiv \sum_{\delta=1}^{p-1} \frac{\{p'\delta\}}{p - \delta} \pmod p,$$

$\{k\}$ being the least positive residue modulo r of k. Whence Sylvester's statement.

J. S. Aladow[13] proved that (1) has at most $(p \mp 1)/4$ roots if $p = 4m \pm 1$.

A. Cunningham[13a] listed 27 cases in which $r^{p-1} \equiv 1$ or $r^l \equiv 1 \pmod{p^l}$, $r < p^{l-1}$, where l is a divisor of $p - 1$. For the 11 cases of the first kind, $p = 5, 7, 17, 19, 29, 37, 43, 71, 487$.

W. Fr. Meyer[14] proved by induction that, if p is a prime, $x^{p-1} - 1$ is divisible by p^k ($1 \leq k < n$), but not by p^{k+1}, for exactly $p^{n-1-k}(p-1)^2$ positive integers $x < p^n$ and prime to p, and is divisible by p^n for the remaining $p - 1$ such integers. Set

$$A = a + \mu_1 p + \ldots + \mu_n p^n \ (1 \leq a < p, \ 0 \leq \mu_i < p), \qquad \lambda_\rho = (a^{p^\rho} - a^{p^{\rho-1}})/p^\rho.$$

[12]Jour. für Math., 115, 1895, 295–300.
[13]St. Petersburg Math. Soc. (Russian), 1899, 40–44.
[13a]Messenger Math., 29, 1899–1900, 158. See Cunningham[128], Ch. VI.
[14]Archiv Math. Phys., (3), 2, 1901, 141–6.

If k is the least index for which $\mu_k \not\equiv \lambda_k$, $\mu_h \equiv \lambda_h$ (mod p) for $h < k$, then $A^{p-1} - 1$ is divisible by p^k, but not by p^{1+k}.

A. Palmström and A. Pollak[15] proved that, if p is a prime and n, m are the exponents to which a belongs modulo p, p^2, respectively, then $a^{np} - 1$ is divisible by p^2, so that m is a multiple of n and a divisor of np, whence $m = n$ or pn. Thus according as a^{p-1} is or is not $\equiv 1$ (mod p^2), $m = n$ or $m = np$.

Worms de Romilly[15a] noted that, if ω is a primitive root of p^2, the incongruent roots of $x^{p-1} \equiv 1$ (mod p^2) are $\omega^{jp}(j = 1, \ldots, p-1)$.

J. W. L. Glaisher[16] proved that if r is a positive integer $< p$, p a prime,

$$r^{p-1} = 1 + g_1 p + \tfrac{1}{2}(g_1^2 - g_2)p^2 + \tfrac{1}{6}(g_1^3 - 3g_1g_2 + 2g_3)p^3 + \ldots,$$

where g_n is the sum of the nth powers of

$$\frac{1}{\sigma}, \frac{2}{[2\sigma]}, \ldots, \frac{r-1}{[(r-1)\sigma]}; \frac{1}{r+\sigma}, \frac{2}{r+[2\sigma]}, \ldots, \frac{r-1}{r+[(r-1)\sigma]}; \frac{1}{2r+\sigma}, \ldots,$$

σ being the least positive residue modulo r of $-p$. If μ_i is the least positive solution of $\sigma\mu_i \equiv i$ (mod r), viz., $p\mu_i + i \equiv 0$, then

$$g_1 = \frac{\mu_1}{1} + \frac{\mu_2}{2} + \ldots + \frac{\mu_{r-1}}{r-1} + \frac{\mu_1}{r+1} + \frac{\mu_2}{r+2} + \ldots + \frac{\mu_{r-1}}{2r-1} + \frac{\mu_1}{2r+1} + \ldots.$$

Set $\mu_r = 0$, $\mu_{i+jr} = \mu_i$. Then

$$g_n = \sum_{i=1}^{p-1} \left(\frac{\mu_i}{i}\right)^n, \qquad \sum_{i=1}^{p-1} \frac{\mu_i}{i^2} \equiv 0 \pmod{p}.$$

Sylvester's corrected results are proved. From $(1+1)^p$,

$$\frac{2^p - 2}{p} \equiv 1 - \tfrac{1}{2} + \tfrac{1}{3} - \ldots - \frac{1}{p-1} \equiv 2\left(1 + \tfrac{1}{3} + \ldots + \frac{1}{p-2}\right) \pmod{p}.$$

For $r' = r + kp$, let μ_i' be the positive root of $p\mu_i' + i \equiv 0$ (mod r'). Then

$$r'^{p-1} = 1 + h_1 p + \tfrac{1}{2}(h_1^2 - h_2)p^2 + \ldots, \qquad h_n = \sum_{i=1}^{p-1} \left(\frac{\mu_i'}{i}\right)^n.$$

It is shown that, for some integer t,

$$h_1 - g_1 + \frac{k}{r} = tp, \qquad h_2 - g_2 \equiv -2\frac{k}{r} - \frac{k^2}{r^2} + 2t \equiv \frac{2k}{r}g_1 - \frac{k^2}{r^2} \pmod{p},$$

$$r^{p^i(p-1)} \equiv 1 + g_1 p^{i+1} - \tfrac{1}{2}g_2 p^{i+2} \pmod{p^{i+3}}.$$

Glaisher,[17] using the same notations, gave

$$r^{p-1} \equiv 1 + p\left(\frac{\mu_1}{1} + \frac{\mu_2}{2} + \ldots + \frac{\mu_{p-1}}{p-1}\right) \pmod{p^2}.$$

[15]L'intermédiaire des math., 8, 1901, 122, 205–6 (7, 1900, 357).
[15a]Ibid., 214–5.
[16]Quar. Jour. Math., 32, 1901, 1–27, 240–251.
[17]Messenger Math., 30, 1900–1, 78.

Glaisher[18] considered q_u in connection with Bernoullian numbers and gave

$$\frac{3^{p-1}-1}{p} \equiv -\tfrac{2}{3}\left(1+\frac{1}{2}+\ldots+\frac{1}{k}\right) \pmod{p=3k+1}.$$

A. Pleskot[19] duplicated the work of Plana.[6]

P. Bachmann[20] gave an exposition of the work by Sylvester,[5] Stern,[8] Mirimanoff.[12]

M. Lerch[21] set, for any odd integer p and for u prime to p,

$$q_u = \frac{1}{p}(u^{\varphi(p)}-1).$$

Then,* as a generalization of (2),

$$q_{uv} \equiv q_u + q_v, \qquad q_{u+pv} \equiv q_u + \frac{v\phi(p)}{u} \pmod{p},$$

$$q_u \equiv \sum_v \frac{1}{uv}\left[\frac{uv}{p}\right], \qquad 2q_2 \equiv \Sigma\frac{1}{\lambda} \equiv -\Sigma\frac{1}{\mu} \pmod{p},$$

where ν ranges over the positive integers $<p$ and prime to p; λ over those $>p/2$; μ over those $<p/2$. Henceforth, let p be an odd prime and set $N = \{(p-1)!+1\}/p$. Then $N \equiv q_1 + \ldots + q_{p-1}$,

$$q_2 \equiv -\tfrac{1}{3}\sum_{\nu=1}^{[p/4]}\frac{1}{\nu}, \qquad 3q_3 \equiv -2\sum_{\nu=1}^{[p/3]}\frac{1}{\nu}, \qquad 5q_5 \equiv -2\sum_{a=1}^{[p/5]}\frac{1}{a} - 2\sum_{b=1}^{[2p/5]}\frac{1}{b},$$

modulo p. If $\psi(n)$ is the number of sets of positive solutions $<p$ of $\mu\nu = n$ and hence the number of divisors between n/p and p of n,

$$N \equiv \sum_{n=1}^{(p-1)^2} \frac{\psi(n)}{n}\left[\frac{n}{p}\right] \pmod{p}.$$

Employing Legendre's symbol and Bernoullian numbers, we have

$$A = \sum_{\nu=1}^{p-1}\left(\frac{\nu}{p}\right)q_\nu \equiv 0 \text{ or } (-1)^{n-1}2B_n \pmod{p},$$

according as $p = 4n+3$ or $4n+1$. In the respective cases,

$$\sum_{\nu=1}^{p-1}\left(\frac{\nu}{p}\right)pq_\nu \equiv Cl(-p) \text{ or } 0 \pmod{p},$$

where $Cl(-\Delta)$ is the number of classes of positive primitive forms $ax^2+bxy+cy^2$ of negative discriminant $b^2-4ac=-\Delta$. Also, modulo p,

$$A-N \equiv \sum_{\nu=1}^{p-1}\frac{1}{\nu^2}\left[\frac{\nu^2}{p}\right], \qquad q_a \equiv 2\Sigma\frac{1}{aa}\left[\frac{aa}{p}\right],$$

$$q_a \equiv 2\Sigma_b\frac{1}{ab}\left[\frac{ab}{p}\right], \qquad q_b \equiv 2\Sigma_a q_a - 2\Sigma_\beta q_\beta + 2\Sigma_a\frac{1}{ab}\left[\frac{ab}{p}\right],$$

where a, a are quadratic residues of p, and b, β non-residues.

---

[18]Proc. London Math. Soc., 33, 1900–1, 49–50.
[19]Zeitschrift für das Realschulwesen, Wien, 27, 1902, 471–2.
[20]Niedere Zahlentheorie, I, 1902, 159–169.   *The greatest integer $\leqq x$ is denoted by $[x]$.
[21]Math. Annalen, 60, 1905, 471–490.

H. F. Baker[22] extended Sylvester's theorem to any modulus N:

$$\frac{r^{\varphi(N)}-1}{N} \equiv \sum_{i=1}^{\varphi(N)} \frac{\{N'm_i\}}{N-m_i} \pmod{N},$$

where the m_i denote the integers $<N$ and prime to N, $N'N\equiv1$ (mod r), and $\{k\}$ is the least positive residue modulo r of k.

Lerch[23] extended Mirimanoff's[12] formula to the case of a composite modulus m. Set

$$q(a, m) = \frac{1}{m}(a^{\varphi(m)}-1).$$

Let a belong to the exponent $\phi(m)/e$. Then $q(a, m)\equiv e\Sigma a/\beta$ (mod m), where β ranges over the residues of the incongruent powers of a, and $ma+\beta\equiv0$ (mod a), $0\leqq a<a$. As an extension of Sylvester's theorem,

$$q(a, m)\equiv\sum_\nu\frac{r_\nu}{\nu}\equiv-\sum_\nu\frac{r_\nu'}{\nu} \pmod{m},$$

where ν ranges over the integers $<m$ and prime to m, while

$$mr_\nu+\nu\equiv0, \qquad mr_\nu'-\nu\equiv0 \pmod{a}, \qquad 0\leqq r_\nu<a, \qquad 0\leqq r_\nu'<a.$$

For $m=m_1\ldots m_k$, where the m_j are relatively prime,

$$q(a, m) \equiv \sum_{j=1}^k n_j n_j' \phi(n_j) q(a, m_j) \pmod{m},$$

where $m=m_j n_j$, $n_j^2 n_j'\equiv1$ (mod m_j).

H. Hertzer[24] verified that, for $a<p<307$, $a^{p-1}-1$ is divisible by p^3 only for $a=68$, $p=113$; $a=3$, 9, $p=11$. He examined all the primes between 307 and 751, but only for a and $p-a$ when $a<\sqrt{p}$, finding only $p=113$, $a=68$. Removing the restriction $a<\sqrt{p}$, he found only the solutions

$$p=11, a=3; \qquad p=331, a=18, 71; \qquad p=353, a=14;$$
$$p=487, a=10, 175; \quad p=673, a=22,$$

together with the square of each a.

A. Friedmann and J. Tamarkine[25] gave formulas connecting q_u with Bernoullian numbers and $[u/p]$.

A. Wieferich[26] proved that if $x^p+y^p+z^p=0$ is satisfied by integers x, y, z prime to p, where p is an odd prime, then $2^{p-1}\equiv1$ (mod p^2). Shorter proofs were given by D. Mirimanoff[27] and G. Frobenius.[28]

D. A. Grave[29] gave the residue of q_2 for each prime $p<1000$ and thought he could prove that 2^p-2 is never divisible by p^2 (error, Meissner[33]).

A. Cunningham[30] verified that 2^p-2 is not divisible by p^2 for any prime $p<1000$, and[31] that 3^p-3 is not divisible by p^2 for a prime $p=2^a3^b+1<100$.

W. H. L. Janssen van Raay[32] noted that 2^p-2 is not divisible by p^2 in general.

[22]Proc. London Math. Soc., (2), 4, 1906, 131–5.   [23]Comptes Rendus Paris, 142, 1906, 35–38.
[24]Archiv Math. Phys., (3), 13, 1908, 107.   [25]Jour. für Math., 135, 1909, 146–156.
[26]Jour. für Math., 136, 1909, 293–302.   [27]L'enseignement math., 11, 1909, 455–9.
[28]Sitzungsber. Ak. Wiss. Berlin, 1909, 1222–4; reprinted in Jour. für Math., 137, 1910, 314.
[29]An elementary text on the theory of numbers (in Russian), Kiev, 1909, p. 315; Kiev Izv. Univ.,
1909, Nos. 2–10.
[30]Report British Assoc. for 1910, 530.   L'intermédiaire des math., 18, 1911, 47; 19, 1912, 159.
Proc. London Math. Soc., (2), 8, 1910, xiii.
[31]L'intermédiaire des math., 18, 1911, 47.   Cf., 20, 1913, 206.
[32]Nieuw Archief voor Wiskunde, (2), 10, 1912, 172–7.

L. Bastien[32a] verified that (1) holds for $p<50$ only for $p=43$, $a=19$, and for Jacobi's[2] cases.   He stated that, if $p=4p\pm1$ is a prime,

$$-\tfrac{1}{2}q_2\equiv1+1/3+1/5+\ldots+1/(2h-1)\ (\mathrm{mod}\ p).$$

W. Meissner[33] gave a table showing the least positive residue of $(2^t-1)/p$ modulo p for each prime $p<2000$, where t is the exponent to which 2 belongs modulo p.   In particular, 2^p-2 is divisible by the square of the prime $p=1093$, contrary to Proth[7] and Grave,[29] but for no other $p<2000$.

In the chapter on Fermat's last theorem will be given not only the condition $q_2\equiv0\ (\mathrm{mod}\ p)$ of Wieferich[26] but also $q_3\equiv0\ (\mathrm{mod}\ p)$, etc., with citations to D. Mirimanoff, Comptes Rendus Paris, 150, 1910, 204-6, and Jour. für Math., 139, 1911, 309-324; H. S. Vandiver, ibid., 144, 1914, 314-8; G. Frobenius, Sitzungsber. Ak. Wiss. Berlin, 1910, 200-8; 1914, 653-81. These papers give further properties of q_u.

P. Bachmann[34] employed the identity

$$(a+b+c)^p-(a+b-c)^p+(a-b-c)^p-(a-b+c)^p$$

$$=2\binom{p}{1}c\{(a+b)^{p-1}-(a-b)^{p-1}\}+2\binom{p}{3}c^3\{(a+b)^{p-3}-(a-b)^{p-3}\}+\ldots$$

for $a=b=1$, $c=2$ or 1 to get expressions for q_2 or q_3, whence

$$\frac{3^p-3}{p}\equiv2\left\{\frac{2^{p-1}}{1}+\frac{2^{p-3}}{3}+\frac{2^{p-5}}{5}+\ldots+\frac{2^2}{p-2}\right\}\ (\mathrm{mod}\ p),$$

for an odd prime p.   Comparing this with the value of $(3^p-3)/p$ obtained by expanding $(2+1)^p$, we see that

$$\frac{2^p-2}{p}\equiv2^{p-1}+\tfrac{1}{2}\cdot2^{p-2}+\tfrac{1}{3}\cdot2^{p-3}+\ldots+\frac{1}{p-1}\cdot2\ (\mathrm{mod}\ p).$$

Again,

$$q_2\equiv2-\left(\frac{p-1}{2}\right)^2+\Sigma sgn.(s-t)\ (-1)^{s+t+1}s\ (\mathrm{mod}\ p),$$

summed for all sets of solutions of $s^2\equiv t^2+1\ (\mathrm{mod}\ p)$.   Finally,

$$q_2\equiv\sum_{h=1}^{p-1}\left\{(r^h-r^{-h})\Sigma_g(r^{2gh}-1)^{-1}\right\},$$

where r is a primitive pth root of unity.

*H. Brocard[35] commented on $a^{p-1}\equiv1\ (\mathrm{mod}\ p^n)$.   *H. G. A. Verkaart[36] treated the divisibility of a^p-a by p.   E. Fauquembergue[37] checked that $2^p\equiv2\ (\mathrm{mod}\ p^2)$ for $p=1093$.

N. G. W. H. Beeger[38] tabulated all roots of $x^{p-1}\equiv1\ (\mathrm{mod}\ p^2)$ for each prime $p<200$.   If ω is a primitive root of p^2, the absolutely least residue

[32a]Sphinx-Oedipe, 7, 1912, 4-6. It is stated that G. Tarry had verified in 1911 that 2^p-2 is not divisible by a prime $p<1013$.

[33]Sitzungsber. Ak. Wiss. Berlin, 1913, 663-7.

[34]Jour. für Math., 142, 1913, 41-50.

[35]Revista de la Sociedad Mat. Española, 3, 1913-4, 113-4.

[36]Wiskundig Tijdschrift, vol. 2, 1906, 238-240.

[37]L'intermédiaire des math., 1914, 33.

[38]Messenger Math., 43, 1913-4, 72-84.

$\pm x_1$ modulo p^2 of ω^p is a root, that $(\pm x_2)$ of x_1^2 is a second root, that $(\pm x_3)$ of $x_1 x_2$ is a third root, etc., until the root $\pm x_s$ is reached, where $s = (p-1)/2$. The remaining roots are $p^2 - x_i (i = 1, \ldots, s)$. He proved that

$$(x_1 \ldots x_s)^2 \equiv (-1)^{\frac{p+1}{2}} \pmod{p^2}.$$

Hence $x_1 \ldots x_s \equiv \pm 1$ if $p = 4n+1$.

W. Meissner[39] wrote h_m for the residue $< p^m$ of $h^{p^{m-1}}$ modulo p^m. When h varies from 1 to $p-1$, we get $p-1$ roots h_m of $x^{p-1} \equiv 1 \pmod{p^m}$. The product of the roots given by $h = 1, \ldots, (p-1)/2$, is $\equiv (-1)^z$ or $(-1)^z \sigma$ $\pmod{p^m}$, according as $p = 4n-1$ or $4n+1$, where z is the number of pairs of integers $< p/2$ whose product is $\equiv -1 \pmod{p}$, and σ is the smaller of the two roots of $x^2 \equiv -1 \pmod{p}$. No number $< p$ which belongs to one of the exponents 2, 3, 4, 6, modulo p, can be a root of $x^{p-1} \equiv 1 \pmod{p^2}$. A root of the latter is given for each prime $p < 300$, and a root modulo p^3 for each $p < 200$; also the exponent to which each root belongs.

N. Nielsen[40] noted that, if we select $2r$ distinct integers a_s, b_s $(s = 1, \ldots, r)$ from $1, \ldots, p-1$, such that $a_s + b_s = p$, then

$$\frac{a_1 \ldots a_r}{b_1 \ldots b_r} = (-1)^r (1 - pA), \quad A \equiv \sum_{s=1}^{r} \frac{1}{b_s} \equiv \sum_{s=1}^{r} \left(q_{a_s} - q_{b_s} \right) \pmod{p}.$$

Proof is given of various results by Lerch,[21] also of simple relations between q_a and Bernoullian numbers, and of the final formula by Plana,[6] here attributed to Euler.[41]

H. S. Vandiver[42] proved that there are not fewer than $[\sqrt{p}]$ and not more than $p - (1 + \sqrt{2p-5})/2$ incongruent least positive residues of $1, 2^{p-1}, \ldots, (p-1)^{p-1}$, modulo p^2.

N. Nielsen[43] noted that, if a is not divisible by the odd prime p,

$$q_a \equiv \frac{a-1}{2a} + \sum_{s=1}^{(p-3)/2} \frac{1}{2s} (-1)^{s-1} B_s (a^{p-2s-1} - 1) \pmod{p},$$

$$q_1 + q_2 + \ldots + q_{p-1} \equiv (-1)^{n-1} B_n + \frac{1}{p} - 1 \pmod{p^2}, \quad n = (p-1)/2.$$

W. Meissner[44] gave various expressions for q_2 and q_3.

A. Gérardin[45] found all primes $p < 2000$, including those of the form $2^n - 1$, for which q_2 is symmetrical when written to the base 2.

H. S. Vandiver[46] proved that $q_2 \equiv 0 \pmod{p^2}$ if and only if

$$1 + \frac{1}{3} + \frac{1}{5} + \ldots + \frac{1}{p-2} \equiv 0 \pmod{p^2}.$$

He gave various expressions for $(n^k - 1)/m$.

[39]Sitzungsber. Berlin Math. Gesell., 13, 1914, 96–107.
[40]Ann. sc. l'école norm. sup., (3), 31, 1914, 171–9.
[41]Euler, Institutiones Calculi Diff., 1755, 406. Proof, Math. Quest. Educ. Times, 48, 1888, 48.
[42]Bull. Amer. Math. Soc., 22, 1915, 61–7.
[43]Oversigt Danske Vidensk. Selsk. Forhandlinger, 1915, 518–9, 177–180; cf. Lerch's[21] N.
[44]Mitt. Math. Gesell. Hamburg, 5, 1915, 172–6, 180.
[45]Nouv. Ann. Math., (4), 17, 1917, 102–8.
[46]Annals of Math., 18, 1917, 112.

CHAPTER V.

EULER'S ϕ-FUNCTION, GENERALIZATIONS: FAREY SERIES.

Number $\phi(n)$ of Integers $<n$ and Prime to n.

L. Euler,[1] in connection with his generalization of Fermat's theorem, investigated the number $\phi(n)$ of positive integers not exceeding n which are relatively prime to n, without then using a functional notation for $\phi(n)$. He began with the theorem that, if the n terms a, $a+d,\ldots$, $a+(n-1)d$ in arithmetical progression are divided by n, the remainders are 0, 1,$\ldots$, $n-1$ in some order, provided d is prime to n; in fact, no two of the terms have the same remainder.

If p is a prime, $\phi(p^m)=p^{m-1}(p-1)$, since p, $2p,\ldots$, $p^{m-1}\cdot p$ are the only ones of the p^m positive integers $\leqq p^m$ not prime to p^m. To prove that

$$(1) \qquad \phi(AB)=\phi(A)\phi(B) \qquad (A, B \text{ relatively prime}),$$

let 1, $a,\ldots$, ω be the integers $<A$ and prime to A. Then the integers $<AB$ and prime to A are

$$
\begin{array}{cccc}
1 & a & \ldots & \omega \\
A+1 & A+a & \ldots & A+\omega \\
2A+1 & 2A+a & \ldots & 2A+\omega \\
\ldots\ldots\ldots & \ldots\ldots\ldots & \ldots & \ldots\ldots\ldots \\
(B-1)A+1 & (B-1)A+a & \ldots & (B-1)A+\omega.
\end{array}
$$

The terms in any column form an arithmetical progression whose difference A is prime to B, and hence include $\phi(B)$ integers prime to B. The number of columns is $\phi(A)$. Hence there are $\phi(A)\phi(B)$ positive integers $<AB$, prime to both A and B, and hence prime to AB. If $p,\ldots$, s are distinct primes, the two theorems give

$$(2) \qquad \phi(p^\lambda\ldots s^\epsilon)=p^{\lambda-1}(p-1)\ldots s^{\epsilon-1}(s-1).$$

Euler[2] later used πN to denote $\phi(N)$ and gave a different proof of (2). First, let $N=p^n q$, where p, q are distinct primes. Among the $N-1$ integers $<N$ there are p^n-1 multiples of q, and $p^{n-1}q-1$ multiples of p, these sets having in common the $p^{n-1}-1$ multiples of pq. Hence

$$\phi(N)=N-1-(p^n-1)-(p^{n-1}q-1)+p^{n-1}-1=p^{n-1}(p-1)(q-1).$$

A simpler proof is then given for the modified form of (2):

$$(3) \qquad \phi(N)=\frac{N(p-1)(q-1)\ldots(s-1)}{pq\ldots s},$$

where p, q, $r,\ldots$, s are the distinct primes dividing N. There are N/p multiples $<N$ of p and hence $N'=N(p-1)/p$ integers $<N$ and prime to p. Of these, N'/q are divisible by q; excluding them, we have $N''=N'(q-1)/q$ numbers $<N$ and prime to both p and q. The rth part of these are said

[1]Novi Comm. Ac. Petrop., 8, 1760–1, 74; Comm. Arith., 1, 274. Opera postuma, I, 492–3.
[2]Acta Ac. Petrop., 4 II (or 8), 1780 (1755), 18; Comm. Arith., 2, 127–133. He took $\phi(1)=0$.

[cf. Poinsot[16]] to be divisible by r; after excluding them we get $N''(r-1)/r$ numbers; etc.

Euler[3] noted in a posthumous paper that, if p, q, r are distinct primes, there are r multiples $\leq pqr$ of pq, and qr multiples of p, and a single multiple of pqr, whence

$$\phi(pqr) = pqr - qr - pr - pq + r + p + q - 1 = (p-1)(q-1)(r-1).$$

In general, if M is any number not divisible by the prime p, and if μ denotes the number of integers $\leq M$ and prime to M, there are $M - \mu$ integers $\leq M$ and not prime to M and hence $p^n(M-\mu)$ integers $\leq Mp^n$ and not prime to M and therefore not prime to Mp^n. Of the Mp^{n-1} multiples $\leq Mp^n$ of p, exclude the $p^{n-1}(M-\mu)$ which are not prime to M; we obtain $p^{n-1}\mu$ multiples of p which are prime to M. Hence

$$\phi(p^nM) = p^nM - p^n(M-\mu) - p^{n-1}\mu = p^{n-1}(p-1)\mu.$$

A. M. Legendre[4] noted that, if $\theta, \ldots, \omega$ are any odd primes not dividing A, the number of terms of the progression $A+B, 2A+B, \ldots, nA+B$ which are divisible by no one of the primes $\theta, \ldots, \omega$ is approximately $n(1-1/\theta) \ldots (1-1/\omega)$, and exactly that number if n is divisible by $\theta, \ldots, \omega$.

C. F. Gauss[5] introduced the symbol $\phi(N)$. He expressed Euler's[1] proof of (1) in a different form. Let a be any one of the $\phi(A)$ integers $< A$ and prime to A, while β is any one of the $\phi(B)$ integers $< B$ and prime to B. There is one and but one positive integer $x < AB$ such that $x \equiv a \pmod{A}$, $x \equiv \beta \pmod{B}$. Since this x is prime to A and to B, it is prime to AB. Making the agreement that $\phi(1) = 1$, Gauss proved

(4) $\Sigma\phi(d) = N$ (d ranging over the divisors of N).

For each d, multiply the integers $\leq d$ and prime to d by N/d; we obtain $\Sigma\phi(d)$ integers $\leq N$, proved to be distinct and to include $1, 2, \ldots, N$.

A. M. Legendre[6] proved (3) as follows: First, let $N = pM$, where p is a prime which may or may not divide M; then $Mp - M$ of the numbers $1, \ldots, N$ are not divisible by p. Second, let $N = pqM$, where p and q are distinct primes. Then $1, \ldots, N$ include M numbers divisible by both p and q; $Mp - M$ numbers divisible by q and not by p; $Mq - M$ numbers divisible by p and not by q. Hence there remain $N(1-1/p)(1-1/q)$ numbers divisible by neither p nor q. Third, a like argument is said to apply to $N = pqrM$, etc.

Legendre (p. 412) proved that if A, C are relatively prime and if $\theta, \lambda, \mu, \ldots,$ ω are odd primes not dividing A, the number of terms $kA - C$ ($k = 1, \ldots, n$), which are divisible by no one of $\theta, \ldots, \omega$, is

$$n - \Sigma\left[\frac{n+\theta_0}{\theta}\right] + \Sigma\left[\frac{n+(\theta\lambda)_0}{\theta\lambda}\right] - \Sigma\left[\frac{n+(\theta\lambda\mu)_0}{\theta\lambda\mu}\right] + \ldots,$$

[3]Tractatus de numerorum, Comm. Arith., 2, 515–8. Opera postuma, I, 1862, 16–17.
[4]Essai sur la théorie des nombres, 1798, p. 14.
[5]Disquisitiones Arithmeticæ, 1801, Arts. 38, 39.
[6]Théorie des nombres, ed. 2, 1808, 7–8; German trans. of ed. 3 by Maser, 8–10.

where the summations extend over the combinations of $\theta, \ldots, \omega$ taken $1, 2, \ldots$, at a time, while Δ_0 is a positive integer $<\Delta$ for which $A\Delta_0 + C$ is divisible by Δ, and. $[x]$ is the greatest integer $\leq x$. We thus derive the approximation stated by Legendre.[4] Taking $A = 1$, $C = 0$ (p. 420), we see that the number of integers $\leq n$, which are divisible by no one of the distinct primes $\theta, \lambda, \ldots, \omega$ is

(5) $$n - \Sigma\left[\frac{n}{\theta}\right] + \Sigma\left[\frac{n}{\theta\lambda}\right] - \Sigma\left[\frac{n}{\theta\lambda\mu}\right] + \ldots.$$

A. von Ettingshausen[7] reproduced without reference Euler's[2] proof of (3) and gave an obscurely expressed proof of (4). Let $N = p^\alpha q^\beta \ldots$, where $p, q, \ldots$ are distinct primes. Consider first only the divisors $d = p^\mu q^\nu$, where $\mu > 0$, $\nu > 0$, so that d involves the primes p and q, but no others. By (3),

$$\phi(d) = d\left(1 - \frac{1}{p}\right)\left(1 - \frac{1}{q}\right), \qquad \overset{\alpha}{\underset{\mu=1}{\Sigma}}\,\overset{\beta}{\underset{\nu=1}{\Sigma}}\, p^\mu q^\nu = (p + p^2 + \ldots + p^\alpha)(q + \ldots + q^\beta),$$

$$\Sigma\phi(p^\mu q^\nu) = (p^\alpha - 1)(q^\beta - 1).$$

Similarly, $\Sigma\phi(p^\mu) = p^\alpha - 1$. In this way we treat together the divisors of N which involve the same prime factors. Hence when d ranges over all the divisors of N,

$$\Sigma\phi(d) = 1 + \underset{p}{\Sigma}(p^\alpha - 1) + \underset{p,\,q}{\Sigma}\,(p^\alpha - 1)(q^\beta - 1) + \underset{p,\,q,\,r}{\Sigma}\,(p^\alpha - 1)(q^\beta - 1)(r^\gamma - 1) + \ldots$$

$$= \underset{p}{\Pi}\{1 + (p^\alpha - 1)\} = \Pi p^\alpha = N,$$

where the summation indices range over the combinations of all the prime factors of N taken $1, 2, \ldots$ at a time. [Cf. Sylvester.[32]]

A. L. Crelle[8] considered the number z_j of integers, chosen from $n_1, \ldots, n_a$, which are divisible by exactly j of the distinct primes $p_1, \ldots, p_m$; and the number s_j of the integers, chosen from $n_1, \ldots, n_a$, which are divisible by at least j of the primes p_i. Then

$$z_1 + z_2 + \ldots + z_m = s_1 - s_2 + s_3 - \ldots \pm s_m.$$

Let ν be the number of the integers $n_1, \ldots, n_a$ which are divisible by no one of the primes p_i. Then

$$a = \Sigma z_i + \nu, \qquad \nu = a - s_1 + s_2 - \ldots \mp s_m.$$

In particular, take $n_1, \ldots, n_a$ to be $1, 2, \ldots, N$, where $N = p^\alpha q^\beta r^\gamma \ldots$, and take $p_1, \ldots, p_m$ to be $p, q, r, \ldots$. Then

$$s_1 = \frac{N}{p} + \frac{N}{q} + \ldots, \qquad s_2 = \frac{N}{pq} + \frac{N}{pr} + \ldots, \qquad s_3 = \frac{N}{pqr} + \ldots,$$

$$\phi(N) = N - s_1 + s_2 - \ldots = N\left(1 - \frac{1}{p}\right)\left(1 - \frac{1}{q}\right)\ldots.$$

He proved (1) for $B = a^\alpha$, where a is a prime not dividing A (p. 40). By Euler's[1] table there are $B\phi(A)$ integers $< AB$ and prime to A. In Euler's

[7] Zeitschrift für Physik u. Math. (eds., Baumgartner and Ettingshausen),Wien, 5, 1829, 287–292.
[8] Abh. Akad. Wiss. Berlin (Math.), 1832, 37–50.

notation, $a(kA+1)$, $a(kA+a)$, ..., $a(kA+\omega)$ give all the numbers between kaA and $(k+1)aA$ which are divisible by a and are prime to A. Taking $k=0, 1, ..., a^{a-1}-1$, we see that there are exactly $a^{a-1}\phi(A)$ multiples of a which are $<AB$ and prime to A. Hence

$$\phi(a^a A) = a^a \phi(A) - a^{a-1}\phi(A) = \phi(a^a)\phi(A).$$

F. Minding[9] proved Legendre's formula (5). The number of integers $\leq n$, not divisible by the prime θ, is $n - [n/\theta]$. To make the general step by induction, let $p_1, ..., p_k$ be distinct primes, and denote by $(B; p_1, ..., p_k)$ the number of integers $\leq B$ which are divisible by no one of the primes $p_1, ..., p_k$. Then, if p is a new prime,

$$(B; p_1, ..., p_k, p) = (B; p_1, ..., p_k) - ([B/p]; p_1, ..., p_k).$$

The truth of (4) for the special case $N = p-1$, where p is a prime, follows (p. 41) from the fact that $\phi(d)$ numbers belong to the exponent d modulo p if d is any divisor of $p-1$.

N. Druckenmüller[10] evaluated $\phi(b)$, first for the case in which b is a product $cd...kl$ of distinct primes. Set $b = \beta l$ and denote by $\psi(b)$ the number of integers $<b$ having a factor in common with b. There are $l\psi(\beta)$ numbers $<b$ which are divisible by one of the primes $c, ..., k$, since there are $\psi(\beta)$ in each of the sets

$$1, 2, ..., \beta; \quad \beta+1, ..., 2\beta; \quad ...; \quad (l-1)\beta+1, ..., l\beta.$$

Again, $l, 2l, ..., \beta l$ are the integers $<b$ with the factor l. Of these, $\phi(\beta)$ are prime to β, while the others have one of the factors $c, ..., k$ and occur among the above $l\psi(\beta)$. Hence $\psi(b) = l\psi(\beta) + \phi(\beta)$. But $\psi(\beta) + \phi(\beta) = \beta$. Hence

$$\phi(b) = (l-1)\phi(\beta) = (c-1)...(l-1).$$

Next, let b be a product of powers of $c, d, ..., l$, and set $b = L\beta$, $\beta = cd...l$. By considering L sets as before, we get

$$\psi(b) = L\psi(\beta), \qquad \phi(b) = L\phi(\beta).$$

E. Catalan[11] proved (4) by noting that

$$\Sigma\phi(p^u q^v ...) = \Pi\{1+\phi(p)+...+\phi(p^u)\} = \Pi p^u = N,$$

where there are as many factors in each product as there are distinct prime factors of N.

A. Cauchy[12] gave without reference Gauss'[5] proof of (1).

E. Catalan[13] evaluated $\phi(N)$ by Euler's[2] second method.

C. F. Arndt[14] gave an obscure proof of (4), apparently intended for Catalan's.[11] It was reproduced by Desmarest, Théorie des nombres, 1852, p. 230.

[9]Anfangsgründe der Höheren Arith., 1832, 13–15.
[10]Theorie der Kettenreihen...Trier, 1837, 21.
[11]Jour. de Mathématiques, 4, 1839, 7–8.
[12]Comptes Rendus Paris, 12, 1841, 819–821; Exercices d'analyse et de phys. math., Paris, 2, 1841, 9; Oeuvres, (2), 12.
[13]Nouv. Ann. Math., 1, 1842, 466–7.
[14]Archiv Math. Phys., 2, 1842, 6–7.

J. A. Grunert[15] examined in a very elementary way the sets

$$jk+1, \quad jk+2, \ldots, \quad jk+k-1, \quad (j+1)k \quad (j=0, 1, \ldots, p-1)$$

and proved that $\phi(pk) = p\phi(k)$ if the prime p divides k, while $\phi(pk) = (p-1)\phi(k)$ if the prime p does not divide k. From these results, (2) is easily deduced [cf. Crelle[17] on $\phi(Z)$].

L. Poinsot[16] gave Catalan's[11] proof of (4) and proved the statements made by Euler[2] in his proof of (3). Thus to show that, of the $N' = N(1-1/p)$ integers $< N$ and prime to p, exactly N'/q are divisible by q, note that the set $1, \ldots, N$ contains N/q multiples of q and the set $p, 2p, \ldots$ contains $(N/p)/q$ multiples of q, while the difference is N'/q.

If P, Q, $R, \ldots$ are relatively prime in pairs, any number prime to $N = PQR \ldots$ can be expressed in the form

$$pQR \ldots + qPR \ldots + rPQ \ldots + \ldots,$$

where p is prime to P, q to Q, etc. If also $p < P$, $q < Q$, etc., no two of these sums are equal. Thus there are $\phi(P)\phi(Q) \ldots$ such sums [certain of which may exceed N].

To prove (4), take (pp. 70–71) a prime p of the form $kN+1$ and any one of the N roots ρ of $x^N \equiv 1 \pmod{p}$. Then there is a least integer d, a divisor of N, such that $\rho^d \equiv 1 \pmod{p}$. The latter has $\phi(d)$ such roots. Also ρ is a primitive root of the last congruence and of no other such congruence whose degree is a divisor of N.

A. L. Crelle[17] considered the product $E = e_1 e_2 \ldots e_n$ of integers relatively prime in pairs, and set $E_j = E/e_j$. When x ranges over the values $1, \ldots, e_i$, the least positive residue modulo E of $E_1 x_1 + \ldots + E_n x_n$ takes each of the values $1, \ldots, E$ once and but once. In case x_i is prime to e_i for $i = 1, \ldots, n$, the residue of $\Sigma E_i x_i$ is prime to E and conversely. Let $d_{i1}, d_{i2}, \ldots$ be any chosen divisors > 1 of e_i which are relatively prime in pairs. Let $\psi(e_i)$ denote the number of integers $\leq e_i$ which are divisible by no one of the $d_{i1}, d_{i2}, \ldots$. Let $\psi(E)$ be the number of integers $\leq E$ which are divisible by no one of the $d_{11}, d_{12}, d_{21}, \ldots$, including now all the d's. Then $\psi(E) = \psi(e_1) \ldots \psi(e_n)$. In case $d_{i1}, d_{i2}, \ldots$ include all the prime divisors > 1 of e_i, $\psi(e_i)$ becomes $\phi(e_i)$. Of the two proofs (pp. 69–73), one is based on the first result quoted, while the other is like that by Gauss.[5]

As before, let $\psi(y)$ be the number of integers $\leq y$ which are divisible by no one of certain chosen relatively prime divisors $d_1, \ldots, d_m$ of y. By considering the xy numbers $ny+r$ ($0 \leq n < x$, $1 \leq r \leq y$), it is proved (p. 74) that, when x and y are relatively prime,

$$\psi(xy) = x\psi(y), \qquad \psi_2(xy) = (x-1)\psi(y),$$

where $\psi_2(xy)$ is the number of integers $\leq xy$ which are divisible neither by x nor by any one of the d's. These formulas lead (pp. 79–83) to the value of $\phi(Z)$. Set

$$Z = p_1^{e_1} \ldots p_\mu^{e_\mu}, \qquad z = p_1 \ldots p_\mu, \qquad n = Z/z,$$

[15]Archiv. Math. Phys., 3, 1843, 196–203.
[16]Jour. de Mathématiques, 10, 1845, 37–43.
[17]Encyklopädie der Zahlentheorie, Jour. für Math., 29, 1845, 58–95.

where $p_1, \ldots, p_\mu$ are distinct primes. For a prime p, not dividing y, we have $\phi(py) = (p-1)\phi(y)$. Take $y = p_1$, $p = p_2$; then

$$\phi(p_1 p_2) = (p_1 - 1)(p_2 - 1).$$

Next, take $y = p_1 p_2$, $p = p_3$, and use also the last result; thus

$$\phi(p_1 p_2 p_3) = (p_1 - 1)(p_2 - 1)(p_3 - 1),$$

and similarly for $\phi(z)$. When ζ ranges over the integers $< z$ and prime to z, the numbers $\nu z + \zeta$ ($\nu = 0, 1, \ldots, n-1$) give without repetition all the integers $< Z$ and prime to Z. Hence $\phi(Z) = n\phi(z)$, which leads to (2). [Cf. Guilmin,[23] Steggall.[78]]

The proofs of (4) by Gauss[5] and Catalan[11] are reproduced without references (pp. 87–90). A third proof is given. Set $N = a^\alpha b^\beta c^\gamma \ldots$, where a, b, $c, \ldots$ are distinct primes. Consider any divisor $\epsilon = b^{\beta_1} c^{\gamma_1} \ldots$ of N such that ϵ is not divisible by a. Then

$$\phi(\epsilon a^k) = a^{k-1}(a-1)\phi(\epsilon).$$

Sum for $k = 0, 1, \ldots, a$; we get $a^\alpha \phi(\epsilon)$. When k ranges over its values and β_1 over the values $0, 1, \ldots, \beta$, and γ_1 over the values $0, 1, \ldots, \gamma$, etc., ϵa^k ranges over all the divisors d of N. Hence $\Sigma \phi(d) = a^\alpha \Sigma \phi(\epsilon)$. Similarly, if ϵ_1 range over the divisors not divisible by a or b,

$$\Sigma \phi(\epsilon) = b^\beta \Sigma \phi(\epsilon_1), \ldots, \qquad \Sigma \phi(d) = a^\alpha b^\beta \ldots = N.$$

E. Prouhet[18] proposed the name indicator and symbol $i(N)$ for $\phi(N)$. He gave Gauss' proof of (1) and Catalan's proof of (4). If δ is the product of the distinct prime factors common to a and b,

$$\phi(ab) = \phi(a)\phi(b)\delta/\phi(\delta).$$

As a generalization, let δ_i be the product of the distinct primes common to i of the numbers $a_1, \ldots, a_n$; then

$$\phi(a_1 \ldots a_n) = \phi(a_1) \ldots \phi(a_n) \frac{\delta_2}{\phi(\delta_2)} \frac{\delta_3^2}{\phi^2(\delta_3)} \cdots \frac{\delta_n^{n-1}}{\phi^{n-1}(\delta_n)}.$$

Friderico Arndt[19] proved (1) by showing that, if x ranges over the integers $< A$ and prime to A, while y ranges over the integers $< B$ and prime to B, then $Ay + Bx$ gives only incongruent residues modulo AB, each prime to AB, and they include every integer $< AB$ and prime to AB. [Crelle's[17] first theorem for $n = 2$.]

V. A. Lebesgue[20] used Euler's[2] argument to show that there are

$$\frac{N(p-1)(q-1) \ldots (k-1)}{p \cdot q \ldots k}$$

integers $< N$ and prime to $p, q, \ldots, k$, the latter being certain prime divisors of N [Legendre,[4] Minding[9]].

[18]Nouv. Ann. Math., 4, 1845, 75–80.
[19]Jour. für Math., 31, 1846, 246–8.
[20]Nouv. Ann. Math., 8, 1849, 347.

G. L. Dirichlet[21] added equations (4) for $N = n, \ldots, 2, 1$, noting that, if $s \leqq n$, $\phi(s)$ occurs in the new left member as often as there are multiples $\leqq n$ of s. Hence

$$\sum_{s=1}^{n} \left[\frac{n}{s}\right] \phi(s) = \tfrac{1}{2}(n^2 + n).$$

The left member is proved equal to $\Sigma \psi[n/s]$, where

$$\psi(x) = \phi(1) + \ldots + \phi(x).$$

It is then shown that $\psi(n) - 3n^2/\pi^2$ is of an order of magnitude not exceeding that of n^δ, where $2 > \delta > \gamma > 1$, γ being such that

$$\sum_{s=2} \frac{1}{s^\gamma} = 1.$$

P. L. Tchebychef[22] evaluated $\phi(n)$ by showing that, if p is a prime not dividing A, the ratio of the number of integers $\leqq pAN$ which are prime to A to the number which are prime to both A and p is $p : p - 1$.

A. Guilmin[23] gave Crelle's[17] argument leading to $\phi(Z)$.

F. Landry[24] proved (3). First, reject from $1, \ldots, N$ the N/p multiples of p; there remain $N(1 - 1/p)$ numbers prime to p. Next, to find how many of the multiples $q, 2q, \ldots, N$ of q are prime to p, note that the coefficients $1, 2, \ldots, N/q$ contain $N/q \cdot (1 - 1/p)$ integers prime to p by the first result, applied to the multiple N/q of p in place of N.

Daniel Augusto da Silva[25] considered any set S of numbers and denoted by $S(a)$ the subset possessing the property a, by $S(ab)$ the subset with the properties a and b simultaneously, by $(a)S$ the subset of numbers in S not having property a; etc. Then

$$(a)S = S - S(a) = S\{1 - (a)\},$$

symbolically. Hence

$$(ba)S = (b)\{(a)S\} = S\{1 - (a)\}\{1 - (b)\},$$
$$(\ldots cba)S = S\{1 - (a)\}\{1 - (b)\}\{1 - (c)\} \ldots.$$

A proof of the latter symbolic formula was given by F. Horta.[25a]

With Silva, let S be the set $1, 2, \ldots, n$, and let $A, B, \ldots$ be the distinct prime factors of n. Let properties $a, b, \ldots$ be divisibility by $A, B, \ldots$. Then there are n/A terms in $S(a)$, $n/(AB)$ terms in $S(ab), \ldots$, and $\phi(n)$ terms in $(\ldots cba)S$. Hence our symbolic formula gives

$$\phi(n) = n\left(1 - \frac{1}{A}\right)\left(1 - \frac{1}{B}\right) \ldots.$$

[21]Abhand. Ak. Wiss. Berlin (Math.), 1849, 78–81; Werke, 2, 60–64.
[22]Theorie der Congruenzen, 1889, §7; in Russian, 1849.
[23]Nouv. Ann. Math., 10, 1851, 23.
[24]Troisième mémoire sur la théorie des nombres, 1854, 23–24.
[25]Proprietades geraes et resoluçao directa das Congruencias binomias, Lisbon, 1854. Report on same by C. Alasia, Rivista di Fisica, Mat. e Sc. Nat., Pavia, 4, 1903, 13–17; reprinted in Annaes Scientificos Acad. Polyt. do Porto, Coimbra, 4, 1909, 166–192.
[25a]Annaes de Sciencias e Lettras, Lisbon, 1, 1857, 705.

E. Betti[26] evaluated $\phi(m)$, where m is a product of powers of the distinct primes $a_1, a_2, \ldots$. Consider the set C_i of the products of the a's taken i at a time and their multiples $\leq m$. Thus C_0 is $1, \ldots, m$, while C_2 is

$$a_1a_2, 2a_1a_2, \ldots, \frac{m}{a_1a_2} \cdot a_1a_2; \quad a_1a_3, 2a_1a_3, \ldots, \frac{m}{a_1a_3} \cdot a_1a_3; \ldots$$

Let x be an integer $< m$ divisible by $a_1, \ldots, a_\alpha$. Then x occurs

$$1 + \binom{a}{2} + \binom{a}{4} + \ldots = 2^{\alpha-1}$$

times in the sets $C_0, C_2, C_4, \ldots$; and $2^{\alpha-1}$ times in $C_1, C_3, \ldots$. Summing

$$1 - \binom{a}{1} + \binom{a}{2} - \binom{a}{3} + \ldots = 0$$

for each of the $m - \phi(m)$ integers $\leq m$ having factors in common with m, we get

$$m - \phi(m) - \Sigma \binom{a}{1} + \Sigma \binom{a}{2} - \ldots = 0.$$

But $\Sigma \binom{a}{1}$ is the number of integers having in common with m one of the factors $a_1, a_2, \ldots$, and hence equals $\Sigma \dfrac{m}{a_1}$. Next, $\Sigma \binom{a}{2}$ is the number of integers having in common with m one of the factors $a_1a_2, a_1a_3, \ldots$, and hence equals $\Sigma\{m/(a_1a_2)\}$. Thus

$$\phi(m) = m - \Sigma\frac{m}{a_1} + \Sigma\frac{m}{a_1a_2} - \ldots.$$

R. Dedekind[27] gave a general theorem on the inversion of functions (to be explained in the chapter on that subject), which for the special case of $\phi(n)$ becomes a proof like Betti's. Cf. Chrystal's Algebra, II, 1889, 511; Mathews' Theory of Numbers, 1892, 5; Borel and Drach,[81] p. 27.

J. B. Sturm[28] evaluated $\phi(N)$ by a method which will be illustrated for the case $N = 15$. From $1, \ldots, 15$ delete the five multiples of 3. Among the remaining ten numbers there are as many multiples of 5 as there are multiples of 5 among the first ten numbers. Hence $\phi(15) = 10 - 2 = 8$. The theorem involved is the following. From the three sets

$$1, 2, 3,^* 4, 5; \qquad 6,^* 7, 8, 9,^* 10; \qquad 11, 12,^* 13, 14, 15^*$$

delete (by marking with an asterisk) the multiples of 3. The numbers 11, 13, 14 which remain in the final set are congruent modulo 5 to the numbers 6, 3, 9 deleted from the earlier sets.

J. Liouville[29] proved by use of (4) that, for $|x| < 1$,

$$\sum_{m=1}^{\infty} \frac{\phi(m)x^m}{1-x^m} = \frac{x}{(1-x)^2}, \qquad \Sigma\frac{\phi(m)x^m}{1-x^{2m}} = \sum_{m=1}^{\infty} \frac{\phi(m)x^m}{1+x^m} = \frac{x(1+x^2)}{(1-x^2)^2},$$

[26]Bertrand's Algèbre, Ital. transl. with notes by Betti, Firenze, 1856, note 5. Proof reproduced by Fontebasso[34], pp. 74–77.
[27]Jour. für Math., 54, 1857, 21. Dirichlet-Dedekind, Zahlentheorie, §138.
[28]Archiv Math. Phys., 29, 1857, 448–452.
[29]Jour. de mathématiques, (2), 2, 1857, 433–440.

where m in Σ' ranges only over the positive odd integers. The final fraction equals $x+3x^3+5x^5+\ldots$. From the coefficient of x^n in the expansion of the third sum, we conclude that, if n is even,

$$\Sigma(-1)^{\delta-1}\phi(d)=0 \qquad (\delta=n/d),$$

where d ranges over all the divisors of n. Let δ_1 range over the odd values of δ, and δ_2 over the even values of δ; then

$$\Sigma\phi\left(\frac{n}{\delta_1}\right)=\Sigma\phi\left(\frac{n}{\delta_2}\right)=\frac{n}{2},$$

the value $n/2$ following from (4). Another, purely arithmetical, proof is given. Finally, by use of (4), it is proved that, if $s>2$,

$$\sum_{n=1}^{\infty}\frac{\phi(n)}{n^s}=S_{s-1}/S_s, \qquad S_m=\sum_{n=1}^{\infty}\frac{1}{n^m}.$$

A. Cayley[30] discussed the solution for N of $\phi(N)=N'$. Set $N=a^\alpha b^\beta\ldots$, where $a, b,\ldots$ are distinct primes. Multiply

$$1+(a-1)\{a\}+a(a-1)\{a^2\}+\ldots+a^{\alpha-1}(a-1)\{a^\alpha\}+\ldots$$

by the analogous series in b, etc.; the bracketed terms are to be multiplied together by enclosing their product in a bracket. The general term of the product is evidently

$$\Pi a^{\alpha-1}(a-1)\cdot\{a^\alpha b^\beta\ldots\}=\phi(N)\{N\}.$$

Hence in the product first mentioned each of the bracketed numbers which are multiplied by the coefficient N' will be a solution N of $\phi(N)=N'$. We need use only the primes a for which $a-1$ divides N', and continue each series only so far as it gives a divisor of N' for the coefficient of $a^{\alpha-1}(a-1)$.

V. A. Lebesgue[31] proved $\phi(Z)=n\phi(z)$ as had Crelle[17] and then $\phi(z)=\Pi(p_i-1)$ by the usual method of excluding multiples of $p_1,\ldots, p_n$ in turn. By the last method he proved (pp. 125-8) Legendre's (5), and the more general formula preceding (5).

J. J. Sylvester[32] proved (4) by the method of Ettingshausen,[7] using (2) instead of (3). By means of (4) he gave a simple proof of the first formula of Dirichlet;[21] call the left member u_n; since $[n/r]-[(n-1)/r]=1$ or 0, according as n is or is not divisible by r,

$$u_n-u_{n-1}=\Sigma\phi(d)=n, \qquad u_n=\frac{n(n+1)}{2}+c.$$

The constant c is zero since $u_1=1$. He stated the generalization

$$\sum_{i=1}^{n}\left\{\phi(i^r)\left(1^{r-1}+2^{r-1}+\ldots+\left[\frac{n}{i}\right]^{r-1}\right)\right\}=1^r+2^r+\ldots+n^r.$$

He remarked that the theorem in its simplest form is

$$n^r=\Sigma\{\phi(i_1{}^{r-1})\phi(i_2{}^{r-2})\ldots\phi(i_{r-1})\cdot n/(i_1 i_2\ldots i_r)\},$$

[30]London Ed. and Dublin Phil. Mag., (4), 14, 1857, 539-540.
[31]Exercices d'analyse numérique, 1859, 43-45.
[32]Quar. Jour. Math., 3, 1860, 186-190; Coll. Math. Papers, 2, 225-8.

the example given being $r=2$, $n=4$, whence the divisors of n are $1\cdot1, 2\cdot1, 4\cdot1$, $1\cdot2, 2\cdot2, 1\cdot4$ and the above terms are

$$1\cdot1\cdot1, \qquad 1\cdot1\cdot1, \qquad 1\cdot1\cdot2, \qquad 2\cdot1\cdot1, \qquad 2\cdot1\cdot1, \qquad 4\cdot2\cdot1,$$

with the sum 4^2. [With this obscure result contrast that by Cantor.[49]]

G. L. Dirichlet[33] completed by induction Euler's[2] method of proving (3), obtaining at the same time the generalization that, if p, q, . . ., s are divisors, relatively prime in pairs, of N, the number of integers $\leq N$ which are divisible by no one of p, . . ., s is

$$N\left(1-\frac{1}{p}\right)\left(1-\frac{1}{q}\right)\cdots\left(1-\frac{1}{s}\right).$$

A proof (§13) of (4) follows from the fact that, if d is a divisor of N, there are exactly $\phi(d)$ integers $\leq N$ having with N the g. c. d. N/d.

P. A. Fontebasso[34] repeated the last remark and gave Gauss' proof of (1).

E. Laguerre[35] employed any real number k and integer m and wrote $(m, m/k)$ for the number of integers $\leq m/k$ which are prime to m. By continuous variation of k he proved that

$$\Sigma(d, d/k)=[m/k],$$

where d ranges over the divisors of m. For $k=1$, this reduces to (4).

F. Mertens[36] obtained an asymptotic value for $\phi(1)+\ldots+\phi(G)$ for G large. He employed the function $\mu(n)$ [see Ch. XIX] and proved that

$$\sum_{m=1}^{G} \phi(m) = \tfrac{1}{2} \sum_{n=1}^{G} \mu(n)\left\{\left[\frac{G}{n}\right]^2+\left[\frac{G}{n}\right]\right\} = \frac{3}{\pi^2}G^2+\Delta$$

$$|\Delta| < G(\tfrac{1}{2} \log_e G+\tfrac{1}{2}C+\tfrac{5}{8})+1,$$

where C is Euler's constant $0.57721\ldots.$ This upper limit for Δ is more exact than that by Dirichlet.[21]

T. Pepin[37] stated that, if $n=a^\alpha b^\beta\ldots(a, b,\ldots$ distinct primes),

$$n=\phi(n)+\Sigma a^{\alpha-1}\phi\left(\frac{n}{a^\alpha}\right)+\Sigma a^{\alpha-1}b^{\beta-1}\phi\left(\frac{n}{a^\alpha b^\beta}\right)+\ldots+a^{\alpha-1}b^{\beta-1}\ldots.$$

Moret-Blanc[38] proved the latter by noting that the first sum is the number of integers $<n$ which are divisible by a single one of the primes $a, b,\ldots$, the second sum is the number of integers $<n$ divisible by two of the primes, . . ., while $a^{\alpha-1}b^{\beta-1}\ldots$ is the number of integers $<n$ divisible by all those primes.

H. J. S. Smith[39] considered the m-rowed determinant Δ_m having as the element in the ith row and jth column the g. c. d. (i, j) of i, j. Let $l_1=m$,

[33]Zahlentheorie, §11, 1863; ed. 2, 1871; ed. 3, 1879; ed. 4, 1894.

[34]Saggio di una introd. all'arit. trascendente, Treviso, 1867, 23–26.

[35]Bull. Soc. Math. France, 1, 1872–3, 77.

[36]Jour. für Math., 77, 1874, 289–91.

[37]Nouv. Ann. Math., (2), 14, 1875, 276.

[38]Ibid., p. 374. L. Gegenbauer, Monatsh. Math. Phys., 4, 1893, 184, gave a generalization to primary complex numbers.

[39]Proc. London Math. Soc., 7, 1875–6, 208–212; Coll. Papers, 2, 161.

$l_2, l_3, \ldots$ be those divisors of $m = p^{\alpha}q^{\beta} \ldots t^{\tau}$ which are given by the expansion of the product

$$\phi(m) = (p^{\alpha} - p^{\alpha-1}) \ldots (t^{\tau} - t^{\tau-1}) = l_1 - l_2 + l_3 - \ldots - l_s.$$

It is proved that

$$\phi(m, k) \equiv (l_1, k) - (l_2, k) + \ldots - (l_s, k)$$

[called Smith's function by Lucas,[72] p. 407] is zero if $k < m$, but equals $\phi(m)$ if $k = m$. Hence if to the mth column of Δ_m we add the columns with indices $l_3, l_5, \ldots$ and subtract the columns with indices $l_2, l_4, \ldots$, we obtain an equal determinant in which the elements of the mth column are zero with the exception of the element $\phi(m)$. Hence $\Delta_m = \Delta_{m-1}\phi(m)$, so that

(6) $$\Delta_m = \phi(1)\phi(2) \ldots \phi(m).$$

If we replace the element $\delta = (i, j)$ by any function $f(\delta)$ of δ, we obtain a determinant equal to $F(1) \ldots F(m)$, where

$$F(m) = f(m) - \Sigma f\left(\frac{m}{p}\right) + \Sigma f\left(\frac{m}{pq}\right) - \ldots.$$

Particular cases are noted. For $f(\delta) = \delta^k$, $F(m)$ becomes Jordan's[200] function $J_k(m)$. Next, if $f(\delta)$ is the sum of the kth powers of the divisors of δ, then $F(m) = m^k$. Finally, if $f(\delta) = 1^k + 2^k + \ldots + \delta^k$, it is stated erroneously that $F(m)$ is the sum $\phi_k(m)$ of the kth powers of the integers $\leqq m$ and prime to m. [Smith overlooked the factors $a^k, a^k b^k, \ldots$ in Thacker's[150] first expression for $\phi_k(n)$, which is otherwise of the desired form $F(n)$. The determinant is not equal to $\phi_k(1) \ldots \phi_k(m)$, as the simple case $k = 1$, $m = 2$, shows.]

In the main theorem we may replace $1, \ldots, m$ by any set of distinct numbers $\mu_1, \ldots, \mu_m$ such that every divisor of each μ_i is a number of the set; the determinant whose element in the ith row and jth column is $f(\delta)$, where $\delta = (\mu_i, \mu_j)$, equals $F(\mu_1) \ldots F(\mu_m)$. Examples of sets of μ's are the numbers in their natural order with the multiples of given primes rejected; the numbers composed of given primes; and the numbers without square factors.

R. Dedekind[40] proved that, if n be decomposed in every way into a product ad, and if e is the g. c. d. of a, d, then

$$\Sigma \frac{a}{e}\phi(e) = n\Pi\left(1 + \frac{1}{p}\right),$$

where a ranges over all divisors of n, and p over the prime divisors of n.

P. Mansion[41] stated that Smith's relation (6) yields a true relation if we replace the elements $1, 2, \ldots$ of the determinant Δ_m by any symbols $x_1, x_2, \ldots$, and replace $\phi(m)$ by $x_{l_1} - x_{l_2} + x_{l_3} - \ldots$. [But the latter is only another form of Smith's $F(m)$ when we write x_{δ} for Smith's $f(\delta)$, so that the generalization is the same as Smith's.]

[40]Jour. für Math., 83, 1877, 288. Cf. H. Weber, Elliptische Functionen, 1891, 244–5; ed. 2, 1908 (Algebra III), 234–5.
[41]Messenger Math., 7, 1877–8, 81–2.

P. Mansion[42] proved (6), showing that $\phi(m, k)$ equals $\phi(m)$ or 0, according as m is or is not a divisor of k. [Cf. Bachmann, Niedere Zahlentheorie, I, 1902, 97–8.] He repeated his[41] "generalization." He stated that if a and b are relatively prime, the products of the $\phi(a)$ numbers $<a$ and prime to a by the numbers $<b$ and prime to b give the numbers $<ab$ and prime to ab [false for $a=4$, $b=3$; cf. Mansion[44]]. His proof of (4) should have been credited to Catalan.[11]

E. Catalan[43] gave a condensation and slight modification of Mansion's[42] paper. C. Le Paige (ibid., pp. 176–8) proved Mansion's[44] theorem that every product equals a determinant formed from the factors.

P. Mansion[44] proved that the determinant $|c_{ij}|$ of order n equals $x_1 x_2 \ldots x_n$ if $c_{ij} = \Sigma x_p$, where p ranges over the divisors of the g. c. d. of i, j. To obtain a "generalization" of Smith's theorem, set $z_1 = x_1$, $z_2 = x_1 + x_2, \ldots, z_i = \Sigma x_d$, where d ranges over all the divisors of i. Solving, we get

$$x_m = z_m - z_{l_2} + z_{l_3} - \ldots,$$

where the l's are defined above.[39] Thus each c_{ij} is a z. For example, if $n = 4$,

$$|c_{ij}| = \begin{vmatrix} z_1 & z_1 & z_1 & z_1 \\ z_1 & z_2 & z_1 & z_2 \\ z_1 & z_1 & z_3 & z_1 \\ z_1 & z_2 & z_1 & z_4 \end{vmatrix}.$$

For $z_i = i$, x_i becomes $\phi(i)$ and we get (6). [As explained in connection with Mansion's[41] first paper, the generalization is due to Smith.]

J. J. Sylvester[45] called $\phi(n)$ the totient $\tau(n)$ of n, and defined the totitives of n to be the integers $<n$ and prime to n.

F. de Rocquigny[46] stated that, if $\phi^2(N)$ denotes $\phi\{\phi(N)\}$, etc.,

$$\phi^p(N^m) = \phi^{p-2}(N^{m-2}) \cdot \phi^{p-1}\{(N-1)^2\},$$

if N is a prime and $m > 2$, $p > 2$. He stated incorrectly (ibid., 50, 1879, 604) that the number of integers $\leq P$ which are prime to $N = a^\alpha b^\beta \ldots$ is $P(1 - 1/a)$ $(1 - 1/b) \ldots$.

A. Minine[47] noted that the last result is correct for the case in which P is divisible by each prime factor $a, b, \ldots$ of N. He wrote symbolically $nE\dfrac{(1)}{x}$ for $[n/x]$, the greatest integer $\leq n/x$. By deleting from $1, \ldots, P$ the $[P/a]$ numbers divisible by a, then the multiples of b, etc., we obtain for the number of integers $\leq P$ which are prime to N the expression

$$\phi(N)_P = P\left\{1 - E\frac{(1)}{a}\right\}\left\{1 - E\frac{(1)}{b}\right\} \ldots$$

[equivalent to (5)]. If $N, N', N'', \ldots$ are relatively prime by twos,

[42]Annales de la Soc. Sc., Bruxelles, 2, II, 1877–8, 211–224. Reprinted in Mansion's Sur la théorie des nombres, Gand, 1878, §3, pp. 3–16.
[43]Nouv. Corresp. Math., 4, 1878, 103–112.
[44]Bull. Acad. R. Sc. de Belgique, (2), 46, 1878, 892–9.
[45]Amer. Jour. Math., 2, 1879, 361, 378; Coll. Papers, 3, 321, 337. Nature, 37, 1888, 152–3.
[46]Les Mondes, Revue Hebdom. des Sciences, 48, 1879, 327.
[47]Ibid., 51, 1880, 333. Math. Soc. of Moscow, 1880. Jour. de math. élém. et spéc., 1880, 278.

$$\phi(N)_P \cdot \phi(N')_{P'} \cdot \phi(N'')_{P''} \ldots = \phi(NN'N'' \ldots)_{P \cdot P'P''} \ldots$$

E. Lucas[48] stated and Radicke proved that

$$\sum_{a=1}^{n} \psi(a,\, n) = \sum_{k=2}^{n} \phi(k), \qquad \sum_{a=1}^{n-1} a\psi(a,\, n) = \tfrac{1}{2} \sum_{k=2}^{n} k\phi(k),$$

if $\psi(a,\, n)$ is the number of integers $>a$, prime to a and $\leqq n$.

H. G. Cantor[49] proved by use of ζ-functions that

$$\Sigma \nu_0^{\rho-1} \nu_1^{\rho-2} \ldots \nu_{\rho-2}^1 \phi(\nu_0) \phi(\nu_1) \ldots \phi(\nu_{\rho-1}) = n^\rho,$$

summed for all distinct sets of positive integral solutions $\nu_0, \ldots, \nu_{\rho-1}$ of $\nu_0 \ldots \nu_\rho = n$, and noted that this result can be derived from the special case (4).

O. H. Mitchell[50] defined the a-totient $\tau_a(k)$ of $k = a^t b^u \ldots$ (where $a, b, \ldots$ are distinct primes) to be the number of integers $<k$ which are divisible by a, but by no one of the remaining prime factors $b, c, \ldots$ of k. Similarly, the ab-totient $\tau_{ab}(k)$ of k is the number of integers $<k$ which are divisible by a and b, but not by $c, \ldots$; etc. If $k = a^t b^u c^v$,

$$\tau_a(k) = a^{t-1}\phi(b^u c^v), \qquad \tau_{ab}(k) = a^{t-1}b^{u-1}\phi(c^v), \qquad \tau_{abc}(k) = a^{t-1}b^{u-1}c^{v-1},$$

$$\phi(k) + \underset{3}{\Sigma}\tau_a(k) + \underset{3}{\Sigma}\tau_{ab}(k) + \tau_{abc}(k) = k.$$

If σ contains the same primes as s, but with the same exponents as in k, so that $\sigma = a^t$ if $s = a$, it is stated (p. 302) that

$$\tau_s(k) = \frac{\sigma}{s}\phi\left(\frac{k}{\sigma}\right).$$

C. Crone[51] evaluated $\phi(n)$ by an argument valid only when n is a product of distinct primes $p_1, \ldots, p_q$. The number of integers $<n$ having a factor in common with n is then

$$A = \Sigma\left(\frac{n}{p_1} - 1\right) - \Sigma\left(\frac{n}{p_1 p_2} - 1\right) + \ldots + (-1)^q \Sigma\left(\frac{n}{p_1 \ldots p_{q-1}} - 1\right).$$

The sum of the second terms of each sum is

$$-\binom{q}{1} + \binom{q}{2} - \ldots - (-1)^q\binom{q}{q-1} = -1 - (-1)^q.$$

Hence the number of integers $<n$ and prime to n is

$$n - 1 - A = n - \Sigma\frac{n}{p_1} + \Sigma\frac{n}{p_1 p_2} - \ldots - (-1)^q\Sigma\frac{n}{p_1 \ldots p_{q-1}} + (-1)^q$$

$$= n\left(1 - \frac{1}{p_1}\right) \ldots \left(1 - \frac{1}{p_q}\right),$$

provided $n = p_1 \ldots p_q$. [To modify the proof to make it valid for any n, we need only add to A the term

$$-(-1)^q\left(\frac{n}{p_1 \ldots p_q} - 1\right)$$

and hence replace $(-1)^q$ by $(-1)^q n/(p_1 \ldots p_q)$ in $n - 1 - A$.]

[48]Nouv. Corresp. Math., 6, 1880, 267–9. Also Lucas,[72] p. 403.
[49]Göttingen Nachrichten, 1880, 161; Math. Ann., 16, 1880, 583–8.
[50]Amer. Jour. Math., 3, 1880, 294.
[51]Tidsskrift for Mathematik, (4), 4, 1880, 158–9.

Franz Walla[52] considered the product P of the first n primes >1. Let $x_1, \ldots, x_\nu$ be the integers $<P/2$ and prime to P, so that $\nu = \phi(P)/2$. Then, if $n>2$, half of the x's are $\equiv 1 \pmod 4$ and the others are $\equiv 3 \pmod 4$. Also, the absolute values of $\frac{1}{2}P - 2x_j$ ($j = 1, \ldots, \nu$) are the $\dot{x}$'s in some order. Half of the x's are $<P/4$.

J. Perott[53] proved that

$$\Phi(N) \equiv \sum_{h=1}^{N} \phi(h) = \frac{1}{2} + \frac{1}{2}\left\{N^2 - \Sigma\left[\frac{N}{p_i}\right]^2 + \Sigma\left[\frac{N}{p_i p_j}\right]^2 - \ldots\right\},$$

the context showing that the summations extend over all the primes p_i for which $1 < p_i \leq N$ [Lucas[72]]. He proved that

$$\lim_{N=\infty} \frac{\Phi(N)}{N^2} = \frac{3}{\pi^2}$$

and gave a table showing the approximation of $3N^2/\pi^2$ to $\Phi(N)$ for $N \leq 100$. The last formula, proved earlier by Dirichlet[21] and Mertens,[36] was proved by G. H. Halphen[54] by the use of integrals and ζ-functions.

Sylvester[54a] defined the frequency δ of a divisor d of one or more given integers $a, b, \ldots, l$ to be the number of the latter which are divisible by d. By use of (4) he proved the generalization

$$\sum_d \delta \, \phi(d) = a + b + \ldots + l.$$

J. J. Sylvester[55] stated that the number of [irreducible proper] fractions whose numerator and denominator are $\leq j$ is $T(j) = \phi(1) + \ldots + \phi(j)$, and that

$$\sum_{k=1}^{j} T\left[\frac{j}{k}\right] \equiv \sum_{k=1}^{j} \sum_{i=1}^{[j/k]} \phi(i) = \frac{j^2 + j}{2},$$

whence $T(j)/j^2$ approximates $3/\pi^2$ as j increases indefinitely.

If $u(x)$ denotes the sum of all the integers $<x$ and prime to x, and if $U(j) = u(1) + \ldots + u(j)$, then $U(j)$ is the sum of the numerators in the above set of fractions, and*

$$\sum_{k=1}^{j} kU\left[\frac{j}{k}\right] = \frac{1}{6}j(j+1)(j+2).$$

When j increases indefinitely, $U(j)/j^3$ approximates $1/\pi^2$. For each integer $n \leq 1000$ the values of $\phi(n)$, $T(n)$, $3n^2/\pi^2$ are tabulated.

Sylvester[56] stated the preceding results and noted that the first formula is equivalent to

$$\sum_{r=1}^{\infty} \left[\frac{j}{r}\right] \phi(r) = \frac{1}{2}(j^2 + j).$$

[52]Archiv Math. Phys., 66, 1881, 353–7.

[53]Bull. des Sc. Math. et Astr., (2), 5, I, 1881, 37–40.

[54]Comptes Rendus Paris, 96, 1883, 634–7.

[54a]Amer. Jour. Math., 5, 1882, 124; Coll. Math. Papers, 3, 611.

[55]Phil. Mag., 15, 1883, 251–7; 16, 1883, 230–3; Coll. Math. Papers, 4, 101–9. Cf. Sylvester.[54]

[56]Comptes Rendus Paris, 96, 1883, 409–13, 463–5; Coll. Math. Papers, 4, 84–90. Proofs by F. Rogel and H. W. Curjel, Math. Quest. Educ. Times, 66, 1897, 62–4; 70, 1899, 56.

*With denominator 3, but corrected to 6 by Sylvester,[56] which accords with Cesàro.[65] The editor of Sylvester's Papers stated in both places that the second member should be $j(j+1)(2j+1)/12$, evidently wrong for $j = 2$.

E. Cesàro[57] proved that, if f is any function,

$$\sum_{n=1}\frac{x^n f(n)}{1-x^n} = \sum_{n=1}^{\infty} x^n F(n), \qquad F(n) \equiv \Sigma f(d),$$

where d ranges over the divisors of n. For $f = \phi$, we have $F(x) = x$ and obtain Liouville's[29] first formula. By the same specialization (p. 64) of another formula (given in Chapter X on sums of divisors[61]), Cesàro derived the final formula of Liouville.[29] If (n, j) is the g. c. d. of n and j, then (p. 77, p. 80)

$$\sum_{j=1}^{n}(n, j) = \Sigma d\phi\left(\frac{n}{d}\right), \qquad \Sigma\frac{1}{(n, j)} = \frac{1}{n}\Sigma d\phi(d), \qquad \Sigma\phi(n, j) = \Sigma\phi(d)\phi\left(\frac{n}{-}\right).$$

If (p. 94) p is one of the integers $a, \beta, \ldots \leqq n$ and prime to n,

$$\Sigma g(a)F(a) = \Sigma G(a)f(a), \qquad F(x) \equiv \Sigma f(d), \qquad G(p) \equiv \Sigma_a g(pa),$$

where d ranges over the divisors of x. For $g(x) = 1$, this gives

$$\Sigma_a f(a)\phi(n, n/a) = \Sigma F(a),$$

where (p. 96) $\phi(n, x)$ is the number of integers $\leqq x$ and prime to n. Cesàro (pp. 144–151, 302–3) discussed and modified Perott's[53] proof of his first formula, criticizing his replacement of $[n/k]$ by n/k for n large. He gave (pp. 153–6) a simple proof that the mean[57] of $\phi(n)$ is $6n/\pi^2$ and reproduced the proofs by Dirichlet[21] and Mertens,[36] the last essentially the same as Perott's. For $\zeta(m) = 1 + 1/2^m + 1/3^m + \ldots$,

$$\Sigma\frac{1}{a^m}\ (m>1), \qquad \Sigma\frac{1}{a}, \qquad \Sigma\frac{1}{a^m\phi(a)}\ (m>1), \qquad \Sigma\frac{1}{\phi(a)}$$

equal asymptotically (pp. 167–9)

$$\zeta(m)/\zeta(m+1), \qquad (6\log n)/\pi^2, \qquad \zeta(m+1), \qquad \log n.$$

As a corollary (p. 251) to Mansion's[41] generalization of Smith's theorem we have the result that the determinant of order n^2, each element being 1 or 0 according as the g. c. d. of its two indices is or is not a perfect square, equals $(-1)^{a+b+\cdots}$, where $p^a q^b \ldots$ is the value of $n!$ expressed in terms of its prime factors.

Cesàro[58] considered any function $F(x, y)$ of the g. c. d. of x, y, and the determinant Δ_n of order n having the element $F(u_i, u_j)$ in the ith row and jth column, where $u_1, \ldots, u_n$ are integers in ascending order such that each divisor of every u_i is a u. Employing the function $\mu(n)$ [see Ch. XIX], he noted that

$$\sum_{i=1}^{n}\mu\left(\frac{u_n}{u_i}\right)F(u_r, u_i) = f(u_n) \text{ or } 0,$$

according as u_r is or is not divisible by u_n, while

$$f(x) = \mu(x)F(1) + \mu\left(\frac{x}{2}\right)F(2) + \mu\left(\frac{x}{3}\right)F(3) + \dots$$

Hence if we multiply the elements of the ith column of Δ_n by $\mu(u_n/u_i)$ and add the products to the last column for $i = 1, \dots, n-1$, the new elements of the last column are zero except the final element, which is $f(u_n)$. Thus

$$\Delta_n = f(u_n)\Delta_{n-1} = f(u_1)f(u_2)\dots f(u_n).$$

[These results are due to Smith,[39] not merely the case $u_i = i$ as stated.]

Cesàro[59] noted that $|u_{ij}| = f(1)\dots f(n)$ if

$$u_{ij} = \sum_{\nu=1}^{n} f(\nu)h\left(\frac{i}{\nu}\right)h_1\left(\frac{j}{\nu}\right),$$

where the function h has the property that the determinant with the general element $h(i/j)$ is unity, and similarly for h_1.

Cesàro[60] gave the last result for the case in which $h(x) = h_1(x) = 1$ or 0 according as x is or is not an integer. P. Mansion (p. 250) stated that he[44] had employed a similar proof.

Cesàro[61] duplicated his paper[58] and transformed its final result into

$$\left|\frac{1}{F[i, j]}\right|_n = \frac{f(1)f(2)\dots f(n)}{F^2(n!)},$$

where $[i, j] = ij/(i, j)$ is the l. c. m. of i, j, and $F(x)$ is a function such that $F(xy) = F(x)F(y)$. In particular, if $F(x) = 1/x$, then $f(x) = \phi(x)\pi(x)/x^2$, where $\pi(n)$ is the product of the negatives of the distinct prime factors of n. Hence

$$|[i, j]|_n = \phi(1)\dots\phi(n)\pi(1)\dots\pi(n).$$

Cesàro[62] investigated the r-rowed minors of the n-rowed determinant whose general element is $F(\delta) = F(i, j)$, where δ is the g. c. d. of i, j. It is shown that the $(n-\nu)$-rowed determinant whose general element is $F(i+\nu, j+\nu)$ is equal to the sum of certain products of $f(1), \dots, f(n)$ taken $n-\nu$ at a time, the case $\nu = 0$ being Smith's theorem. Here

$$f(x) = \sum_j \mu\left(\frac{x}{j}\right)F(j), \qquad F(x) = \Sigma f(d) \qquad (d \text{ divisor of } x).$$

Cesàro[63] stated that the $(n-1)$-rowed determinant, whose general element u_{ij} equals the number of divisors common to $i+1$ and $j+1$, equals the number of integers $\leq n$ deprived of square factors > 1.

[59]Atti. Reale Accad. Lincei, (4), 1, 1884–5, 711–5.
[60]Mathesis, 5, 1885, 248–9.
[61]Giornale di Mat., 23, 1885, 182–197.
[62]Annales de l'école normale sup., (3), 2, 1885, 425–435.
[63]Nouv. Ann. Math., (3), 4, 1885, 56.

Cesàro[64] employed $F(n) = \Sigma f(d)$, $G(n) = \Sigma g(d)$, where d ranges over the divisors of n, and proved that

$$\begin{vmatrix} 0 & G(1) & G(2) & \ldots & G(n) \\ G(1) & F(1,1) & F(1,2) & \ldots & F(1,n) \\ \ldots & \ldots & \ldots & \ldots & \ldots \\ G(n) & F(n,1) & F(n,2) & & F(n,n) \end{vmatrix} = -f(1)\ldots f(n) \sum_{\nu=1}^{n} \frac{g^2(\nu)}{f(\nu)}.$$

In particular, if $F(n)$ is the number of divisors of n and if $G(n)$ is the number of prime divisors of n, the determinant, apart from signs, equals the number of primes $\leq n$.

E. Cesàro[65] wrote (a, b) for the g. c. d. of a, b. If $F(n) = \Sigma f(d)$, where d ranges over the divisors of n, then

$$\sum_{i=1}^{N} F\{(n, i)\} = \Sigma f(d) N / d.$$

In particular, if $I_\epsilon(n)$ is the number of irreducible fractions $\leq \epsilon$ of denominator n,

$$I_\epsilon(n) = \Sigma \left[\frac{n\epsilon}{d}\right] \mu(d), \qquad \Sigma I_\epsilon(d) = [n\epsilon].$$

The last formula, due to Laguerre,[35] follows by inversion (Ch. XIX), and directly from the fact that $I_\epsilon(d)$ is the number of the first $[n\epsilon]$ integers which with n have the g. c. d. n/d. The number of irreducible fractions $\leq \epsilon$ of denominator $\leq n$ is $\Phi_\epsilon(n) = I_\epsilon(1) + \ldots + I_\epsilon(n)$. We have

$$\Phi_\epsilon(n) = \sum_{j=1}^{\infty} \mu(j) \sum_{i=1}^{[n/j]} [i\epsilon], \qquad \lim_{n=\infty} \Phi_\epsilon(n)/n^2 = \frac{3\epsilon}{\pi^2} \quad (\epsilon > 0),$$

due to Sylvester[55] for $\epsilon = 1$. Let $\phi_\epsilon^{(\nu)}(n)$ be the sum of the νth powers of the numerators of the irreducible fractions $\leq \epsilon$ of denominator n. Set

$$\Phi_\epsilon^{(\nu)}(n) = \sum_{i=1}^{n} \phi_\epsilon^{(\nu)}(i), \qquad s_\nu(n) = \sum_{i=1}^{n} i^\nu.$$

Then

$$\sum_{i=1}^{n} i^\nu \Phi_\epsilon^{(\nu)}\left[\frac{n}{i}\right] = \sum_{i=1}^{n} s_\nu[i\epsilon],$$

which generalizes the two formulas of Sylvester.[55]   Also,

$$\Phi_\epsilon^{(\nu)}(n) = \frac{6}{\pi^2} \cdot \frac{\epsilon^{\nu+1}}{\nu+1} \cdot \frac{n^{\nu+2}}{\nu+2}, \text{ asymptotically.}$$

Cesàro[65a] factored determinants of the type in his paper,[58] the function F now being such that $F(xy)/\{F(x)F(y)\}$ is a function of the g. c. d. of x, y.

L. Gegenbauer[65b] gave a complicated theorem involving several general functions, special cases of which give Sylvester's[55] two summation formulas.

[64]Nouv. Ann. Math., (3), 5, 1886, 44–47.
[65]Annali di Mat., (2), 14, 1886–7, 143–6.
[65a]Giornale di Mat., 25, 1887, 18–19.
[65b]Sitzungsber. Ak. Wiss. Wien (Math.), 94, 1886, II, 757–762.

P. S. Poretzky[66] gave a formula for the function $\psi(m)$ whose values are the $\phi(m)$ integers $<m$ and prime to m. For the case $m=2\cdot3\cdot5\ldots p$, where p is a prime,

$$\psi(m)=m\left\{\sum_{p_i=2}^{p}\frac{\psi(p_i)}{p_i}-K\right\},$$

where K is an integer. Application is made to the finding of a prime exceeding a given number, and to a generalization of the sieve of Erastosthenes.

E. Cesàro[67] gave a very simple proof of the known fact that

$$\lim_{n=\infty}\frac{\phi(1)+\ldots+\phi(n)}{n^2}=\frac{3}{\pi^2},$$

which he expressed in words by saying that $\phi(n)$ is asymptotic to $6n/\pi^2$ (not meaning that the limit of $\phi(n)/n$ is $6/\pi^2$). On the distinction between asymptotic mean and median value, see Encyclopédie des sc. math., I, 17 (vol. 3), p. 347.

Cesàro[68] noted that if $F(i, j)$ is a function of the g. c. d. of i, j, then $Q=\Sigma F(i, j)\, x_ix_j$ $(i, j=1,\ldots, n)$ becomes $q=\Sigma f(i)y_i^2$ by the substitution $y_k=x_k+x_{2k}+x_{3k}+\ldots$, provided $F(n)=\Sigma f(d)$, d ranging over the divisors of n. Since the determinant of the substitution is unity, the discriminants of Q and q are equal. Hence we have the theorem of Smith.[39] A generalization is obtained by use of $\Sigma F(\epsilon_i, \epsilon_j)x_ix_j$, where the numbers $\epsilon_1, \epsilon_2,\ldots$ include the divisors of each ϵ.

E. Catalan[69] proved that, if d ranges over the divisors of $N=a^\alpha b^\beta\ldots$,

$$\Sigma\frac{\phi(d)}{d}=\Pi\left\{1+\frac{a(a-1)}{a}\right\},\qquad \Sigma\frac{d}{\phi(d)}=\Pi\left(1+\frac{aa}{a-1}\right).$$

E. Busche[70] derived at once from Dirichlet's[21] formula the result

$$\sum_{x=1}^{\infty}\phi(x)\left\{\rho\left(\frac{n}{x}\right)+\rho\left(\frac{n'}{x}\right)+\ldots\right\}=\Sigma nn',$$

where $\rho(a)=a-[a]$. The case $n=n'=n''=\ldots$ leads to

$$\Sigma\phi(x)=(\nu-1)n^2,$$

where x takes all values for which $\rho(n/x)>\rho(\nu n/x)$. If we take $n=1$ and add $\phi(1)=1$, we get (4) for $N=\nu$. Next, $\Sigma\phi(x)=rr'\delta^2$, where x takes all values for which

$$\frac{y+y'-1}{r+r'}\leq\rho\left(\frac{\delta}{x}\right)<\frac{y}{r},\frac{y'}{r}\qquad (y=1,\ldots, r; y'=1,\ldots, r').$$

[66]Math. phys. soc. Kasan, 6, 1888, 52–142 (in Russian).

[67]Comptes Rendus Paris, 106, 1888, 1651; 107, 1888, 81, 426; Annali di Mat., (2), 16, 1888–9, 178 (discussion with Jensen on terminology).

[68]Atti Reale Accad. Lincei, Rendiconti, 2, 1888, II, 56–61.

[69]Mém. Soc. Sc. Liège, (2), 15, 1888, No. 1, pp. 21–22; Mélanges Math., III, No. 222, dated 1882.

[70]Math. Annalen, 31, 1888, 70–74.

For $\delta = n$, $r' = 1$, $r = \nu - 1$, this becomes the former result; for $r = r' = 1$, $\delta = n$, it becomes $\Sigma\phi(x) = n^2$, where x takes the values for which $\rho(n/x) \geq 1/2$.

H. W. Lloyd Tanner[71] studied the group G of the totitives of n (the integers $< n$ and prime to n), finding all its subgroups and the simple groups whose direct product is G.

E. Lucas[72] proved that, in an arithmetical progression of n terms whose common difference is prime to n, there are $\phi(d)$ terms having with n the g. c. d. n/d. If, when d ranges over the divisors of n, $\Sigma\psi(d) = n$ for every integer n, then (p. 401) $\psi(n) = \phi(n)$, as proved by using $n = 1$, a, $a^2, \ldots$, and $n = ab$, $a^2 b, \ldots$, where $a, b, \ldots$ are distinct primes. He gave (pp. 500–1) a proof of Perott's[53] first formula by induction from $N - 1$ to N, communicated to him by J. Hammond. The name "indicateur" of n is given (preface, xv) to $\phi(n)$ [Prouhet[18]].

C. Moreau (cf. Lucas,[72] 501–3) considered the $C(n)$ circular permutations of n objects of which a are alike, β alike, $\ldots$, λ alike. Thus, if $a = 2$, $\beta = 4$, the $C(6) = 3$ distinct circular permutations are $aabbbb, ababbb, abbabb$. In general,

$$C(n) = \frac{1}{n}\Sigma\phi(d)\frac{(n/d)!}{(a/d)!\ldots(\lambda/d)!},$$

where d ranges over the divisors of the g. c. d. of $a, \beta, \ldots, \lambda$. In the example, $d = 1$ or 2, and the terms of the sum are 15 and 3.

P. A. MacMahon[73] noted that $C(n) = 1$ if $n = a$, so that we have formula (4). His expression for the number of circular permutations of p things n at a time is quoted in Chapter III on Fermat's theorem.

A. Berger[73a] evaluated $\Sigma_{k=1}^{k=n} k^{a-2}\phi(k)$. For $a = 2$ the result is $3n^2/\pi^2 + \lambda n \log n$, where λ is finite for all values of n.

E. Jablonski[74] considered rectilinear permutations of indices $a, \ldots, \lambda$, with the g. c. d. D. Set $a = a'D, \ldots, \lambda = \lambda'D$, $a + \ldots + \lambda = m = m'D$. Then the number of complete rectilinear permutations of indices $a'n, \ldots, \lambda'n$ is

$$P(n) = \frac{(m'n)!}{(a'n)!\ldots(\lambda'n)!}.$$

The number of complete circular permutations is

$$\frac{1}{m}\Sigma\phi(d)P\left(\frac{D}{d}\right),$$

where d ranges over the divisors of D. If $Q(D/d)$ is the number of rectilinear permutations of indices $a, \ldots, \lambda$ which can be decomposed into d identical portions, $\Sigma Q(D/d) = P(D)$. Also

[71]Proc. London Math. Soc., 20, 1888–9, 63–83.

[72]Théorie des nombres, 1891, 396–7. The first theorem was proved also by U. Concina, Il Boll. di Matematica, 1913, 9.

[73]Proc. London Math. Soc., 23, 1891–2, 305–313.

[73a]Nova Acta Regiae Soc. Sc. Upsaliensis, (3), 14, 1891, No. 2, 113.

[74]Comptes Rendus Paris, 114, 1892, 904–7; Jour. de Math., (4), 8, 1892, 331–349. He proved Moreau's[72] formula for $C(n)$.

$$\Sigma Q\left(\frac{D}{d}\right)d^t = \Sigma P\left(\frac{D}{d}\right)J_t(d),$$

where $J_t(d)$ is Jordan's[200] function.

S. Schatunowsky[75] proved that 30 is the largest number such that all smaller numbers relatively prime to it are primes. He employed Tchebychef's[261] theorem of Ch. XVIII that, if $a > 1$, there exists at least one prime between a and $2a$. Cf. Wolfskehl,[91] Landau,[92, 113] Maillet,[93] Bonse,[106] Remak.[112]

E. W. Davis[76] used points with integral coordinates $\geqq 0$ to visualize and prove (1) and (4).

K. Zsigmondy[77] wrote r_s for the greatest integer $\leqq r/s$ and proved that, if a takes those positive integral values $\leqq r$ which are divisible by no one of the given positive integers $n_1, \ldots, n_\rho$ which are relatively prime in pairs,

$$\Sigma f(a) = \sum_{k=1}^{r} f(k) - \sum_{n} \sum_{k=1}^{r_n} f(kn) + \sum_{n,\,n'} \sum_{k=1}^{r_{nn'}} f(knn') - \ldots,$$

$n, n', \ldots$ ranging over the combinations of $n_1, \ldots, n_\rho$ taken $1, 2, \ldots$ at a time. Taking $f(k) = 1$, we obtain for the number $\phi(r; n_1, \ldots, n_\rho)$ of integers $\leqq r$, which are divisible by no one of $n_1, \ldots, n_\rho$, the expression (5) obtained by Legendre for the case in which the n's are all primes. By induction from ρ to $\rho+1$, we get

$$\phi(r; n_1, \ldots, n_\rho, \nu_1, \ldots, \nu_\rho) = \phi(r; n_1, \ldots, n_\rho) - \Sigma_\nu \phi(r_\nu; n_1, \ldots, n_\rho)$$
$$+ \sum_{\nu,\,\nu'} \phi(r_{\nu\nu'}; n_1, \ldots, n_\rho) - \ldots,$$

$$r = \phi(r; n_1, \ldots, n_\rho) + \sum_{i=1}^{\rho} \phi(r_{n_i}; n_1, \ldots, n_{i-1}, n_{i+1}, \ldots, n_\rho)$$
$$+ \sum_{n,\,n'} \phi(r_{nn'}; n_i\text{'s} \neq n, n') + \ldots,$$

$$r = \sum_c \phi(r_c; n_1, \ldots, n_\rho),$$

where c ranges over all combinations of powers $\leqq r$ of the n's. The last becomes (4) when $n_1, \ldots, n_\rho$ are the different primes dividing r. These formulas for r were deduced by him in 1896 as special cases of his inversion formula (see Ch. XIX).

J. E. Steggall[78] evaluated $\phi(n)$ by the second method of Crelle.[17]

P. Bachmann[79] gave an exposition of the work of Dirichlet,[21] Mertens,[36] Halphen[54] and Sylvester[55] on the mean of $\phi(n)$, and (p. 319) a proof of (5).

L. Goldschmidt[80] gave an evaluation of $\phi(n)$ by successive steps which may be combined as follows. Let p be a prime not dividing k. Each of

---

[75]Spaczinskis Bote (phys. math.), 14, 1893, No. 159, p. 65; 15, 1893, No. 180, pp. 276–8 (Russian).
[76]Amer. Jour. Math., 15, 1893, 84.
[77]Jour. für Math., 111, 1893, 344–6.
[78]Proc. Edinburgh Math. Soc., 12, 1893–4, 23–24.
[79]Die Analytische Zahlentheorie, 1894, 422–430, 481–4.
[80]Zeitschrift Math. Phys., 39, 1894, 203–4.

the $\phi(k)$ integers $\leq k$ and prime to k occurs just once among the residues modulo k of the integers from lk to $(l+1)k$; taking $l = 0, 1, \ldots, p-1$, we obtain this residue p times. Hence there are $p\phi(k)$ numbers $\leq pk$ and prime to k. These include $\phi(k)$ multiples of p, whence $\phi(pk) = (p-1)\phi(k)$. For, if r is one of the above residues, then $r, r+k, \ldots, r+(p-1)k$ form a complete set of residues modulo p and hence include a single multiple of p. Hence

$$\phi(abc\ldots) = (a-1)(b-1)(c-1)\ldots,$$

if $a, b, c, \ldots$ are distinct primes. Next, for $n = a^a b^\beta \ldots$, we use the sets of numbers from $lab \ldots$ to $(l+1)ab\ldots$, for $l = 0, 1, \ldots, a^{a-1}b^{\beta-1}\ldots - 1$.

Borel and Drach[81] noted that the period of the least residues of $0, a, 2a, \ldots$ modulo N, contains N/δ terms, if δ is the g. c. d. of a, N; conversely, if d is any divisor of N, there exist integers such that the period has d terms. Taking $a = 0, 1, \ldots, N-1$, we get (4).

H. Weber[82] defined $\phi(n)$ to be the number of primitive nth roots of unity. If a is a primitive ath root of unity and β a primitive bth root, and if a, b are relatively prime, $a\beta$ is a primitive abth root of unity and all of the latter are found in this way. Hence $\phi(ab) = \phi(a)\phi(b)$. This is also proved for relatively prime divisors a, b of $n-1$, where n is a prime, by use of integers a and β belonging to the exponents a and b respectively, modulo n, whence $a\beta$ belongs to the exponent ab.

K. Th. Vahlen[83] proved that, if $I_{a,\beta}(n)$ is the number of irreducible fractions between the limits a and β, $a > \beta \geq 0$, with the denominator n,

$$\Sigma I_{a,\beta}(d) = [(a-\beta)n], \qquad \sum_{k=1}^{n} \left[\frac{n}{k}\right] I_{a,\beta}(k) = \sum_{k=1}^{n} [(a-\beta)k],$$

where d ranges over the divisors of n. For $\beta = 0$, the first was given by Laguerre.[35] Since $I_{1,0}(n) = \phi(n)$, these formulas include (4) of Gauss and that by Dirichlet.[21]

J. J. Sylvester[84] corrected his[55] first formula to read

$$\sum_{k=1}^{\infty} T\left[\frac{j}{k}\right] = \tfrac{1}{2}\{[j]^2 + [j]\} \equiv \Phi(j), \qquad T[n] = \phi(1) + \ldots + \phi([n]),$$

and proved it. By the usual formula for reversion,

$$T[j] = \Phi(j) - \Phi(\tfrac{1}{2}j) + \Phi(\tfrac{1}{3}j) - \Phi(\tfrac{1}{5}j) + \Phi(\tfrac{1}{6}j) - \ldots.$$

A. P. Minin[85] solved $\tfrac{1}{2}\phi(m) = R$ for m when R has certain values. The equation determines the number of regular star polygons of m sides.

Fr. Rogel[86] gave the formula of Dirichlet.[21]

[81]Introd. théorie des nombres, 1895, 23.
[82]Lehrbuch der Algebra, I, 1895, 412, 429; ed. 2, 1898, 456, 470.
[83]Zeitschrift Math. Phys., 40, 1895, 126–7.
[84]Messenger Math., 27, 1897–8, 1–5; Coll. Math. Papers, 4, 738–742.
[85]Report of Phys. Sec. Roy. Soc. of Friends of Nat. Sc., Anthropology, etc. (in Russian), Moscow, 9, 1897, 30–33. Cf. Hammond.[123]
[86]Educat. Times, 66, 1897, 62.

Rogel[87] considered the number of integers $\nu < n$ such that ν and n are not both divisible by the rth power of a prime. Also the number when each prime factor common to ν and n occurs in them exactly to the rth power.

I. T. Kaplan published at Odessa in 1897 a pamphlet in Russian on the distribution of the numbers relatively prime to a given number.

M. Bauer[88] proved that, for x prime to m, $kx+l$ represents

$$\frac{\psi(m)}{\psi(d_1 d_2)} \cdot \frac{\phi(d_1 d_2)}{\phi(m)} \cdot \phi\left(\frac{m}{d_1}\right)$$

integers relatively prime to m and incongruent modulo m, where d_1 is the g. c. d. (k, m) of k, m, and $d_2 = (l, m)$, $(d_1, d_2) = 1$, while

$$\psi(m) = \phi(m) \prod_{i=1}^{s} \left\{ 1 - \frac{1}{\phi(p_i)} \right\}$$

is the number of incongruent integers prime to $m = p_1^{e_1} \ldots p_s^{e_s}$ which are represented by $kx+l$ when k, l, x are prime to m. Of those integers, $\psi(m)/\psi(p_1 \ldots p_r)$ are divisible only by the special prime factors $p_1, \ldots, p_r$ of m.

J. de Vries[88a] proved the first formula of Dirichlet's.[21]

C. Moreau[89] evaluated $\phi(n)$ by the method of Grunert.[15]

E. Landau[90] proved that

$$\sum_{n=1}^{x} \frac{1}{\phi(n)} = \frac{315\zeta(3)}{2\pi^4} \left(\log x + C - \sum_{p} \frac{\log p}{p^2 - p + 1} \right) + \epsilon,$$

where ϵ is of the order of magnitude of $x^{-1} \log x$, C is Euler's constant, and ζ is Riemann's ζ-function.

P. Wolfskehl[91] proved by Tchebychef's theorem that the $\phi(n)$ integers $< n$ and prime to n are all primes only when $n = 1, 2, 3, 4, 6, 8, 12, 18, 24, 30$. [Schatunowsky.[75]]

E. Landau[92] gave a proof, without the use of Tchebychef's theorem, by finding a lower limit to the number of integers k having no square factor > 1, where $t \geq k > 5t/8$.

E. Maillet,[93] by use of Tchebychef's theorem, proved the same result and the generalization: Given any integer r, there exist only a finite number of integers N such that the $\phi(N)$ integers $< N$ and relatively prime to N contain at most r equal or distinct prime factors.

Alois Pichler[94] noted that $\phi(x) = n$ has no solution if n is odd and > 1; while $\phi(x) = 2^n$ has the solutions $x = 2^a bc \ldots (a = 0, 1, \ldots, n+1)$ if

---

[87]Sitzungsber. Böhm. Gesell., Prag, 1897; 1900, No. 30.

[88]Math. Natur. Berichte aus Ungarn, 15, 1897, 41–6.

[88a]K. Akad. Wetenschappen te Amsterdam, Verslagen, 5, 1897, 222.

[89]Nouv. Ann. Math., (3), 17, 1898, 293–5.

[90]Göttingen Nachrichten, 1900, 184.

[91]L'intermédiaire des math., 7, 1900, 253–4; Math. Ann., 54, 1901, 503–4.

[92]Archiv Math. Phys., (3), 1, 1901, 138–142.

[93]L'intermédiaire des math., 7, 1900, 254.

[94]Ueber die Auflösung der Gl. $\varphi(x) = n \ldots$, Jahres-Bericht Maximilians-Gymn. in Wien, 1900-1, 3–17.

$$b = 2^{2^\beta} + 1, \qquad c = 2^{2^\gamma} + 1, \ldots$$

are distinct primes and $2^\beta + 2^\gamma + \ldots = n$ or $n - a + 1$ according as $a = 0$ or $a > 0$. When q is a prime > 3, $\phi(x) = 2q^n$ is impossible if $p = 2q^n + 1$ is not prime; while if p is prime it has the two solutions p, $2p$. If $q = 3$ and p is prime, it has the additional solutions 3^{n+1}, $2 \cdot 3^{n+1}$. Next, $\phi(x) = 2^n q$ is impossible if no one of $p_\nu = 2^{n-\nu} q + 1 (\nu = 0, 1, \ldots, n-1)$ is prime and q is not a prime of the form $2^s + 1$, $s = 2^\lambda \leq n$; but if q is such a prime or if at least one p_ν is prime, the equation has solutions of the respective forms bq^2, where $\phi(b) = 2^{n-s}$; ap_ν, where $\phi(a) = 2^\nu$. Finally, $\phi(x) = 2qr$ has no solution if $p = 2qr + 1$ is not prime and $r \neq 2q + 1$. If p is a prime, but $r \neq 2q + 1$, the two solutions are p, $2p$. If p is not prime, but $r = 2q + 1$, the two solutions are r^2, $2r^2$. If p is prime and $r = 2q + 1$, all four solutions occur. There is a table of the values $n < 200$ for which $\phi(x) = n$ has solutions.

L. Kronecker[95] considered two fractions with the denominator m as equivalent if their numerators are congruent modulo m. The number of non-equivalent reduced fractions with the denominator m is therefore $\phi(m)$. If $m = m'm''$, where m', m'' are relatively prime, each reduced fraction r/m can be expressed in a single way as a sum of two reduced partial fractions r'/m', r''/m''. Conversely, if the latter are reduced fractions, their sum r/m is reduced. Hence $\phi(m) = \phi(m')\phi(m'')$. The latter is also derived (pp. 245–6, added by Hensel) from (4), which is proved (pp. 243–4) by considering the g. c. d. of n with any integer $\leq n$, and also (pp. 266–7) by use of infinite series and products. Proof is given (pp. 300–1) of (5). The Gaussian median value (p. 334) of $\phi(n)/n$ is $6/\pi^2$ with an error whose order of magnitude is $1/\sqrt{n}$, provided we take as the auxiliary number of values of $\phi(n)/n$ a value of the order of magnitude $\sqrt{n} \log_e n$.

E. B. Elliott[96] considered monomials $n = p^a q^b \ldots$ in the independent variables $p, q, \ldots$. In the expansion of $n(1 - 1/p)^m (1 - 1/q)^m \ldots$, the aggregate of those monomial terms whose exponents are all ≥ 0 is denoted by $F_m(n)$. Define $\mu(p^r q^s \ldots)$ to be zero if any one of $r, s, \ldots$ exceeds 1, but to be $(-1)^t$ if no one of them exceeds 1, and t of them equal 1. Then

$$(7) \qquad F_{m-1}(n) = \Sigma F_m(d), \qquad F_{m+1}(n) = \Sigma \mu\left(\frac{n}{d}\right) F_m(d),$$

where d ranges over the monomials $p^\alpha q^\beta \ldots$ with $0 \leq \alpha \leq a$, $0 \leq \beta \leq b, \ldots$. Henceforth, let $p, q, \ldots$ be distinct primes. Then $F_1(n) = \phi(n)$, while $F_{-1}(n)$ is the sum $\sigma(n)$ of the divisors of n. In (7), d now ranges over all the divisors of n, and $\mu(k)$ is Merten's function [Inversion]. For $m = 0$, (7_2) gives the usual expression for $\phi(n)$, while (7_1) defines $\sigma(n)$. For $m = 1$, (7_1) becomes (4).

If $\tau^{(1)}(n) \equiv \tau(n)$ is the number of divisors d of n, write

$$\tau^{(2)}(n) = \Sigma \tau(d), \ldots, \tau^{(k)}(n) = \Sigma \tau^{(k-1)}(d).$$

[95]Vorlesungen über Zahlentheorie, I, 1901, 125–6.
[96]Proc. London Math. Soc., 34, 1901, 3–15.

Then

$$\tau^{(k)}(n) = \prod_a \frac{(a+k)!}{a!k!}, \qquad F_{-k}(n) = \Sigma d\tau^{(k-1)}\left(\frac{n}{d}\right).$$

Generalizing $\mu(s)$, let $\mu^{(k)}(s)$ be zero if the expansion of the product $\Pi(1-p)^k$, extended over all primes p, does not contain a term equal to s, but let it equal the coefficient of s if s occurs in the expansion. Then

$$F_k(n) = \Sigma d\mu^{(k)}\left(\frac{n}{d}\right).$$

The n-rowed determinant in which the element in the rth row and sth column is $F_{m-1}(\delta)$, where δ is the g. c. d. of r, s, is proved equal to $F_m(1)$ $F_m(2)\ldots F_m(n)$, a generalization of Smith's[39] theorem. Finally,

$$\frac{1}{n}\Sigma F_{k+r}\left(\frac{n}{d}\right)F_{-k}(d) = \Sigma \frac{1}{d}F_r(d),$$

the right member being $\tau(n)$, $\Sigma\phi(d)/d$, $\Sigma\sigma(d)/d$ for $r = 0, 1, -1$.

G. Landsberg[96a] gave a simple proof of Moreau's[72] formula for the number of circular permutations.

L. Carlini[97] proved Dirichlet's[21] formula by noting that

(8)
$$\prod_{h=1}^{n}(x^h-1)=0$$

has unity as an n-fold root, while a root $\neq 1$ of x^h-1 is a root of $[n/h]$ factors $x^{th}-1$. Hence the $\phi(h)$ primitive roots of $x^h=1$ furnish $\phi(h)[n/h]$ roots of (8).

M. Lerch[98] found the number N of positive integers $\leq m$ which have no one of the divisors $a, b, \ldots, k, l$, the latter being relatively prime in pairs and having m as their product. Let $F(x)=1$ or 0, according as x is fractional or integral. Let $L=ab\ldots k$. Then [Dirichlet[33]]

$$N = \frac{m(l-1)}{L}\sum_{\rho=1}^{L} F\left(\frac{\rho}{a}\right)\ldots F\left(\frac{\rho}{k}\right) = m\left(1-\frac{1}{a}\right)\ldots\left(1-\frac{1}{l}\right).$$

E. Landau[99] proved that the inferior limit for $x=\infty$ of

$$\frac{1}{x}\phi(x)\log_e\log_e x$$

is e^{-C}, where C is Euler's constant. Hence $\phi(x)$ is comprised between this inferior limit and the maximum $x-1$.

R. Occhipinti[100] proved that, if a_j is an nth root of unity, and if $d_{i1}, \ldots, d_{a_i}$ are the divisors of i,

$$\prod_{j=1}^{n}\left\{\sum_{i=1}^{k_1}\phi(d_{1i})+a_j\sum_{i=1}^{k_2}\phi(d_{2i})+\ldots+a_j^{n-1}\sum_{i=1}^{k_n}\phi(d_{ni})\right\} = \tfrac{1}{2}(-1)^{n-1}n(n+1)n^{n-2}.$$

[96a]Archiv Math. Phys., (3), 3, 1902, 152–4.   [97]Periodico di Mat., 17, 1902, 329.
[98]Prag Sitzungsber., 1903, II.   [99]Archiv Math. Phys., (3), 5, 1903, 86–91.
[100]Periodico di Mat., 19, 1904, 93.   Handbuch,[113] I, 217–9.

G. A. Miller[101] proved (4) by noting that in a cyclic group G of order N there is a single cyclic subgroup of order d, a divisor of N, and it contains $\phi(d)$ operators of order d, while the order of any operator of G is a divisor of N. Thus (4) states merely that the order of G equals the sum of the numbers of the operators of the various possible orders. Next, (1) follows from an enumeration of the operators of highest period AB in a cyclic group of order AB, which is the direct product of its cyclic subgroups of orders A and B. Finally, if p is a prime, all the subgroups of a cyclic group of order p^n are contained in its subgroup of order p^{n-1}, whence $\phi(p^n) = p^n - p^{n-1}$.

G. A. Miller[102] proved the last three theorems and the fact that $\phi(l)$ is even if $l > 2$ by means of the properties of the abelian group whose elements are the integers $< m$ which have with m g. c. d. equal to k.

K. P. Nordlund[103] proved $\phi(mn \ldots) = (m-1)(n-1) \ldots$, where $m, n, \ldots$ are distinct primes, by writing down the multiples $< mnp$ of m, the multiples of mn, etc., whence the number of integers $< mnp$ and not prime to it is $mnp - 1 - (m-1)(n-1)(p-1)$.

E. Busche[104] treated geometrically systems $\begin{pmatrix} ac \\ bd \end{pmatrix}$ of four integers such that $ad - bc > 0$, evaluated the number $\Phi(S)$ of systems incongruent modulo S and prime to S, and generalized (4) to $\Sigma\Phi(S)$.

L. Orlando[105] showed that $\phi(m)$ is determined by (4) [Lucas[72]].

H. Bonse[106] proved Maillet's[93] theorem for $r = 1, 2, 3$ without using Tchebychef's theorem. His lemma was generalized by T. Suzuki.[106a]

J. Sommer[107] gave without reference Crelle's[8] final evaluation of $\phi(n)$.

R. D. Carmichael[108] proved that if n is such that $\phi(x) = n$ is solvable there are at least two solutions x. He found solutions of $\phi(x) = 2^n$ [in accord with Pichler[94]] and proved that there are just $n+2$ solutions (a single one being odd) when $n \leq 31$ and just 33 solutions when $32 \leq n \leq 255$. All the solutions of $\phi(x) = 4n - 2 > 2$ are of the form $p^a, 2p^a$, where p is a prime of the form $4s - 1$; for example, if $n = 5$, the solutions are 19, 27 and their doubles.

Carmichael[109] gave a table showing every value of m for which $\phi(m)$ has any given value ≤ 1000.

A. Ranum[109a] would solve $\phi(x) = n$ by resolving n in every possible way into factors $n_0, ., n_r$, capable of being taken as the values of $\phi(2^{a_0}), \phi(p_1^{a_1})$, $\ldots, \phi(p_r^{a_r})$, where $2, p_1, \ldots, p_r$ are distinct primes. Then $2^{a_0}p_1^{a_1} \ldots p_r^{a_r}$ is a value of x.

Carmichael[110] gave a method of solving $\phi(x) = a$, based on the testing of the equation for each factor x of a definite function of a.

M. Fekete[111] considered the determinant ρ_{kn} obtained by deleting the last row and last column of Sylvester's eliminant for $x^k - 1 = 0$ and $x^n - 1 = 0$

[101]Amer. Math. Monthly, 12, 1905, 41–43.
[102]Amer. Jour. Math., 27, 1905, 315.
[103]Nyt Tidsskrift for Mat., 16A, 1905, 15–29.
[104]Jour. für Math., 131, 1906, 113–135.
[105]Periodico di Mat., 22, 1907, 134–6.
[106]Archiv Math. Phys., (3), 12, 1907, 292–5.
[106a]Tôhoku Math. Jour., 3, 1913, 83–6.
[107]Vorlesungen über Zahlentheorie, 1907, 5.

[108]Bull. Amer. Math. Soc., 13, 1907, 241–3.
[109]Amer. Jour. Math., 30, 1908, 394–400.
[109a]Trans. Amer. Math. Soc., 9, 1908, 193–4.
[110]Bull. Amer. Math. Soc., 15, 1909, 223.
[111]Math. és Phys. Lapok (Math. Phys. Soc.), Budapest, 18, 1909, 349-370. German transl., Math. Naturwiss. Berichte aus Ungarn, 26, 1913 (1908), 196.

$(k<n)$. Thus $|\rho_{kn}| = 1$ or 0 according as k and n are relatively prime or not. Hence

$$\phi(n) = \sum_{k=1}^{n} |\rho_{kn}|, \qquad \phi_1(n) = \sum_{k=1}^{n} k|\rho_{kn}|,$$

where $\phi_1(n)$ is the sum of the integers $\leq n$ and prime to n.

R. Remak[112] proved Maillet's[93] theorem without using Tchebychef's.

E. Landau[113] proved (5), Wolfskehl's[91] theorem and Maillet's[93] generalization.

C. Orlandi[114] proved that, if x ranges over all the positive integers for which $[m/x]$ is odd, then $\Sigma\phi(x) = (m/2)^2$ for m even (Cesàro, p. 144 of this History), while $\Sigma\phi(x) = k^2$ for $m = 2k - 1$.

A. Axer[115] considered the system (P) of all integers relatively prime to the product P of a finite number of given primes and obtained formulas and asymptotic theorems concerning the number of integers $\leq x$ of (P) which are prime to x. Application is made to the probability that two numbers $\leq n$ of (P) are relatively prime and to the asymptotic values of the number (i) of positive irreducible fractions with numerator and denominator in (P) and $\leq n$ and (ii) of regular continued fractions representing positive fractions in (P) with numerator and denominator $\leq n$.

G. A. Miller[116] defined the order of a modulo m to be the least positive integer b such that $ab \equiv 0 \pmod{m}$. If p^a is the highest power of a prime p dividing m, the numbers $\leq m$ whose orders are powers of p are km/p^a ($k = 1, 2, \ldots, p^a$). Hence $\Sigma k_i m/p_i^{a_i}$ ($k_i = 1, \ldots, p_i^{a_i}$) form a complete set of residues modulo $m = \Pi p_i^{a_i}$. If the orders of two integers are relatively prime, the order of their sum is congruent modulo m to the product of their orders. But the number of integers $\leq m$ whose orders equal m is $\phi(m)$. Hence $\phi(\Pi p^a) = \Pi\phi(p^a)$. Since all numbers $\leq m$ whose orders divide d, a divisor of m, are multiples of m/d, there are exactly d numbers $\leq m$ whose orders divide d, and $\phi(d)$ of them are of order d. Hence $m = \Sigma\phi(d)$.

S. Composto[117] employed distinct primes m, n, r, and the $\nu = \phi(mn)$ integers $p_1, \ldots, p_\nu$ prime to mn and $\leq mn$, and proved that

$$p_i, \; p_i+mn, \; p_i+2mn, \ldots, \; p_i+(r-1)mn \qquad (i = 1, \ldots, \nu)$$

include all and only the numbers $rp_1, \ldots, rp_\nu$ and the numbers not exceeding and prime to mnr. Hence $\phi(mnr) = \phi(mn)\cdot(r-1)$. A like theorem is proved for two primes and stated for any number of primes. [The proof is essentially Euler's[1] proof of (1) for the case in which B is a prime not dividing a product A of distinct primes.] Next, if d is a prime factor of n, the integers not exceeding and prime to dn are the numbers $\leq n$ and prime to n, together with the integers obtained by adding to each of them $n, 2n, \ldots,$

[112]Archiv Math. Phys., (3), 15, 1909, 186–193.
[113]Handbuch...Verteilung der Primzahlen, I, 1909, 67–9, 229–234.
[114]Periodico di Mat., 24, 1909, 176–8.
[115]Monatshefte Math. Phys., 22, 1911, 3–25.
[116]Amer. Math. Monthly, 18, 1911, 204–9.
[117]Il Boll. di Matematica Gior. Sc.-Didat., 11, 1912, 12–33.

$(d-1)n$; whence $\phi(dn)=d\phi(n)$. Finally, let $p_1,\ldots,\ p_\nu$ be the $\nu=\phi(n)$ integers $<n$ and prime to n. Then $p_i+kn\ (i=1,\ldots,\nu;\ k=0,\ 1,\ldots)$ give all integers prime to n; let $P_h(n)$ denote the hth one of them arranged in order of magnitude. Then

$$P_{k\nu}(n)=kn-1\ (k\geqq 1),\qquad P_{k\nu+r}(n)=kn+p_r\ (1\leqq r\leqq \nu-1,\ k\geqq 0).$$

If $h=k\nu+r$, $r<\nu$, the sum of the first h numbers prime to n is

$$kn\left\{\frac{k\nu}{2}+r\right\}+p_1+\cdots+p_r,$$

where $p_1,\ldots,\ p_r$ are the first r integers $<n$ and prime to n.

K. Hensel[118] evaluated $\phi(n)$ by the first remark of Crelle.[17]

J. G. van der Corput and J. C. Kuyver[119] proved that the number $I(a/4)$ of integers $\leqq a/4$ and prime to a is $N=\frac{1}{4}a\Pi(1-1/p)$ if a has a prime factor $4m+1$, where p ranges over the distinct prime factors of a; but is $N-2^{k-2}$ if a is a product of powers of k prime factors all of the form $4m-1$. Also $I(a/6)$ is evaluated.

U. Scarpis[120] noted that $\phi(p^n-1)$ is divisible by n if p is a prime.

Several writers[121] discussed the solution of $\phi(x)=\phi(y)$, where x, y are powers of primes. Several[122] proved that $\phi(xy)>\phi(x)\phi(y)$ if x, y have a common factor.

J. Hammond[123] proved that there are $\frac{1}{2}\phi(n)-1$ regular star n-gons.

H. Hancock[124] denoted by $\Phi(i,k)$ the number of triples $(i,k,1)$, $(i,k,2)$, $\ldots$, (i,k,i) whose g. c. d. is unity. Let $i=i_1d$, $k=k_1d$, where i_1, k_1 are relatively prime. Then $\Phi(i,k)=i_1\phi(d)$, $\Phi(k,i)=k_1\phi(d)$.

A. Fleck[125] considered the function, of $m=\Pi p^a$,

$$\phi_k(m)=\Pi_p\left\{\phi(p^a)-\binom{k}{1}\phi(p^{a-1})+\ldots+(-1)^a\binom{k}{a}\phi(p^{a-a})\right\}.$$

Thus $\phi_0(m)=\phi(m)$, $\phi_{-1}(m)=m$, $\phi_{-2}(m)$ is the sum of the divisors of m. Also

$$\underset{d:m}{\Sigma}\ \phi_k(d)=\phi_{k-1}(m),\qquad \phi_k(mn)=\phi_k(m)\phi_k(n),$$

if m, n are relatively prime. For $\zeta(s)=\Sigma m^{-s}$,

$$\sum_{m=1}^{\infty}\frac{\phi_{k-1}(m)}{m^s}=\zeta(s)\sum_{m=1}^{\infty}\frac{\phi_k(m)}{m^s},$$

$$\phi_k(p)=p-\binom{k+1}{1},\quad \phi_k(p^2)=p^2-\binom{k+1}{1}p+\binom{k+1}{2},\ \ldots,$$

$$\phi_k(p^{k+1+\mu})=p^\mu(p-1)^{k+1}.$$

[118]Zahlentheorie, 1913, 97.
[119]Wiskundige Opgaven, 11, 1912–14, 483–8.
[120]Periodico di Mat., 29, 1913, 138.
[121]Amer. Math. Monthly, 20, 1913, 227–8 (incomplete); 309–10.
[122]Math. Quest. Educat. Times, 24, 1913, 72, 106.
[123]Ibid., 25, 1914, 69–70.
[124]Comptes Rendus Paris, 158, 1914, 469–470.
[125]Sitzungsber. Berlin Math. Gesell., 13, 1914, 161–9.

E. Cahen[126] gave F. Arndt's[19] proof without reference.

A. Cunningham[127] tabulated all solutions N of $\phi(N) = 2^r$ for $r = 4, 6, 8,$ 9, 10, 11, 12, 16, each solution being a product of a power of 2 by distinct primes $2^{2^n} + 1$.

J. Hammond[128] noted that, if $\Sigma f(k/n) = F(n)$ or $\Phi(n)$, according as the summation extends over all positive integers k from 1 to n or only over such of them as are prime to n, then $\Sigma \Phi(d) = F(n)$. This becomes (4) when f is constant.

R. Ratat[129] noted that $\phi(n) = \phi(n+1)$ for $n = 1, 3, 15, 104$. For $n < 125$, $2n \neq 2, 4, 16, 104$, he verified that $\phi(2n \pm 1) > \phi(2n)$.

R. Goormaghtigh[130] noted that $\phi(n) = \phi(n+1)$ also for $n = 164, 194, 255$ and 495. He gave very special results on the solution of $\phi(x) = 2a$.

Formulas involving ϕ are cited under Lipschitz,[50, 56] Cesàro,[61] Hammond,[111] and Knopp[160] of Ch. X, Hammond[43] of Ch. XI, and Rogel[243] of Ch. XVIII. Cunningham[95] of Ch. VII gave the factors of $\phi(p^k)$. Dedekind[71] of Ch. VIII generalized ϕ to a double modulus. Minin[120] of Ch. X solved $\phi(N) = \tau(N)$.

SUM $\phi_k(n)$ OF THE kTH POWERS OF THE INTEGERS $\leq n$ AND PRIME TO n.

A. Cauchy[149] noted that $\phi_1(n)$ is divisible by n if $n > 2$, since the integers $< n$ and prime to n may be paired so that the sum of the two of any pair is n.

A. L. Crelle[17] (p. 80, p. 84) noted that $\phi_1(n) = \frac{1}{2}n\phi(n)$. The proof follows from the remark by Cauchy.

A. Thacker[150] defined $\phi_k(n)$ and noted that it reduces for $k = 0$ to Euler's $\phi(n)$. Set $s_k(z) = 1^k + 2^k + \ldots + z^k$, $n = a^\alpha b^\beta c^\gamma \ldots$, where $a, b, \ldots$ are distinct primes. By deleting the multiples of a, then the remaining multiples of b, etc., he proved that

$$\phi_k(n) = s_k(n) - \Sigma_a a^k s_k\left(\frac{n}{a}\right) + \Sigma_{a,b} a^k b^k s_k\left(\frac{n}{ab}\right) - \Sigma_{a,b,c} a^k b^k c^k s_k\left(\frac{n}{abc}\right) + \ldots,$$

where the summation indices range over the combinations of $a, b, c, \ldots$ one, two, ... at a time. In the second paper, he proved Bernoulli's[150a] formula

$$s_k(z) = \frac{z^{k+1}}{k+1} + \frac{1}{2}z^k + \frac{1}{2}\binom{k}{1}B_1 z^{k-1} - \frac{1}{4}\binom{k}{3}B_3 z^{k-3} + \frac{1}{6}\binom{k}{5}B_5 z^{k-5} - \ldots,$$

where $B_1, B_3, \ldots$ are the Bernoullian numbers. Then, by substitution,

$$\phi_k(n) = \frac{n^{k+1}}{k+1}\Pi\left(1 - \frac{1}{a}\right) + \frac{1}{2}\binom{k}{1}B_1 n^{k-1}\Pi(1-a) - \frac{1}{4}\binom{k}{3}B_3 n^{k-3}\Pi(1-a^3)$$

[126]Théorie des nombres, I, 1914, 393.
[127]Math. Quest. Educ. Times, 27, 1915, 103–6.
[128]Ibid., 29, 1916, 53.
[129]L'intermédiaire des math., 24, 1917, 101–2.
[130]Ibid., 25, 1918, 42–4.
[149]Mém. Ac. Sc. de l'Institut de France, 17, 1840, 565; Oeuvres, (1), 3, 272.
[150]Jour. für Math., 40, 1850, 89–92; Cambridge and Dublin Math. Jour., 5, 1850, 243.   Reproduced, with errors as to signs, by Zerr, Amer. Math. Monthly, 5, 1898, 93–5.   Cf. E. Prouhet, Nouv. Ann. Math., 10, 1851, 324–330.
[150a]Jacques Bernoulli, Ars conjectandi, 1713, 95–7.

$$+\tfrac{1}{6}\binom{k}{5}B_5n^{k-5}\Pi(1-a^5)-\ldots,$$

where $\Pi(1-a^i)$ denotes $(1-a^i)(1-b^i)\ldots$

J. Binet[151] wrote $\eta_1,\ldots,\eta_n$ for the integers $<N$ and prime to $N=p^\lambda q^\mu\ldots$ Then, if $B_1,\ -B_3,\ B_5,\ldots$ are the Bernoullian numbers $1/6,\ 1/30,\ 1/42,\ldots,$ and $P_\varrho=(1-p^\varrho)(1-q^\varrho)\ldots,$

$$\sum_{i=1}^{n}\frac{1}{(x+\eta_i)^2}=\Big(\frac{1}{x}-\frac{1}{x+N}\Big)P_{-1}+\Big(\frac{1}{x^3}-\frac{1}{(x+N)^3}\Big)B_1P_1$$

$$+\Big(\frac{1}{x^5}-\frac{1}{(x+N)^5}\Big)B_3P_3+\ldots,$$

for x sufficiently small to insure convergence. Expanding each member into negative powers of x and comparing coefficients, we get

$$n=\Sigma\eta_i^\varrho=P_{-1}N,\quad 2\Sigma\eta_i=P_{-1}N^2,\quad 3\Sigma\eta_i^2=P_{-1}N^3+3B_1P_1N,$$

$$4\Sigma\eta_i^3=P_{-1}N^4+6B_1P_1N^2,\ldots$$

the first being equivalent to the usual formula for $\phi(N)$. The general law can be represented symbolically by

$$g\sum_{i=1}^{n}\eta_i^{\varrho-1}=\frac{1}{2BP}\{(N+BP)^\varrho+(N-BP)^\varrho\},$$

where, after expanding the binomials, we are to replace $N^\varrho/(BP)$ by $P_{-1}N^\varrho$ and any other term $(BP)^{2h-1}$ by $B_{2h-1}P_{2h-1}$. It is easily shown that, if k is odd, $\Sigma\eta^k$ is divisible by N.

Silva[25] used his symbolic formula, taking S to be the sum of $1,\ldots,\ n$, whence $S(a)$ is the sum $\tfrac{1}{2}n(1+n/A)$ of the multiples $\leqq n$ of A. Thus $\phi_1(n)=\tfrac{1}{2}n\phi(n)$. This proof of Crelle's result is thus like that by Brennecke.[152]

W. Brennecke[152] proved Crelle's result by means of

$$1+\ldots+n-\Big\{a\Big(1+2+\ldots+\frac{n}{a}\Big)+b\Big(1+\ldots+\frac{n}{b}\Big)+\ldots\Big\}$$

$$+\Big\{ab\Big(1+\ldots+\frac{n}{ab}\Big)+\ldots\Big\}+\ldots.$$

Set $\mu=\phi(n)$, $a=abc\ldots$ He proved that

$$\phi_2(n)=\tfrac{1}{3}\mu(n^2\pm a/2),\qquad \phi_3(n)=\tfrac{1}{4}\mu n(n^2\pm a),$$

$$\phi_4(n)=\tfrac{1}{5}\mu n^4\pm\tfrac{1}{3}a\mu n^2-\tfrac{1}{30}n(1-a^3)(1-b^3)\ldots,$$

the signs being $+$ or $-$ according as the number of the distinct prime factors $a,\ b,\ldots$ of n is even or odd.

---

[151]Comptes Rendus Paris, 32, 1851, 918–921.
[152]Programm Realschule, Posen, 1855, §§5–6.

G. Oltramare[153] obtained for the sum, sum of squares, sum of cubes, and sum of biquadrates, of the integers $< ma$ and relatively prime to a the respective values

$$\tfrac{1}{2}m^2a\phi(a), \qquad \tfrac{1}{3}m^3a^2\phi(a)+(-1)^a\frac{m}{2\cdot3}a\phi(a_1),$$

$$\tfrac{1}{4}m^4a^3\phi(a)+(-1)^a\frac{m^2}{4}a^2\phi(a_1),$$

$$\tfrac{1}{5}m^5a^4\phi(a)+(-1)^a\frac{m^3}{3}a^3\phi(a_1)-(-1)^a\frac{m}{2\cdot3\cdot5}a\xi(a_1),$$

where a is the number and a_1 the product of the distinct prime factors $\mu, \nu, \ldots$ of a, while $\xi(a_1)=(\mu^3-1)(\nu^3-1)\ldots$ The number of integers $<n$ which are prime to a is $\phi(a)n/a$.

J. Liouville[154] stated that Gauss' proof of $\Sigma\phi(d)=N$ may be extended to the generalization

$$\Sigma\binom{N}{d}^k\phi_k(d)=1^k+2^k+\ldots+N^k,$$

where d ranges over the divisors of N. He remarked that Binet's[151] results are readily proved in various ways. Also,

$$\Sigma\left(\frac{m}{d}\right)^3\phi_3(d)=\left\{\Sigma\frac{m}{d}\phi(d)\right\}^2.$$

N. V. Bougaief[155] stated that, if $\xi(n)$ is the number of distinct prime factors of $n>1$, and $\xi_1(n)$ is their product,

$$6\phi_2(n)=2\phi(n)n^2+(-1)^{\xi(n)}\xi_1(n)\phi(n);$$

also a result quoted below with Gegenbauer's[170] generalization.

August Blind[156] reproduced without reference the formulas and proofs by Thacker,[150] and gave

$$\phi_\nu(m)=m^\nu\phi_0(m)-\binom{\nu}{1}m^{\nu-1}\phi_1(m)+\binom{\nu}{2}m^{\nu-2}\phi_2(m)-\ldots+(-1)^\nu\phi_\nu(m).$$

E. Lucas[157] indicated a proof that $n\phi_{n-1}(x)$ is given symbolically by $(x+Q)^n-Q^n$, where, if $n=a^\alpha b^\beta\ldots$, $Q_k=B_k(1-a^{k-1})(1-b^{k-1})\ldots$ Thus, if π is the product of the negatives of the primes $a, b, \ldots$,

$$2\phi_1(x)=x\phi(x), \qquad 3\phi_2(x)=\phi(x)\left(x^2+\tfrac{1}{2}\pi\right), \qquad 4\phi_3(x)=x\phi(x)(x^2+\pi).$$

---

[153]Mémoires de l'Institut Nat. Génevois, 4, 1856, 1–10.
[154]Comptes Rendus Paris, 44, 1857, 753–4; Jour. de Math., (2), 2, 1857, 393–6.
[155]Nouv. Ann. Math., (2), 13, 1874, 381–3; Bull. Sc. Math. Astr., 10, I, 1876, 18.
[156]Ueber die Potenzsummen der unter einer Zahl m liegenden und zu ihr relativ primen Zahlen, Diss., Bonn, 1876, 37 pp.
[157]Nouv. Ann. Math., (2), 16, 1877, 159; Théorie des nombres, 1891, 394.

Several[157a] found expressions for $\phi_n = \phi_n(N)$ and proved that

$$\phi_0 x^n + n\phi_1 x^{n-1} + \tfrac{1}{2}n(n-1)\,\phi_2 x^{n-2} + \ldots + \phi_n = 0 \quad (n \text{ odd})$$

has the root $-\phi_1/\phi_0$, while the remaining roots can be paired so that the sum of the two of any pair is $-2\phi_1/\phi_0$. If $n = 3$ the roots are in arithmetical progression.

H. Postula[158] proved Crelle's result by the long method of deleting multiples, used by Brennecke.[152] Catalan (ibid., pp. 208–9) gave Crelle's short proof.

Mennesson[159] stated that, if q is any odd number,

$$\phi_q(n) \equiv \tfrac{1}{2}\phi(n^{q+1}) \pmod{q},$$

and (Ex. 366) that the sum of the products $\phi(n) - 1$ at a time of the integers $\leq n$ and prime to n is a multiple of n.

E. Cesàro[160] proved the generalization: The sum ψ_m of the products m at a time of the integers $\alpha, \beta, \ldots \leq N$ and prime to N is divisible by N if m is odd. For by replacing α by $N - \alpha$, β by $N - \beta$, $\ldots$ and expanding,

$$\psi_m = \binom{\phi}{m} N^m - \binom{\phi-1}{m-1} N^{m-1}\psi_1 + \binom{\phi-2}{m-2} N^{m-2}\psi_2 - \ldots \pm \psi_m,$$

where $\phi = \phi(N)$. Also $\phi_m(N)$ is divisible by N if m is odd.

F. de Rocquigny[161] proved Crelle's result. Later, he[162] employed concentric circles of radii $1, 2, 3, \ldots$ and marked the numbers $(m-1)N+1$, $(m-1)N+2, \ldots, mN$ at points dividing the circle of radius m into N equal parts. The lines joining the center to the $\phi(N)$ points on the unit circle, marked by the numbers $<N$ and prime to N, meet the various circles in points marked by all the numbers prime to N. He stated that the sum of the $\phi(N)$ numbers prime to N appearing on the circle of radius m is $\tfrac{1}{2}(2m-1)\phi(N^2)$, and [the equivalent result] that the sum of the numbers prime to N from 0 to mN is $\tfrac{1}{2}m^2\phi(N^2)$. He later recurred to the subject (*ibid., 54, 1881, 160).

A. Minine[163] noted that, if $P > N > 1$ and k is the remainder obtained by dividing P by N, the sum $s(N, P)$ of the integers $<P$ and prime to N may be computed by use of

$$s(N, mN+k) = s(N, k) + \frac{m^2}{2}\phi(N^2) + mN\phi(N)_k,$$

where (Minine[47]) $\phi(N)_k$ is the number of integers $\leq k$ prime to N.

*A. Minine[164] considered the number and sum of all the integers $<P$ which are prime to N [Legendre's (5) and Minine[163]].

[157a]Math. Quest. Educ. Times, 28, 1878, 45–7, 103–5.
[158]Nouv. Corresp. Math., 4, 1878, 204–7. Likewise, R. A. Harris, Math. Mag., 2, 1904, 272.
[159]Ibid., p. 302.
[160]Ibid., 5, 1879, 56–59.
[161]Les Mondes, Revue Hebdom. des Sciences, 51, 1880, 335–6.
[162]Ibid., 52, 1880, 516–9.
[163]Ibid., 53, 1880, 526–9.
[164]Nouveaux théorèmes de la théorie des nombres, Moscow, 1881.

A. Minine[165] investigated the numbers N which divide the sum of all the integers $< N$ and prime to N.

E. Cesàro[166] proposed his theorems[160] as exercises. Proofs, by associating a with $N-a$, etc., were given by Moret-Blanc (3, 1884, 483–4). Cesàro[57] (p. 82) proved the formula of Liouville.[154] Writing (pp. 158–9) ϕ_m for $\phi_m(N)$ and expanding $\phi_m = \Sigma(N-a)^m$, where $a, \beta, \ldots$ are the integers $\leq N$ and prime to N, we get

$$\phi_m = N^m\phi - \binom{m}{1}N^{m-1}\phi_1 + \binom{m}{2}N^{m-2}\phi_2 - \ldots \pm \phi_m,$$

whence ϕ_m is divisible by N if m is odd, but not if m is even. This is evident (p. 257) since $a^m + (N-a)^m$ is divisible by $a+N-a$ if m is odd. The above formula gives $A^m = (1-A)^m$, symbolically, where

$$A_m = \frac{\phi_m}{\phi} \cdot \frac{1}{N^m}$$

is the arithmetic mean of the mth powers of $a/N, \beta/N, \ldots$. The mean value of $\phi_m(N)$ is $6A_m N^{m+1}/\pi^2$. He reproduced (pp. 161–2) an earlier formula,[160] which shows that $B^m = (1-B)^m$, symbolically, if B_m is the arithmetic mean of the products of $a/N, \beta/N, \ldots$ taken m at a time. We have (p. 165) the approximation

$$\sum_{j=1}^{x} \phi_m(j) = \frac{x^{m+2}}{(m+1)(m+2)} \cdot \frac{6}{\pi^2},$$

whence (p. 261) the mean of $\phi_m(N)$ is $6N^{m+2}/(m+1)\pi^2$. Proof is given (pp. 255–6) of Thacker's[150] formula

$$\phi_m(N) = \frac{(N+B\psi)^{m+1} - (B\psi)^{m+1}}{m+1} = \frac{1}{m+1}\sum_{p=0}^{m}\binom{m+1}{p}B_p N^{m-p+1}\psi_p(N),$$

where

$$\psi_p(N) = \Sigma d^{p-1}\mu(d) = \Pi(1 - u^{p-1}),$$

d ranging over the divisors of N, and u over the prime divisors of N. Here $\mu(x)$ is Merten's function (Ch. XIX). It is proved (pp. 258–9) that

$$\Sigma d^{p-1}\psi_p\left(\frac{N}{d}\right) = 1, \qquad \Sigma d^k\psi_s\left(\frac{N}{d}\right) = \Sigma d^k\psi_{s-k}(d),$$

the first characterizing the function $\psi_p(N)$, and reducing to (4) for $p=0$. If a ranges over the integers for which $[2n/a]$ is odd, then (p. 293)

$$\Sigma\phi_m(a)/a^m = \frac{n^2}{m+1} - \frac{m}{12}\Delta_n,$$

exactly if $m=0, 1, 2, 3$, approximately if $m>3$, where Δ_n is the excess of the sum of the inverses of $1, \ldots, n$ over that of $n+1, \ldots, 2n$. In particular, $\Sigma\phi(a) = n^2$.

---

[165]Math. Soc. Moscow (in Russian), 10, 1882–3, 87–101.
[166]Nouv. Ann. Math., (3), 2, 1883, 288.

P. Nazimov[167] (Nasimof) noted that, when x ranges over the integers $\leqq m$ and prime to n, the sum of the values taken by any function $f(x)$ equals

$$\sum_d \mu(d) \sum_{x=1}^{[m/d]} f(dx),$$

where d ranges over all divisors of n. The case $f(x) \equiv 1$ yields Legendre's formula (5). The case $f(x) \equiv x$ yields a result equivalent to that of Minine.[163-4] A generalization was given by Zsigmondy[77] and Gegenbauer.[173]

E. Cesàro[168] noted that, if A_m is the arithmetic mean of the mth powers of the integers $\leqq N$ and prime to N, and B_m that of their products m at a time, we have the symbolic relations

$$A^m = (N-A)^m, \qquad B^m = (N-B)^m.$$

Cesàro[169] proved Thacker's[150] formula expressed as

$$\phi_k(n) = \frac{n^k}{n+1} \sum_{i=0}^{k} \binom{k+1}{i} B_i \zeta_i(n) = \frac{n^k}{k+1} \{(1+B\zeta)^{k+1} - (B\zeta)^{k+1}\},$$

the last being symbolic, where ζ_k is a function such that $\Sigma \zeta_k(d) = n^{1-k}$, d ranging over the divisors of n. By inversion

$$\zeta_k(n) = \sum_d \mu\left(\frac{n}{d}\right) d^{1-k} = \frac{1}{n^{k-1}} \Pi (1 - u^{k-1}),$$

where u ranges over the distinct prime factors of n.

L. Gegenbauer[170] proved that, if $\nu = \left[\sqrt[\rho]{n}\right]$,

$$\sum_{x=1}^{n} \{1^k + 2^k + \ldots + (g_\rho(x))^k\} = \sum_{n=1}^{\nu} \left[\frac{n}{x^\rho}\right] \phi_k(x), \qquad g_\rho(p_1^{\alpha_1} \ldots p_s^{\alpha_s}) \equiv \prod_{i=1}^{s} p_i^{\left[\frac{\alpha_i}{\rho}\right]}.$$

For the case $k=0$, $\rho=2$, this becomes Bougaief's[155] formula

$$\sum_{x=1}^{n} g_2(x) = \sum_{x=1}^{\nu} \left[\frac{n}{x^2}\right] \phi(x), \qquad \nu = [\sqrt{n}].$$

C. Leudesdorf[171] considered for μ odd the sum $\psi_\mu(N)$ of the inverses of the μth powers of the integers $<N$ and prime to N. Then

$$\psi_\mu(N) = \tfrac{1}{2}kN^2 - \tfrac{1}{2}\mu N \psi_{\mu+1}(N),$$

where k is an integer. Thus, if $N=p^l q$, where q is not divisible by the prime $p>3$, $\psi_\mu(N)$ is divisible by p^{2l} unless μ is prime to p, and $\mu+1$ is divisible by $p-1$; for example, $\psi_\mu(p)$ is divisible by p^2. If $p=3$, $\psi_\mu(N)$ is divisible by p^{2l} if μ is an odd multiple of 3. If $p=2$, it is divisible by 2^{2l-1} except when $q=1$.

Cesàro[172] inverted his[57] symbolic form of Thacker's formula for $\phi_m(N)$ in terms of ψ's and obtained

$$nB_p \psi_p(n) = (\phi - nB)^p.$$

[167]Matem. Sbornik (Math. Soc. Moscow), 11, 1883–4, 603–10 (Russian).
[168]Mathesis, 5, 1885, 81.
[169]Giornale di Mat., 23, 1885, 172–4.
[170]Sitzungsber. Ak. Wiss. Wien (Math.), 95, II, 1887, 219–224.
[171]Proc. London Math. Soc., 20, 1889, 199–212.
[172]Periodico di Mat., 7, 1892, 3–6.   See p. 144 of this history.

Hence if a ranges over the integers $\leq n$ and prime to n,

$$\Sigma(a-nB)^p = 0 \text{ or a multiple of } n\psi_p$$

according as p is odd or even.   By this recursion formula,

$$\phi_1 = \tfrac{1}{2}n\phi, \qquad \phi_3 = \tfrac{3}{2}n\phi_2 - \tfrac{1}{4}n^3\phi, \dots$$

L. Gegenbauer[173] gave a formula including those of Nazimov[167] and Zsigmondy.[77]   For any functions $\chi(d)$, $\chi_1(d)$, $f(x_1, \dots, x_s)$,

$$\overset{1,\dots,m}{\underset{x_1,\dots,x_s}{\Sigma}} f(\kappa x_1, \dots, \kappa x_s) \Sigma_\delta \chi(\delta)\chi_1\left(\frac{n}{\delta}\right) = \Sigma_d \chi(d)\chi_1\left(\frac{n}{d}\right) \overset{1,\dots,[m/d]}{\underset{x_1,\dots,x_s}{\Sigma}} f(d\kappa x_1, \dots, d\kappa x_s),$$

where d ranges over all divisors of n which have some definite property P, while δ ranges over those common divisors of n, $x_1, \dots, x_s$ which have property P.   Various special choices are made for χ, χ_1, f and P.   For instance, property P may be that d is an exact ρth power, whence, if $\rho = 1$, d is any divisor of n.   The special results obtained relate mainly to new number-theoretic functions without great interest and suggested apparently by the topic in hand.

T. del Beccaro[174] noted that $\phi_k(n)$ is divisible by n if k is odd [Binet[151]]. When n is a power of 2,

$$1^k + 2^k + \dots + (n-1)^k \equiv 0 \text{ or } \phi(n) \pmod{n},$$

according as k is odd or even.   His proof of (1) is due to Euler.

J. W. L. Glaisher[175] proved that, if a, b, $\dots$ are any divisors of x such that their product is also a divisor, the sum of the nth powers of the integers $< x$ and not divisible by a or b, $\dots$, is

$$\frac{1}{n+1}\left[x^{n+1}\Pi\left(1-\frac{1}{a}\right) + (-1)^s\left\{\binom{n+1}{2}F_1 x^{n-1} - \binom{n+1}{4}F_2 x^{n-3} - \dots\right\}\right],$$

where s is the number of the divisors a, b, $\dots$, and

$$F_n = B_n(a^{2n-1}-1)(b^{2n-1}-1)\dots.$$

If a, b, $\dots$ are all the prime factors of x, this result becomes Thacker's.[150]

N. Nielsen[176] proved by induction on γ that the sum of the nth powers of the positive integers $< mM$ and prime to $M = p_1^{r_1}\dots p_\gamma^{r_\gamma}$ is

$$\frac{m^{n+1}M^n\phi(M)}{n+1} + (-1)^\gamma \overset{[n/2]}{\underset{s=1}{\Sigma}}\frac{(-1)^{s-1}}{n+1}\binom{n+1}{2s}B_s\,(mM)^{n-2s+1}\overset{\gamma}{\underset{i=1}{\Pi}}(p_i^{2s-1}-1).$$

The case $m = 1$ gives Thacker's[150] result.   That result shows (*ibid.*, p. 179) that $\phi_{2n}(m)$ and $\phi_{2n+1}(m)$ are divisible by m and m^2 respectively, for $1 \leq n \leq (p_1-3)/2$, where p_1 is the least prime factor of m, and also gives the residues of the quotients modulo m.   Corresponding theorems therefore hold for the sum of the products of the integers $< m$ and prime to m, taken t at a time.

[173]Sitzungsberichte Ak. Wiss. Wien (Math.), 102, 1893, IIa, 1265–94.
[174]Atti R. Accad. Lincei, Mem. Cl. Fis. Mat., 1, 1894, 344–371.
[175]Messenger Math., 28, 1898–9, 39–41.
[176]Oversigt Danske Vidensk. Selsk. Förhandlinger, 1915, 509–12; cf. 178–9.

SCHEMMEL'S GENERALIZATION OF EULER'S ϕ-FUNCTION.

V. Schemmel[190] considered the $\Phi_n(m)$ sets of n consecutive numbers each $<m$ and relatively prime to m. If $m=a^\alpha b^\beta\ldots$, where a, b, $\ldots$ are distinct primes, and m, m' are relatively prime, he stated that

$$\Phi_n(m)=a^{\alpha-1}(a-n)b^{\beta-1}(b-n)\ldots, \qquad \Phi_n(mm')=\Phi_n(m)\Phi_n(m'),$$

$$\sum_\delta n^{a-a'}n^{\beta-\beta'}\ldots\Phi_n(\delta)=m, \qquad \delta=a^{a'}b^{\beta'}\ldots, \qquad a'\leqq a,\ \beta'\leqq\beta,\ldots,$$

the third formula being a generalization of Gauss' (4). If k is a fixed integer prime to m, $\Phi_n(m)$ is the number of sets of n integers $<m$ and prime to m such that each term of a set exceeds by k the preceding term modulo m. Consider the product P of the λth terms of the $\Phi_n(m)$ sets. If $n=1$, $P\equiv\pm1$ (mod m) by Wilson's theorem. If $n>1$,

$$P^{n-1}\equiv\left\{(-1)^{\lambda-1}k^{n-1}(\lambda-1)!(n-\lambda)!\right\}^{\Phi_n(m)}\ (\text{mod }m).$$

For the case $k=\lambda=1$, $n=2$, we see that the product of those integers $<m$ and prime to m, which if increased by unity give integers prime to m, is $\equiv1$ (mod m).

E. Lucas[191] gave a generalization of Schemmel's function, without mention of the latter. Let $e_1,\ldots,e_k$ be any integers. Let $\Psi(n)$ denote the number of those integers h, chosen from $0, 1,\ldots, n-1$, such that

$$h-e_1,\ h-e_2,\ldots,\ h-e_k$$

are prime to n. For $k<n$, $e_1=0$, $e_2=-1,\ldots,e_k=-(k-1)$, we have k consecutive integers h, $h+1,\ldots,h+k-1$ each prime to n, and the number of such sets is $\Phi_k(n)$. Lucas noted that $\Psi(p)\Psi(q)=\Psi(pq)$ if p and q are relatively prime. Let $n=a^\alpha b^\beta\ldots$, where a, b, $\ldots$ are distinct primes. Let λ be the number of distinct residues of $e_1,\ldots,e_k$ modulo a; μ the number of their distinct residues modulo b; etc. Then

$$\Psi(n)=a^{\alpha-1}(a-\lambda)b^{\beta-1}(b-\mu)\ldots.$$

L. Goldschmidt[192] proved the theorems stated by Schemmel, and himself stated the further generalization: Select any $a-A$ positive integers $<a$, any $b-B$ positive integers $<b$, etc.; there are exactly

$$a^{\alpha-1}(a-A)b^{\beta-1}(b-B)\ldots$$

integers $<m$ which are congruent modulo a to one of the $a-A$ numbers selected and congruent modulo b to one of the $b-B$ numbers selected, etc.

P. Bachmann[193] proved the theorems due to Schemmel and Lucas.

JORDAN'S GENERALIZATION OF EULER'S ϕ-FUNCTION.

C. Jordan,[200] in connection with his study of linear congruence groups, proved that the number of different sets of k (equal or distinct) positive integers $\leqq n$, whose g. c. d. is prime to n, is[*]

$$(10) \qquad J_k(n)=n^k\left(1-\frac{1}{p_1^k}\right)\ldots\left(1-\frac{1}{p_q^k}\right)$$

[190]Jour. für Math., 70, 1869, 191–2. [191]Théorie des nombres, 1891, p. 402.
[192]Zeitschrift Math. Phys., 39, 1894, 205–212. [193]Niedere Zahlentheorie, I, 1902, 91–94, 174–5.
[200]Traité des substitutions, Paris, 1870, 95–97.
[*]He used the symbol [n, k]. Several of the writers mentioned later used the symbol $\phi_k(n)$, which, however, conflicts with that by Thacker.[150]

if $p_1, \ldots, p_q$ are the distinct prime factors of n. In fact, there are n^k sets of k integers $\leqq n$, while $(n/p_1)^k$ of these sets have the common divisor p_1, etc., whence

$$J_k(n) = n^k - \left(\frac{n}{p_1}\right)^k - \left(\frac{n}{p_2}\right)^k - \ldots + \left(\frac{n}{p_1 p_2}\right)^k + \ldots$$

Jordan noted the corollary: if n and n' are relatively prime,

(11) $$J_k(nn') = J_k(n) J_k(n').$$

A. Blind[156] defined the function (10) also for negative values of k, proved (11), and the following generalization of (4):

(12) $$\Sigma J_k(d) = n^k \quad (d \text{ ranging over the divisors of } n).$$

W. E. Story[201] employed the symbol $\tau^k(n)$ for $J_k(n)$ and called it one of the two kinds of kth totients. The second kind is the number $\phi^k(n)$ of sets of k integers $\leqq n$ and not all divisible by any factor of n, such that we do not distinguish between two sets differing only by a permutation of their numbers. He stated that

$$\phi^k(n) = \frac{1}{k!} \{ \tau^k(n) + t_1{}^k \tau^{k-1}(n) + t_2{}^k \tau^{k-2}(n) + \ldots + t^k_{k-1} \tau(n) \},$$

where $1, t_1{}^k, t_2{}^k, \ldots$ are the coefficients of the successive descending powers of x in the expansion of $(x+1)(x+2) \ldots (x+k-1)$.

Story[202] defined "the kth totient of n to the condition κ to be the number of sets of k numbers $\leqq n$ which satisfy condition κ. The number of sets of k numbers $\leqq n$, all containing some common divisor of n satisfying the condition κ, but not all containing any one divisor of n satisfying the condition χ is (if different permutations of k numbers count as different sets)

$$n^k \frac{1}{\delta^k} \frac{1}{\delta'^k} \ldots \left(1 - \frac{1}{\delta_1{}^k}\right) \left(1 - \frac{1}{\delta_1'^k}\right) \ldots,$$

where $\delta, \delta', \ldots$ are the least divisors of n satisfying condition κ, while $\delta_1, \delta_1', \ldots$ are the least divisors of n satisfying condition χ. Here a set of least divisors is a set of divisors no one of which is a multiple of any other."

E. Cesàro[57] (p. 345) stated that, if $\Phi_k(x)$ is the number of sets of k integers $\leqq x$ whose g. c. d. is prime to x, then

$$\Sigma \Phi_k(d) = \binom{n+k-1}{k}, \qquad \Phi_k(n) = \binom{J+k-1}{k},$$

where J^s is to be replaced by $J_s(n)$, and d ranges over the divisors of n.

J. W. L. Glaisher[203] proved (12) by means of a symbolic expression for the infinite series $\Sigma J_k(n) f(x^n)$. If $\mu(n)$ is Merten's function,

$$J_k(n) - \Sigma p_1{}^k J_k\left(\frac{n}{p_1}\right) + \Sigma p_1{}^k p_2{}^k J_k\left(\frac{n}{p_1 p_2}\right) - \ldots = \mu(n),$$

where the summations relate to the distinct prime factors p_i of n. Using

[201]Johns Hopkins University Circulars, 1, 1881, 132.
[202]*Ibid.*, p. 151. Cf. Amer. Jour. Math., 3, 1880, 382–7.
[203]London, Ed. Dublin Phil. Mag., (5), 18, 1884, 531, 537–8.

these formulas for $n = 1, 2, \ldots, n$, we obtain two determinants of order n, each equal to $(-1)^{n-1}J_k(n)$:

$$\begin{vmatrix} 1^k & 2^k & 3^k & 4^k & \cdots \\ 1 & 1 & 1 & 1 & \cdots \\ 0 & 1 & 0 & 1 & \cdots \\ 0 & 0 & 1 & 0 & \cdots \\ \cdots & \cdots & \cdots & \cdots & \cdots \end{vmatrix}, \qquad \begin{vmatrix} 1 & -1 & -1 & 0 & -1 & 1 & \cdots \\ 1 & -2^k & -3^k & 0 & -5^k & 6^k & \cdots \\ 0 & 1^k & 0 & -2^k & 0 & -3^k & \cdots \\ 0 & 0 & 1^k & 0 & 0 & -2^k & \cdots \\ \cdots & \cdots & \cdots & \cdots & \cdots & \cdots & \cdots \end{vmatrix}.$$

L. Gegenbauer[204] proved (12). For $n = p_1^{\nu_1} \ldots p_q^{\nu_q}$, set

$$\pi(n) = (-1)^q p_1 \ldots p_q, \qquad \lambda(n) = (-1)^{\nu_1 + \cdots + \nu_q},$$

$$F(d) = (-1)^{(k+1)w(r/d)} \pi^k\left(\frac{r}{d}\right) J_k\left(\frac{r}{d}\right),$$

where $w(n)$ denotes the number of distinct prime factors of n. By means of the series $\zeta(s) = \Sigma n^{-s}$, he proved that, when d ranges over the divisors of r,

$$\Sigma F(d)d^{2k} = r^k, \qquad \Sigma F(d)d^{3k} = r^k \Sigma d^k J_k(d),$$

$$\Sigma F(d)J_k(d)d^k = 0, \qquad \Sigma(-1)^{(k+1)w(r/d)} d^k \pi^k\left(\frac{r}{d}\right) = 0,$$

the last holding if r has no square factor and following from the third in view of (11),

$$J_k(r) = \Sigma d\mu\left(\frac{r}{d}\right), \qquad \Sigma F(d)d^k\mu(d) = r^{2k}\mu(r), \qquad \Sigma\lambda(d)J_k(d)J_k\left(\frac{r}{d}\right) = 0 \text{ or } J_{2k}(\sqrt{r}),$$

according as r is or. is not a square,

$$\sum_{m,\,n} (-1)^{(k+1)w(m)} \pi^k(m)J_k(m)J_{2k}(n)n^{2k} = r^k\lambda(r)J_k(r) \qquad (mn^2 = r),$$

$$\Sigma F(d)J_{k+t}(d)d^k = r^{k+t}J_k(r), \qquad \Sigma d^t J_k(d)J_t\left(\frac{r}{d}\right) = J_{k+t}(r) \qquad (t > 0),$$

$$\Sigma J_k(n_1) \ldots J_k(n_t)n_1^{(t-1)k} n_2^{(t-2)k} \ldots n_{t-1}^k = r^{tk},$$

where $n_1, \ldots, n_t$ range over all sets of solutions of $n_1 n_2 \ldots n_{t+1} = n$, the case $k = 1$ being due to H. G. Cantor.[49]

E. Cesàro[169] derived (10) from (12), writing ζ_{1-k} for J_k.

E. Cesàro[205] denoted $J_k(n)$ by $\psi^k(n)$ and gave (12).

L. Gegenbauer[170] gave the further generalization

$$\sum_{x=1}^{n} (g_\rho(x))^k = \sum_{x=1}^{\nu} \left[\frac{n}{x^\rho}\right] J_k(x), \qquad \nu = [\sqrt[\rho]{n}].$$

J. Hammond[206] wrote $\psi(n, d)$ for $\Sigma f(\delta)$, where f is an arbitrary function and δ ranges over all multiples $\leq n$ of the fixed divisor d of n. Then

(13) $$\Sigma f(t) = \psi(n, 1) - \Sigma\psi(n, p_1) + \Sigma\psi(n, p_1 p_2) - \ldots,$$

[204]Sitzungsber. Ak. Wiss. Wien (Math.), 89 II, 1884, 37–46. Cf. p. 841. See Gegenbauer[71] of Ch. X.

[205]Annali di Mat., (2), 14, 1886–7, 142–6.

[206]Messenger Math., 20, 1890–1, 182–190.

where t ranges over the integers $\leq n$ which are prime to n, while $p_1, p_2, \ldots$ denote the distinct prime factors of n. If $f(t) \equiv 1$, then $\psi(n, d) = n/d$ and (13) becomes

$$\phi(n) = n - \Sigma \frac{n}{p_1} + \Sigma \frac{n}{p_1 p_2} - \ldots = n\left(1 - \frac{1}{p_1}\right)\left(1 - \frac{1}{p_2}\right)\ldots$$

Next, take $f(t) = a_0 + a_1 t + a_2 t^2 + \ldots$. Using hyperbolic functions,

$$\Sigma f(t) = \tfrac{1}{2} \coth (z/2) = \frac{1}{z} + \frac{z}{12} - \frac{z^3}{720} + \ldots,$$

provided z^r be replaced by $n^r f_r(n) J_{-r}(n)$, where

$$f_1(n) = f'(n) - a_1, \qquad f_2(n) = f''(n) - 2a_2, \ldots, \qquad f_{-1}(n) = \int f(n) dn.$$

Hence, since $J_1(n) = \phi(n)$,

$$\Sigma f(t) = \frac{\phi(n)}{n} f_{-1}(n) + \frac{n}{12} J_{-1}(n) f_1(n) - \frac{1}{720} n^3 J_{-3}(n) f_3(n) + \ldots.$$

In particular, for $f(t) = t^k$, we get $\phi_k(n)$. In Prouhet's[18] first formula, δ may be replaced by the g. c. d. $\Delta_{a,b}$ of a and b. The generalization

$$J_k(ab) = J_k(a) J_k(b) \frac{\Delta_{a,b}^k}{J_k(\Delta_{a,b})}$$

is proved. From (12) we get by addition*

(14) $$\sum_{j=1}^{n} \left[\frac{n}{j}\right] J_k(j) = 1^k + 2^k + \ldots + n^k.$$

Taking $n = 1, 2, \ldots, n$, we obtain equations whose solution gives $J_k(n)$ expressed as a determinant of order n in which the elements of the last column are $1, 1 + 2^k, 1 + 2^k + 3^k, \ldots$, while for $s < n$ the sth column consists of $s - 1$ zeros followed by s units, then s twos, etc. For $s > 0$, the element in the $(s+1)$th row and rth column in Glaisher's[203] first determinant is 1 or 0 according as r/s is integral or fractional.

J. Vályi[207] used $J_2(n) \div \phi(n)$ in his enumeration of the n-fold perspective polygons of n sides inscribed in a cubic curve.

H. Weber[208] proved (10) for $k = 2$.

L. Carlini[209] gave without references (10), (11), (12), with $\phi\left(\frac{k}{n}\right)$ for $J_n(k)$.

E. Cesàro[210] noted that (12) implies (10). For, if $\Sigma f(d) = F(n)$, we have by inversion (Ch. XIX), $f(n) = \Sigma \mu(d) F(n/d)$. The case $f = J_l$ gives

$$\frac{J_l(n)}{n^l} = \Sigma \frac{\mu(d)}{d^l}.$$

The latter is a case of $G(n) = \Sigma g(d)$ and hence, with (12) and

$$\Sigma f(d) G\left(\frac{n}{d}\right) \equiv \Sigma g(d) F\left(\frac{n}{d}\right),$$

---

*This work, Mess. Math., 20, 1890-1, p. 161, for $k = 1$, is really due to Dirichlet.[21] Formula (14) is the case $\rho = 1$ of Gegenbauer's, p. 217.

[207]Math. Nat. Berichte aus Ungarn, 9, 1890, 148; 10, 1891, 171.
[208]Elliptische Functionen, 1891, 225; ed. 2, 1908 (Algebra III), 215.
[209]Periodico di Mat., 6, 1891, 119-122.
[210]Ibid., 7, 1892, 1-6.

yields
$$\Sigma J_k(d)J_l\left(\frac{n}{d}\right)\left(\frac{d}{n}\right)^l = \Sigma\frac{\mu(d)}{d^l}\left(\frac{n}{d}\right)^k = n^k\frac{J_{l+k}(n)}{n^{l+k}},$$
or
$$J_{l+k}(n) = \Sigma d^l J_k(d)J_l\left(\frac{n}{d}\right),$$

which is next to the last formula of Gegenbauer's.[204]   Similarly,
$$\sigma_{l+k}(n) = \Sigma d^k J_l(d)\sigma_k\left(\frac{n}{d}\right),$$

which is the case $t=1$ of Gegenbauer's[72] fifth formula in Ch. X, $\sigma_k(n)$ being the sum of the kth powers of the divisors of n.

E. Weyr[211] interpreted $J_2(n)$ in connection with involutions on loci of genus 1. From the same standpoint, L. Gegenbauer[212] proved (12) for $k=2$ and noted that the value (10) of $J_2(n)$ then follows by the usual method of number-theoretic derivatives.

L. Gegenbauer[212a] wrote $\phi_k(m, n)$ for the number of sets of k positive integers $\leq m$ whose g. c. d. is prime to $n=p_1^{\tau_1}\ldots p_r^{\tau_r}$ and proved a formula including
$$[m]^k = \phi_k(m, n) + \sum_{\sigma=1}^{r}\sum_{\lambda_1,..,\lambda_\sigma=1}^{r}(\lambda_1,\ldots,\lambda_\sigma)^2\,\phi_k\left(\frac{m}{p_{\lambda_1}\ldots p_{\lambda_\sigma}}\,,\,\frac{n}{p_{\lambda 1}^{a\lambda_1}\ldots p_{\lambda_\sigma}^{a\lambda_\sigma}}\right),$$

where $(\lambda_1,\ldots,\lambda_\sigma)$ is the determinant derived from that with unity throughout the main diagonal and zeros elsewhere by replacing the γth row by the λ_γth row for $\gamma=1,\ldots,\sigma$. The case $m=n$, $k=1$, is due to Pepin.[37] There is an analogous formula involving the sum of the kth powers of the positive integers $\leq m$ and prime to n.

E. Jablonski[74] used $J_k(n)$ in connection with permutations.

G. Arnoux[213] proved (10) in connection with modular space.

*J. J. Tschistiakow[214] (or Cistiakov) treated the function $J_k(n)$.

R. D. von Sterneck[215] proved that
$$J_k(n) = \Sigma J_r(\lambda_1)J_{k-r}(\lambda_2) = \Sigma\phi(\lambda_1)\ldots\phi(\lambda_k),$$

the λ's ranging over all sets of integers $\leq n$ whose l. c. m. is n. To generalize this, let $J_k(n; m_1,\ldots,m_k)$ be the number of sets of integers $i_1,\ldots,i_k$, whose g. c. d. is prime to n, while $i_j \leq n/m_j$ for $j=1,\ldots,k$. Then
$$J_k(n; m_1,\ldots,m_k) = \Sigma J_r(\lambda_1; m'_1,\ldots,m'_r)J_{k-r}(\lambda_2; m'_{r+1},\ldots,m'_k)$$
$$= \Sigma J_1(\lambda_1; m_1)\ldots J_1(\lambda_k; m_k),$$

the λ's ranging over all sets of integers $\leq n$ whose l. c. m. is n, while $m'_1,\ldots,$ m'_k form any fixed permutation of $m_1,\ldots,m_k$, and $J_1(n; m)$, designated $\phi^{(m)}(n)$ by the author, is the number of integers $\leq n/m$ which are prime to n. Also,

[211]Sitzungsberichte Ak. Wiss. Wien (Math.), 101, IIa, 1892, 1729–1741.
[212]Monatshefte Math. Phys., 4, 1893, 330.
[212a]Denkschr. Ak. Wiss. Wien (Math.), 60, 1893, 25–47.
[213]Arithmétique graphique; espaces arith. hypermagiques, 1894, 93.
[214]Math. Soc. Moscow, 17, 1894, 530–7 (in Russian).
[215]Monatshefte Math. Phys., 5, 1894, 255–266.

$$\Sigma J_k(d; m_1, \ldots, m_k) = \left[\frac{n}{m_1}\right]\left[\frac{n}{m_2}\right] \cdots \left[\frac{n}{m_k}\right],$$

where d ranges over the divisors of n, the case $k=1$ being due to Laguerre.[35] In the latter case, take $n=1, \ldots, n$ and add. Thus

$$\sum_{k=1}^{n} J_1(k; m)\left[\frac{n}{k}\right] = \sum_{j=1}^{n} \left[\frac{j}{m}\right] = \tfrac{1}{2}\left\{\frac{n^2}{m}+\frac{n}{m}-n+\left(n, \frac{n}{m}\right)\right\},$$

the last equality, in which (n, b) is the g. c. d. of n, b, following from expressions for (n, b) given by Hacks[42] of Ch. XI. In the present paper the above double equation was proved geometrically. For $m=1$, we get Dirichlet's[21] formula. The g. c. d. of three numbers is expressed in terms of them and $[x]$.

The initial formulas were proved geometrically, but were recognized to be special cases of a more general theorem. Let

$$\Sigma f_i(d) = F_i(n) \qquad (i=1, \ldots, k),$$

where d ranges over all divisors of n. Then the function

$$\psi(n) = \Sigma f_1(\lambda_1) \ldots f_k(\lambda_k) \qquad \text{(l. c. m. of } \lambda_1, \ldots, \lambda_k \text{ is } n)$$

has the property

$$\Sigma\psi(d) = F_1(n) \ldots F_k(n).$$

Hence in the terminology of Bougaief (Ch. XIX) the number-theoretic derivative $\psi(n)$ of $F_1(n) \ldots F_k(n)$ equals the sum of the products of the derivatives f_i of the factors F_i, the arguments ranging over all sets of k numbers having n as their g. c. d.

L. Gegenbauer[215a] proved easily that, if $[n, \ldots, t]$ is the g. c. d. of $n, \ldots, t$

$$\sum_{x_1, \ldots, x_s=1}^{n} F([n, x_1, \ldots, x_s]) = \Sigma F(d) \, J_s\left(\frac{n}{d}\right),$$

where d ranges over all divisors of n, and F is any function.

K. Zsigmondy[216] considered any abelian (commutative) group G with the independent generators $g_1, \ldots, g_s$ of periods $n_1, \ldots, n_s$, respectively. Any element $g_1^{h_1} \ldots g_s^{h_s}$ of G is of period δ if and only if δ is the least positive value of x for which $xh_1, \ldots, xh_s$ are multiples of $n_1, \ldots, n_s$, respectively. The number of elements of period δ of G is thus the number of sets of positive integers $h_1, \ldots, h_s$ ($h_1 \leqq n_1, \ldots, h_s \leqq n_s$) such that δ is the least value of x for which $xh_1, \ldots, xh_s$ are divisible by $n_1, \ldots, n_s$, respectively. The number of sets is shown to be

$$\psi(\delta; n_1, \ldots, n_s) = \prod_{j=1}^{s} \delta_j \prod_{i=1}^{r} (1-1/q_i^{l_i}),$$

where δ_j is the g. c. d. of δ and n_j; $q_1, \ldots, q_r$ are the distinct prime factors of δ; while l_i is the number of those integers $n_1, \ldots, n_s$ which contain q_i at least as often as δ contains it. If δ and δ' are relatively prime,

$$\psi(\delta; n_1, \ldots, n_s)\psi(\delta'; n_1, \ldots, n_s) = \psi(\delta\delta'; n_1 \ldots, n_s).$$

[215a]Sitzungsber. Akad. Wiss. Wien (Math.), 103, IIa, 1894, 115.
[216]Monatshefte Math. Phys., 7, 1896, 227–233. For his ϕ we write ψ, as did Carmichael.[11]

If d ranges over all divisors of the product $n_1 \ldots n_s$,

$$\sum_d \psi(\delta; n_1, \ldots, n_s) = n_1 n_2 \ldots n_s.$$

In case δ divides each $n_i (i=1, \ldots, s)$, ψ becomes Jordan's $J_s(\delta)$.

As a generalization (pp. 237-9) consider sets of positive integers $a_1, \ldots, a_s$, where $a_j = 1, 2, \ldots, \gamma_j$ for $j = 1, 2, \ldots, s$. Counting the sets not of the form

$$n_i^{(1)} a_1, \ n_i^{(2)} a_2, \ldots, n_i^{(s)} a_s \qquad (i=1, \ldots, r),$$

we get the number

$$\prod_{j=1}^s \gamma_j - \sum_i \prod_{j=1}^s \left[\frac{\gamma_j}{n_i^{(j)}} \right] + \sum_{i, i'} \prod_{j=1}^s \left[\frac{\gamma_j}{(n_i^{(j)}, \ n_{i'}^{(j)})} \right] - \ldots$$

where $(n_1, n_2, \ldots)$ is the l. c. m. of $n_1, n_2, \ldots$. In particular, take

$$n_i^{(1)} = \ldots = n_i^{(s)} = n_i \qquad (i=1, \ldots, r),$$

where $n_1, \ldots, n_r$ are relatively prime in pairs, and let N be a positive multiple of $n_1, \ldots, n_r$ such that

$$N < m_j, \qquad \frac{N}{\gamma} \geq m_j > \frac{N}{1+\gamma_j} \qquad (j=1, \ldots, s).$$

Then the above expression equals

$$J_s'(N; m_1, \ldots, m_s) = \prod_{j=1}^s \left[\frac{N}{m_j} \right] - \sum_i \prod_{j=1}^s \left[\frac{N}{m_j n_i} \right] + \sum_{i, i'} \prod_{j=1}^s \left[\frac{N}{m_j n_i n_{i'}} \right] - \ldots,$$

which determines the number of sets

$$a_1, \ldots, a_s \qquad \left(a_j = 1, 2, \ldots, \left[\frac{N}{m_j} \right]; j = 1, \ldots, s \right)$$

whose g. c. d. is divisible by no one of $n_1, n_2, \ldots, n_s$. By inversion,

$$\sum J_s' \left(\frac{N}{d}; m_1, \ldots, m_s \right) = \prod_{j=1}^s \left[\frac{N}{m_j} \right],$$

where d ranges over the divisors of N which are products of powers of $n_1, \ldots, n_r$. When $n_1, \ldots, n_s$ are the distinct prime factors of $N, J_s'(N; m_1, \ldots, m_s)$ becomes the function $J_s(N; m_1, \ldots, m_s)$ of von Sterneck.[215] As in the case of the latter function, we have

$$J_s'(N; m_1, \ldots, m_s) = \sum J_1'(\lambda_1; m_1) \ldots J_1'(\lambda_s; m_s),$$

the λ's ranging over all sets whose l. c. m. is N.

L. Carlini[217] proved that if a ranges over the integers for which $[2n/a] = 2k+1$, then

$$\sum J_k(a) = s_{2n}^{(k)} - 2s_n^{(k)}, \qquad s_m^{(k)} \equiv 1^k + \ldots + m^k.$$

For $k=1$, this becomes $\sum \phi(a) = n^2$ [E. Cesàro, p. 144 of this History].

D. N. Lehmer[218] called $J_m(n)$ the m-fold totient of n or multiple totient of n of multiplicity m. He proved that, if $k = p_1^{a_1} \ldots p_r^{a_r}$,

$$J_m(k^n) = k^{m(n-1)} J_m(k), \qquad J_m(ky) = J_m(y) \prod_{i=1}^r \left\{ p_i^{ma_i} - p_i^{m(a_i-1)} \lambda(y, p_i) \right\},$$

where $\lambda(y, p_i) = 0$ or 1 according as p_i is or is not a divisor of y. In the

[217]Periodico di Mat., 12, 1897, 137-9.

[218]Amer. Jour. Math., 22, 1900, 293-335.

second formula the product equals the similar function of y' if y and y' are congruent modulo $p_1 p_2 \ldots p_r$. Consider the function

$$\Phi_m(x, n, k) = \sum_{i=1}^{[x/k]} J_m(i^n k^n),$$

where m, n, k are positive integers and x is a positive number. Then if $S(x, k)$ denotes $1^k + 2^k + \ldots + [x]^k$, it is proved that

$$\sum_{j=1}^{[x]} j^{m(n-1)} \Phi_m\left(\frac{x}{j}, n, 1\right) = S(x, mn),$$

which for $m = n = 1$ becomes Sylvester's[55] formula. By inversion,

$$\Phi_m(x, n, 1) = \sum_{i=1}^{[x]} \mu(i) i^{m(n-1)} S\left(\frac{x}{i}, mn\right),$$

where $\mu(i)$ is Merten's function. For k as above and $k' = k/p_r^{a_r}$,

$$\Phi_m(x, n, k) = p_r^{m(a_r n-1)} \left\{ (p_r - 1) \Phi_m\left(\frac{x}{p_r^{a_r}}, n, k'\right) + \Phi_m\left(\frac{x}{p_r^{a_r}}, n, p_r k'\right) \right\}$$

$$= p_r^{m(a_r n-1)}(p_r - 1) \sum_{j=0}^{l} p_r^{m(n-1)j} \Phi_m\left(\frac{x}{p_r^{a_r+j}}, n, k'\right),$$

where l is the least value of j for which $[x/p_r^{a_r+j}] = 0$. Hence $\Phi_m(x, n, k)$ can be expressed in terms of functions $\Phi_m(y, n, 1)$. True relations are derived from the last four equations by replacing n by $1 - n$ and $\Phi_m(x, 1 - n, k)$ by

$$\Omega_m(x, n, k) = \sum_{i=1}^{[x/k]} J_m(ik) \cdot (ik)^{-nm}.$$

Proof is given of the asymptotic formula

$$\Phi_m(x, n, k) = \frac{x^{mn+1}}{mn+1} \frac{P_{m,k}}{D_{m+1}} + \epsilon, \qquad |\epsilon| \leq A x^{mn} \log x,$$

where A is finite and independent of x, m, n, while

$$D_{m+1} = \sum_{j=1}^{\infty} \frac{1}{j^{m+1}}, \qquad P_{m,k} = \prod_{i=1}^{r} \frac{p_i - 1}{p_i^{a_i-1}(p_i^{m+1} - 1)}, \qquad P_{m,1} = 1.$$

For $m = n = k = 1$, this result becomes that of Mertens[36] (and Dirichlet[21]). The asymptotic expressions found for $\Omega_m(x, n, k)$ are different for the cases $n = 1$, $n = 2$, $n > 2$.

A set of m integers (not necessarily positive) having no common divisor > 1 is said to define a totient point. Let one coordinate, as x_m, have a fixed integral value $\neq 0$, while $x_1, \ldots, x_{m-1}$ take integral values such that $[x_1/x_m], \ldots, [x_{m-1}/x_m]$ have prescribed values; we obtain a compartment in space of m dimensions which contains $J_{m-1}(x_m)$ totient points. For example, if $m = 3$, $x_3 = 6$, and the two prescribed values are zero, there are 24 totient points $(x_1, x_2, 6)$ for which $0 \leq x_1 < 6$, $0 \leq x_2 < 6$, while x_1 and x_2 have no common divisor dividing 6. For $x_1 = 1$ or 5, x_2 has 6 values; for $x_1 = 2$ or 4, $x_2 = 1$, 3 or 5; for $x_1 = 3$, $x_2 = 1$, 2, 5; for $x_1 = 0$, $x_2 = 0$, 1, 5.

Given a closed curve $r = f(\theta)$, decomposable into a finite number of segments for each of which $f(\theta)$ is a single-valued, continuous function. Let

K be the area of the region bounded by this curve, and N the number of points (x, y) within it or on its boundary such that x is a multiple of k and is prime to y. Then

$$\lim_{k=\infty} \frac{N}{K} = \frac{6}{\pi^2} P_{1, k},$$

where K increases by uniform stretching of the figure from the origin. In particular, consider the number N of irreducible fractions $x/y \leqq 1$ whose denominators are $\leqq n$. Since $x \leqq y$, the area K of the triangular region is $n^2/2$. Hence $N = (n^2/2)(6/\pi^2)$, approximately (Sylvester[55]). Again, the number of irreducible fractions whose numerators lie between l and $l+m$, and denominators between l' and $l'+m'$, is $6mm'/\pi^2$, approximately.

There is a similar theorem in which the points are such that y is divisible by k', while three new constants obey conditions of relative primality to each other or to x, y, k, k'.

Extensions are stated for m-dimensional space.

E. Cahen[219] called $J_k(n)$ the indicateur of kth order of n.

G. A. Miller[220] evaluated $J_k(m)$ by noting that it is the number of operators of period m in the abelian group with k independent generators of period m.

G. A. Miller[221] proved (10) and (11) by using the same abelian group.

E. Busche[222] indicated a proof of (10) and (12) by an extension to space of $k+1$ dimensions of Kronecker's[223] plane, in which every point whose rectangular coordinates x, y are integers is associated with the g. c. d. of x, y.

A. P. Minin[224] proved (14) and some results due to Gegenbauer.[204]

R. D. Carmichael[225] gave a simple proof of Zsigmondy's[216] formula for ψ.

G. Métrod[226] stated that the number of incongruent sets of solutions of $xy' - x'y \equiv a \pmod{m}$ is $\Sigma dm J_2(m/d)$, where d ranges over the common divisors of m and a. When a takes its m values, the total number of sets of solutions is

$$m^4 = \sum_{D:m} \phi\left(\frac{m}{D}\right) \sum_{d:D} dm J_2\left(\frac{m}{d}\right).$$

It is asked if like relations hold for J_k, $k > 2$.

Cordone[91] and Sanderson[115] (of Ch. VIII) used Jordan's function in giving a generalization of Fermat's theorem to a double modulus.

FAREY SERIES.

Flitcon[248] gave the number of irreducible fractions < 1 with each denominator < 100, stating in effect the value of Euler's $\phi(n)$ when n is a product of four or fewer primes.

[219] Théorie des nombres, 1900, p. 36; I, 1914, 396–400.
[220] Amer. Math. Monthly, 11, 1904, 129–130.
[221] Amer. Jour. Math., 27, 1905, 321–2.
[222] Math. Annalen, 60, 1905, 292.
[223] Vorlesungen über Zahlentheorie, 1901, I, p. 242.
[224] Matem. Sbornik (Moscow Math. Soc.), 27, 1910, 340–5.
[225] Quart. Jour. Math., 44, 1913, 94–104.
[226] L'intermédiaire des math., 20, 1913, 148. Proof, Sphinx-Oedipe, 9, 1914, 4.
[248] Ladies' Diary, 1751. Reply to Question 281, 1747–8. T. Leybourn's Math. Quest. proposed in Ladies' Diary, 1, 1817, 397–400.

C. Haros[249] proved the results rediscovered by Farey[250] and Cauchy.[252]

J. Farey[250] stated that if all the proper vulgar fractions in their lowest terms, having both numerator and denominator not exceeding a given number n, be arranged in order of magnitude, each fraction equals a fraction whose numerator and denominator equal respectively the sum of the numerators and sum of the denominators of the two fractions adjacent to it in the series. Thus, for $n=5$, the series is

$$\tfrac{1}{5}, \tfrac{1}{4}, \tfrac{1}{3}, \tfrac{2}{5}, \tfrac{1}{2}, \tfrac{3}{5}, \tfrac{2}{3}, \tfrac{3}{4}, \tfrac{4}{5},$$

and

$$\frac{1}{4}=\frac{1+1}{5+3}, \qquad \frac{2}{5}=\frac{1+1}{3+2}.$$

Henry Goodwyn mentioned this property on page 5 of the introduction to his "tabular series of decimal quotients" of 1818, published in 1816 for private circulation (see Goodwyn,[21, 22] Ch. VI), and is apparently to be credited with the theorem. It was ascribed to Goodwyn by C. W. Merrifield.[251]

A. L. Cauchy[252] proved that, if a/b, a'/b', a''/b'' are any three consecutive fractions of a Farey series, b and b' are relatively prime and $a'b-ab'=1$ (so that $a'/b'-a/b=1/bb'$). Similarly, $a''b'-a'b''=1$, so that $a+a''$: $b+b''$ $=a'$: b', as stated by Farey.

Stouvenel[253] proved that, in a Farey series of order n, if two fractions a/b and c/b are complementary (i. e., have the sum unity), the same is true of the fraction preceding a/b and that following c/b. The two fractions adjacent to $1/2$ are complementary and their common denominator is the greatest odd integer $\leqq n$. Hence $1/2$ is the middle term of the series and two fractions equidistant from $1/2$ are complementary. To find the third of three consecutive fractions a/b, a'/b', x/y, we have $a+x=a'z$, $b+y=b'z$ (Farey), and we easily see that z is the greatest integer $\leqq(n+b)/b'$.

M. A. Stern[254] studied the sets m, n, and m, $m+n$, n, and m, $2m+n$, $m+n$, $m+2n$, n, etc., obtained by interpolating the sum of consecutive terms. G. Eisenstein[254a] briefly considered such sets.

*A. Brocot[255] considered the sets obtained by mediation [Farey] from $0/1$, $1/0$:

$$\tfrac{0}{1}, \tfrac{1}{1}, \tfrac{1}{0}; \qquad \tfrac{0}{1}, \tfrac{1}{2}, \tfrac{1}{1}, \tfrac{2}{1}, \tfrac{1}{0}; \dots$$

Herzer[256] and Hrabak[257] gave tables with the limits 57 and 50.

G. H. Halphen[258] considered a series of irreducible fractions, arranged in order of magnitude, chosen according to a law such that if any fraction f is excluded then also every fraction is excluded if its two terms are at least

[249]Jour. de l'école polyt., cah. 11, t. 4, 1802, 364-8.
[250]Philos. Mag. and Journal, London, 47, 1816, 385-6; [48, 1816, 204]; Bull. Sc. Soc. Philomatique de Paris, (3), 3, 1816, 112.
[251]Math. Quest. Educat. Times, 9, 1868, 92-5.
[252]Bull. Sc. Soc. Philomatique de Paris, (3), 3, 1816, 133-5. Reproduced in Exercices de Math., 1, 1826, 114-6; Oeuvres, (2), 6, 1887, 146-8.
[253]Jour. de mathématiques, 5, 1840, 265-275.
[254]Jour. für Math., 55, 1858, 193-220.   [254a]Bericht Ak. Wiss. Berlin, 1850, 41-42.
[255]Calcul des rouages par approximation, Paris, 1862. Lucas.[173]
[256]Tabellen, Basle, 1864.   [257]Tabellen-Werk, Leipzig, 1876.
[258]Bull. Soc. Math. France, 5, 1876-7, 170-5.

equal to the corresponding terms of f. Such a series has the properties noted by Farey and Cauchy for Farey series.

E. Lucas[259] considered series 1, 1 and 1, 2, 1, etc., formed as by Stern. For the nth series it is stated that the number of terms is $2^{n-1}+1$, their sum is $3^{n-1}+1$, the greatest two terms (of rank $2^{n-2}+1\pm2^{n-1}$) are

$$\frac{(1+\sqrt{5})^{n+1}-(1-\sqrt{5})^{n+1}}{2^{n+1}\sqrt{5}}.$$

Changing n to p, we obtain the value of certain other terms.

J. W. L. Glaisher[260] gave some of the above facts on the history of Farey series. Glaisher[261] treated the history more fully and proved (p. 328) that the properties noted by Farey and Cauchy hold also for the series of irreducible fractions of numerators $\leqq m$ and denominators $\leqq n$.

Edward Sang[262] proved that any fraction between A/a and C/γ is of the form $(pA+qC)/(pa+q\gamma)$, where p and q are integers, and is irreducible if p, q are relatively prime.

A. Minine[263] considered the number $S(a, N)$ of irreducible fractions a/b such that $b+aa\leqq N$. Let $\phi(b)_p$ denote the number of integers $\leqq p$ which are prime to b. Then, for $a>0$,

$$S(a, N) = \sum_{b=1}^{N-a} \phi(b)_p, \qquad p = \left[\frac{N-b}{a}\right],$$

since for each denominator b there are $\phi(b)_p$ integers prime to b for which $b+aa\leqq N$ and hence that number of fractions.

A. F. Pullich[264] proved Farey's theorem by induction, using continued fractions.

G. Airy[265] gave the 3043 irreducible fractions with numerator and denominator $\leqq 100$.

J. J. Sylvester[266] showed how to deduce the number of fractions in a Farey series by means of a functional equation.

Sylvester,[55, 56] Cesàro,[65] Vahlen,[83] Axer,[115] and Lehmer[218] investigated the number of fractions in a Farey series.

Sylvester[266a] discussed the fractions x/y for which $x<n$, $y<n$, $x+y\leqq n$.

M. d'Ocagne[267] prolonged Farey's series by adding 1/1 in the pth place, where $p=\phi(1)+\ldots+\phi(n)$. From the first p terms we obtain the next p by adding unity, then the next p by adding unity, etc. Consider a series $S(a, N)$ of irreducible fractions a_i/b_i in order of magnitude such that $b_i+aa_i\leqq N$, where a is any fixed integer called the characteristic. All the series $S(a, N)$ with a given base N may be derived from Farey's series

[259]Bull. Soc. Math. France, 6, 1877–8, 118–9.   [260]Proc. Cambr. Phil. Soc., 3, 1878, 194.

[261]London Ed. Dub. Phil. Mag., (5), 7, 1879, 321–336.

[262]Trans. Roy. Soc. Edinburgh, 28, 1879, 287.

[263]Jour. de math. élém. et spéc., 1880, 278.   Math. Soc. Moscow, 1880.

[264]Mathesis, 1, 1881, 161–3.   [265]Trans. Inst. Civil Engineers; cf. Phil. Mag., 1881, 175.

[266]Johns Hopkins Univ. Circulars, 2, 1883, 44–5, 143; Coll. Math. Papers, 3, 672–6, 687–8.

[266a]Amer. Jour. Math., 5, 1882, 303–7, 327–330; Coll. Math. Papers, IV, 55–9, 78–81.

[267]Annales Soc. Sc. Bruxelles, 10, 1885–6, II, 90.   Extract in Bull. Soc. Math. France, 14, 1885–6, 93–7.

$S(0, N)$ by use of

$$a_i(a, N) = a_i(0, N), \qquad b_i(a, N) = b_i(0, N) - aa_i(0, N).$$

Thus $a_i b_{i-1} - a_{i-1} b_i = 1$, so that the area of $OA_i A_{i-1}$ is $1/2$ if the point A_i has the coordinates a_i, b_i. All points representing terms of the same rank in all the series of the same base lie at equally spaced intervals on a parallel to the x-axis, and the distance between adjacent points is the number of units between this parallel and the x-axis.

A. Hurwitz[268] applied Farey series to the approximation of numbers by rational fractions and to the reduction of binary quadratic forms.

J. Hermes[269] designated as numbers of Farey the numbers $\tau_1 = 1$, $\tau_2 = 2$, $\tau_3 = \tau_4 = 3$, $\tau_5 = 4$, $\tau_6 = \tau_7 = 5$, $\tau_8 = 4, \ldots$ with the recursion formula

$$\tau_n = \tau_{n-2^\nu} + \tau_{2^{\nu+1}-n+1}, \qquad 2^\nu < n \leqq 2^{\nu+1},$$

and connected with the representation of numbers to base 2. The ratios of the τ's give the Farey fractions.

K. Th. Vahlen[269a] noted that the formation of the convergents to a fraction w by Farey's series coincides with the development of w into a continued fraction whose numerators are ± 1, and made an application to the composition of linear fractional substitutions.

H. Made[270] applied Hurwitz's method to numbers $a + bi$.

E. Busche[271] applied geometrically the series of irreducible fractions of denominators $\leqq a$ and numerators $\leqq b$, and noted that the properties of Farey series ($a = b$) hold [Glaisher[261]].

W. Sierpinski[272] used consecutive fractions of Farey series of order m to show that, if x is irrational,

$$\lim_{x = \infty} \left\{ \sum_{k=1}^{n} [kx] - \frac{xn(n+1)}{2} + \frac{n}{2} \right\} = 0.$$

Expositions of the theory of Farey series were given by E. Lucas,[273] E. Cahen,[274] Bachmann.[275]

An anonymous writer,[276] starting with the irreducible fractions < 1, arranged in order of magnitude, with the denominators $\leqq 10$, inserted the fractions with denominator 11 by listing the pairs of fractions $0/1$, $1/10$; $1/6$, $1/5$; $1/4$, $2/7$; ..., the sum of whose denominators is 11, and noting that between the two of each pair lies a fraction with denominator 11 and numerator equal the sum of their numerators.

[268]Math. Annalen, 44, 1894, 417–436; 39, 1891, 279; 45, 1894, 85; Math. Papers of the Chicago Congress, 1896, 125. Cf. F. Klein, Ausgewählte Kapitel der Zahlentheorie, I, 1896, 196–210. Cf. G. Humbert, Jour. de Math., (7), 2, 1916, 116–7.
[269]Math. Annalen, 45, 1894, 371. Cf. L. von Schrutka, 71, 1912, 574, 583.
[269a]Jour. für Math., 115, 1895, 221–233.
[270]Ueber Fareysche Doppelreihen, Diss. Giessen, Darmstadt, 1903.
[271]Math. Annalen, 60, 1905, 288.
[272]Bull. Inter. Acad. Sc. Cracovie, 1909, II, 725–7.
[273]Théorie des nombres, 1891, 467–475, 508–9.
[274]Éléments de la théorie des nombres, 1900, 331–5.
[275]Niedere Zahlentheorie, 1, 1902, 121–150; 2, 1910, 55–96.
[276]Zeitschrift Math. Naturw. Unterricht, 45, 1914, 559–562.

CHAPTER VI.

PERIODIC DECIMAL FRACTIONS; PERIODIC FRACTIONS; FACTORS OF $10^n \pm 1$.

Ibn-el-Banna[1] (Albanna) in the thirteenth century factored $10^n - 1$ for small values of n. The Arab Sibt el-Mâridini[1a] in the fifteenth century noted that in the sexagesimal division of $47° 50'$ by $1° 25'$ the quotient has a period of eight terms.

G. W. Leibniz[2] in 1677 noted that $1/n$ gives rise to a purely periodic fraction to any base b, later adding the correction that n and b must be relatively prime. The length of the period of the decimal fraction for $1/n$, where n is prime to 10, is a divisor of $n-1$ [erroneous for $n=21$; cf. Wallis[3]].

John Wallis[3] noted that, if N has a prime factor other than 2 and 5, the reduced fraction M/N equals an unending decimal fraction with a repetend of at most $N-1$ digits. If N is not divisible by 2 or 5, the period has two digits if N divides 99, but not 9; three digits if N divides 999, but not 99. The period of $1/21$ has six digits and 6 is not a divisor of $21-1$. The length of the period for the reciprocal of a product equals the l. c. m. of the lengths of the periods of the reciprocals of the factors [cf. Bernoulli[8]]. Similar results hold for base 60 in place of 10.

J. H. Lambert[4] noted that all periodic decimal fractions arise from rational fractions; if the period p has n digits and is preceded by a decimal with m digits, we have

$$a + \frac{b}{10^m} + r, \qquad r = \frac{p}{10^m 10^n} + \frac{p}{10^m 10^{2n}} + \cdots = \frac{p}{10^m (10^n - 1)}.$$

John Robertson[5] noted that a pure periodic decimal with a period P of k digits equals $P/9 \ldots 9$, where there are k digits 9.

J. H. Lambert[6] concluded from Fermat's theorem that, if a is a prime other than 2 and 5, the number of terms in the period of $1/a$ is a divisor of $a-1$. If g is odd and $1/g$ has a period of $g-1$ terms, then g is a prime. If $1/g$ has a period of m terms, but $g-1$ is not divisible by m, g is composite. Let $1/a$ have a period of $2m$ terms; if a is prime, $k = 10^m + 1$ is divisible by a; if a is composite, k and a have a common factor; if k is divisible by a and if m is prime, each factor other than $2^p 5^q$ of a is of period $2m$.

Let a be a composite number not divisible by 2, 3 or 5. If $1/a$ has a period of m terms, where m is a prime, each factor of a produces a period

---

[1]Cf. E. Lucas, Arithmétique amusante, 1895, 63–9; Brocard.[103]
[1a]Carra de Vaux, Bibliotheca Math., (2), 13, 1899, 33–4.
[2]Manuscript in Bibliothek Hannover, vol. III, 24; XII, 2, Blatt 4; also, III, 25, Blatt 1, seq., 10, Jan., 1687. Cf. D. Mahnke, Bibliotheca Math., (3), 13, 1912–3, 45–48.
[3]Treatise of Algebra both historical & practical, London, 1685, ch. 89, 326–8 (in manuscript, 1676).
[4]Acta Helvetica, 3, 1758, 128–132.
[5]Phil. Trans., London, 58, 1768, 207–213.
[6]Nova Acta Eruditorum, Lipsiæ, 1769, 107–128.

of m terms. If $1/a$ has a period of mn terms, where m and n are primes, while no factor has such a period, one factor of a divides $10^m - 1$ and another divides $10^n - 1$. If $1/a$ has a period of mnp terms, where m, n, p are primes, but no factor has such a period, any factor of a divides $10^m - 1, \ldots$, or $10^{np} - 1$. These theorems aid in factoring a.

L. Euler[7] gave numerical examples of the conversion of ordinary fractions into decimal fractions and the converse problem.

Euler[7a] noted that if $2p+1$ is a prime $40n \pm 1$, ± 3, ± 9, ± 13, it divides $10^p - 1$; if $2p+1$ is a prime $40n \pm 7$, ± 11, ± 17, ± 19, it divides $10^p + 1$.

Jean Bernoulli[8] gave a résumé of the work by Wallis,[3] Robertson,[5] Lambert[6] and Euler,[7] and gave a table showing the full period for $1/D$ for each odd prime $D < 200$, and a like table when D is a product of two equal or distinct primes < 25. When the two primes are distinct, the table confirms Wallis' assertion that the length of the period for $1/D$ is the l. c. m. of the lengths of the periods for the reciprocals of the factors. But for $1/D^2$, where D is a prime > 3, the length of the period equals D times that for $1/D$. If the period for $1/D$, where D is a prime, has $D-1$ digits, the period for m/D has the same digits permuted cyclically to begin with m. He gave (p. 310) a device communicated to him by Lambert: to find the period for $1/D$, where $D = 181$, we find the remainder 7 after obtaining the part p composed of the first 15 digits of the period; multiply $1/D = p + 7/D$ by 7; thus the next 15 digits of the period are given by $7p$; since $7^3 = D + 162$, the third set of 15 digits is found by adding unity to 7^2p, etc.; since 7 belongs to the exponent 12 modulo D, the period for $1/D$ contains $15 \cdot 12$ digits.

Jean Bernoulli[9] made use of various theorems due to Euler which give the possible linear forms of the divisors of $10^k \pm 1$, and obtained factors of $(10^k - 1)/9$ when $k \leqq 30$, except for $k = 11$, 17, 19, 23, 29, with doubt as to the primality of the largest factor when $k = 13$, 15 or $\geqq 19$. He stated (p. 325) erroneously[10] that $(10^{11}+1)/11 \cdot 23$ has no factor < 3000. Also,

$$10^{15}+1 = 7 \cdot 11 \cdot 13 \cdot 211 \cdot 9091 \cdot 52081.$$

He gave part of the periods for the reciprocals of various primes $\leqq 601$.

L. Euler[11] wrote to Bernoulli concerning the latter's[9] paper and stated criteria for the divisibility of $10^p \pm 1$ by a prime $2p+1 = 4n \pm 1$. If both 2 and 5 or neither occur among the divisors of n, $n \mp 2$, $n \mp 6$, then $10^p - 1$ is divisible by $2p+1$. But if only one of 2 and 5 occurs, then $10^p + 1$ is divisible by $2p+1$ [cf. Genocchi[39]].

Henry Clarke[12] discussed the conversion of ordinary fractions into decimals without dealing with theoretical principles.

[7]Algebra, I, Ch. 12, 1770; French transl., 1774.
[7a]Opusc. anal., 1, 1773, 242; Comm. Arith. Coll., 2, p. 10, p. 25.
[8]Nouv. mém. acad. roy. Berlin, année 1771 (1773), 273–317.
[9]Ibid., 318–337.
[10]P. Seelhoff, Zeitschrift Math. Phys., 31, 1886, 63. Reprinted, Sphinx-Oedipe, 5, 1910, 77–8.
[11]Nouv. mém. acad. roy. Berlin, année 1772 (1774), Histoire, pp. 35–36; Comm. Arith., 1, 584.
[12]The rationale of circulating numbers, London, 1777, 1794.

Anton Felkel[13] showed how to convert directly a periodic fraction written to one base into one to another base. He gave all primes <1000 which can ·divide a period with a prime number of digits <30, as $29m+1$ $=59, 233, \ldots$ Oberreit[14] extended Bernoulli's[9] table of factors of $10^k \doteq 1$.

C. F. Gauss[15] gave a table showing the period of the decimal fraction for k/p^n, $p^n<467$, p a prime, and the period for $1/p^n$, $467 \leqq p^n \leqq 997$.

W. F. Wucherer[16] gave five places of the decimal fraction for n/d, $d<1000$, $n<d$ for $d<50$, $n \leqq 10$ for $d \geqq 50$.

Schröter published at Helmstadt in 1799 a table for converting ordinary fractions into decimal fractions.

C. F. Gauss[17] proved that, if a is not divisible by the prime p ($p \neq 2$, 5), the length of the period for a/p^n is the exponent e to which 10 belongs modulo p^n. If we set $\phi(p^n)=ef$ and choose a primitive root r of p^n such that the index of 10 is f, we can easily deduce from the periods for k/p^n, where $k=1$, $r, \ldots, r^{f-1}$, the period for m/p^n, where m is any integer not divisible by p. For, if i be the index of m to the base r, and if $i=af+\beta$, where $0 \leqq \beta<f$, we obtain the period for m/p^n from that for r^β/p^n by carrying the first a digits to the end. He computed[15] the necessary periods for each $p^n<1000$, but published here the table only to 100. By using partial fractions, we may employ the table to obtain the period for a/b, where b is a product of powers of primes within the limits of the table.

H. Goodwyn[18] noted that, if $a<17$, the period for $a/17$ is derived from the period for $1/17$ by a cyclic permutation of the digits. Thus we may print in a double line the periods for $1/17, \ldots, 16/17$ by showing the period for $1/17$ and, above each digit d of the latter, showing the value of a such that the period for $a/17$ begins with the digit d, while the rest of the period is to be read cyclically from that for $1/17$.

Goodwyn[19] noted that when $1/p$ is converted into a decimal fraction, p being prime, the sum of corresponding quotients in the two half periods is 9, and that for remainders is p, if $p \geqq 7$.

J. C. Burckhardt[20] gave the length of the period for $1/p$ for each prime $p \leqq 2543$ and for 22 higher primes. It follows that 10 is a primitive root of 148 of the 365 primes p, $5<p<2500$.

[13]Abhand. Böhmischen Gesell. Wiss., Prag, 1, 1785, 135–174.

[14]J. H. Lambert's Deutscher Gelehrter Briefwechsel, pub. by J. Bernoulli, Leipzig, vol. 5, 1787, 480–1. The part (464–479) relating to periodic decimals is mainly from Bernoulli's[9] paper.

[15]Posthumous manuscript, dated Oct., 1795; Werke, 2, 1863, 412–434.

[16]Beyträge zum allgemeinern Gebrauch der Decimal Brüche...., Carlsruhe, 1796.

[17]Disq. Arith., 1801, Arts. 312–8. A part was reproduced by Wertheim, Elemente der Zahlentheorie, 1887, 153–6.

[18]Jour. Nat. Phil. Chem. Arts (ed., Nicholson), London, 4, 1801, 402–3.

[19]Ibid., new series, 1, 1802, 314–6. Cf. R. Law, Ladies' Diary, 1824, 44–45, Quest. 1418.

[20]Tables des diviseurs pour tous les nombres du premier million, Paris, 1817, p. 114. For errata see Shanks,[61] Kessler,[92] Cunningham,[124] and Gérardin.[131]

H. Goodwyn[21] gave for each integer $d \leq 100$ a table of the periods for n/d, for the various integers $n < d$ and prime to d. Also, a table giving the first eight digits of the decimal equivalent to every irreducible vulgar fraction $< 1/2$, whose numerator and denominator are both ≤ 100, arranged in order of magnitude, up to $1/2$.

Goodwyn[22, 23] was without doubt the author of two tables, which refer to the preceding "short specimen" by the same author. The first gives the first eight digits of the decimal equivalent to every irreducible vulgar fraction, whose numerator and denominator are both ≤ 1000, from $1/1000$ to $99/991$ arranged in order of magnitude. In the second volume, the "table of circles" occupies 107 pages and contains all the periods (circles) of every denominator prime to 10 up to 1024; there is added a two-page table showing the quotient of each number ≤ 1024 by its largest factor $2^a 5^b$.

For example, the entry in the "tabular series" under $\frac{57}{656}$ is .08689024. The entry in the two-page table under 656 is 41. Of the various entries under 41 in the "table of circles," the one containing the digits 9024 gives the complete period $\overset{\circ}{9}024\overset{\circ}{3}$. Hence $\frac{57}{656} = .0868\overset{\circ}{9}024\overset{\circ}{3}$.

Glaisher[78] gave a detailed account of Goodwyn's tables and checks on them. They are described in the British Assoc. Report, 1873, pp. 31–34, along with tables showing seven figures of the reciprocals of numbers < 100000.

F. T. Poselger[24] considered the quotients $0, a, b, \ldots$ and the remainders $1, a, \beta, \ldots$ obtained by dividing $1, A, A^2, \ldots$ by the prime p; thus

$$\frac{A}{p} = a + \frac{a}{p}, \qquad \frac{A^2}{p} = aA + b + \frac{\beta}{p}, \ldots$$

Adding, we see that the sum $1 + a + \beta + \ldots$ of the remainders of the period is a multiple mp of p; also, $m(A-1) = a + b + \ldots$. Set

$$M = k + \ldots + bA^{t-2} + aA^{t-1},$$

where A belongs to the exponent t modulo p. Then

$$\frac{A^{nt}}{p} = \frac{1}{p} + MS, \qquad S = 1 + A^t + \ldots + A^{(n-1)t}.$$

[21]The first centenary of a series of concise and useful tables of all the complete decimal quotients which can arise from dividing a unit, or any whole number less than each divisor, by all integers from 1 to 1024. To which is now added a tabular series of complete decimal quotients for all the proper vulgar fractions of which, when in their lowest terms, neither the numerator nor the denominator is greater than 100; with the equivalent vulgar fractions prefixed. By Henry Goodwyn, London, 1818, pp. xiv+18; vii+30. The first part was printed in 1816 for private circulation and cited by J. Farey in Philos. Mag. and Journal, London, 47, 1816, 385.

[22]A tabular series of decimal quotients for all the proper vulgar fractions of which, when in their lowest terms, neither the numerator nor the denominator is greater than 1000, London, 1823, pp. v+153.

[23]A table of the circles arising from the division of a unit, or any other whole number, by all the integers from 1 to 1024; being all the pure decimal quotients that can arise from this source, London, 1823, pp. v+118.

[24]Abhand. Ak. Wiss. Berlin (Math.), 1827, 21–36.

If M is divisible by p, we may take $n=1$ and conclude that A^t/p^2 differs from $1/p^2$ by an integer. If M is not divisible by p, S must be, so that n is divisible by p and the length of the period is pt. In general, for the denominator p^λ, we have $n=1$ if M is divisible by $p^{\lambda-1}$, but in the contrary case n is a multiple of $p^{\lambda-1}$. If the period for a prime p has an even number of digits, the sum of corresponding quotients in the two half periods is p.

An anonymous writer[25] noted that, if we add the digits of the period of a circulating decimal, then add the digits of the new sum, etc., we finally get 9. From a number subtract that obtained by reversing its digits; add the digits of the difference; repeat for the sum, etc.; we get 9.

Bredow[26] gave the periods for a/p, where p is a prime or power of a prime between 100 and 200. He gave certain factors of 10^n-1 for $n=6$–10, 12–16, 18, 21, 22, 28, 33, 35, 41, 44, 46, 58, 60, 96.

E. Midy[27] noted that, if a^n, $a^{n_1}, \ldots$ are the least powers of a which, diminished by unity, give remainders divisible by q^h, $q_1{}^{h_1}, \ldots$, respectively (q, $q_1, \ldots$ being distinct primes), and if the quotients are not divisible by q, $q_1, \ldots$, respectively, and if t is the l. c. m. of n, $n_1, \ldots$, then a belongs to the exponent t modulo $p=q^h q_1{}^{h_1} \ldots$, and a^t-1 is divisible by q only h times.

Let the period of the pure decimal fraction for a/b have $2n$ digits. If b is prime to 10^n-1, the sum of corresponding digits in the half periods is always 9, and the sum of corresponding remainders is b. Next, let b and 10^n-1 have $d>1$ as their g. c. d. and set $b'=b/d$. Let a_n be the nth remainder in finding the decimal fraction. Then $a+a_n=b'k$, $a_1+a_{n+1}=b'k_1$, etc. The sums $q+q_n$, $q_1+q_{n+1}, \ldots$ of corresponding digits in the half periods equal

$$(10k-k_1)/d, \qquad (10k_1-k_2)/d, \ldots, \qquad (10k_{n-1}-k)/d.$$

Similar results hold when the period of mn digits is divided into n parts of m digits each. For example, in the period

$$0\ 0\ 2\ 4\ 8\ 1\ 3\ 8\ 9\ 5\ 7\ 8\ 1\ 6\ 3\ 7\ 7\ 1\ 7\ 1\ 2\ 1\ 5\ 8\ 8\ 0\ 8\ 9\ 3\ 3$$

for $1/403$, the two halves are not complementary ($10^{15}-1$ being divisible by 31); for $i=1, 2, 3$, the sum of the digits of rank i, $i+3$, $i+6, \ldots, i+27$ is always 45, while the corresponding sums of the remainders are 2015.

N. Druckenmüller[27a] noted that any fraction can be expressed as $a/x+a_1/x^2+\ldots$.

J. Westerberg[28] gave in 1838 factors of $10^n \pm 1$ for $n \leq 15$.

G. R. Perkins[29] considered the remainder r_x when N^x is divided by P, and the quotient q in $Nr_{x-1}=Pq_x+r_x$. If $r_k=P-1$, there are $2k$ terms in the period of remainders, and

$$r_{k+x}+r_x=P, \qquad q_{k+x}+q_x=N-1.$$

[These results relate to $1/P$ written to the base N.]

[25]Polytechnisches Journal (ed., J. G. Dingler), Stuttgart, 34, 1829, 68; extract from Mechanics' Magazine, N. 313, p. 411.

[26]Von den Perioden der Decimalbrüche, Progr., Oels, 1834.

[27]De quelques propriétés des nombres et des fractions décimales périodiques, Nantes, 1836, 21 pp.

[27a]Theorie der Kettenreihen…, Trier, 1837.

[28]See Chapter on Perfect Numbers.[104]

[29]Amer. Jour. Sc. Arts, 40, 1841, 112–7.

E. Catalan[30] converted periodic decimals into ordinary fractions without using infinite progressions. When $1/13$ is converted into a decimal, the period of remainders is 1, 10, 9, 12, 3, 4; repeat the period; starting in the series of 12 terms with any term (as 10), take the fourth term (4) after it, the fourth term (12) after that, etc.; then the sum 26 of the three is a multiple of 13. In general, if D is a prime and $D-1=mn$, the sum of n terms taken m by m in the period for N/D is a multiple of D [cf. Thibault[31]].

If the sum of two terms of the period of remainders for N/D is D, the same is true of the terms following them. Hence the sum of corresponding terms of the two half periods is D. This happens if the number of terms of the period is $\phi(D)$.

Thibault[31] denoted the numbers of digits in the periods for $1/d$ and $1/d'$ by m and m'. If d' is divisible by d, m' is divisible by m. If d and d' have no common prime factor other than 2 or 5, the number of digits in the period for $1/dd'$ is the l. c. m. of m, m'. Hence it suffices to know the length of the period for $1/p^a$, where p is a prime. If $1/p$ has a period of m digits and if $1/p^n$ is the last one of the series $1/p$, $1/p^2, \ldots$ which has a period of m digits, then the period for $1/p^a$ for $a > n$ has mp^{a-n} digits. For $p=3$, we have $n=2$; hence $1/3^r$ for $r \geq 2$ has a period of 3^{r-2} digits. For any prime p for which $7 \leq p \leq 101$, we have $n=1$, so that $1/p^a$ has a period of mp^{a-1} digits. Note that $1/p$ and $1/p^2$ have periods of the same length to base b if and only if $b^{p-1} \equiv 1 \pmod{p^2}$. Proof is given of Catalan's[30] first theorem, which holds only when $10^m \not\equiv 1 \pmod{D}$, i. e., when m is not a multiple of the number of digits in the period. For example, the sum of the kth and $(6+k)$th remainders for $1/13$ is not a multiple of 13.

E. Prouhet[32] proved Thibault's[31] theorem on the period for $1/p^n$. He[32a] noted that multiples of 142857 have the same digits permuted.

P. Lafitte[33] proved Midy's[27] theorem that, if p is a prime not dividing m and if the period for m/p has an even number of digits, the sum of the two halves of the period is $9 \ldots 9$.

J. Sornin[34] investigated the number m of digits in the period for $1/D$, where D is prime to 10. The period is $x = (10^m - 1)/D$. First, let $D = 10k+1$. Then $x = 10y - 1$, where

$$y = \frac{10^{m-1}+k}{D} = 10z+k, \qquad z = \frac{10^{m-2}-k^2}{D}.$$

Finally, we reach $v = \{1 - (-k)^m\}/D$, and x is an integer if and only if v is. Hence if we form the powers of the number k of tens in D, add 1 to the odd powers, but subtract 1 from the even powers of k, the first exponent giving a result divisible by D is the number m of digits in the period.

[30]Nouv. Ann. Math., 1, 1842, 464–5, 467–9.
[31]Ibid., 2, 1843, 80–89.
[32]Ibid., 5, 1846, 661.
[32a]Ibid., 3, 1844, 376; 1851, 147–152.
[33]Ibid., 397–9. Cf. Amer. Math. Monthly, 19, 1912, 130–2.
[34]Ibid., 8, 1849, 50–57.

Next, if $D = 10k - 1$, we have a like rule to be applied only to the $k^m - 1$. If $D = 10k \pm 3$, $1/(3D)$ has a denominator $10l \pm 1$, and the length of its period, found as above, is shown to be not less than that for $1/D$.

Th. Bertram[35] gave certain numbers p for which $1/p$ has a given length k of period for $k \leqq 100$. Cf. Shanks.[62]

J. R. Young[36] took a part of a periodic decimal, as $.1428571\ 428$ for $1/7$, and marked off from the end a certain number (three) of digits. We can find a multiplier (as 6) such that the product, with the proper carrying (here 2) from the part marked off, has all the digits of the abridged number in the same cyclic order, except certain of the leading digits. In the special case the product is $.8571428$.

W. Loof[37] gave the primes p for which the period for $1/p$ has a given number n of digits, $n \leqq 60$, with no entry for $n = 17, 19, 37-40, 47, 49, 57, 59$, and with doubt as to the primality of large numbers entered for various other n's.

E. Desmarest[38] gave the primes $P < 10000$ for which 10 belongs to the exponent $(P-1)/t$ for successive values of t. The table thus gives the length of the period for $1/P$. He stated (pp. 294-5) that if P is a prime < 1000, and if p is the length of the period for A/P, then except for $P = 3$ and $P = 487$ the length of the period for A/P^2 is pP.

A. Genocchi[39] proved Euler's[11] rule by use of the quadratic reciprocity law. Thus 5 is a quadratic residue or non-residue of N according as $N = 5m \pm 1$ or $5m \pm 3$; for $4n + 1 = 5m \pm 1$, n or $n - 2$ is divisible by 5; for $4n - 1 = 5m \pm 1$, n or $n + 2$ is divisible by 5. Also, 2 is a residue of $4n \pm 1$ for n even, a non-residue for n odd. Hence 10 is a residue of $N = 4n \pm 1$ for n even if n or $n \mp 2$ is divisible by 5, and for n odd if neither is. Thus Euler's inclusion of $n \mp 6$ is superfluous. By a similar proof, 10 is quadratic non-residue of $N = 4n \pm 1$ if both 2 and 5 occur among the divisors of $n \pm 2$, $n \pm 6$, or if neither occurs; a residue if a single one of them occurs.

A. P. Reyer[39a] noted that the period for $a/3^p$ has 3^{p-2} digits and gave the length of the period for a/p for each prime $p < 150$.

*F. van Henekeler[39b] treated decimal fractions.

C. G. Reuschle[40] gave for each prime $p < 15000$ the exponent e to which 10 belongs modulo p. Thus e is the length of the period for $1/p$. He gave all the prime factors of $10^n - 1$ for $n \leqq 16$, $n = 18, 20, 21, 22, 24, 26, 28, 30, 32, 36, 42$; those of $10^n + 1$ for $n \leqq 18$, $n = 21$; also cases up to $n = 243$ of the factors of the quotient obtained by excluding analytic factors.

[35]Einige Sätze aus der Zahlenlehre, Progr. Cöln, Berlin, 1849, 14–15.

[36]London, Ed. Dublin Phil. Mag., 36, 1850, 15–20.

[37]Archiv Math. Phys., 16, 1851, 54–57. French transl. in Nouv. Ann. Math., 14, 1855, 115-7. Quoted by Brocard, Mathesis, 4, 1884, 38.

[38]Théorie des nombres, Paris, 1852, 308. For errata, see Shanks[61] and Gérardin.[131]

[39]Bull. Acad. Roy. Sc. Belgique, 20, II, 1853, 397–400.

[39a]Archiv Math. Phys., 25, 1855, 190–6.

[39b]Ueber die primitiven Wurzeln der Zahlen und ihre Anwendung auf Dezimalbrüche, Leyden, 1855 (Dutch).

[40]Math. Abhandlung...Tabellen, Progr. Stuttgart, 1856. Full title in Ch. I.[108]   Errata, Bork,[105] Hertzer,[119] Cunningham.[121]

W. Stammer[41] noted that $n/p = 0.\dot{a}_1 \ldots \dot{a}_x$ implies

$$\frac{n}{p}(10^x - 1) = a_1 \ldots a_x.$$

J. B. Sturm[42] used this result to explain the conversion of decimal into ordinary fractions without the use of series.

M. Collins[43] stated that, if we multiply any decimal fraction having m digits in its period by one with n digits, we obtain a product with $9mn$ digits in its period if m is prime to n, but with $n(10^m - 1)$ digits if n is divisible by m.

J. E. Oliver[44] proved the last theorem. If x'/x gives a periodic fraction to the base a with a period of ξ figures, then $a^\xi \equiv 1 \pmod{x}$ and conversely. The product of the periodic fractions for $x'/x, \ldots, z'/z$ with period lengths $\xi, \ldots, \zeta$ has the period length

$$\frac{x \ldots z}{M(x, \ldots, z)} \cdot M(\xi, \ldots, \zeta),$$

where $M(x, \ldots, z)$ is the l. c. m. of $x, \ldots, z$. He examined the cases in which the first factor in the formula is expressible in terms of $\xi, \ldots, \zeta$.

Fr. Heime[45] and M. Pokorny[46] gave expositions without novelty.

Suffield[47] gave the more important rules for periodic decimals and indicated the close connection with the method of synthetic division.

W. H. H. Hudson[48] called d a proper prime if the period for n/d has $d-1$ digits. If the period for r/p has $n = (p-1)/\lambda$ digits, there are λ periods for p. The sum of the digits in the period for a proper prime p is $9(p-1)/2$. If $1/p$ has a period of $2n$ digits, the sum of corresponding digits in the two half periods is 9, and this holds also if p is composite but has no factor dividing $10^n - 1$ [Midy[27]]. If $10p+1$ is a proper prime, each digit $0, 1, \ldots, 9$ occurs p times in its period. If a, b are distinct primes with periods of α, β digits, the number of digits in the period for ab is the l. c. m. of α, β [Bernoulli[8]]. Let p have a period of n digits and $1/p = k/(10^n - 1)$. Let m be the least integer for which

$$\binom{m}{1}\frac{k}{p^{x-1}} + \binom{m}{2}\frac{k^2}{p^{x-2}} + \cdots + \binom{m}{x-1}\frac{k^{x-1}}{p}$$

is an integer; then $1/p^x$ has a period of mn digits.

[41]Archiv Math. Phys., 27, 1856, 124.
[42]Ibid., 33, 1859, 94–95.
[43]Math. Monthly (ed., Runkle), Cambridge, Mass., 1, 1859, 295.
[44]Ibid., 345–9.
[45]Ueber relative Prim- und correspondirende Zahlen, primitive und sekundäre Wurzeln und periodische Decimalbrüche, Progr., Berlin, 1860, 18 pp.
[46]Ueber einige Eigenschaften periodischer Dezimalbrüche, Prag, 1864.
[47]Synthetic division in arithmetic, with some introductory remarks on the period of circulating decimals, 1863, pp. iv+19.
[48]Oxford, Cambridge and Dublin Messenger of Math., 2, 1864, 1–6. Glaisher[78] atrributed this useful anonymous paper to Hudson.

V. A. Lebesgue[49] gave for $N \leqq 347$ the periods for $1/N$, r/N,...[cf. Gauss[17]].

Sanio[50] stated that, if m, n,... are distinct primes and $1/m$, $1/n$,... have periods of length q, q',..., then $1/(m^a n^b$...) has the period length $m^{a-1} n^{b-1}$...qq'.... He gave the length of the period for $1/p$ for each prime $p \leqq 700$, and the factors of $10^n - 1$, $n \leqq 18$.

F. J. E. Lionnet[51] stated that, if the period for a/b has n digits, that for any irreducible fraction whose denominator is a multiple of b has a multiple of n digits. If the periods for the irreducible fractions a/b, a'/b',... have n, n',... digits, every irreducible fraction whose denominator is the l. c. m. of b, b',... has a period whose length is the l. c. m. of n, n',.... If the period for $1/p$ has n digits and if p^a is the highest power of the prime p which divides $10^n - 1$, any irreducible fraction with the denominator $p^{a+\beta}$ has a period of np^β digits.

C. A. Laisant and E. Beaujeux[52] proved that if q is a prime and the period for $1/q$ to the base B is $P = ab...h$, with $q-1$ digits, then

$$P - (a+b+...+h) = (B-1)\sigma, \qquad q\left(\sigma + \frac{q-1}{2}\right) = \frac{B^{q-1}-1}{B-1},$$

and stated that a like result holds for a composite number q if we replace $q-1$ by $f = \phi(q)$. Their proof of the generalized Fermat theorem $B^f \equiv 1$ (mod q) is quoted under that topic.

C. Sardi[53] noted that if 10 is a primitive root of a prime $p = 10n+1$, the period for $1/p$ contains each digit $0,...$, 9 exactly n times [Hudson[48]]. For $p = 10n+3$, this is true of the digits other than 3 and 6, which occur $n+1$ times. Analogous results are given for $10n+7$, $10n+9$.

Ferdinand Meyer[54] proved an immediate generalization from 10 to any base k prime to b, b',... of the statements by Lionnet.[51]

Lehmann[54a] gave a clear exposition of the theory.

C. A. Laisant and E. Beaujeux[55] considered the residues r_0, r_1,... when A, AB, AB^2,... are divided by D_1. Let $r_{i-1}B = Q_i D_1 + r_i$. When written to the base B, let $D_1 = a_p...a_2 a_1$, and set $D_i = a_p...a_i$. Then

$$a_1 r_1 + ... + a_p r_p = D_1(r_1 - Q_2 D_2 - ... - Q_p D_p).$$

The further results are either evident or not novel.

For G. Barillari[60a] on the length of the period, see Ch. VII.

[49]Mém. soc. sc. phys. et nat. de Bordeaux, 3, 1864, 245.
[50]Ueber die periodischen Decimalbrüche, Progr., Memel, 1866.
[51]Algèbre élém., ed. 3, 1868. Nouv. Ann. Math., (2), 7, 1868, 239. Proofs by Morel and Pellet, (2), 10, 1871, 39–42, 92–95.
[52]Nouv. Ann. Math., (2), 7, 1868, 289–304.
[53]Giornale di Mat., 7, 1869, 24–27.
[54]Archiv Math. Phys., 49, 1869, 168–178.
[54a]Ueber Dezimalbrüche, welche aus gewöhnlichen Brüchen abgeleitet sind, Progr., Leipzig, 1869.
[55]Nouv. Ann. Math., (2), 9, 1870, 221–9, 271–281, 302–7, 354–360.

*Th. Schröder[56] and J. Hartmann[57] treated periodic decimals.

W. Shanks[58] gave Lambert's method (Bernoulli,[8] end) for shortening the work of finding the length of the period for $1/N$.

G. Salmon[59] remarked that the number n of digits in the period is known if we find two remainders which are powers of 2, since $10^a \equiv 2^p$ and $10^b \equiv 2^q$ imply $10^{aq-bp} \equiv 1$; also if we find three remainders which are products of powers of 2 and 3. Muir[71] noted that it is here implied that $aq - bp$ equals n, whereas it is merely a multiple of n.

J. W. L. Glaisher[60] proved that, for any base r,

$$\frac{1}{(r-1)^2} = .012 \ldots \dot{\overline{r-3}}\ \dot{\overline{r-1}},$$

a generalization of $1/81 = .0\dot{1}2345679\dot{}$.

W. Shanks[61] gave the length of the period for $1/p$, when p is a prime < 30000, and a list of 69 errors or misprints in the table by Desmarest,[38] and 11 in that by Burckhardt.[20]

Shanks[62] gave primes p for which the length n of the period for $1/p$ is a given number ≤ 100, naturally incomplete. Shanks[63] gave additional entries p for $n = 26$, $n = 99$; noted corrections to his former table and stated that he had extended the table to 40000. Shanks[64] mentioned an extension in manuscript from 40000 to 60000. An extension to 120000 in manuscript was made by Shanks, 1875–1880. The manuscript, described by Cunningham,[124] who gave a list of errata, is in the Archives of the Royal Society of London.

Shanks[65] stated that if a is the length of the period for $1/p$, where p is a prime > 5, that for $1/p^n$ is ap^{n-1} [without the restriction by Thibault,[31] Muir[71]].

G. de Coninck[66] stated that, if the last digit (at the right) of A is 1 or 9, the last digit of the period for $1/A$ is 9 or 1; while, if A is a prime not ending in 1 or 9, its last digit is the same as the last in the period.

Moret-Blanc[67] noted that the last property holds for any A not divisible by 2 or 5. For, if a is the integer defined by the period for $1/A$, that for $(A-1)/A$ is $(A-1)a$, whence $a + (A-1)a = 10^n - 1$, if n is the length of the periods. He noted corrections to the remaining nine laws stated by Coninck and implied that when corrected they become trivial or else known facts.

---

[56]Progr. Ansbach, 1872.
[57]Progr. Rinteln, 1872.
[58]Messenger Math., 2, 1873, 41–43.
[59]Ibid., pp. 49–51, 80.
[60]Ibid., p. 188.
[61]Proc. Roy. Soc. London, 22, 1873–4, 200–10, 384–8. Corrections by Workman.[117]
[62]Ibid., pp. 381–4. Cf. Bertram[35], Loof.[37]
[63]Ibid., 23, 1874–5, 260–1.
[64]Ibid., 24, 1875–6, 392.
[65]Messenger Math., 3, 1874, 52–55.
[66]Nouv. Ann. Math., (2), 13, 1874, 569–71; errata, 14, 1875, 191–2.
[67]Ibid., (2), 14, 1875, 229–231.

Karl Broda[68] considered a periodic decimal fraction F having an even number r of digits in the period and a number m of p digits preceding the period. Let x be the first half of the period, y the second half. Then

$$F=\frac{p}{10^m}+\frac{x}{10^{m+r}}+\frac{y}{10^{m+2r}}+\frac{x}{10^{m+3r}}+\cdots=\frac{p}{10^m}+\frac{10^r x+y}{10^m(10^{2r}-1)}$$

$$=\frac{9(p\cdot10^r+x+p)+a}{9\cdot10^m(10^r+1)}$$

if $x+y=a(10^r-1)/9=a\ldots a$ (to r terms). The first paper treated the case $p=m=0$, and gave the generalization to base a in place of 10:

$$\frac{x}{a^r}+\frac{y}{a^{2r}}+\frac{x}{a^{3r}}+\cdots=\frac{a+(a-1)x}{(a-1)(a^r+1)}\quad\text{if }x+y=a\frac{a^r-1}{a-1}.$$

The case $a=a-1$ shows that a purely periodic fraction to the base a equals $(x+1)/(a^r+1)$ if the sum of the half periods has all its digits (to base a) equal to $a-1$. Returning to the base 10, and taking $N=9(10^r+1)$, $Z=9x+a$, where each digit of x is $\leqq a$, we see that Z/N equals a decimal fraction in which x is the first half of the period of r digits, while the second half is such that the sum of corresponding digits in it and x is a. If R is the remainder after r digits of the period have been obtained, $R+Z=a\,(10^r+1)$.

C. G. Reuschle[69] gave tables which serve to find numbers belonging to a given exponent <100 with respect to a given prime modulus <1000.

P. Mansion[70] gave a detailed proof that, if n is prime to 2, 3, 5, and if the period for $1/n$ has $n-1$ digits, the sum of corresponding digits in the half periods is 9.

T. Muir[71] proved that, if p is a prime, either of

$$N^x\equiv1\ (\mathrm{mod}\ p^s),\qquad N^{xp^n}\equiv1\ (\mathrm{mod}\ p^{s+n})$$

follows from the other. If x_1 is the least positive integer x for which the first holds and if p^s is the highest power of p dividing $N^{x_1}-1$, then $x_1 p^n$ is the least positive integer y for which $N^y\equiv1$ (mod p^{s+n}). Hence the known theorem: If $N=\Pi p_i^{m_i}$, where $p_1,p_2,\ldots$ are distinct primes, and if the period for $1/p_i$ has m_i digits, and if $p_i^{b_i}$ is the highest power of p_i dividing $10^{m_i}-1$, the number of digits in the period for $1/N$ is the l. c. m. of the $m_i p_i^{n_i-b_i}$. He asked if $b=1$ when $p>3$, as affirmed by Shanks.[65]

Mansion's proof (*ibid.*, 5, 1876, 33) by use of periodic decimals of the generalized Fermat theorem is quoted under that topic.

D. M. Sensenig[72] noted that a prime $p\neq2,5$, divides N if it divides the sum of the digits of N taken in sets of as many figures each as there are digits in the period for $1/p$.

[68]Archiv Math. Phys., 56, 1874, 85–98; 57, 1875, 297–301.
[69]Tafeln complexer Primzahlen, Berlin, 1875. Errata by Cunningham, Mess. Math., 46, 1916, 60–1.
[70]Nouv. Corresp. Math., 1, 1874–5, 8–12.
[71]Messenger Math., 4, 1875, 1–5.
[72]The Analyst, Des Moines, Iowa, 3, 1876, 25.

*A. J. M. Brogtrop[73] treated periodic decimals.

G. Bellavitis[74] noted that the use of base 2 renders much more compact and convenient Gauss'[15] table and hence constructed such a table.

W. Shanks[75] found that the period for $1/p$, where $p=487$, is divisible by p, so that the period for $1/p^2$ has $p-1$ digits.

J. W. L. Glaisher[76] formed the period 05263... for $1/19$ as follows: List 5; divide it by 2 and list the quotient 2; since the remainder is 1, divide 12 by 2 and list the quotient 6; divide it by 2 and list the quotient, etc. To get the period for $1/199$, start with 50. To get the period, apart from the prefixed zero, for $1/49$, start with 20 and divide always by 5; for $1/499$, start with 200.

Glaisher[77] noted that, if we regard as the same periods those in which the digits and their cyclic order are the same, even if commencing at different places, a number q prime to 10 will have f periods each of a digits, where $af=\phi(q)$. This was used to check Goodwyn's table.[23] If $q=39$, there are four periods each of six digits. If $q-1$ belongs to the period for $1/q$, the two halves of every period are complementary; if not, the periods form pairs and the periods in each pair are complementary. For each prime $N<1000$, except 3 and 487, the period for $1/N^k$ has nN^{k-1} digits if that for $1/N$ has n digits.

Glaisher[78] collected various known results on periodic decimals and gave an account of the tables relating thereto. If q is prime to 10 and if the period for $1/q$ has $\phi(q)$ digits, the products of the period by the $\phi(q)$ integers $<q$ and prime to q have the same digits in the same cyclic order; for example, if $q=49$. He gave (pp. 204–6) for each $q<1024$ and prime to 10 the number a of digits in the period for $1/q$, the number n of periods of irreducible fractions p/q, not regarding as distinct two periods having the same digits in the same cyclic order, and, finally Euler's $\phi(q)$. The values of a and n were obtained by mere counting from the entries in Goodwyn's[23] "table of circles"; in every case, $an=\phi(q)$. For the prime $p=487$, he gave the full periods for $1/p$ and $1/p^2$, each of 486 digits, thus verifying Desmarest's[38] statement of the exceptional character of this p [cf. Shanks[75]].

Glaisher[79] again stated the chief rules for the lengths of periods.

The problem was proposed[80] to find a number whose products by $2, \ldots, 6$ have the same digits, but in a new order.

Birger Hausted[81] solved this problem. Start with any number a of one digit, multiply it by any number p and let b be the digit in the units

[73]Nieuw Archief voor Wiskunde, Amsterdam, 3, 1877, 58–9.
[74]Atti Accad. Lincei, Mem. Sc. Fis. Mat., (3), 1, 1877, 778–800. Transunti, 206. See 62a of Ch. VII.
[75]Proc. Roy. Soc. London, 25, 1877, 551–3.
[76]Messenger Math., 7, 1878, 190–1. Cf. Desmarest.[38]
[77]Report British Assoc., 1878, 471–3.
[78]Proc. Cambridge Phil. Soc., 3, 1878, 185–206.
[79]Solutions of the Cambridge Senate-House Problems and Riders for 1878, pp. 8–9.
[80]Tidsskrift for Math., Kjobenhavn, 2, 1878, 28.
[81]Ibid., pp. 180–3. Jornal de Sc. Math. e Ast., 2, 1878, 154–6.

place of the product ap, β the digit in the tens place. Write the digit b to the left of digit a to form the last two digits of the required number P. The number c in the units place in $bp+\beta$ is written to the left of digit b in P. To cp add the digit in the tens place of bp and place the unit digit of the sum to the left of c in P. The process stops with the kth digit t if the next digit would give a. Then $P=t\ldots cba$ and its products by k integers or fractions has the same k digits in the same cyclic order. For $a=2$, $p=3$, we get $k=28$ and see that P is the period of $2/27$, and the k multipliers are $m/2$, $m=1,\ldots, 28$. [To have an example simpler than the author's, take $a=7$, $p=5$; then $P=142857$, the period of $1/7$; the multipliers are $1,\ldots, 6$.] For proof, we have

$$P=10^{k-1}t+\ldots+10^2c+10b+a, \qquad pP=10^{k-1}a+10^{k-2}t+\ldots+10c+b,$$
$$pP=10^{k-1}a+\frac{P-a}{10}, \qquad \frac{a}{10p-1}=\frac{P}{10^k-1},$$

so that P is the period with k digits for $a/(10p-1)$.

E. Lucas[82] gave the prime factors of $10^{13}\pm1$, $10^{17}\pm1$, $10^{21}\pm1$, $10^{16}+1$, $10^{18}+1$, communicated to him by W. Loof, with the remark that $(10^{19}-1)/9$ has no prime factor <3035479. Lucas gave the factors of $10^{30}+1$.

J. W. L. Glaisher[83] proved his[76] earlier statements, repeated his[77] earlier remarks, and noted that, if q is a prime such that the period for $1/q$ has $q-1$ digits, the products of the period for $1/q$ by 1, 2,..., $q-1$ have the same digits in the same cyclic order. This property, well known for $q=7$, holds also for $q=17$, 19, 23, 29, 47, 59, 61, 97 and for $q=7^2$.

O. Schlömilch[84] stated that, to find every N for which the period for $1/N$ has $2k$ digits such that the sum of the sth and $(k+s)$th digits is 9 for $s=1,\ldots, k$, we must take an integer $N=(10^k+1)/T$; then the first k digits of the period are the k digits of $T-1$.

C. A. Laisant[85] extended his investigations with Beaujeux[52,55] and gave a summary of known properties of periodic fractions; also his[86] process to find the period of simple periodic fractions without making divisions.

V. Bouniakowsky[87] noted that the property of the period of $1/N$, observed by Schlömilch[84] for $N=7$, 11, 13, 77, 91, 143, holds also for the periods of k/N, for $k=N-1$ and $(N-1)/2$, with the same values of N. Consider the decimal fraction $0.y_1y_2\ldots$ with $y_m\equiv y_{m-1}+y_{m-2}$ (mod 9), replacing any residue zero by 9, and taking $y_1>0$, $y_2>0$. The fraction is purely periodic and is either $0.\dot{9}$ or $0.\dot{3}369663\dot{9}$ or has the same digits permuted cyclically, or else has a period of 24 digits and begins with 1, 1 or 2, 2 or 4, 4, or has the same 24 digits permuted cyclically or by the interchange of the two halves

[82]Nouv. Corresp. Math., 5, 1879, 138–9.
[83]Nature, 19, 1879, 208–9.
[84]Zeitschrift Math. Phys., 25, 1880, 416.
[85]Mém. Soc. Sc. Phys. et Nat. de Bordeaux, (2), 3, 1880, 213–34.
[86]Les Mondes, 19, 1869, 331.
[87]Bull. Acad. Sc. St. Pétersbourg, 27, 1881, 362–9.

of the period. The property of Schlömilch holds for these and the generalization to any base, as well as for those with the law $y_m \equiv 2y_{m-1} + y_{m-2}$. But if

$$y_m \equiv 3y_{m-1} - 2y_{m-2} \pmod 9, \qquad y_m \equiv (2^{m-1} - 1)(y_2 - y_1) + y_1 \pmod 9,$$

the fact that $2^6 \equiv 1 \pmod 9$ shows that the period has at most six digits. Those with six reduce by cyclic permutation to nine periods:

$$167943, \quad 235986, \quad 278154, \quad 197346,$$
$$265389, \quad 218457, \quad 764913, \quad 329568, \quad 751248.$$

In the kth of these the sum of corresponding digits in the two half periods is always $\equiv k \pmod 9$.

Karl Broda[88] examined for small values of r and certain primes p the solutions x of $x^r \equiv 1 \pmod p$ to obtain a base x for which the periodic fraction for $1/p$ has a period of r digits, and similarly the condition $x^r \equiv -1 \pmod p$ for an even number of digits in the period (Broda[68]).

F. Kessler[89] factored $10^n - 1$ for $n = 11, 20, 22, 30$.

W. W. Johnson[90] formed the period for $1/19$ by placing 1 at the extreme right, next its double, etc., marking with a star a digit when there is 1 to carry:

$$\overset{*}{0}\,\overset{*}{5}\,2\,6\,\overset{*}{3}\,1\,\overset{*}{5}\,\overset{*}{7}\,\overset{*}{8}\,9\,4\,\overset{*}{7}\,3\,6\,\overset{*}{8}\,\overset{*}{4}\,2\,1.$$

To deduce the value of $1/19$ written on the base 2, use 1 for each digit starred and 0 for the others, reversing the order:

$$\overset{.}{0}\,0\,0\,0\,1\,1\,0\,1\,0\,1\,1\,1\,1\,0\,0\,1\,0\,\overset{.}{1}.$$

If we apply the first process with the multiplier m, we get the period for the reciprocal of $10m - 1$.

E. Lucas[91] gave the prime factors of $10^n - 1$ for n odd, $n \leqq 17$, $n = 21$, and certain factors for $n = 19, \ldots, 41$; those of $10^n + 1$ for $n \leqq 18$ and $n = 21$. He stated that the majority of the results were given by Loof and published by Reuschle. In 1886, Le Lasseur gave

$$10^{17} - 1 = 3^2 \cdot 2071723 \cdot 5363222[3]57,$$

said by Loof to have no divisor $< 400,000$ other than $3, 9$. On the omission of the digit 3, see Cunningham.[123]

F. Kessler[92] listed nine errors in Burckhardt's[20] table and described his own manuscript of a table to $p = 12553$, i. e., for the first 1500 primes.

Van den Broeck[93] stated that $10^{3^n} - 1$ is divisible by 3^{n+2}.

A. Lugli[94] proved that, if p is a prime $\neq 2, 5$, the length of the period of $1/p$ is a divisor of $p - 1$. If the number of digits in the period of a/p is an even number $2t$, the tth remainder on dividing a by p is $p - 1$, and conversely. Hence, if r_h is the hth remainder, $r_h + r_{h+t} = p$ $(h = 1, \ldots, t)$, and the sum of all the r's is tp. If the period of $1/p$ has s digits, $s < p - 1$, then

[88]Archiv Math. Phys., 68, 1882, 85–99.
[89]Zeitschrift Math. Naturw. Unterricht, 15, 1884, 29.
[90]Messenger of Math., 14, 1884–5, 14–18.
[91]Jour. de math. élém., (2), 10, 1886, 160. Cf. l'intermédiaire des math., 10, 1903, 183. Quoted by Brocard, Mathesis, 6, 1886, 153; 7, 1887, 73 (correction, 1889, 110).
[92]Archiv Math. Phys., (2), 3, 1886, 99–102.
[93]Mathesis, 6, 1886, 70. Proofs, 235–6, and Math. Quest. Educ. Times, 54, 1891, 117.
[94]Periodico di Mat., 2, 1887, 161–174.

$p-1=sh$ and we have h sets of s fractions whose periods differ only by the cyclic permu.ation of the digits. If p is a product of distinct primes $p_1, p_2 \ldots$ and if the lengths of the periods of $1/p$, $1/p_1$, $1/p_2, \ldots$ are s, s_1, $s_2, \ldots$, then s is the l. c. m. of s_1, $s_2, \ldots$. If $p = p_1{}^\alpha p_2{}^\beta \ldots$, and s, s_1, s' are the lengths of the periods of $1/p$, $1/p_1$, $1/p_1{}^\alpha$, then s' is one of the numbers s_1, $s_1 p, \ldots$, $s_1 p_1{}^{a-1}$ and hence divides $(p_1-1)p_1{}^{a-1}$; and s is a divisor of $\phi(p)$. Thus p divides $10^{\varphi(p)}-1$.

C. A. Laisant[95] used a lattice of points, whose abscissas are $a+r$, $a+2r, \ldots$, $a+pr$ and ordinates are their residues $<p$ modulo p, to represent graphically periodic decimal fractions and to expand fractions into a difference of two series of ascending powers of fixed fractions.

*A. Rieke[96] noted that a periodic decimal with a period of $2m$ digits equals $(A+1)/(10^m+1)$, where A is the first half of the period. He discussed the period length for any base.

W. E. Heal[97] noted that, if B contains all the prime factors of N, the number of digits in the fraction to the base B for M/N is the greatest integer in $(n+n'-1)/n'$, where $n-n'$ is the greatest difference found by subtracting the exponent of each prime factor of N from the exponent of the same prime factor of B. If B contains no prime factor of N, the fraction for M/N is purely periodic, with a period of $\phi(N)$ digits. If B contains some, but not all, of the prime factors of N, the number of digits preceding the period is the same as in the first theorem. The proofs are obscure. There is given the period for $1/p$ when $p<100$ and has 10 as a primitive root [the same p's as by Glaisher[83]]. Likewise for base 12, with $p<50$.

R. W. Genese[98] noted that, if we multiply the period for $1/81$ [Glaisher[60]] by m, where $m<81$ and prime to it, we get a period containing the digits $0, 1, \ldots, 9$ except $9n-m$, where $9n$ is the multiple of 9 just exceeding m.

Jos. Mayer[99] investigated the moduli with respect to which 10 belongs to a given exponent, and gave the factors of 10^n-1, $n<12$. He discussed the determination of the exponent to which 10 belongs for a given modulus by use of the theory of indices and by the methods of quadratic, cubic, biquadratic,... residues. He used also the fact that there are $(a-a')$ $(\beta-\beta')\ldots$ divisors of $p_1{}^\alpha p_2{}^\beta p_3{}^\gamma \ldots$ which divide no one of the fixed factors $p_1{}^a p_2{}^b p_3{}^\gamma \ldots, p_1{}^a p_2{}^b p_3{}^\gamma, \ldots$, where $a<\alpha$, $b<\beta, \ldots$, and p_1, $p_2, \ldots$ are distinct primes. He gave the length of the period for $1/p$, for each prime $p \leq 2543$ and 22 higher primes [Burckhardt[20]].

L. Contejean[100] proved that, in the conversion of an irreducible fraction a/b into a decimal fraction, if the remainders a_r and a_m are congruent modulo b, so that $10^r a \equiv 10^m a$, then $10^{m-r}-1$ is divisible by the quotient b' of b by the highest factor $2^t 5^s$ of b. Thus the length of the period is

[95]Assoc. franç. avanc. sc., 16, 1887, II, 228–235.
[96]Versuch über die periodischen Brüche, Progr., Riga, 1887.
[97]Annals of Math., 3, 1887, 97–103.
[98]Report British Assoc., 1888, 580–1.
[99]Ueber die Grösse der Periode eines unendlichen Dezimalbruches, oder die Congruence $10^x \equiv 1 \pmod{P}$. Progr. K. Studienanstalt Burghausen, München, 1888, 52 pp.
[100]Bull. soc. philomathique de Paris, (8), 4, 1891–2, 64–70.

$m-r$, while r digits precede the period. The condition that the length of the period be the maximum $\phi(b')$ is that 10 be a primitive root of b', whence $b'=p^n$, since $b'\neq4$ or $2p^n$, p being an odd prime.

P. Bachmann[101] used a primitive root g of the prime p and set

$$\frac{g^{p-1}-1}{p}=Q=c_{p-2}g^{p-2}+\ldots+c_1g+c_0,$$

to the base g. We get the multiples $Q, 2Q, \ldots, (p-1)Q$ by cyclic permutation of the digits of Q. For $p=7$, $g=10$, $Q=142857$.

J. Kraus[102] generalized the last result. When r_1/n is converted into a periodic fraction to base g, prime to n, let $a_1, \ldots, a_k$ be the quotients and $r_1, \ldots, r_k$ the remainders. Then

$$\frac{g^k-1}{n}\, r_\lambda = a_\lambda\, g^{k-1} + a_{\lambda+1}g^{k-2} + \ldots + a_{\lambda-1} \qquad (\lambda=1,\ldots, k),$$

whence

$$r_\lambda(a_1g^{k-1}+\ldots+a_k)=r_1(a_\lambda g^{k-1}+\ldots+a_{\lambda-1}).$$

In particular, let n be such that it has a primitive root g, and take $r_1=1$. Then

$$\frac{g^{\varphi(n)}-1}{n}=Q=a_1g^{\varphi(n)-1}+a_2g^{\varphi(n-2)}+\ldots+a_{\varphi(n)},$$

and if r_λ is prime to n, the product $r_\lambda Q$ has the same digits as Q permuted cyclically and beginning with a_λ.

H. Brocard[103] gave a tentative method of factoring 10^n-1.

J. Mayer[104] gave conditions under which the period of z/P to base a, where z and a are relatively prime to P, shall be complete, i. e., corresponding digits of the two halves of the period have the sum $a-1$.

Heinrich Bork[105] gave an exposition, without use of the theory of numbers, of known results on decimal fractions. There is here first published (pp. 36–41) a table, computed by Friedrich Kessler, showing for each prime $p<100000$ the value of $q=(p-1)/e$, where e is the length of the period for $1/p$. The cases in which $q=1$ or 2 were omitted for brevity. He stated that there are many errors in the table to 15000 by Reuschle.[40] Cunningham[124] listed errata in Kessler's table.

L. E. Dickson[106] proved, without the use of the concept of periodic fractions, that every integer of D digits written to the base N, which is such that its products by D distinct integers have the same D digits in the same cyclic order, is of the form $A(N^D-1)/P$, where A and P are relatively prime. A number of this form is an integer only when P is prime

[101]Zeitschrift Math. Phys., 36, 1891, 381–3; Die Elemente der Zahlentheorie, 1892, 95–97.
    A like discussion occurs in l'intermédiaire des math., 5, 1898, 57–8; 10, 1903, 91–3.
[102]Zeitschrift Math. Phys., 37, 1892, 190–1.
[103]El Progreso Matematico, 1892, 25–27, 89–93, 114–9. Cf. l'intermédiaire des math., 2, 1895, 323–4.
[104]Zeitschrift Math. Phys., 39, 1894, 376–382.
[105]Periodische Dezimalbrüche, Progr. 67, Prinz Heinrichs-Gymn., Berlin, 1895, 41 pp.
[106]Quart. Jour. Math., 27, 1895, 366–77.

to N, and D is a multiple of the exponent d to which N belongs modulo P. The further discussion is limited to the case $D=d$, to exclude repetitions of the period of digits. Then the multipliers which cause a cyclic permutation of the digits are the least residues of N, $N^2, \ldots, N^D$ modulo P. For $A = 1$, we have a solution for any N and any P prime to N. There are listed the 19 possible solutions with $A > 1$, $N \leq 63$, and having the first digit > 0. The only one with $N = 10$ is 142857. General properties are noted.

A like form is obtained (pp. 375-7) for an integer of D digits written to the base N, such that its quotients by D distinct integers have the same D digits in the same cyclic order. The divisors are the least residues of N^D, $N^{D-1}, \ldots, N$ modulo P. For example, if $N = 11$, $P = 7$, $A = 4$, we get $4(11^3 - 1)/7$, or 631 to base 11, whose quotients by 2 and 4 are 316 and 163, to base 11. Another example is 512 to base 9.

E. Lucas[1] gave all the prime factors of $10^n - 1$ for $n \leq 18$.

F. W. Lawrence[107] proved that the large factors of $10^{25} - 1$ and $10^{29} - 1$ are primes.

C. E. Bickmore[108] gave the factors of $10^n - 1$, $n \leq 100$. Here $(10^{23} - 1)/9$ is marked prime on the authority of Loof, whereas the latter regarded its composition as unknown [Cunningham[123]]. There is a misprint for 43037 in $10^{29} - 1$.

B. Bettini[109] considered the number n of digits in the period of the decimal fraction for a/b, i. e., the exponent to which 10 belongs modulo b. If 10 is a quadratic non-residue of a prime b, n is even, but not conversely (p. 48). There is a table of values of n for each prime $b \leq 277$.

V. Murer[110] considered the $n = mq$ remainders obtained when a/b is converted into a decimal fraction with a period of length n, separated them into sets of m, starting with a given remainder, and proved that the sum of the sets is a multiple of $9 \ldots 9$ (to m digits). Further theorems are found when $q = 1$, 2 or 3.

J. Sachs[110a] tabulated all proper fractions with denominators < 250 and their decimal equivalents.

B. Reynolds[111] repeated the rules given by Glaisher[78, 79] for the length of periods. He extended the rules by Sardi[53] and gave the number of times a given digit occurs in the various periods belonging to a denominator N, both for base 10 and other bases.

Reynolds[112] gave numerical results on periodic fractions for various bases the lengths of whose period is 3 or 6, and on the length of the period for $1/N$ for every base $< N - 1$, when N is a prime.

A. Cunningham[113] applied to the question of the length of the period of a periodic fraction to any base the theory of binomial congruences [see

[107]Proc. London Math. Soc., 28, 1896-7, 465.   Cf. Bickmore[49] of Ch. XVI.
[108]Nouv. Ann. Math., (3), 15, 1896, 222-7.
[109]Periodico di Mat., 12, 1897, 43-50.   [110]Ibid., 142-150
[110a]Progr. 632, Baden-Baden, Leipzig, 1898.
[111]Messenger Math., 27, 1897-8, 177-87.
[112]Ibid., 28, 1898-9, 33-36, 88-91.
[113]Ibid., 29, 1899-1900, 145-179.   Errata.[118]

201 of Ch. VII]. He gave extensive tables, and references to papers on higher residues and to tables relating to period lengths.

O. Fujimaki[114] noted that if $10^m - 1$ is exactly divisible by n, and the quotient is $a_1 \ldots a_m$ of m digits, the numbers obtained from the latter by cyclic permutations of the digits are all multiples of $a_1 \ldots a_m$.

J. Cullen, D. Biddle, and A. Cunningham[115] proved that the large factor of 14 digits of $(10^{25}+1)/(10^5+1)$ is a prime.

L. Kronecker[116] treated periodic fractions to any base.

W. P. Workman[117] corrected three errors in Shanks'[61] table.

D. Biddle[118] concluded erroneously that $(10^{17}-1)/9$ is a prime.

H. Hertzer[119] extended Kessler's[105] table from 100000 to 112400, noted Reuschle's[40] error on the conditions that 10 be a biquadratic residue of a prime p and gave the conditions that 10 be a residue of an 8th power modulo p. For errata in the table, see Cunningham.[124]

P. Bachmann[120] proved the chief results on periodic fractions and cyclic numbers to any base g.

A. Tagiuri[121] proved theorems [F. Meyer,[54] Perkins[29]] on purely periodic fractions to any base and on mixed fractions.

E. B. Escott[122] noted a misprint in Bickmore's[108] table and two omissions in Lucas'[91] table, but described inaccurately the latter table, as noted by A. Cunningham.[123]

A. Cunningham[124] described various tables (cited above) which give the exponent to which 10 belongs, and listed many errata.

J. R. Akerlund[125] gave the prime factors of $11 \ldots 1$ (to n digits) for $n \leq 16$, $n = 18$.

K. P. Nordlund[126] applied to periodic fractions the theorem that, if $n_1, \ldots, n_r$ are distinct odd primes, no one dividing a, then $N = n_1{}^{m_1} \ldots n_r{}^{m_r}$ divides $a^k - 1$, where $k = \phi(N)/2^{r-1}$. He gave the period of $1/p$ for p a prime < 100 and of certain a/p.

T. H. Miller,[127] generalizing the fact that the successive pairs of digits in the period for $1/7$ are 14, 28, ..., investigated numbers n to the base r for which

$$\frac{1}{n} = \frac{2n}{r^2} + \frac{4n}{r^4} + \frac{8n}{r^6} + \ldots ,$$

[114]Jour. of the Physics School in Tokio, 7, 1897, 16–21; Abh. Gesch. Math. Wiss., 28, 1910, 22.
[115]Math. Quest. Educat. Times, 72, 1900, 99–101.
[116]Vorlesungen über Zahlentheorie, I, 1901, 428–437.
[117]Messenger Math., 31, 1901–2, 115.
[118]Ibid., p. 34; corrected, ibid., 33, 1903–4, 126 (p. 95).
[119]Archiv Math. Phys., (3), 2, 1902, 249–252.
[120]Niedere Zahlentheorie, I, 1902, 351–363.
[121]Periodico di Mat., 18, 1903, 43–58.
[122]Nouv. Ann. Math., (4), 3, 1903, 136; Messenger Math., 33, 1903–4, 49.
[123]Messenger Math., 33, 1903–4, 95–96.
[124]Ibid., 145–155.
[125]Nyt Tidsskrift for Mat., Kjobenhavn, 16 A, 1905, 97–103.
[126]Göteborgs Kungl. Vetenskaps-Handlingar, (4), VII–VIII, 1905.
[127]Proc. Edinburgh Math. Soc., 26, 1907–8, 95–6.

whence $r^2 - 2n^2 = 2$. Besides the case $r = 10$, $n = 7$, he found $r = 58$, $n = 41$, etc.

A. Cunningham[128] noted two errors in his paper[113] and added

$$252^{12} \equiv 1 (\bmod 997^2), \qquad 390112^4 \equiv 1 \ (\bmod 17^6)$$

and cases modulo p^2, where $p = 103$, 487, attributed to Th. Gosset.

A. Cunningham[129] gave tables of the periods of $1/N$ to the bases 2, 3, 5 for $N \leq 100$.

H. Hertzer[130] noted three errors in Bickmore's[108] table.

A. Gérardin[131] gave factors of $10^n - 1$, $n < 100$, and a table of the exponents to which 10 belongs modulo p, a prime < 10000, with a list of errors in the tables by Burckhardt and Desmarest.

A. Filippov[132] gave two methods of determining the generating factor for the periodic fraction for $1/b$ (cf. Lucas, Théorie des nombres, p. 178).

G. C. Cicioni[133] treated the subject.

E. R. Bennett[134] proved the standard theorems by means of group theory.

W. H. Jackson[135] noted that, if a is prime to 10 and if b is chosen so that $b < 10$, $ab = 10m - 1$, the period for $1/a$ may be written as

$$b\{1 + 10m + (10m)^2 + \ldots + (10m)^{s-1}\} - k \cdot 10^s,$$

where s is the exponent to which 10 belongs modulo a, and k is a positive integer. Thus for $a = 39$, $b = 1$, we have $m = 4$, $s = 6$, and the period is

$$1 + 40 + \ldots + (40)^5 - k \cdot 10^6, \qquad \tfrac{1}{39} = .\dot{0}2564\dot{1}.$$

G. Mignosi[136] discussed the logic underlying the identification of an unending decimal with its generator p/q.

A. Cunningham[137] treated periodic decimals with multiples having the same digits permuted cyclically.

F. Schuh[138] considered the length q_a of the period for $1/p^a$ for the base g, where p is a prime. He proved that q_a is of the form $q_1 p^c$, where $0 \leq c \leq a - 2$ when $p = 2$, $a > 2$, while $0 \leq c \leq a - 1$ in all other cases. For $a > 2$,

$$q_{a-1} = q_1 p^{c-1}, \ldots, \qquad q_{a-c+1} = q_1 p, \qquad q_{a-c} = \ldots = q_2 = q,$$

where $q = q_1$ except when $p = 2$, $g = 4m - 1$, and then $q = 2$. Equality of periods for moduli p^a and p^b can occur for an odd prime p only when this period is q_1, and for $p = 2$ only when it is 1 or 2. It is shown how to find the numbers g which give equal periods for p^a and p, and the odd numbers g which give the period 2 for 2^a.

---

[128]Math. Gazette, 4, 1907–8, 209–210. Sphinx-Oedipe, 8, 1913, 131.

[129]Math. Gazette, 4, 1907–8, 259–267; 6, 1911–12, 63–7, 108–116.

[130]Archiv Math. Phys., (3), 13, 1908, 107.

[131]Sphinx-Oedipe, Nancy, 1908–9, 101–112.

[132]Spaczinskis Bote, 1908, pp. 252–263, 321–2 (Russian).

[133]La divisibilità dei numeri e la teoria delle decimali periodiche, Perugia, 1908, 150 pp.

[134]Amer. Math. Monthly, 16, 1909, 79–82.

[135]Annals of Math., (2), 11, 1909–10, 166–8.

[136]Il Boll. Matematica Gior. Sc.-Didat., 9, 1910, 128–138.

[137]Math. Quest. Educat. Times, (2), 18, 1910, 25–26.

[138]Nieuw Archief Wiskunde, (2), 9, 1911, 408–439. Cf. Schuh,[124–4], Ch. VII.

T. Ghezzi[139] considered a proper irreducible fraction m/p with p prime to the base b of numeration. Let b belong to the exponent n modulo p. In

$$mb = pq_1 + r_1, \qquad r_1 b = pq_2 + r_2, \ldots, \qquad 0 < r_1 < p, \; 0 < r_2 < p, \ldots,$$

$r_1, \ldots, r_n$ are distinct and $r_n = m$. Multiply the respective equations by $b^{n-1}, b^{n-2}, \ldots$ and add; we see that

$$\frac{m}{p} = \frac{q_1 b^{n-1} + \ldots + q_n}{b^n - 1}.$$

A similar proof shows that m/p equals a fraction with the denominator $b^t(b^n - 1)$ when $b = a_1 a_2 a_3$, $p = p_1 a_1{}^r a_2{}^s a_3{}^t$, the a's being primes and p_1 relatively prime to b, while b^t is the least power of b having the divisor $a_1{}^r a_2{}^s a_3{}^t$, and n is the exponent to which b belongs modulo p_1.

F. Stasi[140] gave a long proof showing that the length of the period for b/a does not exceed that for $1/a$. If the period A for $1/p$ has m digits and $n = pq$ is prime to 10, the length of the period for $1/n$ is m if A is divisible by q; is mi if A is prime to q and if the least $A(10^{m(k-1)} + \ldots + 1)$ divisible by q has $m = i$; and is mj if $A = A'a$, $q = aq'$, with A', q' relatively prime, while the least $A'(10^{m(k-1)} + \ldots + 1)$ divisible by q' has $k = j$. For a prime $p \neq 2, 5$, let

$$\frac{1}{p^h} = \frac{A_h}{10^m - 1},$$

and let A_h be the first of the periods of successive powers of $1/p$ not divisible by p; then the period for $1/p^{h+k}$ has mp^k digits. If p_i is a prime $\neq 2, 5$, and r_i is the length of the period for $1/p_i$, and if $1/p_i{}^{\beta_i}$ is the highest power of $1/p_i$ with a period of r_i digits, the length of the period for $1/p_i{}^{\alpha_i}$ is $r_i' = r_i p_i{}^{\alpha_i - \beta_i}$ and that for $1/\Pi p_i{}^{\alpha_i}$ is a multiple of the l. c. m. of the r_i'.

If n is prime to 10 and if $r_1, \ldots, r_m = 1$ are the successive remainders on reducing $1/n$ to a decimal, then $r_i^2 \equiv r_{2i} \pmod{n}$. Hence if $1/n$ has a period of $2i$ digits, $r_i^2 \equiv 1 \pmod{n}$ and conversely. But if it has a period of $2i+1$ digits, $r_{i+1}^2 \equiv 10$ and conversely.

*K. W. Lichtenecker[141] gave the length of the period for $1/p$, when p is a prime ≤ 307, and the factors of $10^n - 1$, $n \leq 10$.

L. Pasternak[142] noted that, after multiplying the terms of a fraction by 9, 3 or 7, we may assume the denominator $N = 10m - 1$. To convert R_0/N into a decimal, we have $10R_{k-1} = Ny_k + R_k$ ($k = 1, 2, \ldots$). Set $R_k = 10z_k + e_k$, $e_k \leq 9$. Since $y_k \leq 9$, $e_k = y_k$ and $R_{k-1} = me_k + z_k$. Hence the successive digits of the period are the unit digits of the successive remainders.

E. Maillet[143] defined a unique development $a_0 + a_1/n + a_2/n^2 + \ldots$ of an arbitrary number, where the a_i are integers satisfying certain conditions. He studied the conditions that the development be limited or periodic.

[139]Il Boll. Matematica Gior. Sc.-Didat., 9, 1910, 263-9.
[140]Ibid., 11, 1912, 226-246.
[141]Zeitschr. für das Realschulwesen, 37, 1912, 338-349.
[142]L'enseignement math., 14, 1912, 285-9.
[143]L'intermédiaire des math., 20, 1913, 202-6.

Welsch[144] discussed briefly the length of the period of a decimal fraction. B. Howarth[145] noted that D^2 is not a factor of $(10^{Dn}-1)/(10^n-1)$ if D is a prime and n is not a multiple of the length of the period for $1/D$. Again,[146] $(10^{mnp^2}-1)/9$ is not divisible by $(10^{mp}-1)(10^{np}-1)/81$. A. Cunningham[147] factored $10^{99}\pm 1$. Known factors of $10^n\pm 1$ are given. Cunningham[148] gave factors of $10^{mpn}-1$.

A. Leman[149] gave an elementary exposition and inserted proofs of Fermat's theorem and related facts, with the aim to afford a concrete introduction to the more elementary facts of the theory of numbers.

S. Weixer[150] would compute the period P for $1/p$ by multiplication, beginning at the right. Let c be the final digit of P, whence $pc=10z-1$. Then c is the first digit of the period P^1 for z/p. The units digit c_1 of $cz=10z_1+c_1$ is the tens digit of P and the units digit of P^1. In $c_1z+z_1=10z_2+c_2$, c_2 is the hundreds digit of P and the tens digit of P^1, etc.

A. Leman[151] discussed the preceding paper.

Problems[152] on decimal fractions may be cited here.

O. Hoppe[153] proved that $(10^{19}-1)/9$ is a prime.

M. Jenkins[154] noted that if $N=a^pb^q\ldots$, where $a, b,\ldots$ are distinct primes $\neq 2, 5$, the period for $1/N$ is complementary (sum of corresponding digits of the half periods is 9) if and only if the lengths of the periods for $1/a, 1/b,\ldots$ contain the same power of 2.

Kraitchik[125] of Ch. VII and Levänen[37] of Ch. XII gave tables of exponents to which 10 belongs. Bickmore and Cullen[115] of Ch. XIV factored $10^{25}+1$.

FURTHER PAPERS INVOLVING NO THEORY OF NUMBERS.

J. L. Lagrange, Leçons élém. à l'école normale en 1795, Oeuvres 7, 200.
James Adams, Annals Phil., Mag. Chem. (Thompson), (2), 2, 1821, 16–18.
C. R. Telosius and S. Mörck, Disquisitio. . . . Acad. Carolina, Lundae, 1838 (in Meditationum Math. . . . Publice Defendent C. J. D. Hill, 1831, Pt. II).
J. A. Arndt, Archiv Math. Phys., 1, 1841, 101–4.
J. Dienger, ibid., 11, 1848, 232; Jour. für Math., 39, 1850, 67.
Wm. Wiley, Math. Magazine, 1, 1882, 7–8.
A. V. Filippov, Kagans Bote, 1910, 214–221 (pedagogic).

---

[144]L'intermédiaire des math., 21, 1914, 10.
[145]Math. Quest. Educat. Times, 28, 1915, 101–4.
[146]Ibid., 27, 1915, 33–4.
[147]Ibid., 29, 1916, 76, 88–9.
[148]Math. Quest. and Solutions, 3, 1917, 59.
[149]Vom Periodischen Dezimalbruch zur Zahlentheorie, Leipzig, 1916, 59 pp.
[150]Zeitschrift Math. Naturw. Unterricht, 47, 1916, 228–230.
[151]Ibid., 230–1.
[152]Zeitschrift Math. Naturw. Unterricht, 12, 1881, 431; 20, 188; 23, 584.
[153]Proc. London Math. Soc., Records of Meeting, Dec. 6, 1917, and Feb. 14, 1918, for a revised proof.
[154]Math. Quest. Educ. Times, 7, 1867, 31–2. Minor results, 32, 1880, 69; 34, 1881, 97–8; 37, 1882, 44; 41, 1884, 113–4; 58, 1893, 108–9; 60, 1894, 128; 63, 1895, 34; 72, 1900, 75–6; 74, 1901, 35; (2), 2, 1902, 65–6, 84–5; 4, 1903, 29, 65–7, 95; 7, 1905, 97, 106, 109–10; 8, 1905, 57; 9, 1906, 73. Math. Quest. and Solutions, 3, 1917, 72 (table); 4, 1917, 22.

CHAPTER VII.

PRIMITIVE ROOTS, BINOMIAL CONGRUENCES.

PRIMITIVE ROOTS, EXPONENTS, INDICES.

J. H. Lambert[1] stated without proof that there exists a primitive root g of any given prime p, so that $g^e - 1$ is divisible by p for $e = p - 1$, but not for $0 < e < p - 1$.

L. Euler[2] gave a proof which is defective. He introduced the term primitive root and proved (art. 28) that at most n integers $x < p$ make $x^n - 1$ divisible by p, the proof applying equally well to any polynomial of degree n with integral coefficients. He stated (art. 29) that, for $n < p$, $x^n - 1$ has all n solutions "real" if and only if n is a divisor of $p - 1$; in particular, $x^{p-1} - 1$ has $p - 1$ solutions (referring to arts. 22, 23, where he repeated his earlier proof of Fermat's theorem). Very likely Euler had in mind the algebraic identity $x^{p-1} - 1 = (x^n - 1)Q$, from which he was in a position to conclude that Q has at most $n - p + 1$ solutions, and hence $x^n - 1$ exactly n. By an incomplete induction (arts. 32–34), he inferred that there are exactly $\phi(n)$ integers $x < p$ for which $x^n - 1$ is divisible by p, but $x^l - 1$ not divisible by p for $0 < l < n$, n being a divisor of $p - 1$ (as the context indicates). In particular, there exist $\phi(p-1)$ primitive roots of p (art. 46). He listed all the primitive roots of each prime ≤ 37.

J. L. Lagrange[3] proved that, if p is an odd prime and

$$x^{p-1} - 1 = X\xi + pF,$$

where X, ξ, F are polynomials in x with integral coefficients, and if x^m and x^μ are the highest powers of x in X and ξ with coefficients not divisible by p, there are m integral values, numerically $< p/2$, of x which make X a multiple of p, and μ values making ξ a multiple of p. For, by Fermat's theorem, the left member is a multiple of p for $x = \pm 1, \pm 2, \ldots, \pm (p-1)/2$, while at most m of these values make X a multiple of p and at most μ make ξ a multiple of p.

L. Euler[4] stated that he knew no rule for finding a primitive root and gave a table of all the primitive roots of each prime ≤ 41.

Euler[5] investigated the least exponent x (when it exists) for which $fa^x + g$ is divisible by N. Find λ such that $-g = \lambda N$ is a multiple, say $a^\alpha r$, of a. Then $fa^{x-\alpha} - r$ is divisible by N. Set $r \mp \lambda' N = a^\beta s$, $\beta \geq 1$. Then $fa^{x-\alpha-\beta} - s$ is divisible by N; etc. If the problem is possible, we finally get f as the residue of $fa^{x-\alpha-\cdots-\zeta}$, whence $x = \alpha + \ldots + \zeta$. For example, to find the least x for which $2^x - 1$ is divisible by $N = 23$, we have

$$1 + 23 = 2^3 3, \quad 3 - 23 = -2^2 5, \quad -5 - 23 = -2^2 7, \quad -7 + 23 = 2^4 1,$$

whence $x = 3 + 2 + 2 + 4 = 11$.

[1]Nova Acta Eruditorum, Leipzig, 1769, p. 127.
[2]Novi Comm. Acad. Petrop., 18, 1773, 85; Comm. Arith., 1, 516–537.
[3]Nouv. Mém. Ac. Roy. Berlin, année 1775 (1777), p. 339; Oeuvres 3, 777.
[4]Opusc. Anal., 1, 1783 (1772), 121; Comm. Arith., 1, 506.
[5]Opusc. Anal., 1, 1783 (1773), 242; Comm. Arith., 2, p. 1; Opera postuma, I, 172–4.

A. M. Legendre[6] started with Lagrange's[3] result that, if p is a prime and n is a divisor of $p-1$,

$$(1) \qquad\qquad x^n \equiv 1 \;(\text{mod } p)$$

has n incongruent integral roots. Let $n = \nu^a \nu'^b \ldots$, where ν, ν', ... are distinct primes. A root a of (1) *belongs** to the exponent n if no one of $a^{n/\nu}$, $a^{n/\nu'}$, ... is congruent to unity modulo p. For, if $a^\theta \equiv 1$, $0 < \theta < n$, let σ be the g. c. d. of θ, n, so that $\sigma = ny - \theta z$ for integers y, z; then

$$a^{\theta z} \equiv 1, \qquad a^\sigma \equiv a^{\theta z + \sigma} \equiv a^{ny} \equiv 1,$$

contrary to hypothesis. Next, of the n roots of (1), n/ν satisfy $x^{n/\nu} \equiv 1$ (mod p), and $n(1-1/\nu)$ do not. Likewise, $n(1-1/\nu')$ do not satisfy $x^{n/\nu'} \equiv 1$; etc. It is said to follow that there are

$$\phi(n) = n\left(1 - \frac{1}{\nu}\right)\left(1 - \frac{1}{\nu'}\right) \ldots$$

numbers belonging to the exponent n modulo p. If

$$\beta^{\nu^a} \equiv 1, \qquad \beta^{\nu^{a-1}} \not\equiv 1 \;(\text{mod } p),$$

β belongs to the exponent ν^a. If β' belongs to the exponent ν'^b, etc., the product $\beta\beta' \ldots$ is stated to belong to the exponent n.

C. F. Gauss[7] gave two proofs of the existence of primitive roots of a prime p. If d is a divisor of $p-1$, and a^d is the lowest power of a congruent to unity modulo p, a is said to belong to the exponent d modulo p. Let $\psi(d)$ of the integers $1, 2, \ldots, p-1$ belong to the exponent d, a given divisor of $p-1$. Gauss showed that $\psi(d) = 0$ or $\phi(d)$, $\Sigma\psi(d) = p-1 = \Sigma\phi(d)$, whence $\psi(d) = \phi(d)$. In his second proof, Gauss set $p-1 = a^\alpha b^\beta \ldots$, where a, $b, \ldots$ are distinct primes, proved the existence of numbers A, $B, \ldots$ belonging to the respective exponents a^α, $b^\beta, \ldots$, and showed that $AB \ldots$ belongs to the exponent $p-1$ and hence is a primitive root of p.

Let a be a primitive root of p, b any integer not divisible by p, and e the integer, uniquely determined modulo $p-1$, for which $a^e \equiv b$ (mod p). Gauss (arts. 57–59) called e the index of b for the modulus p relative to the base a, and wrote $e = \text{ind } b$. Thus

$$a^{\text{ind } b} \equiv b \;(\text{mod } p), \qquad \text{ind } bb' \equiv \text{ind } b + \text{ind } b' \;(\text{mod } p-1).$$

Gauss (arts. 69–72) discussed the relations between indices for different bases and the choice of the most convenient base.

In articles 73–74, he gave a convenient tentative method for finding a primitive root of p. Form the period of 2 (the distinct least positive residues of the successive powers of 2); if 2 belongs to an exponent $t < p-1$, select a number $b < p$ not in the period of 2, and form the period of b; etc.

If a belongs to the exponent t modulo p, the product of the terms in the period of a is $\equiv (-1)^{t+1}$ (mod p), while the sum of the terms is $\equiv 0$ unless $a \equiv 1$ (arts. 75, 79).

---

[6]Mém. Ac. R. Sc., Paris, 1785, 471–3. Théorie des nombres, 1798, 413–4; ed. 3, 1830, Nos. 341–2; German transl. by Maser, 2, pp. 17–18.

*This term was introduced later by Gauss.[7]

[7]Disquisitiones Arith., 1801, arts. 52–55.

The product of all the primitive roots of a prime $p \neq 3$ is $\equiv 1 \pmod{p}$; the sum of the primitive roots of p is $\equiv 0$ if $p-1$ is divisible by a square, but is $\equiv (-1)^n$ if $p-1$ is the product of n distinct primes (arts. 80, 81).

If p is an odd prime and e is the g. c. d. of $\phi(p^n) = p^{n-1}(p-1)$ and t, then $x^t \equiv 1 \pmod{p^n}$ has exactly e incongruent roots. It follows that there exist primitive roots of p^n, i. e., numbers belonging to the exponent $\phi(p^n)$ (arts. 85–89).

For $n > 2$, every odd number belongs modulo 2^n to an exponent which divides 2^{n-2}, so that primitive roots of 2^n are lacking; however, a modified method of employing indices to the base 5 may be used (arts. 90, 91).

If $m = A^a B^b \ldots$, where A, $B,\ldots$ are distinct primes, and $a = \phi(A^a)$, $\beta = \phi(B^b),\ldots$, and if μ is the l. c. m. of a, $\beta,\ldots$, then $z^\mu \equiv 1 \pmod{m}$ for z prime to m. Now $\mu < a \cdot \beta \ldots = \phi(m)$ except when $m = 2^n$, p^n or $2p^n$, where p is an odd prime. Thus there exist primitive roots of m only when $m = 2$, 4, p^n or $2p^n$ (art. 92).

Table I, at the end of Disq. Arith., gives on one page the indices of each prime $< p$ for each prime and power of prime modulus < 100. Gauss gave no direct table to determine the number corresponding to a given index, but indicated (end of art. 316) how his Table III for the conversion of ordinary into decimal fractions leads to the number having a given index (cf. Gauss,[15, 17] Ch. VI).

S. F. Lacroix[8] reproduced Gauss' second proof of the existence of primitive roots of a prime, without a reference.

L. Poinsot[9] argued that the primitive roots of a prime p may be obtained from the algebraic expressions for the imaginary $(p-1)$th roots of unity by increasing the numbers under the radical signs by such multiples of p that the radicals become integral. The $\phi(p-1)$ primitive roots of p may be obtained by excluding from $1,\ldots, p-1$ the residues of the powers whose exponents are the distinct prime factors of $p-1$; while symmetrical, this method is unpractical for large p.

Frégier[10] proved that the 2^nth power of any odd number has the remainder unity when divided by 2^{n+2}, if $n > 0$.

Poinsot[11] developed the first point of his preceding paper. The equation for the primitive 18th roots of unity is $x^6 - x^3 + 1 = 0$. The roots are

$$x = a^i \sqrt[3]{\tfrac{1}{2}(1 + \sqrt{-3})}, \qquad (a^3 = 1).$$

But $\sqrt{-3} \equiv \pm 4$, $\sqrt[3]{-7} \equiv 4$, $\sqrt[3]{-11} \equiv 2 \pmod{19}$. Thus the six primitive roots of 19 are $x = -4, 2, -9, -5, -6, 3$. In general, the algebraic expressions for the nth roots of unity represent the different integral roots of $x^n \equiv 1 \pmod{p}$, where p is a prime $kn+1$, after suitable integers are added to the numbers under the radical signs. Since unity is the only (integral)

[8]Complément des élémens d'algèbre, Paris, ed. 3, 1804, 303–7; ed. 4, 1817, 317–321.
[9]Mém. Sc. Math. et Phys. de l'Institut de France, 14, 1813–5, 381–392.
[10]Annales de Math. (ed., Gergonne), 9, 1818–9, 285–8.
[11]Mém. Ac. Sc. de l'Institut de France, 4, 1819–20, 99–183.

root of $x^p \equiv 1 \pmod{p}$, if p is a prime >2, he concluded (p. 165) that p is a factor of the numbers under the radical signs in the formula for a primitive pth root of unity. Cf. Smith[159] of Ch. VIII.

Poinsot[11a] again treated the same subject.

J. Ivory[12] stated that a primitive root of a prime p satisfies $x^{(p-1)/2} \equiv -1$, but no one of the congruences $x^t \equiv -1 \pmod{p}$, $t = (p-1)/(2a)$, where a ranges over the odd prime factors of $p-1$; while a number not a primitive root satisfies at least one of the $x^t \equiv -1$. Hence if each $a^t \not\equiv -1$ and $a^{(p-1)/2} \equiv -1$, then a is a primitive root.

V. A. Lebesgue[13] stated that prior to 1829 he had given in the Bulletin du Nord, Moscow, the congruence $X \equiv 0$ of Cauchy[14] for the integers belonging to the exponent n modulo p.

A. Cauchy[14] proved the existence of primitive roots of a prime p, essentially as in Gauss' second proof. If $p-1$ is divisible by $n = a^\alpha b^\beta c^\gamma \dots$, where $a, b, c, \dots$ are distinct primes, he proved that the integers belonging to the exponent n modulo p coincide with the roots of

$$X = \frac{(x^n - 1)(x^{n/ab} - 1)(x^{n/ac} - 1)\dots}{(x^{n/a} - 1)(x^{n/b} - 1)\dots(x^{n/abc} - 1)\dots} \equiv 0 \pmod{p}.$$

The roots of the equation $X = 0$ are the primitive nth roots of unity. For the above divisor n of $p-1$, the sum of the lth powers of the primitive roots of $x^n \equiv 1 \pmod{p}$ is divisible by p if l is divisible by no one of the numbers

$$n, n/a, n/b, \dots, n/ab, \dots, n/abc, \dots$$

But if several of these are divisors of l, and if we replace $n, a, b, \dots$ by $\phi(n), 1-a, 1-b, \dots$ in the largest of these divisors in fractional form, we get a fraction congruent to the sum of the lth powers. In case $x^m \equiv 1 \pmod{p}$ has m distinct integral roots, the sum of the lth powers of all the roots is congruent modulo p to m or 0, according as l is or is not a multiple of m.

M. A. Stern[15] proved that the product of all the numbers belonging to an exponent d is $\equiv 1 \pmod{p}$, while their sum is divisible by p if d is divisible by a square, but is $\equiv (-1)^n$ if d is a product of n distinct primes (generalizations of Gauss, D. A., arts. 80, 81). If $p = 2n+1$ and a belongs to the exponent n, the product of two numbers, which do not occur in the period of a, occurs in the period of a. To find a primitive root of p when $p-1 = 2ab\dots$, where $a, b, \dots$ are distinct odd primes, raise any number as 2 to the powers $(p-1)/a, (p-1)/b, \dots$; if no one of the residues modulo p is 1, the negative of the product of these residues is a primitive root of p; in case one of the residues is 1, use 3 or 5 in place of 2. If $p = 2q+1$ and q are odd primes, 2 or -2 is a primitive root of p according as $p = 8n+3$ or $8n+7$. If $p = 4q+1$

[11a]Jour. de l'école polytechnique, cah. 18, t. 11, 1820, 345–410.
[15]Supplement to Encyclopædia Britannica, 4, 1824, 698.
[13]Jour. de Math., 2, 1837, 258.
[14]Exercices de Math., 1829, 231; Oeuvres, (2), 9, 266, 278–90.
[11]Jour. für Math., 6, 1830, 147–153.

and q are primes, 2 and -2 are primitive roots of p. If $p=4q+1$ and $q=3n+1$ are primes, 3 and -3 are primitive roots of p.

F. Minding[16] gave without reference Gauss' second proof of the existence of primitive roots of a prime.

F. J. Richelot[17] proved that, if $p=2^m+1$ is a prime, every quadratic non-residue (in particular, 3) is a primitive root of p.

A. L. Crelle[18] gave a table showing all prime numbers $\leqq 101$ having a given primitive root; also a table of the residues of the powers of the natural numbers when divided by the primes $3,\ldots, 101$. His device[19] for finding the residues modulo p of the powers of a will be clear from the example $p=7$, $a=3$. Write under the natural numbers <7 the residues of the successive multiples of 3 formed by successive additions of 3; we get

$$\begin{array}{cccccc} 1 & 2 & 3 & 4 & 5 & 6 \\ 3 & 6 & 2 & 5 & 1 & 4. \end{array}$$

Then the residues $3, 2, 6,\ldots$ of $3, 3^2, 3^3,\ldots$ modulo 7 are found as follows: after 3 comes the number 2 below 3 in the above table; after 2 comes the number 6 below 2 in the table; etc.

Crelle[20] proved that, if p is a prime and λ is prime to $p-1$ and $<p-1$, the residues modulo p of z^λ range with z over the integers $1, 2,\ldots, p-1$. His proof that there exist $\phi(n)$ numbers belonging to the exponent n modulo p, if n divides $p-1$, is like that by Legendre.[6]

G. L. Dirichlet[21] employed $\phi(k)$ systems of indices for a modulus $k=2^\lambda p^\pi p'^{\pi'}\ldots$, where $p, p',\ldots$ are distinct primes, and $\lambda\geqq 3$. Given any integer n prime to k, and primitive roots $c, c',\ldots$ of $p^\pi, p'^{\pi'},\ldots$, we can determine indices $a, \beta, \gamma, \gamma',\ldots$ such that

$$n\equiv(-1)^a 5^\beta \pmod{2^\lambda}, \qquad n\equiv c^\gamma \pmod{p^\pi}, \qquad n\equiv c'^{\gamma'} \pmod{p'^{\pi'}},\ldots.$$

Michel Ostrogradsky[22] gave for each prime $p<200$ all the primitive roots of p and companion tables of the indices and corresponding numbers. (See Jacobi[23] and Tchebychef.[34])

C. G. J. Jacobi[23] gave for each prime and power of a prime <1000 two companion tables showing the numbers with given indices and the index of each given number. In the introduction, he reproduced the table by Burckhardt, 1817, of the length of the period of the decimal fraction for $1/p$, for each prime $p\leqq 2543$, and 22 higher primes. Of the 365 primes <2500, we therefore have 148 having 10 as a primitive root, and 73 of the form $4m+3$ having -10 as a primitive root. Use is made also of the primes for which 10 or -10 is the square or cube of a primitive root.

[16]Anfangsgründe der höheren Arith., 1832, 36–37.

[17]Jour. für Math., 9, 1832, p. 5.

[18]Ibid., 27–53.

[19]Also, ibid., 28, 1844, 166.

[20]Abh. Ak. Wiss. Berlin, 1832, Math., p. 57, p. 65.

[21]Ibid., 1837, Math., 45; Werke, 1, 1889, 333.

[22]Lectures on alg. and transc. analysis, I–II, St. Pétersbourg, 1837; Mém. Ac. Sc. St. Péters-bourg, sér. 6, sc. math. et phys., 1, 1838, 359–85.

[23]Canon Arithmeticus, Berlin, 1839, xl+248 pp. Errata, Cunningham.[110,135]

To find a primitive root g of p, select any convenient integer a and form the residues of a, a^2, a^3, ... [as by Crelle[18]]. Let n be the exponent to which a belongs. Set $nn'=p-1$. If $n<p-1$, select an integer b not in the period a, ..., a^n. The residue of $b^{n'}$ is in this period of a. If b^f is the least power of b whose residue is in the period of a, then f divides n', say $n'=ff'$ (p. xxiii). Since $a\equiv g^{n'}$, $b^f\equiv a^i$, we have

$$b^f\equiv g^{ff'i}\equiv g^{ff'i+nff'k}, \qquad b\equiv g^{f'(i+nk)} \pmod{p},$$

for some value 0, 1, ..., $f-1$ of k. But k must be chosen so that $i+nk$ is prime to f. For, if $i+nk=du$, where d is a divisor of f, we would have $b^{f/d}\equiv a^u$. The nf residues of $a^r b^s$ ($r=0$, ..., $n-1$; $s=0$, ..., $f-1$) are distinct; their indices to base g are f', $2f'$, ..., nff' in some order and are known. If $nf'<p-1$, we employ an integer not in the set $a^r b^s$ and proceed similarly. Ultimately we obtain a primitive root and at the same time the index of every number. This method was used for the primes between 200 and 1000.

For primes <200, the tables by Ostrogradsky[22] were reprinted with the same errors (noted at the end of the Canon).

Jacobi proved that, if n is an odd prime, any primitive root of n^2 is a primitive root of any higher power of n (p. xxxv).

For the modulus 2^μ, $4\leq\mu\leq 9$, the final tables give the index I of any positive odd number to base 3, where

$$(-1)^{(N-1)(N-3)/8}N\equiv 3^I \pmod{2^\mu}.$$

Robert Murphy[24] stated the empirical theorem that every prime an^2+p has a as a primitive root if $p>a/2$, p is a prime $<a$, and if a is a primitive root of p. For example, a prime $10n^2+7$ has 10 as a primitive root.

H. G. Erlerus[25] considered two odd primes p and p' and a number m such that $m\equiv a$ (mod p), $m\equiv a'$ (mod p'). Let a belong to the exponent e modulo p, and a' to the exponent e' modulo p'. If δ is the g. c. d. of e and e', then m belongs to the exponent ee'/δ modulo pp'. He discussed at length the number of integers belonging to the exponent n for a composite modulus.

A. Cauchy[26] called the least positive integer i for which $m^i\equiv 1$ (mod n) the indicator relative (or corresponding) to the base m and modulus n, which are assumed relatively prime. If the base m is constant, and i_1, i_2 are the indicators corresponding to moduli n_1, n_2, and if $n=n_1 n_2$ is prime to m, then the l. c. m. of i_1 and i_2 is the indicator corresponding to modulus n. If the modulus n is constant, and i_1, i_2 are the indicators corresponding to bases m_1, m_2, and if i_1, i_2 are relatively prime, then $i_1 i_2$ is the indicator corresponding to the base $m_1 m_2$.

Let i_1, i_2 be the indicators corresponding to the bases m_1, m_2 and same modulus n. The g. c. d. ω of i_1, i_2 can be expressed (often in several ways) as a product uv such that i_1/u, i_2/v are relatively prime. For, if $\omega=\alpha\beta...$,

[24]Phil. Mag., (3), 19, 1841, 369.
[25]Elementa Doctrinæ Numerorum, Diss., Halis, 1841, 18–43.
[26]Comptes Rendus Paris, 12, 1841, 824–845; Oeuvres, (1), 6, 124–146.

where $a, \beta, \ldots$ are powers of distinct primes, use a as a factor in forming u in case a is prime to i_1/a, but as a factor of v in case a is prime to i_2/a, and as a factor of either u or v indifferently in case a is prime to both i_1/a and i_2/a. Since i_1/u and i_2/v are relatively prime indicators corresponding to bases $m_1{}^u$ and $m_2{}^v$, it follows from the preceding theorem that the indicator corresponding to base $m_1{}^u \cdot m_2{}^v$ and modulus n is

$$\frac{i_1}{u} \cdot \frac{i_2}{v} = \frac{i_1 i_2}{\omega} = \text{l. c. m. of } i_1, i_2.$$

Hence, given several bases $m_1, m_2, \ldots$ and a single modulus n, we can find a new base relative to which the indicator is the l. c. m. of the indicators corresponding to $m_1, m_2, \ldots$. If the latter bases include all the integers $< n$ and prime to n, the corresponding indicators give all indicators which can correspond to modulus n, so that all of them divide a certain maximum indicator I. Then for every integer m relatively prime to n, $m^I \equiv 1 \pmod{n}$. If $n = \nu^a$, where ν is an odd prime, or if $n = 2$ or 4, $I = \phi(n)$. If $n = 2^k$, $k > 2$, $I = \phi(n)/2$. If I_j is the maximum indicator corresponding to a power n_j of a prime, and if $n = \Pi n_j$, then I is the l. c. m. of $I_1, I_2, \ldots$. The equation $mx - ny = 1$ has the solution $x \equiv m^{I-1} \pmod{n}$.

Cauchy[27] republished the preceding paper, but with an extension from the limit $n = 100$ to the limit $n = 1000$ for his table of the maximum indicator I.

C. F. Arndt[28] gave (without reference) Gauss' second proof of the existence of a primitive root of an odd prime p, and proved the existence of the $\phi(p^n)$ primitive roots of p^n or $2p^n$, and that there are no primitive roots for moduli other than these and 4. If t is a divisor of 2^{n-2}, $n > 2$, exactly t numbers belong to the exponent t modulo 2^n (p. 18). If, for a modulus p^n, $2p^n$, a belongs to the exponent t, then $a \cdot a^2 \ldots a^t$ is congruent to $(-1)^{t+1}$ (pp. 26–27), while the product of the numbers belonging to the exponent t is congruent to $+1$ if $t \neq 2$ (pp. 37–38). He proved also Stern's[15] theorem on the sum of these numbers. He gave the same two theorems also in a later paper.[29]

L. Poinsot[30] used the method of Legendre[6] to prove the existence of $\phi(n)$ integers belonging to the exponent n, a divisor of $p - 1$, where p is a prime. He gave (pp. 71–75) essentially Gauss' first proof, and gave his own[9] method of finding primitive roots of a prime. The existence of primitive roots of p^n, $2p^n$, 4, but of no further moduli, is established by use of the number of roots of binomial congruences (pp. 87–101).

C. F. Arndt[31] noted that if a belongs to an even exponent t modulo 2^n, then $\pm a$, $\pm a^3, \ldots, \pm a^{t-1}$ give the t incongruent numbers belonging to the exponent t, and are congruent to $k \cdot 2^m \mp 1 (k = 1, 3, 5, \ldots)$. The product of the numbers belonging to the exponent t modulo 2^n, $n > 2$, is $\equiv +1$.

[27]Exercices d'Analyse et de Phys. Math., 2, 1841, 1–40; Oeuvres, (2), 12.
[28]Archiv Math. Phys., 2, 1842, 9, 15–16.
[29]Jour. für Math., 31, 1846, 326–8.
[30]Jour. de Mathématiques, (1), 10, 1845, 65–70, 72.
[31]Archiv Math. Phys., 6, 1845, 395, 399.

E. Prouhet[32] gave, without reference, Crelle's[18] method of forming the residues of the powers of a number. The object of the paper is to give a uniform method of proof of theorems, given in various places in Legendre's text, relating to the residues of the first n powers of an integer belonging to the exponent n modulo P, especially when P is a prime or a power of a prime, and the existence of primitive roots. He gave (p. 658) the usual proof that ± 2 is a primitive root of a prime $2q+1$ if q is a prime $4k\pm 1$ (with a misprint).

C. F. Arndt[33] proved that if g is a primitive root of the odd prime p and if p^λ ($\lambda < n$) is the highest power of p dividing $G = g^{p-1} - 1$, then g belongs to the exponent $p^{n-\lambda}(p-1)$ modulo p^n. Conversely, if the last is true of a primitive root g of p, then G is divisible by p^λ and not by $p^{\lambda+1}$. The first result with $\lambda = 1$ shows that any primitive root of p^2 is a primitive root of p^n, $n > 2$. Let g be a primitive root of p; if G is not divisible by p^2, g is a primitive root of p^2; but if G is divisible by p^2, and h is not divisible by p, then $g + hp$ is a primitive root of p^2. Any odd primitive root of p^n is a primitive root of $2p^n$. If g is a primitive root of p^n or $2p^n$, and t is a divisor of $p^{n-1}(p-1)$, then if a ranges over the integers $< t$ and prime to t, the $\phi(t)$ integers belonging to the exponent t modulo p^n or $2p^n$ are g^e, where $e = p^{n-1}(p-1)a/t$. The numbers belonging to the exponent 2^{n-m} modulo 2^n are found more simply than by Gauss[7] and Jacobi[23] (p. 37).

P. L. Tchebychef[34] proved that if $a, \beta, \dots$ are the distinct prime factors of $p-1$, where p is a prime, then a is a primitive root of p if and only if no one of the congruences $x^a \equiv a$, $x^\beta \equiv a, \dots$ (mod p) has an integral root. This furnishes a method (usually impracticable) of finding all primitive roots of p. A second method uses a number a belonging to the exponent n, and a number b not congruent to a power of a, and deduces a number belonging to an exponent $> n$. In the second supplement, he proved that 3 is a primitive root of any prime $2^{2n}+1$; that ± 2 is a primitive root of any prime $2a+1$ such that a is a prime $4k\pm 1$; 3 is a primitive root of $4N2^m+1$ if $m > 0$ and N is a prime $> 9^{2^m}/(4\cdot 2^m)$; 2 is a primitive root of any prime $4N+1$ such that N is an odd prime. The last result was later proposed[35] as a question for solution (with reference to this text). There is given the table of primitive roots and indices for primes < 200, due to Ostrogradsky[22]. Schapira (p. 314) noted that in the list of errata in Jacobi's[23] Canon (p. 222) there is omitted the error 8 for 6 in ind 14 for $p=25$.

V. A. Lebesgue[36] remarked that Cauchy's[14] congruence $X \equiv 0$ shows the existence of $\phi(n)$ integers belonging to the exponent n modulo p, a prime.

[32]Nouv. Ann. Math., 5, 1846, 175–87, 659–62, 675–83.

[33]Jour. für Math., 31, 1846, 259–68.

[34]Theory of Congruences (in Russian), 1849. German translation by Schapira, Berlin, 1889, p. 192. Italian translation by Mlle. Massarini, Rome, 1895, with an extension of the tables of indices to 353.

[35]Nouv. Ann. Math., 15, 1856, 353. Solved by use of Euler's criterion by P. H. Rochette, ibid., 16, 1857, 159. Also proved by Desmarest,[37] p. 278.

[36]Nouv. Ann. Math., 8, 1849, 352; 11, 1852, 420.

E. Desmarest[37] devoted the last 86 pages of his book to primitive roots; the 70 pages claimed to be new might well have been reduced to five by the omission of trivial matters and the use of standard notations. To find (pp. 267–8) a primitive root of the prime $P = 6q+1$, where q is an odd prime, seek an odd solution of $u^2+3 \equiv 0 \pmod{P}$ and set $u = 2R-1$; then $R^3 \equiv -1$ and R belongs to the exponent 6; thus we know the solutions of $x^6 \equiv 1$; let a be any integer prime to P and not such a solution; if $a^q \equiv \pm 1$, then $\pm a$ belongs to the exponent q, and $\pm aR$ is a primitive root of P; but, if $a^{2q} \not\equiv 1$, then $a^{3q} \equiv \mp 1 \pmod{P}$, and $\pm a$ is a primitive root of P. If $P = 8Q+1$ and Q are primes, then $P \equiv 5 \pmod{12}$ and 3 is a quadratic non-residue and hence a primitive root of P.

Let P be a prime of the form $5q \pm 2$. Then $u^2 \equiv 5 \pmod{P}$ is not solvable. Thus, if a is a primitive root of P, $5 \equiv a^e$, where e is odd. Thus if e is prime to $P-1$, 5 is a primitive root of P. It is recommended that 5 be the first number used in seeking by trial a primitive root. And yet he announced the theorem (p. 283) that 5 is in general a primitive root. If P is a prime $5q \pm 2$ also of the form $2^n Q+1$, where Q is an odd prime including 1, then (pp. 284–6) 5 is a primitive root of P provided P is not a factor of $5^{2^n}-1$. He gave the factors of the latter and of $10^{2^n}-1$ for $n = 1, \ldots, 5$.

Results, corresponding to those just quoted for 5, are stated for $\rho = 7, -7, 10, 17$. What is really given is a list of the linear forms of the primes P for which ρ is a quadratic non-residue. If, in addition, $P = 2^n Q+1$, where Q is an odd prime, then ρ is a primitive root, provided $\rho^{2^n} \not\equiv 1 \pmod{P}$. The last condition is ignored in his statement of his results and again in his collection (pp. 297–8) of "principles which give primitive roots" entered in his table (pp. 298–300) giving a primitive root of each prime < 10000.

V. A. Lebesgue[38] proved that, if a and $p = 2^i a+1$ are primes, any quadratic non-residue x of p is a primitive root of p if
$$x^{2^{i-1}}+1 \not\equiv 0 \pmod{p}.$$

J. P. Kulik[39] gave for each prime p between 103 and 353 the indices and all the primitive roots of p. His manuscript extended to 1000. There is an initial table giving the least primitive root of the primes from 103 to 1009.

G. Oltramare[40] called x a root of order or index m of a prime p if x belongs to the exponent $(p-1)/m$ modulo p. Let $X_m(x) \equiv 0 \pmod{p}$ be the congruence whose roots are exclusively the roots of order m of p. By changing x to $x^{1/n}$, we obtain $X_{mn} = \phi(x) \equiv 0$. If $n_1, n_2, \ldots, n$ are the divisors > 1 of n,
$$X_m = \frac{\phi(x^n)}{X_{mn_1} \cdots X_{mn}}.$$

[37]Théorie des nombres. Traité de l'analyse indéterminée du second degré à deux inconnues suivi de l'application de cette analyse à la recherche des racines primitives avec une table de ces racines pour tous les nombres premiers compris entre 1 et 10000, Paris, 1852, 308 pp. For errata, see Cunningham, Mess. Math., 33, 1903, 145.

[38]Nouv. Ann. Math., 11, 1852, 422–4.

[39]Jour. für Math., 45, 1853, 55–81.

[40]Ibid., 303–9.

V. A. Lebesgue[41] noted that, given a primitive root g ($g<p$) of the prime p, we can find at once the primitive roots of p^n. Let g' be the positive residue $<p^2$ when g^p is divided by p^2 and set $h=(g'-g)/p$. Then

$$g+px+p^2y \quad (y=0,\ldots, p^{n-2}-1; \; x=0,\ldots, p-1; \; x\neq h)$$

give $p^{n-2}(p-1)$ primitive roots. Replacing g by g^i, where i is less than and prime to $p-1$, we obtain $\phi\{\phi(p^n)\}$ primitive roots of p^n. In particular, a primitive root of p^2 is a primitive root of p^n (Jacobi[23]). But, if $h=0$, g is not a primitive root of p^2. Since

$$g^{\text{ind } a+e}\equiv p-a \pmod{p^n}, \qquad e=\tfrac{1}{2}p^{n-1}(p-1),$$

we can reduce by half the size of Jacobi's Canon.

D. A. da Silva[42] gave two proofs that $x^d\equiv 1 \pmod{p}$ has $\phi(d)$ primitive roots, if d divides $p-1$, and perfected the method of Poinsot[9,30] for finding the primitive roots of a prime.

F. Landry[42a] was led to the same conclusion as Ivory.[12]   In particular, if $p=2^k+1$, or if $p=2n+1$ (n an odd prime) and $a\neq p-1$, any quadratic non-residue a of p is a primitive root. For each prime $p<10000$, at least one prime ≤ 19 is a quadratic non-residue of p. Cauchy's[14] congruence for the primitive roots is derived and proved.

G. Oltramare[43] proved that $-3^a2^{2\beta}$ is a primitive root of the prime $p=2a\beta+1$, if $a\neq 3$, $\beta\neq 3$, $3^{2a}\not\equiv 1$, $2^{2\beta}\not\equiv 1 \pmod{p}$; that, if

$$p=3\cdot2^m+1=q^2+3r^2, \qquad qx+ry=1,$$

$(-1+qy-3rx)5^3/2$ is a primitive root of p; and analogous theorems. If a and $2a+1$ are primes, 2 or a is a primitive root of $2a+1$, according as a is of the form $4n+1$ or $4n+3$. If a is a prime $\neq 3$ and if $p=2a+1$ is a prime and $m>1$, then 3 is a primitive root of p unless $3^{2m-1}+1\equiv 0 \pmod{p}$. [Cf. Smith.[47]]

P. Buttel[44] attributed to Scheffler (Die unbestimmte Analytik, 1854, §142) the method of Crelle[18] for finding the residues of powers.

C. G. Reuschle's[45] table C gives the Haupt-exponent (i. e., exponent to which the number belongs) (a) of 10, 2, 3, 5, 6, 7 with respect to all primes $p<1000$, and the least primitive root of p; (b) of 10 and 2 for $1000<p<5000$ and a convenient primitive root; (c) of 10 for $5000<p<15000$ (no primitive root given). Numerous errata have been listed by Cunningham.[110]

Allegret[46] stated that if n is odd, n is not a primitive root of a prime $2^{2\lambda}n+1$, $\lambda>0$; proof can be made as in Lebesgue.[38]

[41]Comptes Rendus Paris, 39, 1854, 1069–71; same in Jour. de Math., 19, 1854, 334–6.
[42]Proprietades geraes et resoluçao directa das Congruencias binomias, Lisbon, 1854. Report by C. Alasia, Rivista di Fisica, Mat. e Sc. Nat., Pavia, 4, 1903, 25, 27–28; and Annaes Scientificos Acad. Polyt. do Porto, Coimbra, 4, 1909, 163–192.
[42a]Troisième mémoire sur la théorie des nombres, Paris, 1854, 24 pp.
[43]Jour. für Math., 49, 1855, 161–86.
[44]Archiv Math. Phys., 26, 1856, 247.
[45]Math. Abhandlung...Tabellen, Prog. Stuttgart, 1856; full title in the chapter on perfect numbers.[108]
[46]Nouv. Ann. Math., 16, 1857, 309–310.

H. J. S. Smith[47] stated that some of Oltramare's[43] general results are erroneous at least in expression, and gave a simple proof that $x^d \equiv 1 \pmod{p^n}$ has exactly d roots if d divides $\phi(p^n)$.

V. A. Lebesgue[48] proved that, if p is an odd prime and a, b belong to exponents α, β, there exist numbers belonging to the l. c. m. m of α, β, as exponent. Hence if neither α nor β is a multiple of the other, m exceeds α and β. If $d < p-1$ is the greatest of the exponents to which $1, \ldots, p-1$ belong, the latter do not all belong to exponents dividing d, since otherwise they would give more than d roots of $x^d \equiv 1 \pmod p$. Hence there exist primitive roots of p. If a is odd, $\pm 1 + 2^a a$ belongs to the exponent 2^{m-a} modulo 2^m (p. 87). If h belongs to the exponent k modulo p, a prime, then $h + Pz$ belongs modulo p^n to an exponent which divides kp^{n-1} (p. 101). If f is a primitive root of p, and $f^{p-1} - 1 = pz$, then f is a primitive root of p^n if and only if z is not divisible by p (p. 102).

G. L. Dirichlet[49] proved the last theorem and explained his[21] system of indices for a composite modulus.

V. A. Lebesgue[50] published tables, constructed by J. Hoüel,[51] of indices and corresponding numbers for each prime and power of prime modulus < 200, which differ from Jacobi's[23] only in the choice of the least primitive root. There is an auxiliary table of the indices of $x!$ for prime moduli < 200.

V. A. Lebesgue[52] stated that, if $g < p$ is a primitive root of the prime p and if $g' \equiv g^{p-2} \pmod p$, then g' is a primitive root of p; at least one of g and g' is a primitive root of p^n for n arbitrary.

V. Bouniakowsky[53] proved in a new way the theorems of Tchebychef[34] that 2 is a primitive root of $p = 8n+3$ if p and $4n+1$ are primes, and of $p = 4n+1$ if p and n are primes. He gave a method to find the exponent to which 2 or 10 belongs modulo p.

A. Cayley[54] gave a specimen table showing the indices $\alpha, \beta, \ldots$ for every number $M \equiv a^\alpha b^\beta \ldots \pmod N$, where $M < N$ and prime to N, for $N = 1, \ldots, 50$. There is no apparent way of forming another single table for all N's analogous to Jacobi's tables (one for each N) of numbers corresponding to given indices.

F. W. A. Heime[55] gave the least primitive root of each prime < 1000. His other results are not new. A secondary root of a prime p is one belonging to an exponent $< p-1$ modulo p.

[47]British Assoc. Report, 1859, 228; 1860, 120, §73; Coll. Math. Papers, 1, 50, 158 (Report on theory of numbers).

[48]Introd. théorie des nombres, 1862, 94–96.

[49]Zahlentheorie, §§128–131, 1863; ed. 2, 1871; ed. 3, 1879; ed. 4, 1894.

[50]Mém. soc. sc. phys. et nat. de Bordeaux, 3, cah. 2, 1864–5, 231–274.

[51]Formules et tables numér., Paris, 1866. For moduli ≤ 347.

[52]Comptes Rendus Paris, 64, 1867, 1268–9.

[53]Bull. Ac. Sc. St. Pétersbourg, 11, 1867, 97–123.

[54]Quart. Jour. Math., 9, 1868, 95–96.

[55]Untersuchungen, besonders in Bezug auf relative Primzahlen, primitive u. secundäre Wurzeln, quadratische Reste u. Nichtreste; nebst Berechnung der kleinsten primitiven Wurzeln von allen Primzahlen zwischen 1 und 1000. Berlin, 1868; ed. 2, 1869.

C. J. D. Hill[56] noted that his tables of indices for the moduli 2^n and 5^n ($n \leqq 5$) give the residues of numbers modulo 10^n, i. e., the last n digits. Using also tables for the moduli 9091 and 9901, as well as a table of logarithms, we are able to determine the last 22 digits.

B. M. Goldberg[57] gave the least primitive root of each prime < 10160.

V. Bouniakowsky[58] proved that 3 is a primitive root of p if $p = 24n + 5$ and $(p-1)/4$ are primes; -3 is a primitive root of p if $p = 12n + 11$ and $(p-1)/2$ are primes; if ρ is a primitive root of the prime $p = 4n + 1$, one (or both) of ρ, $p - \rho$ is a primitive root of p^m and of $2p^m$; 5 is a primitive root of $p = 20n + 3$ or $20n + 7$ if p and $(p-1)/2$ are primes, and of $p = 40n + 13$ or $40n + 37$ if p and $(p-1)/4$ are primes; 6 is a primitive root of a prime $24n + 11$ and -6 of $24n + 23$ if $(p-1)/2$ is a prime; 10 is a primitive root of $p = 40n + 7$, 19, 23, and -10 of $p = 40n + 3$, 27, 39, if $(p-1)/2$ is a prime; 10 is a primitive root of a prime $80n + 73$, $n > 0$, or $80n + 57$, $n > 1$, if $(p-1)/8$ is a prime. If $p = 8an + 2a - 1$ or $8an + a - 2$ and $(p-1)/4$ are primes, and if $a^2 + 1$ is not divisible by p, a is a primitive root of p.

V. A. Lebesgue[59] proved certain theorems due to Jacobi[23] and the following theorem which gives a method different from Jacobi's for forming a table of indices for a prime modulus p: If a belongs to the exponent n, and if b is not in the period of a, and if f is the least positive exponent for which $b^f \equiv a^i$, then $x^f \equiv a$ has the root $a^t b^u$, where $ft + iu - 1 = nv$; the root belongs to the exponent nf if and only if u is prime to f.

Consider the congruence $x^m \equiv a \pmod{p}$, where a belongs to the exponent $n = (p-1)/n'$, and m is a divisor of n'. Every root r has a period of mn terms if no one of the residues of $r, r^2, \ldots, r^{m-1}$ is in the period of a. If all the prime divisors of m divide n, the m roots have a period of mn terms; but if m has prime divisors $q, r, \ldots,$ not dividing n, there are only

$$m \left(\frac{q-1}{q} \right) \left(\frac{r-1}{r} \right) \cdots$$

roots having a period of mn terms. The existence of primitive roots follows; this is already the case if $m = n'$.

Mention is made of companion tables in manuscript giving indices of numbers, and numbers corresponding to indices, constructed by J. Ch. Dupain in full for $p < 200$, but from 200 to 1500 with reduction to one-half in view of ind $p - a \equiv$ ind $a \pm (p-1)/2$ modulo $p - 1$.

L. Kronecker[60] proved the existence of two series of positive integers g_j, m_j ($j = 1, \ldots, \rho$) such that the least positive residues modulo $k > 2$ of $g_1^{i_1} g_2^{i_2} \ldots g_\rho^{i_\rho}$ give all the $\phi(k)$ positive integers $< k$ and prime to k, if $i_1 = 0, 1, \ldots, m_1 - 1$; $i_2 = 0, 1, \ldots, m_2 - 1$; etc. [cf. Mertens[92]].

G. Barillari[60a] proved that, if a is prime to b and belongs to the exponent

[56]Jour. für Math., 70, 1869, 282–8; Acta Univ. Lundensis, Lund, 1, 1864 (Math.), No. 6, 18 pp.
[57]Rest- und Quotient-Rechnung, Hamburg, 1869, 97–138.
[58]Bull. Ac. Sc. St. Pétersbourg, 14, 1869, 375–81.
[59]Comptes Rendus Paris, 70, 1870, 1243–1251.
[60]Monatsber. Ak. Berlin, 1870, 881. Cf. Traub, Archiv Math. Phys., 37, 1861, 278–94.
[60a]Giornale di Mat., 9, 1871, 125–135.

m modulo b, and if b^h is the highest power of b which divides $a^m - 1$, and if $n \geq h$, then b^n divides $a^e - 1$ where $e = mb^{n-h}$. Further, if b is a prime, a belongs to the exponent e modulo b^n. For a new prime b', let m', n', h' have the corresponding properties. Then the exponent to which a belongs modulo $B = b^n b'^{n'} \ldots$ is the l. c. m. L of mb^{n-h}, $m'b'^{n'-h'}, \ldots$ For $a = 10$, we see that L is the length of the period for the irreducible fraction N/B.

L. Sancery[61] proved that if p is a prime and $a < p$ belongs to the exponent θ modulo p, there exists an infinitude of numbers $a + px = A$ such that $A^\theta - 1$ is divisible by p^k, but not by p^{k+1}, where k is any assigned positive integer. If A belongs to the exponent θ modulo $p > 2$, A will belong to the exponent θ modulo p^r if the highest power of p which divides $A^\theta - 1$ is $\geq p^r$; but if it be $p^{r-\delta}$, A belongs to the exponent θp^δ modulo p^r [Barillari[60a]]. Hence A is a primitive root of p^r if a primitive root of p and if $A^{p-1} - 1$ is not divisible by p^2, and there are $\phi\{\phi(p^r)\}$ primitive roots of p^r or $2p^r$. [Generalization of Arndt.[33]]

C. A. Laisant[62] noted that if a belongs to the exponent 3 modulo p, a prime, then $a + 1$ belongs to the exponent 6, and conversely. If a belongs to the exponent 6, $a + 1$ will not belong to the exponent 3 unless $p = 7$, $a = 3$. Hence if p is a prime $6m + 1$, there are two numbers a, b belonging to the exponent 3, and two numbers $a + 1$, $b + 1$ belonging to the exponent 6; also, $a + b = p - 1$. If (p. 399) $p + q$ is an odd prime and p is even, then $p^p q^q \equiv q$, $p^q q^p \equiv p \pmod{p+q}$.

G. Bellavitis[62a] gave, for each power $p^i \leq 383$ of a prime p, the periodic fraction for $1/p^i$ to the base 2 and showed how to deduce the indices of numbers for the modulus p^i. Let $q = p^{i-1}(p-1)$ and let 2 belong to the exponent q/r modulo p^i. A root b of $b^r \equiv 2 \pmod{p^i}$ is the base of the system of indices.

G. Frattini[63] proved by the theory of roots of unity that, if p is a prime, the number of interchanges necessary to pass from $1, 2, \ldots, p-2$ to ind 2, ind 3, $\ldots$, ind $(p-1)$ and to

$$\text{ind } 1 - \text{ind } 2, \quad \text{ind } 2 - \text{ind } 3, \ldots, \quad \text{ind } (p-2) - \text{ind } (p-1)$$

are both even or both odd.

Fritz Hofmann[64] used rotations of regular polygons to prove theorems on the sum of the primitive roots of a prime (Gauss[7]).

A. R. Forsyth[65] found the sum of the cth powers of the primitive roots of a prime p. The sum is divisible by p if $p - 1$ contains the square of a prime not dividing c or if it contains a prime dividing c but with an exponent exceeding by at least 2 its exponent in c. If neither of these conditions is satisfied, the result is not so simple.

[61]Bull. Soc. Math. de France, 4, 1875–6, 23–29.
[62]Mém. Soc. Sc. Phys. et Nat. de Bordeaux, (2), 1, 1876, 400–2.
[62a]Atti Accad. Lincei, Mem. Sc. Fis. Mat., (3), 1, 1876–7, 778–800.
[63]Giornale di Mat., 18, 1880, 369–76.
[64]Math. Annalen, 20, 1882, 471–86.
[65]Messenger of Math., 13, 1883–4, 180–5.

J. Perott[66] gave a simple proof that $x^{p^k} \equiv 1 \pmod{p^n}$ has p^k roots. Thus there exists an integer b belonging to the exponent p^{n-1} modulo p^n. Assuming the existence of a primitive root of p, we employ a power of it and obtain a number a belonging to the exponent $p-1$ modulo p^n. Hence ab is a primitive root of p^n.

Schwartz[67] stated, and Hacken proved, the final theorem of Cauchy.[14]

L. Gegenbauer[68] stated 19 theorems of which a specimen is the following: If $p = 8a(8\beta+1) + 24\beta + 5$ and $(p-1)/4$ are primes and if $64a^2 + 48a + 10$ is relatively prime to p, then $8a+3$ is a primitive root of p.

G. Wertheim[69] gave the least primitive root of each prime < 1000 and companion tables of indices and numbers for primes < 100. He reproduced (pp. 125–130) arts. 80–81 of Gauss[7] and stated the generalization by Stern.[15]

H. Keferstein[70] would obtain all primitive roots of a prime p by excluding all residues of powers with exponents dividing $p-1$ [Poinsot[9]].

M. F. Daniëls[71] gave a proof like Legendre's[6] that there are $\phi(n)$ numbers belonging to the exponent n modulo p, a prime, if n divides $p-1$.

*K. Szily[72] discussed the "comparative number" of primitive roots.

E. Lucas[73] gave the name reduced indicator of n to Cauchy's[26] maximum indicator of n, and noted that it is a divisor $< \phi(n)$ of $\phi(n)$ except when $n = 2$, 4, p^k or $2p^k$, where p is an odd prime, and then equals $\phi(n)$. The exponent to which a belongs modulo m is called the "gaussien" of a modulo m (preface, xv, and p. 440).

H. Scheffler[74] gave, without reference, the theorem due to Richelot[17] and the final one by Prouhet.[32] To test if a proposed number a is a primitive root of a prime p, note whether p is of one of the linear forms of primes for which a is a quadratic non-residue, and, if so, raise a to the powers whose exponents divide $(p-1)/2$.

L. Contejean[75] noted that the argument in Serret's Algèbre, 2, No. 318, leads to the following result [for the case $a = 10$]: If p is an odd prime and a belongs to the exponent $e = (p-1)/q$ modulo p, it belongs to the exponent $p^{r-1}e$ modulo p^r when $(a^e - 1)/p$ is not divisible by p, but to a smaller exponent if it is divisible by p [Sancery[61]].

P. Bachmann[76] proved the existence of a primitive root of a prime p by use of the group of the residues $1, \ldots, p-1$ under multiplication.

[66]Bull. des Sc. Math., 9, I, 1885, 21–24. For $k = n-1$ the theorem is contained implicitly in a posthumous fragment by Gauss, Werke, 2, 266.
[67]Mathesis, 6, 1886, 280; 7, 1887, 124–5.
[68]Sitzungsber. Ak. Wiss. Wien (Math.), 95, II, 1887, 843–5.
[69]Elemente der Zahlentheorie, 1887, 116, 375–381.
[70]Mitt. Math. Gesell. Hamburg, 1, 1889, 256.
[71]Lineaire Congruenties, Diss., Amsterdam, 1890, 92–99.
[72]Math. és termes értesitö (Memoirs Hungarian Ac. Sc.), 9, 1891, 264; 10, 1892, 19.   Magyar Tudom. Ak. Ertcsitoje (Report of Hungarian Ac. Sc.), 2, 1891, 478.
[73]Théorie des nombres, 1891, 429.
[74]Beiträge zur Zahlentheorie, 1891, 135–143.
[75]Bull. Soc. Philomathique de Paris, (8), 4, 1891–2, 66–70.
[76]Die Elemente der Zahlentheorie, 1892, 89.

G. B. Mathews[77] reproduced art. 81 of Gauss[7] and gave a second proof by use of Cauchy's[14] congruence $X \equiv 0$ for $n = p - 1$.

K. Zsigmondy[78] treated the problem to find all integers K, relatively prime to given integers a and b, such that $a^\sigma \equiv b^\sigma$ (mod K) holds for the given integral value $\sigma = \gamma$, but for no smaller value. For $b = 1$, it is a question of the moduli K with respect to which a belongs to the exponent γ. Set $\gamma = \Pi q_i^{s_i}$, where the q's are distinct primes and q_1 the greatest. Then all the primes K for which $a^\sigma \equiv b^\sigma$ (mod K) holds for $\sigma = \gamma$, but for no smaller σ, coincide with the prime factors of

$$\Delta = \frac{(a^\gamma - b^\gamma)\Pi(a^{\frac{\gamma}{qq'}} - b^{\frac{\gamma}{qq'}}) \cdots}{\Pi(a^{\gamma/q} - b^{\gamma/q}) \cdots},$$

in which the products extend over the combinations of $q_1, q_2, \ldots$ one, two, $\ldots$ at a time, provided that, if $a^\sigma \equiv b^\sigma$ (mod q_1) for $\sigma = \gamma/q_1^{s_1}$, but for no smaller σ, we do not include among the K's the prime q_1, which then occurs in Δ to the first power only. If the prime p is a K and if p^e is the highest power of p dividing Δ, then p^e is the highest power of p giving a K. The composite K's are now easily found. If a and b are not both numerically equal to unity, it is shown that there is at least one prime K except in the following cases: $\gamma = 1$, $a - b = 1$; $\gamma = 2$, $a + b = \pm 2^\mu$ ($\mu \geqq 1$); $\gamma = 3$, $a = \pm 2$, $b = \mp 1$; $\gamma = 6$, $a = \pm 2$, $b = \pm 1$. The case $b = 1$ shows that, apart from the corresponding exceptions, there exists a prime with respect to which the given integer $a \neq \pm 1$ belongs to the given exponent γ. As a corollary, every arithmetical progression of the type $\mu\gamma + 1$ ($\mu = 1, 2, \ldots$) contains an infinitude of primes.

Zsigmondy[79] considered the function $\Delta_\gamma(a)$ obtained from the above Δ by setting $b = 1$. If a is a primitive root of the prime $p = 1 + \gamma$, the main theorem of the last paper shows that p divides $\Delta_\gamma(a)$. Conversely, $1 + \gamma$ is a prime if it divides Δ. Thus, if all the primes of a set of integers possess the same primitive root a, any integer p of the set is a prime if and only if $\Delta_{p-1}(a)$ is divisible by p. Hence theorems due to Tchebychef[34] imply criteria for primes. For example, a prime $2^{2n} + 1$ has the primitive root 3 implies that $2^{2n} + 1$ is a prime if and only if it divides $3^k + 1$, where $k = 2^{2n}$. Since ± 2 is a primitive root of any prime $2q + 1$ such that q is a prime $4k \pm 1$, we infer that, if q is a prime $4k \pm 1$, then $2q + 1$ is a prime if and only if it divides $(2^q \pm 1)/(2 \pm 1)$. Since 2 is a primitive root of a prime $4N + 1$ such that N is an odd prime, we infer that, if N is an odd prime, $4N + 1$ is a prime if and only if it divides $(2^{2N} + 1)/5$.

G. F. Bennett[80] proved (pp. 196–7) the first theorem of Cauchy,[26] and (pp. 199–201) the results of Sancery.[61] If a and a' belong to exponents t and t' which contain no prime factor raised to the same power in each, then the exponent to which aa' belongs is the l. c. m. of t and t' (p. 194).

[77]Theory of Numbers, 1892, 23–25.
[78]Monatshefte Math. Phys., 3, 1892, 265–284.
[79]Ibid., 4, 1893, 79–80.
[80]Phil. Trans. R. Soc. London, 184 A, 1893, 189–245.

If 2^{s+1} is the highest power of 2 dividing $a^2 - 1$, where a is odd, the exponent to which a belongs modulo 2^λ is $2^{\lambda-s}$ if $\lambda > s$, but, if $\lambda \leqq s$, is 1 if $a \equiv 1$, 2 if $a^2 \equiv 1$, $a \not\equiv 1 \pmod{2^\lambda}$; the result of Lebesgue[48] (p. 87) now follows (pp. 202–6). In case a is not prime to the modulus, there is an evident theorem on the earliest power of a congruent to a higher power (p. 209). If e is a given divisor of $\phi(m)$, there is determined the number of integers belonging to exponent e modulo m [cf. Erlerus[25]]. If $a, a', \ldots$ belong to the exponents $t, t', \ldots$ and if no two of the $tt' \ldots$ numbers $a^r a'^{r'} \ldots$ ($0 \leqq r < t, 0 \leqq r' < t', \ldots$) are congruent modulo m, then $a, a', \ldots$ are called independent generators of the $\phi(m)$ integers $< m$ and prime to m (p. 195); a particular set of generators is given and the most general set is investigated (pp. 220–241) [a special problem on abelian groups].

J. Perott[81] found a number belonging to an exponent which is the l. c. m. of the exponents to which given numbers belong. If, for a prime modulus p, a belongs to an exponent $t > 1$, and b to an exponent which divides t, then b is congruent to a power of a (proof by use of Newton's relations between the sums of like powers of $a, \ldots, a^t$ and their elementary symmetric functions). Hence there exists a primitive root of p.

M. Frolov[82] noted that all the quadratic non-residues of a prime modulus m are primitive roots of m if $m = 2^{2c} + 1$, $m = 2n + 1$ or $4n + 1$ with n an odd prime [Tchebychef[34]]. To find primitive roots of m "without any trial," separate the $m - 1$ integers $< m$ into sets of fours $a, b, -a, -b$, where $ab \equiv 1 \pmod m$. Begin with one such set, say 1, 1, -1, -1. Either a or $m - a$ is even; divide the even one by 2 and multiply the corresponding $\pm b$ by 2; we get another set of four. Repeat the process. If the resulting series of sets contains all $m - 1$ integers $< m$, 2 and -2 are primitive roots if $m = 4h + 1$, and one of them is a primitive root if $m = 4h - 1$. If the sets just obtained do not include all $m - 1$ integers $< m$, further theorems are proved.

G. Wertheim[83] gave the least primitive root of each prime $p < 3000$.

L. Gegenbauer[83a] gave two expressions for the sum s_k of those terms of a complete set of residues modulo p which belong to the exponent k, and evaluated $\Sigma s_{k/t} f(t)$ with t ranging over the divisors of k.

G. Wertheim[84] proved that any prime $2^{4n} + 1$ has the primitive root 7. If $p = 2^n q + 1$ is a prime and q is a prime > 2, any quadratic non-residue m of p is a primitive root of p if $m^{2^n} - 1$ is not divisible by p. As corollaries, we get primes q of certain linear forms for which 2, 5, 7 are primitive roots of a prime $2q + 1$ or $4q + 1$; also, 3 is a primitive root of all primes $8q + 1$ or $16q + 1$ except 41; and cases when 5 or 7 is a primitive root of primes $8q + 1$, $16q + 1$. There is given a table showing the least primitive root of each prime between 3000 and 3500.

[81]Bull. des Sc. Math., (2), 17, I, 1893, 66–83.
[82]Bull. Soc. Math. de France, 21, 1893, 113–128; 22, 1894, 241–5.
[83]Acta Mathematica, 17, 1893, 315–20; correction, 22, 1899, 200 (10 for $p = 1021$).
[83a]Denkschr. Ak. Wiss. Wien (Math.), 60, 1893, 48–60.
[84]Zeitschrift Math. Naturw. Unterricht, 25, 1894, 81–97.

J. Perott[85] employed the sum s_k of the kth powers of $1, 2, \ldots, p-1$, and gave a new proof that $s_1 \equiv 0, \ldots, s_{p-2} \equiv 0, s_{p-1} \equiv -1 \pmod{p}$. If m is the l. c. m. of the exponents to which $1, 2, \ldots, p-1$ belong, evidently $s_m \equiv p-1$, whence $m > p-2$. If A belongs to the exponent m, then $A, A^2, \ldots, A^m$ are incongruent, whence $m \leq p-1$. Thus A is a primitive root.

N. Amici[86] proved that, if $\nu > 2$, a number belongs to the exponent $2^{\nu-2}$ modulo 2^ν if and only if it is of the form $8h \pm 3$, and called such numbers quasi primitive roots of 2^ν. For a base $8h \pm 3$, numbers of the two forms $8k+1$ or $8k \pm 3$, and no others, have indices. The product of two numbers having indices has an index which is congruent modulo $2^{\nu-2}$ to the sum of the indices of the factors. The product of two numbers b_1 and b_2, neither with an index, has an index congruent modulo $2^{\nu-2}$ to the sum of the indices of $-b_1$ and $-b_2$. The product of a number with an index by one without an index has no index.

K. Zsigmondy[87] proved by use of abelian groups that, if $\delta = q_1^{k_1} \ldots q_r^{k_r}$, $m = p_1^{\pi_1} \ldots p_s^{\pi_s}$, where $q_1, \ldots, q_r$ are distinct primes, and $p_1, \ldots, p_s$ are distinct primes, the number of incongruent integers belonging to the exponent δ modulo m is

$$\delta_1 \ldots \delta_s \prod_{i=1}^{r} (1 - 1/q_i^{l_i}),$$

where δ_j is the g. c. d. of δ and $t_j = \phi(p_j^{\pi_j})$, while l_i is the number of the integers $t_1, \ldots, t_s$ which contain the factor $q_i^{k_i}$.

E. de Jonquières[88] proved that the product of an even number of primitive roots of a prime p is never a primitive root, while the product of an odd number of them is either a primitive root or belongs to an exponent not dividing $(p-1)/2$. Similar results hold for products of numbers belonging to like exponents. Certain of the n integers r, for which r^n is a given number belonging to the exponent $e = (p-1)/n$, belong to the exponent ne, while the others (if any are left) belong to an exponent ke, where k divides n. He conjectured that 2 is not a primitive root of a prime $p \equiv 1, 7, 17$ or 23 (mod 24); 3 not of $p \equiv 1, 11, 13$ or 23 (mod 24); 5 not of $p \equiv 1, 11, 19,$ or 29 (mod 30). These results and analogous ones for 7 and 11 were shown by him and T. Pepin[89] to follow from the quadratic reciprocity law and Gauss' theorems on the divisors of $x^2 \pm A$.

G. Wertheim[90] added to his[84] corollaries cases when 6, 10, 11, 13 are primitive roots of primes $2q+1, 4q+1$; also, 10 is a primitive root of all primes $8q+1 \neq 137$ for which q is a prime $10k+7$ or $10k+9$, and of primes $16q+1$ for which q is a prime $10k+1$ or $10k+7$.

Wertheim[91] gave the least primitive root of each prime between 3000 and 5000 and of certain higher primes. He noted errata in his[83] table to 3000.

[85]Bull. des Sc. Mathématiques, 18, I, 1894, 64–66.
[86]Rendiconti Circolo Mat. di Palermo, 8, 1894, 187–201.
[87]Monatshefte Math. Phys., 7, 1896, 271–2.
[88]Comptes Rendus Paris, 122, 1896, p. 1451; 124, 1897, p. 334, p. 428.
[89]Comptes Rendus Paris, 123, 1896, pp. 374, 405, 683, 737.
[90]Acta Math., 20, 1896, 143–152.
[91]Ibid., 153–7; corrections, 22, 1899, 200.

F. Mertens[92] called $i_1, \ldots, i_\rho$ the system of indices of n modulo k if $n \equiv g_1^{i_1} \ldots q_\rho^{i_\rho}$ (mod k) for the g's of Kronecker.[60] Such systems of indices differ from Dirichlet's.

C. Moreau[93] set $N = p^k q^h \ldots, \nu = p^{k-1} q^{h-1} \ldots$, where $p, q, \ldots$ are distinct primes. Take $\epsilon = 1$ if N is not divisible by 4 or if $N = 4$, but $\epsilon = 2$ if N is divisible by 4 and $N > 4$. Let $\psi(N)$ denote the l. c. m. of $\nu/\epsilon, p-1, q-1, \ldots$ [equivalent to Cauchy's[26] maximum indicator for modulus N]. For A prime to N, $A^{\psi(N)} \equiv 1$ (mod N). If $N = p^k$, $2p^k$ or 4 (so that N has primitive roots), $\psi(N) = \phi(N)$ [Lucas[73]]. There is a table of values of $N < 1000$ and certain higher values for which $\psi(N)$ has a given value < 100.

A. Cunningham[94] noted that we may often abbreviate Gauss' method of finding a primitive root of a prime p by testing whether or not the trial root a is a primitive root before computing the residues of all powers of a. The tests are the simple rules to decide whether or not a is a quadratic or cubic residue of p. If a is both a quadratic non-residue and a cubic non-residue of $p = 3\omega + 1$, and if $a^f \not\equiv 1$ for every f dividing $p-1$ except $f = p-1$, then a is a primitive root.

A. Cunningham[95] gave tables showing the residues of the successive powers of 2 when divided by each prime or power of prime < 1000, also companion tables showing the indices x of 2^x whose residues modulo p^k are 1, 2, 3, ... The tables are more convenient than Jacobi's Canon[23] (errata given here) for the problem to find the residue of a given number with respect to a given power of a prime, but less convenient for finding all roots of a given order of a given prime. There are given (p. 172) for each power $p^k < 1000$ of a prime p the factors of $\phi(p^k)$, the exponent ξ to which 2 belongs modulo p^k, and the quotient ϕ/ξ.

E. Cahen[96] proved that if p is a prime $> (3^{2^{m+1}} - 1)/2^{m+3}$ and if $q = 2^{m+2}p + 1$ ($m > 0$) is a prime, then 3 is a primitive root of q, whereas Tchebychef[34] had the less advantageous condition $p > 3^{2^{m+1}}/2^{m+2}$. Other related theorems by Tchebychef are proved. There are companion tables of indices for primes < 200.

G. A. Miller[97] applied the theory of groups to prove the existence of primitive roots of p^n, to show that the primitive roots of p^2 are primitive roots of p^n, and to determine primitive roots of the prime p.

L. Kronecker[98] discussed the existence of primitive roots, defined systems of indices and applied them to the decomposition of fractions into partial fractions. He developed (pp. 375–388) the theory of exponents to which numbers belong modulo p, a prime, by use of the primitive factor

[92]Sitzungsber. Ak. Wien (Math.), 106, II a, 1897, 259.
[93]Nouv. Ann. Math., (3), 17, 1898, 303.
[94]Math. Quest. Educat. Times, 73, 1900, 45, 47.
[95]A Binary Canon, showing residues of powers of 2 for divisors under 1000, and indices to residues, London, 1900, 172 pp. Manuscript was described by author, Report British Assoc., 1895, 613. Errata, Cunningham.[115]
[96]Éléments de la théorie des nombres, 1900, 335–9, 375–390.
[97]Bull. Amer. Math. Soc., 7, 1901, 350.
[98]Vorlesungen über Zahlentheorie, I, 1901, 416–428.

$F_d(x)$ of x^d-1 (dividing the last but not x^t-1 for $t<d$). To every divisor d of $p-1$ belong exactly $\phi(d)$ numbers which are the roots of $F_d(x)\equiv0$ (mod p).

P. G. Foglini[99] gave an exposition of known results on primitive roots, indices, linear congruences, etc. In applying (p. 322) Poinsot's[9] method of finding the primitive roots of a prime p to the case $p=13$, it suffices to exclude the residues of the cubes of the numbers which remain after excluding the residues of squares; for, if x is a residue of a square, $(x^3)^6\equiv1$ and x^3 is the residue of a square.

R. W. D. Christie[100] noted that, if γ is a primitive root of a prime $p=4k-1$,the remaining primitive roots are congruent to $p-\gamma^{2^n}$ $(n=1,2,\ldots)$

A. Cunningham[101] noted that 3, 5, 6, 7, 10 and 12 are primitive roots of any prime $F_n=2^{2^n}+1>5$. Also $F_n^{2^{n+2}}+1\equiv0$ (mod $F_{n+1}>5$).

E. I. Grigoriev[102] noted that a primitive root of a prime p can not equal a product of an even number of primitive roots [evident].

G. Wertheim[103] treated the problem to find the numbers belonging to the exponent equal to the l. c. m. of m, n, given the numbers belonging to the exponents m and n, and proved the first theorem of Stern.[15] He discussed (pp. 251–3) the relation between indices to two bases and proved (pp. 258, 402–3) that the sum of the indices of a number for the various primitive roots of $m=p^n$ or $2p^n$ equals $\frac{1}{2}\phi(m)\phi\{\phi(m)\}$. If a belongs to the exponent 4δ modulo p, the same is true of $p-a$ (p. 266). He gave a table showing the least primitive root of each prime <6200 and for certain larger primes; also tables of indices for primes <100.

P. Bachmann[104] gave a generalization (corrected on p. 402) of Stern's[15] first theorem.

G. Arnoux[105] constructed tables of residues of powers and tables of indices for low composite moduli.

A. Bindoni[106] noted that a table showing the exponent to which a belongs modulo p, a prime, can be extended to a table modulo N by means of the following theorems. Let $a, b_1,\ldots, b_n$ be relatively prime by twos. A number belonging to the exponent t_i modulo b_i belongs modulo $b_1b_2\ldots b_n$ to the l. c. m. of $t_1,\ldots, t_n$ as exponent. If t_i is the least exponent for which $a^{t_i}+1\equiv0$ (mod b_i) and if the t_i are all odd, the least t for which a^t+1 is divisible by $b_1,\ldots, b_n$ is the l. c. m. of $t_1,\ldots, t_n$. If p is an odd prime not dividing a and if a belongs to the exponent t modulo p, and $a^t=pq+1$, and if p^u is the highest power of p dividing q, then a belongs to the exponent tp^{n-1-u} modulo p^n. Hence if a is a primitive root of p, it is one of p^n if

[99]Memorie Pont. Ac. Nuovi Lincei, 18, 1901, 261–348.
[100]Math. Quest. Educat. Times, 1, 1902, 90.
[101]Ibid., pp. 108, 116.
[102]Kazani Izv. fiz. mat. obsc., Bull. Phys. Math. Soc. Kasan, (2), 12,.1902, No. 1, 7–10.
[103]Anfangsgründe der Zahlenlehre, 1902, 236–7, 259–262.
[104]Niedere Zahlentheorie, 1, 1902, 333–6.
[105]Assoc. franç. av. sc., 32, 1903, II, 65–114.
[106]Il Boll. di Matematica Giorn. Sc. Didat., Bologna, 4, 1905, 88–92.

and only if $a^{p-1}-1$ is not divisible by p^2. If t is even, the least x for which $a^x+1\equiv0 \pmod{p^n}$ is $\frac{1}{2}tp^{n-1-u}$.

M. Cipolla[107] gave a historical report on congruences (especially binomial), primitive roots, exponents, indices (in Peano's symbolism).

K. P. Nordlund[108] proved by use of Fermat's theorem that, if $n_1, \ldots, n_r$ are distinct odd primes, no one dividing a, then $N=n_1{}^{m_1}\ldots n_r{}^{m_r}$ divides a^k-1, where $k=\phi(N)/2^{r-1}$.

R. D. Carmichael[109] proved that the maximum indicator of any odd number is even; that of a number, whose least prime factor is of the form $4l+1$, is a multiple of 4; that of $p(2p-1)$ is a multiple of 4 if p and $2p-1$ are odd primes.

A. Cunningham[110] gave a table of the values of ν, where $(p-1)/\nu$ is the exponent to which 2 belongs modulo $p^n<10000$, the omitted values of p being those for which $\nu=1$ or 2 and hence are immediately distinguished by the quadratic character of 2 (extension of his Binary Canon[95]). A list is given of errata in the table by Reuschle.[45] An announcement is made of the manuscript of tables of the exponents to which 3, 5, 6, 7, 10, 11, 12 belong modulo $p^n<10000$, and the least positive and negative primitive roots of each prime <10000 [now in type and extended in manuscript to $p^n<22000$].

A. Cunningham[111] defined the sub-Haupt-exponent ξ_1 of a base q to modulus $m=q^{a_0}\eta_0$ (where η_0 is prime to q, and $a_0\geqq0$) to be the exponent to which q belongs modulo η_0. Similarly, let ξ_2 be the exponent to which q belongs modulo η_1, where $\xi_1=q^{a_1}\eta_1$; etc. Then the ξ's are the successive sub-Haupt-exponents, and the train ends with $\xi_{r+1}=1$, corresponding to $\eta_r=1$. His table I gives these ξ_k for bases $q=2, 3, 5$ and for various moduli including the primes <100.

Paul Epstein[112] desired a function $\psi(m)$, called the Haupt-exponent for modulus m, such that $a^{\psi(m)}\equiv1 \pmod{m}$ for every integer a prime to m and such that this will not hold for an exponent $<\psi(m)$. Thus $\psi(m)$ is merely Cauchy's[26] maximum indicator. Although reference is made to Lucas,[73] who gave the correct value of $\psi(m)$, Epstein's formula requires modification when $m=4$ or 8 since it then gives $\psi=1$, whereas $\psi=2$. The number $\chi(m, \mu)$ of roots of $x^\mu\equiv1 \pmod{m}$ is $2d_0d_1\ldots d_n$ if m is divisible by 4 and if μ is odd, but is $d_1\ldots d_n$ in the remaining cases, where, for $m=2^{a_0}p_1{}^{a_1}\ldots p_n{}^{a_n}$, d_i is the g. c. d. of μ and $\phi(p_i{}^{a_i})$, and d_0 the g. c. d. of μ and 2^{a_0-2}, when $a_0>1$. The number of integers belonging to the exponent $\mu=p^a q^\beta\ldots$ modulo m is

$$\{\chi(m, p^a)-\chi(m, p^{a-1})\}\{\chi(m, q^\beta)-\chi(m, q^{\beta-1})\}\ldots.$$

[107]Revue de Math. (Peano), Turin, 8, 1905, 89–117.
[108]Göteborgs Kungl. Vetenskaps-Handlingar, (4), 7–8, 1905, 12–14.
[109]Amer. Math. Monthly, 13, 1906, 110.
[110]Quar. Jour. Math., 37, 1906, 122–145. Manuscript announced in Mess. Math., 33, 1903–4, 145–155 (with list of errata in earlier tables); British Assoc. Report, 1904, 443; l'inter-médiaire des math., 16, 1909, 240; 17, 1910, 71. Cf. Cunningham.[133]
[111]Proc. London Math. Soc., 5, 1907, 237–274.
[112]Archiv Math. Phys., (3), 12, 1907, 134–150.

This formula is simplified in the case $\mu = \psi(m)$ and the numbers belonging to this Haupt-exponent are called primitive roots of m. The primitive roots of m divide into families of $\phi(\psi(m))$ each, such that any two of one family are powers of each other modulo m, while no two of different families are powers of each other. Each family is subdivided. In general, not every integer prime to m occurs among the residues modulo m of the powers of the various primitive roots of m.

A. Cunningham[113] considered the exponent ξ to which an odd number q belongs modulo 2^m; and gave the values of ξ when $m \leqq 3$, and when $q = 2^x \Omega \doteq 1$ (Ω odd), $m > 3$. When $q = 2^x \mp 1$ and $m > x+1$, the residue of $q^{\xi/2j}$ can usually be expressed in one of the forms 1 ∓ 2^a, $1 \mp 2^a \mp 2^\beta$.

G. Fontené[114] determined the numbers N which belong to a given exponent $p^{m-h}\delta$ modulo p^m, where δ is a given divisor of $p-1$, and $h \geqq 1$, without employing a primitive root of p^m. If $p > 2$, the conditions are that N shall belong to the exponent δ modulo p and that the highest power of p dividing $N^\delta - 1$ shall be p^h, $1 \leqq h \leqq m$.

*M. Demeczky[115] discussed primitive roots.

E. Landau[116] proved the existence of primitive roots of powers of odd primes, discussed systems of indices for any modulus n, and treated the characters of n.

G. A. Miller[117] noted that the determination of primitive roots of g corresponds to the problem of finding operators of highest order in the cyclic group G of order g. By use of the group of isomorphisms of G it is shown that the primitive roots of g which belong to an exponent $2q$, where q is an odd prime, are given by $-a^2$, when a ranges over those integers between 1 and $g/2$ which are prime to g. As a corollary, the primitive roots of a prime $2q+1$, where q is an odd prime, are $-a^2$, $1 < a < q+1$.

A. N. Korkine[118] gave a table showing for each prime $p < 4000$ a primitive root g and certain characters which serve to solve any solvable congruence $x^q \equiv a \pmod{p}$, where q is a prime dividing $p-1$. Let q^a be the highest power of q dividing $p-1$. The characters of degree q are the solutions of

$$u^q \equiv 1, \qquad u'^q \equiv u, \qquad u''^q \equiv u', \ldots, \qquad (u^{(a-1)})^q \equiv u^{(a-2)} \pmod{p}$$

and hence are the residues of the powers of $g^{(p-1)/q^k}$ for $k = 1, \ldots, a$. There are noted some errors in the Canon of Jacobi[23] and the table of Burckhardt. Korkine stated that if p is a prime and a belongs to the exponent $e = (p-1)/\delta$, exactly $\phi(p-1)/\phi(e)$ of the roots of $x^\delta \equiv a \pmod{p}$ are primitive roots of p.

K. A. Posse[119] remarked that Korkine constructed his table without knowing of the table by Wertheim,[91] and extended Korkine's tables to 10000.

[113]Messenger of Math., 37, 1907–8, 162–4.
[114]Nouv. Ann. Math., (4), 8, 1908, 193–216.
[115]Math. és Phys. Lapok, Budapest, 17, 1908, 79–86.
[116]Handbuch . . .Verteilung der Primzahlen, I, 1909, 391–414, 478–486.
[117]Amer. Jour. Math., 31, 1909, 42–4.
[118]Matem. Sborn. Moskva (Math. Soc. Moscow), 27, 1909, 28–115, 120–137 (in Russian). Cf. D. A. Grave, 29, 1913, 7–11. The table was reprinted by Posse.[128]
[119]Ibid., 116–120, 175–9, 238–257. Reprinted by Posse.[129]

R. D. Carmichael[120] called a number a primitive λ-root modulo n if it belongs to the exponent $\lambda(n)$, defined in Ch. III, Lucas.[110] The existence of primitive λ-roots g is proved. The product of those powers of g which are primitive λ-roots is $\equiv 1 \pmod{n}$ if $\lambda(n) > 2$. A method is given to solve $\lambda(x) = a$, and the solutions tabulated for $a \leq 24$.

C. Posse[121] noted that in Wertheim's[83, 91] table, the primitive root 14 of 2161 should be replaced by 23, while 10 is not a primitive root of 3851.

E. Maillet[122] described the manuscript table by Chabanel, deposited in the library of the University of Paris, giving the indices for primes under 10000 and data to determine the number having a given index.

F. Schuh[123] showed how to form the congruence for the primitive roots of a prime and gave two further proofs of the existence of primitive roots. He treated binomial congruences, quadratic residues and made applications to periodic fractions to any base. For any modulus n, he found the least m for which $x^m \equiv 1 \pmod{n}$ holds for every x prime to n, and derived the solutions n of $\phi(n) = m$, i. e., n's having primitive roots.

F. Schuh[124] discussed the solution of $x^q \equiv 1 \pmod{p^a}$ with the least computation. If x belongs to the exponent q modulo n, the powers of x give a cycle of $\phi(q)$ numbers each with the "period" q. The numbers prime to n and having the period q may form several such cycles—more than one if n has no primitive root and q is the maximum period. If $n = 2^a \ (a > 2)$, then $q = 2^s \ (s \leq a - 2)$ and the number of cycles is 1, 3 or 2 according as $s = 0$, $s = 1$ or $s > 1$. In the last case, the cycles are formed by $2^{a-s}(2k+1) \mp 1$.

When q is even, x is said to be of the first or second kind according as $x^{q/2} \equiv -1 \pmod{n}$ or not. If the numbers of a cycle are of the second kind, we get a new cycle of the second kind by changing the signs of the numbers of the first cycle. While for moduli n having primitive roots there exist no numbers of the second kind, when n has no primitive roots and q is a possible even period, there exist at least two cycles of the second kind and of period q. Finally, there is given a table showing the number of cycles of each kind for moduli ≤ 150.

M. Kraitchik[125] gave a table showing for each prime $p < 10000$ a primitive root of p and the least solutions of $2^x \equiv 1$, $10^y \equiv 1 \pmod{p}$.

*J. Schumacher[126] discussed indices.

L. von Schrutka[127] noted that, if $q, r, \ldots$ are the distinct primes dividing $p - 1$, where p is a prime, all non-primitive roots of p satisfy

$$\left(x^{\frac{p-1}{q}} - 1 \right)\left(x^{\frac{p-1}{r}} - 1 \right) \ldots \equiv 0 \pmod{p}.$$

[120]Bull. Amer. Math. Soc., 16, 1909–10, 232–7. Also, Theory of Numbers, pp. 71–4.
[121]Acta Math., 33, 1910, 405–6.
[122]L'intermédiaire des math., 17, 1910, 19–20.
[123]Supplement de Vriend der Wiskunde, Culemborg, 22, 1910, 34–114, 166–199, 252–9; 25, 1913, 33–59, 143–159, 228–259.
[124]Ibid., 23, 1911, 39–70, 130–159, 230–247.
[125]Sphinx-Oedipe, May, 1911, Numéro Spécial, pp. 1–10; errata listed p. 122 by Cunningham and Woodall. Extension to 25000, 1912, 25–9, 39–42, 52–5; errata, 93–4, by Cunningham.
[126]Blätter Gymnasial-Schulwesen, München, 47, 1911, 217–9.
[127]Monatshefte Math. Phys., 22, 1911, 177–186.

To this congruence he applied Hurwitz's[42] method (Ch. VIII) of finding the number of roots and concluded that there are $p-1-\phi(p-1)$ roots. Hence there exist $\phi(p-1)$ primitive roots of p.

A. Cunningham and H. J. Woodall[128] continued to $p < 100000$ the table of Cunningham[110] of the maximum residue indices ν of 2 modulo p.

C. Posse[129] reproduced Korkine's[118] and his own[119] tables and explained their use in the solution of binomial congruences.

C. Krediet[130] treated $x^\rho \equiv 1 \pmod{n}$ of Lucas,[110] Ch. III, and called x a primitive root if it belongs to the exponent φ. The powers of such a root are placed at equal intervals on a circle for various n's.

G. A. Miller[131] proved by use of group theory that, if m is arbitrary, the sum of those integers $< m$ and prime to m which belong to an exponent divisible by 4 is $\equiv 0 \pmod{m}$, and the sum of those belonging to the exponent 2 is $\equiv -1 \pmod{m}$, and proved the corresponding theorem by Stern[15] for a prime modulus.

A. Cunningham[132] tabulated the number of primes $p < 10^4$ for which y belongs to the same exponent modulo p, for $y = 2, 3, 5, 6, 7, 10, 11, 12$; and the number of primes p in each 10000 to 10^5 for which y ($y = 2$ or 10) belongs to the same exponent modulo p. Also, for the same ranges on p and y, the number of primes p for which $y^k \equiv 1 \pmod{p}$ is solvable, where k is a given divisor of $p - 1$.

A. Cunningham[133] stated that he had finished the manuscript of a table of Haupt-exponents to bases 3, 5, 6, 7, 11, 12 for all prime powers < 15000; also canons giving at sight the residues of z^r modulo $p^k < 10000$ for $z = 2$, $r \le 100$; $z = 3, 5, 7, 10, 11$, $r \le 30$.

J. Barinaga[134] considered a number a belonging to the exponent g modulo p, a prime. If a is not divisible by g, the sum of the ath powers of the numbers forming the period of a modulo p is divisible by p. The sum of their products n at a time is congruent to zero modulo p if $n < g$, but to ∓ 1 if $n = g$, according as g is even or odd.

A. Cunningham[135] listed errata in his Binary Canon[95] and Jacobi's Canon.[23]

G. A. Miller[136] employed the group formed by the integers $< m$ and prime to m, combined by multiplication modulo m, to show that, if a number is $\equiv \pm 1 \pmod{2^\gamma}$, but not modulo $2^{\gamma+1}$, where $1 < \gamma < \beta$, it belongs to the exponent $2^{\beta-\gamma}$ modulo 2^β. Also, if p is an odd prime, and $N \equiv 1 \pmod{p}$, N belongs to the exponent $p^{\beta-\gamma}$ modulo p^β if and only if $N - 1$ is divisible by p^γ, but not by $p^{\gamma+1}$, where $\beta > \gamma \ge 1$.

---

[128]Quar. Jour. Math., 42, 1911, 241–250; 44, 1913, 41–48, 237–242; 45, 1914, 114–125.
[129]Acta Math., 35, 1912, 193–231, 233–252.
[130]Wiskundig Tijdskrift, Haarlem, 8, 1912, 177–188; 9, 1912, 14–38; 10, 1913, 40–46, 87–97. (Dutch.)
[131]Amer. Math. Monthly, 19, 1912, 41–6.
[132]Proc. London Math. Soc., (2), 13, 1914, 258–272.
[133]Messenger Math., 45, 1915, 69.   Cf. Cunningham.[110]
[134]Annaes Sc. Acad. Polyt. do Porto, 10, 1915, 74–6.
[135]Messenger Math., 46, 1916, 57–9, 67–8.
[136]Ibid., 101–3.

A. Cunningham[137] gave five primes p for which there is a maximum number of exponents to which the various numbers belong modulo p.

On exponents and indices, see Lebesgue[174-6] and Bouniakowsky[179]; also Reuschle[69] of Ch. VI, Bouniakowsky[111] of Ch. XIV, and Calvitti[48] of Ch. XX.

Binomial Congruences.

Bháscara Achárya[149] (1150 A. D.) found y such that y^2-30 is divisible by 7 by solving $y^2=7c+30$. Changing 30 by multiples of 7, we reach a perfect square 16 with the root 4. Hence set

$$7c+30 = (7n+4)^2, \qquad c=7n^2+8n-2, \qquad y=7n+4.$$

Taking $n=1$, we get $y=11$. Such a problem is impossible if, after abrading the absolute term (30 above) by the divisor (7 above) and the addition of multiples of the divisor, we do not reach a square.

Similarly for the case of a cube, with corresponding conditions for impossibility (§206, p. 265). For $y^3=5e+6$, abrade 6 by the divisor 5 to get the cube 1; adding 43·5, we get $216=6^3$. Hence set $y=5n+6$.

An anonymous Japanese manuscript[150] of the first part of the eighteenth century gave a solution of $x^n-ky=a$ by trial. The residues $a_1, \ldots, a_{k-1}$ of $1^n, \ldots, (k-1)^n$ modulo k are formed; if $a_r=a$, then $x=r$. It was noted that $a_{k-r}=a_r$ or $k-a_r$ according as k is even or odd, and that the residue of r^n is r times that of r^{n-1}.

Matsunaga,[150a] in the first half of the eighteenth century, solved $a^2+bx=y^2$ by expressing b as a product mn and finding p, q and A so that $mp-nq=1$, $2pa\equiv A \pmod{n}$. Then $x=(Am-2a)A/n$ [and $y=a-mb$]. But if $Am=2a$, write $A+n$ in place of A and proceed as before. Or write $2a+b$ in the form $bQ+R$, whence $x=2a+b-(Q+1)R$. To solve $69+11x=y^2$, consider the successive squares until we reach $5^2\equiv3 \pmod{11}$. Write $2·5+11$ in the form $1·11+10$. Then for $a=5$, $b=11$, $Q=1$, $R=10$, the preceding expression for x becomes 1, whence $5^2+11·1=6^2$. Then write $2·6+11$ in the form $2·11+1$. Then $23-(2+1)·1=20$ gives $6^2+20·11=16^2$, and $x=(256-69)/11=17$.

L. Euler[151] proved that, if n divides $p-1$, where p is a prime, and if $a=c^n+kp$, then (by powering and using Fermat's theorem), $a^{(p-1)/n}-1$ is divisible by p. Conversely, if a^m-1 is divisible by the prime $p=mn+1$, we can find an integer y such that $a-y^n$ is divisible by p. For,

$$a^m-y^{mn} = (a-y^n)Q(y),$$

and the differences of order $mn-n$ of $Q(1)$, $Q(2), \ldots, Q(mn)$ are the same

[137]Math. Quest. and Solutions (Ed. Times), 3, 1917, 61–2; corrections, p. 65.

[149]Vija-gañita, §§ 204–5; Algebra, with arith. and mensuration, from the Sanscrit of Brahmegupta and Bháscara, transl. by H. T. Colebrooke, London, 1817, pp. 263–4.

[150]Abhand. Geschichte Math. Wiss., 30, 1912, 237.

[150a]Ibid., 234–5.

[151]Novi Comm. Acad. Petrop., 7, 1758–9 (1755), p. 49, seq., §64, §72, §77; Comm. Arith., 1, 270–1, 273. In Novi Comm., 1, 1747–8, p. 20; Comm. Arith., 1, p. 60, he proved the first statement and stated the converse

as those of the term y^{mn-n} for $y=1,\ldots,mn$, and hence equal $(mn-n)!$, so that $Q(y)$ is not divisible by p for some values $1,\ldots,mn$ of y.

Euler[152] recurred to the subject. The main conclusion here and from his former paper is the criterion that, if $p=mn+1$ is a prime, $x^n\equiv a$ (mod p) has exactly n roots or no root, according as $a^m\equiv1$ (mod p) or not. In particular, there are just m roots of $a^m\equiv1$, and each root a is a residue of an nth power.

Euler[152a] stated that, if $aq+b=p^2$, all the values of x making $ax+b$ a square are given by $x=ay^2\pm2py+q$.

J. L. Lagrange[153] gave the criterion of Euler, and noted that if p is a prime $4n+3$, $B^{(p-1)/2}-1$ is divisible by p, so that $x\equiv B^{n+1}$ is a root of $x^2\equiv B$ (mod p). Given a root ξ of the latter, where now p is any odd prime not dividing B, we can find a root of $x^2\equiv B$ (mod p^2) by setting $x=\xi+\lambda p$, $\xi^2-B=p\omega$. Then $x^2-B=(\lambda^2+\mu)p^2$ if $2\xi\lambda+\omega=\mu p$. The latter can be satisfied by integers λ, μ since 2ξ and p are relatively prime. We can proceed similarly and solve $x^2\equiv B$ (mod p^n).

Next, consider $\xi^2\equiv B$ (mod 2^n), for $n>2$ and B odd (since the case B even reduces to the former). Then $\xi=2z+1$, $\xi^2-B=Z+1-B$, where $Z=4z(z+1)$ is a multiple of 8. Thus $1-B$ must be a multiple of 8. Let $n>3$ and $1-B=2^r\beta$, $r>3$. If $r\geqq n$, it suffices to take $z=2^{n-2}\zeta$, where ζ is arbitrary. If $r<n$, Z must be divisible by 2^r, whence $z=2^{r-2}\zeta$ or $2^{r-2}\zeta-1$. Hence $w\equiv\zeta(2^{r-2}\zeta\pm1)+\beta$ must be divisible by 2^{n-r}. If $n-r\leqq r-2$, it suffices to take $\zeta\pm\beta$ divisible by 2^{n-r}. The latter is a necessary condition if $n-r>r-2$. Thus $\zeta=2^{r-2}\rho\mp\beta$, $w=2^{r-2}(\zeta^2\pm\rho)$. Hence $\zeta^2\pm\rho$ must be divisible by 2^{n-2r+2}. We have two sub-cases according as the exponent of 2 is $\leqq$ or $>r-1$; etc.

Finally, the solution of $x^2\equiv B$ (mod m) reduces to the case of the powers of primes dividing m. For, if f and g are relatively prime and ξ^2-B is divisible by f, and ψ^2-B by g, then x^2-B is divisible by fg if $x=\mu f\pm\xi=\nu g\pm\psi$. But the final equality can be satisfied by integers μ, ν since f is prime to g.

A. M. Legendre[154] proved that if p is a prime and ω is the g. c. d. of n and $p-1=\omega p'$, there is no integral root of

$$(1)\qquad\qquad x^n\equiv B\ (\text{mod }p)$$

unless $B^{p'}\equiv1$ (mod p); if the last condition is satisfied, there are ω roots of (1) and they satisfy

$$(2)\qquad\qquad x^\omega\equiv B^l\ (\text{mod }p),$$

where l is the least positive integer for which

$$(3)\qquad\qquad ln-q(p-1)=\omega.$$

For, from (1) and $x^{p-1}\equiv1$, we get $x^{ln}\equiv B^l$, $x^{q(p-1)}\equiv1$, and hence (2), by use of (3). Set $n=\omega n'$. Then, by (2) and (1),

$$B^{n'l}\equiv x^n\equiv B,\qquad B^{p'l}\equiv x^{p'\omega}\equiv x^{p-1}\equiv1\ (\text{mod }p).$$

[152]Novi Comm..Petrop., 8, 1760–1, 74; Opusc. Anal. 1, 1772, 121; Comm. Arith., 1, 274, **487**.
[152a]Opera postuma, I, 1862, 213–4 (about 1771).
[153]Mém. Acad. R. Sc. Berlin, 23, année 1767, 1769; Oeuvres, 2, 497–504.
[154]Mém. Ac. R. Sc. Paris, 1785, 468, 476–481. (Cf. Legendre.[155])

Since $ln' - qp' = 1$, the first gives $B^{qp'} \equiv 1$. Hence

$$B^{p'} = B^{p'(ln'-qp')} = (B^{p'l})^{n'}/(B^{qp'})^{p'} \equiv 1 \ (\text{mod } p).$$

Conversely, if $B^{p'} \equiv 1$,

$$x^{p-1} - 1 \equiv x^{p'\omega} - B^{p'l} \ (\text{mod } p)$$

has the factor $x^\omega - B^l$, so that (Lagrange[3]) congruence (2) has ω roots.

If $4n$ divides $p-1$, the roots of $x^{2n} \equiv -1$ (mod p) are the odd powers of an integer belonging to the exponent $4n$ modulo p.

Let n divide $p-1$, and m divide $(p-1)/n$. Let ω be the g. c. d. of m, n and set $n = \omega\nu$. Determine positive integers l and q such that $l\nu - qm = 1$. If $B^m \equiv \pm 1$ (mod p), (1) is satisfied by the roots of $x^\omega \equiv B^l y$ (mod p), where y ranges over the roots of $y^\nu \equiv (\pm 1)^q$ (mod p). For, the last two congruences give

$$x^n = x^{r\omega} \equiv B^{rl}y^r \equiv B^{qm+1}(\pm 1)^q \equiv B \ (\text{mod } p).$$

Hence by means of the roots of $y^\nu \equiv \pm 1$, we reduce the solution of (1) to binomial congruences of lower degrees. In particular, let $n=2$, $m=(p-1)/2$, and let 2 be prime to m, so that $p = 4a-1$, $l=a$, $q=1$. Then $x^2 \equiv B$ (mod p) requires that $B^m \equiv 1$, so that we have the solutions $x \equiv \pm B^a$ without trial (Lagrange[153]). Next, if $n=2$ and $B^{2k+1} \equiv -1$, the theorem gives $x \equiv B^{k+1}y$, where $y^2 \equiv -1$. But we may generalize the last result. Consider $x^2 + c^2 \equiv 0$ (mod p). Since p must have the form $4a+1$, we have $p = f^2 + g^2$. Determine u and z so that $c = gu - pz$. Then $x \equiv fu$ (mod p).

Let a belong to the exponent nw modulo p, where w divides $(p-1)/n$. Then the roots of $B^w \equiv 1$ (mod p) are $B \equiv a^{n\mu}$ $(\mu=1, \ldots, w-1)$, and, for a fixed B, the roots of (1) are $x = a^{mw+\mu}$ $(m=0, 1, \ldots, n-1)$. For, a^n belongs to the exponent w, whence $B \equiv a^{n\mu}$.

Legendre[155] gave the same theorems in his text. He added that, knowing a root θ of (1), it is easy to find a root of $x^n \equiv B$ (mod p^a), with the possible exception of the case in which n is divisible by p. Let $\theta^n - B = Mp$ and set $x = \theta + Ap$. Then $x^n - B$ is divisible by p^2 if

$$M + n\theta^{n-1}A = pM',$$

which can be satisfied by integers A, M' if n is not divisible by p. To solve (1) when p is composite, $p = a^\alpha b^\beta \ldots$, where $a, b, \ldots$ are distinct primes, determine all the roots λ of $\lambda^n \equiv B$ (mod a^α), all the roots μ of $\mu^n \equiv B$ (mod b^β),.... Then if $x \equiv \lambda$ (mod a^α), $x \equiv \mu$ (mod b^β),..., x will range over all the roots of (1).

Legendre[156] noted that if p is a prime $8n+5$ we can give explicitly the solutions of $x^2 + a \equiv 0$ (mod p) when it is solvable, viz., when $a^{4n+2} \equiv 1$. For, either $a^{2n+1} + 1 \equiv 0$ and $x = a^{n+1}$ is a solution, or $a^{2n+1} - 1 \equiv 0$ and $\theta = a^{n+1}$ satisfies $\theta^2 - a \equiv 0$ (mod p), so that it remains only to solve $x^2 + \theta^2 \equiv 0$, which was done at the end of his[154] memoir. For $p = 8n+1$, let $n = \alpha\beta$, where α is a power of 2 and β is odd; if $a^\beta \equiv \pm 1$, $x^2 + a \equiv 0$ can be solved as in the

[155]Théorie des nombres, 1798, 411–8; ed. 2, 1808, 349–357; ed. 3, 1830, Nos. 339–351; German transl. by Maser, 1893, 2, pp. 15–22.

[156]Ibid., 231–8; ed. 2, 1808, pp. 211–219; Maser, I, pp. 246–7.

case $p=8n+5$; but in general no such direct solution is known, and it is best to represent some multiple of p by the form y^2+az^2.

If we have found θ such that θ^2+a is divisible by the prime p, not dividing a, we readily solve $x^2+a\equiv0$ (mod p^n). For, from

$$(\theta\pm\sqrt{-a})^n=r\pm s\sqrt{-a}, \qquad (\theta^2+a)^n=r^2+as^2,$$

r^2+as^2 is divisible by p^n. Now s is not divisible by p. Thus we may take $r=sx+p^ny$, whence x^2+a is divisible by p^n. [Cf. Tchebychef, Theorie der Congruenzen, §30.]

The case of any composite modulus N is easily reduced to the preceding (end of Lagrange's[153] paper). Legendre proved that, if N is odd and prime to a, the number of solutions of $x^2+a\equiv0$ (mod N) is 2^{i-1} where i is the number of distinct prime factors of N; the same is true for modulus $2N$. Henceforth let N be odd or the double of an odd number and let d be the g. c. d. of N and a. If d has no square factor, the congruence has 2^{i-1} roots, where i is the number of distinct odd prime factors of N not dividing a. But if $d=\omega\psi^2$, where ω has no square factor, the congruence has $2^{i-1}\psi$ roots where i is the number of distinct odd prime factors of N/d.

C. F. Gauss[157] treated congruence (1) by the use of indices. However, we can give a direct solution (arts. 66–68) when a root is known to be congruent to a power of B. For, by (1) and $x\equiv B^k$, $B\equiv B^{kn}$. If therefore a relation of the last type is known, a root of (1) is B^k. The condition for the relation is $1\equiv kn$ (mod t), where t is the exponent to which B belongs modulo p. It is shown that t must divide $m=(p-1)/n$. We may discard from m any factor of n; if the resulting number is m/q, the unique solution k of $1\equiv kn$ (mod m/q) is the desired k. [Cf. Poinsot[165].]

Gauss (arts. 101–5) gave the usual method of reducing the solution of $x^2\equiv A$ (mod m) for any composite modulus to the case of a prime modulus and gave the number of roots modulo p^n in the various possible subcases. His well-known and practical "method of exclusion" (arts. 319–322) employs successive small powers of primes as moduli. Another method (arts. 327–8) is based on the theory of binary quadratic forms [cf. Smith[170]].

The congruence $ax^2+bx+c\equiv0$ (mod m) is reduced (art. 152) to $y^2\equiv b^2-4ac$ (mod $4am$). For each root y, it remains to solve $2ax+b\equiv y$ (mod $4am$).

Gauss[158] showed in a somewhat incomplete posthumous paper that, if t is a prime and $t^{r-1}(t-1)=a^\alpha b^\beta\ldots$, where $a, b,\ldots$ are distinct primes, the solution of $x^n\equiv1$ (mod t^r) may be made to depend upon the solution of a congruences of degree a, β congruences of degree b, etc. Use is made of the periods formed of the primitive roots of the congruence, as in Gauss' theory of roots of unity.

Legendre[159] solved $x^2+a\equiv0$ (mod 2^m) when a is of the form $-1\mp8a$ by

[157]Disquis. Arith., 1801, Arts. 60–65.

[158]Werke, 2, 1863, 199–211. Maser's German transl. of Gauss' Disq. Arith., etc., 1889, 589–601 (comments, p. 683).

[159]Théorie des nombres, ed. 2, 1808, pp. 358–60 (Nos. 350–2). Maser, 2, 1893, 25–7.

use of the expansion of $(1+z)^{1/2}$:

$$\sqrt{1\pm 8a}=1\pm\tfrac{1}{2}2^3a-\frac{1\cdot 1}{2\cdot 4}2^6a^2\pm\frac{1\cdot 1\cdot 3}{2\cdot 4\cdot 6}2^9a^3-\ldots\pm N2^{3n}a^n+\ldots,$$

$$N=\frac{1\cdot 1\cdot 3\cdot 5\ldots(2n-3)}{2\cdot 4\cdot 6\cdot 8\ldots 2n}.$$

The coefficient of a^n is an integer divisible by 2^{n+1}. Retain only the terms whose coefficients are not divisible by 2^{m-1} and call their sum θ. Hence every term of θ^2+a is divisible by 2^m. Thus the general solution of the proposed congruence is $x\equiv 2^{m-1}x'\pm\theta$.

P. S. Laplace[160] attempted to prove that, if p is a prime and $p-1=ae$, there exists an integer $x<e$ such that x^e-1 is not divisible by p. For, if $x=e$ and all earlier values of x make x^e-1 divisible by p,

$$f\equiv(e^e-1)-e\{(e-1)^e-1\}+\binom{e}{2}\{(e-2)^e-1\}-\ldots$$

would be divisible by p. The sum of the second terms of the binomials is

$$-1+e-\binom{e}{2}+\ldots=-(1-1)^e=0,$$

while the sum of the first terms of the binomials is $e!$ by the theory of differences, and is not divisible by p since $e<p$. [But the former equality implies that the last term of f is $(-1)^e(0-1)$, whereas the theorem is trivial if x is allowed to take the value 0. Again, nothing in the proof given prevents a from being unity; then the statement that there is a positive integer $x<p-1$ such that $x^{p-1}-1$ is not divisible by p contradicts Fermat's theorem.]

L. Poinsot[11] deduced roots of $x^n\equiv 1$ (mod p) from roots of unity.

M. A. Stern[15] (p. 152) proved that if n is odd and p is a prime, $x^n\equiv -1$ (mod p) is solvable and the number of roots is the g. c. d. of n and $p-1$; while, if n is even, it is solvable if and only if the factor 2 occurs in $p-1$ to a higher power than in n.

G. Libri[161] gave a long formula, involving sums of trigonometric functions, for the number of roots of $x^2+c\equiv 0$ (mod p).

V. A. Lebesgue[13] applied a theorem on $f(x_1,\ldots,x_k)\equiv 0$ to derive Legendre's[154] condition $B^{p'}\equiv 1$ for the existence of roots of (1), and the number of roots. Cf. Lebesgue[17] of Ch. VIII.

Erlerus[25] (pp. 9-13) proved that, if $p_1,\ldots,p_\mu$ are distinct odd primes,

$$x^2\equiv 1\ (\text{mod } 2^r p_1^{\epsilon_1}\ldots p_\mu^\epsilon\ )$$

has 2^μ, 2^μ, $2^{\mu+1}$ or $2^{\mu+2}$ roots according as $\nu=0$, 1, 2 or >2.

For the last result and the like number of roots of $x^2\equiv a$, see the reports, in Ch. III on Fermat's theorem, of the papers by Brennecke[57] and Crelle[58] of 1839, Crelle,[66] Poinsot[67] (erroneous) and Prouhet[69] of 1845, and Schering[102] of 1882.

C. F. Arndt[162] proved that the number of roots of $x^t\equiv 1$ (mod p^n) for

---

[160]Communication to Lacroix, Traité Calcul Diff. Int., ed. 2, vol. III, 1818, 723.

[161]Jour. für Math., 9, 1832, 175-7. See Libri,[16] Ch. VIII.

[162]Archiv Math. Phys., 2, 1842, 10-14, 21-22.

p an odd prime is the g. c. d. of t and $\phi(p^n)$; the same holds for modulus $2p^n$. He found the number of roots of $x^2 \equiv r \pmod{m}$, m arbitrary. By using $\Sigma\phi(t) = \delta$, if t ranges over the divisors of δ, he proved (pp. 25–26) the known result that the number of roots of $x^n \equiv 1 \pmod{p}$ is the g. c. d. δ of n and $p-1$. The product of the roots of the latter is congruent to $(-1)^{\delta+1}$; the sum of the roots is divisible by p; the sum of the squares of the roots is divisible by p if $\delta > 2$.

P. F. Arndt[163] used indices to find the number of roots of $x^3 \equiv a$.

A. L. Crelle[164] gave an exposition of known results on binomial congruences.

L. Poinsot[165] considered the direct solution of $x^n \equiv A \pmod{p}$, where p is a prime and n is a divisor of $p-1 = nm$ (to which the contrary case reduces). Let the necessary condition $A^m \equiv 1$ be satisfied. Hence we may replace A by A^{1+mk} and obtain the root $x \equiv A^e$ if $1+mk = ne$ is solvable for integers k, e, which is the case if m and n are relatively prime [cf. Gauss[157]]. The fact that we obtain a single root $x \equiv A^e$ is explained by the remark that it is a root common to $x^n \equiv A$ and $x^m \equiv 1$, which have a single common root when n is prime to m. Next, let n and m be not relatively prime. Then there is no root A^e if A belongs to the exponent m modulo p. But if A belongs to a smaller exponent m' and if m' is prime to n, there exists as before a root $A^{e'}$, where $1+m'k = ne'$. The number of roots of $x^n \equiv 1 \pmod{N}$ is found (pp. 87–101).

C. F. Arndt[166] proved that $x^t \equiv 1 \pmod{2^n}$, $n>2$, has the single root 1 if t is odd; while for t even the number of roots is double the g. c. d. of t and 2^{n-2}. The sum of the kth powers of the roots of $x^t \equiv 1 \pmod{p}$ is divisible by the prime p if k is not a multiple of t. By means of Newton's identities it is shown that the sum, sum of products by twos, threes, etc., of the roots of $x^t \equiv 1 \pmod{p}$ is divisible by the prime p, while their product is $\equiv +1$ or -1 according as the number of roots is odd or even. If the sum, sum of products by twos, threes, etc., of m integers is divisible by the prime p, while their product is $\equiv -(-1)^m$, the m integers are the roots of $x^m \equiv 1 \pmod{p}$.

A. Cauchy[167] stated that if $I = p^\lambda q^\mu \ldots$, where p, q, $\ldots$ are m distinct primes, and if n is an odd prime, $x^n \equiv 1 \pmod{I}$ has n^m distinct roots, including primitive roots, i. e., numbers belonging to the exponent n. [But $x^3 \equiv 1 \pmod{5}$ has a single root.]

Cauchy[168] later restricted p, q, $\ldots$ to be primes $\equiv 1 \pmod{n}$. Then $x^n \equiv 1 \pmod{p^\lambda}$ has a primitive root r_1, and $x^n \equiv 1 \pmod{q^\mu}$ has a primitive root r_2, so that $x^n \equiv 1 \pmod{I}$ has a primitive root, viz., an integer $\equiv r_1 \pmod{p^\lambda}$ and $\equiv r_2 \pmod{q^\mu}$, etc.; but no primitive root if p, q, $\ldots$ are not all $\equiv 1 \pmod{n}$.

[163]Von den Kubischen Resten, Torgau, 1842, 12 pp.
[164]Jour. für Math., 28, 1844, 111–154.
[165]Jour. de Mathématiques, (1), 10, 1845, 77–87.
[166]Archiv Math. Phys., 6, 1845, 380, 396–9.
[167]Comptes Rendus Paris, 24, 1847, 996; Oeuvres, (1), 10, 299.
[168]Comptes Rendus Paris, 25, 1847, 37; Oeuvres, (1), 10, 331.

Hoëné Wronski[169] stated without proof that, if $x^m \equiv a \pmod{M}$,

$$a = (-1)^{\omega+1} \{hK + (-1)^{k+1}\}^m A[M/K, \omega]^{\omega-2} + Mi,$$
$$x = h + (-1)^{\tau+k} A[M/K, \pi]^{\tau-1} + Mj,$$

and that M must be a factor of $aK^m - \{hK - (-1)^{k+1}\}^m$. Here the "alephs" $A[M/K, \omega]^r$, for $r = 0, 1, \ldots$, are the numerators of the reduced fractions obtained in the development of M/K as a continued fraction. In place of K, Wronski wrote the square of $1^{k/1} = k!$. Concerning these formulas, see Hanegraeff,[171] Bukaty,[180] Dickstein.[194] Cf. Wronski[151] of Ch. VIII.

E. Desmarest[37] noted that, if $x^2 + D \equiv 0 \pmod{p}$ is solvable, $x^2 + Dy^2 = mp$ can be satisfied by a value of $m < 3 + p/16$ and a value of $y \leq 3$. His proof is not satisfactory.

D. A. da Silva[42] (Alasia, p. 31) noted that $x^D \equiv 1 \pmod{m}$, where $m = p_1^{e_1} p_2^{e_2} \ldots$, has the roots $\Sigma x_i q_i m / p_i^{e_i}$ where x_i is a root of $x^{D_i} \equiv 1 \pmod{p_i^{e_i}}$, D_i being the g. c. d. of D and $\phi(p_i^{e_i})$, while the q's are integers such that $\Sigma q_i m / p_i^{e_i} \equiv 1 \pmod{m}$.

Da Silva[169a] proved that a solvable congruence $x^n \equiv r \pmod{m}$ can be reduced to the case r prime to m and then to the case $m = p^a$, p a prime > 2. Then, if δ is the g. c. d. of n and $\phi(p^a) = \delta \delta_1$, there is a root if and only if $r^{\delta_1} \equiv 1 \pmod{p^a}$ and hence if and only if $r^d \equiv 1 \pmod{p^{a'+1}}$, where $p^{a'}$ is the g. c. d. of n and p^{a-1}, while d is the quotient of $p-1$ by its g. c. d. with n.

H. J. S. Smith[170] indicated a simplification in Gauss'[157] second method of solving $x^2 \equiv A$. If $r^2 + D \equiv 0 \pmod{P}$ is solvable, $mP = x^2 + Dy^2$ is solvable for some value of $m < 2\sqrt{D/3}$. Employing all values of m under that limit for which also

$$\left(\frac{m}{D}\right) = \left(\frac{P}{D}\right),$$

and finding with Gauss all prime representations of the resulting products by the form $x^2 + Dy^2$, we get $\pm r \equiv x'/y'$, $x''/y'', \ldots \pmod{P}$, where x', y'; x'', y''; ... denote the sets of solutions of $mP = x^2 + Dy^2$.

Eg. Hanegraeff[171] reduced $x^m \equiv r$ to $\theta^m r \equiv 1 \pmod{p}$ by use of $\theta x \equiv 1$. When p/θ is developed into a continued fraction, let μ and $P_{\mu-1}$ be the number of quotients and number of convergents preceding the last. Let ν, $P_{\nu-1}$ be the corresponding numbers for p/θ^m. Then

$$x \equiv (-1)^{\mu-1} P_{\mu-1}, \qquad r \equiv (-1)^{\nu-1} P_{\nu-1} \pmod{p}.$$

For p a prime, we get all roots by taking $\theta = 1, \ldots, (p-1)/2$. By starting with $\theta(x-h) \equiv 1$ in place of $\theta x \equiv 1$, we get

[169]Réforme des Mathématiques, being Vol. i of Réforme du savoir humain, 1847. Wronski's mathematical discoveries have been discussed by S. Dickstein, Bibliotheca Math., (2), 6, 1892, 48–52, 85–90; 7, 1893, 9–14 [on analysis, (2), 8, 1894, 49, 85; (2), 10, 1896, 5]. Bull. Int. Ac. Sc. Cracovie, 1896; Rozprawy, Krakow, 4, 1913, 73, 396. Cf. l'intermédiaire des math., 22, 1915, 68; 23, 1916, 113, 164–7, 181–3, 199, 231–4; 25, 1918, 55–7.

[169a]C. Alasia, Annaes Sc. Acad. Polyt. do Porto, 9, 1914, 65–95. There are many confusing misprints; for example, five at the top of p. 76.

[170]British Assoc. Report, 1860, 120–, §68; Coll. M. Papers, 1, 148–9.

[171]Note sur l'équation de congruence $x^m \equiv r \pmod{p}$, Paris, 1860.

$$x-h\equiv(-1)^{\mu-1}P_{\mu-1}, \qquad r\equiv(-1)^{\nu-1}(\theta h+1)^m P_{\nu-1} \pmod{p}.$$

By taking $\theta=(1^{k/1})^2$ and replacing 1 by $(-1)^{k+1}$ in $\theta(x-h)\equiv 1$, the last results become the fundamental formula given without proof by Wronski[169] in his Réforme des Mathématiques.

G. L. Dirichlet[172] discussed the solution of $x^2\equiv D$ for any modulus.

G. F. Meyer[173] gave an elementary discussion of the solution of $x^3\equiv b$ (mod k), for k a prime, power of prime, or any integer.

V. A. Lebesgue[174] employed a prime p, a divisor n of $p-1=nn'$, and a number a belonging to the exponent n' modulo p. Then the roots of $x^n\equiv a$ (mod p) are $a^\alpha b^\beta$, where b is not in the period of a, and b is a quadratic non-residue of p if a is a quadratic residue, and b^n is the least power of b congruent to a term of the period of a. If we set $b^n\equiv a^r$ (mod p), then must $na+\nu\beta\equiv 1$ (mod n'). The roots x are primitive roots of p. In the construction of a table of indices, his method is to seek a primitive root giving to ± 2 the minimum index (rather than to ± 10, used by Jacobi); thus we use the theorem for $a=\pm 2$.

Lebesgue[175] gave reasons why the conditions imposed on b in his preceding paper are necessary. He added that when we have found that $x^n\equiv a$ (mod p) leads to a primitive root $x=g$ of p, it is easy to solve $x^m\equiv r$ (mod p) when m divides $p-1$, by expressing r as a power of g by the equivalent of an abridged table of indices.

Lebesgue[176] noted that the usual method of solution by indices leads to the theorem: If a belongs to the exponent e modulo p, and if n divides $p-1$, and we set $n=e'm$, where e' has only prime factors which divide e, while m is prime to e, then, for every divisor M of m, $x^n\equiv a$ (mod p) has $e'\phi(M)$ roots belonging to the exponent M.

If a belongs to the exponent e modulo p, there are $e\phi(n)$ numbers b, not in the period of a, for which $b^n\equiv a^i$ (mod p), with n a minimum. A common divisor of n and i does not divide e. Then the n roots of $x^n\equiv a$ (mod p) are $a^t b^u$, where $nt-iu-1=ev$, $t<e$, $u<n$. This generalization of his[174] earlier theorem is used to find the period of a primitive root of p from the period of 2.

R. Gorgas[177] stated that, if ρ is the residue modulo M of the pth term of $\{(M-1)/2\}^2,\ldots,2^2,1^2$, then $p(p-1)=\rho\pm m+Ma$, according as $M=4m\pm 1$. Take the lower signs and solve for p; we get

$$2p=1\pm b, \qquad b^2=M(4a-1)+4\rho.$$

Set $4\rho=Mc+\rho'$. Hence the initial equation $x^2=My+\rho$ has been replaced by $b^2=M(4a+c-1)+\rho'$ of like form. Let ρ' be the p'th place from the end. The process may be repeated until we reach an equation $P(P-1)=MA+\rho_m-m$ solvable by inspection.

[172]Zahlentheorie, 1863, §§32-7; ed. 2, 1871; ed. 3, 1879; ed. 4, 1894.
[173]Archiv Math. Phys., 43, 1865, 413-36.
[174]Comptes Rendus Paris, 61, 1865, 1041-4.
[175]Ibid., 62, 1866, 20-23.
[176]Ibid., 63, 1866, 1100-3.
[177]Ueber Lösung dioph. Gl. 2. Gr., Progr., Magdeburg, 1867.

Ladrasch[178] obtained known results on $x^3 \equiv a$ for any modulus.

V. Bouniakowsky[179] gave a method of solving $q \cdot 3^x \equiv \pm r \pmod{P}$, where P is odd. His first illustration is $3^x \equiv \pm 1 \pmod{25}$. Write the integers $\leq (25-1)/2$ in a line. Under the first four write in order the integers $\equiv 0 \pmod 3$; under the next four write in reverse order those $\equiv 1$; under the last four write in order those $\equiv 2$.

$$\begin{array}{cccc|cccc|cccc}
1^* & 2^* & 3^* & 4^* & 5 & 6^* & 7^* & 8^* & 9^* & 10 & 11^* & 12^* \\
3 & 6 & 9 & 12 & 10 & 7 & 4 & 1 & 2 & 5 & 8 & 11
\end{array}$$

Mark with an asterisk 1 in the first line; below it lies 3; mark with an asterisk 3 in the first line; etc. The number 10 of the integers marked with an asterisk is the least solution x of $3^x \equiv -1 \pmod{25}$. The sign is determined by the number of integers in the second set marked by an asterisk. The method applies to any $P = 6n+1$. But for $P = 6n+5$, we use for the second set of numbers in the second line those $\equiv 2 \pmod 3$ in reverse order, and for the third set those $\equiv 1$ in order. If $P = 23$, we see that each of the 11 numbers in the first line are marked with an asterisk, whence $3^{11} \equiv -1 \pmod{23}$. A like marking occurs for $P = 5, 11, 17, 29$. For $P = 35$, 12 numbers are marked, whence 12 is the least x for which $3^x \equiv 1 \pmod{35}$. Starting with the unmarked number 5, we get the cycle 5, 15, 10, whence $3^3 \equiv -1 \pmod 7$; similarly, the cycle 7, 14 gives $3^2 \equiv -1 \pmod 5$.

For $q \cdot 3^x \equiv \pm 4 \pmod{25}$, we begin with 4 in the second row. Since it lies below 7, we mark 7 with an asterisk in the second row; etc. Wê use an affix n on the number which is the nth marked by an asterisk.

$$\begin{array}{cccc|cccc|cccc}
1 & 2 & 3 & 4 & 5 & 6 & 7 & 8 & 9 & 10 & 11 & 12 \\
3^{*6} & 6^{*3} & 9^{*5} & 12^{*10} & 10 & 7^{*2} & 4^{*1} & 1^{*7} & 2^{*4} & 5 & 8^{*8} & 11^{*9}
\end{array}$$

For $q = 11$, we have the entry 8^{*8} below 11; hence $11 \cdot 3^8 \equiv -4$, the sign following from the number of entries ≤ 8 in the second set which are marked with an asterisk. Similarly for any $q \leq 12$, except $q = 5, 10$.

Bukaty[180] discussed the formula of Wronski.[169]

T. N. Thiele[181] used a mosaic (empty and filled squares on cross-section paper) to test $y^2 \equiv d \pmod c$, where c is an integer or Gauss complex integer $a + b\sqrt{-1}$, employing the graph of $y^2 - cx = d$.

Dittmar[182] discussed $x^3 \equiv r \pmod p$. Using Cauchy's[14] explicit congruence for the numbers belonging to a given exponent, he gave the expanded form of the congruence with the roots belonging to the successive exponents $1, \ldots, 21$.

---

[178]Von den Kubischen Resten u. Nichtresten, Progr., Dortmund, 1870.

[179]Bull. Ac. Sc. St. Pétersbourg, 14, 1870, 356–375.

[180]Déduction et démonstration de trois lois primordiales de la congruence des nombres, Paris, 1873.

[181]"Om Talmonstre," Forhandl. Skandinaviske Naturforskeres, Kjobenhavn, 11, 1873, 192–5.

[182]Die Theorie der Reste, insbesondere derer vom 3. Grade, nebst einer Tafel der Kubischen Reste aller Primzahlen der Form $6n+1$ zwischen den Grenzen 1 und 100. Progr. Köln Gym., Berlin, 1873.

L. Sancery[61] (pp. 17–23) employed the modulus $M = p^r$ or $2p^r$, where p is an odd prime. Let a belong to the exponent n modulo M. Let Δ be the g. c. d. of m and $\phi(M)/n$. Set $\Delta = \Delta_1\Delta_2$ where $\Delta_1 = p_1{}^{e_1}p_2{}^{e_2}\ldots$, and p_i is a prime dividing both Δ and n, and $p_i{}^{e_i}$ is the power of p_i dividing Δ. Let δ be any divisor of Δ_2. Then $x^m \equiv a \pmod{M}$ has $\phi(n\Delta_1\delta)/\phi(n)$ roots belonging to the exponent $n\Delta_1\delta$; the power $a\Delta_1\delta$ of such a root is congruent to a, where a can be found by means of a linear congruence. Given a number belonging to the exponent $n\Delta_1\delta$, we can find $\Delta_1\delta$ roots of the congruence.

C. G. Reuschle[182a] tabulated the roots of $f \equiv 0 \pmod{p}$, where $p = m\lambda + 1$ and λ are primes and f is the maximum irreducible algebraic prime factor of $a^\lambda - 1$; also the roots of

$$\eta^2 + c \equiv 0, \quad \eta^4 + c^2 \equiv 0, \quad \eta^8 + c^4 \equiv 0, \quad \eta^2 \pm \eta + d \equiv 0,$$

for $c < 13$, $d = -1$ to -26, $d = +2$ to $+21$, and for various cubic and quartic congruences.

A. Kunerth's method for $y^2 \equiv c \pmod{b}$ will be given in Vol. 2, Ch. XII.

E. Lucas[182b] treated $x^2 + 1 \equiv 0 \pmod{p^m}$, where p is a prime > 2, for use in the question of the number of satins. Given $a^2 + 1 \equiv 0 \pmod{p}$, set

$$(a+i)^m = A + Bi, \qquad \beta B \equiv 1 \pmod{p^m}.$$

Then $A\beta$ is a root x of the proposed congruence.

B. Stankewitsch[183] proved that if $x^2 \equiv q \pmod{p}$ is solvable, p being an odd prime, the positive root $< p/2$ is $\equiv B/A \pmod{p}$, where

$$A = S_{i-1} + qS_{i-3} + q^2S_{i-5} + \ldots + q^{\frac{i-2}{2}}S_1, \qquad B = S_i + qS_{i-2} + \ldots + q^{\frac{i}{2}},$$

where $i = (p-1)/2$ and S_k denotes the sum of the products of $1, 2, \ldots, i$ taken k at a time. Let n be a divisor of $p-1$. Let $F(x)$ be the g. c. d. modulo p of $x^n - 1$ and $\Pi(x^{n/a} - 1)$, where a ranges over the distinct prime factors of n. Call $f(x)$ the quotient of $x^n - 1$ by $F(x)$. Then the roots of $f(x) \equiv 0 \pmod{p}$ are the primitive roots of $x^n \equiv 1 \pmod{p}$. [Cf. Cauchy.[14]]

N. V. Bougaief[184] noted that if $p = 8n + 5$ is a prime and if $x^2 \equiv q \pmod{p}$ is solvable, it has the root $q^{(p+3)/8}$ or $\left(\frac{p-1}{2}\right)! \, q^{(p+3)/8}$ according as $q^{2n+1} \equiv 1$ or -1. If $p = 2^\lambda l + 1$, l odd, and $q^l \equiv 1$, it has the root $x \equiv q^{(l+1)/2}$. [Legendre.[156]]

T. Pepin[185] treated $x^3 \equiv 2$ by tables of indices.

P. Gazzaniga[186] gave a generalization of Gauss' lemma (the case $n = \delta = 2$,

[182a] Tafeln Complexer Primzahlen..., Berlin, 1875. Errata, Cunningham.[135]
[182b] Géométrie des tissus, Assoc. franç., 40, 1911, 83–6; French transl. of his Italian paper in l'Ingegnere Civile, 1880, Turin.
[183] Moscow Math. Soc., 10, 1882–3, I, 112 (in Russian).
[184] Ibid., p. 103.
[185] Atti Accad. Pont. Nuovi Lincei, 38, 1884–5, 201.
[186] Atti Reale Istituto Veneto, (6), 4, 1885–6, 1271–9.

$\nu = 0$). Separate the residues modulo p of kq, for $k = 1, 2, \ldots, (p-1)/\delta$, into three sets:

$$0 < r_1, \ldots, r_\rho < \frac{p}{\delta} < s_1, \ldots, s_\nu < \frac{\delta - 1}{\delta} p < t_1, \ldots, t_\mu < p$$

and form the differences $m_i = p - t_i$. From the set $1, \ldots, (p-1)/\delta$, delete the r_i and m_i; there remain ν numbers v_i. If y_i is a root of $s_i y_i \equiv v_i$ (mod p), then $x^n \equiv q$ (mod p) is solvable if and only if $(-1)^\nu y_1 \ldots y_\nu \equiv 1$ (mod p), where δ is the g. c. d. of n and $p-1$.

P. Seelhoff[187] gave the known cases in which $x^2 \equiv r$ (mod p) can be solved explicitly [Lagrange,[153] Legendre[156]]. In the remaining cases, one uses Gauss' method of exclusion, the process of Desmarest,[37] or, with Seelhoff, use various quadratic residues of p (*ibid.*, p. 306). Here $x^2 \equiv 41$ (mod 120097) is treated.

A. Berger[188] considered a quadratic congruence reducible to $x^2 \equiv D$ (mod $4n$), where $D \equiv 0$ or 1 (mod 4). If D is prime to n, the number of roots is

$$\psi(D, 4n) = 2\Pi\left\{1 + \left(\frac{D}{p}\right)\right\} = 2\Sigma\left(\frac{D}{d}\right)\zeta_d = 2\Sigma\left(\frac{D}{d}\right)\zeta_{d_1},$$

where p ranges over the distinct prime factors of n, while d and d_1 range over the pairs of complementary divisors of n, and $\zeta_d = 0$ or 1 according as d has a square factor or not. If $g(nm) = g(n)g(m)$ for all integers n, m, and $g(1) = 1$,

$$\Sigma\left(\frac{D^2}{n}\right)\psi(D, 4n)g(n) = 2\Sigma\left(\frac{D^2}{n}\right)g(n)\cdot\Sigma\left(\frac{D}{n}\right)g(n) \div \Sigma\left(\frac{D^2}{n}\right)g(n)^2,$$

where n ranges over all positive integers. Mean values are found:

$$\sum_{k=1}^{n}\left(\frac{D^2}{k}\right)\psi(D, 4k) = \frac{12n}{\pi^2\Pi(1+1/p)}\sum_{h=1}^{\infty}\left(\frac{D}{h}\right)\frac{1}{h} + \lambda n^{2/3},$$

$$\sum_{k=1}^{n}\psi(\Delta, 4k) = \frac{12n}{\pi^2}\sum_{h=1}^{\infty}\left(\frac{\Delta}{h}\right)\frac{1}{h} + \lambda_1 n^{2/3},$$

where Δ is a fundamental discriminant according to Kronecker, λ, λ_1 are finite for all n's, and p ranges over all primes.

G. Wertheim[189] presented the theory of $x^2 \equiv a$ (mod m).

R. Marcolongo[190] treated $x^2 + P \equiv 0$ (mod p) in the usual manner when explicit solutions are known. Next, from a particular set of solutions x, y of $x^2 + p^m y + P = 0$, where p is a prime > 2, we get the solution

$$\pm x_1 \equiv x - p^m y[a_1 \ldots a_{n-1}] \text{ (mod } p^{m+1})$$

of $x_1^2 + p^{m+1}y_1 + P = 0$, where $[a_1 \ldots a_{n-1}]$ is the numerator of next to the last convergent to the continued fraction for $p^m/(2x)$. The method is Serret's, Alg. Supér., II. For $p = 2$ the results obtained are the same as in Dirichlet's Zahlentheorie, §36.

---

[187]Zeitschrift Math. Phys., 31, 1886, 378–80.

[188]Öfversigt K. Vetenskaps-Ak. Förhandlingar, Stockholm, 44, 1887, 127–153. Nova Acta regiæ soc. sc. Upsalensis, (3), 12, 1884.

[189]Elemente der Zahlentheorie, 1887, 182–3, 207–217.   [190]Giornale di Mat., 25, 1887, 161–173.

F. J. Studnička[191] treated at length the solution in integers x, y ($y < b$) of $bx + 1 = y^2$, discussed by Leibniz in 1716.

L. Gegenbauer[192] gave a new derivation of the equations of Berger[188] leading to asymptotic expressions for the number of solutions of $x^2 \equiv D$.

A. Tonelli[193] gave a method of solving $x^2 \equiv c$ (mod p), when p is a prime $4h + 1$ and some quadratic non-residue g of p is known. Set $p = 2^s \gamma + 1$, where γ is odd. By Euler's criterion, the power $\gamma 2^{s-1}$ of c and g are congruent to $+1$, -1. Set $\epsilon_0 = 0$ or 1, according as the power $\gamma 2^{s-2}$ of c is congruent to $+1$ or -1. Then

$$g^{\epsilon_0 \gamma 2^{s-1}} c^{\gamma 2^{s-2}} \equiv +1 (\text{mod } p).$$

For $s \geq 3$, set $\epsilon_1 = 0$ or 1 according as the square root of the left member is $\equiv +1$ or -1. Then

$$g^{\epsilon_1 \gamma 2^{s-1} + \epsilon_0 \gamma 2^{s-2}} c^{\gamma 2^{s-3}} \equiv +1 \ (\text{mod } p).$$

Proceeding similarly, we ultimately get

$$g^{2e\gamma} c^\gamma \equiv +1 \ (\text{mod } p), \qquad e = \epsilon_0 + 2\epsilon_1 + \ldots + 2^{s-2}\epsilon_{s-2}.$$

Thus $x \equiv \pm g^{e\gamma} c^{(\gamma+1)/2}$ (mod p). Then $X^2 \equiv c$ (mod p^λ) has the root

$$X \equiv x^{p^{\lambda-1}} c^{(p^\lambda - 2p^{\lambda-1} + 1)/2} \ (\text{mod } p^\lambda).$$

G. B. Mathews[77] (p. 53) treated the cases in which $x^2 \equiv a$ (mod p) is solvable by formulas. Cf. Legendre.[156]

S. Dickstein[194] noted that H. Wronski[169] gave the solution

$$y = hK + (-1)^{k+1} + Mi, \qquad z = h + (-1)^{\pi+k} A\left[\frac{M}{K}, \pi\right]^{(\pi-1)} + Mj$$

of $z^n - ay^n \equiv 0$ (mod M) with $(1^k/^1)^2$ in place of K, and gave, as the condition for solvability,

$$a(1^k/^1)^{2n} - 1 \equiv 0 \ (\text{mod } M).$$

But there may be solutions when the last condition is satisfied by no integer k. This is due to the fact that the value assigned to y imposes a limitation, which may be avoided by using the same expressions for y, z in a parameter K, subject to the condition $aK^n - 1 \equiv 0$ (mod M).

M. F. J. Mann[194a] proved that, if $n = 2^k \lambda^a \mu^b \ldots$, where λ, μ, $\ldots$ are distinct odd primes, the number of solutions of $x^p \equiv 1$(mod n) is $GG_1G_2\ldots$ $g_1 g_2 \ldots$, where $G = 1$ if n or p is odd, otherwise G is the g. c. d. of $2p$ and 2^{k-1}, and where G_1, G_2, .., g_1, g_2, .. are the g. c. d.'s of p with λ^{a-1}, $\mu^{b-1}, \ldots$, $\lambda - 1$, $\mu - 1, \ldots$, respectively.

A. Tonelli[195] gave an explicit formula for the roots of $x^2 \equiv c$ (mod p^λ),

[191]Casopis, Prag, 18, 1889, 97; cf. Fortschritte Math., 1889, 30.
[192]Denkschriften Ak. Wiss. Wien (Math.), 57, 1890, 520.
[193]Göttingen Nachrichten, 1891, 344–6.
[194]Bull. Internat. de l'Acad. Sc. de Cracovie, 1892, 372 (64–65); Berichte Krakauer Ak. Wiss., 26, 1893, 155–9.
[194a]Math. Quest. Educ. Times, 56, 1892, 24–7.
[195]Atti R. Accad. Lincei, Rendiconti, (5), 1, 1892, 116–120.

when p is an odd prime, and a quadratic non-residue g of p is known. Set $p = 2^s a + 1$, where $s \geq 1$ and a is odd. Then $\gamma = a p^{\lambda-1}$ is odd, and $\phi(p^\lambda) = 2^s \gamma$. Tonelli's earlier work for modulus p now holds for modulus p^λ and we get $x \equiv \pm g^{c\gamma} c^{(\gamma+1)/2}$. If $s = 1$, then $e = 0$ and the root is that given by Lagrange if $\lambda = 1$. If $s = 2$, whence $p = 4a + 1 = 8l + 5$, the expression for x is given a form free of $e = \epsilon_0$:

$$x \equiv \pm (c^a + 3)^\gamma c^{(\gamma+1)/2}, \qquad \gamma = a p^{\lambda-1}.$$

A. Tonelli[196] expressed the root x in a form free of e for every s:

$$x \equiv \pm v_0^{\gamma 2^{s-2}} v_1^{\gamma 2^{s-3}} \ldots v_{s-3}^{2\gamma} v_{s-2}^\gamma c^{\frac{\gamma+1}{2}},$$

where the v's are given by the recursion formula

$$v_{s-h} = c^{2^{s-h}a} v_{s-2}^{2^{s-h+1}a} \ldots v_{s-h+1}^{2^{s-2}a} + k \qquad (h = 2, 3, \ldots).$$

Here k is an existing integer such that $k+1$ is a quadratic residue of p, and $k-1$ a non-residue. Thus, if $s = 3$,

$$x \equiv \pm (c^{2a} + k)^\gamma \{(c^{2a} + k)^{2a} c^a + k\}^{2\gamma} c^{\frac{\gamma+1}{2}},$$

where we may take $k = -2$ if a is not divisible by 3, but $k = -4$ if a is divisible by 3, while neither a nor $4a + 1$ are divisible by 5.

N. Amici[86] proved that $x^{2^k} \equiv b \pmod{2^\nu}$, b odd, $k \leq \nu - 2$, is solvable only when b is of the form $2^{k+2}h + 1$ and then has 2^{k+1} roots, as shown by use of indices. For $(x^m)^{2^k} \equiv b$, the same condition on b is necessary; thus it remains to solve $x^m \equiv \beta \pmod{2^\nu}$ when m is odd. If $\beta = 8k + 1$ or $8k + 3$, it has an index to the base $8h + 3$ and we get an unique root. If $\beta = 8k - 3$ or $8k - 1$, then $x^m \equiv -\beta$ has a root a by the preceding case, and $-a$ is a root of the proposed congruence.

Jos. Mayer[197] found the number of roots of $x^3 \equiv a \pmod{p^n}$, for the primes 2, 3, $p = 6m \pm 1$. If $a_1, a_2, \ldots$ are residues of nth powers modulo p, and if q is the g. c. d. of n and $p-1$, then $a_1 a_2 \ldots \equiv +1$ or $-1 \pmod{p}$, according as $p' = (p-1)/q$ is odd or even. If p' is even, we can pair the numbers belonging to the exponent p' so that the sum of a pair is 0 or p; hence there exists a residue of an nth power $\equiv -1 \pmod{p}$; but none if p' is odd.

K. Zsigmondy[87] obtained by the use of abelian groups known theorems on the number, product and sum of the roots of $x^\delta \equiv 1 \pmod{m}$.

G. Speckmann[198] considered $x^2 \equiv a \pmod{p}$, where p is an odd prime. Set $P = (p-1)/2$. When they exist, the roots may be designated $P - k$, $P + 1 + k$, whose sum is p. The successive differences of P^2, $(P+1)^2$, $(P+2)^2, \ldots$ are p, $p+2$, $p+4, \ldots$. The sum of $z = s+1$ terms of $2, 4, 6, \ldots$ is $s^2 + 3s + 2 = z^2 + z$. Adding to the latter the remainder r obtained by dividing P^2 by p, we must get $pn + a$. Hence in $pn + a - r$ we give to n the values

[196] Atti R. Accad. Lincei, Rendiconti, (5), 2, 1893, 259–265.
[197] Ueber $n$te Potenzreste und binomische Congruenzen dritten Grades, Progr., Freising, 1895.
[198] Archiv Math. Phys., (2), 14, 1896, 445–8; 15, 1897, 335–6.

0, 1, 2,... until we reach a number of the form z^2+z (found by extracting the square root). Then $k=z$, so that the roots $P-k$, $P+1+k$ are found.

N. Amici[199] proved that if neither m nor b is divisible by the prime p, and if a is a given root of $x^m \equiv b$ (mod p), and if β, q are (existing) integers such that

$$\beta\phi(p^\lambda)-p^{\lambda-1}+1=mq,$$

then $a^{p^{\lambda-1}}b^q$ is a root of $x^m \equiv b$ (mod p^λ). Hence we limit attention to the case $\lambda=1$. Consider henceforth $x^{2^k} \equiv b$ (mod p), where $p=2^s h+1$ is an odd prime, h being odd, and b not divisible by p. First, let $k \geqq s$. Then $b^h \equiv 1$ (mod p) is a necessary and sufficient condition for solvability and $x \equiv \pm b^q$ are roots, where q is such that $2^k q-1$ is divisible by h. If g is a quadratic non-residue of p, all 2^s roots are given by $\pm b^q g^{he}$, where $e=\epsilon_1+2\epsilon_2 +...+2^{s-2}\epsilon_{s-1}$, the ϵ_i taking the values 0 and 1 independently. Finally, let $k<s$. Then two roots $\pm\beta$ are determined by the method of Tonelli, while all the roots are given by

$$x \equiv \pm\beta g^{ht}, \qquad t=\epsilon_1+2\epsilon_2+...+2^{k-2}\epsilon_{k-1}, \qquad \epsilon_i=0 \text{ or } 1.$$

R. Alagna[200] considered a prime $p=4k+1$ for which k is a prime. Since 2 is known to be a primitive root of p, it is easy to write down those powers of 2 which give all the roots of $x^d \equiv 1$ (mod p), where d is one of the six divisors 2^i or $2^i k$ of $p-1$, likewise of $x^d \equiv N$, since N must be congruent to an even power of 2. For the modulus p^λ, we may apply the first theorem of Amici or proceed directly. The same questions are treated for a prime $4k+3$ for which $2k+1$ is a prime.

A. Cunningham[201] treated at length the solution of $x^l \equiv 1$ (mod N^t), where N is a prime, and gave tables showing all incongruent roots when $t=1, 2, N \leqq 101$, l any admissible divisor of $N-1$; also for a few additional t's when N is small.

Cunningham[201a] treated $a^p \equiv 1$ (mod q^2) and $3.2^x \equiv \pm 1$ (mod p). He[201b] treated the problem to find $b^\eta \equiv +1$ or $\pm a$, given $a^t \equiv 1$, $a^x \equiv \pm b$ (mod p), where ξ is odd and ξ, x, η are the least values of their kind; also given $a^\xi \equiv 1$, $a^x \equiv \pm b$, $a^z \equiv \pm c$, to find the least β and γ such that $b^\beta \equiv c$, $c^\gamma \equiv b$ (mod p).

W. H. Besant[202] would solve $y^2 = ax+b$ by finding the roots s of $s^2 \equiv b$ (mod a). Then $y=ar+s$, $x=ar^2+2rs+(s^2-b)/a$.

G. Speckmann[203] replaced $x^n \equiv k$ (mod p) by the pair of congruences $x^{n-1} \equiv r$, $xr \equiv k$ (mod p). In $np+k$ give to n the values 0, 1, 2,... until we find one for which $np+k=rx$ such that, by trial, $x^{n-1} \equiv r$. The method is, of course, impractical.

[199]Rendiconti Circolo Mat. di Palermo, 11, 1897, 43–57.
[200]Rendiconti Circolo Mat. di Palermo, 13, 1899, 99–129.
[201]Messenger of Math., 29, 1899–1900, 145–179. Errata, Cunningham[226], p. 155. See 13a of Ch. IV.
[201a]Math. Quest. Educ. Times, 71, 1899, 43–4; 75, 1901, 52–4.
[201b]Ibid., (2), 1, 1902, 70–2.
[202]Math. Gazette, 1, 1900, 130.
[203]Archiv Math. Phys., (2), 17, 1900, 110–2, 120–1.

G. Picou[204] applied to the case $n=2$ Wronski's[169] formula for the residues of nth powers modulo M, M arbitrary. For example, if $M=16a\pm1$,

$$(h\pm8a)^2\equiv\mp a(4h-1)^2\ (\mathrm{mod}\ M).$$

[If $8a$ were replaced by $4a$, we would have an identity in h.]

P. Bachmann[104] (pp. 344–351) discussed $x^m\equiv a$ (mod p^a), $p>2$, $p=2$.

G. Arnoux[205] solved $x^{14}\equiv79$ (mod $3\cdot5\cdot7$) by getting the residue 2 of 79 modulo 7 and that of 14 modulo $\phi(7)=6$ and solving $x^2\equiv2$ (mod 7) by use of a table of residues of powers modulo 7. Similarly for moduli 3, 5. Take the product of the roots as usual.

M. Cipolla[206] generalized the results of Alagna[200] to the case of a prime $p=2^mq+1$, $m>0$, q an odd prime, including unity. For any divisor d of $p-1$, the roots of $x^d\equiv N$ (mod p) are expressed as given powers of a primitive root a of p. If 2 belongs to the exponent $2^r\omega$ modulo p, where ω is odd, then $q^q\equiv1$ (mod p) if and only if 2^{p-1} is the highest power of 2 dividing m.

Cunningham[206a] found the sum of the roots of $(y^n\pm1)/(y\pm1)\equiv0$ (mod p).

M. Cipolla[207] proved the existence of an integer k such that k^2-q is a quadratic non-residue of the prime p not dividing the given integer q. Let

$$u_n=\tfrac12\sqrt{q}\{(k+\sqrt{q})^n-(k-\sqrt{q})^n\},$$
$$v_n=\tfrac12\{(k+\sqrt{k^2-q})^n+(k-\sqrt{k^2-q})^n\}.$$

By expansion of the binomials it is shown that the roots of $x^2\equiv q$ (mod p) are given by $\pm u_{(p-1)/2}$ and by $\pm v_{(p+1)/2}$. These may be computed by use of

$$w_n\equiv2kw_{n-1}-qw_{n-2}\ (\mathrm{mod}\ p)\qquad(w=u\ \text{or}\ v),$$

with the initial values $u_0=1$, $u_1=p$; $v_0=1$, $v_1=k$. Although u_n, v_n are the functions of Lucas, the exposition is here simple and independent of the theory of Lucas (Ch. XVII).

M. Cipolla[208] proved that if q is a quadratic residue and k^2-q is a quadratic non-residue of an odd prime p, $z^2\equiv q$ (mod p^λ) has the roots

$$\pm\tfrac12\sqrt{q}\{(k+\sqrt{q})^r-(k-\sqrt{q})^r\},$$

where $r=p^{\lambda-1}(p-1)/2$. Other expressions for the roots are

$$\pm\tfrac12q^t\{(k+\sqrt{k^2-q})^s+(k-\sqrt{k^2-q})^s\},$$
$$t=(p^\lambda-2p^{\lambda-1}+1)/2,\qquad s=p^{\lambda-1}(p+1)/2.$$

Thus if $z_1^2\equiv q$ (mod p), the roots modulo p^λ are $\pm q^t z_1^{p^{\lambda-1}}$ (Tonelli[193]). Finally, let $n=\Pi p_i^{\lambda_i}$, where the p's are primes >3; take $\epsilon_i=\pm1$ when $p_i\equiv\mp1$ (mod 4). There exists a number Δ of the form k^2-q such that

[204]L'intermédiaire des math., 8, 1901, 162.
[205]Assoc. franç. av. sc., 31, 1902, II, 185–201.
[206]Periodico di Mat., 18, 1903, 330–5.
[206a]Math. Quest. Educ. Times, (2), 4, 1903, 115–6; 5, 1904, 80–1.
[207]Rendiconto Accad. Sc. Fis. e Mat. Napoli, (3), 9, 1903, 154–163.
[208]Ibid., (3), 10, 1904, 144–150.

$(\Delta/p_1) = \epsilon_1, \ldots, (\Delta/p_\nu) = \epsilon_\nu$, where the symbols are Legendre's. Call M the l. c. m. of $p_i^{\lambda-1}(p_i - \epsilon_i)/2$ for $i = 1, \ldots, \nu$. Then $z^2 \equiv q \pmod{n}$ has the root

$$\tfrac{1}{2}q^{\{\varphi(n) - M + 1\}/2}\{(k + \sqrt{\Delta})^M + (k - \sqrt{\Delta})^M\}.$$

A. Cunningham[209] indicated how his tables may be used to solve directly $x^n \equiv -1 \pmod{p}$ for $n = 2, 3, 4, 6, 12$. From $p = a^2 + b^2$, we get the roots $x \equiv \pm a/b$ of $x^2 \equiv -1 \pmod{p}$. Also $p = a^2 + b^2 = c^2 + 2d^2$ gives the roots $\pm d(a+b)/(ce)$ and $\pm c(a \pm b)/(2de)$ of $x^4 \equiv -1 \pmod{p}$, where $e = a$ or b. Again, $p = A^2 + 3B^2$ gives the roots $(A-B)/(2B)$, $(B+A)/(B-A)$, and their reciprocals, of $x^3 \equiv 1 \pmod{p}$.

M. Cipolla[107] gave a report (in Peano's symbolism) on binomial congruences.

M. Cipolla[210] proved that if p is an odd prime not dividing q and if $z^2 \equiv q \pmod{p}$ is solvable, the roots are

$$z \equiv \pm 2(qs_1 + q^2 s_3 + q^3 s_5 + \ldots + q^{(p-3)/2}s_{p-4} + s_{p-2})$$

where

$$s_r = 1^r + 2^r + \ldots + \left(\frac{p-1}{2}\right)^r.$$

Then $x^2 \equiv q \pmod{p^\lambda}$ has the root $z^{p^{\lambda-1}}q^e$, $e = (p^\lambda - 2p^{\lambda-1} + 1)/2$. For $p \equiv 1 \pmod 4$, $x^4 \equiv q \pmod{p}$ has the root

$$4\sum_{i=1}^{2l} q^i s_{2i-1} \cdot \sum_{j=1}^{l} q^{j-1} s_{4j-3} + 2\sum_{i=1}^{l} q^i s_{4i-1} \qquad \left(l = \frac{p-1}{4}\right).$$

M. Cipolla[211] extended the method of Legendre[159] and proved that

$$x^{2^m} \equiv 1 + 2^s A \pmod{2^k},$$

for A odd and $s \geqq m + 2$, has a root

$$x = 1 + 2^s A c_1 - 2^{2s} A^2 c_2 + \ldots + (-1)^{n-1} 2^{ns} A^n c_n, \qquad n = \left[\frac{k-2}{s-m-1}\right],$$

where

$$c_1 = \frac{1}{2^m}, \qquad c_n = \frac{(2^m - 1)(2 \cdot 2^m - 1) \ldots (\overline{n-1} \cdot 2^m - 1)}{2^{mn} n!}$$

are the coefficients in

$$(1+z)^{1/2^m} = 1 + c_1 z - c_2 z^2 + c_3 z^3 - \ldots - (-1)^n c_n z^n + \ldots.$$

O. Meissner[212] gave for a prime $p = 8n + 5$ the known root

$$\xi = D^{\frac{p+3}{8}} \text{ of } x^2 \equiv D \pmod{p}, \qquad D^{\frac{p-1}{4}} \equiv 1 \pmod{p}.$$

But if $D^{(p-1)/4} \equiv -1 \pmod{p}$, a root is $\xi\{(p-1)/2\}!$, since the square of the last factor is congruent to $(-1)^{(p+1)/2}$ by Wilson's theorem.

Tamarkine and Friedmann[213] expressed the roots of $z^2 \equiv q \pmod{p}$ by a formula, equivalent to Cipolla's,[210]

[209]Quadratic Partitions, 1904, Introd., xvi–xvii. Math. Quest. Educ. Times, 6, 1904, 84–5; 7, 1905, 38–9; 8, 1905, 18–9.
[210]Rendiconto Accad. Sc. Fis. e Mat. Napoli, (3), 11, 1905, 13–19.
[211]Ibid., 304–9.
[212]Archiv Math. Phys. (3), 9, 1905, 96.
[213]Math. Annalen, 62, 1906, 409.

$$z \equiv \pm 2 \sum_{m=0}^{(p-3)/2} q^{\frac{1}{2}(p-1)-m} s_{2m+1}.$$

For, according as y^2 is or is not $\equiv q \pmod{p}$, we have

$$y\{1-(y^2-q)^{p-1}\} \equiv y \text{ or } 0 \pmod{p}.$$

We can express s_{2m+1} in terms of Bernoullian numbers.

A. Cunningham[214] gave a tentative method of solving $x^2 \equiv a \pmod{p}$. He[214a] noted that a root $Y = 2\eta^2$ of $Y^4 \equiv -1$ leads to the roots of $y^8 \equiv -1 \pmod{p}$.

M. Cipolla[215] employed an odd prime p and a divisor n of $p-1 = n\nu$. If $r_1, \ldots, r_\nu$ form a set of residues of p whose nth powers are incongruent, and if $q^\nu \equiv 1 \pmod{p}$, then $x^n \equiv q \pmod{p}$ has the root

$$x \equiv \sum_{k=0}^{\nu-1} A_k q^k, \qquad A_k = -n \sum_{j=1}^{\nu} r_j^{nk-1}.$$

For $n=2$, this becomes his[210] earlier formula by taking $1, 2, \ldots, (p-1)/2$ as the r's. Next, let $p-1 = m\mu$, where m and μ are relatively prime and m is a multiple of n. If γ and δ belong to the exponents m and μ modulo p, the products $\gamma^r \delta^s$ ($r < m/n$, $s < \mu$) may be taken as $r_1, \ldots, r_\nu$. According as $nk \equiv 1$ or not $\pmod{\mu}$, we have

$$A_k \equiv -n\mu \frac{\gamma^{(nk-1)m/n}-1}{\gamma^{nk-1}-1} \text{ or } A_k \equiv 0 \pmod{p}.$$

If n is a prime and n^r is its highest power dividing $p-1$, there exists a number ω not an nth power modulo p and we may set $m = n^r$, $\gamma \equiv \omega^\mu \pmod{p}$. In particular, if $n=2$, $x^2 \equiv q$ has the root

$$x \equiv \frac{-1}{2^{r-2}} q^{\frac{p+2^r-1}{2^{r+1}}} \sum_{s=0}^{2^{r-1}-1} q^{s(p-1)/2^r}/(\omega^{(2s+1)(p-1)/2^r}-1),$$

where ω is a quadratic non-residue of p. If $p \equiv 5 \pmod{8}$, we may take $\omega = 2$ and get

$$\tfrac{1}{2}q^{\frac{p+3}{8}}\{2^t+1-(2^t-1)q^t\}, \qquad t = \frac{p-1}{4}.$$

M. Cipolla[216] considered the congruence, with p an odd prime,

$$x^{p^r} \equiv a \pmod{p^m}, \qquad r < m,$$

a necessary condition for which is that $h = (a^{p^r}-a)/p^{r+1}$ be an integer. Determine A by $a^{p^r}A \equiv h \pmod{p^m}$. Then the given congruence has the root ax_0 if x_0 is a root of

$$x^{p^r} \equiv 1 - Ap^{r+1} \pmod{p^m}.$$

This is proved to have the root

[214]Math. Quest. Educ. Times, (2), 13, 1908, 19–20.
[214a]Ibid., 10, 1906, 52–3.
[215]Math. Annalen, 63, 1907, 54–61.
[216]Atti R. Accad. Lincei, Rendiconti, (5), 16, I, 1907, 603–8.

$$x_0 = 1 - \sum_{i=1}^{k} c_i A^i p^{i(r+1)}, \qquad k = m+1+\left[\frac{m-2}{p-2}\right],$$

where $c_1 = 1/p^r, \ldots$ are given by the expansion

$$\sqrt[p^r]{1-z} = 1 - c_1 z - c_2 z^2 - \cdots .$$

M. Cipolla[217] treated $x^n \equiv a \pmod{p^m}$ where n divides $\phi(p^m)$. We may set $n = p^r \nu$, where ν divides $p-1$. Determine integers a, β such that

$$a p^r + \nu \beta \equiv 1 \ \{\text{mod } p^{m-r-1}(p-1)\}.$$

Then the initial congruence has the root $y x_1^a$ if $y^{p^r} \equiv a^\beta \pmod{p^m}$, solved as in his preceding paper, and if x_1 is a root of $x^r \equiv a \pmod{p^m}$. The latter has the root

$$\frac{1}{t} a^{(p^m - 2p^{m-1}+1)/\nu} \sum_{k=0}^{t-1} a^{kp^{m-1}} \sum_{i=1}^{t} \rho_i^{\nu k-1},$$

where $t = (p-1)/\nu$, $\rho_i \equiv r_i^{p^{m-1}} \pmod{p^m}$, $r_1, \ldots, r_t$ being integers prime to p such that their νth powers are incongruent and form a group modulo p^m.

K. A. Posse[218] gave a simplified exposition of Korkine's[118] method of solving binomial congruences. Cf. Posse,[129] Schuh.[123-4]

F. Stasi[219] proved that we obtain all solutions of $x^2 \equiv a^2 \pmod{n}$, where n is odd and prime to a, by expressing n as a product of two relatively prime factors P and Q in all ways, setting $x - a = Pz$ and finding z from $Pz + 2a \equiv 0 \pmod{Q}$. [Instead of his very long proof, it may be shown at once that we may take $x - a$, $x + a$ divisible by P, Q, respectively.]

L. Grosschmid[220] gave for the incongruent roots of $x^2 \equiv r \pmod{M}$ an explicit formula obtained by means of the ideal factors of M in a quadratic number-field.

L. Grosschmid[221] treated the roots of quadratic binomial congruences.

A. Cunningham[222] solved $x^2 \equiv -1 \pmod{p}$, where $p = 616318177$ is a prime factor of $2^{37} - 1$; by using various small moduli, he obtained $p = 24561^2 + 3616^2$.

L. von Schrutka[222a] used a correspondence between the integers and certain rational numbers to treat quadratic congruences without novelty as to results. The method will be given under the topic Fields in a later volume of this History.

Grosschmid[223] employed the products R and N of all the quadratic residues and non-residues, respectively, $\leq 2n$, of a prime $p = 4n+1$. Then

$$R^2 \equiv (-1)^{n+1}, \qquad N^2 \equiv (-1)^n \pmod{p}.$$

[217]Atti R. Accad. Lincei, Rendiconti, (5), 16, I, 1907, 732–741.
[218]Charlkov Soobšč. Mat. Obšč (Report Math. Soc. Charkov), (2), 11, 1910, 249–268 (Russian).
[219]Il Boll. Matematica Gior. Sc.-Didat., 9, 1910, 296–300.
[220]Jour. für Math., 139, 1911, 101–5.
[221]Math. és Phys. Lapok, Budapest, 20, 1911, 47–72 (Hungarian).
[222]Math. Questions Educat. Times, (2), 20, 1911, 33–4 (76).
[222a]Monatshefte Math. Phys., 23, 1912, 92–105.
[223]Archiv Math. Phys., (3), 21, 1913, 363; 23, 1914–5, 187–8.

Hence $\pm R$ and $\pm N$ are the roots of $x^2 \equiv -1$ (mod p) according as $p = 8m + 1$ or $8m + 5$.

U. Concina[224] proved the first result by Legendre.[154]

A. Cunningham[225] tabulated the roots of $y^4 \equiv \pm 2$, $2y^4 \equiv \pm 1$ (mod p), for each prime $p < 1000$.

Cunningham[226] listed the roots of $y^l \equiv \pm 1$ (mod p^r), where $l = qp^a$, p being an odd prime $\leqq 19$, $p^r < 10^4$, $a = 1$ and often also $a = 2$, q a factor of $p - 1$.

A. Gérardin and L. Valroff[227] solved $2y^4 \equiv 1$ (mod p), $1000 < p < 5300$.

Cunningham[228] announced the completion of tables giving all proper roots of $y^m \equiv 1$ (mod p^k) for m odd $\leqq 15$, and of $y^m \equiv -1$ (mod p^k) for m even $\leqq 14$. These tables have since been completed up to $p^k < 100000$ and are now nearly all in type.

T. G. Creak[228] announced the completion of like tables for $m = 16$ to 50; 52, 54, 56, 63, 64, 72, 75, and $10^3 < p^k < 10^4$.

H. C. Pocklington[229] noted that if p is a prime $8m + 5$ and $a^{2m+1} \equiv -1$, $x^2 \equiv a$ (mod p) has the roots $\pm \frac{1}{2}(4a)^{m+1}$. He showed how to use $(t + u\sqrt{D})^n$ to solve $x^2 \equiv -D$ (mod $p = 4k + 1$), and treated $x^3 \equiv a$.

*J. Maximoff[230] treated binomial congruences and primitive roots.

*G. Rados[231] gave a new proof of known criteria for the solvability of $x^2 \equiv D$ (mod p). He[232] gave a new exposition of the theory of binomial congruences without using indices.

Congruences $x^{p-1} \equiv 1$ (mod p^n) are treated in Chapter IV. Euler[4, 7] of Ch. XVI solved $x^2 \equiv -1$ (mod p). Lazzarini[172] of Ch. I erred on the number of roots of $z^2 \equiv -3$ (mod n). Many papers in Ch. XX treat $x^k \equiv x$ (mod 10^n). The following papers from the first part of Ch. VII treat also binomial congruences: Euler,[2] Lagrange,[3] Poinsot,[11] Cauchy,[14] Lebesgue,[59] Epstein,[112] Korkine.[118]

[224]Periodico di Mat., 28, 1913, 212–6.
[225]Messenger Math., 43, 1913–4, 52–3.
[226]Ibid., 148–163. Cf. Cunningham.[201]
[227]Sphinx-Oedipe, 1913, 34; 1914, 18–37, 73.
[228]Messenger Math., 45, 1915–6, 69.
[229]Proc. Cambridge Phil. Soc., 19, 1917, 57–9.
[230]Bull. Soc. Phys.–Math. Kasan, (2), XXI.
[231]Math. és Termés Értesitö, 33, 1915, 758–62.
[232]Ibid., 34, 1916, 641–55.

CHAPTER VIII.

HIGHER CONGRUENCES.

A CONGRUENCE OF DEGREE n HAS AT MOST n ROOTS IF THE MODULUS p IS A PRIME.

J. L. Lagrange[1] proved that, if a is not divisible by the prime p, $ax^n + bx^{n-1} + \ldots$ is divisible by p for at most n integers x between $p/2$ and $-p/2$. For, let $a, \beta, \ldots, \rho, \sigma$ be $n+1$ such distinct integers. Then the quotient of

$$a(x^n - a^n) + b(x^{n-1} - a^{n-1}) + \ldots$$

by $x - a$ is a polynomial $ax^{n-1} + \ldots$ which is divisible by p when $x = \beta, \ldots, \sigma$. Proceeding as before, we finally have $a(\rho - \sigma)$ divisible by p, which is impossible.

L. Euler[2] noted that $x^n - 1$ is divisible by a prime p for not more than n integers x, $0 < x < p$. For, if $x = a$, is such an integer, then $x - a$ divides $x^n - 1 - mp$, where m is a suitable integer; the quotient f is of degree $n-1$. If $x = b$ is a second such integer, $x - b$ divides $f - m'p$. Proceeding as in algebra, we obtain the theorem stated. [The argument is applicable to any polynomial of degree n in x.]

A. M. Legendre[3] noted that $P \equiv (x-a)Q + pA$ has only one more root than Q.

C. F. Gauss[4] proved the theorem by assuming that there is a congruence $ax^n + \ldots \equiv 0 \pmod p$ with more than n roots $a, \ldots$, and that every congruence of degree l, $l < n$, has at most l roots. Substituting $y + a$ for x, we obtain a congruence $ay^n + \ldots \equiv 0$ with more than n roots, one of which is zero. Removing the factor y, we obtain $ay^{n-1} + \ldots \equiv 0$ with more than $n-1$ roots, contrary to hypothesis.

Gauss[5] noted that if a is a root of $\xi \equiv 0 \pmod p$, then ξ is divisible by $x - a$ modulo p. If $a, b, \ldots$ are incongruent roots, ξ is divisible modulo p by the product $(x-a)(x-b) \ldots$. Hence the number of roots does not exceed the degree of ξ.

A. Cauchy[6] made the proof by use of $X \equiv (x-a)X_1 \pmod p$, identically in x, where the degree of X_1 is one less than the degree of X.

A. L. Crelle[7] and S. Earnshaw[8] gave Lagrange's proof.

Crelle[9] proved that if $e_1, \ldots, e_n$ are n distinct roots,

$$ax^n + \ldots \equiv a(x-e_1)\ldots(x-e_n) + pN.$$

[1]Mém. Ac. Berlin, 24, année 1768 (1770), p. 192; Oeuvres, 2, 1868, 667–9.
[2]Novi Comm. Ac. Petrop., 18, 1773, p. 93; Comm. Arith., 1, 519–20.
[3]Mém. Ac. Roy. Sc., Paris, 1785, 466; Théorie des nombres, 1798, 184.
[4]Disq. Arith., 1801, Art. 43.
[5]Posthumous paper, Werke, 2, p. 217, Art. 338 (p. 214, Art. 333). Maser's German translation of Gauss' Disq. Arith., etc., 1889, p. 607 (p. 604).
[6]Exercices de Math., 4, 1829, 219; Oeuvres, (2), 9, 261; Comptes Rendus Paris, 12, 1841, 831–2; Exercices d'Analyse et de Phys. Math., 2, 1841, 1–40, Oeuvres, (2), 12.
[7]Berlin Abhand., Math., 1832, p. 34.
[8]Cambridge Math. Jour., 2, 1841, 79.
[9]Berlin Abhand., Math., 1843, 50–54.

L. Poinsot[10] gave the proof due to Crelle.[9]

J. A. Grunert[11] proceeded by induction from $n-1$ to n, making use of the first part of Lagrange's proof.

D. A. da Silva[12] gave a proof.

Number of Roots of Higher Congruences.

G. Libri[16] found that $f(x, y, \ldots) \equiv 0 \pmod{m}$ has

$$\frac{1}{m} \sum_{x=a}^{b} \sum_{y=c}^{d} \ldots \left\{ \sum_{k=0}^{m-1} \cos\frac{2k\pi f}{m} + i \sin\frac{2k\pi f}{m} \right\}$$

sets of solutions such that $a \leq x \leq b$, $c \leq y \leq d$, ... The total number of sets of solutions is

$$\frac{1}{m} \sum_{x=0}^{m} \sum_{y=0}^{m} \ldots \left\{ 1 + \cos\frac{2\pi f}{m} + \cos\frac{4\pi f}{m} + \ldots + \cos 2\frac{(m-1)\pi f}{m} \right\}.$$

V. A. Lebesgue[17] proved that if p is a prime we obtain as follows the residue modulo p of the number S_k of sets of solutions of $F(x_1, \ldots, x_n) \equiv 0 \pmod{p}$, in which each x_i is chosen from $0, 1, \ldots, p-1$, and F is a polynomial with integral coefficients. Let ΣA be the sum of the coefficients of the terms $A x_1^a \ldots x_k^g$ of the expansion of F^{p-1} in which each of the exponents $a, \ldots, g$ is a multiple >0 of $p-1$. Then $S_k \equiv (-1)^{k+1} \Sigma A \pmod{p}$.

Henceforth, let $p = hm+1$. First, let $F = x^m - a$. In F^{p-1} the coefficient of $x^{m(p-1-n)}$ is $\binom{p-1}{n}(-a)^n \equiv a^n \pmod{p}$. The exponent of x will be a multiple >0 of $p-1$ only when $n = k(p-1)/d$, for $k = 0, 1, \ldots, d-1$, where d is the g. c. d. of m and $p-1$. Thus $S_1 \equiv \Sigma a^{k(p-1)/d} \pmod{p}$, while evidently $S_1 < p$. According as $a^{(p-1)/d} \equiv 1$ or not, we get $S_1 = d$ or 0.

Next, let $F = x^m - ay^m - b$. Set $c = ay^m + b$. In $(x^m - c)^{p-1}$ we omit the terms in which the exponent of x is not a multiple >0 of $p-1$ and also the $x^{m(p-1)}$ not containing y. Since the arithmetical coefficient is $\equiv 1$ as in the first case, we get

$$c^h x^{m(p-1-h)} + c^{2h} x^{m(p-1-2h)} + \ldots + c^{(m-1)h} x^{mk}.$$

In this, we replace c^{kh} by those terms of $(ay^m + b)^{kh}$ in which the exponents are multiples >0 of $p-1$, viz.,

$$\sum_{l=0}^{k-1} \binom{kh}{lh}(ay^m)^{kh-lh}b^{lh}.$$

Set $y = 1$, and sum for $k = 1, \ldots, m-1$; we get $-S_2 \pmod{p}$. It is shown otherwise that S_2 is a multiple $<mp$ of m.

To these two cases is reduced the solution of

(1) $$F = a_1 x_1^m + \ldots + a_k x_k^m \equiv a \pmod{p = hm+1}.$$

[10]Jour. de Mathématiques, 10, 1845, 12–15.

[11]Klügel's Math. Wörterbuch, 5, 1831, 1069–71.

[12]Proprietades...Congruencias binomias, Lisbon, 1854.   Cf. C. Alasia, Rivista di fisica, mat. e sc. nat., 4, 1903, p. 9.

[16]Mém. divers Savants Ac. Sc. de l'Institut de France (Math.), 5, 1838, 32 (read 1825).   Jour. für Math., 9, 1832, 54.   To be considered in vol. ii.

[17]Jour. de Math., 2, 1837, 253–292.   Cf. vol. 3, 113; vol. 4, 366.

Denote by P the sum of the first f terms of F and by Q the sum of the last $k-f$ terms. ·Let g be a primitive root of p. Let P^0 be the number of sets of solutions of $P\equiv0$ (mod p); $P^{(i)}$ the number for $P\equiv g^i$ (mod p); Q^0 and $Q^{(i)}$ the corresponding numbers for $Q\equiv0$, $Q\equiv g^i$. Then the number of sets of solution of $P\equiv Q$ (mod p) is $P^0Q^0+h\Sigma_{i=1}^{i=m}P^{(i)}Q^{(i)}$. Hence we may deduce the number of sets of solutions of $F\equiv0$ from the numbers for $P\equiv A$ and $Q\equiv -A$. For $F\equiv a$, we employ $P=F$, $Q=g^kx^m$ and get $F^0=P^0$ $+(p-1)P^{(k)}$, which determines the desired $P^{(k)}$.

The theory is applied in detail to (1) for $m=2$, k arbitrary, and for $m=3, 4, k=2$. Finally, the method of Libri[16] is amplified.

Th. Schönemann[18] noted that, if S_k is the sum of the kth powers of the roots of an equation $x^n+\ldots=0$ with integral coefficients, that of x^n being unity, and if $S_{(p-1)t}\equiv n$ (mod p) for $t=1, 2,\ldots, n$, where p is a prime $>n$, the corresponding congruence $x^n+\ldots\equiv0$ (mod p) has n real roots.

A. L. Cauchy[19] considered $F(x)\equiv0$ (mod M), with $M=AB\ldots$, where $A, B\ldots$ are powers of distinct primes. If $F(x)\equiv0$ (mod A) has a roots, $F(x)\equiv0$ (mod B) has β roots, etc., the proposed congruence has $a\beta\ldots$ roots in all. For, if $a, b,\ldots$ are roots for the moduli $A, B,\ldots$ and $X\equiv a$ (mod A), $X\equiv b$ (mod B),$\ldots$, then X is a root for modulus M.

P. L. Tchebychef[20] proved that, if p is a prime, a congruence $f(x)\equiv0$ (mod p) of degree $m<p$ has m roots if and only if the coefficients of the remainder obtained by dividing x^p-x by $f(x)$ are all divisible by p.

Ch. Hermite[21] proved the theorem: If μ and μ' are the numbers of sets of solutions of $\phi(x, y)\equiv0$ for the respective moduli M and M', which are relatively prime, the number of sets of solutions modulo MM' is $\mu\mu'$. If $\phi\equiv0$ is solvable for a prime modulus p, it will be solvable modulo p^n if

$$\phi\equiv0, \qquad \frac{\partial\phi}{\partial x}\equiv0, \qquad \frac{\partial\phi}{\partial y}\equiv0 \text{ (mod } p)$$

have no common sets of solutions. In this case, the number of sets of solutions modulo p^n is $p^{n-1}\pi$ if π is the number for modulus p. Similar results are said to hold for any number k of unknowns. If M is a product of powers of the distinct primes $p_1,\ldots, p_n$, and if π_i is the number of sets of solutions of the congruence modulo p_i, then the number of sets for modulus M is

$$M^{k-1}\frac{\pi_1\ldots\pi_n}{(p_1\ldots p_n)^{k-1}}.$$

For $x^2+Ay^2\equiv\Delta$ (mod M), we have $\pi_i=p_i-(-A/p_i)$, where (a/p) is ±1 according as a is a quadratic residue or non-residue of p.

Julius König gave a theorem in a seminar at the Technische Hochschule in Budapest during the winter, 1881-2, which was published in the following paper and that by Rados.[24]

[18]Jour. für Math., 19, 1839, 293.
[19]Comptes Rendus Paris, 25, 1847, 36; Oeuvres, (1), 10, 324.
[20]Theorie der Congruenzen, in Russian, 1849; in German, 1889, §21.
[21]Jour. für Math., 47, 1854, 351-7; Oeuvres, 1, 243-250.

G. Raussnitz[23] proved the theorem, due to König: Let

(2) $f(x) = a_0 x^{p-2} + a_1 x^{p-3} + \ldots + a_{p-2}$,

where the a's are integers and a_{p-2} is not divisible by the prime p.   Then $f(x) \equiv 0 \pmod{p}$ has real roots if and only if the cyclic determinant

(3) $$D = \begin{vmatrix} a_0 & a_1 & a_2 & \ldots & a_{p-3} & a_{p-2} \\ a_1 & a_2 & a_3 & \ldots & a_{p-2} & a_0 \\ \cdots\cdots\cdots\cdots\cdots\cdots\cdots\cdots \\ a_{p-2} a_0 & a_1 & \ldots & a_{p-4} & a_{p-3} \end{vmatrix}$$

is divisible by p.   In order that it have at least k distinct real roots it is necessary that all $p-k$ rowed minors of D be divisible by p.   If also not all $p-k-1$ rowed minors are divisible by p, the congruence has exactly k distinct real roots.

The theorem is applicable to any congruence not having the root zero, since we may then reduce the degree to $p-2$ by Fermat's theorem.

Gustav Rados[24] proved König's theorem, using the fact that a system of $p-1$ linear homogeneous congruences modulo p in $p-1$ unknowns has at least k sets of solutions linearly independent modulo p if and only if the $p-k$ rowed minors are divisible by p.

L. Kronecker[25] noted that, if p is a prime, the condition for the existence of exactly $p-m-1$ roots of (2), distinct from one another and from zero, is that the rank of the system

(3′) (a_{i+k}) $(i, k = 0, 1, \ldots, p-2)$

modulo p is exactly m, where $a_{s+p-1} = a_s$.   The same is the condition for the existence of a $(p-m-1)$-fold manifold of sets of solutions of the system of linear congruences

$$\sum_{k=0}^{p-2} a_{h+k}\phi_k \equiv 0 \pmod{p} \qquad (h = 0, 1, \ldots, p-2).$$

L. Kronecker[26] gave a detailed proof of his preceding results, noted that the rank is m if not all principal m-rowed minors are divisible by p while all $m+1$ rowed minors are, and added that $c_0 + c_1 x + \ldots + c_{p-2} x^{p-2} \equiv 0$ $\pmod{p}$ has exactly s roots $\neq 0$ if one and the same linear homogeneous congruence holds between every set of $p-s$ (but not fewer) successive terms of the periodic series $c_0, c_1, \ldots, c_{p-2}, c_0, c_1, \ldots$.

L. Gegenbauer[27] proved Kronecker's version of König's theorem.

Gegenbauer[28] noted that Kronecker's theorems imply the corollary:

[23]Math. und Naturw. Berichte aus Ungarn, 1, 1882–3, 266–75.
[24]Jour. für Math., 99, 1886, 258–60; Math. Termes Ertesito, Magyar Tudon Ak., Budapest, 1, 1883, 296; 3, 1885, 178.
[25]Jour. für Math., 99, 1886, 363, 366.
[26]Vorlesungen über Zahlentheorie, 1, 1901, 389–415, including several additions by Hensel (pp. 393, 399, 402–3).
[27]Sitzungsber. Ak. Wiss. Wien (Math.), 95, II, 1887, 165–9, 610–2.
[28]Ibid., 98, IIa, 1889, p. 32, foot-note. Cf. Gegenbauer.[25]

There exist exactly $p-m-2$ roots of (2), distinct from one another and from zero, if and only if there exist exactly $p-m-2$ distinct linear homogeneous functions

$$\sum_{h=0}^{p-2} a_{k,\,h} a_h \qquad (k=1,\ldots,p-m-2)$$

which remain divisible by p after applying all cyclic permutations of the a_h, so that

$$\sum_{h=0}^{p-2} a_{k,\,h} a_{i+h} \equiv 0 \pmod{p} \qquad \binom{k=1,\ldots,\,p-m-2}{i=0,1,\ldots,\,p-2}.$$

A simple proof of this corollary is given.

L. Gegenbauer[29] noted that the number of roots of $f(x)\equiv 0 \pmod{k}$ is

$$\{f(x),\,k\} = \sum_{x=0}^{k-1} D(k), \qquad D(k) = \left[\frac{|f(x)|}{k}\right] - \left[\frac{|f(x)|-1}{k}\right],$$

since $D(k)=1$ or 0 according as $f(x)$ is divisible by k or not. Let $k_1,\ldots,k_\delta$ be a series of increasing positive integers and $g(x)$ any function. In the first equation take $k=k_l$, multiply by $g(k_l)$ and sum for $l=1,\ldots,\delta$. Reversing the order of the summation indices l, x in the new right-hand member, we get

$$\sum_{l=1}^{\delta} \{f(x),\,k_l\} g(k_l) = \sum_{x=0}^{k_\delta-1} G, \qquad G=\Sigma D(\mu)g(\mu),$$

where in G the summation index μ takes those of the values $k_1,\ldots,k_\delta$ which exceed x. Thus G represents the sum $G(f(x);\,k_1,\ldots,k_\delta;\,x)$ of the values of $f(\mu)$ when μ ranges over those of the numbers $k_1,\ldots,k_\delta$ which exceed x and are divisors of $f(x)$. In particular, if $g(x)=1$, G becomes the number ψ of the k's which exceed x and divide $f(x)$.

Let $f(x)=m\pm nx$. Then $f(x)\equiv 0 \pmod{k}$ has (k,n) roots or no root according as m is or is not divisible by the g. c. d. (k,n) of k and n; let $(k,n;m)$ denote (k,n) or 0 in the respective cases. Then

$$\sum_{l=1}^{\delta} (k_l,\,n;\,m)\,g(k_l) = \sum_{x=0}^{k_\delta-1} G(m\pm nx;\,k_1,\ldots,k_\delta;\,x).$$

Let $G(a,b)$ denote the sum of the values of $g(\mu)$ when μ ranges over all the divisors $>b$ of a; $\psi(a,b)$ the number of divisors $>b$ of a. Taking $k_l=l$ for $l=1,\ldots,\delta$, we deduce

$$\sum_{l=1}^{\delta} (l,\,n;\,m)g(l) = \sum_{x=0}^{\delta-1} \{G(m\pm nx,\,x) - G(m\pm nx,\,\delta)\}.$$

For $g(l)=1$, this reduces to Lerch's[100] relation (16) in Ch. X. Again,

$$\sum_{x=1}^{a} \{G(m+nx,\,x-1) - G(m+nx,\,b+x)\} = \sum_{\mu=0}^{b} \{G(m-n\mu,\,\mu) - G(m-n\mu,\,\mu+a)\},$$

[29]Sitzungsberichte Ak. Wiss. Wien (Math.), 98, IIa, 1889, 28–36.

which for $g(x) = 1$ yields the first formula of Lerch. Next, if the k's are primes and q is a prime distinct from them,

$$\sum_{x=0}^{k_\delta-1} G(x^n - q; k_1, \ldots, k_\delta; x) = \sum_{l=1}^{\delta} (k_l - 1, n; q)g(k_l).$$

Finally, he treated $f(x)$ of degree $d = k_\delta - 2$, whose constant term is prime to each k_i and coefficient of x^{d-i} is divisible by the prime k_μ if $i < k_\delta - k_\mu$.

Gegenbauer[30] noted that, if $p - 1 - \mu$ is the rank of the system (3) modulo p, the congruence, satisfied by the distinct roots $\neq 0$ of (2) and by these only, is given symbolically by

$$\left(\frac{\partial}{\partial a_1}x - \frac{\partial}{\partial a_0}\right)^\mu \mid a_{i+k} \mid \equiv 0 \ (\text{mod } p) \qquad (i, k = 0, \ldots, p-2).$$

He obtained easily Kronecker's[25] form of the last congruence. He gave necessary and sufficient conditions, expressed in terms of a complicated determinant and its $\mu - 1$ successive derivatives with respect to a_{p-2}, in order that (2) and a second congruence of degree $p - 2$ shall have μ common roots $\neq 0$, and found the congruence satisfied by these μ common roots. He deduced determinantal expressions for the sum σ_r of the rth powers of the roots of (2), and for the coefficients in terms of the σ's.

Michael Demeczky[31] would employ Euclid's process to find the g. c. d. $G(x)$ modulo p of (2) and $x^p - x$. If $G(x) \equiv 0 \ (\text{mod } p)$ is of degree ν it has ν real roots and these give all the real roots of (2). Multiple roots are then treated. The case of any composite modulus is known to reduce to the case of p^π, p a prime. If (2) has λ distinct real roots, not multiple roots, we can derive λ real roots of $f(x) \equiv 0 \ (\text{mod } p^\pi)$. If $p_1, \ldots, p_n$ are distinct primes and if $f(x) \equiv 0 \ (\text{mod } p_i)$ has λ_i real roots, then $f(x) \equiv 0 \ (\text{mod } p_1 \ldots p_n)$ has $\lambda_1 \ldots \lambda_n$ real roots, and is satisfied by every integer x if the former are. Various sets of necessary and sufficient conditions are found that $f(x) \equiv 0 \ (\text{mod } m = \Pi p_i^{\pi_i})$ shall have m distinct real roots; one set is that $f(x) \equiv 0 \ (\text{mod } p_i^{\pi_i})$ identically for each i.

L. Gegenbauer[32] proved that a congruence modulo p, a prime, of degree $p - 2$ in each of n variables has a set of solutions each $\neq 0$ if and only if p divides the determinant of a cyclic matrix

$$\begin{Bmatrix} A^0 & A^1 & \ldots & A^{r-1} \\ A^{r-1} & A^0 & \ldots & A^{r-2} \\ \cdots & \cdots & \cdots & \cdots \\ A^1 & A^2 & \ldots & A^0 \end{Bmatrix},$$

where A^μ is itself a cyclic matrix in $B^0, \ldots, B^{r_1-1}$; etc., until we reach matrices in the coefficients of the congruence. An upper limit is found for

---

[30]Sitzungsber. Ak. Wiss. Wien (Math.), 98, IIa, 1889, 652-72.

[31]Math. u. Naturw. Berichte aus Ungarn, 8, 1889-90, 50-59. Math. és Termés Ertesitō, 7, 1889, 131-8.

[32]Sitzungsber. Ak. Wiss. Wien (Math.), 99, IIa, 1890, 799-813.

the number of sets of solutions each not divisible by p.   He proved that

$$\sum_{j=1}^{s} a_j x_j^{\frac{p-1}{2}} + \sum_{j=1}^{n} a_{s+j} x_{s+j} + b \equiv 0 \pmod{p}$$

has p^{n+s-1} sets of solutions.   Of these,

$$\frac{1}{p}\left(\frac{p-1}{2}\right)^s \left\{ 2^s[(p-1)^n - (-1)^n] - (-1)^{n-1} pr \right\}$$

have each $x \not\equiv 0$, where r is the number of the 2^s integers

$$b \pm a_1 \pm a_2 \pm \ldots \pm a_s$$

which are divisible by p.   The number of sets of solutions of

$$\sum_{j=1}^{s} a_j x_j^{\frac{p-1}{2}} + \sum_{j=1}^{n} a_{s+j} x_{s+j}^2 + b \equiv 0 \pmod{p}$$

is expressed in terms of the functions used for quadratic congruences.

*E. Snopek[33] gave a generalization of König's criterion for the solvability of a congruence modulo p.

L. Gegenbauer[34] proved that if the ρ congruences

$$\sum_{k=0}^{p-2} z_{k\lambda} x^{p-2-k} \equiv 0 \pmod{p} \qquad (\lambda = 0, 1, \ldots, \rho-1)$$

have in common at least $p-\rho$ distinct roots not divisible by p then all ρ-rowed determinants in the matrix $(z_{k\lambda})$ are divisible by p.   The converse is proved when a certain condition holds.   By specialization, König's theorem is obtained.

Gegenbauer[35] proved that, if r is less than the prime p and if $z_0, \ldots, z_{r-1}$ are incongruent and not divisible by p, the system of linear congruences

(4) $$\sum_{k=0}^{p-2} b_{k+\rho} y_k \equiv 0 \pmod{p} \qquad (\rho = 0, 1, \ldots, p-2)$$

has all its sets of solutions of the form

(5) $$y_k \equiv \sum_{\lambda=0}^{r-1} a_\lambda z_\lambda^k \qquad (k = 0, 1, \ldots, p-2)$$

or not, according as the matrix $(b_{k+\rho})$, $k = r, r+1, \ldots, p-2; \rho = 0, \ldots, p-2$, has a $p-r-1$ rowed determinant prime to p or not.   Next, if

(6) $$\sum_{k=0}^{p-2} b_k x^k \equiv 0 \pmod{p}$$

has exactly r distinct roots $z_0, \ldots, z_{r-1}$ each not divisible by p, every system of solutions of (4) is given by (5), and conversely.   By combining this theorem of Kronecker's with the former, we obtain Kronecker's form of König's theorem.

[33]Prace Mat. Fiz., Warsaw, 4, 1893, 63–70 (in Polish).
[34]Sitzungsber. Ak. Wiss. Wien (Math.), 102, IIa, 1893, 549–64.
[35]Monatshefte Math. Phys., 5, 1894, 230–2.   Cf. Gegenbauer.[14]

K. Zsigmondy[36] proved that, if p is a prime, there are exactly

$$\psi(n,\,k) = p^n - \binom{k}{1}p^{n-1} + \binom{k}{2}p^{n-2} - \ldots + (-1)^n\binom{k}{n}$$

congruences $x^n + \ldots \equiv 0 \pmod{p}$ not having as roots k given distinct numbers. Also,

$$\psi(n,\,k) = p\psi(n-1,\,k) + (-1)^n\binom{k}{n}, \qquad \psi(n,\,k+1) = \psi(n,\,k) - \psi(n-1,\,k).$$

If $n \geq k$, $\psi(n,\,k) = p^{n-k}(p-1)^k$. For $n = k$, $\psi(n,\,k)$ is the number $\psi(n)$ of congruences of degree n with no root. The number with exactly i roots is $\binom{n}{i}\psi(n-i)$. There are $\binom{p-1}{i}\psi(i-r)$ distinct matrices (3) of rank i such that a_{r-1} is the first one of $a_0, a_1, \ldots$ not divisible by p.

K. Zsigmondy[37] considered a function $\Phi(f)$ of a polynomial $f(x)$ such that Φ is unaltered when the coefficients of $f(x)$ are increased by integral multiples of the prime p. Let $f_k^{(i)}(x)$, $i = 1, \ldots, p^k$, denote the polynomials of degree k which are distinct modulo p and have unity as the coefficient of x^k. It is stated that

$$\sum_a \Phi\{f_n^{(a)}(x)\} = \sum_{j=1}^{p^n} \Phi\{f_n^{(j)}(x)\} - \sum_i \sum_{j=1}^{p^{n-1}} \Phi\{(x-a_i)f_{n-1}^{(j)}(x)\}$$

$$+ \sum_{i,\,i'} \sum_{j=1}^{p^{n-2}} \Phi\{(x-a_i)(x-a_{i'})f_{n-2}^{(j)}(x)\} - \ldots,$$

where a takes those values $1, 2, \ldots, p^n$ for which $f_n^{(a)}(x) \equiv 0 \pmod{p}$ does not have as a root one of the given incongruent numbers $a_1, \ldots, a_s$; while, in the outer sums on the right, $i, i', \ldots$ range over the combinations of $1, \ldots, s$ without repetitions.

Zsigmondy[38] had earlier given the preceding formula for the case in which $a_1, \ldots, a_s$ denote $0, 1, \ldots, p-1$. Then taking $\Phi(f) = 1$, we get the number of congruences of degree n with no root (Zsigmondy[36]). Taking $\Phi(f) = f$, we see that the sum of the congruences of degree n with no root is $\equiv 0 \pmod{p}$, aside from specified exceptions. Taking $\Phi(f) = \omega^f$, where ω is a pth root of unity, and $n \geq p$, we see that the system $f_n^{(a)}(x)$ takes each of the values $1, \ldots, p-1 \pmod{p}$ equally often.

Zsigmondy[39] proved his[36,37] earlier formulas, obtained for an integral value of x the number of complete sets of residues modulo p into which fall the values of the $f_n^{(a)}(x)$ not having prescribed roots, and investigated the system B_n of the least positive residues modulo p of the left members of all congruences of degree n having no root. In particular, he found how often the system B_n contains each residue, or non-residue, of a qth power. He investigated (pp. 19–36) the number of polynomials in x which take k prescribed residues modulo p for k given values of x.

---

[36]Sitzungsber. Ak. Wiss. Wien (Math.), 103, IIa, 1894, 135–144.
[37]Monatshefte Math. Phys., 7, 1896, 192–3.
[38]Jahresbericht d. Deutschen Math. Verein., 4, 1894–5, 109–111.
[39]Monatshefte Math. Phys., 8, 1897, 1–42.

L. Gegenbauer[40] proved that (2) has as a root a quadratic residue or non-residue of the prime p if and only if the respective determinant

$$P = \mid a_{\mu+i} + a_{\mu+i+\pi} \mid, \quad N = \mid a_{\mu+i} - a_{\mu+i+\pi} \mid \quad (i, \; \mu = 0, \ldots, \pi-1)$$

be divisible by p, where $\pi = (p-1)/2$. From this it is proved that (2) has exactly $\pi - r$ distinct quadratic residues (or non-residues) of p as roots if and only if P (or N) and its $\pi-1-r$ successive derivatives with respect to $a_{\pi-1} + a_{p-2}$ have the factor p, while the derivative of order $\pi-r$ is prime to p. These residues satisfy the congruence

$$\left\{ x \frac{\partial}{\partial(a_1 + a_{\pi+1})} - \frac{\partial}{\partial(a_0 + a_\pi)} \right\}^{(\pi-r)} K \equiv 0 \pmod{p},$$

where $K = P$ or N, while the νth power of the sign of differentiation represents the νth derivative. A second set of conditions is obtained. Congruence (2) has exactly $\pi-1-\kappa$ distinct quadratic residues as roots if and only if the determinants of type P with now $i = 0, \ldots, \kappa, \kappa+1$ and $\mu = 0, \ldots, \kappa, \tau$, are divisible by p for $\tau = \kappa+1, \ldots, \pi-1$; while p is not a factor of the determinant of type P with now $i, \mu = 0, \ldots, \kappa$. These residues are the roots of

$$\sum_{\tau=\kappa}^{\pi} \mid a_{\mu+i} + a_{\mu+i+\pi} \mid x^{\pi-1-\tau} \equiv 0 \pmod{p},$$

where $i = 0, \ldots, \kappa$, and $\mu = 0, \ldots, \kappa-1$, τ in the determinants. For non-residues we have only to use the differences of a's in place of sums.

S. O. Satunovskij[41] noted that, for a prime modulus p, a congruence of degree n ($n < p$) has n distinct roots if and only if its discriminant is not divisible by p and $S_{p+q} \equiv S_{q+1} \pmod{p}$ for $q = 1, \ldots, n-1$, where S_k is the sum of the kth powers of the n roots.

A. Hurwitz[42] gave an expression for the number N of real roots of

$$f(x) = a_0 + a_1 x + \ldots + a_r x^r \equiv 0 \pmod{p},$$

where p is a prime. By Fermat's theorem,

$$N \equiv \sum_{x=1}^{p-1} \{1 - f(x)^{p-1}\} \pmod{p}.$$

Let $f(x)^{p-1} = C_0 + C_1 x + \ldots$. Then N is determined by

$$N + 1 \equiv C_0 + C_{p-1} + C_{2(p-1)} + \ldots \pmod{p}.$$

Let $f(x_1, x_2)$ be the homogeneous form of $f(x)$. Let A be the number of sets of solutions of $f(x_1, x_2) \equiv 0 \pmod{p}$, regarding (x_1, x_2) and (x_1', x_2') as the same solution if $x_1' \equiv \rho x_1$, $x_2' \equiv \rho x_2 \pmod{p}$ for an integer ρ. Then

$$A - 1 \equiv -a_0^{p-1} - a_r^{p-1} + \Sigma \frac{(p-1)!}{a_0! \ldots a_r!} a_0^{a_0} \ldots a_r^{a_r} \pmod{p},$$

[40]Sitzungsber. Ak. Wiss. Wien (Math.), 110, IIa, 1901, 140–7.
[41]Kazanî Izv. fiz. mat. Obsc. (Math. Soc. Kasan), (2), 12, 1902, No. 3, 33–49. Zap. mat. otd. Obsc., 20, 1902, I–II.
[42]Archiv Math. Phys., (3), 5, 1903, 17–27.

where the summation extends over the sets of solutions ≥ 0 of

$$a_0 + a_1 + \ldots + a_r = p - 1, \qquad a_1 + 2a_2 + \ldots + ra_r \equiv 0 \pmod{p-1}.$$

The right member is an invariant modulo p of $f(x_1, x_2)$ with respect to all linear homogeneous transformations on x_1, x_2 with integral coefficients whose determinant is not divisible by p. The final sum in the expression for $A - 1$ is congruent to $N + 1$. If $r = 2$, $p > 2$, the invariant is congruent to the power $(p-1)/2$ of the discriminant $a_1^2 - 4a_0 a_2$ of f.

*E. Stephan[43] investigated the number of roots of linear congruences and systems of congruences.

H. Kühne[44] considered $f(x) = x^m + \ldots + a_m$ with no multiple irreducible factor and with a_m not a multiple of the prime p. For $n < m$, let $g = x^n + \ldots + b_n$ have arbitrary coefficients. The resultant $R(f, g)$ is zero modulo p if and only if f and g have a common factor modulo p. Thus the number of all g's of degree n which have no common factor with f modulo p is ρ_n, where

$$\rho_n \equiv \Sigma \{R(f, g)\}^\omega \pmod{p^n}, \qquad \omega = p^{n-1}(p-1),$$

the summation extending over the p^n possible g's. He expressed ρ_n as a sum of binomial coefficients. For any two binary forms ϕ, ψ of degrees m, n, it is shown that

$$J_n = \underset{\psi}{\Sigma} \{R(\phi, \psi)\}^\omega$$

is invariant modulo p^n under linear transformations with integral coefficients of determinant prime to p; J_1 is Hurwitz's[42] invariant.

M. Cipolla[45] used the method of Hurwitz[42] to find the sum of the kth powers of the roots of a congruence, and extended the method to show that the number of common roots of $f(x) \equiv 0$, $g(x) \equiv 0 \pmod{p}$, of degrees r, s, is congruent to $-\Sigma C_j K_i$, where i, j take the values for which

$$0 < i \leq s(p-1), \qquad 0 < j \leq r(p-1), \qquad i + j \equiv 0 \pmod{p},$$

the C's being as with Hurwitz, and similarly

$$g(x)^{p-1} = K_0 + K_1 x + \ldots.$$

The number of roots common to n congruences is given by a sum.

L. E. Dickson[46] gave a two-fold generalization of Hurwitz's[42] formula for the number of integral roots of $f(x) \equiv 0 \pmod{p}$. The first generalization is to the residue modulo p of the number of roots which are rational in a root of an irreducible congruence of a given degree. A further generalization is obtained by taking the coefficients a_i of $f(x)$ to be elements in the Galois field of order p^n (cf. Galois[62], etc.). Then let N be the number of roots of $f(x) = 0$ which belong to the Galois field of order $P = p^{nm}$. Then

[43]Jahresber. Staatsoberrealsch. Steyer, 34, 1903–4, 3–40.
[44]Archiv Math. Phys., (3), 6, 1904, 174–6.
[45]Periodico di Mat., 22, 1907, 36–41.
[46]Bull. Amer. Math. Soc., 14, 1907–8, 313.

$N \equiv N^*$ (mod p), where N^*+1 is derived from either of Hurwitz's two sums for $N+1$ by replacing p by P. The same replacement in Hurwitz's expression for $A-1$ leads to the invariant A^*-1, where A^* is congruent modulo p to the number of distinct sets of solutions in the Galois field of order p^{nm} of the equation $f(x_1, x_2) = 0$.

G. Rados[47] considered the sets of solutions of

$$f(x, y) = \sum_{k=0}^{p-2} (a_0^{(k)} x^{p-2} + a_1^{(k)} x^{p-3} + \ldots + a_{p-2}^{(k)}) y^{p-k-2} \equiv 0 \pmod{p}$$

for a prime p. Let A_k denote the matrix of D, in (3), with a_i replaced by $a_i^{(k)}$. Let C denote the determinant of order $(p-1)^2$ obtained from D by replacing a_k by matrix A_k. Then $f \equiv 0$ has a solution other than $x \equiv y \equiv 0$ if and only if C is divisible by p; it has exactly r sets of solutions other than $x \equiv y \equiv 0$ if and only if C is of rank $(p-1)^2-r$.

To obtain theorems including the possible solution $x \equiv y \equiv 0$, use

$$\phi(x, y) = \sum_{k=0}^{p-1} (a_0^{(k)} x^{p-1} + a_1^{(k)} x^{p-2} + \ldots + a_{p-1}^{(k)}) y^{p-k-1} \equiv 0 \pmod{p},$$

$$a = \begin{pmatrix} a_0 & a_1 & \ldots & a_{p-3} & a_{p-2} & a_{p-1} \\ a_1 & a_2 & \ldots & a_{p-2} & a_{p-1}+a_0 & 0 \\ a_2 & a_3 & \ldots & a_{p-1}+a_0 & a_1 & 0 \\ \ldots & \ldots\ldots\ldots & \ldots & \ldots & \ldots \\ a_{p-1}+a_0 & a_1 & \ldots & a_{p-3} & a_{p-2} & 0 \end{pmatrix},$$

and a_k derived from a by replacing a_i by $a_i^{(k)}$. Let γ be the determinant derived from $|a|$ by replacing a_k by matrix a_k and 0 by a matrix whose p^2 elements are zeros. Then $\phi \equiv 0$ has a set of real solutions if and only if $\gamma \equiv 0$ (mod p); it has r sets of solutions if and only if γ is of rank p^2-r.

*P. B. Schwacha[48] discussed the number of roots of congruences.

*G. Rados[49] treated higher congruences.

Theory of Higher Congruences, Galois Imaginaries.

C. F. Gauss,[50] in a posthumous paper, remarked that "the solution of congruences is only a part of a much higher investigation, viz., that of the factorization of functions modulo p. Even when $\xi(x) \equiv 0$ has no real root, ξ may be a product of factors of degrees ≥ 2, each of which could be said to have imaginary roots. If use had been made of a similar freedom which younger mathematicians have permitted themselves, and such imaginary roots had been introduced, the following investigation could be greatly condensed." As the later work of Serret[74] shows, such imaginaries can be

[47] Ann. Sc. École Normale Sup., (3), 27, 1910, 217–231. Math. és Termés Értesitö (Report of Hungarian Ac.), Budapest, 27, 1909, 255–272.

[48] Ueber die Existenz und Anzahl der Wurzeln der Kongruenz $\Sigma c_i x^i \equiv 0$ (mod m), Progr. Wilhering, 1911, 30 pp.

[49] Math. és. Termés Ertesitö, Budapest, 29, 1911, 810–826.

[50] Werke, 2, 1863, 212–240. Maser's German translation of Gauss' Disq. Arith., etc., 1889, 604–629.

introduced in a way free from any logical objections. Avoiding their use, Gauss began his investigation by showing that two polynomials in x with integral coefficients have a greatest common divisor modulo p, which can be found by Euclid's process. It is understood throughout that p is a prime (cf. Maser, p. 627). Hence if A and B are relatively prime polynomials modulo p, there exist two polynomials P and Q such that

$$PA + QB \equiv 1 \pmod{p}.$$

Thus if A has no factor in common with B or C modulo p, we find by multiplying the preceding congruence by C that A has no factor in common with the product BC modulo p. If a polynomial is divisible by A, B, C, ..., no two of which have a common factor modulo p, it is divisible by their product.

A polynomial is called prime modulo p if it has no factor of lower degree modulo p. Any polynomial is either prime or is expressible in a single way as a product of prime polynomials modulo p. The number of distinct polynomials $x^n + a x^{n-1} + \ldots$ modulo p is evidently p^n. Let (n) of these be prime functions. Then $p^n = \Sigma d(d)$, where d ranges over all the divisors of n (only a fragment of the proof is preserved). It is said to follow easily from this relation that, if n is a product of powers of the distinct primes a, b, ..., then

$$n(n) = p^n - \Sigma p^{n/a} + \Sigma p^{n/ab} - \ldots .$$

The τth powers of the roots of an equation $P = 0$ with integral coefficients are the roots of an equation $P_\tau = 0$ of the same degree with integral coefficients. If τ is a prime, $P_\tau \equiv P \pmod{\tau}$.

A prime function P of degree m, other than x itself, divides $x^\nu - 1$ for some value of $\nu < p^m$. If ν is the least such integer, ν is a divisor of $p^m - 1$. Hence P divides

$$(1) \hspace{4cm} x^{p^m - 1} - 1.$$

The latter is congruent modulo p to the product of the prime functions, other than x, whose degrees are the various divisors of m.

If $P = x^m - A x^{m-1} + B x^{m-2} - \ldots$ is a prime function modulo p, the remainders by dividing the sum, the sum of the products by twos, etc., of

$$x, \ x^p, \ x^{p^2}, \ldots, \ x^{p^{m-1}}$$

by P are congruent to A, B, etc., respectively.

If ν is not divisible by p and if m is the least positive integer for which $p^m \equiv 1 \pmod{\nu}$, each prime function dividing $x^\nu - 1$ modulo p is a divisor of (1) and its degree is therefore a divisor of m. Let δ be a divisor of m, and δ', δ'', ... the divisors $< \delta$ of δ; let μ be the g. c. d. of ν and $p^\delta - 1$, μ' the g. c. d. of ν and $p^{\delta'} - 1$, ... and set $\lambda' = \mu/\mu'$, $\lambda'' = \mu/\mu''$, ... Then the number of prime divisors modulo p of degree δ of $x^\nu - 1$ is N/δ, if N is the number of integers $< \mu$ which are divisible by no one of λ', λ'', ... A method of finding all prime functions dividing $x^\nu - 1$ is based on periods of powers of x with exponents $< \nu$ and prime to ν (pp. 620-2).

If X has been expressed as a product of relatively prime factors modulo p, we can express X as a product of a like number of factors mod p^n congruent to the former factors modulo p. There is a fragment on the case of multiple factors.

C. G. J. Jacobi[61] noted that, if q is a prime $6n-1$, $x^{q+1}\equiv1$ (mod q) has $q-1$ imaginary roots $a+b\sqrt{-3}$, where $a+3b^2\equiv1$ (mod q), besides the roots ±1.

E. Galois[62] employed imaginary roots of any irreducible congruence $F(x)\equiv0$ (mod p), where p is a prime. Let i be one imaginary root of this congruence of degree ν. Let a be one of the $p^\nu-1$ expressions

$$a+a_1i+a_2i^2+\ldots+a_{\nu-1}i^{\nu-1}$$

in which the a's are integers $<p$, not all zero. Since each power of a can be expressed as such a polynomial, we have $a^n=1$ for some positive integer n. Let n be a minimum. Then $1, a, \ldots, a^{n-1}$ are distinct. Multiply them by a new polynomial β in i; we get n products distinct from each other and from the preceding powers of a. If $2n<p^\nu-1$, we use a new multiplier, etc. Hence n divides $p^\nu-1$, and

(2) $$a^{p^\nu-1}=1.$$

[This is known as Galois's generalization of Fermat's theorem.] It follows that there exist primitive roots a such that $a^e\not\equiv1$ if $e<p^\nu-1$. Any primitive root satisfies a congruence of degree ν irreducible modulo p.

Every irreducible function $F(x)$ of degree ν divides $x^{p^\nu}-x$ modulo p. Since $\{F(x)\}^{p^n}\equiv F(x^{p^n})$ modulo p, the roots of $F(x)\equiv0$ are

$$i, i^p, i^{p^2}, \ldots, i^{p^{\nu-1}}.$$

All the roots of $x^{p^\nu}=x$ are polynomials in a certain root i, which satisfies an irreducible congruence of degree ν. To find all irreducible congruences of degree ν modulo p, delete from $x^{p^\nu}-x$ all factors which it has in common with $x^{p^\mu}-x$, $\mu<\nu$. The resulting congruence is the product of the desired ones; the factors may be obtained by the method of Gauss, since each of their roots is expressible in terms of a single root. In practice, we find by trial one irreducible congruence of degree ν, and then a primitive root of (2); this is done for $p=7$, $\nu=3$.

Any congruence of degree n has n real or imaginary roots. To find them, we may assume that there is no multiple root. The integral roots are found from the g. c. d. of $F(x)$ and $x^{p-1}-1$. The imaginary roots of the second degree are found from the g. c. d. of $F(x)$ and $x^{p^2-1}-1$; etc.

V. A. Lebesgue[63] noted that, if p is a prime, the roots of all quadratic

[61]Jour. für Math., 2, 1827, 67; Werke, 6, 235.
[62]Sur la théorie des nombres, Bulletin des Sciences Mathématiques de M. Férussac, 13, 1830, 428. Reprinted in Jour. de Mathématiques, 11, 1846, 381; Oeuvres Math. d'Evariste Galois, Paris, 1897, 15–23; Abhand. Alg. Gleich. Abel u. Galois, Maser, 1889, 100.
[63]Jour. de Mathématiques, 4, 1839, 9–12.

congruences modulo p are of the form $a+b\sqrt{n}$, where n is a fixed quadratic non-residue of p, while a, b are integers. But the cube root of a non-cubic residue is not reducible to this form $a+b\sqrt{n}$. The $p+1$ sets of integral solutions of $y^2-nz^2\equiv a$ (mod p) yield the $p+1$ real or imaginary roots $x=y+z\sqrt{n}$ of $x^{p+1}\equiv a$ (mod p). The latter congruence has primitive roots if $a=1$.

Th. Schönemann[64] built a theory of congruences without the use of Euclid's g. c. d. process. He began with a proof by induction that if a function is irreducible modulo p and divides a product AB modulo p, it divides A or B. Much use is made of the concept norm Nf_φ of $f(x)$ with respect to $\phi(x)$, i. e., the product $f(\beta_1)\ldots f(\beta_m)$, where $\beta_1,\ldots,\beta_m$ are the roots of $\phi(x)=0$; the norm is thus essentially the resultant of f and ϕ. The norm of an irreducible function with respect to a function of lower degree is shown by induction to be not divisible by p. Hence if f is irreducible and $Nf_\varphi\equiv 0$ (mod p), then f is a divisor of ϕ modulo p. A long discussion shows that if $a_1,\ldots,a_n$ are the roots of an algebraic equation $f(x)=x^n+\ldots=0$ and if $f(x)$ is irreducible modulo p, then $\Pi_{i=1}^{i=n}\{z-\phi(a_i)\}$ is a power of an irreducible function modulo p.

If a is a root of $f(x)$ and $f(x)$ is irreducible modulo p, and if $\phi(a)=\psi(a)+pR(a)$, we write $\phi\equiv\psi$ (mod p, a); then $\phi(x)-\psi(x)$ is divisible by $f(x)$ modulo p. If the product of two functions of a is $\equiv 0$ (mod p, a), one of the functions is $\equiv 0$.

If $f(x)=x^n+\ldots$ is irreducible modulo p and if $f(a)=0$, then

$$f(x)\equiv(x-a)(x-a^p)\ldots(x-a^{p^{n-1}}),\qquad a^{p^{n-1}}\equiv 1\ (\text{mod }p,a),$$

$$x^{p^{n-1}}-1=\prod_{i=1}^{p^{n-1}}\{x-\phi_i(a)\}\ (\text{mod }p,a),$$

where ϕ_i is a polynomial of degree $n-1$ in a with coefficients chosen from $0,1,\ldots,p-1$, such that not all are zero. There exist $\phi(p^n-1)$ primitive roots modulis p, a, i. e., functions of a belonging to the exponent p^n-1.

Let $F(x)$ be irreducible modulis p, a, i. e., have no divisor of degree ≥ 1 modulis p, a. Let $F(\beta)=0$, algebraically. Two functions of β with coefficients involving a are called congruent modulis p, a, β if their difference is the product of p by a polynomial in a, β. It is proved that

$$F(x)\equiv(x-\beta)(x-\beta^{p^n})\ldots(x-\beta^{p^{(m-1)n}}),\qquad \beta^{p^{mn}}\equiv 1\ (\text{mod }p,a,\beta).$$

If $\nu<n$, n being the degree of $f(x)$, and if the function whose roots are the $(p^\nu-1)$th powers of the roots of $f(x)$ is $\not\equiv 0$ (mod p) for $x=1$, then $f(x)$ is irreducible modulo p. Hence if m is a divisor of $p-1$ and if g is a primitive root of p, and if k is prime to m, then x^m-g^k is irreducible modulo p.

If $\nu<m$, m being the degree of $F(x)$, and if the function whose roots are the (p^m-1)th powers of the roots of $F(x)$ is $\not\equiv 0$ (mod p, a) for $x=1$, then

[64]Grundzüge einer allgemeinen Theorie der höhern Congruenzen, deren Modul eine reelle Primzahl ist, Progr., Brandenburg, 1844. Same in Jour. für Math., 31, 1846, 269–325.

$F(x)$ is irreducible modulis p, a. Hence if m is a divisor of p^n-1, and if $g(a)$ is a primitive root of

$$x^{p^n-1} \equiv 1 \pmod{p, a},$$

and if k is prime to m, then $x^m - g^k$ is irreducible modulis p, a.

If $F(x, a)$ is irreducible modulis p, a, and if at least one coefficient satisfies

$$x^{p^\nu-1} \equiv 1 \pmod{p, a}$$

if and only if ν is a multiple of n, then

$$\psi(x) \equiv \prod_{j=0}^{n-1} F(x, a^{p^j}) \pmod{p, a}$$

has integral coefficients and is irreducible modulo p.

If $G(x)$ is of degree mn and is irreducible modulo p, and $G(a)=0$, algebraically, and if $r(a)$ is a primitive root of $x^{p^{mn}} \equiv 1 \pmod{p, a}$, then

$$\chi(x) \equiv \prod_{j=0}^{m-1} (x-t^{p^j}), \qquad t=r^e, \qquad e=\frac{p^{mn}-1}{p^m-1},$$

has integral coefficients and is irreducible modulo p.

The last two theorems enable us to prove the existence of irreducible congruences modulo p of any degree. First,

$$(x^{p^{p^n-1}}-1)/(x^{p^{p^{n-1}-1}}-1)$$

is the product of the irreducible functions of degree p^n modulo p. To prove the existence of an irreducible function of degree lp^n, where l is any integer prime to p, assume that there exists an irreducible function of each degree $< lp^n$, and hence for the degree $a = Ap^n$, where $A = \phi(l) < l$. Let a be a root of the latter, and r a primitive root of $x^{P-1} \equiv 1 \pmod{p, a}$, where $P = p^a$. Since l divides $P-1$ by Euler's generalization of Fermat's theorem, $x^l - r$ is irreducible modulis p, a. Hence by the theorem preceding the last, $\prod_{j=0}^{l-a}(x^l - r^{p^j})$ is irreducible modulo p. Since its degree is $lp^n A$, the last theorem gives an irreducible congruence of degree lp^n.

Every irreducible factor modulo p of $x^{p^n-1}-1$ is of degree a divisor of n. Conversely, every irreducible function of degree a divisor of n is a factor of that binomial. If n is a prime, the number of irreducible functions modulo p of degree n^ν is $(p^{n^\nu} - p^{n^{\nu-1}})/n^\nu$. If n is a product of powers of distinct primes $A, B, \ldots$, say four, the number of irreducible congruences of degree n modulo p is

$$\frac{1}{n}\{P^{ABCD} - P^{ABC} - \ldots - P^{BCD} + P^{AB} + \ldots + P^{CD} - P^A - \ldots - P^D\},$$

where $P = p^{n/(ABCD)}$. Replacing p by p^m, we get the number of irreducible congruences of degree n modulis p, a, where a is a root of an irreducible congruence of degree m.

If n is a prime and p belongs to the exponent e modulo n, $f = (x^n-1)/(x-1)$ is congruent modulo p to the product of $(n-1)/e$ irreducible functions of degree e modulo p. Hence if p is a primitive root of n, f is irreducible modulo p, and therefore with respect to each of the infinitude of primes $p + \gamma n$. Thus f is algebraically irreducible.

Schönemann[65] considered congruences modulo p^m. If $g(x)$ is not divisible by p, and $f = x^n + \ldots$ is irreducible modulo p^m and if $A(x)$ is not divisible by f modulo p, then $fg \equiv AB \pmod{p^m}$ implies that $B(x)$ is divisible by f modulo p^m. If $f \equiv f_1$, $g \equiv g_1 \pmod{p}$ and the leading coefficients of the four functions are unity, while f and g have no common factor modulo p, then $fg \equiv f_1 g_1 \pmod{p^m}$ implies $f \equiv f_1$, $g \equiv g_1 \pmod{p^m}$. He proved the final theorem of Gauss.[60] Next, $(x-a)^n + pF(x)$ is irreducible modulo p^2 if and only if $F(a) \not\equiv 0 \pmod{p}$; an example is

$$\frac{x^p - 1}{x - 1} = (x-1)^{p-1} + pF(x), \qquad F(1) = 1.$$

Henceforth, let $f(x)$ be irreducible modulo p and of degree n. If $f(x)^n + pF(x)$ is reducible modulo p^2, then (p. 101) $f(x)$ is a factor of $F(x)$ modulo p. If $f(a) = 0$ and $g(a) \not\equiv 0 \pmod{p, a}$, then $g^e \equiv 1 \pmod{p^m, a}$, where $e = p^{m-1}(p^n - 1)$. If the roots of $G(z)$ are the (p^{m-1})th powers of the roots of $f(x)$, then

$$G(z) \equiv (z - \beta)(z - \beta^p) \ldots (z - \beta^{p^{n-1}}) \pmod{p^m, a}.$$

If M is any integer and if $F(x)$ has the leading coefficient unity, we can find z and w such that $(x^z - 1)^w$ is divisible by $F(x)$ modulo M.

A. Cauchy[66] noted the uniqueness of the factorization of a function $f(x)$ with integral coefficients into irreducible factors modulo p, a prime. An irreducible function divides a product only when it divides one factor modulo p. A common divisor of two functions divides their g. c. d. modulo p.

Cauchy[67] employed an indeterminate quantity or symbol i and defined $f(i)$ to be not the value of the polynomial $f(x)$ for $x = i$, but to be $a + bi$ if $a + bx$ is the remainder obtained by dividing $f(x)$ by $x^2 + 1$. In particular, if $f(x)$ is $x^2 + 1$ itself, we have $i^2 + 1 = 0$.

Similarly, if $\omega(x) \equiv 0$ is an irreducible congruence modulo p, a prime, let i denote a symbolic root. Then $\phi(i)\psi(i) \equiv 0$ implies either $\phi(i) \equiv 0$ or $\psi(i) \equiv 0 \pmod{p}$. At most n integral functions of i satisfy $f(x, i) \equiv 0 \pmod{p}$, if the degree of f in x is $n < p$. If our $\omega(x)$ divides $x^n - 1$, but not $x^m - 1$, $m < n$, modulo p, where n is not a divisor of $p - 1$, call i a symbolic primitive root of $x^n \equiv 1 \pmod{p}$. Then $x^n - 1 \equiv (x-1)(x-i) \ldots (x - i^{n-1})$.) If s is a primitive root of n and if $n - 1 = gh$, and $p^g \equiv 1 \pmod{n}$,

$$\prod_{j=0}^{g-1} (x - i^{s^{k+jh}})$$

equals a function of x with integral coefficients, while every factor of $x^n - 1$ modulo p with integral coefficients equals such a product.

---

[65]Jour. für Math., 32, 1846, 93–105.
[66]Comptes Rendus Paris, 24, 1847, 1117; Oeuvres, (1). 10, 308–12.
[67]Comptes Rendus Paris, 24, 1847, 1120; Oeuvres, (1), 10, 312–23.

G. Eisenstein[68] stated that if $f(x) \equiv 0$ is irreducible modulo p, and a is a root of the equation $f(x) = 0$ of degree n, and if $a_0, a_1, \ldots$ are any integers,

$$K = a_0 + a_1 a + \ldots + a_{n-1} a^{n-1}$$

is congruent modulis p, a to one and but one expression

$$B = b_0 \beta + b_1 \beta^p + b_2 \beta^{p^2} + \ldots + b_{n-1} \beta^{p^{n-1}},$$

where the b's are integers and β is a suitably chosen function of a. Hence the p^n numbers B form a complete set of residues modulis p, a. If ω is a primitive nth root of unity, and if

$$\phi(\lambda) = a + \omega^\lambda a^p + \omega^{2\lambda} a^{p^2} + \ldots + \omega^{(n-1)\lambda} a^{p^{n-1}},$$

the product $\phi(\lambda)\phi(\lambda') \ldots$ is independent of a if $\lambda + \lambda' + \ldots$ is divisible by n. Th. Schönemann[69] proved the last statement in case n is not divisible by p. To make $K = B$, raise it to the powers p, $p^2, \ldots, p^{n-1}$ and reduce by $\beta^{p^n} \equiv \beta$ (mod p, a). This system of n congruences determines β uniquely if the cyclic determinant of order n with the elements b_i is not divisible by p; in the contrary case there may not exist a β. The statement that the expressions B form p^n distinct residues is false if β is a root of a congruence of degree $< n$ irreducible modulo p; it is true if β is a root of such a congruence of degree n and if

$$\beta + \beta^p + \ldots + \beta^{p^{n-1}} \not\equiv 0 \pmod{p, a}.$$

J. A. Serret[70] made use of the g. c. d. process to prove that if an irreducible function $F(x)$ divides a product modulo p, a prime, it divides one factor modulo p. Then, following Galois, he introduced an imaginary quantity i verifying the congruence $F(i) \equiv 0$ (mod p) of degree $\nu > 1$, but gave no formal justification of their use, such as he gave in his later writings. However, he recognized the interpretation that may be given to results obtained from their use. For example, after proving that any polynomial $a(i)$ with integral coefficients is a root of $a^{p^\nu} \equiv a$ (mod p), he noted that this result, for the case $a = i$, may be translated into the following theorem, free from the consideration of imaginaries: If $F(x)$ is of degree ν, has integral coefficients, and is irreducible modulo p, there exist polynomials $f(x)$ and $\chi(x)$ with integral coefficients such that

$$x^{p^\nu} - x = f(x)F(x) + p\chi(x).$$

The existence of an irreducible congruence of any given degree and any prime modulus is called the chief theorem of the subject. After remarking that Galois had given no satisfactory proof, Serret gave a simple and ingenious argument; but as he made use of imaginary roots of congruences without giving an adequate basis to their theory, the proof is not conclusive.

[68] Jour. für Math., 39, 1850, 182.
[69] Jour. für Math., 40, 1850, 185–7.
[70] Cours d'algèbre supérieure, ed. 2, Paris, 1854, 343–370.

R. Dedekind[71] developed the subject of higher congruences by the methods of elementary number theory without the use of algebraic principles. As by Gauss[60] he developed the theory of the g. c. d. of functions modulo p, a prime, and their unique factorization into prime (or irreducible) functions, apart from integral factors. Two functions A and B are called congruent modulis p, M, if $A - B$ is divisible by the function M modulo p. We may add or multiply such congruences. If the g. c. d. of A and B is of degree d, $Ay \equiv B \pmod{p, M}$ has p^d incongruent roots $y(x)$ modulis p, M.

Let $\phi(M)$ denote the number of functions which are prime to M modulo p and are incongruent modulis p, M. Let μ be the degree of M. A primary function of degree a is one in which the coefficient of x^a is $\equiv 1 \pmod{p}$. If D ranges over the incongruent primary divisors of M, then $\Sigma\phi(D) = p^\mu$. If M and N are relatively prime modulo p, then $\phi(MN) = \phi(M)\phi(N)$. If A is a prime function of degree a, $\phi(A^a) = p^{aa}(1 - 1/p^a)$. If M is a product of powers of incongruent primary prime functions $a, \ldots, \rho$,

$$\phi(M) = p^\mu \left(1 - \frac{1}{p^a}\right) \ldots \left(1 - \frac{1}{p^\rho}\right).$$

If F is prime to M modulo p, $F^{\phi(M)} \equiv 1 \pmod{p, M}$, which is the generalization of Fermat's theorem. Hence if A is prime to M, the above linear congruence has the solution $y \equiv BA^{\phi-1}$.

If P is a prime function of degree π, a congruence of degree n modulis p, P has at most n incongruent roots. Also

(3) $y^{p^{\pi}-1} - 1 \equiv \Pi(y - F) \pmod{p, P}$,

identically in y, where F ranges over a complete set of functions incongruent modulis p, P and not divisible by P. In particular, $1 + \Pi F \equiv 0 \pmod{p, P}$, the generalization of Wilson's theorem.

There are $\phi(p^\pi - 1)$ primitive roots modulis p, P. Hence we may employ indices in the usual manner, and obtain the condition for solutions of $y^n \equiv A \pmod{p, P}$, where A is not divisible by P. In particular, A is a quadratic residue or non-residue of P according as

$$A^{(p^\pi - 1)/2} \equiv +1 \text{ or } -1 \pmod{p, P}.$$

His extension of the quadratic reciprocity law will be cited under that topic.

A function A belongs to the exponent ρ with respect to the prime function P of degree π if ρ is the least positive integer for which $A^{p^\rho} \equiv A \pmod{p, P}$. Evidently ρ is a divisor of π. Let $N(\rho)$ be the number of incongruent functions which belong to an exponent ρ which divides π. Then $p^\rho = \Sigma N(d)$, where d ranges over the divisors of ρ. By the principle of inversion (Ch. XIX),

$$N(\rho) = p^\rho - \Sigma p^{\rho/a} + \Sigma p^{\rho/ab} - \Sigma p^{\rho/abc} + \ldots,$$

where $a, b, \ldots$ are the distinct primes dividing ρ. Since the quotient of this sum by its last term is not divisible by p, we have $N(\rho) > 0$.

---

[71]Jour. für Math., 54, 1857, 1-26.

The product of the incongruent primary prime functions modulo p whose degree divides π is congruent modulo p to

$$\{\pi\} = x^{p^\pi} - x.$$

Then, if $\psi(\rho)$ is the number of primary prime functions of any degree ρ, $\Sigma d\psi(d) = p^\pi$, where the summation extends over all divisors d of π. A comparison of this with $\Sigma N(d) = p^\rho$ above shows that $N(\rho) = \rho\psi(\rho)$. Another proof is based on the fact that

$$(y-A)(y-A^p)\ldots(y-A^{p^{\rho-1}})$$

is congruent modulis p, P to a polynomial in y with integral coefficients which is a prime function. Moreover, if in (3) we associate the linear factors in which the F's belong to the same exponent, we obtain a factor of the left member which is irreducible modulo p.

The product of the incongruent primary prime functions of degree m (m being divisible by no primes other than a, b, ...) is congruent modulo p to

$$\frac{\{m\}\cdot\Pi\{m/ab\}\ldots}{\Pi\{m/a\}\cdot\Pi\{m/abc\}\ldots}.$$

H. J. S. Smith[72] gave an exposition of the theory.

E. Mathieu,[73] in his famous paper on multiply transitive groups, gave without proof the factorization (p. 301; for $m=1$, p. 275)

$$h(z^{p^{mn}} - z) \equiv \Pi_a \{(hz)^{p^{m(n-1)}} + (hz)^{p^{m(n-2)}} + \ldots + (hz)^{p^m} + \dot{h}z + a\},$$

where a ranges over the roots of $a^{p^m} \equiv a$, while $h^{p^{mn}} \equiv h$; and (p. 302; for $m=1$, p. 280)

$$h(z^{p^{mn}} - z) \equiv \Pi_\beta (h^{p^m} z^{p^m} - hz - \beta),$$

where β ranges over the roots of

$$z^{p^{m(n-1)}} + z^{p^{m(n-2)}} + \ldots + z^{p^m} + z \equiv 0.$$

If Ω is a root of a congruence of degree n whose coefficients are roots of $z^{p^m} \equiv z$ and whose first member is prime to $z^{p^m} - z$, then (p. 303) all the roots of $z^{p^{mn}} \equiv z$ are given by $A_0 + A_1\Omega + \ldots + A_{n-1}\Omega^{n-1}$, where the A's satisfy $z^{p^m} \equiv z$.

J. A. Serret,[74] in contrast to his[70] earlier exposition, here avoided at the outset the use of Galois imaginaries. An irreducible function of degree ν modulo p divides $x^{p^\mu} - x$ modulo p if and only if ν divides μ. A simple

[72]British Assoc. Reports, 1860, 120, §§69–71; Coll. M. Papers, 1, 149–155.

[73]Jour. de Mathématiques, (2), 6, 1861, 241–323.

[74]Mém. Ac. Sc. de l'Institut de France, 35, 1866, 617–688. Same in Cours d'algèbre supérieure, ed. 4, vol. 2, 1879, 122–189; ed. 5, 1885.

proof is given for Dedekind's[71] final theorem on the product of all irreducible functions of degree m modulo p.

A function $F(x)$ of degree ν, irreducible modulo p, is said to belong to the exponent n if n is the least positive integer such that $x^n - 1$ is divisible by $F(x)$ modulo p. Then n is a divisor of $p^\nu - 1$, and a proper divisor of it, since it does not divide $p^\mu - 1$ for $\mu < \nu$. Let n be a product of powers of the distinct primes $a, b, \ldots$. Then the product of all functions of degree ν, irreducible modulo p, which belong to an exponent n which is a proper divisor of $p^\nu - 1$, is congruent modulo p to

$$\frac{(x^n - 1) \cdot \Pi(x^{n/ab} - 1) \ldots}{\Pi(x^{n/a} - 1) \cdot \Pi(x^{n/abc} - 1) \ldots}$$

and their number is therefore $\phi(n)/\nu$.

By a skillful analysis, Serret obtained theorems of practical importance for the determination of irreducible congruences of given degrees. If we know the N irreducible functions of degree μ modulo p, which belong to the exponent $l = (p^\mu - 1)/d$, then if we replace x by x^λ, where λ is prime to d and has no prime factor different from those which divide $p^\mu - 1$, we obtain the N irreducible functions of degree $\lambda\mu$ which belong to the exponent λl, exception being made of the case when p is of the form $4h - 1$, μ is odd, and λ is divisible by 4. In this exceptional case, we may set $p = 2^i t - 1$, $i \geq 2$, t odd; $\lambda = 2^j s$, $j \geq 2$, s odd. Let k be the least of i, j. Then if we know the $N/2^{k-1}$ irreducible functions of odd degree μ modulo p which belong to the exponent l and if we replace x by x^λ, where λ is of the form indicated, is prime to d and contains only primes dividing $p^\mu - 1$, we obtain $N/2^{k-1}$ functions of degree $\lambda\mu$ each decomposable into 2^{k-1} irreducible factors, thus giving N irreducible functions of degree $\lambda\mu/2^{k-1}$ which belong to the exponent λl. Apply these theorems to $x - g^e$, which belongs to the exponent $(p-1)/d$ if g is a primitive root of p and if d is the g. c. d. of e and $p-1$; we see that $x^\lambda - g^e$ is irreducible unless the exceptional case arises, and is then a product of 2^{k-1} irreducible functions. In that case, irreducible trinomials of degree λ are found by decomposing $x^\nu - g^e$, where $\nu = 2^{i-1}\lambda$.

If a is not divisible by p, $x^p - x - a$ is irreducible modulo p.

There is a development of Dedekind's theory of functions modulis p and $F(x)$, where $F(x)$ is irreducible modulo p. Finally, that theory is considered from the point of view of Galois. Just as in the theory of congruences of integers modulo p we treat all multiples of p as if they were zero, so in congruences in the unknown X,

$$G(X, x) \equiv 0 \pmod{p, F(x)},$$

we operate as if all multiples of $F(x)$ vanish. There is here an indeterminate x which we can make use of to cause the multiples of $F(x)$ to vanish if we agree that this indeterminate x is an imaginary root i of the irreducible congruence $F(x) \equiv 0 \pmod{p}$. From the theorems of the theory of functions modulis p, $F(x)$, we may read off briefer theorems involving i (cf. Galois[62]).

Harald Schütz[75] considered a congruence

$$X^n + a_1 X^{n-1} + \ldots + a_n \equiv 0 \pmod{M(x)}$$

in which the a's and the coefficients of M are any complex integers (cf. Cauchy,[67] for real coefficients). Let $a_1, \ldots, a_n$ be the roots of the corresponding algebraic equation. Let $M = 0$ have the distinct roots $\mu_1, \ldots, \mu_m$. Then the congruence has n^m distinct roots. For, let $X - a_p$ $= f_i(x)$ have the factor $x - \mu_i$, for $i = 1, \ldots, m$. Taking $i > 1$, we have

$$f_i(x) = f_1(x) + a_{p_1} - a_{p_i}.$$

Set $x = \mu_i$. Then the right member must vanish. Using these and $f_1(\mu_1) = 0$, we have m independent linear relations for the coefficients of $f_1(x)$.

C. Jordan[76] followed Galois in employing from the outset a symbol for an imaginary root of an irreducible congruence, proved the theorems of Galois, and that, if $j, j_1, \ldots$ are roots of irreducible congruences of degrees $p^\alpha, q^\beta, \ldots$ where $p, q, \ldots$ are distinct primes, their product $jj_1 \ldots$ is a root of an irreducible congruence of degree $p^\alpha q^\beta \ldots$.

A. E. Pellet[77] stated that, if i is a root of an irreducible congruence of degree ν modulo p, a prime, the number of irreducible congruences of degree ν_1 whose coefficients are polynomials in i is

$$\frac{1}{\nu_1} \{ p^{\nu\nu_1} - \Sigma p^{\nu\nu_1/q_1} + \Sigma p^{\nu\nu_1/q_1 q_2} - \ldots + (-1)^m p^{\nu\nu_1/q_1 \ldots q_m} \}$$

if $q_1, \ldots, q_m$ are the distinct primes dividing ν_1. Of these congruences, $\phi(n)/\nu_1$ belong to the exponent n if n is a proper divisor of $(p^\nu)^{\nu_1} - 1$.

Any irreducible function of degree μ modulo p with integral coefficients is a product of δ irreducible factors of degree μ/δ with coefficients rational in i, where δ is the g. c. d. of μ, ν.

In an irreducible function of degree ν_1 and belonging to the exponent n and having as coefficients rational functions of i, replace x by x^λ, where λ contains only prime factors dividing n; the resulting function is a product of $2^{k-1}D/n$ irreducible functions of degree $\lambda n\nu_1/(2^{k-1}D)$ belonging to the exponent λn, where D is the g. c. d. of λn and $p^{\nu_1} - 1$, and 2^{k-1} is the highest power of 2 dividing the numerators of each of the fractions $(p^{\nu_1} + 1)/2$ and $\lambda n/(2D)$ when reduced to their lowest terms.

Let g be a rational function of i, and m the number of distinct values among $g, g^p, g^{p^2}, \ldots$. If neither $g + g^p + \ldots + g^{p^{m-1}}$ nor ν/m is divisible by p, then $x^p - x - g$ is irreducible; in the contrary case it is a product of linear functions.

Hence if we replace x by $x^p - x$ in an irreducible function of degree μ having as coefficients rational functions of i, we get a new irreducible function provided the coefficient of $x^{\mu-1}$ in the given function is not zero.

[75]Untersuchungen über Functionale Congruenzen, Diss. Göttingen, Frankfurt, 1867.
[76]Traité des substitutions, 1870, 14–18.
[77]Comptes Rendus Paris, 70, 1870, 328–330.

[Proof in Pellet.[88]] In particular, if p is a primitive root of a prime n, we have the irreducible function, modulo p,

$$\frac{(x^p-x)^n-1}{x^p-x-1}.$$

C. Jordan[78] listed irreducible functions [errata, Dickson,[102] p. 44].

J. A. Serret[79] determined the product V_n of all functions of degree p^n irreducible modulo p, a prime. In the expansion of $(\xi-1)^\mu$ replace each power ξ^k by x^{p^k}; denote the resulting polynomial in x by X_μ. Then

$$X_{\nu p^m}\equiv \{(\xi-1)^{p^m}\}^\nu\equiv(\xi^{p^m}-1)^\nu, \qquad X_{p^m}\equiv x^{p^{p^m}}-x \pmod{p}.$$

Hence $V_n=X_{p^n}/X_{p^{n-1}}$. Moreover,

$$X_{\mu+1}=(\xi-1)^{\mu+1}=\xi(\xi-1)^\mu-(\xi-1)^\mu\equiv X_\mu{}^p-X_\mu \pmod{p}.$$

Multiply this by the relations obtained by replacing μ by $\mu+1,\ldots,\mu+\nu-1$. Thus

$$X_{\mu+\nu}\equiv X_\mu(X_\mu{}^{p-1}-1)(X_{\mu+1}^{p-1}-1)\ldots(X_{\mu+\nu-1}^{p-1}-1) \pmod{p}.$$

Take $\mu=p^{n-1}$, $\mu+\nu=p^n$. Hence

$$V_n\equiv \prod_{\lambda=1}^{p^n-p^{n-1}} f_\lambda \pmod{p}, \qquad f_\lambda=X_{p^{n-1}+\lambda-1}^{p-1}-1.$$

Each f_λ decomposes into $p-1$ factors $X-g$ where $g=1,\ldots,p-1$. The irreducible functions of degree p^n whose product is f_λ are said to belong to the λth class. When x is replaced by x^p-x, X_μ is replaced by $X_{\mu+1}$ since ξ^i is replaced by $\xi^i(\xi-1)$ and hence $(\xi-1)^\mu$ by $(\xi-1)^{\mu+1}$; thus f_λ is replaced by $f_{\lambda+1}$, while the last factor in $V_n=\mathrm{H}f_\lambda$ is replaced by $X_{p^n}^{p-1}-1$, which is the first factor in V_{n+1}. Hence if $F(x)$ is of degree p^n and is irreducible modulo p and belongs to the λth class, $F(x^p-x)$ is irreducible or the product of p irreducible functions of degree p^n according as $\lambda=$ or $<p^n-p^{n-1}$.

For $n=1$, the irreducible functions of the λth class have as roots polynomials of degree λ in a root of $i^p-i\equiv1$, which is irreducible modulo p. Hence if we eliminate i between the latter and $f(i)=x$, where $f(i)$ is the general polynomial of degree λ in i, we obtain the general irreducible function of degree p of the λth class.

For any n, the determination of the irreducible functions of degree p^n of the first class is made to depend upon a problem of elimination (Algèbre, p. 205) and the relation to these of the functions of the λth class, $\lambda>1$, is investigated.

G. Bellavitis[79a] tabulated the indices of Galois imaginaries of order 2 for each prime modulus $p=4n+3\leq63$.

Th. Pepin[80] proved that $x^2-ny^2\equiv1 \pmod{p}$ has $p+1$ sets of solutions

[78]Comptes Rendus Paris, 72, 1871, 283–290.
[79]Jour. de Mathématiques, (2), 18, 1873, 301–4, 437–451. Same as in Cours d'algèbre supérieure, ed. 4, vol. 2, 1879, 190–211.
[79a]Atti Accad. Lincei. Mem. Sc. Fis. Mat., (3), 1, 1876-7, 778–800.
[80]Atti Accad. Pont. Nuovi Lincei, 31, 1877-8, 43–52.

x, y selected from $0, 1, \ldots, p-1$, provided n is a quadratic non-residue of the prime p. Then $x+y\sqrt{n}$ is a root of $\xi^{p+1}\equiv 1 \pmod{p}$, which therefore has $p+1$ complex roots, all a power of one root. There is a table of indices for these roots when $p=29$ and $p=41$. [Lebesgue.[63]]

A. E. Pellet[81] considered the product Δ of the squares of the differences of the roots of a congruence $f(x)\equiv 0 \pmod{p}$ having no equal roots. Then Δ is a quadratic non-residue of p if $f(x)$ has an odd number of irreducible factors of even degree, a quadratic residue if $f(x)$ has no irreducible factor of even degree or has an even number of them. For, if $\delta_1, \ldots, \delta_i$ are the values of Δ for the various irreducible factors of $f(x)$, then $\Delta\equiv a^2\delta_1\ldots\delta_i$ $\pmod{p}$, where a is an integer. Hence it suffices to consider an irreducible congruence $f(x)\equiv 0 \pmod{p}$. Let ν be its degree and i a root. In

$$y = \prod_{l=1}^{\nu-1} \prod_{k=0}^{l-1} (x^{p^k} - x^{p^l})$$

replace x by the ν roots; we get two distinct values if ν is even, one if ν is odd. In the respective cases, $y^2\equiv\Delta \pmod{p}$ is irreducible or reducible.

R. Dedekind[82] noted that, if $P(x)$ is a prime function of degree f modulo p, a prime, a congruence $F(x)\equiv 0 \pmod{p, P}$ is equivalent to the congruence $F(\alpha)\equiv 0 \pmod{\pi}$, where π is a prime ideal factor of p of norm p^f, and α is a root of $P(\alpha)\equiv 0 \pmod{\pi}$.

A. E. Pellet[83] denoted by $f(x)=0$ the equation of degree $\phi(k)$ having as its roots the primitive kth roots of unity, and by $f_1(y)=0$ the equation derived by setting $y=x+1/x$. If p is a prime not dividing k, $f(x)$ is congruent modulo p to a product of $\phi(k)/\nu$ irreducible factors whose degree ν is the least integer for which $p^\nu-1$ is divisible by k. If $f_1(y)\equiv 0 \pmod{p}$ has an integral root a, $f(x)$ is divisible modulo p by $x^2-2ax+1$. Either the latter has two real roots and $f(x)$ and $f_1(y)$ have all their roots real and $p-1$ is divisible by k, or it is irreducible and $f(x)$ is a product of quadratic factors modulo p and the roots of $f_1(y)$ are all real and $p+1$ is divisible by k. If k divides neither $p+1$ nor $p-1$, $f_1(y)$ is a product of factors of equal degree modulo p. [Cf. Sylvester,[29] etc., Ch. XVI.]

Let k be a divisor $\neq 2$ of $p+1$. Let λ be an odd number divisible by no prime not a factor of k, and relatively prime to $(p+1)/k$. Then $x^{2\lambda}-2ax^\lambda+1$ is irreducible modulo p [Serret,[74] No. 355]. Also, if b is not divisible by p

$$F = (x+b)^{2\lambda} - 2a(x^2-b^2)^\lambda + (x-b)^{2\lambda}$$

is irreducible modulo p; replacing x^2 by y, we obtain a function of degree λ irreducible modulo p. If k is a divisor $\neq 2$ of $p-1$ and if λ is odd, prime to $(p-1)/k$ and divisible by no prime not a factor of k, F decomposes modulo p into two irreducible functions of degree λ.

The function $f(x^2)$ is either irreducible or the product of two irreducible factors of degree ν. In the respective cases, the product Δ of the squares of

[81]Comptes Rendus Paris, 86, 1878, 1071–2.
[82]Abhand. K. Gesell. Wiss. Göttingen, 23, 1878, p. 25. Dirichlet-Dedekind, Zahlentheorie, ed. 4, 1894, 571–2.
[83]Comptes Rendus Paris, 90, 1880, 1339–41.

the differences of the roots of $f(x^2) \equiv 0$ is a quadratic non-residue or residue of p [Pellet[31]]. Let Δ_1 be the like product for $f(x)$. Then $\Delta = (-1)^{\nu} 2^{2\nu} f(0) \Delta_1^2$. Hence $f(ax^2 + b)$ is irreducible if $(-1)^{\nu} f(b)/a^{\nu}$ is a quadratic non-residue and then $f(ax^{2^i} + b)$ is irreducible modulo p for every i and even ν.

O. H. Mitchell[84] gave analogues of Fermat's and Wilson's theorems modulis p (a prime) and a function of x.

A. E. Pellet[85] considered the exponent n to which belongs the product P of the roots of a congruence $F(x) \equiv 0$ of degree ν irreducible modulo p. If q is a prime factor of n, $F(x^q)$ is irreducible or the product of q irreducible factors of degree ν modulo p according as q is not or is a divisor of $(p-1)/n$. In particular, $F(x^{\lambda})$ is irreducible modulo p if, for ν even, λ contains only prime factors of n not dividing $(p-1)/n$; for ν odd, we can use the factor 2 in λ only once if $p = 4m + 1$. Let i be a root of $F(x) \equiv 0$, i_1 a root of an irreducible congruence $F_1(x) \equiv 0 \pmod{p}$ of degree ν_1 prime to ν. Then ii_1 is a root of an irreducible congruence $G(x) \equiv 0 \pmod{p}$ of degree $\nu\nu_1$. $F(x)$ belongs to the exponent Nn modulo p, where n is prime to $(p^{\nu} - 1) \div \{(p-1)N\}$. Let q_1 be a prime factor of N not dividing $p-1$. Then $G(x^{q_1})$ is irreducible or decomposes into q_1 irreducible factors of degree $\nu\nu_1$ according as q_1 is not or is a divisor of $(p^{\nu} - 1)/N$. Thus $G(x^{\lambda})$ is irreducible if λ contains only prime factors of N dividing neither $p-1$ nor $(p^{\nu} - 1)/N$.

O. H. Mitchell[86] defined the prime totient of $f(x)$ to mean the number of polynomials in x, incongruent modulo p, of degree less than the degree of (x) and having no factor in common with f modulo p. Those which contain S, but no prime factor of f not contained in S, are called S-totitives of f.

C. Dina[87] proved known results on congruences modulis p and $F(x)$.

A. E. Pellet[88] proved that, if μ distinct values are obtained from a rational function of x with integral coefficients by replacing x successively by the m roots of an irreducible congruence modulo p, then μ is a divisor of m and these μ values are the roots of an irreducible congruence. Thus if A is a rational function of any number of roots of congruences irreducible modulo p, and ν is the number of distinct values among A, A^p, $A^{p^2}, \ldots$, these values satisfy an irreducible congruence modulo p. If A belongs to the exponent n modulo p, then ν is the least positive integer for which $p^{\nu} \equiv 1 \pmod{n}$. He proved a result of Serret's[74] stated in the following form: If, in an irreducible function $F(x)$ modulo p of degree ν and exponent n, x is replaced by x^{λ}, where λ contains only primes dividing n, then $F(x^{\lambda})$ is a product of irreducible factors of degree νq and exponent $n\lambda$, where q is the least integer for which $p^{\nu q} \equiv 1 \pmod{n\lambda}$. He proved the first theorem of Pellet[85] and the last one of Pellet.[77]

[84]Johns Hopkins University Circulars, 1, 1880–1, 132.
[85]Comptes Rendus Paris, 93, 1881, 1065–6. Cf. Pellet.[88]
[86]Amer. Jour. Math., 4, 1881, 25–38.
[87]Giornale di Mat., 21, 1883, 234–263. For comments on 263–9, see the chapter on quadratic reciprocity law.
[88]Bull. Soc. Math. France, 17, 1888–9, 156–167.

E. H. Moore[89] stated that every finite field (Körper) is, apart from notations, a Galois field composed of the p^n polynomials in a root of an irreducible congruence of degree n modulo p, a prime.

E. H. Moore[90] proved the last theorem and others on finite fields.

K. Zsigmondy[36] noted that the number of congruences of degree n modulo p, having no irreducible factor of degree i, is

$$p^n - \binom{I}{1}p^{n-i} + \binom{I}{2}p^{n-2i} - \ldots,$$

where I is the number of functions of degree i irreducible modulo p.

G. Cordone[91] noted that if a function is prime to each of its derivatives with respect to each prime modulus $p_1, \ldots, p_n$ and is irreducible modulo $M = p_1^{e_1} \ldots p_n^{e_n}$, it is irreducible with respect to at least one of $p_1, \ldots, p_n$. If $F(x)$ is not identically $\equiv 0$ modulo p_1, nor modulo p_2, etc., and if it divides a product modulo M and is prime to one factor according to each modulus $p_1, \ldots, p_n$, then $F(x)$ divides the other factor modulo M.

Let $F(x)$ be a function of degree r irreducible with respect to each prime $p_1, \ldots, p_n$, while $f(x)$ is not divisible by $F(x)$ with respect to any one of the p's, then (pp. 281-8)

$$\{f(x)\}^{\phi_r(M)} \equiv 1 \pmod{M, F(x)}, \qquad \phi_r(M) = M^r\left(1 - \frac{1}{p_1^r}\right) \ldots \left(1 - \frac{1}{p_n^r}\right),$$

$\phi_r(M)$ being the number of functions $c_1 x^{r-1} + \ldots + c_r$, in which the c's take such values $0, 1, \ldots, M-1$ whose g. c. d. is prime to M. Let A be the product of these reduced functions modulis $M, F(x)$. Then (pp. 316-8), $A \equiv -1 \pmod{M, F}$ if $M = p^k, 2p^k$ or 4, where p is an odd prime, while $A \equiv +1$ in all other cases.

Borel and Drach[92] gave an exposition of the theory of Galois imaginaries from the standpoint of Galois himself.

H. Weber[93] considered the finite field (Congruenz Körper) formed of the p^n classes of residues modulo p of the polynomials, with integral coefficients, in a root of an irreducible equation of degree n. He proved the generalization of Fermat's theorem, the existence of primitive roots, and the fact that every element is a square or a sum of the squares of two elements.

Ivar Damm[94] gave known facts about the roots of congruences modulis $p, f(x)$, where $f(x)$ is irreducible modulo p, without exhibiting the second modulus and without making it clear that it is not a question of ordinary congruences modulo p. Let e be a fixed primitive root of the prime p. Then the roots of every irreducible quadratic congruence are of the form $a \pm b\omega$, where $\omega^2 = e$. Let $k^{p+1} = e$, $k_1 = k^p$.

[89]Bull. New York Math. Soc., 3, 1893-4, 73-8.
[90]Math. Papers Chicago Congress of 1893, 1896, 208-226; University of Chicago Decennial Publications, (1), 9, 1904, 7-19.
[91]El Progreso Matemático, 4, 1894, 265-9.
[92]Introd. théorie des nombres, 1895, 42-50, 58-62, 343-350.
[93]Lehrbuch der Algebra, II, 1896, 242-50, 259-261; ed. 2, 1899, 302-10, 320-2.
[94]Bidrag till Läran om Kongruenser med Primtalsmodyl, Diss., Upsala, 1896, 86 pp.

Analogous to the definition of trigonometric functions in terms of exponentials, he defined quasi cosines and sines by

$$Cqx = \frac{1}{2}(k^x + k_1{}^x), \qquad Sqx = \frac{1}{2\omega}(k^x - k_1{}^x),$$

and Tqx as their quotient. Their relations are discussed. He defined pseudo cosines and sines by

$$Cpx = Cq[(p-1)x] = e^{-x}Cq2x, \qquad Spx = -e^{-x}Sq2x.$$

For each prime $p < 100$, he gave (pp. 65-86) the (integral) values of

$$e^x, \text{ ind } x, Cqx, Sqx, Tqx, Cpx, Spx$$

for $x = 1, 2, \ldots, p+1$.

L. E. Dickson[95] extended the results of Serret[74] to the more general case in which the coefficients of the functions are polynomials in a given Galois imaginary (i. e., are in a Galois field of order p^n). For the corresponding generalization of the results of Serret[79] on irreducible congruences modulo p of degree a power of p, additional developments were necessary. To obtain the irreducible functions of degree p in the $GF[p^n]$ which are of the first class, we need the complete factorization, in the field,

$$h(z^{p^n} - z - \nu) = \Pi(h^p z^p - hz - \beta)$$

where $h\nu$ is an integer and β ranges over the roots of

$$B = \beta^{p^{n-1}} + \beta^{p^{n-2}} + \ldots + \beta^p + \beta = h\nu,$$

all of whose roots are in the field. For the case $\nu = 0$ this factorization is due to Mathieu.[73] Thus $h^p z^p - hz - \beta$ is irreducible in the field if and only if $B \neq 0$. In particular, if β is an integer not divisible by p, $z^p - z - \beta$ is irreducible in the $GF[p^n]$ if and only if n is not divisible by p.

R. Le Vavasseur[96] employed Galois imaginaries to express in brief notation the groups of isomorphisms of certain types of groups, for example, that of the abelian group G generated by n independent operators $a_1, \ldots, a_n$, each of period a prime p. If i is a root of an irreducible congruence of degree n modulo p, and if $j = a_1 + ia_2 + \ldots + i^{n-1}a_n$, he defined a^j to be $a_1{}^{a_1} \ldots a_n{}^{a_n}$. Then the operators of G are represented by the real and imaginary powers of a.

A. Guldberg[97] considered linear differential forms

$$Ay = a_k \frac{d^k y}{dx^k} + \ldots + a_1 \frac{dy}{dx} + a_0 y,$$

with integral coefficients. The product of two such forms is defined by Boole's symbolic method to be

$$Ay \cdot By = (a_k \frac{d^k}{dx^k} + \ldots + a_0)(b_l \frac{d^l}{dx^l} + \ldots + b_1 \frac{d}{dx} + b_0)y.$$

[95]Bull. Amer. Math. Soc., 3, 1896-7, 381-9.
[96]Mém. Ac. Sc. Toulouse, (9), 9, 1897, 247-256.
[97]Comptes Rendus Paris, 125, 1897, 489.

If the product is $\equiv Cy$ (mod p), Ay and By are called divisors modulo p of Cy. Let Δy be of order n and irreducible modulo p. Then Ay is congruent modulis p, Δy to one and but one of the p^n forms

$$(4) \qquad \sum_{i=0}^{n-1} c_i \frac{d^i y}{dx_i} \qquad (c_i = 0, 1, \ldots, p-1).$$

If u is any one of these forms (4) and if $e = p^n - 1$, Guldberg stated the analogue of Fermat's theorem

$$\frac{d^e u}{dx^e} = u \text{ (mod } p, \Delta y),$$

but incorrectly gave the right member to be unity [cf. Epsteen,[106] Dickson[107]].

L. Stickelberger[98] considered $F(x) = x^n + a_1 x^{n-1} + \ldots$ with integral coefficients, such that the product D of the squares of the differences of the roots is not zero. Let p be any prime not dividing D. Let ν be the number of factors of $F(x)$ which are irreducible modulo p. He proved by the use of prime ideals that

$$\left(\frac{D}{p}\right) = (-1)^{n-\nu},$$

where the symbol in the left member is that of Legendre [see quadratic residues].

L. E. Dickson[99] proved the existence of the Galois field $GF[p^r]$ of order p^n by induction from $r = n$ to $r = qn$, by showing that

$$(x^{p^{nq}} - x)/(x^{p^n} - x)$$

is a product of factors of degree q belonging to and irreducible in the $GF[p^n]$. Any such factor defines the $GF[p^{nq}]$.

L. Kronecker[100] treated congruences modulis p, $P(x)$ from the standpoint of modular systems.

F. S. Carey[101] gave for each prime $p < 100$ a table of the residues of the first $p + 1$ powers of a primitive root $a + bj$ of $z^{p^2-1} \equiv 1$ (mod p) where $j^2 \equiv \nu$ (mod p), ν being an integral quadratic non-residue of p. The higher powers are readily derived. While only the single modulus p is exhibited, it is really a question of a double modulus p and $x^2 - \nu$. Methods of "solving" $z^{p^2-1} \equiv 1$ are discussed. In particular, for $n = 3$, there is given a primitive root for each prime $p < 100$.

L. E. Dickson[102] gave a systematic introductory exposition of the theory, with generalizations and extensions.

M. Bauer[103] proved that, if $f(x) = 0$ is an irreducible equation with integral coefficients and leading coefficient unity, w a root, D its discriminant, $d = D/k^2$ that of the domain defined by w, p a prime not dividing k, $x > 1$,

[98]Verhand. I. Internat. Math. Kongress, 1897, 186.
[99]Bull. Amer. Math. Soc., 6, 1900, 203–4.
[100]Vorlesungen über Zahlentheorie, I, 1901, 212–225 (expanded by Hensel, p. 506).
[101]Proc. London Math. Soc., 33, 1900–1, 294–310.
[102]Linear groups with an exposition of the Galois field theory, Leipzig, 1901, pp. 1–71.
[103]Math. Naturw. Berichte aus Ungarn, 20, 1902, 39–42; Math. és Phys. Lapok, 10, 1902, 28–33.

then $f(x)$ is congruent modulo p^a to a product of $F_1(x), \ldots, F_s(x)$, each irreducible modulo p^a, such that $F_i(x) \equiv f_i(x)^{e_i} \pmod{p}$, where $f(x) \equiv \Pi f_i(x)^{e_i}$ $\pmod{p}$, and $f_i(x)$ is irreducible modulo p. There is an example of an irreducible cyclotomic function reducible with respect to every prime power modulus.

P. Bachmann[104] gave an exposition of the general theory.

G. Arnoux[105] exhibited in the form of tables the work of finding a primitive root of the $GF[7^3]$ and of the $GF[5^4]$, and tabulated the reducible and irreducible congruences of degrees 1, 2, 3, modulo 5.

S. Epsteen[106] proved the result of Guldberg,[97] and developed the theory of residues of linear differential forms parallel to the theory of finite fields, as presented by Dickson.[102]

L. E. Dickson[107] noted that the last mentioned subjects are identical abstractly. Let the irreducible form Δy be

$$\delta_n \frac{d^n y}{dx^n} + \ldots + \delta_1 \frac{dy}{dx} + \delta_0 y.$$

To the element (4) we make correspond the element $\Sigma c_i z^i$ of the Galois field of order p^n, where z is a root of the irreducible congruence

$$\delta_n z^n + \ldots + \delta_1 z + \delta_0 \equiv 0 \pmod{p}.$$

Since product relations are preserved by this correspondence, the p^n residues (4) define a field abstractly identical with our Galois field.

Dickson[107a] proved that $x^m \equiv x \pmod{m = p^n}$ has p and only p roots if p is a prime and hence does not define the Galois field of order m as occasionally stated.

A. Guldberg[107b] employed the notation of finite differences and wrote

$$Fy_x = \sum_{i=0}^{n} a_i \theta^i y_x, \quad Gy_x = \sum_{i=0}^{m} b_i \theta^i y_x, \quad Fy_x.Gy_x = \sum_{i=0}^{n} a_i \theta^i . \sum_{i=0}^{m} b_i \theta^i y_x,$$

where $\theta y_x = y_{x+1}$, $\theta^2 y_x = y_{x+2}, \ldots$, symbolically. To these linear forms with integral coefficients taken modulo p, a prime, we may apply Euclid's g. c. d. process and prove that factorization is unique. Next, let b_m be not divisible by p, so that Gy_x is of order m. With respect to the two moduli p, Gy_x, a complete set of p^m residues of linear forms is $a_{m-1}y_{x+m-1} + \ldots + a_0 y_x$ ($a_i = 0$, $1, \ldots, p-1$). Amongst these occur $\phi(Gy_x) = p^m(1 - 1/p^{m_1}) \ldots (1 - 1/p^{m_q})$ forms Fy_x prime to Gy_x if $m_1, \ldots, m_q$ are the orders of the irreducible factors of Gy_x modulo p, and

$$Fy_x^{\phi(Gy_x)} \equiv y_x \pmod{p, Gy_x}$$

In particular, if Gy_x is irreducible and of order m,

$$Fy_x^{p^m-1} \equiv y_x \pmod{p, Gy_x}.$$

[104]Niedere Zahlentheorie, 1, 1902, 363–399.
[105]Assoc. franç. av. sc., 31, 1902, II, 202–227.
[106]Bull. Amer. Math. Soc., 10, 1903–4, 23–30.
[107]Ibid., pp. 30–1.
[107a]Amer. Math. Monthly, 11, 1904, 39–40.
[107b]Annali di Mat., (3), 10, 1904, 201–9.

W. H. Bussey[108] gave for each Galois field of order <1000 companion tables showing the residues of the successive powers of a primitive root, and the powers corresponding to the residues arranged in a natural order. These tables serve the same purposes in computations with Galois fields that tables of indices serve in computations with integers modulo p^n, where p is a prime.

G. Voronoï[109] proved the theorem of Stickelberger.[98] Thus, for $n=3$, $(D/p) = -1$ only when $v=2$. Hence a cubic congruence has a single root if $(D/p) = -1$, and three real roots or none if $(D/p) = +1$.

P. Bachmann[110] developed the general theory from the standpoint of Kronecker's modular systems and considered its relation to ideals (p. 241).

M. Bauer[111] employed a polynomial $f(z)$ of degree n irreducible modulo p, and another one $M(z)$ of degree less than that of $f'(z)$ and not divisible by $f(z)$ modulo p. Then if $(t, a) = 1$, the equation

$$f'(z) + p^a M(z) = 0$$

is irreducible. The case $a=1$ is due to Schönemann[65] (p. 101).

G. Arnoux,[112] starting with any prime m and integer n, introduced a symbol i such that $i^s \equiv 1 \pmod{m}$ and such that $i, i^2, \ldots, i^s$ are distinct, where $s = m^n - 1$, without attempting a logical foundation. If $f(x)$ is irreducible modulo m and of degree n, there is only a finite number of distinct residues of powers of x modulis $f(x)$, m; let x^k and x^{k+p} have the same residue. Thus $x^p - 1$ is divisible by $f(x)$ modulo m. It is stated (p. 95) without proof that p divides s. "Call a a root of $f(x) \equiv 0$. To make a coincide with the primitive root i of $x^s = 1$, we must take $p = s$, whence every such primitive root is a root of an irreducible congruence of degree n modulo m." Following this inadequate basis is an exposition (pp. 117–136) of known properties of Galois imaginaries.

L. I. Neikirk[113] represented geometrically the elements of the Galois field of order p^n defined by an irreducible congruence

$$f(x) = x^n + a_1 x^{n-1} + \ldots + a_n \equiv 0 \pmod{p}.$$

Let j be a root of the equation $f(x) = 0$ and represent

$$c_1 j^{n-1} + \ldots + c_{n-1} j + c_n \qquad (c\text{'s integers})$$

by a point in the complex plane. The p^n points for which the c's are chosen from $0, 1, \ldots, p-1$ represent the elements of the Galois field.

G. A. Miller[114] listed all possible modular systems p, $\phi(x)$, where p is a prime and the coefficient of the highest power of x is unity, in regard to which a complete set of prime residues forms a group of order ≤ 12. If $\phi(x)$ is the product of k distinct irreducible functions $\phi_1, \ldots, \phi_k$ modulo p,

[108]Bull. Amer. Math. Soc., 12, 1905, 21–38; 16, 1909–10, 188–206.
[109]Verhand. III. Internat. Math. Kongress, 1905, 186–9.
[110]Allgemeine Arith. d. Zahlenkörper, 1905, 81–111.
[111]Jour. für Math., 128, 1905, 87–9.
[112]Arithmétique Graphique, Fonctions Arith., 1906, 91–5.
[113]Bull. Amer. Math. Soc., 14, 1907–8, 323–5.
[114]Archiv Math. Phys., (3), 15, 1909–10, 115–121.

the residues prime to p, $\phi(x)$ constitute the direct product of the groups with respect to the various p, $\phi_i(x)$.   Not every abelian group can be represented as a congruence group composed of a complete set of prime residues with respect to $F_1, \ldots, F_\lambda$, where the F's are functions of a single variable.

Mildred Sanderson[115] employed two moduli m and $P(y)$, the first being any integer and the second any polynomial in y with integral coefficients. Such a polynomial $f(y)$ is said to have an inverse $f_1(y)$ if $ff_1 \equiv 1$ (mod m, P). If $P(y)$ is of degree r and is irreducible with respect to each prime factor of m, a function $f(y)$, whose degree is $< r$, has an inverse modulis m, $P(y)$, if and only if the g. c. d. of the coefficients of $f(y)$ is prime to m.   For such an f, $f^n \equiv 1$ (mod m, P), where n is Jordan's function $J_r(m)$ [Jordan,[200] Ch. V].   In case m is a prime, this result becomes Galois'[62] generalization of Fermat's theorem.   The product of the n distinct residues having inverses modulis m, $P(y)$, is congruent to -1 when m is a power of an odd prime or the double of such a power or when $r=1$, $m=4$; but congruent to $+1$ in all other cases—a two-fold generalization of Wilson's theorem. There exists a polynomial $P(y)$ of degree r which is irreducible with respect to each prime factor of m.   Then if $A(y)$, $B(y)$ are of degrees $< r$ and their coefficients are not all divisible by a factor of m, there exist polynomials $\alpha(y)$, $\beta(y)$, such that $\alpha A + \beta B \equiv 1$ (mod m, P).

Several writers[116] discussed the irreducible quadratic factors modulo p of $(x^a - 1)/(x^k - 1)$, where $k = 1$ or 2, p is a prime, a a divisor of $p+1$.

G. Tarry[117] noted that, if $j^2 \equiv q$ (mod m), where q is a quadratic non-residue of the prime m, the Galois imaginary $a + bj$ is a primitive root if its norm $(a+bj)(a-bj)$ is a primitive root of m and if the ratio $a:b$ and the analogous ratios of the coordinates of the first m powers of $a+bj$ are incongruent.

L. E. Dickson[118] proved that two polynomials in two variables with integral coefficients have a unique g. c. d. modulo p, a prime.   Thus the unique factorization theorem holds.

G. Tarry[119] stated that $A\rho$ is a primitive root of the $GF[p^2]$ if the norm of $A = a + bj$ is a primitive root of p and if the imaginary ρ belongs to the exponent $p+1$.   The $\phi(p+1)$ numbers ρ are found by the usual process to obtain the primitive roots of a prime.

U. Scarpis[120] proved that an equation of degree ν irreducible in the Galois field of order p^n has in the field of order p^{mn} either ν roots or no root according as ν is or is not a divisor of m [Dickson,[102] p. 19, lines 7–9].

CUBIC CONGRUENCES.

A. Cauchy[130] solved $y^3 + By + C \equiv 0$ (mod p) when it has three distinct

[115]Annals of Math., (2), 13, 1911, 36–9.
[116]L'intermédiaire des math., 18, 1911, 195, 246; 19, 1912, 61–69, 95–6; 21, 1914, 158–161; 22, 1915, 77–8. Sphinx-Oedipe, 7, 1912, 2–3.
[117]Assoc. franç. av. sc., 40, 1911, 12–24.   [118]Bull. Amer. Math. Soc., (2), 17, 1911, 293–4.
[119]Sphinx-Oedipe, 7, 1912, 43–4, 49–50.   [120]Annali di Mat., (3), 23, 1914, 45.
[130]Exercices de Math., 4, 1829, 279–292; Oeuvres, (2), 9, 326–333.

integral roots y_1, y_2, y_3, and p is a prime $\equiv 1 \pmod 3$, and $B \not\equiv 0 \pmod p$. Set

$$3v_1 = y_1 + ry_2 + r^2y_3, \qquad 3v_2 = y_1 + r^2y_2 + ry_3, \qquad r^2 + r + 1 \equiv 0 \pmod p.$$

The roots of $u^2 + Cu - B^3/27 \equiv 0 \pmod p$ are $u_1 = v_1^3$, $u_2 = v_2^3$. After finding v_1 from $v_1^3 \equiv u_1 \pmod p$, we get $v_2 \equiv -B/(3v_1)$, and determine the y's from $\Sigma y_i \equiv 0$ and the expressions for $3v_1$, $3v_2$. Thus

$$y_1 \equiv v_1 + v_2, \qquad y_2 \equiv r^2v_1 + rv_2, \qquad y_3 \equiv rv_1 + r^2v_2 \pmod p.$$

Since by hypothesis the cubic congruence has three distinct integral roots, the quadratic has two distinct integral roots, whence

$$u_i^{\frac{p-1}{3}} \equiv 1, \qquad D^{\frac{p-1}{2}} \equiv 1 \pmod p, \qquad D = \frac{C^2}{4} + \frac{B^3}{27},$$

$$\left(-\frac{C}{2} - D^{\frac{1}{2}}\right)^{\frac{p-1}{3}} + \left(-\frac{C}{2} + D^{\frac{1}{2}}\right)^{\frac{p-1}{3}} \equiv 2, \qquad D^{\frac{p-1}{2}} \equiv 1 \pmod p.$$

Conversely, if the last two conditions are satisfied, the cubic congruence has three distinct real roots provided $p \equiv 1 \pmod 3$, $B \not\equiv 0 \pmod p$.

G. Oltramare[131] found the conditions that one or all of the roots of $x^3 + 3px + 2q \equiv 0 \pmod \mu$ given by Cardan's formula become integral modulo μ, a prime. Set

$$D = q^2 + p^3, \qquad \sigma = -q + \sqrt{D}, \qquad \tau = -q - \sqrt{D},$$

First, let μ be a prime $6n - 1$. If D is a quadratic residue of μ, there is a single rational root $-2q/(p + \sigma^{2n} + \tau^{2n})$. If D is a quadratic non-residue of μ, there are three rational roots or no root according as the rational part M of the development of σ^{2n-1} by the binomial theorem satisfies or does not satisfy $Mp^2 + q \equiv 0 \pmod \mu$; if also $\mu = 18m + 11$ and there are three rational roots, they are

$$2M\sqrt[3]{p^2}, \qquad -\sqrt[3]{p^2}(M \pm N\sqrt{-3D}),$$

if $\sigma^{2m+1} = M + N\sqrt{D}$; with a like result when $\mu = 18m + 5$.

Next, let $\mu = 6n + 1$. If D is a quadratic non-residue of μ, there is one rational root or none according as the rational part M of the development of σ^{2n} is or is not such that

$$(2M - 1)^2(M + 1) \equiv -2q^2/p^3 \pmod \mu,$$

and if a rational root exists it is $2q/\{p(2M - 1)\}$. If D is a quadratic residue of μ, there are three rational roots or none according as $\sigma^{2n} \equiv 1 \pmod \mu$ or not. When there are three, they are given explicitly if $\mu = 18m + 7$ or $18m + 13$, while if $\mu = 18m + 1$ there are sub-cases treated only partially.

G. T. Woronoj[132] (or Voronoï) employed Galois imaginaries $a + bi$, where $i^2 - N \equiv 0 \pmod p$ is irreducible, p being an odd prime, to treat the solution of

$$x^3 - rx - s \equiv 0 \pmod p.$$

---

[131]Jour. für Math., 45, 1853, 314–339.

[132]Integral algebraic numbers depending on a root of a cubic equation (in Russian), St. Petersburg, 1894, Ch. I. Cf. Fortschritte Math., 25, 1893–4, 302–3. Cf. Voronoï.[109]

If $4r^3-27s^2$ is a quadratic non-residue of p, the congruence has one and only one root; but if it is a residue, there are three roots or no root.

G. Cordone[133] gave simpler proofs of Oltramare's[131] theorem II on the case $\mu = 6n-1$, gave theorems to replace VII and VIII, and proved that the condition in IX is sufficient as well as necessary.

Ivar Damm[94] found when Cardan's formula gives three real roots, one or no real root of a cubic congruence, and expressed the roots by use of his quasi sine and cosine functions. For the prime modulus $p=3n+1$, $f=x^3+ax+b$ is irreducible if

$$c=\sqrt{\frac{b^2}{4}+\frac{a^3}{27}} \text{ is real, } \left(-\frac{b}{2}+c\right)^{\frac{p-1}{6}} \equiv \pm 1.$$

If $p=3n-1$, it is irreducible if c and $(-b/2+c)^n$ are both imaginary. There are given (p. 52) explicit expressions for b such that f is irreducible.

J. Iwanow[134] gave another proof of the theorem of Woronoj.[132]

Woronoj[135] gave another proof of the same theorem and stated that the congruence has the same number of roots for all primes representable by a binary quadratic form whose determinant equals $-4r^3+27s^2$.

G. Arnoux[136] gave double-entry tables of the roots of the congruences $x^r+bx^s+a\equiv 0 \pmod{m}$, and solved numerical cubic congruences by interpreting Cardan's formulas.

G. Arnoux[137] treated $x^3+bx+a\equiv 0 \pmod{m}$ by use of Cardan's formula. For $m=11$, he gave a table of the real roots for $a\leqq 10$, $b\leqq 10$, and the residues of

$$R=\frac{a^2}{4}+\frac{b^3}{27}.$$

When R is a quadratic residue, the cube roots of $-a/2\pm \sqrt{R}$ are found by use of a table for the Galois field of order 11^2 defined by $i^2\equiv 2 \pmod{11}$, and the cubic is seen to have a real and two imaginary roots involving i. If R is a quadratic non-residue, there are three real roots or none. Like results are said to hold when $m-1$ is not divisible by 3. If $m\equiv 1 \pmod 3$, there is a single real root if R is a quadratic non-residue; three real or three imaginary roots of the third order if R is a residue.

L. E. Dickson[138] proved that, if p is a prime >3, $x^3+\beta x+b\equiv 0 \pmod p$ has no integral root if and only if $-4\beta^3-27b^2$ is a quadratic residue of p, say $\equiv 81\mu^2$, and if $\frac{1}{2}(-b+\mu\sqrt{-3})$ is not congruent to the cube of any $y+z\sqrt{-3}$, where y and z are integers. The reducible and irreducible cubic congruences are given explicitly. Necessary and sufficient conditions for the irreducibility of a quartic congruence are proved.

---

[133]Rendiconti Circolo Mat. di Palermo, 9, 1895, 221–36.

[134]Bull. Ac. Sc. St. Petersburg, 5, 1896, 137–142 (in Russian).

[135]Natural Sc. (Russian), 10, 1898, 329; cf. Fortschritte Math., 29, 1898, 156.

[136]Assoc. franç. av. sc., 30, 1901, II, 31–50, 51–73; corrections, 31, 1902, II, 202.

[137]Assoc. franç. av. sc., 33, 1904, 199–230 [182–199], and Arnoux[112], 166–202.

[138]Bull. Amer. Math. Soc., 13, 1906, 1–8.

D. Mirimanoff[139] noted that the results by Arnoux[112, 137] may be combined by use of the discriminant $D = -4b^3 - 27a^2 = -3 \cdot 6^2 R$ in place of R, since -3 is a quadratic residue of a prime $p = 3k+1$, non-residue of $p = 3k-1$, and we obtain the result as stated by Voronoï.[132]

To find which of the values 1 or 3 is taken by ν when D is a quadratic residue, apply the theorem that if $f(x) \equiv 0 \pmod{p}$ is an irreducible congruence of degree n and if x_0 is one of its imaginary roots (say one of the roots of the equation $f(x) = 0$), the roots are

$$x_0, \qquad x_1 = x_0^p, \ldots, \qquad x_{n-1} = x_0^{p^{n-1}}.$$

Hence a function unaltered by the cyclic substitution $(x_0 x_1 \ldots x_{n-1})$ has an integral value modulo p. Take $n = 3$, $D \equiv d^2$, a a root $\neq 1$ of $z^3 \equiv 1$ $\pmod{p}$, and let

$$M = (x_0 + ax_1 + a^2 x_2)^3.$$

If $p \equiv 1 \pmod 3$, a is an integer, and M is an integer if $\nu = 1$, while M is the cube of an integer if $\nu = 3$. Thus we have Arnoux's criterion:[112] $\nu = 3$ if M or $\frac{3}{2}(-9a + \sqrt{-3d})$ is a cubic residue modulo p. If $p \equiv -1 \pmod 3$ $\nu = 3$ if and only if $M^k \equiv 1 \pmod{p}$, where $k = (p^2 - 1)/3$.

For quartic congruences, we can use $(x_0 - x_1 + x_2 - x_3)^2$.

R. D. von Sterneck[140] noted that if p is a prime > 3 not dividing A, and if $k = 3AC - B^2 \not\equiv 0 \pmod{p}$, then the number of incongruent values taken by $Ax^3 + Bx^2 + Cx + D$ is $\frac{1}{3}\{2p + (-3/p)\}$; but, if $k \equiv 0$, the number is p if $p = 3n-1$, $(p+2)/3$ if $p = 3n+1$. Generalization by Kantor.[181]

C. Cailler[141] treated $x^3 + px + q \equiv 0 \pmod{l}$, where l is a prime > 3. By the algebraic method leading to Cardan's formula, we write the congruence in the form

(1) $$x^3 - 3abx + ab(a+b) \equiv 0 \pmod{l},$$

where a, b are the roots of $z^2 + 3qz/p - p/3 \equiv 0 \pmod{l}$, whence

$$z \equiv (x_0 + ax_1 + a^2 x_2)^3/(9p), \qquad a^2 + a + 1 \equiv 0 \pmod{l}.$$

Let $\Delta = 4p^3 + 27q^2$. If 3Δ is a quadratic residue of l, a and b are distinct and real. If 3Δ is a non-residue, a and b are Galois imaginaries $r \pm s\sqrt{N}$, where N is any non-residue. For a root x of (1),

$$y^3 \equiv \frac{a}{b} \pmod{l}, \qquad y = \frac{x-a}{x-b}.$$

Use is made of a recurring series S with the scale of relation $[a+b, -ab]$ to get $y_0, y_1, \ldots$. Write $Q = (3\Delta/l)$. If $l = 3m-1$, $Q = 1$, then

$$y \equiv (b/a)^{m-1}, \qquad x \equiv \frac{a^m - b^m}{a^{m-1} - b^{m-1}} = \frac{y_m}{y_{m-1}}.$$

If $l = 3m+1$, $Q = 1$, the congruence is possible only when the real number a/b is a cubic residue, i. e., if $y_m \equiv 0$ in S; let a/b belong to the exponent $3\mu \mp 1$ modulo l, whence

[139]L'enseignement math., 9, 1907, 381–4.
[140]Sitzungsber. Ak. Wiss. Wien (Math.), 116, 1907, IIa, 895–904.
[141]L'enseignement math., 10, 1908, 474–487.

$$y \equiv \left(\frac{a}{b}\right)^{\pm \mu}, \qquad x \equiv \frac{y_{2\mu}}{y_{2\mu-1}} \text{ or } \frac{y_{\mu+1}}{y_\mu},$$

according as the upper or lower sign holds. If $l=3m+1$, $Q=-1$, then

$$y_{3m+2} \equiv 0, \qquad \left(\frac{a}{b}\right)^{3m+3} \equiv \frac{a}{b}, \qquad \text{real } x \equiv \frac{y_{2m+2}}{y_{2m+1}}.$$

If $l=3m\dot-1$, $Q=-1$, there are three real roots if and only if a/b is a cubic residue of l, viz., $y_m \equiv 0$; when real, the roots may be found as in the second case.

Cailler[142] noted that a cubic equation $X=0$ has its roots expressible rationally in one root and $\sqrt{\Delta}$, where Δ is the discriminant (Serret's Algèbre, ed. 5, vol. 2, 466–8). Hence, if p is a prime, $X \equiv 0$ (mod p) has three real roots if one, when and only when Δ is a quadratic residue of p. If $p=9m\pm1$, his[141] test shows that $x^3-3x+1\equiv 0$ (mod p) has three real roots, but no real root for other prime moduli $\neq 3$. The function $F(x)=x^3+x^2-2x-1$ for the three periods of the seventh roots of unity is divisible by the primes $7m\pm1$ (then 3 real roots, Gauss[60], p. 624) and 7, but by no other primes.

E. B. Escott[143] noted that the equation $F(x)=0$ last mentioned has the roots a, $\beta=a^2-2$, $\gamma=\beta^2-2$, so that $F(x)\equiv 0$ (mod p) has three real roots if one real root. To find the most general irreducible cubic equation with roots a, β, γ such that

$$\beta=f(a), \quad \gamma=f(\beta), \quad a=f(\gamma),$$

we may assume that $f(x)$ is of degree 2. For $f(a)=a^2-n$, we get

(2) $$x^3+ax^2-(a^2-2a+3)x-(a^3-2a^2+3a-1)=0,$$

with $\beta=a^2-c$, $\gamma=\beta^2-c$, $a=\gamma^2-c$, $c=a^2-a+2$. The corresponding congruence has three real roots if one. To treat $f(a)=a^2+ka+l$, add $k/2$ to each root. For the new roots, $\beta'=a'^2-n$, as in the former case. To treat $f(a)=ta^2+ga+h$, the products of the roots by t satisfy the preceding relation.

L. E. Dickson[144] determined the values of a for which the congruence corresponding to (2) has three integral roots. Replace x by $z-a$; we get

$$z^3-2az^2+(2a-3)z+1\equiv 0 \text{ (mod } p).$$

If one root is z, the others are $1-1/z$ and $1/(1-z)$. Evidently a is rational in z. If -3 is a quadratic non-residue of p, there are exactly $(p-2)/3$ values of a for which the congruence has three distinct integral roots. If -3 is a residue, the number is $(p+2)/3$. A second method, yielding an explicit congruence for these values of a, is a direct application of his[138] general criteria for the nature of the roots of a cubic congruence.

T. Hayashi[145] treated cyclotomic cubic equations with three real roots by use of Escott's[143] results.

---

[142]L'intermédiaire des math., 16, 1909, 185–7.　　[144]Ibid., (2), 12, 1910–11, 149–152.
[143]Annals of Math., (2), 11, 1909–10, 86–92.　　[145]Ibid., 189–192.

MISCELLANEOUS RESULTS ON CONGRUENCES.

Linear congruences will be treated in Vol. 2 under linear diophantine equations, quadratic congruences in two or more variables, under sums of four squares; $ax^n+by^n+cz^n\equiv0$, under Fermat's last theorem.

Fermat[148] stated that not every prime p divides one of the numbers $a+1, a^2+1, a^3+1,\ldots$. For, if k is the least value for which a^k-1 is divisible by p and if k is odd, no term a^h+1 is divisible by p. But if k is even, $a^{k/2}+1$ is divisible by p.

Fermat[149] stated that no prime $12n\doteq1$ divides 3^x+1, every prime $12n\doteq5$ divides certain 3^x+1, no prime $10n\doteq1$ divides 5^x+1, every prime $10n\doteq3$ divides certain 5^x+1, and intimated that he possessed a rule relating to all primes. See Lipschitz.[166]

A. M. Legendre[150] obtained from a given congruence $x^n\equiv ax^{n-1}+\ldots$ (mod p), p an odd prime, one having the same roots, but with no double roots. Express $x^{(p-1)/2}$ in terms of the powers of x with exponents $<n$, and equate the result to $+1$ and to -1 in turn. The g. c. d. of each and the given congruence is the required congruence. An exception arises if the proposed congruence is satisfied by $0, 1,\ldots, p-1$.

Hoëné de Wronski[151] developed $(n_1+\ldots+n_\omega)^m$, replaced each multinomial coefficient by unity, and denoted the result by $A[n_1+\ldots+n_\omega]^m$. Thus $A[n_1+n_2]^2=n_1^2+n_1n_2+n_2^2$. Set $N_\omega=n_1+\ldots+n_\omega$. Then (pp.65–9),

$$(1) \qquad A[N_\omega-n_p]^m-A[N_\omega-n_q]^m=(n_q-n_p)A[N_\omega]^{m-1}\equiv0 \text{ (mod } n_q-n_p).$$

Let $(n_1\ldots n_\omega)_m$ be the sum of the products of $n_1,\ldots, n_\omega$ taken m at a time. Then (p. 143), if $A[N_\omega]^0=1$,

$$(2) \qquad A[N_\omega]^\mu=(n_1\ldots n_\omega)_1A[N_\omega]^{\mu-1}-(n_1\ldots n_\omega)_2A[N_\omega]^{\mu-2}$$
$$+(n_1\ldots n_\omega)_3A[N_\omega]^{\mu-3}-\ldots+(-1)^{\mu+1}(n_1\ldots n_\omega)_\mu A[N_\omega]^0.$$

He discussed (pp. 146–151) in an obscure manner the solution of $X_1\equiv X_2$ (mod X), where the X's are polynomials in ξ of degree ν. Set $N_\omega=n_1+\ldots+n_{\omega-2}+n_p+n_q$. Let the negatives of $n_1,\ldots, n_{\omega-2}, n_p$ be the roots of $P=P_0+P_1x+\ldots+P_{\omega-2}x^{\omega-2}+x^{\omega-1}=0$; the negatives of $n_1,\ldots, n_{\omega-2}, n_q$ the roots of $Q=Q_0+\ldots+x^{\omega-1}=0$. We may add ζ_1X and ζ_2X to the members of our congruence. It is stated that the new first member may be taken to be $A[N_\omega-n_q]^m$, whence by (2)

$$X_1+\zeta_1X=P_{\omega-2}A[N_\omega-n_q]^{m-1}-P_{\omega-3}A[N_\omega-n_q]^{m-2}+\ldots,$$

and the A's may be expressed in terms of the P's by (2). Similarly, $X_2+\zeta_2X$ may be expressed in terms of the Q's. By (1), $X=n_q-n_p=Q_{\omega-2}-P_{\omega-2}$. Since $P=0$, $Q=0$ have $\omega-2$ roots in common, we have further conditions on the coefficients P_i, Q_i. It is argued that $\omega-3$ of the latter

[148]Oeuvres, 2, 209, letter to Frenicle, Oct. 18, 1640.
[149]Oeuvres, 2, 220, letter to Mersenne, June 15, 1641.
[150]Mém. Ac. Sc. Paris, 1785, 483.
[151]Introduction à la Philosophie des Mathématiques et Technie de l'Argorithmie, Paris, 1811. He used the Hebrew aleph for the A of this report. Cf. Wronski[169] of Ch. VII.

remain arbitrary, and that ξ is a function of them and one of the n's, which has an arbitrary rational value.

A. Cauchy[152] noted that if f and F are polynomials in x, Lagrange's interpolation formula leads to polynomials u and v such that $uf+vF=R$, where R is a constant [provided f and F have no common factor]. If the coefficients are all integers, R is an integer. Hence R is the greatest of the integers dividing both f and F. For $f=x^p-x$, we may express R as a product of trigonometric functions. If also $F(x)=(x^n+1)/(x+1)$, where n and p are primes, $R=0$ or ± 2 according as p is or is not of the form $nx+1$. Hence the latter primes are the only ones dividing x^n+1, but not $x+1$.

Cauchy[153] proved that a congruence $f(x)\equiv 0$ (mod p) of degree $m<p$ is equivalent to $(x-r)^i\phi(x)\equiv 0$, where ϕ is of degree $m-i$, if and only if

$$f(r)\equiv 0, \qquad f'(r)\equiv 0,\ldots, \qquad f^{(i-1)}(r)\equiv 0 \text{ (mod } p),$$

where p is a prime. The theorem fails if $m\geq p$. He gave the method of Libri (Mémoires, I) for solving the problem: Given $f(x)\equiv 0$ (mod p) of degree $m\leq p$ and with exactly m roots, and $f_1(x)$ of degree $l\leq m$, to find a polynomial $\phi(x)$, also with integral coefficients, whose roots are the roots common to f and f_1. He gave the usual theorem on the number of roots of a binomial congruence and noted conditions that a quartic congruence have four roots.

Cauchy[154] stated that if I is an arbitrary modulus and if $r_1,\ldots,r_m$ are roots of $f(x)\equiv 0$ (mod I) such that each difference r_i-r_j is prime to I, then

$$f(x)\equiv (x-r_1)\ldots(x-r_m)Q(x) \text{ (mod } I).$$

If in addition, m exceeds the degree of $f(x)$, then $f(x)\equiv 0$ (mod I) for every x. A congruence of degree n modulo p^λ, where p is a prime, has at most n roots unless every integer is a root. If $f(r)\equiv 0$ (mod I) and if in the irreducible fraction equal to

$$\tau=\frac{f(r)}{If'(r)}$$

the denominator is prime to I, then $r-\tau I$ is a root of $f(x)\equiv 0$ (mod I^2).

V. A. Lebesgue[155] wrote $a/b\equiv c$ (mod p) if b is prime to p and $a\equiv bc$ (mod p), and $a/b\equiv c/d$ (mod p) if b, d are prime to p and $ad\equiv bc$ (mod p).

J. A. Serret[156] stated and A. Genocchi proved that, if p is a prime, the sum of the mth powers of the p^n polynomials in x, of degree $n-1$ and with integral coefficients $<p$, is a multiple of p if $m<p^n-1$, but not if $m=p^n-1$.

J. A. Serret[157] noted that all the real roots of a congruence $f(x)\equiv 0$ (mod p), where p is a prime, satisfy $\psi(x)\equiv 0$, where ψ is the g. c. d. of $f(x)$ and $x^{p-1}-1$.

[152]Exercices de Math., 1, 1826, 160–6; Bull. Soc. Philomatique; Oeuvres, (2), 6, 202–8.
[153]Exercices de Math., 4, 1829, 253–279; Oeuvres, (2), 9, 298–326.
[154]Comptes Rendus Paris, 25, 1847, 37; Oeuvres, (1), 10, 324–30.
[155]Nouv. Ann. Math., 9, 1850, 436.
[156]Nouv. Ann. Math., 13, 1854, 314; 14, 1855, 241–5.
[157]Cours d'algèbre supérieure, ed. 2, 1854, 321–3.

N. H. Abel[158] proved that we can solve by radicals any abelian equation, i. e., one whose roots are r, $\phi(r)$, $\phi^2(r) = \phi[\phi(r)], \ldots$, where ϕ is a rational function. H. J. S. Smith[159] concluded that when the roots of a congruence can be similarly expressed modulo p, its solution can evidently be reduced to the solution of binomial congruences, and the expressions for the roots of the corresponding equation may be interpreted as the roots of the congruence. For the special case $x^n \equiv 1$, this was done by Poinsot in 1813–20 in papers discussed in the chapter on primitive roots.

M. Jenkins[159a] noted that all solutions of $a^x \equiv 1 \pmod{x}$ are $x - U_n - u_1 u_2 \ldots u_n$, where u_1 is any divisor of any power of $a-1$; u_2 any divisor prime to $a-1$, of any power of $a^{u_1} - 1$; $\ldots$; u_n any divisor, prime to $a^{U_{n-2}} - 1$, of any power of $a^{U_{n-1}} - 1$. For $a^x + 1 \equiv 0 \pmod{x}$, modify the preceding by taking odd factors of $a+1$ instead of factors of $a-1$.

J. J. Sylvester[160] proved that if p is a prime and the congruence $f(x) \equiv 0 \pmod{p}$ of degree n has n real roots and if the resultant of $f(x)$ and $g(x)$ is divisible by p, then $g(x) \equiv 0$ has at least one root in common with $f(x) \equiv 0$. There are exactly $p-1$ real roots of $x^{p-1} \equiv 1 \pmod{p^j}$.

A. S. Hathaway[161] noted the known similarity between equations and congruences for a prime modulus. He[162] made abstruse remarks on higher congruences.

G. Frattini[163] proved that $x^2 - Dy^4 \equiv \lambda$ and $x^4 - Dy^2 \equiv \lambda$ are each solvable when the modulus is a prime $p > 5$ and $D \not\equiv 0$. If $d = B^2 - AC \not\equiv 0$, then $Ax^4 + 2Bx^2y + Cy^2 \equiv \lambda \pmod{p}$ is solvable since $dx^4 + \lambda C$ can be made congruent to a square and hence to $(Cy + Bx^2)^2$. Likewise for $ax^2 + 2bx + c \equiv y^4$.

A. Hurwitz[164] discussed the congruence of fractions and the theory of the congruence of infinite series. If $\phi(x) = r_0 + r_1 x + \ldots + r_n x^n / n! + \ldots$ and if $\psi(x)$ is a similar series with the coefficients s_n, then ϕ and ψ are called congruent modulo m if and only if $r_n \equiv s_n \pmod{m}$ for $n = 1, 2, \ldots$.

G. Cordone[165] treated the general quartic congruence for a prime modulus μ by means of a cubic resolvent. The method is similar to Euler's solution of a quartic equation as presented by Giudice in Peano's Rivista di Matematica, vol. 2. For the special case $x^4 + 6Hx^2 + K \equiv 0 \pmod{\mu}$, set $t = (\mu - 1)/2$, $r^2 = 9H^2 - K$; then if K is a quadratic residue of μ, there are four rational roots or none according as $(-3H + r)^t \equiv +1$ or not; but if K is a non-residue, there are two rational roots or none according as one of the congruences

$$(-3H + r)^t \equiv +1, \qquad (-3H - r)^t \equiv -1$$

is satisfied or not.

[158]Jour. für Math., 4, 1829, 131; Oeuvres, 1, 114.
[159]Report British Assoc. 1860, 120 seq., §66: Coll. M. Papers, 1, 141–5.
[159a]Math. Quest. Educ. Times, 6, 1866, 91–3.
[160]Amer. Jour. Math., 2, 1879, 360–1; Johns Hopkins University Circulars, 1, 1881, 131.   Coll. Papers, 3, 320–1.
[161]Johns Hopkins Univ. Circulars, 1, 1881, 97.   [162]Amer. Jour. Math., 6, 1884, 316–330.
[163]Rendiconti Reale Accad. Lincei, Rome, (4), 1, 1885, 140–2.
[164]Acta Mathematica, 19, 1895, 356.
[165]Rendiconti Circolo Mat. di Palermo 9, 1895, 209–243.

R. Lipschitz[166] examined Fermat's[148] statement and proved that the primes p for which $a^x+1 \equiv 0 \pmod{p}$ is impossible are those and only those for which a solution u of $u^{2^k} \equiv a \pmod{p}$ is a quadratic non-residue of p and for which $\lambda \leqq k$, where 2^λ is the highest power of 2 dividing $p-1$. Cases when $a^x+1 \equiv 0$ is impossible and not embraced by Fermat's rule are $a=2$, $p=89$, 337; $a=3$, $p=13$; $a=-2$, $p=281$; etc.

L. Kronecker[167] called $f(x)$ an invariant of the congruence $k \equiv k' \pmod{m}$, if the latter congruence implies the equality $f(k)=f(k')$. If also, conversely, the equality implies the congruence, $f(x)$ is called a proper (or characteristic) invariant, an example being the least positive residue of an integer modulo m. It is shown that every invariant of $k \equiv k' \pmod{m}$ can be represented as a symmetric function of all the integers congruent to k modulo m.

G. Wertheim[168] proved that $a^x+1 \equiv 0 \pmod{p}$ is impossible if a belongs to an odd exponent modulo p [Fermat[148]].

E. L. Bunitzky[169] (Bunickij) noted that, for any integer M, the congruences

$$f(a+kh) \equiv r_k \pmod{M} \qquad (k=0, 1, \ldots, n)$$

hold if and only if the coefficients A_k of $f(x)$ satisfy the conditions

$$k! \, h^k A_k \equiv \Delta^k r_0 \pmod{M} \qquad (k=1, \ldots, n).$$

If k is the least value of x for which $x! \, h^x$ is divisible by M, and if the g. c. d. of M and h is $k < m$, where m is a divisor of M, then if $f(x) \equiv 0 \pmod{M}$ has the roots a, $a+h, \ldots, a+(k-1)h$, it has also the roots $a+jh$ $(j=k, k+1, \ldots, m-1)$.

G. Biase[170] called a similar to b in the ratio $m:n$ modulo k if the remainders on dividing a and b by k are in the ratio $m:n$. Two numbers similar to a third in two given ratios modulo k are similar to each other modulo k in a ratio equal to the quotient of the given ratios.

The problem[171] to find n numbers whose n^2-n differences are incongruent modulo n^2-n+1 is possible for $n=6$, but not for $n=7$.

R. D. von Sterneck[140] proved that, if A is not divisible by the odd prime p, Ax^4+Bx^2+C takes $\psi(2AB, p)$ incongruent values (when x ranges over the set $0, 1, \ldots, p-1$) if B is not divisible by p, while if B is divisible by p, it takes $(p+3)/4$ or $(p+1)/2$ values according as $p=4n+1$ or $p=4n-1$. In terms of Legendre's symbol,

$$\psi(a, p) = \frac{1}{8}\left[3p+4-2\left(\frac{-2a}{p}\right)+\left(\frac{-1}{p}\right)+2\left(\frac{-a}{p}\right)\right].$$

---

[166]Bull. des Sc. Math., (2), 22, I, 1898, 123–8. Extract in Oeuvres de Fermat, 4, 196–7.
[167]Vorlesungen über Zahlentheorie, I, 1901, 131–142.
[168]Anfangsgründe der Zahlenlehre, 1902, 265.
[169]Zap. mat. otd. Obsc. (Soc. of natur.), Odessa, 20, 1902, III–VIII (in Russian); cf. Fortschr. Math., 33, 1902, p. 205.
[170]Il Boll. Matematica Gior. Sc. Didat., Bologna, 4, 1905, 96.
[171]L'intermédiaire des math., 1906, 141; 1908, 64; 19, 1912, 130–1. Amer. Math. Monthly, 13, 1906, 215; 14, 1907, 107–8.

E. Landau[172] proved that, if $f(x) = 0$ is an equation with integral coefficients and at least one root of odd multiplicity, there exist an infinitude of primes $p = 4n - 1$ such that $f(x) \equiv 0 \pmod{p}$ has a root.

R. D. von Sterneck[173] found the number of combinations of the ith class (with or without repetition) of the numbers prime to p of a complete set of residues modulo p^x whose sum is congruent to a given integer modulo p^x, p being a prime.

E. Piccioli[174] gave known theorems on adding and multiplying congruences.

C. Jordan[175] found the number of sets of integers a_{ik} for which the determinant $|a_{ik}|$ of order n is congruent to a given integer modulo M.

C. Krediet[176] gave theorems on congruences of degree n for a prime modulus analogous to those for an algebraic equation of degree n, including the question of multiple roots. The determination of roots is often simplified by seeking first the roots which are quadratic residues and then those which are non-residues. The exposition is not clear or simple.

G. Rados[177] proved that, if p is a prime,

$$f(x) = a_0 x^{p-2} + \ldots + a_{p-2} \equiv 0, \qquad g(x) = b_0 x^{p-2} + \ldots + b_{p-2} \equiv 0 \pmod{p}$$

have a common root if and only if each $R_i \equiv 0 \pmod{p}$, where

$$\Phi(u) = R_0 u^{p-1} + R_1 u^{p-2} + \ldots + R_{p-1}$$

$$= \begin{vmatrix} a_0 u + b_0 & a_1 u + b_1 & \ldots & a_{p-2} u + b_{p-2} \\ a_1 u + b_1 & a_2 u + b_2 & \ldots & a_0 u + b_0 \\ \ldots & \ldots & \ldots & \ldots \\ a_{p-2} u + b_{p-2} & a_0 u + b_0 & \ldots & a_{p-3} u + b_{p-3} \end{vmatrix}$$

For $g = f'$, let $\Phi(u)$ become $D_0 u^{p-2} + \ldots + D_{p-2}$; thus $f(x) \equiv 0 \pmod{p}$ has a multiple root if and only if each $D_i \equiv 0 \pmod{p}$. Each of these theorems is extended to three congruences. Finally, if $f(x)$ and $f'(x)$ are relatively prime algebraically, there is only a finite number of primes p for which the number of roots of $f \equiv 0 \pmod{p^k}$ exceeds the degree of f.

G. Frattini[178] proved that if p and q are primes, q a divisor of $p-1$, every homogeneous symmetric congruence in q variables is solvable modulo p by values of the variables distinct from each other and from zero except when the degree of the congruence is divisible by q.

C. Grötzsch[179] noted that if a is a root of $x^x \equiv a \pmod{p}$, where a is prime to p, then $x \equiv a \pmod{p^2 - p}$ is a root, and proved that if θ is the g. c. d. of ind a and $p-1$ and if ind $a > 0$, it has exactly

[172]Handbuch...Verteilung der Primzahlen, 1, 1909, 440.
[173]Sitzungsber. Ak. Wiss Wien (Math.), 118, 1909, IIa, 119–132.
[174]Il Pitagora, Palermo, 16, 1909–10, 125–7.
[175]Jour. de Math., (6), 7, 1911, 409–416.
[176]Wiskundig Tijdschrift, Haarlem, 7, 1911, 193–202 (Dutch).
[177]Ann. sc. école norm. sup., (3), 30, 1913, 395–412.
[178]Periodico di Mat., 29, 1913, 49–53.
[179]Archiv Math. Phys., (3), 22, 1914, 49–53.

$$N = \phi(p-1) + \Sigma\delta\phi\left(\frac{p-1}{\delta}\right)$$

roots incongruent modulo $p(p-1)$, where δ ranges over all divisors >1 of θ. If ind $a=0$, the number of such roots is $p-1+N$, where now δ ranges over the divisors >1 of $p-1$.

A. Châtelet[180] noted that divergences between congruences and equations are removed by not limiting attention to the given congruence $f(x) \equiv 0$ of degree n, but considering simultaneously all the polynomials $g(x)$ derived from $f(x)$ by a Tschirnhausen transformation $ky = \phi(x)$, where k is an integer and ϕ has integral coefficients and is of degree $n-1$.

*M. Tihanyi[180a] proved a simple congruence.

R. Kantor[181] discussed the number of incongruent values modulo m taken by a polynomial in n variables, and especially for $ax^3 + \ldots + d$ modulo p^r, generalizing von Sterneck.[140]

The solvability of $x^3 + 9x + 6 \equiv 0$ and $x^3 + y(y+1) \equiv 0$ (mod p) has been treated.[182]

A. Cunningham[183] announced the completion, in conjunction with Woodall and Creak, of tables of least solutions (x, a) of the congruences

$$r^x \equiv \pm y^a, \quad r^x y^a \equiv \pm 1 \ (\text{mod } p^k < 10000), \quad r = 2, 10; y = 3, 5, 7, 11.$$

T. A. Pierce[184] gave two proofs that $f(x) \equiv 0$ (mod p) has a real root if and only if the odd prime p divides $\Pi(1 - a_i{}^{p-1})$, where a_i ranges over the roots of the equation $f(x) = 0$.

Christie[185] stated that $t^p(t^p+1) \equiv 1$ (mod p) if $t = 2 \sin 18°$ and p is any odd prime. Cunningham gave a proof and a generalization.

*G. Rados[186] found the congruence of degree r having as its roots the r distinct roots $\neq 0$ of a given congruence of degree $p-2$ modulo p, a prime.

[180]Comptes Rendus Paris, 158, 1914, 250–3.
[180a]Math. és Phys. Lapok, Budapest, 23, 1914, 57–60.
[181]Monatshefte Math. Phys., 26, 1915, 24–39.
[182]Wiskundige Opgaven, 12, 1915, 211–2, 215–7.
[183]Messenger Math., 45, 1915–6, 69.
[184]Annals of Math., (2), 18, 1916, 53–64.
[185]Math. Quest. Educ. Times, 71, 1899, 82–3.
[186]Math. és Termés Értesitö, 33, 1915, 702–10.

CHAPTER IX.

DIVISIBILITY OF FACTORIALS AND MULTINOMIAL COEFFICIENTS.

HIGHEST POWER OF A PRIME DIVIDING $m!$.

Genty[1] noted that the highest power of 2 dividing $(2^n)!$ is 2^{2^n-1}, and the quotient is $3^{n-1}(5\cdot7)^{n-2}(9\cdot11\cdot13\cdot15)^{n-3}(17\dots31)^{n-4}\dots(2^n-1)$. In general if $P=2^{m_1}+2^{m_2}+\dots+2^{m_r}$, where the n's decrease, the highest power of 2 dividing $P!$ is 2^{P-r}.

A. M. Legendre[2] proved that if p^μ is the highest power of the prime p which divides $m!$, and if $[x]$ denotes the greatest integer $\leq x$,

$$(1) \qquad \mu=\left[\frac{m}{p}\right]+\left[\frac{m}{p^2}\right]+\left[\frac{m}{p^3}\right]+\dots=\frac{m-s}{p-1},$$

where $s=a_0+\dots+a_n$ is the sum of the digits of m to the base p:

$$m=a_0p^n+a_1p^{n-1}+\dots+a_n \qquad (0\leq a_i<p).$$

Th. Bertram[3] stated Legendre's result in an equivalent form.

H. Anton[4] proved that, if $n=vp+a$, $a<p$, $v<p$, and p is a prime,

$$\frac{n!}{p^v}\equiv(p-1)^va!v! \pmod{p},$$

while, if $v=v'p+a'$, $a'<p$, $v'<p$,

$$\frac{n!}{p^{v+v'}}\equiv(p-1)^{v+v'}a!a'!v!v'! \pmod{p}.$$

D. André[5] stated that the highest power p^μ of the prime p dividing $n!$ is given explicitly by $\mu=\Sigma_{k=1}^{k=\infty}[n/p^k]$ and claimed that merely the method of finding μ had been given earlier. He applied this result to prove that the product of n consecutive integers is divisible by $n!$.

J. Neuberg[6] determined the least integer m such that $m!$ is divisible by a given power of a prime, but overlooked exceptional cases.

L. Stickelberger[7] and K. Hensel[8] gave the formula [cf. Anton[4]].

$$(2) \qquad \frac{m!}{p^\mu}\equiv(-1)^\mu a_0!a_1!\dots a_n! \pmod{p}.$$

F. de Brun[9] wrote $g[u]$ for the exponent of the highest power of the prime p dividing u. He gave expressions for

$$\psi(n;k)=\prod_{j=1}^{n}j^{j^k}, \qquad g[\psi(n;k)]$$

in terms of the functions $h(a;k)=1^k+2^k+\dots+a^k$. A special case gives (1).

---

[1]Hist. et Mém. Ac. R. Sc. Inscript. et Belles Lettres de Toulouse, 3, 1788, 97–101 (read Dec. 4, 1783).

[2]Théorie des nombres, ed. 2, 1808, p. 8; ed. 3, 1830, I, p. 10.

[3]Einige Sätze aus der Zahlenlehre, Progr. Cöln, Berlin, 1849, 18 pp.

[4]Archiv Math. Phys., 49, 1869, 298–9.

[5]Nouv. Ann. Math., (2), 13, 1874, 185.

[6]Mathesis, 7, 1887, 68–69. Cf. A. J. Kempner, Amer. Math. Monthly, 25, 1918, 204–10.

[7]Math. Annalen, 37, 1890, 321.

[8]Archiv Math. Phys., (3), 2, 1902, 294.

[9]Arkiv för Matematik, Astr., Fysik, 5, 1904, No. 25 (French).

R. D. Carmichael[10] treated the problem to find m, given the prime p and $s = \Sigma a_i$, in Legendre's formula; a given solution m_2 leads to an infinitude of solutions $m_2 p^k$, k arbitrary. Again, to find m such that p^{m-t} is the highest power of $p > 2$ dividing $m!$, we have $m - t = (m-s)/(p-1)$, and see that m has a limited number of values; there is always at least one solution m.

Carmichael[11] used the notation $H\{y\}$ for the index of the highest power of the prime p dividing y, and evaluated

$$h = H\left\{\prod_{x=0}^{n-1}(xa+c)\right\},$$

where a, c are relatively prime positive integers. Set $c_0 = c$ and let i_r be the least integer such that $i_r a + c_{r-1}$ is divisible by p, the quotient being c_r. Let

$$e_1 = \left[\frac{n-1-i_1}{p}\right], \qquad e_r = \left[\frac{e_{r-1}-i_r}{p}\right], \quad r > 1.$$

Then $h = \sum_{r=1}^{t-1}(e_r+1)$, where t is the least subscript for which

$$c_t(a+c_t)(2a+c_t)\ldots(e_t a+c_t)$$

is not divisible by p. It follows that

$$\left[\frac{n}{p}\right] + \left[\frac{n}{p^2}\right] + \ldots \leqq h \leqq \left[\frac{n-1}{p}\right] + \left[\frac{n-1}{p^2}\right] + \ldots + R,$$

where R is the index of the highest power of p not exceeding $n-1$. If n is a power of p, $h = (n-1)/(p-1)$. But if $n = \delta_k p^k + \ldots + \delta_1 p + \delta_0$, $\delta_k \neq 0$, and at least one further δ is not zero,

$$\frac{n-\sigma}{p-1} \leqq h \leqq k + \frac{n-\sigma}{p-1}, \qquad \sigma = \delta_k + \ldots + \delta_0.$$

In case the first x for which $xa+c$ is divisible by p gives c as the quotient, all the c_r are equal and hence all the i_r; then

$$h = \left[\frac{n-1-i+p}{p}\right] + \left[\frac{n-1-i-ip+p^2}{p^2}\right] + \left[\frac{n-1-i-ip-ip^2+p^3}{p^3}\right] + \ldots.$$

The case $a = c = 1$ yields Legendre's[s] result. The case $a = 2$, $c = 1$, gives

$$H\{1\cdot3\cdot5\ldots(2n-1)\} = \left[\frac{2n-1+p}{2p}\right] + \left[\frac{2n-1+p^2}{2p^2}\right] + \ldots.$$

E. Stridsberg[12] wrote H_m for (1) and considered

$$\pi_t = a(a+m)\ldots(a+mt),$$

where a is any integer not divisible by the positive integer m. Let p be a prime not dividing m. Write a_j for the residue of aj modulo m. He noted that, if $pj \equiv 1 \pmod{m}$,

[10]Bull. Amer. Math. Soc., 14, 1907–8, 74–77; Amer. Math. Monthly, 15, 1908, 15–17.
[11]Ibid., 15, 1908–9, 217.
[12]Arkiv för Matematik, Astr., Fysik, 6, 1911, No. 34.

$$(p^{k+1}a_j^{k+1}-a)/m$$

is an integer, and wrote L_k for its residue modulo p^{k+1}. Set

$$t=\sum_{\nu=0}^{k}\tau_\nu p^\nu \quad (0\leqq \tau_\nu \leqq p-1), \qquad T_\mu=\sum_{\nu=0}^{\mu}\tau_\nu p^\nu \quad (\mu\leqq k),$$

$$\Sigma_\mu=\left[\frac{p^{\mu+1}-1+T_\mu/L_\mu}{p^{\mu+1}}\right].$$

He proved that π_t is divisible by p^s, where $s=H_t+\Sigma_{\mu=0}^{\mu=k}\Sigma_\mu$. If τ_σ is the first one of the numbers $\tau_0, \tau_1, \ldots$ which is $<p-1$, π_t is divisible by p^v,

$$v=H_{t+1}+A_t, \quad 0\leqq A_t=\sum_{\mu=\sigma}^{k}\Sigma_\mu\leqq k+1.$$

A. Cunningham[13] proved that if z^ζ is the highest power of the prime z dividing p, the number of times p is a factor of $p^n!$ is the least of the numbers

$$\frac{p^n}{z^{\zeta n}}\cdot\frac{z^{\zeta n-\zeta+1}-1}{z-1},$$

for the various primes z dividing p.

W. Jänichen[14] stated and G. Szegö proved that

$$\Sigma\mu(n/d)\nu(d)=\phi(n)/(p-1),$$

summed for the divisors d of n, where $\nu(d)$ is the exponent of the highest power of p (a prime factor of n) which divides $d!$, for μ as in Ch. XIX.

Integral Quotients Involving Factorials.

Th. Schönemann[18] proved, by use of symmetric functions of pth roots of unity, that if δ is the g. c. d. of $\mu, \nu, \ldots$,

$$\frac{\delta\cdot(m-1)!}{\mu!\nu!\ldots}=\text{integer}, \qquad (m=\mu+\nu+\ldots).$$

He gave (p. 289) an arithmetical proof by showing that the fractions obtained by replacing δ by $\mu, \nu, \ldots$ are integers.

A. Cauchy[19] proved the last theorem and that

$$\frac{(a+2b+\ldots+nk)\cdot(m-1)!}{a!\ldots k!}=\text{integer}, \qquad (m=a+\ldots+k).$$

D. André[20] noted that, except when $n=1$, $a=4$, $n(\dot{n}+1)\ldots(na-1)$ is not or is divisible by a^n according as a is a prime or not.

E. Catalan[21] found by use of elliptic functions that

$$\frac{(m+n-1)!}{m!n!}, \qquad \frac{(2m)!(2n)!}{m!n!(m+n)!}$$

are integers, provided m, n are relatively prime in the first fraction.

[13]L'intermédiaire des math., 19, 1912, 283-5. Text modified at suggestion of E. Maillet.
[14]Archiv Math. Phys., (3), 13, 1908, 361; 24, 1916, 86-7.
[18]Jour. für Math., 19, 1839, 231-243.
[19]Comptes Rendus Paris, 12, 1841, 705-7; Oeuvres, (1), 6, 109.
[20]Nouv. Ann. Math., (2), 11, 1872, 314.
[21]Ibid., (2), 13, 1874, 207, 253. Arith. proofs, Amer. Math. Monthly, 18, 1911, 41-3.

P. Bachmann[22] gave arithmetical proofs of Catalan's results.

D. André[23] proved that, if $a_1, \ldots, a_n$ have the sum N and if k of the a's are not divisible by the integer >1 which divides the greatest number of the a's, then $(N-k)!$ is divisible by $a_1! \ldots a_n!$.

J. Bourguet[24] proved that, if $k \geq 2$,

$$\frac{(km_1)!\,(km_2)!\ldots(km_k)!}{m_1!\ldots m_k!\,(m_1+\ldots+m_k)!} = \text{integer}.$$

M. Weill[25] proved that the multinomial coefficient $(tq)! \div (q!)^t$ is divisible by $t!$.

Weill[26] stated that the following expression is an integer:

$$\frac{(a+\beta+\ldots+pq+p_1q_1+\ldots+rst)!}{a!\beta!\ldots(p!)^q q!(p_1!)^{q_1} q_1!\ldots(r!)^{st}(s!)^t t!}.$$

Weill[27] stated the special case that $(a+\beta+pq+rs)!$ is divisible by $a!\beta!(q!)^p p!(s!)^r r!$.

D. André[28] proved that $(tq)! \div (q!)^t$ is divisible by $(t!)^k$ if for every prime p the sum of the digits of q to base p is $\geq k$.

Ch. Hermite[29] proved that $n!$ divides

$$m(m+k)(m+2k)\ldots\{m+(n-1)k\}k^{n-1}.$$

C. de Polignac[30] gave a simple proof of the theorem by Weill[25] and expressed the generalization by André[28] in another and more general form.

E. Catalan[31] noted that, if s is the number of powers of 2 having the sum $a+b$,

$$\frac{(2a)!\,(2b)!}{a!\,b!\,(a+b)!}$$

is an even integer and the product of 2^s by an odd number.

E. Catalan[32] noted that, if $n = a+b+\ldots+t$,

$$\frac{n!(n+t)}{a!\,b!\ldots t!}$$

is divisible by $a+t, b+t, \ldots, a+b+t, \ldots, a+b+c+t, \ldots$

E. Cesàro[33] stated and Neuberg proved that $\binom{n}{p}$ is divisible by $n(n-1)$ if p is prime to $n(n-1)$, and $p-1$ prime to $n-1$; and divisible by $(p+1) \times (p+2)$ if $p+1$ is prime to $n+1$, and $p+2$ is prime to $(n+1)(n+2)$.

[22]Zeitschrift Math. Phys., 20, 1875, 161–3. Die Elemente der Zahlentheorie, 1892, 37–39.
[23]Bull. Soc. Math. France, 1, 1875, 84.
[24]Nouv. Ann. Math., (2), 14, 1875, 89; he wrote $\Gamma(n)$ incorrectly for $n!$; see p. 179.
[25]Comptes Rendus Paris, 93, 1881, 1066; Mathesis, 2, 1882, 48; 4, 1884, 20; Lucas, Théorie des nombres, 1891, 365, ex. 3. Proof by induction, Amer. M. Monthly, 17, 1910, 147.
[26]Bull. Soc. Math. France, 9, 1880–1, 172. Special case, Amer. M. Monthly, 23, 1916, 352–3.
[27]Mathesis, 2, 1882, 48; proof by Liénard, 4, 1884, 20–23.
[28]Comptes Rendus Paris, 94, 1882, 426.
[29]Faculté des Sc. de Paris, Cours de Hermite, 1882, 138; ed. 3, 1887, 175; ed. 4, 1891, 196. Cf. Catalan, Mém. Soc. Sc. de Liège, (2), 13, 1886, 262–4 (=Mélanges Math.); Heine.[29a]
[30]Comptes Rendus Paris, 96, 1883, 485–7. Cf. Bachmann, Niedere Zahlentheorie, I, 1902, 59–62.
[31]Atti Accad. Pont. Nouvi Lincei, 37, 1883–4, 110–3.
[32]Mathesis, 3, 1883, 48; proof by Cesàro, p. 118.
[33]Ibid., 5, 1885, 84.

E. Catalan[34] noted that

$$\left(\frac{2n-2p}{n-p}\right)^2 \left(\frac{2p}{p}\right)^2 \div \left(\frac{n}{p}\right) = \text{integer}.$$

F. Gomes Teixeira[35] discussed the result due to Weill.[26]
De Presle[36] proved that

$$\frac{(k+1)(k+2)\ldots(k+hl)}{l!\,(h!)^l} = \text{integer},$$

being the product of an evident integer by $(hl)!/\{l!(h!)^l\}$.

E. Catalan[37] noted that, if n is prime to 6,

$$\frac{(2n-4)!}{n!(n-2)!} = \text{integer}.$$

H. W. Lloyd Tanner[38] proved that

$$\frac{\{(\lambda_1+\ldots+\lambda_h)g\}!}{(\lambda_1!\ldots\lambda_h!)^g(g!)^h} = \text{integer}.$$

L. Gegenbauer stated and J. A. Gmeiner[39] proved arithmetically that, if $n = \Sigma_{j=1}^{j=r} a_{j1}a_{j2}\ldots a_{js}$, the product

$$m(m+k)(m+2k)\ldots\{m+(n-1)k\}k^{n-r}$$

is divisible by

$$\prod_{j=1}^{r} \prod_{\nu=1}^{s}(a_{j\nu}!)^{a_{j\nu+1}\ldots a_{js}},$$

where m, k, n, $a_{11}, \ldots, a_{rs}$ are positive integers. This gives Hermite's[29] result by taking $r=s=1$. The case $m=k=1$, $s=2$, is included in the result by Weill.[26]

Heine[39a] and A. Thue[40] proved that a fraction, whose denominator is $k!$ and whose numerator is a product of k consecutive terms of an arithmetical progression, can always be reduced until the new denominator contains only such primes as divide the difference of the progression [a part of Hermite's[29] result].

F. Rogel[41] noted that, if P be the product of the primes between $(p-1)/2$ and $p+1$, while n is any integer not divisible by the prime p,

$$(n-1)(n-2)\ldots(n-p+1)P/p \equiv 0 \pmod{P}.$$

S. Pincherle[42] noted that, if n is a prime,

$$P = (x+1)(x+2)\ldots(x+n-1)$$

is divisible by n and, if x is not divisible by n, by $n!$. If $n = \Pi p^a$, P is divisible

[34]Nouv. Ann. Math., (3), 4, 1885, 487. Proof by Landau, (4), 1, 1901, 282.
[35]Archiv Math. Phys., (2), 2 1885, 265-8.   [36]Bull. Soc. Math. France, 16, 1887-8, 159.
[37]Mém. Soc. Roy. Sc. Liège, (2), 15, 1888, 111 (Mélanges Math. III).   Mathesis, 9, 1889, 170.
[38]Proc. London Math. Soc., 20, 1888-9, 287.   [39]Monatshefte Math. Phys., 1, 1890, 159-162.
[39a]Jour. für Math., 45, 1853, 287-8. Cf. Math. Quest. Educ. Times, 56, 1892, 62-63.
[40]Archiv for Math. og Natur., Kristiania, 14, 1890, 247-250.
[41]Archiv Math. Phys., (2), 10, 1891, 93.
[42]Rendiconto Sess. Accad. Sc. Istituto di Bologna, 1892-3, 17.

by $n!$ if and only if divisible by $\Pi p^{\alpha+\beta}$, where β is the exponent of the power of p dividing $(n-1)!$.

G. Bauer[43] proved that the multinomial coefficient $(n+n_1+n_2+\ldots)!$ $\div \{n!n_1!\ldots\}$ is an integer, and is even if two or more n's are equal.

E. Landau[44] generalized most of the preceding results. For integers a_{ij}, b_{ij}, each $\geqq 0$, and positive integers x_j, set

$$f = \frac{u_1!\ldots u_m!}{v_1!\ldots v_n!}, \qquad u_i = \sum_{j=1}^{r} a_{ij}x_j, \qquad v_i = \sum_{j=1}^{r} b_{ij}x_j.$$

Then f is an integer if and only if

$$\sum_{i=1}^{m} u_i \geqq \sum_{i=1}^{n} v_i$$

for all real values of the x_j for which $0 \leqq x_j \leqq 1$. A new example is

$$\frac{(4m)!\,(4n)!}{m!n!(2m+n)!(m+2n)!} = \text{integer}.$$

P. A. MacMahon[45] treated the problem to find all a's for which

$$\left(\frac{n+1}{1}\right)^{a_1}\left(\frac{n+2}{2}\right)^{a_2}\cdots\left(\frac{n+m}{m}\right)^{a_m}$$

is an integer for all values of n; in particular, to find those "ground forms" from which all the forms may be generated by multiplication. For $m=2$, the ground forms have $(a_1, a_2) = (1, 0)$ or $(1, 1)$. For $m=3$, the additional ground forms are $(1, 1, 1)$, $(1, 2, 1)$, $(1, 3, 1)$. For $m=4$, there are 3 new ground forms; for $m=5$, 13 new.

J. W. L. Glaisher[46] noted that, if $B_p(x)$ is Bernoulli's function, $i.$ $e.$, the polynomial expression in x for $1^{p-1}+2^{p-1}+\ldots+(x-1)^{p-1}$ [Bernoulli[150a] of Ch. V],

$$x(x+1)\ldots(x+p-1)/p \equiv B_p(x) - x \pmod{p}.$$

He gave ($ibid.$, 33, 1901, 29) related congruences involving the left member and $B_{p-1}(x)$.

Glaisher[47] noted that, if r is not divisible by the odd prime p, and $l = kp+t$, $0 \leqq t < p$,

$$l(r+l)(2r+l)\ldots\{(p-1)r+l\}/p \equiv -\left\{\left[\frac{t}{p}\right]_r + k\right\} \pmod{p},$$

where $[t/p]_r$ denotes the least positive root of $px \equiv t \pmod{r}$. The residues mod p^3 of the same product $l(r+l)\ldots$ are found to be complicated.

E. Maillet[48] gave a group of order $t!(q!)^t$ contained in the symmetric group on tq letters, whence follows Weill's[25] result.

[43]Sitzungsber. Ak. Wiss. München (Math.), 24, 1894, 346–8.
[44]Nouv. Ann. Math., (3), 19, 1900, 344–362, 576; (4), 1, 1901, 282; Archiv Math. Phys., (3), 1, 1901, 138. Correction, Landau.[53]
[45]Trans. Cambr. Phil. Soc., 18, 1900, 12–34.
[46]Proc. London Math. Soc., 32, 1900, 172.
[47]Messenger Math., 30, 1900–1, 71–92.
[48]Mém. Prés. Ac. Sc. Paris, (2), 32, 1902, No. 8, p. 19.

M. Jenkins[48a] counted in two ways the arrangements of $n = \phi f + \gamma g + \ldots$ elements in ϕ cycles of f letters each, γ cycles of g letters, ..., where f, g, ... are distinct integers > 1, and obtained the result

$$\frac{n!}{f^\phi \phi! g^\gamma \gamma! \ldots} = n!\left(\frac{1}{2!} - \frac{1}{3!} + \frac{1}{4!} - \cdots + (-1)^n \frac{1}{n!}\right).$$

C. de Polignac[49] investigated at length the highest power of $n!$ dividing $(nx)!/(x!)^n$. Let n_p be the sum of the digits of n to base p. Then

$$(x+n)_p = x_p + n_p - k(p-1), \qquad (xn)_p = x_p \cdot n_p - k'(p-1),$$

where k is the number of units "carried" in making the addition $x+n$, and k' the corresponding number for the multiplication $x \cdot n$.

E. Schönbaum[50] gave a simplified exposition of Landau's first paper.[44]

S. K. Maitra[51] proved that $(n-1)(2n-1)\ldots\{(n-2)n-1\}$ is divisible by $(n-1)!$ if and only if n is a prime.

E. Stridsberg[52] gave a very elementary proof of Hermite's[29] result.

E. Landau[53] corrected an error in his[44] proof of the result in No. III of his paper, no use of which had been made elsewhere.

Birkeland[18] of Ch. XI noted that a product of $2^p k$ consecutive odd integers is $\equiv 1 \pmod{2^p}$.

Among the proofs that binomial coefficients are integers may be cited those by:

G. W. Leibniz, Math. Schriften, pub. by C. I. Gerhardt, 7, 1863, 102.
B. Pascal, Oeuvres, 3, 1908, 278–282.
Gioachino Pessuti, Memorie di Mat. Soc. Italiana, 11, 1804, 446.
W. H. Miller, Jour. für Math., 13, 1835, 257.
S. S. Greatheed, Cambr. Math. Jour., 1, 1839, 102, 112.

Proofs that multinomial coefficients are integers were given by:

C. F. Gauss, Disq. Arith., 1801, art. 41.
Lionnet, Complément des éléments d'arith., Paris, 1857, 52.
V. A. Lebesgue, Nouv. Ann. Math., (2), 1, 1862, 219, 254.

FACTORIALS DIVIDING THE PRODUCT OF DIFFERENCES OF r INTEGERS.

H. W. Segar[60] noted that the product of the differences of any r distinct integers is divisible by $(r-1)!(r-2)!\ldots2!$. For the special case of the integers 1, 2, ..., n, $r+1$, the theorem shows that the product of any n consecutive integers is divisible by $n!$.

A. Cayley[61] used Segar's theorem to prove that

$$m(m-n)\ldots(m-\overline{r-1}n)\cdot n^r$$

is divisible by $r!$ if m, n are relatively prime [a part of Hermite's[29] result]. Segar[62] gave another proof of his theorem. Applying it to the set

[48a]Quar. Jour. Math., 33, 1902, 174–9.   [49]Bull. Soc. Math. France, 32, 1904, 5–43.
[50]Casopis, Prag, 34, 1905, 265–300 (Bohemian).
[51]Math. Quest. Educat. Times, (2), 12, 1907, 84–5.
[52]Acta Math., 33, 1910, 243.   [53]Nouv. Ann. Math., (4), 13, 1913, 353–5.
[60]Messenger Math., 22, 1892–3, 59.   [61]Messenger Math. 22, 1892–3, p. 186. Cf. Hermite.[29]
[62]Ibid., 23, 1893–4, 31.   Results cited in l'intermédiaire des math., 2, 1895, 132–3, 200; 5, 1898, 197; 8, 1901, 145.

$a, a+N, \ldots, a+N^n$, we conclude that the product of their differences is divisible by $n!(n-1)!\ldots2!=\nu$. But the product equals

$$P = (N-1)^{n-1} (N^2-1)^{n-2} \ldots (N^{n-2}-1)^2 (N^{n-1}-1),$$

multiplied by a power of N. Hence, if N is prime to $n!$, P is divisible by ν; in any case a least number λ is found such that $N^\lambda P$ is divisible by ν. It is shown that the product of the differences of $m_1, \ldots, m_k$ is divisible by $k!(k-1)!\ldots2!$ if there be any integer p such that $m_1+p, \ldots, m_k+p$ are relatively prime to each of $1, 2, \ldots, k$. It is proved that the product of any n distinct integers multiplied by the product of all their differences is a multiple of $n!(n-1)!\ldots2!$.

E. de Jonquières[63] and F. J. Studnička[64] proved the last theorem.

E. B. Elliott[65] proved Segar's theorem in the form: The product of the differences of n distinct numbers is divisible by the product of the differences of $0, 1, \ldots, n-1$. He added the new theorems: The product of the differences of n distinct squares is divisible by the product of the differences of $0^2, 1^2, \ldots, (n-1)^2$; that for the squares of n distinct odd numbers, multiplied by the product of the n numbers, is divisible by the product of the differences of the squares of the first n odd numbers, multiplied by their product.

RESIDUES OF MULTINOMIAL COEFFICIENTS.

Leibniz[4, 7] of Ch. III noted that the coefficients in $(\Sigma a)^p - \Sigma a^p$ are divisible by p.

Ch. Babbage[69] proved that, if n is a prime, $\binom{2n-1}{n-1}-1$ is divisible by n^2, while $\binom{p+n}{p}-1$ is divisible by p if and only if p is a prime.

G. Libri[70] noted that, if $m=6p+1$ is a prime,

$$6p - \binom{6p}{3}3 + \binom{6p}{5}3^2 - \ldots \equiv 0,$$

$$2^{6p-2} + 6p - 1 - \binom{6p-1}{3}3 + \binom{6p-1}{5}3^2 - \ldots \equiv 0 \pmod{m}.$$

E. Kummer[71] determined the highest power p^N of a prime p dividing

$$\frac{(A+B)!}{A!\,B!}, \qquad A \equiv a_0 + a_1 p + \ldots + a_l p^l, \qquad B \equiv b_0 + b_1 p + \ldots + b_l p^l,$$

where the a_i and b_i belong to the set $0, 1, \ldots, p-1$. We may determine c_i in this set and $\epsilon_i=0$ or 1 such that

$$(3) \quad a_0+b_0 = \epsilon_0 p + c_0, \qquad \epsilon_0 + a_1 + b_1 = \epsilon_1 p + c_1, \qquad \epsilon_1 + a_2 + b_2 = \epsilon_2 p + c_2, \ldots$$

Multiply the first equation by 1, the second by p, the third by p^2, etc., and add. Thus

$$A+B = c_0 + c_1 p + \ldots + c_l p^l + \epsilon_l p^{l+1}.$$

[63]Comptes Rendus Paris, 120, 1895, 408–10, 534–7.
[64]Vestnik Ceske Ak., 7, 1898, No. 3, 165 (Bohemian).
[65]Messinger Math., 27, 1897–8, 12–15.
[69]Edinburgh Phil. Jour., 1, 1819, 46.
[70]Jour. für Math., 9, 1832, 73. Proofs by Stern, 12, 1834, 288.
[71]Ibid., 44, 1852, 115–6. Cayley, Math. Quest. Educ. Times, 10, 1868, 88–9.

Hence, by Legendre's formula (1),

$$(p-1)N = A+B-\gamma-\epsilon_l-(A-a)-(B-\beta), \quad a=\Sigma a_i, \quad \beta=\Sigma b_i, \quad \gamma=\Sigma c_i.$$

Insert the value of $a+\beta$ obtained by adding equations (3). Thus

$$N = \epsilon_0+\epsilon_1+\ldots+\epsilon_l.$$

A. Genocchi[72] proved that, if m is the sum of n integers $a, b, \ldots, k$, each divisible by $p-1$, and if $m < p^n - 1$, then $m! \div \{a!b!\ldots k!\}$ is divisible by the prime p.

J. Wolstenholme[73] proved that $\binom{2n-1}{n-1} \equiv 1 \pmod{n^3}$ if n is a prime > 3.

H. Anton[4] (303–6) proved that if $n=vp+a$, $r=wp+b$, where a, b, v, w are all less than the prime p,

$$\binom{n}{r} \equiv \binom{a}{b}\binom{v}{w}, \qquad \frac{1}{p}\binom{n}{r} \equiv \frac{1}{p}\binom{p+a}{b}\binom{v-1}{w}v \pmod{p},$$

according as $a \geq b$ or $a < b$.

M. Jenkins[73a] considered for an odd prime p the sum

$$\sigma_r = \Sigma \binom{(m+n)r}{mr+k(p-1)},$$

extended over all the integers k between $nr/(p-1)$ and $-mr/(p-1)$, inclusive, and proved that $\sigma_r \equiv \sigma_\rho \pmod{p}$ if the g. c. d. of $r, p-1$ equals that of $\rho, p-1$.

E. Catalan[74] noted that $\binom{np-1}{p-1} \equiv 1 \pmod{p}$, if p is a prime.

Ch. Hermite[75] proved by use of roots of unity that the odd prime p divides

$$\binom{2n+1}{p-1}+\binom{2n+1}{2p-2}+\binom{2n+1}{3p-3}+\ldots.$$

E. Lucas[76] noted that, if $m=pm_1+\mu$, $n=pn_1+\nu$, $\mu < p$, $\nu < p$, and p is a prime,

$$\binom{n}{m} \equiv \binom{n_1}{m_1}\binom{\nu}{\mu} \pmod{p}.$$

In general, if $\mu_1, \mu_2, \ldots$ denote the residues of m and the integers contained in the fractions $m/p, m/p^2, \ldots$, while the ν's are the residues of $n, [n/p], \ldots$,

$$\binom{n}{m} \equiv \binom{\nu_1}{\mu_1}\binom{\nu_2}{\mu_2}\ldots \pmod{p}.$$

E. Lucas[77] proved the preceding results and

$$\binom{p}{n} \equiv 0, \qquad \binom{p-1}{n} \equiv (-1)^n, \qquad \binom{p+1}{n} \equiv 0 \pmod{p},$$

according as n is between 0 and p, 0 and $p-1$, or 1 and p.

[72]Nouv. Ann. Math., 14, 1855, 241–3.
[73]Quar. Jour. Math., 5, 1862, 35–9. For mod. n^2, Math. Quest. Educ. Times, (2), 3, 1903, 33.
[73a]Math. Quest. Educ. Times, 12, 1869, 29.   [74]Nouv. Corresp. Math., 1, 1874–5, 76.
[75]Jour. für Math., 81, 1876, 94.   [76]Bull. Soc. Math. France, 6, 1877–8, 52.
[77]Amer. Jour. Math., 1, 1878, 229, 230. For the second, anon.[43] of Ch. III (in 1830).

J. Wolstenholme[78] noted that the highest power of 2 dividing $\binom{2m-1}{m}$ is $q-p-1$, where q is the sum of the digits of $2m-1$ to base 2, and 2^p is the highest power of 2 dividing m.

J. Petersen[79] proved by Legendre's formula that $\binom{a+b}{a}$ equals the product of the powers of all primes p, the exponent of p being $(t_a+t_b-t_{a+b}) \div (p-1)$, where t_a is the sum of the digits of a to base p.

E. Cesàro[80] treated Kummer's[71] problem. He stated (Ex. 295) and Van den Broeck[81] proved that the exponent of the highest power of the prime p dividing $\binom{2n}{n}$ is the number of odd integers among $[2n/p]$, $[2n/p^2]$, $[2n/p^3]$, ...

O. Schlömilch[81a] stated in effect that $\binom{kn}{n+1}$ is divisible by n.

E. Catalan[82] proved that if n is odd,

$$\binom{2n}{n} + 10\binom{2n-2}{n-1} \equiv 0 \pmod{n+2}.$$

W. J. C. Sharp[82a] noted that $(p+n)! - p!n!$ is divisible by p^2, if p is a prime $>n$. This follows also from $\binom{p+n}{n} \equiv 1 \pmod{p}$ [Dickson[90]].

L. Gegenbauer[83] noted that, if σ is any integer, τ one of the form $6s$ or $3s$ according as n is odd or even,

$$(\tau-\sigma)\binom{2n}{n} + 5\sigma\binom{2n-2}{n-1} \equiv 0 \pmod{n+2}.$$

The case n odd, $\sigma=2$, $\tau=3$, gives Catalan's result.

E. Catalan[84] proved Hermite's[75] theorem.

Ch. Hermite[85] stated that $\binom{m}{n}$ is divisible by $m-n+1$ if m is divisible by n; by $(m-n+1)/\epsilon$ if ϵ is the g. c. d. of $m+1$ and n; by m/δ, if δ is the g. c. d. of m, n.

E. Lucas[86] noted that, if $n \leq p-1$, $p-2$, $p-3$, respectively,

$$\binom{p-1}{n} \equiv (-1)^n, \qquad \binom{p-2}{n} \equiv (-1)^n(n+1),$$

$$\binom{p-3}{n} \equiv (-1)^n \frac{(n+1)(n+2)}{2} \pmod{p},$$

if p is a prime, and proved Hermite's[75] result (p. 506).

F. Rogel[87] proved Hermite's[75] theorem by use of Fermat's.

[78]Jour. de math. élém. et spéc., 1877–81, ex. 360.
[79]Tidsskrift for Math., (4), 6, 1882, 138–143.
[80]Mathesis, 4, 1884, 109–110.
[81]*Ibid.*, 6, 1886, 179.
[81a]Zeitschrift Math. Naturw. Unterricht, 17, 1886, 281.
[82]Mém. Soc. Roy. Sc. de Liège, (2), 13, 1886, 237–241 (=Mélanges Math.).   Mathesis, 10, 1890, 257–8.
[82a]Math. Quest. Educ. Times, 49, 1888, 74.
[83]Sitzungsber. Ak. Wiss. Wien (Math.), 98, 1889, IIa, 672.
[84]Mém. Soc. Sc. Liège, (2), 15, 1888, 253–4 (Mélanges Math. III).
[85]Jour. de math. spéciales, problems 257–8.   Proofs by Catalan, *ibid.*, 1889, 19–22; 1891, 70; by G. B. Mathews, Math. Quest. Educ. Times, 52, 1890, 63; by H. J. Woodall, 57, 1892, 91.
[86]Théorie des nombres, 1891, 420.   [87]Archiv Math. Phys., (2), 11, 1892, 81–3.

C. Szily[88] noted that no prime $>2a$ divides

$$\sum_{k=0}^{a} \binom{a}{k}^2,$$

and specified the intervals in which its prime factors occur.

F. Morley[89] proved that, if $p=2n+1$ is a prime, $\binom{2n}{n}-(-1)^n 2^{4n}$ is divisible by p^3 if $p>3$. That it is divisible by p^2 was stated as an exercise in Mathews' Theory of Numbers, 1892, p. 318, Ex. 16.

L. E. Dickson[90] extended Kummer's[71] results to a multinomial coefficient M and noted the useful corollary that it is not divisible by a given prime p if and only if the partition of m into $m_1, \ldots, m_t$ arises by the separate partition of each digit of m written to the base p into the corresponding digits of $m_1, \ldots, m_t$. In this case he proved that

$$M \equiv \prod_{i=0}^{n} \frac{a_i!}{a_i^{(1)}! \ldots a_i^{(t)}!} \pmod{p}, \qquad m_k = a_0^{(k)} p^n + \ldots + a_n^{(k)}.$$

This also follows from (2) and from

$$(x_1 + \ldots + x_t)^m \equiv (x_1 + \ldots + x_t)^{a_n}(x_1^p + \ldots + x_t^p)^{a_{n-1}} \ldots (x_1^{p^n} + \ldots + x_t^{p^n})^{a_0}$$
$$\pmod{p}.$$

F. Mertens[91] considered a prime $p \leq n$, the highest powers p^τ and 2^ν of p and 2 which are $\leq n$, and set $n_a = [n/2^a]$. Then $n! \div \{n_1! n_2! \ldots n_\nu!\}$ is divisible by Πp^τ, where p ranges over all the primes p.

J. W. L. Glaisher[92] gave Dickson's[90] result for the case of binomial coefficients. He considered (349–60) their residues modulo p^n, and proved (pp. 361–6) that if $(n)_r$ denotes the number of combinations of n things r at a time, $\Sigma(n)_r \equiv (j)_k \pmod{p}$, where p is any prime, n any integer $\equiv j$ $\pmod{p-1}$, while the summation extends over all positive integers r, $r \leq n$, $r \equiv k \pmod{p-1}$, and j, k are any of the integers $1, \ldots, p-1$. He evaluated $\Sigma[(n)_r \div p]$ when r is any number divisible by $p-1$, and $(n)_r$ is divisible by p, distinguishing three cases to obtain simple results.

Dickson[93] generalized Glaisher's[92] theorem to multinomial coefficients: Let k be that one of the numbers $1, 2, \ldots, p-1$ to which m is congruent modulo $p-1$, and let $k_1, \ldots, k_t$ be fixed numbers of that set such that $k_1 + \ldots + k_t \equiv k \pmod{p-1}$. Then if p is a prime,

$$\sum_{m_1, \ldots, m_t} (m_1, m_2, \ldots, m_t) \equiv \begin{cases} (k_1, k_2, \ldots, k_t) & \text{if } k_1 + \ldots + k_t = k \\ 0 & \text{if } k_1 + \ldots + k_t > k \end{cases} \pmod{p},$$

where

$$(m_1, \ldots, m_t) = \frac{(m_1 + \ldots + m_t)!}{m_1! \ldots m_t!}.$$

The second of the two proofs given is much the simpler.

[88] Nouv. Ann. Math., (3), 12, 1893, Exercices, p. 52.* Proof, (4), 16, 1916, 39–42.
[89] Annals of Math., 9, 1895, 168–170.
[90] Ibid., (1), 11, 1896–7. 75–6: Quart. Jour. Math., 33, 1902, 378–384.
[91] Sitzungsber. Ak. Wiss. Wien (Math.), 106, IIa, 1897, 255–6.
[92] Quar. Jour. Math., 30, 1899, 150–6, 349–366.
[93] Ibid., 33, 1902, 381–4.

Glaisher[94] discussed the residues modulo p^3 of binomial coefficients.

T. Hayashi[95] proved that if p is a prime and $\mu+\nu=p$,

$$\binom{rp+\mu+s-1}{rp+s} \equiv (-1)^s \binom{\nu}{s}, \; 0, \; 1 \pmod{p},$$

according as $0<s\leqq\nu$, $\nu<s<p$, or $s=0$.

T. Hayashi[96] proved that, if l_0 is the least positive residue of l modulo p, and if $\nu=p-\mu$,

$$\binom{l+\mu-1}{l} \equiv (-1)^{l_0} \binom{\nu}{l_0} \equiv \binom{\nu+l-1}{l} + \binom{\nu+l-p-1}{l-p} + \binom{\nu+l-2p-1}{l-2p} + \cdots$$

modulo p. Special cases of the first result had been given by Lucas.[86]

A. Cunningham[97] proved that, if p is a prime,

$$\binom{p-1}{x} \equiv (-1)^x \pmod{p}, \qquad \tfrac{1}{2}\binom{2p}{p} \equiv 1 \pmod{p^3, \; p>3}.$$

B. Ram[98] noted that, if $\binom{n}{m}$, $m=1,\ldots,\,n-1$, have a common factor $a>1$, then a is a prime and $n=a^r$. There is at most one prime $<n$ which does not divide $\Pi\binom{n}{m}$ for $m=1,\ldots,\,n-2$, and then only when $n+1=qa^r$, where a is a prime and $q<a$. For $m=0, 1,\ldots,\,n$, the number of odd $\binom{n}{m}$ is always a power of 2.

P. Bachmann[99] proved that, if $h(p-1)$ is the greatest multiple $<k$ of $p-1$,

$$\binom{k}{p-1} + \binom{k}{2(p-1)} + \cdots + \binom{k}{h(p-1)} \equiv 0 \pmod{p},$$

the case k odd being due to Hermite.[75]

G. Fontené stated and L. Grosschmid[100] proved that

$$\binom{Pk}{P(p-1)} \equiv (-1)^k \pmod{p}, \qquad P=p^a, \; a\geqq 0.$$

A. Fleck[101] proved that, if $0\leqq\rho<p$, $a+b\equiv 0 \pmod{p}$,

$$\sum_{k=0}^{\infty} \binom{m}{\rho+kp} a^{m-\rho-kp} b^{\rho+kp} \equiv 0 \pmod{p^e}, \quad e=\left[\frac{m-1}{p-1}\right].$$

N. Nielsen[102] proved Bachmann's[99] result by use of Bernoulli numbers.

[94]Quar. Jour. Math., 31, 1900, 110–124.

[95]Jour. of the Physics School in Tokio, 10, 1901, 391–2; Abh. Geschichte Math. Wiss., 28, 1910, 26–28.

[96]Archiv Math. Phys., (3), 5, 1903, 67–9.

[97]Math. Quest. Educat. Times, (2), 12, 1907, 94–5.

[98]Jour. of the Indian Math. Club, Madras, 1, 1909, 39–43.

[99]Niedere Zahlentheorie, II, 1910, 46.

[100]Nouv. Ann. Math., (4), 13, 1913, 521–4.

[101]Sitzungs. Berlin Math. Gesell., 13, 1913–4, 2–6.　Cf. H. Kapferer, Archiv Math. Phys. (3), 23, 1915, 122.

[102]Annali di mat., (3), 22, 1914, 253.

A. Fleck[103] proved that

$$\binom{a}{a}\binom{a+1}{a}\binom{a+2}{a}\ldots\binom{p-1}{a} \equiv (-1)^{a(a+1)/2}\binom{a}{1}\binom{a}{2}\ldots\binom{a}{a-1}\pmod{p}$$

if and only if p is a prime. The case $a=1$ is Wilson's theorem.
Guérin[104] asked if Wolstenholme's[73] result is new and added that

$$\binom{kp-1}{p-1} \equiv k-1 \pmod{p^3}, \quad p \text{ prime} > 3.$$

THE CONGRUENCE $1 \cdot 2 \cdot 3 \ldots (p-1)/2 \equiv \pm 1 \pmod{p}$.

J. L. Lagrange[110] noted that $p-1, p-2, \ldots, (p+1)/2$ are congruent modulo p to $-1, -2, \ldots, -(p-1)/2$, respectively, so that Wilson's theorem gives

$$(4) \qquad \left(1 \cdot 2 \cdot 3 \ldots \frac{p-1}{2}\right)^2 \equiv (-1)^{\frac{p+1}{2}} \pmod{p}.$$

For p a prime of the form $4n+3$, he noted that

$$(5) \qquad 1 \cdot 2 \cdot 3 \ldots \frac{p-1}{2} \equiv \pm 1 \pmod{p}.$$

E. Waring[111] and an anonymous writer[112] derived (4) in the same manner.

G. L. Dirichlet[113] noted that, since -1 is a non-residue of $p=4n+3$, the sign in (5) is $+$ or $-$, according as the left member is a quadratic residue or non-residue of p. Hence if m is the number of quadratic non-residues $< p/2$ of p,

$$1 \cdot 2 \cdot 3 \ldots \frac{p-1}{2} \equiv (-1)^m \pmod{p}.$$

C. G. J. Jacobi[114] observed that, for $p>3$, m is of the same parity as N, where $2N-1=(Q-P)/p$, P being the sum of the least positive quadratic residues of p, and Q that of the non-residues. Writing the quadratic residues in the form $\pm k$, $1 \le k \le \frac{1}{2}(p-1)$, let m be the number of negative terms $-k$, and $-T$ their sum. Since -1 is a non-residue, m is the number of non-residues $< \frac{1}{2}p$ and

$$\Sigma(\pm k) = Sp, \qquad P = \Sigma(+k) + \Sigma(p-k) = mp + Sp,$$

$$P+Q = 1 + \ldots + p - 1 = \frac{p(p-1)}{2}, \quad 2N-1 = \frac{P+Q}{p} - \frac{2P}{p} = \frac{p-1}{2} - 2(m+S).$$

Since $p=4n+3$, $N=n+1-m-S$. But $n+1$ and S are of the same parity since

$$pS + 2T = 1 + 2 + \ldots + \tfrac{1}{2}(p-1) = \tfrac{1}{8}(p^2-1) = (2n+1)(n+1).$$

[103]Sitzungs. Berlin Math. Gesell., 15, 1915, 7–8.
[104]L'intermédiaire des math., 23, 1916, 174.
[110]Nouv. Mém. Ac. Berlin, 2, 1773, année 1771, 125; Oeuvres, 3, 432.
[111]Meditat. Algebr., 1770, 218; ed. 3, 1782, 380.
[112]Jour. für Math., 6, 1830, 105.
[113]Ibid., 3, 1828, 407–8; Werke, 1, 107. Cf. Lucas, Théorie des nombres, 438; l'intermédiaire des math., 7, 1900, 347.
[114]Ibid., 9, 1832, 189–92; Werke, 6, 240–4.

He stated empirically that N is the number of reduced forms $ay^2+byz+cz^2$, $4ac-b^2=p$ for b odd, $ac-\frac{1}{4}b^2=p$ for b even, where $b<a$, $b<c$.

C. F. Arndt[115] proved in two ways that the product of all integers relatively prime to $M=p^n$ or $2p^n$, and not exceeding $(M-1)/2$, is $\equiv \pm 1$ (mod M), when p is a prime $4k+3$, the sign being $+$ or $-$ according as the number of residues $>M/2$ of M is even or odd. Again,

$$\{1\cdot 3\cdot 5\cdot 7 \ldots (p-2)\}^2 \equiv \pm 1 \pmod{p},$$

the sign being $+$ or $-$ according as the prime p is of the form $4n+3$ or $4n+1$. In the first case, $1\cdot 3 \ldots (p-2) \equiv \pm 1 \pmod{p}$.

L. Kronecker[116] obtained, for Dirichlet's[113] exponent m, the result $m\equiv \nu$ (mod 2), where ν is the number of positive integers of the form $q^{4l+1}r^2$ in the set $p-2^2$, $p-4^2$, $p-6^2$, ..., and q is a prime not dividing r. Liouville (p. 267) gave $m\equiv k+\nu''$ (mod 2), when $p=8k+3$ and ν'' is the number of positive integers of the form $q^{4l+1}r^2$ in the set $p-4^2$, $p-8^2$, $p-12^2$, ...

J. Liouville[117] gave the result $m\equiv \sigma+\tau$ (mod 2), for the case $p=8k+3$, where τ is the number of positive integers of the form $2q^{4l+1} r^2$ (q a prime not dividing r) in the set $p-1^2$, $p-3^2$, $p-5^2$, ..., and σ is the number of equal or distinct primes $4g+1$ dividing b, where $p=a^2+2b^2$ (uniquely).

A. Korkine[118] stated that, if $[x]$ is the greatest integer $\leqq x$,

$$m\equiv \frac{p-3}{4}+\sum_{k=1}^{(p-3)/4}\left[\sqrt{pk}\right] \pmod{p}.$$

J. Franel[119] proved the last result by use of Legendre's symbol and

$$(-1)^m\equiv \prod_{r=1}^{(p-1)/2}\left(\frac{r}{p}\right), \qquad \left(\frac{r}{p}\right)=(-1)^\mu, \qquad \mu\equiv \sum_{s=1}^{(p-1)/2}\left[\frac{rs}{p}\right] \pmod{2}.$$

M. Lerch[120] obtained Jacobi's[114] result.

H. S. Vandiver[120a] proved Dirichlet's[113] result and that

$$m\equiv \sum_{j=1}^{(p-1)/2}\left[\frac{j^2}{p}\right] \pmod{2}.$$

R. D. Carmichael[121] noted that (4) holds if and only if p is a prime.

E. Malo[122] considered the residue $\pm r$ of $1\cdot 2 \ldots (p-1)/2$ modulo p, where p is a prime $4m+1$, and $0<r<p/2$. Thus $r^2\equiv -1$. The numbers $2, 3, \ldots, (p-1)/2$, with r excluded, may be paired so that the product of the two of a pair is $\equiv \pm 1$ (mod p). If this sign is minus for k pairs, $1\cdot 2 \ldots (p-1)/2 \equiv (-1)^k r \pmod{p}$.

*J. Ouspensky gave a rule to find the sign in (5).

OTHER CONGRUENCES INVOLVING FACTORIALS.

V. Bouniakowsky[129] noted that $(p-1)!=PP'$, $P\pm P'\equiv 0 \pmod{p}$ according as $p=4k\mp 1$. For, if ρ is a primitive root of p, we may set $P=\rho\rho^2$

[115]Archiv Math. Phys., 2, 1842, 32, 34–35.
[116]Jour. de Math., (2), 5, 1860, 127.
[117]Ibid., 128.
[118]L'intermédiaire des math., 1, 1894, 95.
[119]Ibid., 2, 1895, 35–37.
[120]Prag Sitzungsber. (Math.), 1898, No. 2.
[120a]Amer. Math. Monthly, 11, 1904, 51–6.
[121]Ibid., 12, 1905, 106–8.
[122]L'intermédiaire des math., 13, 1906, 131–2
[123]Bull. Soc. Phys. Math. Kasan, (2), 21.
[129]Mém. Ac. Sc. St. Pétersbourg, (6), 1, 1831, 564.

$\ldots\rho^t$, $P'=\rho^{t+1}\ldots\rho^{p-1}$ with $t=(p-1)/2$, when $p=4k-1$; but $P=\rho\rho^{p-1}$ $\rho^3\rho^{p-3}\ldots$, $P'=\rho^2\rho^{p-2}$ $\rho^4\rho^{p-4}\ldots$, when $p=4k+1$.

G. Oltramare[130] gave several algebraic series for the reciprocal of the binomial coefficient $\binom{2m}{m}$ and concluded that, if the moduli are primes,

$$1+(m!)^4\equiv -2\left\{\left(\frac{1}{3}\right)^2+\left(\frac{1\cdot5}{3\cdot7}\right)^2+\left(\frac{1\cdot5\cdot9}{3\cdot7\cdot11}\right)^2+\ldots\right\}\pmod{4m+1},$$

$$2^5+(m!)^4\equiv -2^6\left\{\left(\frac{3}{1}\right)^2+\left(\frac{3\cdot7}{1\cdot5}\right)^2+\left(\frac{3\cdot7\cdot11}{1\cdot5\cdot9}\right)^2+\ldots\right\}\pmod{4m+3}.$$

V. Bouniakowsky[131] considered the integers $q_1,\ldots,q_s$, each $<N$ and prime to N, arranged in ascending order of magnitude. If λ is any chosen integer $\leq s$, multiply

$$q_s=N-q_1, \qquad q_{s-1}=N-q_2,\ldots, \qquad q_{s-\lambda+1}=N-q_\lambda$$

together and multiply the resulting equation by $q_1\ldots q_{s-\lambda}$. Apply the generalized Wilson theorem $q_1\ldots q_s+(-1)^s\equiv 0\pmod{N}$. Hence

$$q_1q_2\ldots q_\lambda\cdot q_1q_2\ldots q_{s-\lambda}+(-1)^{s+\lambda}\equiv 0\pmod{N}.$$

For N a prime, we have $s=N-1$ and

$$\lambda!(N-1-\lambda)!+(-1)^\lambda\equiv 0\pmod{N} \qquad (1\leq\lambda\leq N-1).$$

C. A. Laisant and E. Beaujeux[132] gave the last result and

$$\binom{p-1}{k}\equiv(-1)^k\pmod{p}, \qquad k=\frac{p-1}{2}.$$

F. G. Teixeira[133] proved that if $a=2^{2p-1}p-a$, $a<2p-1$,

$$a(a+1)\ldots(a+2p-1)\equiv 3^2\cdot5^2\ldots(2p-1)^2p$$
$$\pmod{a+a+1+a+2+\ldots+a+2p-1}.$$

Thus, for $p=3$, $a=1$, $a=95$,

$$95\cdot96\cdot97\cdot98\cdot99\cdot100\equiv 3^2\cdot5^2\cdot3\pmod{585=95+\ldots+100}.$$

M. Vecchi[134] noted that the final formula by Bouniakowsky[131] follows by induction. Taking $\lambda=(N-1)/2$, we get Lagrange's formula (4). From the latter, we get

$$\{3\cdot5\cdot7\ldots(2y-1)\}^2\left\{\left(\frac{p-2y-1}{2}\right)!\right\}^2/2^{2y}\equiv(-1)^{\frac{p+1}{2}}\pmod{p}.$$

The case $y=(p-1)/2$ gives Arndt's[115] result

$$(6) \qquad \{3\cdot5\cdot7\ldots(p-2)\}^2\equiv(-1)^{\frac{p+1}{2}}\pmod{p}.$$

Vecchi[135] proved that, if ν is the number of odd quadratic non-residues of a prime $p=4n+3$, then $1\cdot3\cdot5\ldots(p-2)\equiv(-1)^\nu\pmod{p}$. If μ is the number of non-residues $<p/2$, $1\cdot3\cdot5\ldots(p-2)\equiv(-1)^{\mu+1}2^{(p-1)/2}\pmod{p}$.

[130]Mém. de l'Institut Nat. Genevois, 4, 1856, 33–6.
[131]Bull. Ac. Sc. St. Pétersbourg, 15, 1857, 202–5.
[132]Nouv. Corresp. Math., 5, 1879, 156 (177).
[133]Jornal de Sciencias Math. e Astr., 3, 1881, 105–115.
[134]Periodico di Mat., 16, 1901, 22–4.
[135]Ibid., 22, 1907, 285–8.

R. D. Carmichael[121] proved that, if $a+1$ and $2a+1$ are both primes, $(a!)^4-1$ is divisible by $(a+1)(2a+1)$, and conversely.

A. Arévalo[136] proved (6) and Lucas'[77] residues of binomial coefficients.

N. G. W. H. Beeger[137] proved that [if p is a prime]

$$(p-1)!+1 \equiv s-p+1 \pmod{p^2}, \qquad s=1+2^{p-1}+\ldots+(p-1)^{p-1}=ph_{p-1},$$

where h is a Bernoulli number defined by the symbolical equation $(h+1)^n = h^n$, $h_1=1/2$. By use of Adams'[137a] table of h_i, $i<114$, it was verified that $p=5$, $p=13$ are the only $p<114$ for which $(p-1)!+1 \equiv 0 \pmod{p^2}$.

T. E. Mason[138] and J. M. Child[139] noted that, if p is a prime >3,

$$(np)! \equiv n!(p!)^n \pmod{p^{n+3}}.$$

N. Nielsen[140] proved that, if $p=2n+1$, $P=1\cdot3\cdot5\ldots(2n-1)$,

$$P^2 \equiv (-1)^n 2^{2n}(2n)! \pmod{p^2},$$
$$(-1)^n 2^{2n} P^2 \equiv 2^{2n}\cdot3\cdot5\ldots(4n-1) \pmod{16n^2}.$$

If p is a prime >3, $P \equiv (-1)^n 2^{3n} n! \pmod{p^3}$. He gave the last result also elsewhere.[141]

C. I. Marks[142] found the smallest integer x such that $2\cdot4\ldots(2n)x$ is divisible by $3\cdot5\ldots(2n-1)$.

---

[136]Revista de la Sociedad Mat. Española, 2, 1913, 130–1.

[137]Messenger Math., 43, 1913–4, 83–4.

[137a]Jour. für Math., 85, 1878, 269–72.

[138]Tôhoku Math. Jour., 5, 1914, 137.

[139]Math. Quest. Educat. Times, 26, 1914, 19.

[140]Annali di mat., (3), 22, 1914, 81–2.

[141]K. Danske Vidensk. Selsk. Skrifter, (7), 10 1913, 353.

[142]Math. Quest. Educ. Times, 21, 1912, 84–6.

CHAPTER X.

SUM AND NUMBER OF DIVISORS.

The sum of the kth powers of the divisors of n will be designated $\sigma_k(n)$. Often $\sigma(n)$ will be used for $\sigma_1(n)$, and $\tau(n)$ for the number $\sigma_0(n)$ of the divisors of n; also,

$$T(n) = \tau(1) + \tau(2) + \ldots + \tau(n).$$

The early papers in which occur the formulas for $\tau(n)$ and $\sigma(n)$ were cited in Chapter II.

L. Euler[1, 2, 3] applied to the theory of partitions the formula

$$(1) \qquad p(x) \equiv \prod_{k=1}^{\infty} (1-x^k) = s \equiv 1 - x - x^2 + x^5 + x^7 - x^{12} - \ldots$$

Euler[4] verified for $n < 300$ that

$$(2) \qquad \sigma(n) = \sigma(n-1) + \sigma(n-2) - \sigma(n-5) - \sigma(n-7) + \sigma(n-12) + \ldots,$$

in which two successive plus signs alternate with two successive minus signs, while the differences of $1, 2, 5, 7, 12, \ldots$ are $1, 3, 2, 5, 3, 7, \ldots$, the alternate ones being $1, 2, 3, 4, \ldots$ and the others being the successive odd numbers. He stated that (2) can be derived from (1).

Euler[5] noted that the numbers subtracted from n in (2) are pentagonal numbers $(3x^2 - x)/2$ for positive and negative integers x, and that if $\sigma(n-n)$ occurs it is to be replaced by n. He was led to the law of the series s by multiplying together the earlier factors of $p(x)$, but had no proof at that time that $p = s$. Comparing the derivatives of the logarithms of p and s, he found for $-x \, dp/(p \, dx)$ the two expressions equated in

$$(3) \qquad \sum_{n=1}^{\infty} \frac{nx^n}{1-x^n} = \frac{x + 2x^2 - 5x^5 - 7x^7 + 12x^{12} + \ldots}{s}.$$

He verified for a few terms that the expansion of the left member is

$$(4) \qquad \sum_{n=1}^{\infty} x^n \sigma(n).$$

Multiplying the latter by the series s and equating the product to the numerator of the right member of (3), he obtained (2) from the coefficients of x^n.

Euler[6] proved (1) by induction. To prove (2), multiply the left member of (3) by $-dx/x$ and integrate. He obtained $\log p(x)$ and hence $\log s$, and then (3) by differentiation.

[1]Letter to D. Bernoulli, Jan. 28, 1741, Corresp. Math. Phys. (ed. Fuss), II, 1843, 467.
[2]Euler, Introductio in Analysin Infinitorum, 1748, I, ch. 16.
[3]Novi Comm. Ac. Petrop., 3, 1750-1, 125; Comm. Arith., 1, 91.
[4]Letter to Goldbach, Apr. 1, 1747, Corresp. Math. Phys. (ed. Fuss), I, 1843, 407.
[5]Posth. paper of 1747, Comm. Arith., 2, 639; Opera postuma, I, 1862, 76-84. Novi Comm. Ac. Petrop., 5, ad annos 1754-5, 59-74; Comm. Arith., 1, 146-154.
[6]Letter to Goldbach, June 9, 1750, Corresp. Math. Phys. (ed. Fuss), I, 1843, 521-4. Novi Comm. Ac. Petrop., 5, 1754-5, 75-83; Acta Ac. Petrop., 4 I, 1780, 47, 56; Comm. Arith., 1, 234-8; 2, 105. Cf. Bachmann, Die Analytische Zahlentheorie, 1894, 13-29.

Material on (1) will be given in the chapter on partitions in Vol. II.
J. H. Lambert,[7] by expanding the terms by simple division, obtained

$$\sum_{n=1}^{\infty} \frac{x^n}{1-x^n} = x + 2x^2 + 2x^3 + 3x^4 + \ldots,$$

in which the coefficient of x^n is $\tau(n)$. Similarly, he obtained (4) from the left
member of (3).

E. Waring[8] reproduced Euler's[6] proof of (2).

E. Waring[9] employed the identity

$$\prod_{k=1}^{n} (x^k - 1) \equiv x^b - x^{b-1} - x^{b-2} + x^{b-5} + x^{b-7} - - \ldots = A,$$

the coefficient of x^{b-v}, for $v \leq n$, being $(-1)^z$ if $v = (3z^2 \pm z)/2$ and zero if v is
not of that form. If $m \leq n$, the sum of the mth powers of the roots of
$A = 0$ is $\sigma(m)$. Thus (2) follows from Newton's identities between the
coefficients and sums of powers of the roots. He deduced

$$(5) \quad 1 - \frac{m(m-1)}{2}\sigma(2) + \frac{m(m-1)(m-2)}{3}\sigma(3) - \frac{m(m-1)(m-2)(m-3)}{4}\sigma(4)$$

$$+ \ldots + \frac{m(m-1)(m-2)(m-3)}{2 \cdot 2^2}\{\sigma(2)\}^2 - \ldots = c \cdot m!,$$

where $c = \pm 1$ or 0 is the coefficient of x^{b-m} in series A. Let

$$\Pi(x^p - 1) = x^{b'} - x^{b'-1} - x^{b'-2} + x^{b'-4} + x^{b'-8} - \ldots = A',$$

where p ranges over the primes 1, 2, 3, 5,..., n. If $m \leq n$, the sum of the
mth powers of the roots of $A' = 0$ equals the sum $\sigma'(m)$ of the prime divisors
of m. Thus

$$\sigma'(m) = \sigma'(m-1) + \sigma'(m-2) - \sigma'(m-4) - \sigma'(m-8) + \sigma'(m-10) + \sigma'(m-11)$$
$$- \sigma'(m-12) - \sigma'(m-16) + \ldots.$$

We obtain (5) with σ replaced by σ', and c by the coefficient of $x^{b'-m}$ in series
A'. Consider

$$\prod_{j=1}^{n} (x^{jl} - 1) = x^b - x^{b-l} - x^{b-2l} + x^{b-5l} + \ldots = B,$$

with coefficients as in series A. The sum of the (lm)th powers of the roots
of $B = 0$ equals the sum $\sigma^{(l)}(m)$ of those divisors of m which are multiples of l.
Thus

$$\sigma^{(l)}(m) = \sigma^{(l)}(m-l) + \sigma^{(l)}(m-2l) - \sigma^{(l)}(m-5l) - \ldots,$$

with the same laws as (2). The sum of those divisors of m which are divisible

---

[7]Anlage zur Architectonic, oder Theorie des Ersten und des Einfachen in der phil. und math.
Erkenntniss, Riga, 1771, 507. Quoted by Glaisher.[75]
[8]Meditationes Algebraicæ, ed. 3, 1782, 345.
[9]Phil. Trans. Roy. Soc. London, 78, 1788, 388–394.

by the relatively prime numbers $a, b, c, \ldots$ is

$$\Sigma \sigma^{(a)}(m) - \Sigma \sigma^{(ab)}(m) + \Sigma \sigma^{(abc)}(m) - \ldots$$

Waring noted that $\sigma(a\beta) = a\sigma(\beta) + ($sum of those divisors of β which are not divisible by $a)$. Similarly,

$$\sigma(a\beta\gamma \ldots) = a\sigma(\beta\gamma \ldots) + (\text{sum of divisors of } \beta\gamma \ldots \text{ not divisible by } a)$$
$$= a\beta\sigma(\gamma\delta \ldots) + (\text{sum of divisors of } \beta\gamma \ldots \text{ not divisible by } a)$$
$$+ a(\text{sum of divisors of } \gamma\delta \ldots \text{ not divisible by } \beta),$$

etc. Again, $\sigma^{(l)}(a\beta) = a\sigma^{(l)}(\beta) + ($sum of divisors of β divisible by l but not by $a)$. The generalization is similar to that just given for σ.

C. G. J. Jacobi[10] proved for the series s in (1) that

$$s^3 = 1 - 3x + 5x^3 - 7x^6 + \ldots = \sum_{n=0}^{\infty} (-1)^n (2n+1) \, x^{n(n+1)/2}.$$

Jacobi[11] considered the excess $E(n)$ of the number of divisors of the form $4m+1$ of n over the number of divisors of the form $4m+3$ of n. If $n = 2^p uv$, where each prime factor of u is of the form $4m+1$ and each prime factor of v is of the form $4m+3$, he stated that $E(n) = 0$ unless v is a square, and then $E(n) = \tau(u)$.

Jacobi[12] proved the identity

(6) $(1 + x + x^3 + \ldots + x^{k(k+1)/2} + \ldots)^4 = 1 + \sigma(3)x + \ldots + \sigma(2n+1)x^n + \ldots.$

A. M. Legendre[13] proved (1).

G. L. Dirichlet[14] noted that the mean (mittlerer Werth) of $\sigma(n)$ is $\pi^2 n/6 - 1/2$, that of $\tau(n)$ is $\log n + 2C$, where C is Euler's constant $0.57721. \ldots$ He stated the approximations to $T(n)$ and $\psi(n)$, proved later[17], without obtaining the order of magnitude of the error.

Dirichlet[15] expressed m in all ways as a product of a square by a complementary factor ϵ, denoted by ν the number of distinct primes dividing ϵ, and proved that $\Sigma 2^\nu = \tau(m)$.

Stern[15a] proved (2) by expanding the logarithm of (1). If C'_n is the number of all combinations with repetitions with the sum n,

$$\sigma(n) = nC'_n - C'_1 \sigma(n-1) - C'_2 \sigma(n-2) - \ldots.$$

Let $S(n)$ be the sum of the even divisors of n. Then, by (1),

$$S(2n) = S(2n-2) + S(2n-4) - S(2n-10) - S(2n-14) + \ldots, \quad S(0) = 2n.$$

[10]Fundamenta Nova, 1829, § 66, (7); Werke, 1, 237. Jour. für Math., 21, 1840, 13; French transl., Jour. de Math, 7, 1842, 85; Werke, 6, 281. Cf. Bachmann,[6] pp. 31-7.

[11]Ibid., §40; Werke, 1, 1881, 163.

[12]Attributed to Jacobi by Bouniakowsky[19] without reference. See Legendre (1828) and Plana (1863) in the chapter on polygonal numbers, vol. 2.

[13]Théorie des nombres, ed. 3, 1830, vol. 2, 128.

[14]Jour. für Math., 18, 1838, 273; Bericht Berlin Ak., 1838, 13-15; Werke, 1, 373, 351-6.

[15]Ibid., 21, 1840, 4. Zahlentheorie, § 124.

[15a]Ibid., 177-192.

Let $S'(n)$ be the sum of the odd divisors of n, and C_n be the number of all combinations without repetitions with the sum n, so that $C_7 = 5$. Then

$$S'(n) = nC_n - S'(n-1)C_1 - S'(n-2)C_2 + \ldots,$$

$$D(n) = -D(n-1) - D(n-3) - D(n-6) - \ldots, \quad D(n) = S'(n) - S(n).$$

A complicated recursion formula for $\tau(n)$ is derived from

$$\log\{(1-x)(1-x^2)^{\frac{1}{2}}(1-x^3)^{\frac{1}{2}}\ldots\} = -\sum_{n=1}^{\infty} \frac{1}{n}\tau(n)x^n.$$

Complicated recursion formulas are found for the number of integers $< m$ not factors of m, and for the sum of these integers. A recursion formula for the sum $s_r(n)$ of the divisors $\leq r$ of n is obtained by expanding

$$\log\{(1-x)(1-x^2)\ldots(1-x^r)\} = -\sum_{n=1}^{\infty} \frac{1}{n}s_r(n)x^n.$$

Jacobi[16] proved (1).

Dirichlet[17] obtained approximations to $T(n)$. An integer $s \leq n$ occurs in as many terms of this sum as there are multiples of s among $1, 2, \ldots, n$. The number of these multiples is $[n/s]$, the greatest integer $\leq n/s$. Hence

$$T(n) = \sum_{s=1}^{n} \left[\frac{n}{s}\right].$$

This sum is approximately the product of n by

$$\sum_{s=1}^{n} \frac{1}{s} = \log n + C + \frac{1}{2n} + \ldots.$$

Hence $T(n)$ is of the same order of magnitude as $n \log n$.

Let μ be the least integer $\geq \sqrt{n}$ and set $\nu = [n/\mu]$. Then if $g(x)$ is any function and $G(x) = g(1) + g(2) + \ldots + g(x)$,

$$\sum_{s=1}^{n} \left[\frac{n}{s}\right] g(s) = -\nu G(\mu) + \sum_{s=1}^{\mu} \left[\frac{n}{s}\right] g(s) + \sum_{s=1}^{\nu} G\left\{\left[\frac{n}{s}\right]\right\}.$$

In particular, if $g(x) = 1$,

$$T(n) = -\mu\nu + \sum_{s=1}^{\mu} \left[\frac{n}{s}\right] + \sum_{s=1}^{\nu} \left[\frac{n}{s}\right].$$

Giving to $[n/s]$ the approximation n/s, we see that

(7) $$T(n) = n \log_e n + (2C - 1)n + \epsilon,$$

where ϵ is of the same order of magnitude as $\sqrt{n}$.

Let $\rho(n)$ be the number of distinct prime factors > 1 of n. Then $2^{\rho(n)}$ is the number of ways of factoring n into two relatively prime factors, taking

---

[16]Jour. für Math., 32, 1846, 164; 37, 1848, 67, 73.
[17]Abhand. Ak. Wiss. Berlin, 1849, Math., 69–83; Werke, 2, 49–66. French transl., Jour. de Math., (2), 1, 1856, 353–370.

account of the order of the factors. The number of pairs of relatively prime integers ξ, η for which $\xi\eta\leqq n$ is therefore

$$\psi(n)=\sum_{j=1}^{n}2^{\rho(j)}.$$

For the preceding C and $T(n)$, it is proved that

$$T(n)=\sum_{s=1}^{t}\psi\left[\frac{n}{s^2}\right],\qquad t=[\sqrt{n}],$$

$$\psi(n)=\frac{6n}{\pi^2}\left(\log_e n+\frac{12C'}{\pi^2}+2C-1\right)+m,\qquad C'=\sum_{s=2}^{\infty}\frac{\log s}{s^2},$$

where m is of the order of magnitude of n^δ, $\delta>\gamma/2$, while γ is determined by $\Sigma s^{-\gamma}=1$ ($s=2$ to ∞). Moreover, $T(n)$ is the number of pairs of integers x, y for which $xy\leqq n$. He noted that

$$\sigma(1)+\sigma(2)+\ldots+\sigma(n)=\sum_{s=1}^{n}s\left[\frac{n}{s}\right]$$

and that the difference between this sum and $\pi^2 n^2/12$ is of an order of magnitude not exceeding $n\log_e n$.

G. H. Burhenne[18] proved by use of infinite series that

$$\tau(n)=\frac{1}{n!}\sum_{k=1}^{n}f^{(n)}(0),\qquad f(x)\equiv\frac{x^k}{1-x^k},$$

and then expressed the result as a trigonometric series.

V. Bouniakowsky[19] changed x into x^3 in (6), multiplied the result by x^4 and obtained

$$(x^{1^2}+x^{3^2}+x^{5^2}+\ldots)^4=x^4+\sigma(3)x^{12}+\ldots+\sigma(2m+1)x^{8m+4}+\ldots.$$

Thus every number $8m+4$ is a sum of four odd squares in $\sigma(2m+1)$ ways. By comparing coefficients in the logarithmic derivative, we get

$$(8)\quad (1^2-\overline{2m+1})\sigma(2m+1)+(3^2-\overline{2m-1})\sigma(2m-1)+(5^2-\overline{2m-5})\sigma(2m-5)$$
$$+\ldots=0,$$

in which the successive differences of the arguments of σ are 2, 4, 6, 8, ... For any integer N,

$$(9)\quad (1^2-N)\sigma(N)+(3^2-\overline{N-1\cdot2})\sigma(N-1\cdot2)+(5^2-\overline{N-2\cdot3})\sigma(N-2\cdot3)$$
$$+\ldots=0,$$

where $\sigma(0)$, if it occurs, means $N/6$. It is proved (p. 269) by use of Jacobi's[10] result for s^3 that

$$1+x+x^3+x^6+\ldots=\frac{p(x^2)^2}{p(x)}=(1+x)(1+x^2)(1+x^3)\ldots$$
$$(1-x^2)(1-x^4)(1-x^6)\ldots,$$

[18]Archiv Math. Phys., 19, 1852, 442–9.

[19]Mém. Ac. Sc. St. Pétersbourg (Sc. Math. Phys.), (6), 4, 1850, 259–295 (presented, 1848).
        Extract in Bulletin, 7, 170 and 15, 1857, 267–9.

where the exponents in the series are triangular numbers. Hence if we count the number of ways in which n can be formed as a sum of different terms from $1, 2, 3, \ldots$ together with different terms from $2, 4, 6, \ldots$, first taking an even number of the latter and second an odd number, the difference of the counts is 1 or 0 according as n is a triangular number or not. It is proved that

(10) $\sigma(n) + \{\sigma(2) - 4\sigma(1)\}\sigma(n-2) + \sigma(3)\sigma(n-4) + \{\sigma(4) - 4\sigma(2)\}\sigma(n-6)$
   $+ \sigma(5)\sigma(n-8) + \{\sigma(6) - 4\sigma(3)\}\sigma(n-10) + \ldots = \dfrac{n+1}{8}\sigma(n+2).$

The fact that the second member must be an integer is generalized as follows: for n odd, $\sigma(n)$ is even or odd according as n is not or is a square; for n even, $\sigma(n)$ is even if n is not a square or the double of a square, odd in the contrary case. Hence squares and their doubles are the only integers whose sums of divisors are odd.

V. Bouniakowsky[20] proved that $\sigma(N) \equiv 2 \pmod 4$ only when $N = kc^2$ or $2kc^2$, where k is a prime $4l+1$ [corrected by Liouville[30]].

V. A. Lebesgue[21] denoted by $1 + A_1 x + A_2 x^2 + \ldots$ the expansion of the mth power of $p(x)$, given by (1), and proved, by the method used by Euler for the case $m=1$, that

$$\sigma(n) + A_1 \sigma(n-1) + A_2 \sigma(n-2) + \ldots + A_{n-1}\sigma(1) + nA_n/m = 0.$$

This recursion formula gives

$$A_1 = -m, \quad A_2 = \frac{m(m-3)}{1\cdot 2}, \quad A_3 = \frac{-m(m-1)(m-8)}{1\cdot 2\cdot 3}, \quad \ldots$$

The expression for A_k was not found.

E. Meissel[22] proved that (cf. Dirichlet[17])

(11) $T(n) = \displaystyle\sum_{j=1}^{n}\left[\frac{n}{j}\right] = 2\sum_{j=1}^{\nu}\left[\frac{n}{j}\right] - \nu^2 \qquad (\nu = [\sqrt{n}]).$

J. Liouville[23] noted that by taking the derivative of the logarithm of each member of (6) we get the formula, equivalent to (8):

$$\Sigma\left\{n - \frac{5m(m+1)}{2}\right\}\sigma(2n+1-m^2-m) = 0,$$

summed for $m = 0, 1, \ldots$, the argument of σ remaining ≥ 0.

J. Liouville[24] stated that it is easily shown that

$$\Sigma d\sigma(d) = \Sigma\left(\frac{m}{d}\right)^2 \sigma(d),$$

[20]Mém. Ac. Sc. St. Pétersbourg, (6), 5, 1853, 303–322.
[21]Nouv. Ann. Math., 12, 1853, 232–4.
[22]Jour. für Math., 48, 1854, 306.
[23]Jour. de Math., (2), 1, 1856, 349–350 (2, 1857, 412).
[24]Ibid., (2), 2, 1857, 56; Nouv. Ann. Math., 16, 1857, 181; proof by J. J. Hemming, ibid., (2), 4, 1865, 547.

where d ranges over the divisors of m. He proved (p. 411) that

$$\Sigma(-1)^{m/d}d = 2\sigma(m/2) - \sigma(m).$$

J. Liouville[25] stated without proof the following formulas, in which d ranges over all the divisors of m, while $\delta = m/d$:

$$\Sigma\sigma(d) = \Sigma\delta\tau(d), \qquad \Sigma\phi(d)\tau(\delta) = \sigma(m), \qquad \Sigma\theta(d)\tau(\delta) = \{\tau(m)\}^2,$$

$$\Sigma\sigma(d)\sigma(\delta) = \Sigma d\tau(d)\tau(\delta), \qquad \Sigma\tau(d)\tau(\delta) = \Sigma'\left\{\tau\left(\frac{m}{D^2}\right)\right\}^2,$$

where $\phi(d)$ is the number of integers $< d$ and prime to d, $\theta(d)$ is the number of decompositions of d into two relatively prime factors, and the accent on Σ denotes that the summation extends only over the square divisors D^2 of m. He gave (p. 184)

$$\Sigma\theta(d) = \tau(m^2), \qquad \Sigma'\theta\left(\frac{m}{D^2}\right) = \tau(m),$$

the latter being implied in a result due to Dirichlet.[15]

Liouville[26] gave the formulas, numbered $(a), \ldots, (k)$ by him, in which $\lambda(m) = +1$ or -1, according as the total number of equal or distinct prime factors of m is even or odd:

$$\Sigma\tau(d^{2\mu}) = \tau(m)\tau(m^\mu), \qquad \Sigma\tau(d^{2\mu})\tau(\delta) = \Sigma\tau(d)\tau(d^\mu), \qquad \Sigma\phi(\delta)\sigma(d) = m\tau(m),$$

$$\Sigma\delta\sigma(d) = \Sigma d\tau(d), \qquad \Sigma\lambda(d) = 1 \text{ or } 0, \qquad \Sigma\lambda(d)\theta(d)\tau(\delta) = 1 \text{ or } 0,$$

according as m is or is not a square;

$$\Sigma\lambda(d)\theta(d)\tau(\delta^2) = 1, \qquad\qquad \Sigma\lambda(d)\theta(d) = \lambda(m), \quad \Sigma\lambda(d)\theta(\delta) = 1,$$

$$\Sigma\lambda(d)\theta(d)\theta(\delta) = 0, \qquad\qquad \Sigma\lambda(\delta)\sigma(d) = m\Sigma'\frac{1}{D^2}.$$

The number of square divisors D^2 of m is $\Sigma\lambda(d)\tau(\delta)$.

Liouville[27] gave the formulas, numbered I–XVIII by him:

$$\Sigma\tau(\delta^2)\phi(d) = \Sigma\delta\theta(d), \qquad\qquad \Sigma d\tau(\delta^2) = \Sigma\theta(\delta)\sigma(d),$$

$$\Sigma\tau(\delta^2)\lambda(d) = \tau(m), \qquad\qquad \Sigma\{\tau(\delta)\}^2\lambda(d)\theta(d) = \tau(m),$$

$$\Sigma\phi(d)\tau(\delta)\tau(\delta^\mu) = \Sigma d\tau(\delta^{2\mu}), \qquad \Sigma\theta(\delta)\tau(d)\tau(d^\mu) = \Sigma\tau(\delta^2)\tau(d^{2\mu}),$$

$$\Sigma\tau(\delta^{2\mu})\sigma(d) = \Sigma\delta\tau(d)\tau(d^\mu), \qquad \Sigma'\phi(D)\tau\left(\frac{m}{D^2}\right) = \Sigma'D\,\theta\left(\frac{m}{D^2}\right),$$

$$\Sigma'\theta(D)\tau\left(\frac{m}{D^2}\right) = \Sigma'\tau(D^2)\theta\left(\frac{m}{D^2}\right), \qquad \Sigma'\tau(D)\tau(D^\mu)\theta\left(\frac{m}{D^2}\right)\Sigma'\tau(D^{2\mu})\tau = \left(\frac{m}{D^2}\right),$$

$$\Sigma\lambda(\delta)\tau(d)\tau(d^\mu) = \Sigma'\tau\left(\frac{m^{2\mu}}{D^{4\mu}}\right), \qquad \Sigma\lambda(d)\sigma(d) = m\lambda(m)\Sigma'\frac{1}{D^2},$$

[25] Jour. de Mathématiques, (2), 2, 1857, 141–4. "Sur quelques fonctions numériques," 1st article. Here Σabc denotes $\Sigma(abc)$.

[26] Ibid., 244–8, second article of his series.

[27] Ibid., 377–384, third article of his series.

$$\Sigma'\lambda(D)\tau\left(\frac{m}{D^2}\right)=\Sigma''\theta\left(\frac{m}{e^4}\right), \qquad \Sigma\{\theta(d)\}^\mu=\tau(m^{2\mu}),$$

$$\Sigma\phi(\delta)\tau(d^{2\mu})=\Sigma\delta\{\theta(d)\}^\mu, \qquad \Sigma\{\theta(d)\}^\mu\tau(\delta)=\Sigma\tau(d^{2\mu})=\tau(m)\tau(m^{2\mu-1}),$$

$$\Sigma\tau(d^{2\mu})\theta(\delta)=\Sigma\{\theta(d)\}^\mu\tau(\delta^2), \qquad \Sigma\tau(d^{2\mu})\lambda(\delta)=\Sigma'\left\{\theta\left(\frac{m}{D^2}\right)\right\}^\mu,$$

where, in Σ'', e ranges over the biquadrate divisors of m.

Liouville[28] gave the formula

$$\Sigma\{\tau(d)\}^3=\{\Sigma\tau(d)\}^2,$$

which implies that if $2m$ (m odd) has no factor of the form $4\mu+3$ and if we find the number of decompositions of each of its even factors as a sum of two odd squares, the sum of the cubes of the numbers of decompositions found will equal thesquare of their sum. Thus, for $m=25$,

$$50=1^2+7^2=7^2+1^2=5^2+5^2, \qquad 10=3^2+1^2=1^2+3^2, \qquad 2=1+1,$$

whence $3^3+2^3+1^3=6^2$.

Liouville[29] stated that, if $a, b, \ldots$ are relatively prime in pairs,

$$\sigma_n(ab\ldots)=\sigma_n(a)\sigma_n(b)\ldots,$$

while if $p, q, \ldots$ are distinct primes,

$$\sigma_n(p^\alpha q^\beta\ldots)=\frac{p^{n(\alpha+1)}-1}{p^n-1}\cdot\frac{q^{n(\beta+1)}-1}{q^n-1}\cdots$$

He stated the formulas

$$\Sigma\sigma_\mu(d)\phi(\delta)=m\sigma_{\mu-1}(m), \qquad \Sigma d^\mu\sigma_\nu(\delta)=\Sigma d^\nu\sigma_\mu(\delta),$$

$$\Sigma\lambda(d)\tau(d^2)\sigma_\mu(\delta)=\Sigma d^\mu\tau(\delta)\lambda(\delta), \qquad \Sigma d^\mu\sigma_\mu(\delta)=\Sigma d^\mu\tau(d),$$

$$\Sigma d^\mu\sigma_\mu(d)=\Sigma\delta^{2\mu}\sigma_\mu(d), \qquad \Sigma d^\mu\sigma_{3\mu}(\delta)=\Sigma d^\mu\sigma_{2\mu}(d),$$

$$\Sigma d^\mu\sigma_{\nu+t}(d)\sigma_\nu(\delta)=\Sigma d^\nu\sigma_{\mu+t}(d)\sigma_\mu(\delta), \qquad \Sigma\lambda(d)\sigma_\mu(\delta)=\Sigma'\left(\frac{m}{D^2}\right)^\mu,$$

$$\Sigma\tau(d^{2\mu})\sigma_\nu(\delta)=\Sigma d^\nu\tau(\delta)\tau(\delta^\mu), \qquad \Sigma\{\theta(d)\}^\nu\sigma_\mu(\delta)=\Sigma d^\mu\tau(\delta^{2\nu}),$$

and various special cases of them. To the seventh of these Liouville[30] later gave several forms, one being the case $\rho=0$ of

$$\Sigma d^{\mu-\nu}\sigma_{\nu+\tau}(d)\sigma_{\mu+\rho}(\delta)=\Sigma d^{\mu-\nu}\sigma_{\nu+\rho}(d)\sigma_{\mu+\tau}(\delta),$$

and proved (p. 84) the known theorem that $\sigma(m)$ is odd if and only if m is a square or the double of a square [cf. Bouniakowsky,[19] end]. He proved that $\sigma(N)\equiv2$ (mod 4) if and only if N is the product of a prime $4\lambda+1$, raised to the power $4l+1$ ($l\geqq0$), by a square or by the double of a square not divis-

[28]Jour. de Mathématiques, (2), 2, 1857, 393–6; Comptes Rendus Paris, 44, 1857, 753.
[29]Ibid., 425–432, fourth article of his series.
[30]Ibid., (2), 3, 1858, 63.

ible by the prime $4\lambda+1$. The condition given by Bouniakowsky[20] is necessary, but not sufficient. Also,

$$\sigma_3(m) = \sum_{j=1}^{m} \sigma(2j-1)\sigma(2m-2j+1) \qquad (m \text{ odd}).$$

J. Liouville's series of 18 articles, "Sur quelques formules...utiles dans la théorie des nombres," in Jour. de Math., 1858–1865, involve the function σ_n, but will be reported on in volume II of this History in connection with sums of squares. A paper of 1860 by Kronecker will be considered in connection with one by Hermite.[70]

C. Traub[31] investigated the number $(N; M, t)$ of divisors T of N which are $\equiv t \pmod{M}$, where M is prime to t and N. Let $a, b, \ldots, l$ be the integers $< M$ and prime to M; let them belong modulo M to the respective exponents $a', b', \ldots, l'$; let m be a common multiple of the latter. Since any prime factor of N is of the form $Mx+k$, where $k=a, \ldots, l$, any T is congruent to

$$a^A b^B \ldots l^L \equiv t \pmod{M}, \qquad 0 \leqq A < a', \ldots, 0 \leqq L < l'.$$

Let $A', \ldots, L'$ be one of the n sets of exponents satisfying these conditions. If P is a primitive mth root of unity, the function

$$\psi = \frac{1}{a' \ldots l'} \Sigma P^e, \qquad e = (A-A')am/a' + \ldots + (L-L')\lambda m/l',$$

summed for all sets $0 \leqq a < a', \ldots, 0 \leqq \lambda < l'$, has the property that $\psi = 1$ if $A \equiv A' \pmod{a'}, \ldots, L \equiv L' \pmod{l'}$ simultaneously, while $\psi = 0$ in all other cases. Thus $(N; M, t) = \Sigma\Sigma\psi$, where one summation refers to the n sets mentioned, while the other refers to the various divisors T of N. This double sum is simplified.

[The properties found (pp. 278–294) for the set of residues modulo M of the products of powers of $a, \ldots, l$ may be deduced more simply from the modern theory of commutative groups.]

V. Bouniakowsky[32] considered the series

$$\psi(x) = \sum_{n=1}^{\infty} \frac{1}{n^x}, \qquad \{\psi(x)\}^m = \sum_{n=1}^{\infty} \frac{z_{n,m}}{n^x}.$$

By forming the product of $\psi(x)^{m-1}$ by $\psi(x)$, he proved that $z_{n,2}$ is the number $N_0(n) = \tau(n)$ of the divisors of n, and $z_{n,m}$ equals

$$N_{m-2}(n) = \Sigma N_{m-3}(d),$$

where (and below) d ranges over the divisors of n. Also,

$$\psi(x)\psi(x-1) = \sum_{n=1}^{\infty} \frac{\sigma(n)}{n^x}.$$

From $\psi(x)^i \psi(x-1)^j$ for $(i, j) = (2, 1), (2, 2), (1, 2)$, he derived the first and fourth formulas of Liouville's[25] first article and the fourth of his[26] second article. He extended these three formulas to sums of powers of the divisors

[31] Archiv Math. Phys., 37, 1861, 277–345.
[32] Mém. Ac. Sc. St. Pétersbourg, (7), 4, 1862, No. 2, 35 pp.

and proved the second formula in Liouville's first article and the first two summation formulas of Liouville.[29]   He proved

$$\Sigma\sigma(d) = \Sigma N_1(d)\phi\left(\frac{n}{d}\right),$$

$$\sum_{x=1}^{\sigma} \tau(2x-1) = 2\sigma - 1 + \sum_{x=1}^{k}\left[\frac{\sigma-1-x}{2x+1}\right], \qquad k = \left[\frac{\sigma-2}{3}\right],$$

$$\tau(2\sigma-1) = 2 + \eta + \sum_{x=1}^{l}\left\{\left[\frac{\sigma-1-x}{2x+1}\right] - \left[\frac{\sigma-2-x}{2x+1}\right]\right\}, \qquad l = \left[\frac{\sigma}{3}\right] - 1,$$

where $\eta = 1$ or 0 according as $2\sigma - 1$ is divisible by 3 or not.   The last two were later generalized by Gegenbauer.[80]

E. Lionnet[33] proved the first two formulas of Liouville.[29]

J. Liouville[34] noted that, if q is divisible by the prime a,

$$\sigma_\mu(aq) + a^\mu\sigma_\mu\left(\frac{q}{a}\right) = (a^\mu + 1)\sigma_\mu(q).$$

C. Sardi[35] denoted by A_n the coefficient of x^n in Jacobi's[10] series for s^3, so that $A_n = 0$ unless n is a triangular number.   From that series he got

$$\sum_p (-1)^p(2p+1)\sigma\{n - p(p+1)/2\} = (-1)^{(t+1)/2}tn/3 \text{ or } 0 \qquad (t = \sqrt{1+8n}),$$

according as n is or is not a triangular number, and

$$\frac{n}{3}A_n + A_{n-1}\sigma(1) + \ldots + A_1\sigma(n-1) + A_0\sigma(n) = 0.$$

This recursion formula determines A_n in terms of the σ's, or $\sigma(n)$ in terms of the A's.   In each case the values are expressed by means of determinants of order n.

M. A. Andreievsky[36] wrote $N_{4h\pm1}$ for the number of the divisors of the form $4h\pm1$ of $n = a^\alpha b^\beta \ldots$, where $a, b, \ldots$ are distinct primes.   We have

$$N_{4h+1} - N_{4h-1} = \Sigma\left(\frac{-1}{d}\right) = \sum_{a'=0}^{\alpha}\left(\frac{-1}{a}\right)^{a'} \cdot \sum_{\beta'=0}^{\beta}\left(\frac{-1}{b}\right)^{\beta'} \ldots,$$

where d ranges over all the divisors of n and the symbols are Legendre's. Evidently

$$\sum_{a'=0}^{\alpha}\left(\frac{-1}{a}\right)^{a'} = a+1 \text{ if } a = 4l+1,$$

$$= 0 \text{ or } 1 \text{ if } a = 4l-1,$$

according as a is odd or even.   Hence, if any prime factor $4l-1$ of n occurs to an odd power, we have $N_{4h+1} = N_{4h-1}$.   Next, let

$$n = p_1^{a_1}p_2^{a_2}\ldots q_1^{2\beta_1}q_2^{2\beta_2}\ldots,$$

where each p_i is a prime of the form $4l+1$, each q_i of the form $4l-1$.   Then

$$N_{4h+1} - N_{4h-1} = (a_1+1)(a_2+1)\ldots = \tau\left(\frac{n}{D^2}\right), \qquad D = q_1^{\beta_1}q_2^{\beta_2}\ldots.$$

[33]Nouv. Ann. Math., (2), 7, 1868, 68–72.
[34]Jour. de math., (2), 14, 1869, 263–4.
[35]Giornale di Mat., 7, 1869, 112–5.
[36]Mat. Sbornik (Math. Soc. Moscow), 6, 1872–3, 97–106 (Russian).

The sum of the N's is $\tau(n) = \tau(D^2)\tau(n/D^2)$. Hence

$$\frac{N_{4h+1}}{N_{4h-1}} = \frac{\tau(D^2)+1}{\tau(D^2)-1},$$

which is never an integer other than 1 or 2 when n is odd. If it be 2, $\tau(D^2) = 3$ requires that D be a prime. Similarly, for Legendre's symbol $(2/a)$,

$$N_{8h\pm1} - N_{8h\pm3} = \sum_{a'=0}^{a} \left(\frac{2}{a}\right)^{a'} \cdot \sum_{\beta'=0}^{\beta} \left(\frac{2}{b}\right)^{\beta'} \cdots$$

is zero if any prime factor $8l\pm3$ of n occurs to an odd power, but is $\Pi(n_i+1)$ if in n each p_i is a prime $8l\pm1$ and each q_i a prime $8l\pm3$. For n odd, $N_{8h\pm1}/N_{8h\pm3}$ can not be an integer other than 1 or 2; if 2, D is a prime.

F. Mertens[37] proved (11). He considered the number $\nu(n)$ of divisors of n which are not divisible by a square >1. Evidently $\nu(n) = 2^\rho$, where ρ is the number of distinct prime factors of n. If $\mu(n)$ is zero when n has a square factor >1 and is $+1$ or -1 according as n is a product of an even or odd number of distinct primes, $\nu(n) = \Sigma\mu^2(d)$, where d ranges over the divisors of n. Also,

$$\sum_{k=1}^{n}\nu(k) = \sum_{k=1}^{t}\mu(k)T\left(\frac{n}{k^2}\right), \qquad t = [\sqrt{n}].$$

He obtained Dirichlet's[17] expression $\psi(n)$ for this sum, finding for m a limit depending on C and n, of the order of magnitude of $\sqrt{n}\log_e n$.

E. Catalan[37a] noted that $\Sigma\sigma(i)\sigma(j) = 8\sigma_3(n)$ where $i+j = 4n$. Also, if i is odd, $\sigma(i)$ equals the sum of the products two at a time of the E's of the odd numbers whose sum is $2i$, where E denotes the excess of the number of divisors $4\mu+1$ over the number of divisors $4\mu-1$.

H. J. S. Smith[38] proved that, if $m = p_1^{a_1}p_2^{a_2}\cdots$,

$$\sigma_s(m) - \Sigma\sigma_s\left(\frac{m}{p_1}\right) + \Sigma\sigma_s\left(\frac{m}{p_1p_2}\right) - \ldots = m^s.$$

For, if $P = 1+p^s+\ldots+p^{as}$, $P' = 1+p^s+\ldots+p^{(a-1)s}$, then

$$\sigma_s(m) = P_1P_2\ldots, \qquad \sigma_s\left(\frac{m}{p_1}\right) = P_1'P_2\ldots, \qquad \sigma_s\left(\frac{m}{p_1p_2}\right) = P_1'P_2'P_3\ldots,$$

and the initial sum equals $(P_1 - P_1')(P_2 - P_2')\ldots = m^s$.

J. W. L. Glaisher[39] stated that the excess of the sum of the reciprocals of the odd divisors of a number over that for the even divisors is equal to the sum of the reciprocals of the divisors whose complementary divisors are odd. The excess of the sum of the divisors whose complementary divisors are odd over that when they are even equals the sum of the odd divisors.

G. Halphen[40] obtained the recursion formula

$$\sigma(n) = 3\sigma(n-1) - 5\sigma(n-3) + \ldots - (-1)^x(2x+1)\sigma\left\{n - \frac{x(x+1)}{2}\right\} + \ldots,$$

[37]Jour. für Math., 77, 1874, 291–4.

[37a]Recherches sur quelques produits indéfinis, Mém. Ac. Roy. Belgique, 40, 1873, 61–191. Extract in Nouv. Ann. Math., (2), 13, 1874, 518–523.

[38]Proc. London Math. Soc., 7, 1875–6, 211.

[39]Messenger Math., 5, 1876, 52.

[40]Bull. Soc. Math. France, 5, 1877, 158.

where, if n is of the form $x(x+1)/2$, $\sigma(0)$ is to be taken to be $n/3$ [Glaisher[63, 67]]. The proof follows from the logarithmic derivative of Jacobi's[10] expression for s^3, as in Euler's[5] proof of (2).

Halphen[41] formed for an odd function $f(z)$ the sum of

$$(-1)^x f\left(\frac{a-x^2}{y}+x\right),$$

x taking all integral values between the two square roots of a, and y ranging over all positive odd divisors of $a-x^2$. This sum is

$$(-1)^{a+1}\sqrt{a}\,f(\sqrt{a})$$

if a is a square, zero if a is not a square. Taking $f(z)=z$, we get a recursion formula for the sum of those divisors d of x for which x/d is odd [see the topic Sums of Squares in Vol. II of this History]. Taking $f(z)=a^z-a^{-z}$, we get a recursion formula for the number of odd divisors $<a/m$ of a. A generalization of (2) gives a recursion formula for the sum of the divisors of the forms $2nk$, $n(2k+1)\doteq m$, with fixed n, m.

E. Catalan[42] denoted the square of (1) by $1+L_1x+\ldots+L_nx^n+\ldots$ Thus

$$\sigma(n)+L_1\sigma(n-1)+L_2\sigma(n-2)+\ldots+L_{n-1}=-\frac{n}{2}L_n,$$

$$L_n-L_{n-1}-L_{n-2}+L_{n-5}+L_{n-7}-\ldots=0 \text{ or } (2\lambda+1)(-1)^\lambda,$$

according as n is not or is of the form $\lambda(\lambda+1)/2$. In view of the equality of (3) and (4) and the fact that $1/p=\Sigma\psi(n)x^n$, where $\psi(n)$ is the number of partitions of n into equal or distinct positive integers, he concluded that

$$\sigma(n)=\psi(n-1)+2\psi(n-2)-5\psi(n-5)-7\psi(n-7)+12\psi(n-12)+\ldots.$$

J. W. L. Glaisher[43] noted that, if $\theta(n)$ is the excess of the sum of the odd divisors of n over the sum of the even divisors,

$$\theta(n)+\theta(n-1)+\theta(n-3)+\theta(n-6)+\ldots=0,$$

where $1, 3, 6,\ldots$ are the triangular numbers, and $\theta(n-n)=-n$.

E. Cesàro[44] denoted by s_n the sum of the residues obtained by dividing n by each integer $<n$, and stated that

$$s_n+\sigma(1)+\sigma(2)+\ldots+\sigma(n)=n^2.$$

E. Catalan[45] proved the equivalent result that the sum of the divisors of $1,\ldots,n$ equals the sum of the greatest multiples, not $>n$, of these numbers.

Catalan[46] stated that, if $\phi(a, n)$ is the greatest multiple $\leq n$ of a,

$$\sigma(n)=\sum_{a=1}^{n}\{\phi(a, n)-\phi(a, n-1)\}.$$

[41]Bull. Soc. Math. France, 6, 1877–8, 119–120, 173–188.
[42]Assoc. franç. avanc. sc., 6, 1877, 127–8. Cf. Catalan.[57a]
[43]Messenger Math., 7, 1877–8, 66–7.
[44]Nouv. Corresp. Math., 4, 1878, 329; 5, 1879, 22; Nouv. Ann. Math., (3), 2, 1883, 289; 4, 1885, 473.
[45]Ibid., 5, 1879, 296–8; stated, 4, 1879, ex. 447.
[46]Ibid., 6, 1880, 192.

Radicke (p. 280) gave an easy proof and noted that if we take $n=1,\ldots, m$ and add, we get the result by E. Lucas[47]

$$\sigma(1)+\ldots+\sigma(m)=\phi(1, m)+\ldots+\phi(m, m).$$

J. W. L. Glaisher[48] stated that if $f(n)$ is the sum of the odd divisors of n and if $g(n)$ is the sum of the even divisors of n, and $f(0)=0$, $g(0)=n$, then

$$f(n)+f(n-1)+f(n-3)+f(n-6)+f(n-10)+\ldots$$
$$=g(n)+g(n-1)+g(n-3)+\ldots.$$

Chr. Zeller[49] proved (11).

R. Lipschitz[50] wrote $G(t)$ for $\sigma(1)+\ldots+\sigma(t)$, $D(t)$ for $(t^2+t)/2$, and $\Phi(t)$ for $\phi(1)+\ldots+\phi(t)$, using Euler's $\phi(t)$. Then if 2, 3, 5, 6, $\ldots$ are the integers not divisible by a square >1,

$$T(n)-T\left[\frac{n}{2}\right]-T\left[\frac{n}{3}\right]-T\left[\frac{n}{5}\right]+\ldots=n,$$

$$G(n)-2G\left[\frac{n}{2}\right]-3G\left[\frac{n}{3}\right]-5G\left[\frac{n}{5}\right]+\ldots=n,$$

$$D(n)-D\left[\frac{n}{2}\right]-D\left[\frac{n}{3}\right]-D\left[\frac{n}{5}\right]+\ldots=\Phi(n),$$

the sign depending on the number of prime factors of the denominator. He discussed (pp. 985-7) Dirichlet's[17] results on the mean of $\tau(n)$, $\sigma(n)$, $\phi(n)$.

A. Berger[51] proved by use of gamma functions that the mean of the sum of the divisors d of n is $\pi^2 n/6$, that of $\Sigma\ d/2^d$ is 1, that of $\Sigma 1/d$ is $\pi^2/6$.

G. Cantor[51a] gave the second formula of Liouville[25] and his[26] third.

A. Piltz[52] considered the number $T_k(n)$ of sets of positive integral solutions of $u_1\ldots u_k=n$, where differently arranged u's give different sets. Thus $T_1(n)=1$, $T_2(n)=\tau(n)$. If σ is the real part of the complex number s, and n^s denotes $e^{s\log n}$ for the real value of the logarithm, he proved that

$$t_k(x; s)=\sum_{n=1}^{x}\frac{T_k(n)}{n^s}=x^{1-s}\sum_{m=0}^{k}b_m\log^m x+b+O(x^l)+O(x^l\log^{k-2}x),$$

where $l=1-\sigma-1/k$, and the b's are constants, $b_k=0$ for $s\neq 1$; while $O(f)$ is[90] of the order of magnitude of f. Taking $s=0$, we obtain the number $\Sigma T_k(n)$ of sets of positive integral solutions of $u_1\ldots u_k\leqq x$.

H. Ahlborn[53] treated (11).

E. Cesàro[54] noted that the mean of the difference between the number of odd and number of even divisors of any integer is $\log 2$; the limit for

---

[47]Nouv. Corresp. Math., 5, 1879, 296.

[48]Nouv. Corresp. Math., 5, 1879, 176.

[49]Göttingen Nachrichten, 1879, 265.

[50]Comptes Rendus Paris, 89, 1879, 948-50. Cf. Bachmann[20] of Ch. XIX.

[51]Nova Acta Soc. Sc. Upsal., (3), 11, 1883, No. 1 (1880). Extract by Catalan in Nouv. Corresp. Math., 6, 1880, 551-2. Cf. Gram.[54a]

[51a]Göttingen Nachr., 1880, 161; Math. Ann., 16, 1880, 586.

[52]Ueber das Gesetz, nach welchem die mittlere Darstellbarkeit der natürlichen Zahlen als Produkte einer gegebenen Anzahl Faktoren mit der Grösse der Zahlen wächst. Diss., Berlin, 1881.

[53]Progr., Hamburg, 1881.

[54]Mathesis, 1, 1881, 99-102. Nouv. Ann. Math., (3), 1, 1882, 240; 2, 1883, 239, 240. Also Cesàro,[61] 113-123, 133.

$n = \infty$ of $T(n)/(n \log n)$ is 1; cf. (7); the mean of $\Sigma(d+p)^{-1}$ is $(1+1/2+\ldots +1/p)/p$. As generalizations of Berger's[51] results, the mean of $\Sigma d/p^d$ is $1/(p-1)$; the mean of the sum of the rth powers of the divisors of n is $n^r \zeta(r+1)$ and that of the inverses of their rth powers is $\zeta(r+1)$, where

$$(12) \qquad \zeta(s) = \sum_{n=1}^{\infty} 1/n^s.$$

J. W. L. Glaisher[55] proved the last formula of Catalan[42] and

$$\sigma(n) - \sigma(n-4) - \sigma(n-8) + \sigma(n-20) + \sigma(n-28) - \ldots$$
$$= Q(n-1) + 3Q(n-3) - 6Q(n-6) - 10Q(n-10) + \ldots,$$

where $Q(n)$ is the number of partitions of n without repetitions, and 4, 8, 20, ... are the quadruples of the pentagonal numbers. He gave another formula of the latter type.

R. Lipschitz,[56] using his notations,[50] proved that

$$T(n) - \Sigma T \left[\frac{n}{a}\right] + \Sigma T \left[\frac{n}{ab}\right] - \ldots = n + \Sigma \left[\frac{n}{P}\right],$$

$$G(n) - \Sigma a G\left[\frac{n}{a}\right] + \Sigma a b G\left[\frac{n}{ab}\right] - \ldots = n + \Sigma P\left[\frac{n}{P}\right],$$

$$D(n) - \Sigma D \left[\frac{n}{a}\right] + \Sigma D \left[\frac{n}{ab}\right] - \ldots = \Phi(n) + \Sigma \Phi\left[\frac{n}{P}\right],$$

where P ranges over those numbers $\leq n$ which are composed exclusively of primes other than given primes $a, b, \ldots$, each $\leq n$.

Ch. Hermite[57] proved (11) very simply.

R. Lipschitz[58] considered the number $\tau_s(t)$ of those divisors of t which are exact sth powers of integers and proved that

$$\sum_{t=1}^{n} \tau_s(t) = \sum_{j=1}^{p} \left[\frac{n}{j^s}\right] = \sum_{j=1}^{n} \left[\frac{n^{1/s}}{j^{1/s}}\right]$$
$$= -\mu\nu + \sum_{x=1}^{\mu} \left[\frac{n}{x^s}\right] + \sum_{y=1}^{\nu} \left[\frac{n^{1/s}}{y^{1/s}}\right] = -\mu^2 + \sum_{x=1}^{\mu} \left[\frac{n}{x^s}\right] + \sum_{y=1}^{\mu} \left[\frac{n^{1/s}}{y^{1/s}}\right],$$

where p^s is the largest sth power $\leq n$, and $\nu = [n/\mu^s]$. The last expression, found by taking $\mu = [n^{(1+s)^{-1}}]$, gives a generalization of (11).

T. J. Stieltjes[59] proved (7) by use of definite integrals.

E. Cesàro[60] proved (7) arithmetically and (11).

E. Cesàro[61] proved that, if d ranges over the divisors of n, and δ over those of x,

$$(13) \qquad \Sigma G(d) f\left(\frac{n}{d}\right) = \Sigma g(d) F\left(\frac{n}{d}\right), \qquad F(x) \equiv \Sigma f(\delta), \qquad G(x) \equiv \Sigma g(\delta).$$

Taking $g(x) = 1, f(x) = x, \phi(x), 1/x$, we get the first two formulas of Liouville[25]

[55]Messenger Math., 12, 1882–3, 169–170.
[56]Comptes Rendus Paris, 96, 1883, 327–9.
[57]Acta Math., 2, 1883, 299–300.
[58]Ibid., 301–4.
[59]Comptes Rendus Paris, 96, 1883, 764–6.
[60]Ibid., 1029.
[61]Mém. Soc. Sc. Liège, (2), 10, 1883, Mém. 6, pp. 26–34.

and the fourth of Liouville.[26] Taking $g=x$, $f=\phi$, we get the third formula of Liouville.[26] For $g=1/x$, $f=\phi$, we get

$$\Sigma d\phi(d)\sigma\left(\frac{n}{d}\right) = \Sigma d^2.$$

For $g=\phi$ or x^r, $f=x^s$, we get the first two of Liouville's[29] summation formulas. If $\pi(x)$ is the product of the negatives of the prime factors $\neq 1$ of x,

$$\Sigma\pi(d)\phi(d)\tau\left(\frac{n}{d}\right)\frac{1}{d^2}=\frac{\sigma(n)}{n}, \qquad \Sigma\pi\left(\frac{n}{d}\right)\phi\left(\frac{n}{d}\right)\sigma(d)d^2=n\Sigma d^2,$$

$$\Sigma\pi(d)\phi(d)\sigma\left(\frac{n}{d}\right)\frac{1}{d}=\tau(n), \qquad \Sigma\pi(d)\phi(d)\frac{1}{d^3}=\frac{1}{n^2}\Sigma d\phi(d).$$

Further specializations of (13) and of the generalization (p. 47)

$$\Sigma G(d)f\left(\frac{n}{d}\right)=\Sigma F(d)g\left(\frac{n}{d}\right), \qquad F(x)\equiv\Sigma\psi(\delta)f\left(\frac{x}{\delta}\right), \qquad G(x)\equiv\Sigma\psi(\delta)g\left(\frac{x}{\delta}\right),$$

led Cesàro (pp. 36–59) to various formulas of Liouville[25-27] and many similar ones. It is shown (p. 64) that

$$\sum_{n=1}^{\infty}\frac{F(n)}{n^m}=\zeta(m)\sum_{n=1}^{\infty}\frac{f(n)}{n^m},$$

for ζ and F as in (12), (13). For $f(n)=\phi(n)$, we have the result quoted under Cesàro[57] in Ch. V. For $f(n)=1$ and n^k, $m-k>1$,

$$\sum_{n=1}^{\infty}\frac{\tau(n)}{n^m}=\zeta^2(m), \qquad \Sigma_n\frac{\sigma_k(n)}{n^m}=\zeta(m)\zeta(m-k).$$

If (n, j) is the g. c. d. of n, j, then (pp. 77–86)

$$\sum_{j=1}^{n}\frac{n}{(n, j)}=2\Sigma\sigma(d)-1, \qquad n\tau(n)=\Sigma\sigma(n, j), \qquad \sigma(n)=\Sigma\tau(n, j),$$

$$\sum_{j=1}^{n}\sigma_k(n, j)=n\sigma_{k-1}(n), \qquad \Sigma j\sigma(n, j)=\frac{n}{2}\{n\tau(n)+\sigma(n)\}.$$

If in the second formula of Liouville[25] we take $m=1,\ldots, n$ and add, we get

$$\sum_{j=1}^{n}\phi(j)T\left[\frac{n}{j}\right]=\sum_{j=1}^{n}\sigma(j).$$

Similarly (pp. 97–112) we may derive a relation in $[x]$ from any given relation involving all the divisors of x, or any set of numbers defined by x, such as the numbers $a, b,\ldots$ for which $x-a^2$, $x-b^2,\ldots$ are all squares. Formula (7) is proved (pp. 124–8). It is shown (pp. 135–143) that the mean of the sum of the inverses of divisors of n which are multiples of k is $\pi^2/(6k^2)$; the excess of the number of divisors $4\mu+1$ over the number of divisors $4\mu+3$ is in mean $\pi/4$, and that for $4\mu+2$ and 4μ is $\frac{1}{2}\log 2$; the mean of the sum of the inverses of the odd divisors of any integer is $\pi^2/8$; the mean is found of various functions of the divisors. The mean (p. 172) of the number of divisors of an integer which are mth powers is $\zeta(m)$, and hence is $\pi^2/6$ if

$m = 2$. The mean (pp. 216–9) of the number of divisors of the form $a\mu + r$ of n is, for $r > 0$,

$$\frac{1}{r} + \frac{1}{a} \left\{ \log n/a + 2C - \int_0^1 \frac{1 - x^{r/a}}{1 - x} dx \right\}$$

(cf. pp. 341–2 and, for $a = 4, 6$, pp. 136–8), while several proofs (also, p. 134) are given of the known result that the number of divisors of n which are multiples of a is in mean

$$\frac{1}{a} (\log n/a + 2C).$$

If (pp. 291–2) a ranges over the integers for which $[2n/d]$ is odd, the number (sum) of the a's is the excess of the number (sum) of the divisors of $n+1, n+2, \ldots, 2n$ over that of $1, \ldots, n$; the means are $n \log 4$ and $\pi^2 n^2/6$. If (pp. 294–9) k ranges over the integers for which $[n/k]$ is odd, the number of the k's is the excess of the number of odd divisors of $1, \ldots, n$ over the number of their even divisors, and the sum of the k's is the sum of the odd divisors of $1, \ldots, n$; also

$$\Sigma \phi(k) = q^2, \qquad q = \left[\frac{n+1}{2}\right].$$

Several asymptotic evaluations by Cesàro are erroneous. For instance, for the functions $\lambda(n)$ and $\mu(n)$, defined by Liouville[26] and Mertens,[37] Cesàro (p. 307, p. 157) gave as the mean values $6/\pi^2$ and $36/\pi^4$, whereas each is zero.[62]

J. W. L. Glaisher[63] considered the sum $\Delta(n)$ of the odd divisors of n. If $n = 2^r m$ (m odd), $\Delta(n) = \sigma(m)$. The following theorems were proved by use of series for elliptic functions:

$$\Delta(1)\Delta(2n-1) + \Delta(3)\Delta(2n-3) + \Delta(5)\Delta(2n-5) + \ldots + \Delta(2n-1)\Delta(1)$$

equals the sum of the cubes of those divisors of n whose complementary divisors are odd. The sum of the cubes of all divisors of $2n+1$ is

$$\Delta(2n+1) + 12\{\Delta(1)\Delta(2n) + \Delta(2)\Delta(2n-1) + \ldots + \Delta(2n)\Delta(1)\}.$$

If A, B, C are the sums of the cubes of those divisors of $2n$ which are respectively even, odd, with odd complementary divisor,

$$2\Delta(2n) + 24\{\Delta(2)\Delta(2n-2) + \Delta(4)\Delta(2n-4) + \ldots + \Delta(2n-2)\Delta(2)\}$$
$$= \frac{1}{3}(2A - 2B - C) = \frac{1}{7}(3 \cdot 2^{3r} - 10)B$$

if $2n = 2^r m$ (m odd). Halphen's formula[40] is stated on p. 220. Next,

$$n\sigma(2n+1) + (n-5)\sigma(2n-1) + (n-15)\sigma(2n-5)$$
$$+ (n-30)\sigma(2n-11) + \ldots = 0,$$

[62]H. v. Mangoldt, Sitzungsber. Ak. Wiss. Berlin, 1897, 849, 852; E. Landau, Sitzungsber. Ak. Wiss. Wien, 112, IIa, 1903, 537.
[63]Quar. Jour. Math., 19, 1883, 216–223.

in which the differences between the arguments of σ in the successive terms are 2, 4, 6, 8,..., and those between the coefficients are 5, 10, 15,..., while $\sigma(0) = 0$. Finally, there is a similar recursion formula for $\Delta(n)$.

Glaisher[64] proved his[43] recursion formula for $\theta(n)$, gave a more complicated one and the following for $\sigma(n)$:

$$\sigma(n) - 2\{\sigma(n-1) + \sigma(n-2)\} + 3\{\sigma(n-3) + \sigma(n-4) + \sigma(n-5)\} - \ldots$$
$$+ (-1)^{r-1} r\{\ldots + \sigma(1)\} = (-1)^s (s^3 - s)/6,$$

where $s = r$ unless $r\sigma(1)$ is the last term of a group, in which case, $s = r+1$. He proved Jacobi's[11] statement and concluded from the same proof that $E(n) = \Pi E(n_i)$ if $n = \Pi n_i$, the n's being relatively prime. It is evident that $E(p^r) = r+1$ if p is a prime $4m+1$, while $E(p^r) = 1$ or 0 if p is a prime $4m+3$, according as r is even or odd. Also $E(2^r) = 1$. Hence we can at once evaluate $E(n)$. He gave a table of the values of $E(n)$, $n = 1, \ldots, 1000$. By use of elliptic functions he found the recursion formulæ

$$E(n) - 2E(n-4) + 2E(n-16) - 2E(n-36) + \ldots = 0 \text{ or } (-1)^{(\sqrt{n}-1)/2}\sqrt{n},$$

for n odd, according as n is not or is a square; for any n.

$$E(n) - E(n-1) - E(n-3) + E(n-6) + E(n-10) - \ldots$$
$$= 0 \text{ or } (-1)^n\{(-1)^{(t-1)/2}t - 1\}/4, \qquad t \equiv \sqrt{8n+1},$$

according as n is not or is a triangular number 1, 3, 6, 10,.... He gave recursion formulæ for

$$S(2n) = E(2) + E(4) + \ldots + E(2n),$$
$$S(2n-1) = E(1) + E(3) + \ldots + E(2n-1).$$

The functions E, S, θ, σ are expressed as determinants.

J. P. Gram[64a] deduced results of Berger[51] and Cesàro.[54]

Ch. Hermite[65] expressed $\sigma(1) + \sigma(3) + \ldots + \sigma(2n-1)$, $\sigma(3) + \sigma(7) + \ldots + \sigma(4n-1)$ and $\sigma(1) + \sigma(5) + \ldots + \sigma(4n+1)$ as sums of functions
$$E_2(x) = \{[x]^2 + [x]\}/2.$$

Chr. Zeller[66] gave the final formula of Catalan.[42]

J. W. L. Glaisher[67] noted that, if in Halphen's[40] formula, n is a triangular number, $\sigma(n-n)$ is to be given the value $n/3$; if, however, we suppress the undefined term $\sigma(0)$, the formula is

$$\sigma(n) - 3\sigma(n-1) + 5\sigma(n-3) - \ldots = 0 \text{ or } (-1)^{r-1}(1^2 + 2^2 + \ldots + r^2),$$

according as n is not a triangular number or is the triangular number $r(r+1)/2$. He reproduced two of his[63, 64, 76] own recursion formulas for $\sigma(n)$ (with ψ for σ in two) and added

$$\sigma(n) - \{\sigma(n-2) + \sigma(n-3) + \sigma(n-4)\} + \{\sigma(n-7) + \sigma(n-8) + \sigma(n-9)$$
$$+ \sigma(n-10) + \sigma(n-11)\} - \{\sigma(n-15) + \ldots\} + \ldots = A - B,$$

[64]Proc. London Math. Soc., 15, 1883–4, 104–122.
[64a]Det K. Danske Vidensk. Selskabs Skrifter, (6), 2 1881–6 (1884), 215–220 296.
[65]Amer. Jour. Math., 6, 1884, 173–4.
[66]Acta Math., 4, 1884, 415–6.
[67]Proc. Cambr. Phil. Soc., 5, 1884, 108–120.

where A and B denote the number of positive and negative terms respectively, not counting $\sigma(0) = n$ as a term;

$$n\sigma(n) + 2\{(n-2)\sigma(n-2) + (n-4)\sigma(n-4)\}$$
$$+ 3\{(n-6)\sigma(n-6) + (n-8)\sigma(n-8) + (n-10)\sigma(n-10)\} + \ldots$$
$$= \sigma(n) + (1^2 + 3^2)\{\sigma(n-2) + \sigma(n-4)\}$$
$$+ (1^2 + 3^2 + 5^2)\{\sigma(n-6) + \sigma(n-8) + \sigma(n-10)\} + \ldots \qquad (n \text{ odd}).$$

He reproduced his[64] formulas for $\theta(n)$ and $E(n)$. He announced (ibid., p. 86) the completion of tables of the values of $\phi(n)$, $\tau(n)$, $\sigma(n)$ up to $n = 3000$, and inverse tables.

Möbius[68] obtained certain results on the reversion of series which were combined by J. W. L. Glaisher[69] into the general theorem: Let $a, b, \ldots$ be distinct primes; in terms of the undefined quantities $e_a, e_b, \ldots$, let $e_n = e_a{}^\alpha e_b{}^\beta \ldots$ if $n = a^\alpha b^\beta \ldots$, and let $e_1 = 1$. Then, if

$$F(x) = \Sigma e_n f(x^n),$$

where n ranges over all products of powers of $a, b, \ldots$, we have

$$f(x) = \Sigma(-1)^r e_\nu F(x^\nu),$$

where ν ranges over the numbers without square factors and divisible by no prime other than $a, b, \ldots$, while r is the number of the prime factors of ν. Taking

$$e_n = n^r, \qquad f(x) = \frac{x}{1-x},$$

Glaisher obtained the formula of H. J. S. Smith[38] and

$$\sigma_r(n) - \Sigma a^r \sigma_r\left(\frac{n}{a}\right) + \Sigma a^r b^r \sigma_r\left(\frac{n}{ab}\right) - \ldots = 1.$$

Using the same f, but taking $e_2 = 0$, $e_p = p^r$, when p is an odd prime, he proved that, if $\Delta_r(n)$ is the sum of the rth powers of the odd divisors of n,

$$\Delta_r(n) - \Sigma \Delta_r\left(\frac{n}{a}\right) + \Sigma \Delta_r\left(\frac{n}{ab}\right) - \ldots = 0 \text{ or } n^r,$$

according as n is even or odd. In the latter case, it reduces to Smith's.

If $\Delta'_r(n)$ is the sum of the rth powers of those divisors of n whose complementary divisors are odd, while $E_r(n)$ [or $E'_r(n)$] is the excess of the sum of the rth powers of those divisors of n which [whose complementary divisors] are of the form $4m+1$ over the sum of the rth powers of those divisors which [whose complementary divisors] are of the form $4m+3$,

$$\Delta'_r(n) - \Sigma a^r \Delta'_r\left(\frac{n}{a}\right) + \Sigma a^r b^r \Delta'_r\left(\frac{n}{ab}\right) - \ldots = \nu = \tfrac{1}{2}\{1 - (-1)^n\},$$

$$\Delta'_r(n) - \Sigma \Delta'_r\left(\frac{n}{A}\right) + \Sigma \Delta'_r\left(\frac{n}{AB}\right) - \ldots = n^r,$$

[68]Jour. für Math., 9, 1832, 105–123; Werke, 4, 591.
[69]London, Ed. Dublin Phil. Mag., (5), 18, 1884, 518–540.

$$E_r(n) - \Sigma(-1)^{(A-1)/2} A^r E_r\left(\frac{n}{A}\right) + \Sigma(-1)^{(AB-1)/2} A^r B^r E_r\left(\frac{n}{AB}\right) - \ldots = 1,$$

$$E_r(n) - \Sigma E_r\left(\frac{n}{a}\right) + \Sigma E_r\left(\frac{n}{ab}\right) - \ldots = (-1)^{(n-1)/2} n^r \nu,$$

$$E'_r(n) - \Sigma a^r E'_r\left(\frac{n}{a}\right) + \Sigma a^r b^r E'_r\left(\frac{n}{ab}\right) - \ldots = (-1)^{(n-1)/2} \nu,$$

$$E'_r(n) - \Sigma(-1)^{(A-1)/2} E'_r\left(\frac{n}{A}\right) + \Sigma(-1)^{(AB-1)/2} E'_r\left(\frac{n}{AB}\right) - \ldots = n^r,$$

where $A, B, \ldots$ are the odd prime factors of n. Note that $\nu = 0$ or 1 according as n is even or odd. By means of these equations, each of the five functions $\sigma_r(n), \ldots, E'_r(n)$ is expressed in two or more ways as a determinant of order n.

Ch. Hermite[70] quoted five formulas obtained by L. Kronecker[71] from the expansions of elliptic functions and involving as coefficients the functions $\Phi(n) = \sigma(n)$, the sum $X(n)$ of the odd divisors of n, the excess $\Psi(n)$ of the sum of the divisors $> \sqrt{n}$ of n over the sum of those $< \sqrt{n}$, the excess $\Phi'(n)$ of the sum of the divisors of the form $8k \pm 1$ of n over the sum of the divisors of the form $8k \pm 3$, and the excess $\Psi'(n)$ of the sum of the divisors $8k \pm 1$ exceeding $\sqrt{n}$ and the divisors $8k \pm 3$ less than $\sqrt{n}$ over the sum of the divisors $8k \pm 1$ less than $\sqrt{n}$ and the divisors $8k \pm 3$ exceeding $\sqrt{n}$. Hermite found the expansions into series of the right-hand members of the five formulas, employing the notations

$$E_1(x) = [x + \tfrac{1}{2}] - [x], \qquad E_2(x) = [x][x+1]/2,$$
$$a = 1, 3, 5, \ldots; \qquad b = 2, 4, 6, \ldots; \qquad c = 1, 2, 3, \ldots,$$

and A for a number of type a, etc. He obtained

$$X(1) + X(3) + \ldots + X(A) = \Sigma E_2\left(\frac{A+a}{2a}\right),$$

$$\sigma(1) + \sigma(2) + \ldots + \sigma(C) = \Sigma E_2(C/c),$$

$$\Psi(1) + \Psi(2) + \ldots + \Psi(C) = \Sigma E_2\left(\frac{C - c^2}{c}\right),$$

$$X(2) + X(4) + \ldots + X(B) = \tfrac{1}{3}\Sigma\left\{a\left[\frac{B}{2a}\right] + bE_1\left[\frac{B}{2b}\right]\right\},$$

$$\Phi'(1) + \Phi'(3) + \ldots + \Phi'(A) = \Sigma(-1)^{(a^2-1)/8} a\left[\frac{A+a}{2a}\right],$$

$$\Psi'(1) + \Psi'(3) + \ldots + \Psi'(A) = \Sigma(-1)^{(a^2+7)/8} a\left\{\left[\frac{A + 2a - a^2}{2a}\right] + \left[\frac{A - a^2}{2a}\right] - \left[\frac{A+a}{2a}\right]\right\}.$$

[70]Bull. Ac. Sc. St. Pétersbourg, 29, 1884, 340–3; Acta Math., 5, 1884–5, 315–9.
[71]Jour. für Math., 57, 1860, bottom p. 252 and top p. 253.

The first three had been found and proved purely arithmetically by Lipschitz and communicated to Hermite.

Hermite proved (11) by use of series. Also,

$$\sum_{a=1}^{n} F(a) = \sum_{a=1}^{n} \left[\frac{n}{a}\right] f(a), \qquad F(n) \equiv \Sigma f(d),$$

where d ranges over the divisors of n. When $f(d) = 1$, $F(n)$ becomes $\tau(n)$ and the formula becomes the first one by Dirichlet.[17]

L. Gegenbauer[72] considered the sum $\rho_{k,t}(n)$ of the kth powers of those divisors d_t of n whose complementary divisors are exact tth powers, as well as Jordan's function $J_k(n)$ [see Ch. V]. By means of the ζ-function, (12), he proved that

$$\sum_{m,n} \sigma_k(m) \rho_{0,2}(n) = \Sigma \rho_{0,2t}(d) \rho_{k,t}\left(\frac{r}{d}\right),$$

where d ranges over the divisors of r, and m, n over all pairs of integers for which $mn^t = r$;

$$\Sigma J_{tk}(n) \rho_{\nu,t}(m) = r^k \rho_{\nu-k,t}(r), \qquad \Sigma \sigma_{\nu-k}(m) \tau(n) m^k = \Sigma \rho_{k,t}(d) \rho_{\nu,t}\left(\frac{r}{d}\right),$$

the latter for $t = 1$ being Liouville's[29] seventh formula for $\nu = 0$;

$$\Sigma d^\nu \rho_{k,t}\left(\frac{r}{d}\right) = \Sigma d^k \rho_{\nu,t}\left(\frac{r}{d}\right), \qquad \Sigma J_\nu(d) d^k \rho_{k,t}\left(\frac{r}{d}\right) = \rho_{k+\nu,t}(r),$$

the latter for $t = \nu = 1$, $k = 0$, being the second formula of Liouville[25], while for $t = 1$ it is the final formula by Cesàro[210] of Ch. V;

$$\Sigma \lambda(d) d^k \rho_{k,2t}\left(\frac{r}{d}\right) = \Sigma \lambda(d) \rho_{k,t}(d) \rho_{k,t}\left(\frac{r}{d}\right) = 0 \text{ or } \rho_{2k,t}(\sqrt{r}),$$

according as r is not or is a square;

$$\Sigma \lambda(n) \rho_{k,t}(m) = \rho_{k,2t}(r), \qquad \Sigma \lambda(d) \tau(d^2) = \lambda(r) \tau(r),$$

$$\Sigma \tau^2(d) J_k\left(\frac{r}{d}\right) = r^k \Sigma \frac{\tau(d^2)}{d^k}, \qquad \Sigma d^k \tau(d^2) \sigma_k\left(\frac{r}{d}\right) = \Sigma d^k \tau^2(d),$$

$$\sum_{x=1}^{n} \left[\frac{n}{x}\right] \tau(x^2) = \sum_{r=1}^{n} \tau^2(r), \qquad \sum_{x=1}^{n} \left[\frac{n}{x}\right] \lambda(x) \sigma_k(x) = \sum_{r=1}^{n} \rho_{k,2}(r).$$

By changing the sign of the first subscript of ρ, we obtain formulas for the sum $P_{k,t}(n) = n^k \rho_{-k,t}(n)$ of the kth powers of those divisors of n which are tth powers. By taking the second subscript of ρ to be unity, we get formulas for $\sigma_k(n)$. There are given many formulas involving also the number $f_a(n)$ of solutions of $n_1 n_2 \ldots n_a = n$, and the number $\omega(n)$ of ways n can be expressed as a product of two relatively prime factors. Two special cases [(107),(128)] of these are the first formula of Liouville[26] and the ninth summation formula of Liouville,[29] a fact not observed by Gegenbauer. He proved that, if $p \leqq n$,

$$\sum_{x=p+1}^{n} B(x) = - \sum_{x=A+1}^{B} C(x) + Bn - Ap,$$

---

[72]Sitzungsber. Ak. Wiss. Wien (Math.), 89, II, 1884, 47–73, 76–79.

where
$$B(x) = \left[\sqrt[\tau]{\frac{bx^\sigma + \beta}{a}} - \rho\right], \qquad C(x) = \left[\sqrt[\sigma]{\frac{a(x^\tau + \rho) - \beta}{b}}\right],$$

and $B = B(n)$, $A = B(p+1)$; also that
$$\sum_{x=p+1}^{n} D(x) = \sum_{x=D+1}^{E} F(x) + Dn - Ep,$$

where
$$D(x) = \left[\sqrt[\tau]{\frac{a}{bx^\sigma + \beta}} - \rho\right], \qquad F(x) = \left[\sqrt[\sigma]{\frac{a}{b(x^\tau + \rho)}} - \frac{\beta}{b}\right],$$

and $D = D(n)$, $E = D(p+1)$. It is stated that special cases of these two formulas (here reported with greater compactness) were given by Dirichlet, Zeller, Berger and Cesàro. In the second, take $\tau = 1$, $\rho = 0$, and choose the integers a, β, b, n so that
$$bn^\sigma + \beta > a > b(n-1)^\sigma + \beta,$$

whence $D = 0$. If χ_r is the number of divisors of r which are of the form $bx^\sigma + \beta$, we get
$$\sum_{r=1}^{a} \chi_r = \sum_{x=1}^{p} \left[\frac{a}{bx^\sigma + \beta}\right] + \sum_{x=1}^{E} \left[\sqrt[\sigma]{\frac{a}{bx}} - \frac{\beta}{b}\right] - Ep, \qquad E = \left[\frac{a}{b(p+1)^\sigma + \beta}\right].$$

Change n to $n+1$ and set $\beta = 0$, $b = \sigma = 1$, whence $a = n$ [also set $p = [\sqrt{n}]$]; we get Meissel's[22] formula (11). Other specializations give the last one of the formulas by Lipschitz,[58] and
$$\sum_{r=1}^{n} k(r) = \sum_{x=1}^{\nu} \left[\frac{2n}{2x+1}\right] + \sum_{x=1}^{\nu} \left[\frac{n}{x} - \frac{1}{2}\right] - \nu^2 + t,$$

where $\nu = [\sqrt{n}]$, $k(r)$ is the number of odd divisors of r, while $t = 0$ or 1 according as $[n/\nu - \frac{1}{2}] > \nu - 1$ or $= \nu - 1$.

L. Gegenbauer[73] proved by use of ζ-functions many formulas involving his[72] functions ρ, f and divisors d_t. Among the simplest formulas, special cases of the more general ones, are
$$\Sigma \sigma_k(d) d^\lambda = \Sigma \sigma_{k+\lambda}\left(\frac{r}{d}\right) d^\lambda = \Sigma \sigma^\lambda\left(\frac{r}{d}\right) d^{k+\lambda}, \qquad \Sigma \mu^2(d_4) = \Sigma \lambda(h),$$

$$\Sigma \theta(h) \mu^2(d_2) = \Sigma \mu^2(h), \qquad \Sigma \tau(h^2) \mu^2(d_2) = \Sigma \theta(h), \qquad \Sigma \mu^2(d) \tau\left(\frac{r}{d}\right) = \tau(r^2),$$

$$\Sigma \tau(d_2) \mu(h) = \theta(r), \qquad \Sigma \mu^2(d) J_k\left(\frac{r}{d}\right) = \Sigma d_2{}^k \mu(h),$$

summed for d, d_2, d_4, where $h = \sqrt{r/d_2}$. Other special cases are the fourth and sixth formulas of Liouville,[29] the first, third and last of Liouville.[25] Beginning with p. 414, the formulas involve also
$$\omega_k(n) = n^k \prod_{i=1}^{r} (1 + 1/p_i^k), \qquad n = \prod_{i=1}^{r} p_i^{r_i}.$$

---

[73]Sitzungsber. Ak. Wien (Math.), 90, II, 1884, 395–459. The functions used are not defined in the paper. For his ψ_h, ψ, ω, we write σ_h, τ θ, where θ is the notation of Liouville.[25]

Beginning with p. 425 and p. 430 there enter the two functions

$$n^\mu \prod_{i=1}^{r} \left(\frac{\Delta}{p_i}\right)^{n_i-1} \left\{ \left(\frac{\Delta}{p_i}\right) \mp p_i^{k-\mu} \right\},$$

in which (Δ/p) is Legendre's symbol, with the value 1 or -1.

J. W. L. Glaisher[74] investigated the excess $\zeta_r(n)$ of the sum of the rth powers of the odd divisors of n over the sum of the rth powers of the even divisors, the sum $\Delta'_r(n)$ of the rth powers of those divisors of n whose complementary divisors are odd, wrote ζ for ζ_1, and Δ' for Δ'_1, and proved

$$\Delta'_3(n) = n\Delta'(n) + 4\Delta'(1)\Delta'(n-1) + 4\Delta'(2)\Delta'(n-2) + \ldots + 4\Delta'(n-1)\Delta'(1),$$

$$\zeta_3(n) = (2n-1)\zeta(n) - 4\zeta(1)\zeta(n-1) - 4\zeta(2)\zeta(n-2) - \ldots - 4\zeta(n-1)\zeta(1),$$

$$n\Delta'(n) = \Delta'(1)\Delta'(2n-1) - \Delta'(2)\Delta'(2n-2) + \ldots + \Delta'(2n-1)\Delta'(1),$$

$$(-1)^{n-1}n\zeta(n) = \Delta'(n) + 8\zeta(1)\Delta'(n-2) + 8\zeta(2)\Delta'(n-4) + \ldots,$$

$$\Delta'_3(n) = n\Delta'(n) + \Delta'(2)\Delta'(2n-2) + \Delta'(4)\Delta'(2n-4) + \ldots + \Delta'(2n-2)\Delta'(2),$$

$$-\zeta_3(n) = 3\Delta(n) + 4\{\Delta(1)\Delta(n-1) + 9\Delta(2)\Delta(n-2) + \Delta(3)\Delta(n-3)$$
$$+ 9\Delta(4)\Delta(n-4) + \ldots + \Delta(n-1)\Delta(1)\} \qquad (n \text{ even}),$$

$$\frac{2^{2r-1}\Delta'_{2r+1}(n)}{(2r)!} = \frac{[1, 2r-1]}{1!(2r-1)!} + \frac{[3, 2r-3]}{3!(2r-3)!} + \ldots + \frac{[2r-1, 1]}{(2r-1)!1!},$$

where

$$[p, q] = \sigma_p(1)\sigma_q(2n-1) + \sigma_p(3)\sigma_q(2n-3) + \ldots + \sigma_p(2n-1)\sigma_q(1).$$

For n odd, $\zeta(n) = \Delta'(n) = \sigma(n)$ and the fourth formula gives

$$(n-1)\sigma(n) = 8\{\sigma(1)\sigma(n-2) + \zeta(2)\sigma(n-4) + \sigma(3)\sigma(n-6) + \zeta(4)\sigma(n-8) + \ldots\}.$$

Glaisher[75] proved that

$$5\sigma_3(n) - 6n\sigma(n) + \sigma(n)$$
$$= 12\{\sigma(1)\sigma(n-1) + \sigma(2)\sigma(n-2) + \ldots + \sigma(n-1)\sigma(1)\},$$

$$\sigma(1)\sigma(2n-1) + \sigma(3)\sigma(2n-3) + \ldots + \sigma(2n-1)\sigma(1)$$
$$= \Delta'_3(n) = \tfrac{1}{8}\{\sigma_3(2n) - \sigma_3(n)\}.$$

The latter includes the first theorem in his[63] earlier paper.

Glaisher[76] proved for Jacobi's[11] $E(n)$ that

$$\sigma(2m+1) = E(1)E(4m+1) + E(5)E(4m-3) + E(9)E(4m-7) + \ldots$$
$$+ E(4m+1)E(1),$$

$$E(t) - 2E(t-4) + 2E(t-16) - 2E(t-36) + \ldots = 0 \qquad (t = 8n+5),$$

$$\sigma(v) - 2\sigma(v-4) + 2\sigma(v-16) - 2\sigma(v-36) + \ldots = 0 \qquad (v = 8n+7),$$

$$\sigma(u) + \sigma(u-8) + \sigma(u-24) + \sigma(u-48) + \ldots = 4\{\sigma(m) + 2\sigma(m-4)$$
$$+ 2\sigma(m-16) + 2\sigma(m-36) + \ldots\} \qquad (m = 2n+1, \ u = 8n+3),$$

and three formulas analogous to the last (pp. 125, 129). He repeated (p. 158) his[74] expressions for $\Delta'_3(n)$.

---

[74] Messenger Math., 14, 1884–5, 102–8.
[75] Ibid., 156–163.
[76] Quar. Jour. Math., 20, 1885, 109, 116, 121, 118.

L. Gegenbauer[77] considered the number $\tau_1(k)$ of the divisors $\leqq [\sqrt{n}]$ of k
and the number $\tau_2(k)$ of the remaining divisors and proved that

$$\sum_{k=1}^{n} \tau_1(k) = \frac{n}{2}(\log_e n + 2C) + O(\sqrt{n}),$$

$$\Sigma \tau_2(k) = \frac{n}{2}(\log_e n + 2C - 2) + O(\sqrt{n}),$$

$O(s)$ being [90] of the order of magnitude of s. He proved (p. 55) that the mean
of the sum of the reciprocals of the square divisors of any integer is $\pi^4/90$;
that (p. 64) of the reciprocals of the odd divisors is $\pi^2/8$; the mean (p. 65)
of the cubes of the reciprocals of the odd divisors of any integer is $\pi^4/96$,
that of their fifth powers is $\pi^6/960$. The mean (p. 68) of Jacobi's[11] $E(n)$ is
$\pi/4$.

G. L. Dirichlet[78] noted that in (7), p. 282 above, we may take ϵ to be of
lower [unstated] order of magnitude than his former $\sqrt{n}$.

L. Gegenbauer[79] considered the sum $\tau_{r,k,s}(n)$ of the kth powers of those
divisors of n which are rth powers and are divisible by no (sr)th power
except 1; also the number $Q_a(b)$ of integers $\leqq b$ which are divisible by no ath
power except 1. It follows at once that, if $\mu_s(m) = 0$ if m is divisible by an
sth power > 1, but $= 1$ otherwise,

$$\tau_{r,k,s}(n) = \Sigma \mu_s\left(\sqrt[r]{\frac{n}{d_r}}\right)\frac{n^k}{d_r^k},$$

where the summation extends over all the divisors d_r of n whose complementary divisors are rth powers, and that

(14) $$\sum_{x=1}^{n} \tau_{r,k,s}(x) = \sum_{x=1}^{\nu} \left[\frac{n}{x^r}\right] x^{rk} \mu_s(x), \qquad \nu = [\sqrt[r]{n}].$$

From the known formula $Q_r(n) = \Sigma[n/x^r]\mu(x)$, $x = 1, \ldots, \nu$, is deduced

$$\sum_{y=1}^{\nu} Q_r\left(\left[\frac{n}{y^r}\right]\right) y^k = \sum_{x=1}^{\nu} \left[\frac{n}{x^r}\right]\left\{\Sigma \mu\left(\frac{x}{d}\right) d^k\right\},$$

the right member reducing to n for $k = 0$ and thus giving a result due to
Bougaief. From this special result and (14) is derived

$$\sum_{x=1}^{n} \tau_{r,0,s}(x) = \sum_{x=1}^{\nu} Q_{sr}\left(\left[\frac{n}{x^r}\right]\right).$$

From these results he derived various expressions for the mean value of
$\tau_{r,-k,s}(x)$ and of the sum $\tau_{r,k,s}(n)$ of the kth powers of those divisors of n
which are rth powers and are divisible by at least one (sr)th power other
than 1. He obtained theorems of the type: The mean value of the number
of square divisors not divisible by a biquadrate is $15/\pi^2$; the mean value of
the excess of the number of divisors of one of the forms $4r\mu + j(j = 1, 3, \ldots,$
$2r - 1)$ over the number of the remaining odd divisors is

$$\frac{\pi}{4r} \sum_{l=1}^{r} \cot\frac{(2l-1)\pi}{4r}.$$

[77]Denkschr. Akad. Wien (Math.), 49, I, 1885, 24.
[78]Göttingen Nachrichten, 1885, 379; Werke, 2, 407; letter to Kronecker, July 23, 1858.
[79]Sitzungsberichte Ak. Wiss. Wien (Math.), 91, II, 1885, 600–621.

L. Gegenbauer[80] considered the number $A_0(a)$ of those divisors of a which are congruent modulo k and have a complementary divisor $\equiv 1$ (mod k). He proved that, if $\rho < k$,

$$\sum_{x=1}^{\sigma} A_0(kx-\rho) = \sum_{x=1}^{\sigma}\left[\frac{(k-1)x+\sigma-\rho}{kx-\rho}\right] = \sigma + \sum_{x=1}^{a}\left[\frac{\rho x+\sigma}{kx+1}\right]$$

$$= \sigma+a+\sum_{x=1}^{b}\left[\frac{\sigma-1-(k-\rho)x}{kx+1}\right], \quad a\equiv\left[\frac{\sigma-1}{k-\rho}\right], \quad b\equiv\left[\frac{\sigma-2}{2k-\rho}\right].$$

If we replace σ by $\sigma-1$ and subtract, we obtain expressions for $A_0(k\sigma-\rho)$. The above formulas give, for $k=2$, $\rho=1$,

$$\sum_{x=1}^{\sigma}\tau(2x-1) = \sigma+\sum_{x=1}^{\sigma-1}\left[\frac{x+\sigma}{2x+1}\right], \quad \tau(2\sigma-1) = 2+\sum_{x=1}^{\sigma-2}\left\{\left[\frac{x+\sigma}{2x+1}\right]-\left[\frac{x+\sigma-1}{2x+1}\right]\right\}.$$

and formulas of Bouniakowsky.[32] The same developments show that an odd number a is a prime if

$$\left[\frac{a}{2(2x+1)}+\frac{1}{2}\right] = \left[\frac{a-2}{2(2x+1)}+\frac{1}{2}\right]$$

for $x\leqq[(a-3)/2]$; likewise for $a=6k\pm1$ if the same equality holds when $x\leqq[(a-5)/6]$, with similar tests for $a=3n-1$, or $4n-1$.

C. Runge[81] proved that $\tau(n)/n^{\epsilon}$ has the limit zero as n increases indefinitely, for every $\epsilon>0$.

E. Catalan[82] noted that, if x_{np} is the number of ways of decomposing a product of n distinct primes into p factors >1, order being immaterial,

$$x_{np} = px_{n-1\,p}+x_{n-1\,p-1} = \{p^{n-1}-\binom{p-1}{1}(p-1)^{n-1}+\binom{p-1}{2}(p-2)^{n-1}-\ldots\pm1\}$$
$$\div\{(p-1)!\}.$$

E. Cesàro[83] considered the number $F_m(x)$ of integers $\leqq x$ which are not divisible by mth powers, and the number $T_m(x)$ of those divisors of x which are mth powers, evaluated sums involving these and other functions, and determined mean values and probabilities relating to the greatest square divisor of an arbitrary integer.

R. Lipschitz[84] considered the sum $k(m)$ of the odd divisors of m increased by half the sum of the even divisors, and the function $l(m)$ obtained by interchanging the words "even," "odd." He proved that

$$k(m)-2k(m-1)+2k(m-9)-\ldots=(-1)^{m-1}m \text{ or } 0,$$

according as m is a square or is not;

$$l(m)+l(m-1)+l(m-3)+l(m-6)+\ldots=-m \text{ or } 0,$$

according as m is a triangular number or is not;

$$K(m) = k(1)+k(2)+\ldots+k(m) = [m]+\left[\frac{m}{2}\right]+3\left[\frac{m}{3}\right]+2\left[\frac{m}{4}\right]+\ldots+\mu\left[\frac{m}{m}\right],$$

$$L(m) = l(1)+l(2)+\ldots+l(m) = -[m]+2\left[\frac{m}{2}\right]-3\left[\frac{m}{3}\right]+4\left[\frac{m}{4}\right]-\ldots,$$

[80]Sitzungsberichte Ak. Wien. (Math.), 91, II, 1885, 1194–1201.　　[81]Acta Math., 7, 1885, 181–3.
[82]Mém. soc. roy. sc. Liège, (2), 12, 1885, 18–20; Mélanges Math., 1868, 18.
[83]Annali di Mat., (2), 13, 1885, 251–268. Reprint "Excursions arith. à l'infini," 17–34.
[84]Comptes Rendus Paris, 100, 1885, 845. Cf. Glaisher[116], also Fergola[21] of Ch. XI, Vol. II.

where $\mu = m$ or $m/2$ according as m is odd or even.   Cf. Hacks.[96]

M. A. Stern[85] noted that Zeller's[66] formula follows from $B = \rho A$, where

$$\frac{1}{p(x)} = A = \sum_{n=0}^{\infty} \psi(n) x^n, \qquad \frac{\rho}{p(x)} = B = \sum_{n=1}^{\infty} \sigma(n) x^{n-1}, \qquad \rho = 1 + 2x - 5x^4 - 7x^6 + \ldots,$$

where $p(x)$ is defined by (1), $\psi(n)$ is the number of partitions of n, and the second equation follows from the equality of (3) and (4) after removing the factor x.   Next, if $N(n)$ denotes the number of combinations of $1, 2, \ldots, n$ without repetitions producing the sum n,

$$\sum_{n=1}^{\infty} N(n) x^n = (1+x)(1+x^2) \ldots = \frac{(1-x^2)(1-x^4) \ldots}{(1-x)(1-x^2) \ldots},$$

then by the second equation above,

$$B(1 - \dot{x}^2 - x^4 + x^{10} + x^{14} - \ldots) = \rho \Sigma N(n) x^n,$$

$$\sigma(n) - \sigma(n-2) - \sigma(n-4) + \sigma(n-10) + \sigma(n-14) - \ldots$$
$$= N(n-1) + 2N(n-2) - 5N(n-5) - 7N(n-7) + \ldots,$$

where $\sigma(n-n) = 0$, $N(n-n) = 1$.

S. Roberts[86] noted that Euler's[4] formula (2) is identical with Newton's relation $S_{-n} = S_{-n+1} + S_{-n+2} - \ldots$ for obtaining the sum S_{-n} of the $(-n)$th powers of the roots of $s = 0$, where s and p are defined by (2).   In p, the sum of the $(-n)$th powers of the roots of $1 - x^k = 0$ is k or 0 according as k is or is not a divisor of n.   Hence the like sum for p is $\sigma(n)$.   [Cf. Waring[9].]   The process can be applied to products of factors $1 - f(k) x^k$.   His further results may be given the following simpler form.   Let ϕ_n be the sum of the even divisors of n, and ψ_n the sum of the odd divisors of n, and set $s_n = \phi_n + 2\psi_n$ if n is even, $s_n = -2\psi_n$ if n is odd.   By elliptic function expansions,

$$s_{2n} + 8\{ s_{2n-1}\psi_1 + 3s_{2n-2}\psi_2 + s_{2n-3}\psi_3 + 3s_{2n-4}\psi_4 + \ldots + s_1 \psi_{2n-1} \} + 12n\psi_{2n} = 0,$$
$$s_{2n+1} + 8\{ s_{2n}\psi_1 + 3s_{2n-1}\psi_2 + \ldots + 3s_1\psi_{2n} \} + (4n+2)\psi_{2n+1} = 0,$$

the coefficients being 1 and 3 alternately.   He indicated a process for finding a recursion formula involving the sums of the cubes of the even divisors and the sums of the cubes of the odd divisors, but did not give the formula.

N. V. Bougaief[86a] obtained, as special cases of a summation formula,

$$\sum_u \{ 8x + 5 - 5(2u-1)^2 \} \sigma(2x + 1 - u^2 + u) = 0, \qquad \Sigma\{ n - 3\sigma(u) \} P\{ n - \sigma(u) \} = 0,$$

where $P(n)$ is the number of solutions u, v of $\sigma(u) + \sigma(v) = n$.

L. Gegenbauer[86b] proved that the number of odd divisors of $1, 2, \ldots, n$ equals the sum of the greatest integers in $(n+1)/2$, $(n+2)/4$, $(n+3)/6, \ldots$, $(2n)/(2n)$.   The number of divisors of the form $Bx - \gamma$ of $1, \ldots, n$ is expressed as a sum of greatest integers; etc.

J. W. L. Glaisher[87] considered the sum $\Delta_s(n)$ of the sth powers of the odd divisors of n, the like sum $D_s(n)$ for the even divisors, the sum $D'_s(n)$ of the

[85]Acta Mathematica, 6, 1885, 327-8.
[86]Quar. Jour. Math., 20, 1885, 370-8.
[86a]Comptes Rendus. Paris, 100, 1885, 1125, 1160.
[86b]Denkschr. Akad. Wiss Wien (Math.), 49, II, 1885, 111.
[87]Messenger Math., 15, 1885-6, 1-20.

sth powers of the divisors of n whose complementary divisors are even, the excess $\zeta'_s(n)$ of the sum of the sth powers of the divisors whose complementary divisors are odd over that when they are even, and the similar functions[74] Δ'_s, ζ_s, σ_s. The seven functions can be expressed in terms of any two:

$$\Delta_s = \sigma_s - 2^s D'_s, \qquad D_s = 2^s D'_s, \qquad \Delta'_s = \sigma_s - D'_s,$$
$$\zeta_s = \sigma_s - 2^{s+1}D'_s, \qquad \zeta'_s = \sigma_s - 2D'_s,$$

where the arguments are all n. Since $D'_s(2k) = \sigma_s(k)$, we may express all the functions in terms of $\sigma_s(n)$ and $\sigma_s(n/2)$, provided the latter be defined to be zero when n is odd. Employ the abbreviation $\Sigma fF = \Sigma Ff$ for

$$f(1)F(n-1) + f(2)F(n-2) + f(3)F(n-3) + \ldots + f(n-1)F(1).$$

This sum is evaluated when f and F are any two of the above seven functions with $s = 1$ (the subscript 1 is dropped). If

$$f(n) = a\sigma(n) + \beta D'(n), \qquad F(n) = a'\sigma(n) + \beta' D'(n),$$

then

$$\Sigma fF = aa'\Sigma\sigma\sigma + (a\beta' + a'\beta)\Sigma\sigma D' + \beta\beta'\Sigma D'D'.$$

By using the first formula in each of two earlier papers,[74, 75] we get

$$12\Sigma\sigma\sigma = 5\sigma_3(n) - 6n\sigma(n) + \sigma(n),$$
$$12\Sigma D'D' = 5D_3'(n) - 3nD'(n) + D'(n),$$
$$24\Sigma\sigma D' = 2\sigma_3(n) + (1-3n)\sigma(n) + (1-6n)D'(n) + 8D_3'(n).$$

Hence all 21 functions can now be expressed at once linearly in terms of σ_3, D_3', σ and D'. The resulting expressions are tabulated; they give the coefficients in the products of any two of the series $\Sigma_1^\infty f(n)x^n$, where f is any one of our seven functions without subscript.

Glaisher[88] gave the values of $\Sigma\sigma_3\sigma_i$ for $i = 3, 5, 9$ and $\Sigma\sigma_5\sigma_7$, where the notation is that of the preceding paper. Also, if $\rho = n - r$,

$$12\sum_{r=1}^{n} r\rho(r)\sigma(\rho) = n^2\sigma_3(n) - n^3\sigma(n), \qquad \sum_{r=1}^{n} rf(r)F(\rho) = \frac{n}{2}\Sigma fF.$$

L. Gegenbauer[89] gave purely arithmetical proofs of generalizations of theorems obtained by Hermite[70] by use of elliptic function expansions. Let

$$S_k(r) = \sum_{j=1}^{r} j^k, \qquad \sigma = \sum_{x=1}^{\nu} S_k\left(\left[\frac{n}{x}\right]\right) - \nu S_k(\nu), \qquad \nu \equiv [\sqrt{n}].$$

Then (p. 1059),

$$\sum_{x=1}^{n}\left[\frac{n}{x}\right]x^k = \sum_{x=1}^{\nu}\left[\frac{n}{x}\right]x^k + \sigma.$$

The left member is known to equal the sum of the kth powers of all the divisors of $1, 2, \ldots, n$. The first sum on the right is the sum of the kth powers of the divisors $\leqq \sqrt{n}$ of $1, \ldots, n$. Hence if $\Lambda_k(x)$ is the excess of the sum $\psi_k'(x)$ of the kth powers of the divisors $> \sqrt{x}$ of x over the sum of the kth powers of the remaining divisors, it follows at once that

$$\sum_{x=1}^{n} \Lambda_k(x) = - \sum_{x=1}^{\nu} \left[\frac{n}{x}\right] x^k + \sigma.$$

Also

$$\sum_{x=1}^{n} \psi_k'(x) = \sum_{x=1}^{\nu} S_k\left(\left[\frac{n}{x}\right]\right) + S_{k+1}(\nu) - (\nu+1) S_k(\nu),$$

with a similar formula for $\Sigma \Psi_k(x)$, where $\Psi_k(x)$ is the excess of $\psi_k'(x)$ over the sum of the kth powers of the divisors $< \sqrt{x}$ of x. For $k=1$, the last formula reduces to the third one of Hermite's.

Let $\chi_k(x)$ be the sum of the kth powers of the odd divisors of x; $\chi_k''(x)$ that for the odd divisors $> \sqrt{x}$; $X_k''(x)$ the excess of the latter sum over the sum of the kth powers of the odd divisors $< \sqrt{x}$ of x; $\chi_k'''(x)$ the excess of the sum of the kth powers of the divisors $8s \pm 1 > \sqrt{x}$ of x over the sum of the kth powers of the divisors $8s \pm 3 < \sqrt{x}$ of x. For $y=2x$ and $y=2x-1$, the sum from $x=1$ to $x=n$ of $\chi_k(y)$, $\chi_k''(y)$, $X_k''(y)$ and $\chi_k'''(y)$ are expressed as complicated sums involving the functions S_k and $[x]$.

E. Pfeiffer[90] attempted to prove a formula like (7) of Dirichlet,[17] where now ϵ is $O(n^{1/3+k})$ for every $k>0$. Here $Og(T)$ means a function whose quotient by $g(T)$ remains numerically less than a fixed finite value for all sufficiently large values of T. E. Landau[91] noted that the final step in the proof fails from lack of uniform convergence and reconstructed the proof to secure this convergence.

L. Gegenbauer,[92] in continuation of his[80] paper, gave similar but longer expressions for

$$\sum_{x=0}^{\sigma} \tau(y), \qquad \sum_{x=0}^{\sigma} \sigma_k(y) \qquad (y=4x+1, 6x+1, 8x+3, 8x+5, 8x+7)$$

and deduced similar tests for the primality of y.

Gegenbauer[92a] found the mean of the number of divisors $\lambda x + a$ of a number of s digits with a complementary divisor $\mu y + \beta$; also for divisors $ax^2 + by^2$.

Gegenbauer[92b] evaluated $A(1) + \ldots + A(n)$ where $A(x)$ is the sum of the ρth powers of the σth roots of those divisors d of x which are exact σth powers and whose complementary divisors exceed $kd^{T/\sigma}$. A special case gives (11), p. 284 above.

Gegenbauer[92c] gave a formula involving the sum of the kth powers of those divisors of $1, \ldots, m$ whose complementary divisors are divisible by no rth power >1.

[90]Ueber die Periodicität in der Teilbarkeit..., Jahresbericht der Pfeiffer'schen Lehr- und Erziehungs-Anstalt zu Jena, 1885–6, 1–21.

[91]Sitzungsber. Ak. Wiss. Wien (Math.), 121, IIa, 1912, 2195–2332; 124, IIa, 1915, 469–550. Landau.[161]

[92]Ibid., 93, II, 1886, 447–454.

[92a]Sitzungsber. Ak. Wiss. Wien (Math.), 93, 1886, II, 90–105.

[92b]Ibid., 94, 1886, II, 35–40.

[92c]Ibid., 757–762.

Ch. Hermite[93] proved that if $F(N)$ is the number of odd divisors of N,

$$\sum_{n=1}^{\infty} q^{(n^2+n)/2}/(1-q^n) = \Sigma F(N)q^N,$$

and then that

$$F(1)+F(2)+\ldots+F(N) = \frac{1}{2}N \log N + \left(C-\frac{1}{4}\right)N,$$

$$\Phi(1)+\Phi(2)+\ldots+\Phi(N) = \frac{1}{2}N \log N/k + \left(C-\frac{1}{2}\right)N,$$

asymptotically, where $\Phi(N)$ is the number of decompositions of N into two factors d, d', such that $d' > kd$.

E. Catalan[93a] noted that, if $n = i+i' = 2i''d$,

$$\Sigma\sigma(i)\sigma(i') = \Sigma d^3, \qquad \Sigma\{\sigma(i)\sigma(2n-i)\} = 8\Sigma\{\sigma(i)\sigma(n-i)\}.$$

E. Cesàro[94] proved Lambert's[7] result that $\tau(n)$ is the coefficient of x^n in $\Sigma x^k/(1-x^k)$. Let $T_\nu(n)$ be the number of sets of positive integral solutions of

$$\xi_1 + 2\xi_2 + \ldots + \nu\xi_\nu = n$$

and $s_\nu(n)$ the sum of the values taken by ξ_ν. Then

$$s_\nu(n) = T_\nu(n) + T_\nu(n-\nu) + T_\nu(n-2\nu) + \ldots,$$
$$\tau(n) = s_1(n) - s_2(n) + s_3(n) - \ldots.$$

Let

$$t_n(x) = \Sigma(-1)^{d+1}T_d(x-n),$$

summed for the divisors d of n. Then

$$\tau(n) = t_1(n) + t_2(n) + \ldots + T_1(n) - T_2(n) + T_3(n) - \ldots.$$

E. Busche[95] employed two complementary divisors δ_m and δ_m' of m, an arbitrary function f, and a function $y = \Phi(x)$ increasing with x whose inverse function is $x = y\psi(y)$. Then

$$\sum_{x=1}^{a} \{f([\psi(x)], x) - f(0, x)\} = \Sigma\{f(\delta'_m, \delta_m) - f(\delta'_m - 1, \delta_m)\},$$

where in the second member the summation extends over all divisors of all positive integers, and $\Phi(m) \leqq \delta_m \leqq a$. In particular,

$$\sum_{x=1}^{a} f(x)[\psi(x)] = \Sigma f(\delta_m), \qquad \sum_{x=1}^{a} [\psi(x)] = \text{number of } \delta_m,$$

subject to the same inequalities. In the last equation take $\psi(x) = x$, $a = [\sqrt{n}]$; we get (11).

J. Hacks[96] proved that, if m is odd,

$$\mathfrak{F}(m) \equiv \tau(1) + \tau(3) + \tau(5) + \ldots + \tau(m) = \Sigma\left[\frac{m+t}{2t}\right],$$

[93]Jour. für Math., 99, 1886, 324–8.
[93a]Mém. Soc. R. Sc. Liège, (2), 13, 1886, 318 (Mélanges Math., II).
[94]Jornal de sciencias math. e astr., 7, 1886, 3–6.
[95]Jour. für Math., 100, 1887, 459–464. Cf. Busche.[103]
[96]Acta Math., 9, 1887, 177–181. Corrections, Hacks,[97] p. 6, footnote.

$$\mathfrak{G}(m) \equiv \sigma(1) + \sigma(3) + \sigma(5) + \dots + \sigma(m) = \Sigma t \left[\frac{m+t}{2t} \right],$$

where t ranges over the odd integers $\leqq m$. For the K and L of Lipschitz[84] and $G(m) = \sigma(1) + \sigma(2) + \dots + \sigma(m)$, it is shown that

$$\mathfrak{F}(m) \equiv \mathfrak{G}(m) \equiv \left[\frac{\sqrt{m}+1}{2} \right] \equiv K(m) \quad (\text{mod } 2),$$

$$L(m) \equiv G(m) \equiv [\sqrt{m}] + \left[\sqrt{\frac{m}{2}} \right], \qquad T(m) \equiv [\sqrt{m}] \quad (\text{mod } 2).$$

J. Hacks[97] gave a geometrical proof of (11) and of Dirichlet's[17] expression for $T(n)$, just preceding (7). He proved that the sum of all the divisors, which are exact ath powers, of $1, 2, \dots, m$ is

$$\sum_{j=1}^{m} \{1^a + 2^a + \dots + [\sqrt[e]{m/j}]^a\}.$$

He gave (pp. 13–15) several expressions for his[96] $\mathfrak{F}(m)$, $\mathfrak{G}(m)$, $K(m)$.

L. Gegenbauer[97a] gave simple proofs of the congruences of Hacks.[96]

M. Lerch[98] considered the number $\psi(a, b)$ of divisors $> b$ of a and proved that

(15) $$\sum_{\rho=0}^{[n/2]} \psi(n-\rho, \rho) = n, \qquad \sum_{\rho=0}^{n} \psi(n+\rho, \rho) = 2n.$$

A. Strnad[99] considered the same formulas (15).

M. Lerch[100] considered the number $\chi(a, b)$ of the divisors $\leqq b$ of a and proved that

$$\sum_{\sigma=0}^{[(m-1)/a]} \{\psi(m-\sigma a, k+\sigma) - \chi(m-\sigma a, a)\}$$
$$+ \sum_{\lambda=1}^{k} \{\psi(m+\lambda a, \lambda-1) - \chi(m+\lambda a, a)\} = 0.$$

This reduces to his (15) for $a=1$, $k=1$ or $m+1$. Let $(k, n; m)$ denote the g. c. d. (k, n) of k, n or zero, according as (k, n) is or is not a divisor of m. Then

(16) $$\sum_{a=0}^{a-1} \{\psi(m+an, a) - \psi(m+an, a)\} = \sum_{k=1}^{a} (k, n; m).$$

In case m and n are relatively prime, the right member equals the number $\phi(a, n)$ of integers $\leqq a$ which are prime to n. Finally, it is stated that

(17) $$\sum_{a=0}^{c} \psi(m-an, a) = \sum_{a=0}^{c} \chi(m-an, n), \qquad c = \left[\frac{m-1}{n} \right].$$

Gegenbauer,[29] Ch. VIII, proved (16) and the formula preceding it.

[97]Acta Math., 10, 1887, 9–11.
[97a]Sitzungsber. Ak. Wiss. Wien (Math.), 95, 1887, II, 297–8.
[98]Prag Sitzungsberichte (Math.), 1887, 683–8.
[99]Casopis mat. fys., 18, 1888, 204.
[100]Compt. Rend. Paris, 106, 1888, 186. Bull. des sc. math. et astr., (2), 12, I, 1888, 100–108, 121–6.

C. A. Laisant[101] considered the number $n_k(N)$ of ways N can be expressed as a product of k factors (including factors unity), counting $PQ\ldots$ and $QP\ldots$ as distinct decompositions. Then

$$n_k(N) = n_{k-1}(N)\,\Pi\Big(1 + \frac{e_i}{k-1}\Big), \qquad N = \Pi p_i^{e_i}.$$

E. Cesàro[102] proved Gauss' result that the number of divisors, not squares, of n is asymptotic to $6\pi^{-2}\log n$. Hence $\tau(n^2)$ is asymptotic to $3\pi^{-2}\log^2 n$. The number of decompositions of n into two factors whose g. c. d. has a certain property is asymptotic to the product of $\log n$ by the probability that the g. c. d. of two numbers taken at random has the same property.

E. Busche[103] gave a geometric proof of his[95] formula. But if we take $\Phi(x)$ to be a continuous function decreasing as x increases, with $\Phi(0) > 0$, then the number of positive divisors of y which are $\leq \psi(y)$ is $\Sigma[\Phi(x)/x]$, summed for $x = 1, 2, \ldots$, with $\Phi(x) \geq 0$. This result is extended to give the number of non-associated divisors of $y + zi$ whose absolute value is $\leq \phi(y, z)$.

J. W. L. Glaisher[104] considered the excess $H(n)$ of the number of divisors $\equiv 1 \pmod 3$ of n over the number of divisors $\equiv 2 \pmod 3$, proved that $H(pq) = H(p)H(q)$ if p, q are relatively prime, and discussed the relation of $H(n)$ to Jacobi's[11] $E(n)$.

Glaisher[105] gave recursion formulae for $H(n)$ and a table of its values for $n = 1, \ldots, 100$.

L. Gegenbauer[106] found the mean value of the number of divisors of an integer which are relatively prime to given primes $p_1, \ldots, p_\sigma$, and are also $(\rho\tau)$th powers and have a complementary divisor which is divisible by no τth powers. Also the mean of the sum of the reciprocals of the kth powers of those divisors of an integer which are prime to $p_1, \ldots, p_\sigma$ and are rth powers. Also many similar theorems.

Gegenbauer[106a] expressed $\Sigma_{x=0}^{x=n} F(4x+1)$ and $\Sigma F(4x+3)$ in terms of Jacobi's symbols (Δ/y) and greatest integers $[y]$ when $F(x)$ is the sum of the kth powers of those divisors $\leq \sqrt{x}$ of x which are prime to D, or are divisible by no rth power > 1, etc.; and gave asymptotic evaluations of these sums.

J. P. Gram[107] considered the number $D_n(m)$ of divisors $\leq m$ of n, the number $N_{2,3\ldots}(n)$ of integers $\leq n$ which are products of powers of the primes $2, 3, \ldots$, and the sum $L_{2,3\ldots}(n)$ of the values of $\lambda(k)$ whose arguments k are the preceding N numbers, where $\lambda(2^\alpha 3^\beta \ldots) = (-1)^{\alpha+\beta+\cdots}$.
If $p = p_1^{\alpha_1} p_2^{\alpha_2} \ldots$, where the p_i are distinct primes,

$$D_p(n) = N(n) - \Sigma N(n/p_1^{\alpha_1+1}) + \Sigma N(n/p_1^{\alpha_1+1} p_2^{\alpha_2+1}) - \ldots.$$

[101]Bull. Soc. Math. France, 16, 1888, 150.
[102]Atti R. Accad. Lincei, Rendiconti, 4, 1888, I, 452–7.
[103]Jour. für Math., 104, 1889, 32–37.
[104]Proc. London Math. Soc., 21, 1889–90, 198–201, 209.   [105]*Ibid.*, 395–402. See Glaisher.[141]
[106]Denkschriften Ak. Wiss. Wien (Math.), 57, 1890, 497–530.
[106a]Sitzungsber. Ak. Wiss. Wien (Math.), 99, 1890, IIa, 390–9.
[107]Det K. Danske Videnskab. Selskabs Skrifter (natur. og math.), (6), 7, 1890, 1–28, with résumé in French, 29–34.

In particular, if the p_i include all the primes in order, we may replace $N(x)$ by $[x]$, the greatest integer $\leq x$. Since there are as many divisors $> a$ of n as there are divisors $< n/a$,

$$D_p(n) + D_p\left(\frac{p}{n}\right) = \epsilon + \Pi(a_i + 1),$$

where $\epsilon = 1$ or 0 according as n is or is not a divisor of p. These two formulas serve as recursion formulas for the computation of $N(n)$. For the case of two primes $p_1 = 2$, $p_2 = 3$,

$$N(2^x) = x + 1 + \sum_{j=1}^{x}\left[\frac{j \log 2}{\log 3}\right], \qquad N(3^y) = y + 1 + \sum_{j=1}^{y}\left[\frac{j \log 3}{\log 2}\right].$$

The functions L satisfy similar formulas and are computed similarly.

J. W. L. Glaisher[108] stated a theorem, which reduces for $m = 1$ to Halphen's,[40]

$$S \equiv \sigma_m(n) - 3\sigma_m(n-1) + 5\sigma_m(n-3) - 7\sigma_m(n-6) + 9\sigma_m(n-10) - \ldots$$

$$= 2\Sigma\binom{m}{k}\{\sigma_{m-k}(n-1) - (1^k + 2^k)\sigma_{m-k}(n-3) + (1^k + 2^k + 3^k)\sigma_{m-k}(n-6) - \ldots\}$$
$$+ \delta(-1)^{g-1}(1 + 2^{m+1} + 3^{m+1} + \ldots + g^{m+1}),$$

provided m is odd, where k ranges over the even numbers $2, 4, \ldots, m-1$, while $\delta = 0$ or $\delta = 1$ according as n is not or is of the form $g(g+1)/2$. As in Glaisher[67] for $m = 1$, the series are stopped before any term $\sigma_i(n-n)$ is reached; but, if we retain such terms, we must set $\delta = 0$ for every n and define $\sigma_i(0)$ by

$$\sigma(0) = \frac{n}{m+2}, \qquad \binom{m}{3}\sigma_3(0) = \left\{\frac{m}{m+2} - (m+1)B_1\right\}n,$$

$$\binom{m}{5}\sigma_5(0) = \left\{\frac{m}{m+2} - (m+1)B_1 + \binom{m+1}{3}\frac{B_2}{2}\right\}n,$$

$$\binom{m}{7}\sigma_7(0) = \left\{\frac{m}{m+2} - (m+1)B_1 + \binom{m+1}{3}\frac{B_2}{2} - \binom{m+1}{5}\frac{B_3}{3}\right\}n, \ldots,$$

where $B_1, B_2, \ldots$ are the Bernoullian numbers.

Glaisher[109] stated the simpler generalization of Halphen[40]:

$$S + \Sigma_k \frac{1}{2^k(k+1)}\binom{m}{k}\{\sigma_{m-k}(n) - 3^{k+1}\sigma_{m-k}(n-1) + 5^{k+1}\sigma_{m-k}(n-3) - \ldots\}$$

$$= \delta(-1)^{g-1}\frac{1}{2^{m+1}(m+1)}\{g(2g+1)^{m+1} - 1 - 3^{m+1} - 5^{m+1} - \ldots - (2g-1)^{m+1}\},$$

where the summation index k ranges over the even numbers $2, 4, \ldots, m-1$, and m is odd. If we include the terms $\sigma_{2r-1}(0) = (-1)^r B_r/(4r)$ in the left member, the right member is to be replaced by

$$\delta(-1)^{g-1}\frac{(2g+1)^{m+2}}{2^{m+2}(m+2)}.$$

[108] Messenger Math., 20, 1890–1, 129–135.
[109] Ibid., 177–181.

Glaisher[110] considered the set $G_n\{\psi(d), \chi(d), \ldots\}$ of the values of $\psi(d)$, $\chi(d), \ldots$ when d ranges over all the divisors of n, and wrote $-G(\psi, \chi, \ldots)$ for $G(-\psi, -\chi, \ldots)$. By use of the ζ-function (12), he proved (p. 377) that the numbers given by

$$G_n(d) - G_{n-1}(d, d\pm1) + G_{n-3}(d, d\pm1, d\pm2) - G_{n-6}(d, d\pm1, d\pm2, d\pm3) + \ldots$$

all cancel if n is not a triangular number, but reduce to one 1, two 2's, three 3's, $\ldots$, g g's, each taken with the sign $(-)^{g-1}$, if n is the gth triangular number $g(g+1)/2$. For example, if $n=6$, whence $g=3$,

$$\{1, 2, 3, 6\} - \{1, 5; 2, 6; 0, 4\} + \{1, 3; 2, 4; 0, 2; 3, 5; -1, 1\}$$
$$= \{1, 2, 2, 3, 3, 3\}.$$

Let $\psi(d)$ be an odd function, so that $\psi(-d) \equiv -\psi(d)$, and let $\Sigma_r f(d)$ denote the sum of the values of $f(d)$ when d ranges over the divisors of r. Then the above theorem implies that

$$\Sigma_n \psi(d) - \Sigma_{n-1}\{\psi(d) + \psi(d\pm1)\} + \Sigma_{n-1-2}\{\psi(d) + \psi(d\pm1) + \psi(d\pm2)\}$$
$$- \Sigma_{n-1-2-3}\{\psi(d) + \psi(d\pm1) + \psi(d\pm2) + \psi(d\pm3)\} + \ldots$$
$$= \delta(-1)^{g-1}\{\psi(1) + 2\psi(2) + 3\psi(3) + \ldots + g\psi(g)\},$$

where $\delta = 0$ or 1 according as n is not or is of the form $g(g+1)/2$, and where $\psi(d\pm i)$ is to be replaced by $\psi(d+i) + \psi(d-i)$. Taking $\psi(d) = d^m$, where m is odd, we obtain Glaisher's[108] recursion formula for $\sigma_m(n)$, other forms of which are derived. For the function[74] ζ_3, we derive

$$\zeta_3(n) + \zeta_3(n-1) + \zeta_3(n-3) + \ldots + 6\{\zeta(n-1) - (1^2-2^2)\zeta(n-3)$$
$$+ (1^2-2^2+3^2)\zeta(n-6) - \ldots\}$$
$$= (-1)^{g-1}(1^4 - 2^4 + 3^4 - \ldots + (-1)^{g-1}g^4) \text{ or } 0,$$

according as n is of the form $g(g+1)/2$ or not.

Next he proved a companion theorem to the first:

$$G_n\binom{2d+1}{-[2d-1]} - G_{n-1}\binom{2d+3}{-[2d-3]} + G_{n-3}\binom{2d+5}{-[2d-5]} - G_{n-6}\binom{2d+7}{-[2d-7]} + \ldots$$

all cancel if n is not a triangular number, but reduce to 1, 3, 5, $\ldots$, $2g-1$, each taken with the sign $(-)^g$, together with $(-1)^{g+1}(2g+1)$ taken g times, if n is the gth triangular number $g(g+1)/2$. For example, if $n=6$,

$$\left\{\begin{matrix} 3, & 5, & 7, & 13 \\ -1, & -3, & -5, & -11 \end{matrix}\right\} - \left\{\begin{matrix} 5, & 13 \\ -1, & -7 \end{matrix}\right\} + \left\{\begin{matrix} 7, & 11 \\ -3, & -1 \end{matrix}\right\} = \left\{-1, -3, -5, 7, 7, 7\right\}.$$

Hence if $\chi(d)$ be any even function, so that $\chi(-d) = \chi(d)$,

$$\Sigma_n\{\chi(2d+1) - \chi(2d-1)\} - \Sigma_{n-1}\{\chi(2d+3) - \chi(2d-3)\} + \Sigma_{n-3} - \ldots$$
$$= \delta(-1)^{g-1}\{g\chi(2g+1) - \chi(1) - \chi(3) - \ldots - \chi(2g-1)\}.$$

Taking $\chi(k) = k^{m+1}$, where k and m are odd, we get Glaisher's[109] formula.

---

[110]Proc. London Math. Soc., 22, 1890–1, 359–410. Results stated in London, Edinb., Dublin Phil. Mag., (5), 33, 1892, 54–61.

He proved two theorems relating to the divisors of $1, 2, \ldots, n$:

$$G_n\left(\begin{matrix} d+1 \\ -[d-1] \end{matrix}\right) - (G_{n-1}+G_{n-2})\left(\begin{matrix} -d+2 \\ -[d-2] \end{matrix}\right)$$
$$+ (G_{n-3}+G_{n-4}+G_{n-5})\left(\begin{matrix} d+3 \\ -[d-3] \end{matrix}\right) - \ldots$$

all cancel with the exception of $-2, -4, \ldots, -(p-2)$, each taken twice, p taken $p-1$ times and -0, if p be even; but with the exception of $1, 3, \ldots, p-2$, each taken twice, and $-p$ taken $p-1$ times, if p be odd, where $p(p+1)/2$ is the triangular number next $>n$;

$$G_n(d) - (G_{n-1}+G_{n-2})(d \pm 1) + (G_{n-3}+G_{n-4}+G_{n-5})(d, d \pm 2)$$
$$- (G_{n-6}+ \ldots +G_{n-9})(d \pm 1, d \pm 3) + (G_{n-10}+ \ldots +G_{n-14})(d, d \pm 2, d \pm 4) - \ldots$$

all cancel with the exception of k taken k times, for $k = 1, 3, 5, \ldots, p-1$, if p be even; and of $-k$ taken k times, for $k = 2, 4, 6, \ldots, p-1$, if p be odd; here zeros are ignored.

The last two theorems yield (as before) corresponding relations for any even function χ and any odd function ψ. Applying them to $\chi(d+1) = (d+1)^m$ and $\psi(d) = d^m$, where m is odd, and in the first case dividing by $2(m+1)$, and modifying the right members, we get for

$$T \equiv \sigma_m(n) - 2\{\sigma_m(n-1)+\sigma_m(n-2)\} + 3\{\sigma_m(n-3) \\ +\sigma_m(n-4)+\sigma_m(n-5)\} - \ldots$$

the respective relations

$$T + \Sigma_k \frac{1}{k+1}\binom{m}{k}\{\sigma_{m-k}(n) - 2^{k+1}\left(\sigma_{m-k}(n-1)+\sigma_{m-k}(n-2)\right)$$
$$+ 3^{k+1} \text{ (next three)} - \ldots\}$$
$$= (-1)^p\left\{\frac{p^{m+2}}{2(m+2)} - B_1 p^m + \frac{2^2}{3}\binom{m}{2}\frac{B_2}{2}p^{m-2} - \frac{2^4}{5}\binom{m}{4}\frac{B_3}{3}p^{m-4} + \ldots \pm 2^{m-1}\frac{B_s}{s}p\right\},$$

where $s = (m+1)/2$ and $\sigma_i(0)$ terms are suppressed;

$$T = \Sigma_k 2\binom{m}{k}\{\sigma_{m-k}(n-1)+\sigma_{m-k}(n-2) - 2^k \text{ (next three)} + (1^k+3^k)(\text{next four})$$
$$- (2^k+4^k)(\text{next five}) + (1^k+3^k+5^k)(\text{next six}) - (2^k+4^k+6^k)(\text{next seven}) + \ldots\}$$
$$+ \left\{\begin{matrix} 1^{m+1}+3^{m+1}+5^{m+1}+ \ldots +(p-1)^{m+1}, & \text{if } p \text{ be even,} \\ -2^{m+1}-4^{m+1}-6^{m+1}- \ldots -(p-1)^{m+1}, & \text{if } p \text{ be odd,} \end{matrix}\right.$$

where, in each, k takes the values $2, 4, \ldots, m-1$. These sums of like powers of odd or even numbers are expressed by the same function of Bernoullian numbers. For $m=1$, the first formula becomes that by Glaisher,[64] republished.[67] Three further G_n formulas are given, but not applied to σ_n.

J. Hammond[111] wrote $(n; m) = 1$ or 0 according as n/m is integral or fractional, also $\tau(x) = \sigma(x) = 0$ if x is fractional, and stated that

$$\tau(n/m) = \sum_{j=1}^{\infty} (n; jm), \qquad \sigma(n/m) = \sum_{j=1}^{\infty} j(n; jm).$$

[111]Messenger Math., 20, 1890-1, 158-163.

From the sum of Euler's $\phi(d)$ for the divisors d of n, he obtained

$$\sigma(n) = \sum_{j=1}^{n} \tau\left(\frac{n}{j}\right)\phi(j), \qquad n\tau(n) = \sum_{j=1}^{n} \sigma\left(\frac{n}{j}\right)\phi(j).$$

E. Lucas[112] proved the last formulas, the result of Cesàro,[44] and the related one $\sigma(n) + s_n = s_{n-1} + 2n - 1$.

A. Berger[113] considered the mean of the number of decompositions of $1, 2, \ldots, x$ into three or more factors, and gave long expressions for $\psi(1) + \ldots + \psi(n)$, where $\psi(k) = \Sigma d^s d_1{}^{s_1}$, summed for the solutions of $dd_1 = k$. He gave (pp. 116–125) complicated results on the mean value of $\sigma_k(n)$.

D. N. Sokolov and D. T. Egorov[113a] proved, by use of Bougaief's formulas for sums extending over all the divisors of a number, the formulas in Liouville's[25-29] series of four articles.

J. W. L. Glaisher[114] gave Zeller's[66] formula and

$$P(n-1) + 2^2 P(n-2) - 5^2 P(n-5) - 7^2 P(n-7) + \ldots$$
$$= \frac{-1}{12}\{5\sigma_3(n) - (18n-1)\sigma(n)\},$$

where $1, 2, 5, \ldots$ are pentagonal numbers $(3r^2 \pm r)/2$ and $P(0) = 1$.

Glaisher[115] proved formulæ which are greatly shortened by setting

$$a_{ij}(n) = n^i\sigma_j(n) - 3(n-1)^i\sigma_j(n-1) + 5(n-3)^i\sigma_j(n-3) - 7(n-6)^i\sigma_j(n-6) + \ldots.$$

Write a_{ij} for $a_{ij}(n)$. Besides the formula [of Halphen[40]] $a_{01} = 0$, he gave

$$a_{03} - 2a_{11} = 0, \qquad a_{05} - 10a_{13} + \frac{40}{3}a_{21} = 0,$$

$$a_{07} - \frac{126}{5}a_{15} + \frac{756}{5}a_{23} - 168a_{31} = 0,$$

$$a_{09} - 50a_{17} + 720a_{25} - 3360a_{33} + 3360a_{41} = 0,$$

with the agreement that $\sigma(0) = n/3$ and

$$\sigma_3(0) = \frac{-t^2+1}{240}, \qquad \sigma_5(0) = \frac{t^3-1}{504}, \qquad \sigma_7(0) = \frac{-t^4+1}{480}, \qquad \sigma_9(0) = \frac{t^5-1}{264},$$

where $t = 8n+1$, but did not find the general formula of this type. Next, he gave five formulas of another set, the first one being that of his earlier paper,[64] the second involving the same function of σ_3 with added terms in $r\sigma(r)$. Finally, denoting Euler's formula (2) by $E\sigma(n) = 0$, it is shown that

$$5E\sigma_3(n) - 18E\{n\sigma(n)\} = 0.$$

Glaisher[116] showed that his[76] third formula holds for all odd numbers v not expressible as a sum of three squares and hence in particular for the

[112]Théorie des nombres, 1891, 403–6, 374, 388.
[113]Nova Acta Soc. Upsal., (3), 14, 1891 (1886), No. 2, p. 63.
[113a]Math. Soc. Moscow, 16, 1891, 89–112, 236–255.
[114]Messenger Math., 21, 1891–92, 47–8.
[115]Ibid., 49–69.
[116]Ibid., 122, 126. The further results are quoted in the chapter on sums of three squares.

former case $v \equiv 7 \pmod 8$. Also the left member of the third formula equals

$$4\{E(v-1)-3E(v-9)+5E(v-25)-\ldots\}$$

when v is odd, provided $E(0)=1/4$. If $\Delta'(n)$ denotes the sum of those divisors of n whose complementary divisors are odd,

$$\Delta'(n)-2\Delta'(n-1)+2\Delta'(n-4)-2\Delta'(n-9)+\ldots=0 \text{ or } (-1)^{n-1}n,$$

according as n is not or is a square. [Cf. Lipschitz.[84]] Since $\Delta'(n)=\sigma(n)$ for n odd, we deduce a formula involving σ's and Δ''s.

M. Lerch[117] proved (11) and

$$\sum_{k=1}^{n} \sigma_s(k)=\Sigma k^s\left[\frac{n}{k}\right], \qquad \Sigma F(k)=\Sigma f(k)\left[\frac{n}{k}\right],$$

if $F(n)=\Sigma f(d)$, d ranging over the divisors of n.

K. Th. Vahlen[118] proved Liouville's[23] formula and Jacobi's[10] result.

A. P. Minin[119] proved that 2, 8, 9, 12, 18, $8q$ and $12p$ (where q is a prime >2, p a prime >3) are the only numbers such that each is divisible by the number of its divisors and the quotient is a prime. Minin[120] found that 1, 3, 8, 10, 18, 24 and 30 are the only numbers N for which the number of divisors equals the number of integers $<N$ and prime to N.

M. Lerch[121] considered the number $\chi(a, b)$ and sum $X(a, b)$ of the divisors $\leq b$ of a, proved his[100] final formula (17) and

$$\sum_{a=1}^{c} X(m-an, a)= \sum_{a=1}^{c} a\{\chi(m-an, n)-\psi(m-an, a)\},$$

(18) $$\sum_{a=0}^{c} \psi\left(m-an, \frac{a}{r}\right)= \sum_{a=0}^{c} \chi(m-an, rn), \qquad c\equiv\left[\frac{m-1}{n}\right].$$

If δ ranges over the divisors of n,

$$\frac{1}{n}\sum_{a=0}^{n-1} \tau\{(a-am, n)\}=\Sigma\frac{(\delta, m; a)}{\delta}, \qquad \frac{1}{n}\sum_{a=0}^{n-1}\sigma\{(a-am, n)\}=\Sigma(\delta, m; a),$$

$$\sum_{a=1}^{n} (am, n)=n\Sigma\frac{\phi(\delta)}{\delta}\cdot(\delta, (m, n)).$$

He quoted (p. 8) from a letter to him from Chr. Zeller that $\sum_{a=1}^{m-1} a\psi(m-a, a)$ equals the sum of the remainders obtained on dividing m by the integers $<m$.

M. Lerch[122] proved that

$$\Sigma\psi(m+\rho-\sigma n, \sigma)=\Sigma\chi(m+\rho-\sigma n, n)-\Sigma\left[\frac{m+\rho}{n}-\rho\right],$$

$$\Sigma\psi(m-\rho-\rho n, \sigma)=\Sigma\chi(m-\rho-\sigma n, n)-\frac{1}{2}\left[\frac{m-1}{n+1}\right]\cdot\left[\frac{m+n}{n+1}\right],$$

[117]Casopis, Prag, 21, 1892, 90–95, 185–190 (in Bohemian). Cf. Jahrbuch Fortschritte Math., 24, 1892, 186–7.
[118]Jour. für Math., 112, 1893, 29.
[119]Math. Soc. Moscow, 17, 1893, 240–253.
[120]Ibid., 17, 1894, 537–544.
[121]Prag Sitzungsberichte (Math.), 1894, No. 11.
[122]Ibid., No. 32.

summed for ρ, $\sigma = 0, 1, \ldots$, with $\rho \leqq \sigma$. Also,

$$\sum_{a=0}^{m-1} (-1)^a \psi(m-a, a) = 2 \sum_{a=0}^{m-1} (-1)^a \Theta'(m-a) + (-1)^m m,$$

$$\sum_{a=0}^{m-1} \psi'(m-a, a) = \sum_{k=1}^{m} (-1)^{k-1} \left[\frac{m}{k}\right],$$

$$\sum_{a=0}^{m-1} \psi_0(m-a, 2a) = m + \frac{(-1)^{[\sqrt{m}]}-1}{2} + 2 \sum_{\nu=1}^{} (-1)^\nu \left[\frac{m-\nu^2}{2\nu}\right],$$

where $\Theta'(k)$ is the number of odd divisors of k; $\psi'(n, a)$ is the number of divisors $> a$ of n whose complementary divisors are odd; while $\psi_0(k, \mu)$ is the number of even divisors $> \mu$ of k.

In No. 33, he expressed in terms of greatest integer functions

$$\sum_{\rho, \sigma} \{\psi(m-\rho-\sigma n, k+\sigma) - \chi(m-\rho-\sigma n, n)\},$$

$$\sum_{a} \{\psi(m-a, k+a) - (k+a)\psi(m-a, k+a)\}.$$

E. Busche[123] gave a geometrical proof of Meissel's[22] (11).

J. Schröder[124] obtained (11) and the first relation (15) of Lerch[98] as special cases of the theorem that

$$\sum_{\rho_1, \ldots, \rho_m =}^{0, 1, 2, \ldots} \psi_{r\nu+s}(n - r \sum_{i=1}^{m} i\rho_i, \sum_{i=1}^{m} \rho_i)$$

equals the coefficient of x^n in the expansion of

$$\frac{1 - \prod_{i=0}^{m-1} (1 - x^{ri+s})}{(1 - x^{rm}) \prod_{i=0}^{m-1} (1 - x^{ri+s})},$$

where $\psi_{r\nu+s}(a, \beta)$ is the number of divisors of a which are $> \beta$ and have a complementary divisor of the form $r\nu + s (\nu = 0, 1, \ldots)$. He obtained

$$\sum_{\rho=}^{0, 1, \ldots} \psi_{r\nu+1}(n - r\rho, \rho) = \left[\frac{n+r-1}{r}\right].$$

Schröder[125] determined the mentioned coefficient of x^n.

Schröder[126] proved the generalization of (11):

$$\sum_{\rho=1}^{n} \left[\frac{n}{\rho}\right]^r = \sum_{\rho=1}^{\nu} \left[\frac{n}{\rho}\right]^r + \sum_{\rho=2}^{\nu} \{\rho^r - (\rho-1)^r\} \left[\frac{n}{\rho}\right] + n - \nu^{r+1}, \qquad \nu = [\sqrt{n}].$$

For $\sigma(1) + \ldots + \sigma(n)$, Dirichlet,[17] end, he gave the value

$$\sum_{s=1}^{n} s \left[\frac{n}{s}\right] = \frac{1}{2} \sum_{\rho=1}^{\nu} \left\{\left[\frac{n}{\rho}\right]^2 + (2\rho+1)\left[\frac{n}{\rho}\right]\right\} - \frac{1}{2}\nu(\nu+1).$$

E. Busche[127] proved that if $X = \phi(m)$ is an increasing (or decreasing) function whose inverse function is $m = \Phi(X)$, the divisors of the natural

[123]Mittheilungen Math. Gesell. Hamburg, 3, 1894, 167–172.
[124]Ibid., 177–188.
[125]Ibid., 3, 1897, 302–8.
[126]Ibid., 3, 1895, 219–223.
[127]Ibid., 3, 1896, 239–40.

numbers between $\phi(m)$ and a, including the limits, are the numbers x from 1 to a (or those $\geq a$) each taken $\xi = [\Phi(x)/x]$ times, and the numbers within the limits which are multiples of x are $x, 2x, \ldots, \xi x$. For example, if $a = 3$, $\phi(m) = 900/m^2$, then $\Phi(x) = 30/\sqrt{x}$ and it is a question of the divisors of $3 \ldots, 17$; for $x = 3$, $\xi = 5$ and 3 is a divisor of 3, 6, 9, 12, 15. For $\Phi(x) = n$, $a = 1$, the theorem states that among the divisors of $1, \ldots, n$ any one x occurs $[n/x]$ times and that these divisors are $1, \ldots, n$; $1, \ldots, [n/2]$; $1, \ldots$, $[n/3]$; etc.   Hence the sum of the divisors of $1, \ldots, n$ is

$$\sum_{x=1}^{n} x\left[\frac{n}{x}\right] = \tfrac{1}{2} \sum_{x=1}^{n} \left\{ \left[\frac{n}{x}\right]^2 + \left[\frac{n}{x}\right] \right\}$$

and their product is

$$\prod_{x=1}^{n} x^{[n/x]} = \prod_{x=1}^{n} [n/x]!.$$

He proved (pp. 244–6) that the number of divisors $\equiv r \pmod{s}$ of $1, 2, \ldots, n$ equals $A + B$, where A is the number of integers $[n/x]$ for $x = 1, \ldots, n$ which have one of the residues $r, r+1, \ldots, s-1 \pmod{s}$, and B is the number of all divisors of $1, 2, \ldots, [n/s]$. The number of the divisors δ of m, such that

$$\sqrt{\frac{m}{n}} \leq \delta \leq n$$

and such that δ' divides m/δ, equals the number of divisors of $1, 2, \ldots, n$. The number of primes among $n, [n/2], \ldots, [n/n]$ equals the number of those divisors of $1, \ldots, n$ which are primes decreased by the number of divisors which exceed by unity a prime.

P. Bachmann[128] gave an exposition of the work of Dirichlet,[14, 17] Mertens,[37] Hermite,[57] Lipschitz,[58] Cesàro,[60] Gegenbauer,[77] Busche,[123, 127] Schröder.[124, 126]

N. V. Bougaief[129] stated that

$$n + \Sigma d\left[\frac{n}{d-1}\right] = \sum_{j=1}^{n} X\left(n, 1 + \left[\frac{n}{j}\right]\right), \qquad \nu + \Sigma\left[\sqrt{\frac{n}{d-1}}\right] = \sum_{j=1}^{\nu} \chi\left(n, 1 + \left[\frac{n}{j^2}\right]\right),$$

where d ranges over the divisors > 1 of n, and $\nu = [\sqrt{n}]$;

$$\Sigma d\left[\frac{n}{d^2}\right] = \sum_{j=1}^{n} X\left(n, \left[\sqrt{\frac{n}{j}}\right]\right),$$

where d ranges over the divisors of n for which $d^2 < n$. If θ is any function,

$$n \Sigma \frac{n}{d} \theta(d) = \sum_{j=1}^{n} \Sigma_d \theta(d),$$

where, on the left, d ranges over all the divisors of n; on the right, only over those $\leq [n^2/j]$. For $\theta(d) \equiv 1$, this gives

$$n\sigma(n) = \sum_{j=1}^{n} \chi\left(n, \left[\frac{n^2}{j}\right]\right).$$

[128]Die Analytische Zahlentheorie, 1894, 401–422, 431–6, 490–3.
[129]Comptes Rendus Paris, 120, 1895, 432–4. He used $\xi(a, b)$, $\xi_1(a, b)$ with the same meaning as $\chi(b, a)$, $X(b, a)$ of Lerch,[121] and $\xi_1(n)$ for $\sigma(n)$.

M. Lerch[130] proved relations of the type

$$\sum_{k=1}^{m} \chi(k,\,a) = a\left[\frac{m}{a}\right] + \sum_{k=1}^{m} \psi\left(k,\,\frac{m}{a}\right).$$

The number of solutions of $[n/x] = [n/(x+1)]$, $x<n$, is

$$\sum_{r\geq k} \psi(n-r,\,r) + \sum_{\rho<k} \chi(n-\rho,\,\rho) \qquad (k \equiv -\tfrac{1}{2} + \sqrt{n+1/4}\ ).$$

F. Nachtikal[131] gave an elementary proof of (15).

M. Lerch[132] proved that

$$\sum_{\sigma>rsa} \left\{ \psi\left(m-\sigma a,\,\frac{\sigma}{r}\right) + \psi(m-\sigma a,\,ra) \right\}$$

remains unaltered if we interchange r and s. He proved (18) and showed that it also follows from the special case (17). From (17) for $n=2$ he derived

$$\sum_{a=0}^{c} \psi(m-2a,\,a) = \frac{3m+1}{4} + (-1)^m \frac{m-1}{4}, \qquad c \equiv \left[\frac{m-1}{2}\right].$$

L. Gegenbauer[132a] proved a formula which includes as special cases four of the five general formulas by Bougaief.[129] When x ranges over a given set S of n positive integers, the sum $\Sigma f(x)[\chi(x)]$ is expressed as sums of expressions $\Phi(\rho)$ and $\Phi_1(\rho)$, where ρ takes values depending upon χ, while $\Phi(z)$ is the sum of the values of $f(x)$ for x in S and $x \geqq z$, and $\Phi_1(z)$ is the analogous sum with $x \leqq z$.

F. Rogel[133] differentiated repeatedly the relation

$$\prod_{\omega=1}^{\infty} (1-x^\omega)^{\omega^{n-1}} = e^{-T}, \qquad T \equiv \sum_{\omega=1}^{n} \sigma_n(\omega)\frac{x^\omega}{\omega}, \qquad |x|<1,$$

then set $x=0$ and found that

$$\sum_{i=1}^{r} \sum \frac{(-1)^i}{a_1!\dots a_r!} \left\{\frac{\sigma_n(2)}{2}\right\}^{a_2} \dots \left\{\frac{\sigma_n(r)}{r}\right\}^{a_r} = \sum_{i=1}^{r} \sum (-1)^i \binom{1}{a_1}\binom{2^{n-1}}{a_2}\dots\binom{r^{n-1}}{a_r},$$

the summations extending over all sets of a's for which

$$a_1+a_2+\dots+a_r=i, \qquad a_1+2a_2+\dots+ra_r=r.$$

Starting with the reciprocals of the members of the initial relation, he obtained similarly a second formula; subtracting it from the former result, he obtained

$$\sigma_n(r) = r^n + \tfrac{1}{2}\sum_{i=3}^{r}\sum\left\{\prod_{j=2}^{r-3}\binom{a_j+j^{n-1}-1}{a_j} - (-1)^i\prod_{j=1}^{r-3}\binom{j^{n-1}}{a_j}\right\}$$

$$-\Sigma\Sigma'\frac{1}{a_1!\dots a_{r-3}!}\prod_{j=2}^{r-3}\left\{\frac{\sigma_n(j)}{j}\right\}^{a_j},$$

---

[130]Casopis, Prag, 24, 1895, 25–34, 118–124; 25, 1896, 228–30.

[131]Ibid., 25, 1896, 344–6.

[132]Jornal de Sciencias Math. e Astr. (Teixeira), 12, 1896, 129–136.

[132a]Monatshefte Math. Phys., 7, 1896, 26.

[133]Sitzungsber. Gesell. Wiss. (Math.), Prag, 1897, No. 7, 9 pp.

where $i = 3, 5, 7, \ldots$ in Σ', while the a's range over the solutions of

$$a_1 + \ldots + a_{r-3} = i, \qquad a_1 + 2a_2 + \ldots + (r-3)a_{r-3} = r.$$

The case $n = 0$ leads to relations for $\tau(r)$.

J. de Vries[133a] proved the first formula of Lerch's.[117]

A. Berger[134] considered the excess $\psi(k)$ of the sum of the odd divisors of k over the sum of the even divisors and proved that

$$\psi(n) + \psi(n-1) + \psi(n-3) + \psi(n-6) + \psi(n-10) + \ldots = 0 \text{ or } n,$$

according as n is not or is a triangular number; also Euler's (2).

J. Franel[135] employed two arbitrary functions f, g and set

$$\theta(n) = \Sigma f(d) g\left(\frac{n}{d}\right), \qquad F(n) = \sum_{j=1}^{n} f(j), \qquad G(n) = \sum_{j=1}^{n} g(j),$$

where d ranges over the divisors of n. Then

$$\sum_{j=1}^{n} \theta(j) = \sum_{r=1}^{\nu} f(r) G\left[\frac{n}{r}\right] + \sum_{r=1}^{\nu} g(r) F\left[\frac{n}{r}\right] - F(\nu) G(\nu),$$

where $\nu = [\sqrt{n}]$. The case $f(x) = g(x) = 1$ gives Meissel's[22] (11). Next, he evaluated $\Sigma \vartheta(j)$, where $\vartheta(n) = \Sigma f(x) g(y) h(z)$, summed for the sets of positive integral solutions of $xyz = n$. In particular, $\vartheta(n)$ is the number of such sets if $f = g = h = 1$. Using Dirichlet's series, it is shown (p. 386) that

$$\sum_{j=1}^{n} \vartheta(j) = \frac{n}{2}\{(\log n + 3C - 1)^2 - 3C^2 + 6C_1 + 1\} + \epsilon,$$

where ϵ is of the order of magnitude of $n^{2/3} \log n$, C is Euler's constant and $C_1 = 0.0728\ldots$ [Piltz,[52] Landau[137]].

Franel[136] proved that

$$\sum_{r=1}^{p} \frac{\tau(r)}{r} = \frac{1}{2} \log^2 p + 2C \log p + \epsilon + A_0,$$

where A_0 is a coefficient in a certain expansion, and $\epsilon p^{1/2}$ remains in absolute value inferior to a fixed number for every p.

E. Landau[137] gave an immediate proof of (11) and of

$$\sum_{\nu=1}^{x} T_3(\nu) = \sum_{\nu=1}^{x} \tau(\nu) \left[\frac{x}{\nu}\right],$$

where $T_3(\nu)$ is the number of decompositions of ν into three factors. He obtained by elementary methods a formula yielding the final result of Franel[135] on $\Sigma T_3(\nu)$.

R. D. von Sterneck[137a] proved Jacobi's[10] formula for s^3.

[133a] K. Akad. Wetenschappen te Amsterdam, Verslagen, 5, 1897, 223.

[134] Nova Acta Soc. Sc. Upsaliensis, (3), 17, 1898, No. 3, p. 26.

[135] Math. Annalen, 51, 1899, 369–387.

[136] Ibid., 52, 1899, 536–8.

[137] Ibid., 54, 1901, 592–601.

[137a] Sitzungsber. Ak. Wiss. Wien (Math.), 109, IIa, 1900, 31–33.

J. Franel[138] stated that, if $f(n)$ is the number of positive integral solutions of $x^a y^b = n$, where a, b are distinct positive integers,

$$\sum_{r=1}^{n} f(r) = \zeta\left(\frac{b}{a}\right) n^{\frac{1}{a}} + \zeta\left(\frac{a}{b}\right) n^{\frac{1}{b}} + O\left(n^{\frac{1}{a+b}}\right),$$

where[90] $O(s)$ is of the order of magnitude of s. Taking $a=1$, $b=2$, we see that $f(n)$ is the number of divisors of q, where q^2 is the greatest square dividing n, and that the mean of $f(n)$ is $\pi^2/6$.

E. Landau[139] proved the preceding formula of Franel's.

Elliott[96] of Ch. V gave formulas involving $\sigma(n)$ and $\tau(n)$.

L. Kronecker[140] proved that the sum of the odd divisors of a number equals the algebraic sum of all its divisors taken positive or negative according as the complementary divisor is odd or even (attributed to Euler[2]); proved (pp. 267–8) the result of Dirichlet[15] and (p. 345) proved (7) and found the median value (Mittelwert) of $\tau(n)$ to be $\log_e n + 2C$ with an error of the order of magnitude of $n^{-1/4}$ when the number of values employed is of the order of $n^{3/4}$. Calling a divisor of n a smaller or greater divisor according as it is less than or greater than $\sqrt{n}$, he found (pp. 343–369) the mean and median value of the sum of all smaller (or greater) divisors of $1, 2, \ldots, N$ [cf. Gegenbauer[77]], the sum of their reciprocals, and the sum of their logarithms. The mean of Jacobi's[11] $E(n)$ is $\pi/4$ (p. 374).

J. W. L. Glaisher[141] tabulated for $n=1, \ldots, 1000$ the values of the function[104] $H(n)$ and of the excess $J(n)$ of the number of divisors of n which are of the form $8k+1$ or $8k+3$ over the number of divisors of the form $8k+5$ or $8k+7$. When n is odd, $2J(n)$ is the number of representations of n by x^2+2y^2.

J. W. L. Glaisher[142] derived from Dirichlet's[17] formula, and also independently, the simpler formula

$$\sum_{s=1}^{n}\left[\frac{n}{s}\right] g(s) = -\rho G(\rho) + \sum_{s=1}^{\rho}\left[\frac{n}{s}\right] g(s) + \sum_{s=1}^{\rho} G\left\{\left[\frac{n}{s}\right]\right\},$$

where $\rho = [\sqrt{n}]$. The case $g(s) = 1$ gives Meissel's[22] formula (11), which is applied to find asymptotic formulæ involving $n/s - [n/s]$. The error of the approximation (7) is discussed at length (pp. 38–75, 180–2). The first formula above is applied (pp. 183–229) to find exact and asymptotic formulas for $\Sigma f(s)$, when $f(n)$ is Jacobi's[11] $E(n)$, Glaisher's[141] $H(n)$ or $J(n)$, or the excess $D(n)$ of the number of odd divisors of n over the number of even divisors, or more general functions (p. 215, p. 223) involving the number of divisors with specified residues modulo r.

G. Voronoï[143] proved a formula like Dirichlet's[17] (7), but with ϵ now of the same order of magnitude as $\sqrt[3]{n} \log_e n$.

[138]L'intermédiaire des math., 6, 1899, 53; 18, 1911, 52–3.
[139]Ibid., 20, 1913, 155.
[140]Vorlesungen über Zahlentheorie, I, 1901, 54–55.
[141]Messenger Math., 31, 1901–2, 64–72, 82–91.
[142]Quar. Jour. Math., 33, 1902, 1–75, 180–229.
[143]Jour. für Math., 126, 1903, 241–282.

H. Mellin[144] obtained asymptotic expressions for $\Sigma\tau(n)$, $\Sigma\sigma(n)$.

I. Giulini[145] noted that, if m and h are given integers, and $\beta(r)$ is the sum of the divisors $d=mk+h$ of r, then

$$\beta(1)+\ldots+\beta(n)=\Sigma_k d[n/d], \qquad k=0, 1,\ldots, [(n-h)/m].$$

The number and sum of the divisors $d=mk+h$ of $1,\ldots, n$ are

$$\sum_{k=0}^{[(n-h)/m]}\left[\frac{n}{d}\right], \qquad m\sum_{r=1}^{[n/(m+h)]}E_2\left(\frac{n-hr}{mr}\right)+h\sum_{s=1}^{[n/h]}\left[\frac{n-sh}{ms}+1\right],$$

respectively, where $E_2(x)=[x][x+1]/2$.

G. Voronoï[145a] gave for $T(x)$ the precise analytic expression

$$x(\log x+2C-1)+\tfrac{1}{4}+\tfrac{1}{2}\tau(x)-2\int_x^\infty g(t)dt+\int_0^\infty \{g(-x+ti)-g(-x-ti)\}i\,dt,$$

and (p. 515) approximations to these integrals, where

$$g(x)=-\tfrac{1}{4}\log x-\tfrac{1}{2}C-\frac{\log 4\pi^2 x}{4\pi^2 x}+\frac{1}{2\pi^2}\sum_{n=1}^\infty \tau(n)\log\frac{x}{n}\left(\frac{1}{x-n}+\frac{1}{x+n}\right).$$

He discussed at length the function $g(x)$ and (pp. 467, 480-514) the asymptotic value of $\Sigma\tau(n)(x-n)^k/k!$.

J. Schröder[146] proved that the sum of the νth powers of $1,\ldots, n$ is

$$\sum_{\rho=1}^n \rho^\nu\left[\frac{n}{\rho}\right]=n\sigma_{\nu-1}(n)+\sum_{\rho=t+1}^{n-1}\rho^\nu+\sum_{\rho=1}^{t}{}'\rho^\nu\left[\frac{n}{\rho}\right],$$

where $t=[n/2]$, and the accent on the last Σ denotes that the summation extends only over the values $\leq t$ of ρ which are not divisors of n.

E. Busche[147] proved that, if we multiply each divisor of m by each divisor of n, the number of times we obtain a given divisor a of mn is $\tau(\mu\nu/a)$, where μ is the g. c. d. of m, a, and ν is that of n, a. A like theorem is proved for the divisors of $mnp\ldots$. He stated (p. 233; cf. Bachmann[168]) that

$$\sigma_h(m)\sigma_h(n)=\Sigma d^h\sigma_h\left(\frac{mn}{d^2}\right),$$

where d ranges over the common divisors of m, n.

C. Hansen[148] denoted by $T_1(n)$ and $T_3(n)$ the number of divisors of n of the respective forms $4k-1$ and $4k-3$, and set

$$A_n=T_3(4n-3)-T_1(4n-3).$$

By use of Jacobi's $\theta_3(\nu, s)$ for $\nu=1/4$, he proved that

$$\sum_{n=1}^\infty A_n s^{4n-3}=\sum_{n=1}^\infty (-1)^{n+1}\frac{s^{2n-1}}{1-s^{4n-2}}=\frac{s-3s^9+5s^{25}-\cdots}{1-2s^4+2s^{16}+\cdots},$$

[144]Acta Math., 28, 1904, 49.
[145]Giornale di mat., 42, 1904, 103-8.
[145a]Annales sc. l'école norm. sup., (3), 21, 1904, 213-6, 245-9, 258-267, 472-480. Cf. Hardy.[180]
[146]Mitt. Math. Gesell. Hamburg, 4, 1906, 256-8.
[147]Ibid., 4, 1906, 229.
[148]Oversigt K. Danske Videnskabernes Selskabs Forhandlinger, 1906, 19-30 (in French).

and hence deduced the law of a recursion formula for A_n. The law of a recursion formula for $B_n = 4\{T_3(n) - T_1(n)\}$ is obtained from

$$\sum_{n=0}^{\infty} B_n s^{8n} \sum_{n=0}^{\infty} s^{(2n+1)^2} \cos(2n+1)\frac{\pi}{4} = \sum_{n=0}^{\infty} (2n+1)s^{(2n+1)^2} \sin(2n+1)\frac{\pi}{4},$$

with $B_0 = 1$, which was found by use of Jacobi's $\theta(\frac{1}{4}, s)$. Next,

$$\Phi(s) \equiv \sum_{n=1}^{\infty} \frac{s^n}{1+s^{2n}} = \frac{1}{4} \sum_{n=1}^{\infty} B_n s^n$$

is shown to satisfy the functional equation

$$\Phi(is) = \frac{i}{2}\{\Phi(s) - \Phi(-s)\} - \Phi(s^2) + 2\Phi(s^4).$$

If a convergent series $\Sigma c_n s^n$ is a solution $\Phi(s)$ of the latter, the coefficients are uniquely determined by the $c_{4k-3}(k=1, 2, \ldots)$, which are arbitrary. Hence the function B_n is determined for all values of n by its values for $n = 4k - 3$ $(k = 1, 2, \ldots)$.

S. Wigert[149] proved that, for sufficiently large values of n, $\tau(n) < 2^t$, where $t = (1+\epsilon)\log n \div \log \log n$, for every $\epsilon > 0$; while there exist certain values of n above any limit for which $\tau(n) > 2^s$, $s = (1-\epsilon)\log n \div \log \log n$.

J. V. Pexider[150] proved that, if a, n are positive, a an integer,

$$\sum_{k=1}^{[n]} \left[\frac{n}{k}\right] = \sum_{k=1}^{a} \left[\frac{n}{k}\right] + \sum_{k=1}^{[n/a]} \left[\frac{n}{k}\right] - a\left[\frac{n}{a}\right],$$

by the method used, for the case in which n is an integral multiple of a, by E. Cesàro.[60] Taking $a = [\sqrt{n}]$, we have the second equation (11). Proof is given of the first equation (11) and

$$\Sigma k\left[\frac{n}{k}\right] = \Sigma\sigma(k), \qquad \Sigma\left[\frac{n}{d}\right]\left[\frac{n-1}{d}\right] = \Sigma d^2 - \sigma[n],$$

where d ranges over the divisors of $[n]$.

O. Meissner[151] noted that, if $m = p_1^{e_1} \ldots p_n^{e_n}$, where p_1 is the least of the distinct primes $p_1, \ldots, p_n$, then

$$\prod_{i=1}^{n} \frac{p_i}{p_i - 1} > \frac{\sigma(m)}{m} > \prod_{i=2}^{n} \frac{p_i}{p_i - 1}, \qquad 1 < \frac{\sigma(m)}{m \log m} < G,$$

where G is finite and independent of m. If $k > 1$, $\sigma_k(m)/m^k$ is bounded.

W. Sierpinski[152] proved that the mean of the number of integers whose squares divide n, of their sum, and of the greatest of them, are

$$\frac{\pi^2}{6}, \qquad \frac{1}{2}\log n + \frac{3}{2}C, \qquad \frac{3}{\pi^2}\log n + \frac{9C}{\pi^2} + \frac{36}{\pi^4}\sum_{s=1}^{\infty}\frac{\log s}{s^2},$$

respectively, where C is Euler's constant.

J. W. L. Glaisher[153] derived formulas differing from his[110] earlier ones only in the replacement of d by $(-1)^{d-1}d$, i. e., by changing the sign of each

[149]Arkiv för mat., ast., fys., 3, 1906–7, No. 18, 9 pp.
[150]Rendiconti Circolo Mat. Palermo, 24, 1907, 58–63.
[151]Archiv Math. Phys., (3), 12, 1907, 199.
[152]Sprawozdania Tow. Nank. (Proc. Sc. Soc. Warsaw), 1, 1908, 215–226 (Polish).
[153]Proc. London Math. Soc., (2), 6, 1908, 424–467.

even divisor d. In the case of the theorems on the cancellation of actual divisors, the results follow at once from the earlier ones. But the recursion formulæ for σ_n and ζ_n are new and too numerous to quote. Cancellation formulas (pp. 449–467) are proved for the divisors whose complementary divisors are odd, and applied to obtain recursion formulæ for the related function $\Delta_r'(n)$ of Glaisher.[74, 87]

E. Landau[155] proved that log 2 is the superior limit for $x = \infty$ of $\log \tau(x) \cdot \log \log x \div \log x$.

M. Fekete[156] employed the determinant $R_{k,n}^{(t)}$ obtained by deleting the last t rows and last t columns of Sylvester's eliminant of $x^k - 1 = 0$ and $x^n - 1 = 0$. Set, for $k \leqq n$,

$$b_n(k) = 1 - |R_{k,n}^{(k-1)}|, \quad c_n(i,k) = |R_{i,k}^{(1)}|(1 - |R_{k,n}^{(k-1)}|)(1 - |R_{i,n}^{(i-1)}|)(1 - |R_{ik,n}^{(iE-1)}|).$$

Then $b_n(k) = 1$ or 0 according as k is or is not a divisor of n; while $c_n(i, k) = 1$ if $ik = n$ and i is relatively prime to k, but $= 0$ in the contrary cases. Thus

$$\tau(n) = \sum_{k=1}^{n} b_n(k), \qquad \sigma(n) = \sum_{k=1}^{n} k b_n(k),$$

while the number and sum of those divisors d of n, which are relatively prime to the complementary divisors n/d, equal, respectively,

$$\sum_{i,k=1}^{n} c_n(i, k), \qquad \frac{1}{2} \sum_{i,k=1}^{n} (i+k) c_n(i, k).$$

J. Schröder[157] deduced from his[124] final equation the results

$$\sum_{\rho=}^{0,1,\ldots} \psi_{r+s}\left(n - \rho, \left[\frac{\rho}{r}\right]\right) = \left[\frac{n}{s}\right], \qquad \Sigma \psi\left(n - \rho, \left[\frac{\rho}{r}\right]\right) = \sum_{s=1}^{r} \left[\frac{n}{s}\right].$$

The final sum equals $\Sigma_{s=1}^{s=n} \psi(s, [s/(r+1)])$.

P. Bachmann[158] gave an exposition of the work of Euler,[5, 6] Glaisher,[63, 64] Zeller,[66] Stern,[85] Glaisher,[110] Liouville.[30]

E. Landau[159] proved that the number of positive integers $\leqq x$ which have exactly n positive integral divisors is asymptotic to

$$A x^{1/(p-1)} (\log \log x)^{w-1} / \log x,$$

where p is the least prime factor of n, and p occurs exactly w times in n, while A depends only on n.

K. Knopp[160] obtained, by enumerations of lattice points,

$$\sum_{k=1}^{n} f_1(q, k) = \sum_{k=1}^{n} f_2(k, q) = \sum_{k=1}^{w} f_1(q, k) + \sum_{k=1}^{w} f_2(k, q) - F(w, w),$$

where $q = [n/k]$ and

$$f_1(r, k) = \sum_{j=1}^{r} f(j, k), \qquad f_2(k, s) = \sum_{j=1}^{s} f(k, j), \qquad F(r, s) = \sum_{j=1}^{s} f_1(r, j).$$

[155]Handbuch...Verteilung der Primzahlen, I, 1909, 219–222.
[156]Math. és Phys. Lapok (Math. phys. soc.), Budapest, 18, 1909, 349–370. German transl., Math. Naturwiss. Berichte aus Ungarn, 26, 1913 (1908), 196–211.
[157]Mitt. Math. Gesell. Hamburg, 4, 1910, 467–470.
[158]Niedere Zahlentheorie, II, 1910, 268–273, 284–304, 375.
[159]Annaes Sc. Acad. Polyt. do Porto, Coimbra, 6, 1911, 129–137.
[160]Sitzungsber. Berlin Math. Gesell., 11, 1912, 32–9; with Archiv Math. Phys.

Taking $f(h, k) \equiv 1$, we obtain Meissel's[22] (11), a direct proof of which is also given.   Taking $f(h, k) = f(h)g(hk)$, we get

$$\sum_{k=1}^{n} \sum_{j=1}^{q} f(j)g(jk) = \sum_{k=1}^{n} f(k) \sum_{j=1}^{q} g(jk), \qquad q = \left[\frac{n}{k}\right],$$

special cases of which yield many known formulas involving Möbius's function $\mu(n)$ or Euler's function $\phi(n)$.

E. Landau[161] proved the result due to Pfeiffer[90], and a theorem more effective than that by Piltz[52], having the O terms replaced by $O(x^a)$, where, for every $\epsilon > 0$,

$$a = \frac{k-1}{k+1} - \sigma + \epsilon.$$

E. Landau[162] extended the theorem of Piltz[52] to an arbitrary algebraic domain, defining $T_k(n)$ to be the number of representations of n as the norm of a product of k ideals of the domain.

J. W. L. Glaisher[163], generalizing his[142] formula, proved that

$$\sum_{s=1}^{n} F\left[\frac{n}{s}\right]g(s) = \sum_{s=1}^{\rho} F\left[\frac{n}{s}\right]g(s) + \sum_{s=1}^{\rho} G\left[\frac{n}{s}\right]f(s) - F(\rho)G(\rho),$$

where $F(s) = f(1) + \ldots + f(s)$, $G(s) = g(1) + \ldots + g(s)$, $\rho = [\sqrt{n}]$.   A similar generalization of another formula by Dirichlet[17] is proved, also analogous theorems involving only odd arguments.

Glaisher[164] applied the formulas just mentioned to obtain theorems on the number and sum of powers of divisors, which include all or only the even or only the odd divisors.   Among the results are (11) and those of Hacks.[96, 97]   The larger part of the paper relates to asymptotic formulas for the functions mentioned, and the theorems are too numerous to be cited here.

E. Landau[91] gave another proof of the result by Voronoï[143].   He proved (p. 2223) that $\tau(n) < 4n^{1/3}$.

J. W. L. Glaisher[165] stated again many of his[164] results, but without determining the limits of the errors of the asymptotic formulas.

S. Minetola[166] proved that the number of ways a product of m distinct primes can be expressed as a product of n factors is

$$\frac{1}{n!}\left\{n^m - \binom{n}{1}(n-1)^m + \binom{n}{2}(n-2)^m - \ldots \pm \binom{n}{n-1}1^m\right\}.$$

T. H. Gronwall[167] noted that the superior limits for $x = \infty$ of

$$\sigma_a(x)/x^a \quad (a > 1), \qquad \sigma(x)/(x \log \log x)$$

are the zeta function $\zeta(a)$ and e^C, respectively, C being Euler's constant.

[161]Göttingen Nachrichten, 1912, 687–690, 716–731.
[162]Trans. Amer. Math. Soc., 13, 1912, 1–21.
[163]Quar. Jour. Math., 43, 1912, 123–132.
[164]Ibid., 315–377. Summary in Glaisher.[165]
[165]Messenger Math., 42, 1912–13, 1–12.
[166]Il Boll. di Matematica Gior. Sc.-Didat., Roma, 11, 1912, 43–46; cf. Giornale di Mat., 45, 1907, 344–5; 47, 1909, 173, §1, No. 7.
[167]Trans. Amer. Math. Soc., 14, 1913, 113–122.

P. Bachmann[168] proved the final formula of Busche.[147]

K. Knopp[169] studied the convergence of $\Sigma b_n x^n/(1-x^n)$, including the series of Lambert[7], and proved that the function defined in the unit circle by Euler's[1] product (1) can not be continued beyond that circle.

E. T. Bell[170] proved that, if P is the product of all the distinct prime factors of m, and λ is their number, and d ranges over all divisors of m,

$$6^\lambda \Sigma \tau(d)\tau\left(\frac{m}{d}\right) = \tau(m)\,\tau(Pm)\,\tau(P^2m).$$

J. F. Steffensen[171] proved that,[90] if lx denotes $\log x$,

$$\sum_1 \frac{\tau(n)}{n} = \tfrac{1}{2}l^2\nu + O(l\nu), \qquad \sum_1 \frac{\sigma(n)}{n^2} = \frac{\pi^2}{6}l\nu + O(1).$$

S. Wigert[172] proved, for the sum $n \cdot s(n)$ of the divisors of n,

$$(1-\epsilon)e^C \log \log n < s(n) < (1+\epsilon)e^C \log \log n,$$

$$\sum_{n \leq x} s(n) = \frac{\pi^2}{6}x - \psi(x), \qquad \psi(x) = x\sum_{n>x}\frac{1}{n^2} + \sum_{n \leq x}\frac{1}{n}\rho\left(\frac{x}{n}\right),$$

for $\epsilon > 0$ and $\rho(x) = x - [x]$. For x sufficiently large,

$$(\tfrac{1}{4}-\epsilon) \log x < \psi(x) < (\tfrac{3}{4}+\epsilon) \log x.$$

Besides results on $\Sigma s(x)(x-n)^k$, $\Sigma s(n) \log x/n$, he proved that

$$\sum_{n \leq x} ns(n) = \frac{\pi^2 x^2}{12} + x\{\tfrac{1}{2}\log x - \psi(x)\} + O(x).$$

E. Landau[173] gave corrections and simplifications in the proofs by Wigert.[172]

E. T. Bell[174] introduced a function including as special cases the functions treated by Liouville,[25-29] restated his theorems and gave others.

J. G. van der Corput[175] proved, for $\mu(d)$ as in Chapter XIX,

$$\sum_{d=1}^{x} d^n\mu(d) \sum_{k=1}^{x/d} \sigma_n(k) = x.$$

S. Ramanujan[176] proved that $\tau(N)$ is always less than 2^k and 2^t, where[90]

$$k = \frac{\log N}{\log \log N} + O\left\{\frac{\log N}{(\log \log N)^2}\right\}, \qquad t = Li(\log N) + O\{\log Ne^{-a(\log \log N)^{1/2}}\},$$

for $Li(x)$ as in Ch. XVIII, and for a a constant. Also, $\tau(N)$ exceeds 2^k and 2^t for an infinitude of values of N. A highly composite number N is one for which $\tau(N) > \tau(n)$ when $N > n$; if $N = 2^{a_1}3^{a_2}\ldots p^{a_p}$, then $a_2 \geq a_3 \geq a_5 \geq$

[168]Archiv Math. Phys., (3), 21, 1913, 91.
[169]Jour. für Math., 142, 1913, 283–315; minor errata, 143, 1913, 50.
[170]Amer. Math. Monthly, 21, 1914, 130–1.
[171]Acta Math., 37, 1914, 107. Extract from his Danish Diss., "Analytiske Studier med Anvendelser paa Taltheorien," Kopenhagen, 1912.
[172]Ibid., 113–140.
[173]Göttingsche gelehrte Anzeigen, 177, 1915, 377–414.
[174]Univ. of Washington Publications Math. Phys., 1, 1915, 6–8, 38–44.
[175]Wiskundige Opgaven, 12, 1915, 182–4.
[176]Proc. London Math. Soc., (2), 14, 1915, 347–409.

...$\geq a_p$, while $a_p = 1$ except when $N = 4$ or 36. The value of λ for which $a_2 > a_3 > \ldots > a_\lambda$ is investigated at length. The ratio of two consecutive highly composite numbers N tends to unity. There is a table of N's up to $\tau(N) = 10080$. An N is called a superior highly composite number if there exists a positive number ϵ such that

$$\frac{\tau(N_2)}{N_2{}^\epsilon} < \frac{\tau(N)}{N^\epsilon} \geq \frac{\tau(N_1)}{N_1{}^\epsilon}$$

for all values of N_1 and N_2 such that $N_2 > N > N_1$. Properties of $\tau(N)$ are found for (superior) highly composite .numbers.

Ramanujan[177] gave for the zeta function (12) the formula

$$\frac{\zeta(s)\zeta(s-a)\zeta(s-b)\zeta(s-a-b)}{\zeta(2s-a-b)} = \sum_{j=1}^{\infty} j^{-s}\sigma_a(j)\sigma_b(j),$$

and found asymptotic formulæ for

$$\sum_{1}^{n}\tau^s(j), \qquad \sum_{j=1}^{n}\tau(jv+c), \qquad \prod_{j=1}^{n}\sigma_s(j), \qquad \sum_{j=1}^{n}\sigma_a(j)\sigma_1(j), \qquad D_v(n),$$

for $a = 0$ or 1, where

$$D_v(n) = \sum_{j=1}^{n}\tau(jv) = \Sigma\mu(d)\tau\left(\frac{v}{d}\right)D_1\left(\frac{n}{d}\right),$$

summed for the divisors d of v. If δ is a common divisor of u, v,

$$\tau(uv) = \sum_{1}^{\infty}\mu(n)\tau\left(\frac{u}{n}\right)\tau\left(\frac{v}{n}\right) = \Sigma\mu(\delta)\tau\left(\frac{u}{\delta}\right)\tau\left(\frac{v}{\delta}\right).$$

E. Landau[177a] gave another asymptotic formula for the number of decompositions of the numbers $\leq x$ into k factors, $k \geq 2$.

Ramanujan[178] wrote $\sigma_s(0) = \frac{1}{2}\zeta(-s)$ and proved that

$$\sum_{j=0}\sigma_r(j)\sigma_s(n-j) = \frac{\Gamma(r+1)\Gamma(s+1)}{\Gamma(r+s+2)} \cdot \frac{\zeta(r+1)\zeta(s+1)}{\zeta(r+s+2)}\sigma_{r+s+1}(n)$$

$$+ \frac{\zeta(1-r)+\zeta(1-s)}{r+s}n\sigma_{r+s-1}(n) + O(n^{2(r+s+1)/3}),$$

for positive odd integers r, s. Also that there is no error term in the right member if $r = 1$, $s = 1, 3, 5, 7, 11$; $r = 3$, $s = 3, 5, 9$; $r = 5$, $s = 7$.

J. G. van der Corput[179] wrote s for the g. c. d. of the exponents $a_1, a_2, \ldots$ in $m = \Pi p_i{}^{a_i}$ and expressed in terms of zeta function $\zeta(i)$, $i = 2, \ldots, k+1$,

$$\sum_{m=2}^{\infty}\{\sigma_k(s)-1\}/m$$

if $k > 1$; the sum being $1 - C$ if $k = -1$, where C is Euler's constant.

---

[177]Messenger Math., 45, 1915–6, 81–84.

[177a]Sitzungsber. Ak. Wiss. München, 1915, 317–28.

[178]Trans. Cambr. Phil. Soc., 22, 1916, 159–173.

[179]Wiskundige Opgaven, 12, 1916, 116–7.

G. H. Hardy[180] proved that for Dirichlet's[17] formula (7) there exists a constant K such that $\epsilon > Kn^{1/4}$, $\epsilon < -Kn^{1/4}$, for an infinitude of values of n surpassing all limit. In Piltz's[52] formula

$$\sum_{n=1}^{x} T_k(n) = x\{a_{k1}(\log x)^{k-1} + \ldots + a_{kk}\} + \epsilon_k,$$

$\epsilon_k > Kx^t$, $\epsilon_k < -Kx^t$, where $t = (k-1)/(2k)$. He gave two proofs of an equivalent to Voronoï's[145a] explicit expression for $T(x)$.

Hardy[181] wrote $\Delta(n)$ for Dirichlet's ϵ in (7) and proved that,[90] for every positive e, $\Delta(n) = O(n^{e+1/4})$ on the average, i. e.,

$$\frac{1}{n}\int_1^n |\Delta(t)| dt = O(n^{e+1/4}).$$

G. H. Hardy and S. Ramanujan[182] employed the phrase "almost all numbers have a specified property" to mean that the number of the numbers $\leq x$ having this property is asymptotic to x as x increases indefinitely, and proved that if f is a function of n which tends steadily to infinity with n, then almost all numbers have between $a-b$ and $a+b$ different prime factors, where $a = \log \log n$, $b = f\sqrt{a}$. The same result holds also for the total number of prime factors, not necessarily distinct. Also a is the normal order of the number of distinct prime factors of n or of the total number of its prime factors, where the normal order of $g(n)$ is defined to mean $f(n)$ if, for every positive ϵ, $(1-\epsilon)f(n) < g(n) < (1+\epsilon)f(n)$ for almost all values of n.

S. Wigert[183] gave an asymptotic representation for $\Sigma_{n \leq x} \tau(n)(x-n)^k$.

E. T. Bell[184] gave results bearing on this chapter.

F. Rogel[185] expressed the sum of the rth powers of the divisors $\leq q$ of m as an infinite series involving Bernoullian functions.

A. Cunningham[186] found the primes $p < 10^4$ (or 10^5) for which the number of divisors of $p-1$ is a maximum 64 (or 120).

Hammond[43] of Ch. XI and Rogel[243] of Ch. XVIII gave formulas involving σ and τ. Bougaief[59, 62] of Ch. XIX treated the number of divisors $\leq m$ of n. Gegenbauer[60] of Ch. XIX treated the sum of the ρth powers of the divisors $\geq m$ of n.

[180]Proc. London Math. Soc., (2), 15, 1916, 1–25.
[181]*Ibid.*, 192–213.
[182]Quar. Jour. Math., 48, 1917, 76–92.
[183]Acta Math., 41, 1917, 197–218.
[184]Annals of Math., 19, 1918, 210–6.
[185]Math. Quest. Educ. Times, 72, 1900, 125–6.
[186]Math. Quest. and Solutions, 3, 1917, 65.

CHAPTER XI.

MISCELLANEOUS THEOREMS ON DIVISIBILITY, GREATEST COMMON DIVISOR, LEAST COMMON MULTIPLE.

THEOREMS ON DIVISIBILITY.

An anonymous author[1] noted that for n a prime the sum of $1, 2, \ldots, n-1$ taken by twos (as $1+2$, $1+3, \ldots$), by fours, by sixes, etc., when divided by n give equally often the residues $1, 2, \ldots, n-1$, and once oftener the residue 0. The sum by threes, fives, $\ldots$, give equally often the residues $1, \ldots, n-1$ and once fewer the residue 0.

J. Dienger[2] noted that if $m^{2r+1} \doteq 1$ and $(m^{4r+2}-1)/(m^2-1)$ are divisible by the prime p, then the sum of any $2r+1$ consecutive terms of the set $1, m^{2n}, m^{2 \cdot 2n}, m^{3 \cdot 2n}, \ldots$ is divisible by p. The case $m=2$, $r=1$, $p=3$ or 7 was noted by Stifel (Arith. Integra).

G. L. Dirichlet[3] proved that when n is divided by $1, 2, \ldots, n$ in turn the number of cases in which the remainder is less than half the divisor bears to n a ratio which, as n increases, has the limit $2 - \log 4 = 0.6137 \ldots$; the sum of the quotients of the n remainders by the corresponding divisors bears to n a ratio with the limit $0.423 \ldots$

Dirichlet[4] generalized his preceding result. The number h of those divisors $1, 2, \ldots, p$ ($p \leq n$), which yield a remainder whose ratio to the divisor is less than a given proper fraction a, is

$$h = \sum_{s=1}^{p} \left\{ \left[\frac{n}{s} \right] - \left[\frac{n}{s} - a \right] \right\}.$$

Assuming that p^2/n increases indefinitely with n, the limit of h/p is a if n/p increases indefinitely with n, but if n/p remains finite is

$$\frac{n}{p} \int_0^1 \frac{1-\varphi^a}{1-\varphi} d\varphi - \frac{n}{p} \sum_{s=1}^{q} \left(\frac{1}{s} - \frac{1}{s+a} \right) + \left\{ 1 - \frac{n}{p(p+a)} \right\} \left\{ q - \left[\frac{n}{p} - a \right] \right\}, \qquad q = \left[\frac{n}{p} \right].$$

J. J. Sylvester[5] noted that 2^{m+1} is a factor of the integral part of k^{2m+1} and of the integer just exceeding k^{2m}, where $k = 1 + \sqrt{3}$.

N. V. Bougaief[6] called a number primitive if divisible by no square > 1, secondary if divisible by no cube. The number of primitive numbers $\leq n$ is

$$H_1(n) = \sum_{1}^{t_1} q(u) + \sum_{1}^{t_2} q(u) + \ldots, \qquad t_i = [\sqrt{n/i}],$$

where $q(u)$ is zero if u is not primitive, but is $+1$ or -1 for a primitive u, according as u is a product of an even or odd number of prime factors.

[1]Jour. für Math., 6, 1830, 100–4.   [2]Archiv Math. Phys., 12, 1849, 425–9.

[3]Abh. Ak. Wiss. Berlin, 1849, 75–6; Werke, 2, 57–58. Cf. Sylvester, Amer. Jour. Math., 5, 1882, 298–303; Coll. Math. Papers, IV, 49–54.

[4]Jour. für. Math., 47, 1854, 151–4. Berlin Berichte, 1851, 20–25; Werke, 2, 97–104; French transl. by O. Terquem, Nouv. Ann. Math., 13, 1854, 396.

[5]Quar. Journ. Math., 1, 1857, 185. Lady's and Gentleman's Diary, London, 1857, 60–1.

[6]Comptes Rendus Paris, 74, 1872, 449–450. Bull. Sc. Math. Astr., 10, I, 1876, 24. Math. Sbornik (Math. Soc. Moscow), 6, 1872–3, I, 317–9, 323–331.

To obtain the number $H_2(n)$ of secondary numbers $\leqq n$, replace square roots by cube roots in the t_i. We have

$$H_1(n)+H_1\left(\left[\frac{n}{2^2}\right]\right)+H_1\left(\left[\frac{n}{3^2}\right]\right)+\ldots=n, \qquad H_2(n)+H_2\left(\left[\frac{n}{2^3}\right]\right)+\ldots=n,$$

and similarly for $H_{\lambda-1}(n)$ given by (2) below.

J. Grolous[7] considered the probability R_k that a number be divisible by at least one of the integers $Q_1, \ldots, Q_k$, relatively prime by twos, and showed that

$$R_n=\frac{1}{Q_1}+\frac{1}{Q_2}(1-R_1)+\ldots+\frac{1}{Q_n}(1-R_{n-1}).$$

Chr. Zeller[7a] modified Dirichlet's[4] expression for h. The sums

$$\sum_{s=1}^{p}\left[\frac{n}{s}-a\right], \qquad \sum_{s=1}^{p-1}\left[\frac{n}{s+a}\right]$$

are equal. The sum of the terms of the second with $s>\mu=[\sqrt{p}]$ equals the excess of the sum of the first μ terms of the first over μ^2 or μ^2-1, the latter in the case of numbers between μ^2 and $\mu^2+\mu$. Hence we may abbreviate the computation of h.

E. Cesàro[8] obtained Dirichlet's[3, 4] results and similar ones. The mean (p. 174) of the number of decompositions of N into two factors having p as their g. c. d. is $6(\log N)/(p^2\pi^2)$. The mean (p. 230) of the number of divisors common to two positive integers n, n' is $\pi^2/6$, that of the sum of their common divisors is

$$\tfrac{1}{2}\log_e nn'+2C-\frac{\pi^2}{12}+\tfrac{1}{2},$$

where $C=0.57721\ldots$. The sum of the inverses of the nth powers of two positive integers is in mean $\zeta(n+2)$ where ζ is defined by (12) of Ch. X.

E. Cesàro[9] proved the preceding results on mean values; showed that the number of couples of integers whose l. c. m. is n is the number of divisors of n^2, if (a, b) and (b, a) are both counted when $a\neq b$; found the mean of the l. c. m. of two numbers; found the probability that in a random division the quotient is odd, and the mean of the first or last digit of the quotient; the probability that the g.c.d. of several numbers shall have specified properties.

Cesàro[9a] noted that the probability that an integer has no divisor >1 which is an exact rth power is $1/\zeta(r)$.

L. Gegenbauer[10] proved that the number of integers $\leqq x$ and divisible by no square is asymptotic to $6x/\pi^2$, with an error of order inferior to $\sqrt{x}$. He proved the final formulas of Bougaief.[6]

---

[7]Bull. Sc. Soc. Philomatique de Paris, 1872, 119–128.
[7a]Nachrichten Gesell. Wiss. Göttingen, 1879, 265–8.
[8]Mém. Soc. R. Sc. de Liège, (2), 10, 1883, No. 6, 175–191, 219–220 (corrections, p. 343).
[9]Annali di mat., (2), 13, 1885, 235–351, "Excursions arith. à l'infini."
[9a]Nouv. Ann. Math., (3), 4, 1885, 421.
[10]Denkschr. Akad. Wien (Math.), 49, I, 1885, 47–8. Sitzungsber. Akad. Wien, 112, II a, 1903, 562; 115, II a, 1906, 589. Cf. A. Berger, Nova Acta Soc. Upsal., (3), 14, 1891, Mém. 2, p. 110; E. Landau, Bull. Soc. Math. France, 33, 1905, 241. See Gegenbauer,[72, 79] Ch. X.

Gegenbauer[10a] proved that the arithmetical mean of the greatest integers contained in k times the remainders on the division of n by $1, 2, \ldots, n$ approaches

$$k \log k + k - 1 - k \sum_{x=1}^{k-1} 1/x$$

as n increases. The case $k=2$ is due to Dirichlet.

Gegenbauer[11] gave formulas involving the greatest divisor $t_a(n)$, not divisible by a, of the integer n. In particular, he gave the mean value of the greatest divisor not divisible by an ath power.

L. Gegenbauer,[12] employing Merten's function μ (Ch. XIX) and $R(a) = a - |a|$, gave the three general formulas

$$\sum_{x_1} \sum_{y=1}^{r+n} \mu\left(\frac{x_1}{y}\right) f(y) = \sum_{k=1}^{r+n} f(k) - \sum_{k=1}^{r} f(k) - \sum_{k=1}^{n} f(k),$$

$$\sum_{x_2} \sum_{y=1}^{rn} \mu\left(\frac{x_2}{y}\right) f(y) = \sum_{k=(r-1)n+1}^{rn} f(k) - \sum_{k=1}^{n} f(k),$$

$$\sum_{x,\,y=1}^{rn} \mu\left(\frac{x}{y}\right) f(y)\left\{\frac{1}{n}R\left(\frac{n}{x}\right) - \frac{1}{r}R\left(\frac{r}{x}\right)\right\} = \frac{1}{r}\sum_{k=1}^{r} f(k) - \frac{1}{n}\sum_{k=1}^{n} f(k),$$

where x_2 ranges over the divisors $>n$ of $(r-1)n+1$, $(r-1)n+2, \ldots, rn$, while x_1 ranges over all positive integers for which

$$\frac{a+\beta-1}{r+n} \leq \frac{R(g/x_1)}{g} < \frac{a}{r}, \frac{\beta}{n} \qquad \left(a = 1, \ldots, \frac{r}{g}; \beta = 1, \ldots, \frac{n}{g}\right),$$

where g is the g. c. d. of r, n. Take $f(x) = 1$ or 0 according as x is an sth power or not. Then the functions

$$(1) \qquad \sum_{k=1}^{m} f(k), \qquad \sum_{y=1}^{x} \mu\left(\frac{x}{y}\right) f(y)$$

become $[\sqrt[s]{m}]$ and $\lambda_s(x)$, with the value 0 if the exponent of any prime factor of x is $\not\equiv 0$, 1 (mod s), otherwise the value $(-1)^\sigma$, where σ is the number of primes occurring in x to the power $ks+1$. Thus

$$\sum_{x_1} \lambda_s(x_1) = \left[\sqrt[s]{r+n}\right] - \left[\sqrt[s]{r}\right] - \left[\sqrt[s]{n}\right],$$

$$\sum_{x_2} \lambda_s(x_2) = \left[\sqrt[s]{rn}\right] - \left[\sqrt[s]{(r-1)n}\right] - \left[\sqrt[s]{n}\right].$$

If $f(x) = 0$ or 1 according as x is divisible by an sth power or not, the functions (1) become $Q_s(m)$ and $\mu(\sqrt[s]{x})$, the former being the number of integers $\leq m$ divisible by no sth power. If $f(x) = 1$ or 0 according as x is prime or not, the functions (1) become the number of primes $\leq m$ and a simple function $a(x)$; then the third formula shows that the mean density of the primes $\leq r$ is

$$\sum_{y=1}^{r} a(y)\left\{R\left(\frac{1}{y}\right) - \frac{1}{r}R\left(\frac{r}{y}\right)\right\}.$$

[10a]Denkschr. Akad. Wien (Math.), 49, II, 1885, 108.

[11]Sitzungsber. Akad. Wiss. Wien (Math.), 94, 1886, II, 714.

[12]*Ibid.*, 97, 1888, IIa, 420–6.

If $f(x) = \log x$, the second function (1) becomes $\nu(x)$, having the value* $\log p$ when x is a power of the prime p, otherwise the value 0. Besides the resulting formulas, others are found by taking $f(x) = \nu(x)$, Jacobi's symbol (Δ/x) in the theory of quadratic residues, and finally the number of representations of x by the system of quadratic forms of discriminant Δ.

L. Saint-Loup[13] represented graphically the divisors of a number. Write the first 300 odd numbers in a horizontal line; the 300 following numbers are represented by points above the first, etc. Take any prime as 17 and mark all its multiples; we get a rectilinear distribution of these multiples, which are at the points of intersection of two sets of parallel lines.

J. Hacks[14] proved that the number of integers $\leq m$ which are divisible by an nth power > 1 is

$$p_n(m) = \Sigma \left[\frac{m}{k_1{}^n}\right] - \Sigma \left[\frac{m}{k_1{}^n k_2{}^n}\right] + \Sigma \left[\frac{m}{k_1{}^n k_2{}^n k_3{}^n}\right] - \cdots,$$

where the k's range over the primes > 1 [Bougaief[6]]. Then $\psi_2(m) = m - p_2(m)$ is the number of integers $\leq m$ not divisible by a square > 1, and

$$\psi_2(m) + \psi_2\left(\frac{m}{4}\right) + \psi_2\left(\frac{m}{9}\right) + \cdots + \psi_2\left(\frac{m}{[\sqrt{m}]^2}\right) = m.$$

A like formula holds for $\psi_3 = m - p_3(m)$, using quotients of m by cubes.

L. Gegenbauer[14a] found the mean of the sum of the reciprocals of the kth powers of those divisors of a term of an unlimited arithmetical progression which are rth powers; also the probability that a term be divisible by no rth power; and many such results.

L. Gegenbauer[15] noted that the number of integers $1, \ldots, n$ not divisible by a λth power is

$$(2) \qquad Q_\lambda(n) = \sum_{x=1}^{[n^{1/\lambda}]} \left[\frac{n}{x^\lambda}\right] \mu(x).$$

Ch. de la Vallée Poussin[16] proved that, if x is divided by each positive number $ky + b \leq x$, the mean of the fractional parts of the quotients has for $x = \infty$ the limit $1 - C$; if x is divided by the primes $\leq x$, the mean of the fractional parts of the quotients has for $x = \infty$ the limit $1 - C$. Here C is Euler's constant.[8]

L. Gegenbauer[17] proved, concerning Dirichlet's[3] quotients Q of the remainders (found on dividing n by $1, 2, \ldots, n$ in turn) by the corresponding divisors, that the number of Q's between 0 and $1/3$ exceeds the number of Q's between $2/3$ and 1 by approximately $0.179n$, and similar theorems.

*Cf. Bougaief[16] of Ch. XIX.
[13]Comptes Rendus Paris, 107, 1888, 24; École Norm. Sup., 7, 1890, 89.
[14]Acta Math., 14, 1890-1, 329-336.
[14a]Sitzungsber. Ak. Wien (Math.), 100, IIa, 1891, 1018-1053.
[15]Ibid., 100, 1891, IIa, 1054. Denkschr. Akad. Wien (Math.), 49 I, II, 1885; 50 I, 1885. Cf. Gegenbauer[79] of Ch. X.
[16]Annales de la soc. sc. Bruxelles, 22, 1898, 84-90.
[17]Sitzungsberichte Ak. Wiss. Wien (Math.), 110, 1901, IIa, 148-161.

He investigated the related problem of Dirichlet.[4] Finally, he used as divisors all the sth powers $\leq n$ and found the ratio of the number of remainders less than half of the corresponding divisors to the number of the others.

L. E. Dickson[17a] and H. S. Vandiver proved that $2^n > 2(n+1)(n'+1)\ldots$, if $1, n, n',\ldots$ are the divisors of an odd number $n > 3$.

R. Birkeland[18] considered the sum s_q of the qth powers of the roots $a_1,\ldots, a_m$ of $z^m + A_1 z^{m-1} + \ldots + A_m = 0$. If $s_1,\ldots, s_m$ are divisible by the power a^p of a prime a, then A_q is divisible by a^p unless q is divisible by a. If q is divisible by a, and a^{p_1} is the highest power of a dividing q, then A_q is divisible by a^{p-p_1}. Then $(n+aa_1)\ldots(n+aa_m) - n^m$ is divisible by a^p. In particular, the product of m consecutive odd integers is of the form $1 + 2^p t$ if m is divisible by 2^p.

E. Landau[19] reproduced Poussin's[16] proof of the final theorem and added a simplification. He then proved a theorem which includes as special cases the two of Poussin and the final one by Dirichlet[3]. Given an infinite class of positive numbers q without a finite limit point and such that the number of q's $\leq x$ is asymptotic to $x/w(x)$, where $w(x)$ is a non-decreasing positive function having

$$\lim_{x=\infty} \frac{w(2x)}{w(x)} = 1;$$

then if x is divided by all the q's $\leq x$, the mean of the fractional parts of the quotients has for $x = \infty$ the limit $1 - C$.

St. Guzel[20] wrote $\delta(n)$ for the greatest odd divisor of n and proved in an elementary way the asymptotic formulas

$$\sum_{n=1}^{[x]} \delta(n) = \frac{x^2}{3} + O(x), \qquad \sum_{n=1}^{[x]} \frac{\delta(n)}{n} = \tfrac{2}{3}x + O(1),$$

for O as in Pfeiffer[90], Ch. X.

A. Axer[21] considered the $\chi^{\lambda,\nu}(n)$ decompositions of n into such a pair of factors that always the first factor is not divisible by a λth power and the second factor not by a νth power, $\lambda \geq 2$, $\nu \geq 2$. Then $\sum_{n=1}^{n=x} \chi^{\lambda,\nu}(n)$ is given asymptotically by a complicated formula involving the zeta function.

F. Rogel[22] wrote $R_{\lambda,n}$ for the algebraic sum of the partial remainders $t - [t]$ in (2), with n replaced by z, and obtained

$$Q_\lambda(z) = z P_{\lambda,n} + R_{\lambda,n}, \qquad P_{\lambda,n} = \prod_{\nu=2}^{n}\left(1 - \frac{1}{p_\nu}\right),$$

where p_n is the nth prime and $p_n^\lambda \leq z < p_{n+1}^\lambda$. He gave relations between the values of $Q_\lambda(z)$ for various z's and treated sums of such values, and tabulated the values of $Q_2(z)$ and $R_{2,n}$ for $z \leq 288$. He[22a] gave many relations

[17a]Amer. Math. Monthly, 10, 1903, 272; 11, 1904, 38–9.
[18]Archiv Math. og Natur., Kristiania, 26, 1904, No. 10.
[19]Bull. Acad. Roy. Belgique, 1911, 443–472.
[20]Wiadomosci mat., Warsaw, 14, 1910, 171–180.
[21]Prace mat. fiz., 22, 1911, 73–99 (Polish), 99–102 (German). Review in Bull. des sc. math., (2), 38, II, 1914, 11–13.
[22]Sitzungsber. Ak. Wiss. Wien (Math.), 121, IIa, 1912, 2419–52.
[22a]Ibid., 122, IIa, 1913, 669–700. See Rogel[248] of Ch. XVIII.

between the $Q_x(z)$, relations involving the number $A(z)$ of primes $\leqq z$, and relations involving both Q's and A's.

A. Rothe[23] called b a maximal divisor of a if no larger divisor of a contains b as a factor. Then a/b is called the index of b with respect to a. If also c is a maximal divisor of b, etc., $a, b, c, \ldots, 1$ are said to form a series of composition of a. In all series of composition of a, the sets of indices are the same apart from order [a corollary of Jordan's theorem on finite groups applied to the case of a cyclic group of order a].

*Weitbrecht[24] noted tricks on the divisibility of numbers.

*E. Moschietti[25] discussed the product of the divisors of a number.

Each[26] of the consecutive numbers 242, 243, 244, 245 has a square factor >1; likewise for the sets of three consecutive numbers beginning with 48 or 98 or 124.

C. Avery and N. Verson[27] noted that the consecutive numbers 1375, 1376, 1377 are divisible by 5^3, 2^3, 3^3, respectively.

J. G. van der Corput[28] evaluated the sum of the nth powers of all integers, not divisible by a square >1, which are $\leqq x$ and are formed of r prime factors of m.

GREATEST COMMON DIVISOR, LEAST COMMON MULTIPLE.

On the number of divisions in finding the g. c. d. of two integers, see Lamé[11] et seq. in Ch. XVII; also Binet[33] and Dupré[34].

V. A. Lebesgue[35] noted that the l. c. m. of $a, \ldots, k$ is $(p_1 p_3 p_5 \ldots)/(p_2 p_4 p_6 \ldots)$ if p_1 is the product of $a, \ldots, k$, while p_2 is the product of their g. c. d.'s two at a time, and p_3 the product of their g. c. d.'s three at a time, etc. If a, b, c have no common divisor, there exist an infinitude of numbers $ax+b$ relatively prime to c.

V. Bouniakowsky[36] determined the g. c. d. N of all integers represented by a polynomial $f(x)$ with integral coefficients without a common factor. Since N divides the constant term of $f(x)$, it remains to find the highest power p^μ of a prime p which divides $f(x)$ identically, $i.\,e.$, for $x = 1, 2, \ldots, p^\mu$. Divide $f(x)$ by $X_p = (x-1) \ldots (x-p)$ and call the quotient Q and remainder R. Then must $R \equiv 0 \pmod{p^\mu}$ for $x = 1, \ldots, p$, so that each coefficient of R is divisible by p^μ, and $\mu \leqq \mu_1$, where p^{μ_1} is the highest power of p dividing the coefficients of R. If $\mu_1 = 1$, we have $\mu = 1$. Next, let $\mu_1 > 1$. Divide

[23]Zeitschrift Math.-Naturw. Unterricht, 44, 1913, 317–320.
[24]Vom Zahlenkunststück zur Zahlentheorie, Korrespondenz-Blatt d. Schulen Württembergs, Stuttgart, 20, 1913, 200–6.
[25]Suppl. al Periodico di Mat., 17, 1914, 115–6.
[26]Math. Quest. Educ. Times, 36, 1881, 48.
[27]Math. Miscellany, Flushing, N. Y., 1, 1836, 370–1.
[28]Nieuw Archief voor Wiskunde, (2), 12, 1918, 213–27.
[33]Jour. de Math., (1), 6, 1841, 453.
[34]Ibid., (1), 11, 1846, 41.
[35]Nouv. Ann. Math., 8, 1849, 350; Introduction à la théorie des nombres, 1862, 51–53; Exercises d'analyse numérique, 1859, 31–32, 118–9.
[36]Mém. acad. sc. St. Pétersbourg, (6), sc. math. et phys. 6 (sc. math. phys. et nat. 8), 1857 305–329 (read 1854); extract in Bulletin, 13, 149.

Q by $(x-p-1)\ldots(x-2p)$ and call the quotient Q' and remainder R'. Then must $X_pR'+X_{2p}Q'\equiv0$ and hence $X_pR'\equiv0$ (mod p^κ). Thus if μ_2 is the exponent of the highest power of p dividing the coefficients of R', we have $\mu\le\mu_2+1$. In general, if μ_k and λ_{k-1} are the exponents of the highest powers of p dividing the coefficients of the remainder $R^{(k-1)}$ and $X_{(k-1)p}$ identically, then $\mu\le\mu_k+\lambda_{k-1}$. Finally, if $l=[m/p]$, $\mu\le\lambda_l$. The extension to several variables is said to present difficulties. [For simpler methods, see Hensel[44] and Borel.[48]] It is noted (p. 323) that

$$(x^p-x)^n, \qquad (x^{\varphi(p^n)}-1)x^n$$

are identically divisible by p^n. It is conjectured (p. 328) that $f(x)/N$ represents an infinitude of primes when $f(x)$ is irreducible.

E. Cesàro[37] and J. J. Sylvester[38] proved that the probability that two numbers taken at random from $1,\ldots,n$ be relatively prime is $6/\pi^2$ asymptotically.

L. Gegenbauer[39] gave 16 sums involving the g. c. d. of several integers and deduced 37 asymptotic theorems such as the fact that the square of the g. c. d. of four integers has the mean value $15/\pi^2$. He gave the mean of the kth power of the g. c. d. of r integers.

J. Neuberg[39a] noted that, if two numbers be selected at random from $1,\ldots,N$, the probability that their sum is prime to N is $k=\phi(N)$ or $k/(N-1)$ according as N is odd or even.

T. J. Stieltjes,[40] starting with a set of n integers, replaced two of them by their g. c. d. and l. c. m., repeated the same operation on the new set, etc. Finally, we get a set such that one number of every pair divides the other. Such a reduced set is unique. The l. c. m. of $a,\ldots,l$ can be expressed (pp. 14–16) as a product $a'\ldots l'$ of relatively prime factors dividing $a,\ldots,l$, respectively. The l. c. m. (or g. c. d.) of $a, b, \ldots, l$ equals the quotient of $P=ab\ldots l$ by the g. c. d. (or l. c. m.) of $P/a, P/b, \ldots, P/l$.

E. Lucas[41] gave theorems on g. c. d. and l. c. m.

L. Gegenbauer[41a] considered in connection with the theory of primes, the g. c. d. of r numbers with specified properties.

J. Hacks[42] expressed the g. c. d. of m and n in the forms

$$2\sum_{s=1}^{n-1}\left[\frac{sm}{n}\right]-mn+m+n, \qquad 2\sum_{s=1}^{[n/2]}\left[\frac{sm}{n}\right]+2\sum_{s=1}^{[m/2]}\left[\frac{sn}{m}\right]-2\left[\frac{m}{2}\right]\left[\frac{n}{2}\right]+\epsilon,$$

where $\epsilon=0$ or 1 according as m, n are both or not both even.

J. Hammond[43] considered arbitrary functions f and F of p and a, such

[37]Mathesis, 1, 1881, 184; Johns Hopkins Univ. Circ., 2, 1882–3, 85.
[38]Johns Hopkins Univ. Circ., 2, 1883, 45; Comptes Rendus Paris, 96, 1883, 409; Coll. Papers, 3, 675; 4, 86.
[39]Sitzungsberichte Ak. Wiss. Wien (Math.) 92, 1885, II, 1290–1306.
[39a]Math. Quest. Educ. Times, 50, 1889, 113–4.
[40]Sur la théorie des nombres, Annales de la fac. des sciences de Toulouse, 4, 1890, final paper.
[41]Théorie des nombres, 1891, 345–6; 369, exs. 4, 5.
[41a]Monatshefte Math. Phys., 3, 1892, 319–335.
[42]Acta Math., 17, 1893, 208.
[43]Messenger Math. 24 1894–5 17–19.

that $f(p, 0) = 1$, $F(p, 0) = 0$, and any two integers $m = \Pi p^a$, $n = \Pi p^\beta$, where the p's are distinct primes and, for any p, $a \geq 0$, $\beta \geq 0$. Set

$$\psi(m) = \Pi f(p, a), \qquad \Phi = \Sigma F(p, a).$$

By the usual proof that mn equals the product of the g. c. d. M of m and n by their l. c. m. μ, we get

$$\psi(m)\psi(n) = \psi(M)\psi(\mu), \qquad \Phi(m) + \Phi(n) = \Phi(M) + \Phi(\mu).$$

In particular, if m and n are relatively prime,

$$\psi(m)\psi(n) = \psi(mn), \qquad \Phi(m) + \Phi(n) = \Phi(mn).$$

These hold if ψ is Euler's ϕ-function, the sum $\sigma(m)$ of the divisors of m or the number $\tau(m)$ of divisors of m; also, if $\Phi(m)$ is the number of prime factors of m or the sum of the exponents a in $m = \Pi p^a$.

K. Hensel[44] proved that the g. c. d. of all numbers represented by a polynomial $F(u)$ of degree n with integral coefficients equals the g. c. d. of the values of $F(u)$ for any $n+1$ consecutive arguments. For a polynomial of degree n_1 in u_1, n_2 in u_2, $\ldots$ we have only to use n_1+1 consecutive values of u_1, n_2+1 consecutive values of u_2, etc.

F. Klein[45] discussed geometrically Euclid's g. c. d. process.

F. Mertens[46] calls a set of numbers primitive if their g. c. d. is unity. If $m \neq 0$, $k > 1$, and $a_1, \ldots, a_k$, m is a primitive set, we can find integers $x_1, \ldots, x_k$ so that $a_1 + mx_1, \ldots, a_k + mx_k$ is a primitive set. Let d be the g. c. d. of $a_1, \ldots, a_k$ and find δ, μ so that $d\delta + m\mu = 1$. Take integral solutions a of $a_1 a_1 + \ldots + a_k a_k = d$ and primitive solutions β_i not all zero of $a_1\beta_1 + \ldots + a_k\beta_k = 0$. Then $\gamma_i = \beta_i + \delta a_i (i = 1, \ldots, k)$ is a primitive set. Determine integers ξ so that $\gamma_1\xi_1 + \ldots + \gamma_k\xi_k = 1$ and set $x_i = \mu\xi_i$. Then $a_i + mx_i$ form a primitive set.

R. Dedekind[47] employed the g. c. d. d of a, b, c; the g. c. d. $(b, c) = a_1$, $(c, a) = b_1$, $(a, b) = c_1$. Then $a' = a_1/d$, $b' = b_1/d$, $c' = c_1/d$ are relatively prime in pairs. Then $db'c'$ is the l. c. m. of b_1, c_1, and hence is a divisor of a. Thus $a = db'c'a''$, $b = dc'a'b''$, $c = da'b'c''$. The 7 numbers $a', \ldots, a'', \ldots, d$ are called the "Kerne" of a, b, c. The generalization from 3 to n numbers is given.

E. Borel[48] considered the highest power of a prime p which divides a polynomial $P(x, y, \ldots)$ with integral coefficients for all integral values of $x, y, \ldots$. If each exponent is less than p, we have only to find the highest power of p dividing all the coefficients. In the contrary case, reduce all exponents below p by use of $x^p = x + px_1, x_1{}^p = x_1 + px_2, \ldots$ and proceed as above with the new polynomial in x, x_1, x_2, $\ldots$, y, y_1, $\ldots$. Then to find all arithmetical divisors of a polynomial P, take as p in turn each prime less than the highest exponent appearing in P.

L. Kronecker[49] found the number of pairs of integers i, k having t as their g. c. d., where $1 \leq i \leq m$, $1 \leq k \leq n$. The quotient of this number by

[44]Jour. für Math., 116, 1896, 350–6.
[45]Ausgewählte Kapitel der Zahlentheorie, I, 1896.
[46]Sitzungsberichte Ak. Wiss. Wien (Math.), 106, 1897, II a, 132–3.
[47]Ueber Zerlegungen von Zahlen durch d. grössten gemeinsamen Teiler, Braunschweig, 1897.
[48]Bull. Sc. Math. Astr., (2), 24 I, 1900, 75–80. Cf. Borel and Drach[180] of Ch. III.
[49]Vorlesungen über Zahlentheorie, I, 1901, 306–312.

mn is the mean. When m and n increase indefinitely, the mean becomes $6/(\pi^2 t^2)$. The case $t=1$ gives the probability that two arbitrarily chosen integers are relatively prime; the proof in Dirichlet's Zahlentheorie fails to establish the existence of the probability.

E. Dintzl[50] proved that the g. c. d. $\Delta(a, \ldots, e)$ is a linear function of $a, \ldots, e$, and reproduced the proof of Lebesgue's[35] formula as given in Merten's Vorlesungen über Zahlentheorie and by de Jough.[51]

A. Pichler,[50a] given the l. c. m. or g. c. d. of two numbers and one of them, found values of the other number.

J. C. Kluyver[52] constructed several functions z (involving infinite series or definite integrals) which for positive integral values of the two real variables equals their g. c. d. He gave to Stern's[53] function the somewhat different form

$$z = 2 \sum_{\mu=0}^{[x]} P\left(\frac{\mu y}{x}\right), \qquad P(u) \equiv u - [u] - \tfrac{1}{2}.$$

W. Sierpinski[54] stated that the probability that two integers $\leq n$ are relatively prime is

$$\frac{1}{n^2} \sum_{k=1}^{n} \mu(k) \left[\frac{n}{k}\right]^2,$$

contrary to Bachmann, Analyt. Zahlentheorie, 1894, 430.

G. Darbi[55] noted that if $a = (a, N)$ is the g. c. d. of a, N,

$$(N, abc\ldots) = a\left(b, \frac{N}{a}\right)\left(c, \frac{N}{a(b, N/a)}\right)\cdots$$

and gave a method of finding the g. c. d. and l. c. m. of rational fractions without bringing them to a common denominator.

E. Gelin[56] noted that the product of n numbers equals ab, where a is the l. c. m. of their products r at a time, and b is the g. c. d of their products $n-r$ at a time.

B. F. Yanney[57] considered the greatest common divisors $D_1, D_2, \ldots$ of $a_1, \ldots, a_n$ in sets of k, and their l. c. m.'s $L_1, L_2, \ldots$. Then

$$\prod_{i=1}^{b} D_i L_i^{k-1} \geq (a_1 \ldots a_n)^c \geq \prod_{i=1}^{b} D_i^{k-1} L_i, \qquad b = \binom{n}{k}, \quad c = \binom{n-1}{k-1}.$$

The limits coincide if $k=2$. The products have a single term if $k=n$.

P. Bachmann[58] showed how to find the number N obtained by ridding a given number n of its multiple prime factors. Let d be the g. c. d. of n and $\phi(n)$. If $\delta = n/d$ occurs to the rth power, but not to the $(r+1)$th power in n, set $n_1 = n/\delta^r$. From n_1 build δ_1 as before, etc. Then $N = \delta \delta_1 \delta_2 \ldots$

[50]Zeitschrift für das Realschulwesen, Wien, 27. 1902, 654–9, 722.
[50a]Ibid., 26, 1901, 331–8.
[51]Nieuw Archief voor Wiskunde, (2), 5, 1901, 262–7.
[52]K. Ak. Wetenschappen Amsterdam, Proceedings of the Section of Sciences, 5, II 1903, 658–662. (Versl. Ak. Wet., 11, 1903, 782–6.)
[53]Jour. für Math., 102, 1888, 9–19.
[54]Wiadomosci Mat., Warsaw, 11, 1907, 77–80.
[55]Giornale di Mat., 46, 1908, 20–30.
[56]Il Pitagora, Palermo, 16, 1909–10, 26–27.
[57]Amer. Math. Monthly, 19, 1912, 4–6.
[58]Archiv Math. Phys., (3), 19, 1912, 283–5.

Erroneous remarks[59] have been made on the g. c. d. of $2^x - 1$, $3^x - 1$.

M. Lecat[60] noted that, if a_{ij} is the l. c. m. of i and j, the determinant $|a_{ij}|$ was evaluated by L. Gegenbauer,[61] who, however, used a law of multiplication of determinants valid only when the factors are both of odd class.

J. Barinaga[61a] proved that, if δ is prime to $N = nk$, the sum of those terms of the progression N, $N+\delta$, $N+2\delta$, ..., which are between nk and $n(k+h\delta)$ and which have with $n = mp$ the g. c. d. p, is $\frac{1}{2}n\phi(n/p)(2k+h\delta)h$.

R. P. Willaert[62] noted that, if $P(n)$ is a polynomial in n of degree p with integral coefficients, $f(n) = aA^{an} + P(n)$ is divisible by D for every integral value of n if and only if the difference $\Delta^k f(0)$ of the kth order is divisible by D for $k = 0, 1, \ldots, p+1$. Thus, if $p = 1$, the conditions are that $f(0)$, $f(1)$, $f(2)$ be divisible by D.

*H. Verhagen[63] gave theorems on the g. c. d. and l. c. m.

H. H. Mitchell[64] determined the number of pairs of residues a, b modulo λ whose g. c. d. is prime to λ, such that ka, kb is regarded as the same pair as a, b when k is prime to λ, and such that λ and $ax+by$ have a given g. c. d.

W. A. Wijthoff[65] compared the values of the sums

$$\overset{ab}{\underset{m=1}{\Sigma}} (-1)^{m-1} m^s F\{(m, a)\}, \qquad \overset{(ab-1)/2}{\underset{m=1}{\Sigma}} m^s F\{(m, a)\}, \qquad s = 1, 2,$$

where (m, a) is the g. c. d. of m, a, while F is any arithmetical function.

F. G. W. Brown and C. M. Ross[66] wrote $l_1, l_2, \ldots, l_n$ for the l. c. m. of the pairs A_1, A_2; A_2, A_3; ...; A_n, A_1, and $g_1, g_2, \ldots, g_n$ for the g. c. d. of these pairs, respectively. If L, G are the l. c. m. and g. c. d. of $A_1, A_2, \ldots, A_n$, then

$$g_1 g_2 \ldots g_n = G^n, \qquad \frac{l_1 l_2 \ldots l_n}{g_1 g_2 \ldots g_n} = \frac{L^2}{G^2}.$$

C. de Polignac[67] obtained for the g. c. d. (a, b) of a, b results like

$$(a\lambda, b\mu) = (a, b) \cdot (\lambda, \mu) \cdot \left(\frac{a}{(a, b)}, \frac{\mu}{(\lambda, \mu)} \right) \cdot \left(\frac{b}{(a, b)}, \frac{\lambda}{(\lambda, \mu)} \right).$$

Sylvester[68] and others considered the g. c. d. of D_n and D_{n+1} where D_n is the n-rowed determinant whose diagonal elements are 1, 3, 5, 7, ..., and having 1, 2, 3, 4, ... in the line parallel to that diagonal and just above it, and units in the parallel just below it, and zeros elsewhere.

On the g. c. d., see papers 33–88, 215–6, 223 of Ch. V, Cesàro[61] of Ch. X, Cesàro[8, 9] of Ch. XI, and Kronecker[30] of Ch. XIX.

[59]L'intermédiaire des math., 20, 1913, 112, 183–4, 228; 21, 1914, 36–7.
[60]Ibid., 21, 1914, 91–2.
[61]Sitzungs. Ak. Wiss. Wien (Math.), 101, 1892, II a, 425–494;
[61a]Annaes Sc. Acad. Polyt. do Porto, 8, 1913, 248–253.
[62]Mathesis, (4), 4, 1914, 57.
[63]Nieuw Tijdschrift voor Wiskunde, 2, 1915, 143–9.
[64]Annals of Math., (2), 18, 1917, 121–5.
[65]Wiskundige Opgaven, 12, 1917, 249–251.
[66]Math. Quest. and Solutions, 5, 1918, 17–18.
[67]Nouv. Corresp. Math., 4, 1878, 181–3.
[68]Math. Quest. Educ. Times, 36, 1881, 97–8; correction, 117–8.

CHAPTER XII.

CRITERIA FOR DIVISIBILITY BY A GIVEN NUMBER.

In the Talmud[1], $100a+b$ is stated to be divisible by 7 if $2a+b$ is divisible by 7.

Hippolytos[1a], in the third century, examined the remainder on the division of certain sums of digits by 7 or 9, but made no application to checking numerical computation.

Avicenna or Ibn Sînâ (980–1037) is said to have been the discoverer of the familiar rule for casting out nines (cf. Fontès[39]); but it seems to have been of Indian origin.[1b]

Alkarkhi[1c] (about 1015) tested by 9 and 11.

Ibn Mûsâ Alchwarizmî[1d] (first quarter of the ninth century) tested by 9.

Leonardo Pisano[1e] gave in his Liber Abbaci, 1202, a proof of the test for 9, and indicated tests for 7, 11.

Ibn Albannâ[1f] (born about 1252), an Arab, gave tests for 7, 8, 9.

In the fifteenth century, the Arab Sibt el-Mâridini[1c] tested addition by casting out multiples of 7 or 8.

Nicolas Chuquet[1g] in 1484 checked the four operations by casting out 9's.

J. Widmann[1h] tested by 7 and 9.

Luca Paciuolo[2] tested by 7, as well as by 9, the fundamental operations, but gave no rule to calculate rapidly the remainder on division by 7.

Petrus Apianus[2a] tested by 6, 7, 8, 9.

Robert Recorde[2b] tested by 9.

Pierre Forcadel[3] noted that to test by $7 = 10 - 3$ we multiply the first digit by 3, subtract multiples of 7, add the residue to the next digit, then multiply the sum by 3, etc.

Blaise Pascal[4] stated and proved a criterion for the divisibility of any number N by any number A. Let $r_1, r_2, r_3, \ldots$, be the remainders obtained when $10, 10r_1, 10r_2, \ldots$ are divided by A. Then $N = a+10b+100c+\ldots$ is divisible by A if and only if $a+r_1b+r_2c+\ldots$ is divisible by A.

---

[1]Babylonian Talmud, Wilna edition by Romm, Book Aboda Sara, p. 9b.

[1a]M. Cantor, Geschichte der Math., ed. 3, I, 1907, 461.

[1b]Ibid., 511, 611, 756–7, 763–6.

[1c]Cf. Carra de Vaux, Bibliotheca Math., (2), 13, 1899, 33–4.

[1d]M. Cantor, Geschichte der Math., ed. 3, I, 1907, 717.

[1e]Scritti, 1, 1857, 8, 20, 39, 45; Cantor, Geschichte, 2, 1892, 8–10.

[1f]Le Talkhys d'Ibn Albannâ publié et traduit par A. Marre, Atti Accad. Pont. Nuovi Lincei, 17, 1863–4, 297. Cf. M. Cantor, Geschichte Math., I, ed. 2, 757, 759; ed. 3, 805–8.

[1g]Le Triparty en la science de nombres, Bull. Bibl. St. Sc. Math., 13, 1880, 602–3.

[1h]Behēde vnd hubsche Rechnung...., Leipzig, 1489.

[2]Summa de arithmetica geometria proportioni et proportionalita, Venice, 1494, f. 22, r.

[2a]Ein newe...Kauffmans Rechnung, Ingolstadt, 1527, etc.

[2b]The Grovnd of Artes, London, c. 1542, etc.

[3]L'Arithmeticqve de P. Forcadel de Beziers, Paris, 1556, 59–60.

[4]De numeris multiplicibus, presented to the Académie Parisienne, in 1654, first published in 1665; Oeuvres de Pascal, 3, Paris, 1908, 311–339; 5, 1779, 123–134.

D'Alembert[5] noted that if $N = A \cdot 10^m + B \cdot 10^n + \ldots + E$ is divisible by $10 - b$, then $Ab^m + Bb^n + \ldots + E$ is divisible by $10 - b$; if N is divisible by $10 + b$, then $A(-b)^m + B(-b)^n + \ldots + E$ is divisible by $10 + b$. The case $b = 1$ gives the test for divisibility by 9 or 11. By separating N into parts each with an even number of digits, $N = A \cdot 10^m + \ldots + E$, where $m, \ldots$ are even; then if N is divisible by $100 - b$, $Ab^{m/2} + \ldots + E$ is divisible by $100 - b$.

De Fontenelle[6] gave a test for divisibility by 7 which is equivalent to the case $b = 3$ of D'Alembert; to test 3976 multiply the first digit by 3 and add to the second digit; it remains to test 1876. For proof see F. Sanvitali, Hist. Literariae Italiae, vol. 6, and Castelvetri.[8]

G. W. Kraft[7] gave the same test as Pascal for the factor 7.

J. A. A. Castelvetri[8] gave the test for 99: Separate the digits in pairs, add the two-digit components, and see if the sum is a multiple of 99. For 999 use triples of digits.

Castelvetri[9] tested 1375, for example, for the factor 11 by noting that $13 + 75 = 88$ is divisible by 11. If the resulting sum be composed of more than two digits, pair them, add and repeat. To test for the factor 111, separate the digits into triples and add. The proof follows from the fact that 10^{2k} has the remainder 1 when divided by 11.

J. L. Lagrange[10] modified the method of Pascal by using the least residue modulo A (between $-A/2$ and $A/2$) in place of the positive residue. He noted that if a number is written to any base a its remainder on division by $a - 1$ is the same as for the sum of its digits.

J. D. Gergonne[11] noted that on dividing $N = A_0 + A_1 b^m + A_2 b^{2m} + \ldots$, written to base b, by a divisor of $b^m - 1$, the remainder is the same as on dividing the sum $A_0 + A_1 + A_2 + \ldots$ of its sets of m digits. Similarly for $b^m + 1$ and $A_0 - A_1 + A_2 - A_3 + \ldots$.

C. J. D. Hill[12] gave rules for abbreviating the testing for a prime factor p, for $p < 300$ and certain larger primes.

C. F. Liljevalch[12a] noted that if $10^n a - \beta$ is divisible by p then $a - 10^n b$ will be a multiple of p if and only if $a\alpha - \beta b$ is a multiple of p.

J. M. Argardh[13] used Hill's symbols, treating divisors 7, 17, 27, 1429.

F. D. Herter[14] noted that $a + 10b + 100c + \ldots$ is divisible by $10n \pm 1$ if

[5]Manuscript R. 240* 6 (8°), Bibl. Inst. France, 21, ff. 316–330, Sur une propriété des nombres.
[6]Histoire Acad. Paris, année 1728, 51–3.   [7]Comm. Ac. Sc. Petrop, 7, ad annos 1734–5, p. 41.
[8]De Bononiensi Scientiarum et Artium Instituto atque Academia Comm., 4, 1757; commentarii, 113–139; opuscula, 242–260.
[9]De Bononiensi Scientiarum et Artium Instituto atque Academia Comm., vol. 5, 1767, part 1, pp. 134–144; part 2, 108–119.
[10]Leçons élém. sur les math. données à l'école normale en 1795, Jour. de l'école polytechnique, vols. 7, 8, 1812, 194–9; Oeuvres, 7, pp. 203–8.
[11]Annales de math. (ed., Gergonne), 5, 1814–5, 170–2.
[12]Jour. für Math., 11, 1834, 251–261; 12, 1834, 355. Also, De factoribus numerorum compositorum dignoscendis, Lund, 1838.
[12a]De factoribus numerorum compositorum dignoscendis, Lund, 1838.
[13]De residuis ex divisione…, Diss. Lund, 1839.
[14]Ueber die Kennzeichen der Theiler einer Zahl, Progr. Berlin, 1844.

$a \mp b/n + c/n^2 \mp \ldots$ is divisible by $10n \pm 1$, with a like test for $10n \pm 3$ (replacing $1/n$ by $3/n$), and deduced the usual tests for 9, 11, 7, 13, etc.

A. L. Crelle[15] noted that to test $x_m A^m + \ldots + x_1 A + x_0$ for the divisor s we may select any integer n prime to s, take $r \equiv nA$ (mod s), and test
$$x_m r^m + n x_{m-1} r^{m-1} + \ldots + n^m x_0$$
for the divisor s. For example, if $A = 10$, $s = 7$, $10^3 \equiv -1$ (mod 7), so that $x_0 - x_1 + x_2 - \ldots \pm x_m$ is to be tested for the divisor 7, where $x_0, \ldots$ are the three-digit components of the proposed number from right to left. Similarly for $s = 9$, 11, 13, 17, 19.

A. Transon[16] gave a test for the divisibility of a number by any divisor of $10^a \cdot n \pm 1$.

A. Niegemann[17] noted that 354578385 is divisible by 7 since 35457 + 2×8385 is divisible by 7. In general if the number formed by the last m digits of N is multiplied by k, and the product is added to the number derived from N by suppressing those digits, then N is divisible by d if the resulting sum is divisible by d. Here $k(0 < k < d)$ is chosen so that $10^m k - 1$ is divisible by d. Thus $k = 2$ if $m = 4$, $d = 7$.

Many of the subsequent papers are listed at the end of the chapter.

H. Wilbraham[18] considered the exponent p to which 10 belongs modulo m, where m is not divisible by 2 or 5. Then the decimal for $1/m$ has a period of p digits. If any number N be marked off into periods of p digits each, beginning with units, so that $N = a_1 + 10^p a_2 + 10^{2p} a_3 + \ldots$, then $a_1 + a_2 + \ldots \equiv N$ (mod m), and N is divisible by m if and only if $a_1 + a_2 + \ldots$ is divisible by m.

E. B. Elliott[19] let $10^p = MD + r_p$. Thus $N = 10^p n_p + \ldots + 10 n_1 + n_0$ is divisible by D if $N = \Sigma n_j MD + \Sigma n_j r_j$ is divisible by D. The values of the r's are tabulated for $D = 3$, 7, 8, 9, 11, 13, 17.

A. Zbikowski[20] noted that $N = a + 10k$ is divisible by 7 if $k - 2a$ is divisible by 7. If δ is of the form $10n + 1$, $N = a + 10k$ is divisible by δ if $k - na$ is divisible by δ; this holds also if δ is replaced by a divisor of a number $10n + 1$.

V. Zeipel[21] tests for a divisor b by use of $nb = 10d + 1$. Then $10a_2 + a_1$ is divisible by b if $a_2 - a_1 d$ is divisible by b.

J. C. Dupain[22] noted, for use when division by $p - 1$ is easy, that $N = (p - 1)Q + R$ is divisible by p if $R - Q$ is divisible by p.

F. Folie[23] proved that if a, c are such that $ak' \pm ck = mp$ then $AB + C$ is divisible by the prime $p = aB + c$ if $Ak' \pm Ck = m'p$, provided a, c, k, k' are

[15]Jour. für Math., 27, 1844, 125–136.
[16]Nouv. Ann. Math., 4, 1845, 173–4 (cf. 81–82 by O. R.).
[17]Entwickelung u. Begründung neuer Gesetze über die Theilbarkeit der Zahlen. Jahresber. Kath. Gym. Köln, 1847–8.
[18]Cambridge and Dublin Math. Jour., 6, 1851, 32.
[19]The Math. Monthly (ed. Runkle), 1, 1859, 45–49.
[20]Bull. ac. sc. St. Pétersbourg, (3), 3, 1861, 151–3; Mélanges math. astr. ac. St. Pétersbourg, 3, 1859–66, 312.
[21]Öfversigt finska vetensk. forhandl., Stockholm, 18, 1861, 425–432.
[22]Nouv. Ann. Math., (2), 6, 1867, 368–9.
[23]Mém. Soc. Sc. Liège, (2), 3, 1873, 85–96.

not multiples of p. Application is made to the primes $p \leq 37$. Again, if p is a prime and

$$aB^2 + cB + d = ak'' + ck' + dk = Ak'' + Ck' + Dk = mp,$$

where k, k', k'' are prime to p, then $AB^2 + CB + D$ is divisible by p provided $k'^2 - kk''$ is a multiple of p.

C. F. Möller and C. Holten[24] would test the divisibility of n by a given prime p by seeking a such that $ap \equiv \pm 1$ (mod 10) and subtracting from n such a multiple of ap that the difference ends with zero.

L. L. Hommel[25] made remarks on the preceding method.

V. Schlegel[26] noted that if the divisor to be tested ends with 1, 3, 7 or 9, its product by 1, 7, 3 or 9 is of the form $d = 10\lambda + 1$. Then a, with the final digit u, is divisible by d if $a_1 = (a - ud)/10$ is. Then treat a_1 as we did a, etc.

P. Otto[27] would test Z for a given prime factor p by seeking a number n such that if the product by n of the number formed by the last s digits of Z be subtracted from the number represented by the remaining digits, the remainder is divisible by p if and only if Z is. Material is tabulated for the application of the method when $p < 100$.

N. V. Bougaief[27a] noted that $a_\mu \ldots a_1$ to base B is divisible by D if $a_1 \ldots a_\mu$ to base d is divisible by D, where $dB \equiv 1$ (mod D). For $B = 10$ and $D = 10n + 9$, 1, 3, 7, we may take $d = n + 1$, $9n + 1$, $3n + 1$, $7n + 5$, respectively. Again, $kB^2 + aB + b$ is divisible by D if $kB + a + bd$ is divisible.

W. Mantel and G. A. Oskamp[28] proved that, to test the divisibility of a number to any base by a prime, the value of the coefficient required to eliminate one, two, ... digits on subtraction is periodic. Also the number of terms of the period equals the length of the period of the periodic fraction arising on division by the same prime.

G. Dostor[28a] noted that $10t + u$ is divisible by any divisor a of $10A \pm 1$ if $t \mp Au$ is divisible by a. [A case of Liljevalch[12a].]

Hočevar[29] noted that if N, written to base a, is separated into groups G_1, G_2, ... each of q digits, N is divisible by a factor of $a^q + 1$ if $G_1 - G_2 + G_3 - \ldots$ is divisible. Thus, for $a = 2$, $q = 4$, $N = 104533$, or 11001100001010101 to base 2 is divisible by 17 since $0101 - 0101 + 1000 - 1001 + 1 = 0$.

J. Delboeuf[30] stated that if p, q are such that $pa + qb$ is a multiple of D and if $N = Aa + B\beta$ is a multiple of $D = aa + b\beta$, then $pA + qB$ is a multiple of D.

E. Catalan (ibid., p. 508) stated and proved the preceding test in the following form: If a, b and also a', b' are relatively prime, and

$$N = aa' + bb', \qquad Nx = Aa + Bb, \qquad Nx' = A'a' + B'b',$$

then $AA' + BB'$ is a multiple of N (and a sum of 2 squares if N is).

[24]Tidsskrift for Math., (3), 5, 1875, 177–180.   [25]Tidsskrift for Math., (3), 6, 1876, 15–19.
[26]Zeitschrift Math. Phys., 21, 1876, 365–6.   [27]Zeitschrift Math. Phys., 21, 1876, 366–370.
[27a]Mat. Sbornik (Math. Soc. Moscow), 8, 1876, I, 501–5.
[28]Nieuw Archief voor Wiskunde, Amsterdam, 4, 1878, 57–9, 83–94.
[28a]Archiv Math. Phys., 63, 1879, 221–4.
[29]Zur Lehre von der Teilbarkeit..., Prog. Innsbruck, 1881.
[30]La Revue Scientifique de France, (3), 38, 1886, 377–8.

Noël (*ibid.*, 378–9) gave tests for divisors 11, 13, 17,..., 43.
Bougon (*ibid.*, 508) gave several tests for the divisor 7. For example, a number is divisible by 7 if the quadruple of the number of its tens diminished by the units digit is divisible by 7, as 1883 since $188 \cdot 4 - 3 = 749$ is divisible by 7. J. Heilmann (*ibid.*, 187) gave a test for the divisor 7.
P. Breton and Schobbens (*ibid.*, 444–5) gave tests for the divisor 13.
S. Dickstein[31] gave a rule to reduce the question of the divisibility of a number to any base by another to that for a smaller number.
A. Loir[32] gave a rule to test the divisibility of N, having the units digit a, by a prime P. From $(N-a)/10$, subtract the product of a by the number, say $(mP-1)/10$, of tens in such a multiple mP of P that the units digit is 1. To the difference obtained apply the same operation, etc., until we exhaust N. If the final difference be P or 0, N is divisible by P.
R. Tucker[33] started with a number N, say 5443, cut off the last digit 3 and defined $u_2 = 544 - 2 \cdot 3 = 538$, $u_3 = 53 - 2 \cdot 8$, etc. If any one of the u's is divisible by 7, N is divisible by 7. R. W. D. Christie (p. 247) extended the test to the divisors 11, 13, 17, 37, the respective multipliers being 1, 9, 5, 11, provided always the number tested ends with 1, 3, 7 or 9.
R. Perrin[34] would find the minimum residue of N modulo p as follows. Decompose N, written to base x, into any series of digits, each with any number of digits, say A, B_i, C_j,..., where B_i has i digits. Let p be any integer prime to x and find q_1 so that $q_1 x \equiv 1 \pmod{p}$. Let a be any one of the integers prime to p and numerically $< p/2$. Let β be the ith integer following a in that one of the series containing a which are defined thus: as the first series take the residues modulo p of 1, q, q^2,...; as the second series take the products of the preceding residues by any new integer prime to p; etc. Let γ be the jth integer following β in the same series, etc. Then $N' = Aa + B_i\beta + C_j\gamma + ...$ is or is not divisible by p according as N is or not. By repetitions of the process, we get the minimum residue of N modulo p. The special case $A + B_1 q_1$, with p a prime, is due to Loir.[32]
Dietrichkeit[35] would test $Z = 10k + a$ for the divisor n by testing $k - xa$, where $10x + 1$ is some multiple of n. To test Z (pp. 316–7) for the divisor 7, test the sum of the products of the units digit, tens digit,... by 1, 3, 2, 6, 4, 5, taken in cyclic order beginning with any term (the remainders on converting $1/7$ into a decimal fraction). Similarly for $1/n$, when n is prime to 10.
J. Fontès[36] would test N for a divisor M by using a number $< N$ and $\equiv N \pmod{M}$, found as follows. For the base B, let q be the absolutely least residue of B^m modulo M. Commencing at the right, decompose N into sets of m digits, as $\lambda_m,..., a_m$, and set $f(x) = a_m x^n + \beta_m x^{n-1} + ... + \lambda_m$, whence $N = f(B^m)$. By expanding $N = f(q + M\Omega)$, we see that $f(q)$ is the desired number $< N$ and $\equiv N \pmod{M}$.
S. Levänen[37] gave a table showing the exponent to which 10 belongs for

[31]Lemberg Museum (Polish), 1886. [32]Comptes Rendus Paris, 106, 1888, 1070–1; errata, 1194.
[33]Nature, 40, 1889, 115–6. [34]Assoc. franç. avanc. sc., 18, 1889, II, 24–38.
[35]Zeitschr. Math. Phys., 36, 1891, 64. [36]Comptes Rendus Paris, 115, 1892, 1259–61.
[37]Öfversigt af finska vetenskaps-soc. förhandlingar, 34, 1892, 109–162. Cf. Jahrbuch Fortschr.
Math., 24, 1892, 164–5.

primes $b < 200$ and certain larger primes, from which are easily deduced tests for the divisor b.

Several[37a] noted that if 10 belongs to the exponent n modulo d, and if $S_1, S_2, \ldots$ denote the sums of every nth digit of N beginning with the first, second, ... at the right, the remainder on the division of N by d is that of $S_1 + 10S_2 + 10^2 S_3 + \ldots$

J. Fontès[38] would find the least residue of N modulo M. If 10^n has the residue q modulo M, we do not change the least residue of N if we multiply a set of n digits of N by the same power of q as of 10^n. Thus for $M = 19$, $N = 10433 = 10^4 + 4 \cdot 10^2 + 33$, 10^2 has the residue 5 modulo 19 and we may replace N by $5^2 + 4 \cdot 5 + 33$. The method is applied to each prime $M \le 149$.

Fontès[39] gave a history of the tests for divisibility, and an "extension of the method of Pascal," similar to that in his preceding paper.

P. Valerio[40] would test the divisibility of N by 39, for example, by subtracting from N a multiple of 39 with the same ending as N.

F. Bělohlávek[41] noted that $10A + B$ is divisible by $10p \pm 1$ if $A \mp pB$ is.

C. Börgen[42] noted that $Z = a_n \cdot 10^n + \ldots + a_1 \cdot 10 + a_0$ is divisible by N if

$$\sum_{\nu=0}^{n-a+1} (a_{\nu-a+1} \cdot 10^{a-1} + \ldots + a_\nu)(10^a - N)^{\nu/a}$$

is divisible by N. For $N = 7$, take $a = 1$; then $10^a - N = 3$ and Z is divisible by 7 if $a_0 + 3a_1 + 2a_2 - a_3 - 3a_4 - 2a_5 + \ldots$ is divisible by 7.

J. J. Sylvester[42a] noted that, if the r digits of N, read from left to right, be multiplied by the first r terms of the recurring series $1, 4, 3, -1, -4, -3;$ $1, 4, \ldots$ [the residues, in reverse order, of $10, 10^2, \ldots$, modulo 13], the sum of the products is divisible by 13 if and only if N is divisible by 13.

C. L. Dodgson[42b] discussed the quotient and remainder on division by 9 or 11.

L. T. Riess[43] noted that, if p is not divisible by 2 or 5, $10b + a (a < 10)$ is divisible by p if $b - xa$ is divisible by p, where $mp = 10x + a$ $(a < 10)$ and $m = 1, 7, 3, 9$ according as $p \equiv 1, 3, 7, 9$ (mod 10), respectively.

A. Loir[44] gave tests for prime divisors < 100 by uniting them by twos or threes so that the product P ends in 01, as $7 \cdot 43 = 301$. To test N, multiply the number formed of the last two digits of N by the number preceding 01 in P, subtract the product from N, and proceed in the same manner with the difference. Then P is a factor if we finally get a difference which is zero. If a difference is a multiple of a prime factor p of P, then N is divisible by p.

Plakhowo[45] gave the test by Bougaief, but without using congruences.

[37a]Math. Quest. Educ. Times, 57, 1892, 111.
[38]Assoc. franç. avanc. sc., 22, 1893, II, 240–254.
[39]Mém. ac. sc. Toulouse, (9), 5, 1893, 459–475.
[40]La Revue Scientifique de France, (3), 52, 1893, 765.
[41]Casopis, Prag, 23, 1894, 59.　　[42]Nature, 57, 1897–8, 54.
[42a]Educat. Times, March, 1897. Proofs. Math. Quest. Educ. Times, 66, 1897, 108. Cf. W. E. Heal, Amer. Math. Monthly, 4, 1897, 171–2.
[42b]Nature, 56, 1897, 565–6.
[43]Russ. Nat., 1898, 329. Cf. Jahrb. Fortschritte Math., 29, 1898, 137.
[44]Assoc. franç. avanc. sc., 27, 1898, II, 144–6.
[45]Bull. des sc. math. et phys. élémentaires, 4, 1898–9, 241–3.

To test $N = a_0 + a_1B + \ldots + a_nB^n$ for the divisor D prime to B, determine d and x so that $Bd = Dx + 1$. Multiply this equation by a_0 and subtract from N. Thus

$$N = BN' - Da_0x, \qquad N' = a_0d + (a_1 + a_2B + \ldots + a_nB^{n-1})B.$$

Hence N is divisible by D if and only if N' is divisible by D. Now, N' is derived from N by supressing the units digit a_0 and adding to the result the product a_0d. Next operate with N' as we did with N.

J. Malengreau[46] would test N for a factor q prime to 10 by seeking a multiple $11 \ldots 1$ (to m digits) of q, then an exponent t such that the number of digits of $10^t \cdot N$ is a multiple of m. From each set of m digits of $10^t \cdot N$ subtract the nearest multiple of $1 \ldots 1$ (to m digits). The sum of the residues is divisible by q if and only if N is divisible by q.

G. Loria[47] proved that $N = a_0 + ga_1 + \ldots + g^k a_k$ is divisible by a if and only if a divides the sum $a_0 + \ldots + a_k$ of the digits of N written to a base g of the form $ka + 1$; or if a divides $a_0 - a_1 + a_2 - \ldots$ when the base g is of the form $ka - 1$. Taking $g = 10^m$, we have the test, in Gelin's Arithmétique, in terms of groups of m digits. We may select m to be $\frac{1}{2}\phi(a)$ or a number such that $10^m \pm 1$ has the factor a. In place of $a_0 + a_1 + \ldots$ when $g = 10^m$, we may employ

$$\rho a_0 + \lambda a_1 + 10\lambda a_2 + \ldots + 10^{m-2}\lambda a_{m-1}$$
$$+ \sum_{k=1}^{n-1} 10^{km-1}\lambda(a_{km} + 10a_{km+1} + \ldots + 10^{m-1}a_{km+m-1}),$$

where $\lambda = 1$, 2 or 5, and ρ is determined by $10\rho/\lambda \equiv 1 \pmod a$. Taking $a = 7, 13, 17, 19, 23$, special tests for divisors are obtained.

G. Loria[48] proved that, if $a_0, a_1, \ldots$ are successive sets of t digits of N, counted from the right, and $\sigma = a_0 \pm a_1 + a_2 \pm a_3 + \ldots$, then

$$N - \sigma = a_1(10^t \mp 1) + a_2(10^{2t} - 1) + a_3(10^{3t} \mp 1) + \ldots,$$

so that a factor of $10^t \mp 1$ divides N if and only if it divides σ.

A. Tagiuri[49] extended the last result to any base g. We have

$$N = a_0 + ga_1 + \ldots = N_{0m} + g^m N_{1m} + g^{2m} N_{2m} + \ldots$$

if $N_{\rho m} = a_{\rho m} + a_{\rho m+1}g + \ldots + a_{\rho m+m-1}g^{m-1}$. Hence, if $g^m \equiv \pm 1 \pmod a$,

$$N \equiv N_{0m} \pm N_{1m} + N_{2m} \pm \ldots \pmod a.$$

L. Ripert[50] noted that $10D + u$ is divisible by $10\delta + i$ if $Di - \delta u$ is divisible, and gave many tests for small divisors.

G. Biase[51] derived tests that $10d + u$ has the factor 7 or 19 from

$$2(10d + u) \equiv 2u - d \pmod 7, \qquad 2(10d + u) \equiv 2u + d \pmod{19}.$$

O. Meissner[52] reported on certain tests cited above.

[46]Mathesis, (3), 1, 1901, 197–8.
[47]Rendiconti Accad. Lincei (Math.), (5), 10, 1901, sem. 2, 150–8. Mathesis, (3), 2, 1902, 33–39.
[48]Il Boll. Matematica Gior. Sc.-Didat., Bologna, 1, 1902. Cf. A. Bindoni, ibid., 4, 1905, 87.
[49]Periodico di Mat., 18, 1903, 43–45.   [50]L'enseignement math., 6, 1904, 40–46.
[51]Il Boll. Matematica Gior. Sc.-Didat., Bologna, 4, 1905, 92–6.
[52]Math. Naturw. Blätter, 3, 1906, 97–99.

E. Nannei[53] employed $r_1 = a_1 - a_0 x$, $r_2 = a_2 - r_1 x$, ... $(x < 10)$. Then, if $r_n = 0$, $N = 10^n a_n + \ldots + 10 a_1 + a_0$ is divisible by $10x + 1$ and the quotient has the digits r_{n-1}, r_{n-2}, ..., r_1, a_0. The cases $x = 1$, 2 are discussed and several tests for 7 deduced. For $x = 1/3$, we conclude that, if $r_n = 0$, N is divisible by 13 and the digits of the quotient are $r_{n-1}/3$, ..., $r_1/3$, $a_0/3$.

A. Chiari[54] employed D'Alembert's[5] method for $10 + b$, $b = 3$, 7, 9.

G. Bruzzone[55] noted that, to find the remainder R when N is divided by an integer x of r digits, we may choose y such that $x + y = 10^r$, form the groups of r digits counting from the right of N, and multiply the successive groups (from the right) by 1, y, y^2, ... or by their residues modulo x; then R equals the remainder on dividing the sum of the products by x. If we choose $x - y = 10^r$, we must change alternate signs before adding. For practical use, take $y = 1$.

Fr. Schuh[56] gave three methods to determine the residue of large numbers for a given modulus.

Stuyvaert[57] let a, b, ... be the successive sets of n digits of N to the base B, so that $N = a + bB^n + cB^{2n} + \ldots$. Then N is divisible by a factor D of $B^n \mp R^n$ if and only if $a \pm bR^n + cR^{2n} \pm \ldots$ is divisible by D. For $R = 1$, $B = 10$, $n = 1$, 2, ..., we obtain tests for divisors of 9, 99, 11, 101, etc. A divisor, prime to B, of $mB + 1$ divides $N = a + bB$ if and only if it divides $b - ma$.

FURTHER PAPERS GIVING TESTS FOR A GIVEN DIVISOR d.

J. R. Young and Mason for $d = 7$, 13 [Pascal[4]], Ladies' Diary, 1831, 34–5, Quest. 1512.
P. Gorini [Pascal[4]], Annali di Fis., Chim. Mat., (ed., Majocchi), 1,1841, 237.
A. Pinaud for $d = 7$, 13, Mém. Acad. Sc. Toulouse, 1, 1844, 341, 347.
*Dietz and Vincenot, Mém. Acad. Metz, 33, 1851–2, 37.
Anonymous writer for $d = 9$, 11, Jour. für Math., 50, 1855, 187–8.
*H. Wronski, Principes de la phil. des math. Cf. de Montferrier, Encyclopédie math., 2, 1856, p. 95.
O. Terquem for $d \leqq 19$, 23, 37, 101, Nouv. Ann. Math., 14, 1855, 118–120.
A. P. Reyer for $d = 7$, Archiv Math. Phys., 25, 1855, 176–196.
C. F. Lindman for $d = 7$, 13, ibid., 26, 1856, 467–470.
P. Buttel for $d = 7$, 9, 11, 17, 19, ibid., 241–266.
De Lapparent [Herter[14]], Mém. soc. imp. sc. nat. Cherbourg, 4, 1856, 235–258.
Karwowski [Pascal[4]], Ueber die Theilbarkeit..., II, Progr., Lissa, 1856.
*D. van Langeraad, Kenmerken van deelbarheid der geheele getallen, Schoonhoven, 1857.
Flohr, Ueber Theilbarkeit und Reste der Zahlen, Progr., Berlin, 1858.
V. Bouniakowsky for $d = 37$, 989, Nouv. Ann. Math., 18, 1859, 168.
Elefanti for $d = 7 \cdot 13$, Proc. Roy. Soc. London, 10, 1859–60, 208.
A. Niegemann for $d = 10^{m} \cdot n + a$, Archiv Math. Phys., 38, 1862, 384–8.
J. A. Grunert for $d = 7$, 11, 13, ibid., 42, 1864, 478–482.
V. A. Lebesgue, Tables diverses pour la décomposition des nombres, Paris, 1864, p. 13.

[53]Il Pitagora, Palermo, 13, 1906–7, 54–9.
[54]Ibid., 14, 1907–8, 35–7.
[55]Ibid., 15, 1908–9, 119–123.
[56]Supplem. De Vriend der Wiskunde, 24, 1912, 89–103.
[57]Les Nombres Positifs, Gand, 1912, 59–62.

C. M. Ingleby for $d=9$, 11, British Assoc. Report, 35, 1865, 7 (trans.).

M. Jenkins for any prime d, Math. Quest. Educ. Times, 8, 1868, 69, 111.

F. Unferdinger [Gergonne[11]], Sitzungsber. Ak. Wiss. Wien (Math.), 59, 1869, II, 465–6.

H. Anton for $d=9$, 11, 13, 101, Archiv Math. Phys., 49, 1869, 241–308.

W. H. Walenn, British Assoc. Report, 40, 1870, 16–17 (trans.); Phil. Mag., (4), 36, 1868, 346–8; (4), 46, 1873, 36–41; (4), 49,.1875, 346–351; (5), 2, 1876, 345; 4, 1877, 378; 9, 1880, 56, 121, 271.

M. A. X. Stouff for $d<100$, Nouv. Ann. Math., (2), 10, 1871, 104.

J. Lubin, *ibid.*, (2), 12, 1874, 528–30 (trivial).

Szenic for $d=7$, 9, 37, Von der Kongruenz der Z., Progr. Schrimm, 1873.

E. Brooks for $d=7$, Des Moines Analyst, 2, 1875, 129.

W. J. Greenfield and M. Collins for $d=47$, 73, Math. Quest. Educ. Times, 22, 1875, 87.

F. da Ponte Horta for $d=7$, 9, 11, 13, Jornal de Sciencias Mat. Ast., 1, 1877, 57–62.

Mennesson for $d=7$, Nouv. Corresp. Math., 4, 1878, 151; generalization by Cesàro, p. 156.

C. Lange, for $d=7$, 13, 17, 19, Ueber die Teilbarkeit der Zahlen, Progr., Berlin, 1879.

F. Jorcke for $d=7$, 9, 11, Ueber Zahlenkongruenzen..., Progr. Fraustadt, 1878.

K. Broda for any base, Archiv Math. Phys., 63, 1879, 413–428.

A. Badoureau for $d=19$, Nouv. Ann. Math., (2), 18, 1879, 35–6.

S. M. Drach for $d=7$, Math. Quest. Educ. Times, 35, 1881, 71–2.

W. A. Pick for $d=7$, *ibid.*, 38, 1883, 64.

A. Evans for $d=7$, Des Moines Analyst, 10, 1883, 134.

K. Haas, Theilbarkeitsregeln..., Progr., Wien, 1883.

G. Wertheim, Elemente der Zahlentheorie, 1887, 31–33.

B. Adam for $d<100$, Ueber die Teilbarkeit..., Progr. Gym. Clausthal, 1889.

A. Loir for $d<138$, Jour. de math. élém., 1889, 66, 107–10, 121–3.

A. G. Fazio [Schlegel[26]], Sui caratteri..., Palermo, 1889.

E. Gelin, Mathesis, (2), 2, 1892, 65, 93; (2), 12, 1902, 65–74, 93–99 (extract in Mathesis, (3), 10, 1910, Suppl. I); Ann. Soc. Sc. Bruxelles, 34, 1909–10, 66; Recueil de problèmes d'arith., 1896. Extracts by M. Nassò, Revue de Math. (ed., Peano), 7, 1900–1, 42–52.

Speckmann, Dorsten, Haas, Dörr, Zeitschrift Math. Phys., 37, 1892, 58, 63, 128, 192, 383.

Lalbaletrier, Jour. de Math. (ed., de Longchamps), 1894, 54.

H. T. Burgess [Pascal[4]], Nature, 57, 1897–8, 8–9, 30, 55.

A. Conti [Pascal[4]], Periodico di Mat., 13, 1898, 180–6, 207-9.

F. Mariantoni, *ibid.*, 149–151, 191–2, 217–8.

T. Lange for $d<30$, Archiv. Math. Phys., (2), 16, 1898, 220–3.

W. J. Greenstreet, Math. Gazette, 1, 1900, 186–7.

Christie for $d=2^n p, 5^n p$ (p prime), Math. Quest. Educ. Times, 73, 1900, 119.

A. Cunningham and D. Biddle for $d=rp \pm 1$, *ibid.*, 75, 1901, 49–50.

M. Zuccagni for $d=7$, Suppl. al Periodico di Mat., 6, fasc. V.

Calvitti for $d=7$, *ibid.*, 8, fasc. IV.

S. Dickstein, Wiad. Mat., Warsaw, 6, 1902, 253–7 (Polish).

B. Niewenglowski, *ibid.*, 252–3.

Pietzker for $d=7$, 11, 13, 27, 37, Unterrichtsblätter Math. Naturwiss., 9, 1903, 85–110.

A. Church for $d=7$, 13, 17, Amer. Math. Monthly, 12, 1905, 102–3.

E. A. Cazes, Assoc. franç., 36, 1907, 55–63.

A. Gérardin for $d=7$, 13, 17, 37, 43, Sphinx-Oedipe, 1907–8, 2.

M. Morale for $d=7$, Suppl. al Periodico di Mat., 11, 1908, 103.

*T. Ghezzi, *ibid.*, 12, 1908–9, 129–130.

Lenzi, Il Boll. Matematica Gior. Sc.-Didat., 7, 1908.

R. Polpi, *ibid.*, 8, 1909, 281–5.
M. Morale for $d=7$, 13, Suppl. al Periodico di Mat., 13, 1909–10, 38–9.
A. L. Csada, *ibid.*, 56–8.
*A. La Paglia, *ibid.*, 14, 1910–11, 136–7, extension of Morale to any d.
A. V. Filippov, 8 methods for $d=9$, Kagans Bote, 1910, 88–92, No. 520.
P. Cattaneo for $d=11$, Il Boll. Matematica Gior. Sc.-Didat., 9, 1910, 305–6.
*L. Miceli, Condizioni di divisibilità di un numero N per un numero a..., Matera, 1911, 8 pp.
R. Ayza for $d=a\cdot 10^n \pm 1$, Revista sociedad mat. española, Madrid, 1, 1911, 162–6.
*Paoletti, Il Pitagora, Palermo, 18, 1911–12, 128–132.
*R. La Marca, Criterî di congruenza e criterî di divisibilità, Torre del Greco, 1912, 30 pp.
K. W. Lichtenecker, Zeitschr. für Realschulwesen, 37, 1912, 338–49.
R. E. Cicero, Sociedad Cientifica Antonio Alzate, 32, 1912–3, 317–331.
J. G. Galé for $d=7$, Revista sociedad mat. española, 3, 1913–4, 46–7.
C. F. Iodi for $d=7$, 13, 17, 19, Suppl. al Periodico di Mat., 18, 1914, 20–3.
E. Kylla for $d=11$, Unterrichtsblätter Math. Naturwiss., 20, 1914, 156.
R. Krahl for $d=7$, Zeitschrift Math. Naturw. Unterricht, 45, 1914, 562.
P. A. Fontebasso, Il Boll. Matematica, 13, 1914–5.
G. M. Persico, Periodico di Mat., 32, 1917, 105–124.
Sammlung der Aufgaben in Zeitschrift Math. Naturw. Unterricht, 1898: for $d=7$, II, 337; IV, 404, 407; for $d=9$, 11, XXIV, 606; XXV, 587–8; for $d=37$, etc., XXVI, 18, 25–27.
Criteria for divisibility in connection with tables were given by Barlow,[48] Tarry[86] and Lebon[87] of Ch. XIII, and Harmuth[31] of Ch. XIV.

PAPERS ON DIVISIBILITY NOT AVAILABLE FOR REPORT.

Joubin, Jour. Acad. Soc. Sc. France et de l'Etranger, Paris, 2, 1834, 230.
J. Lenthéric, Théorie de la divisibilité des nombres, Paris, 1838.
R. Volterrani, Saggio sulla divisione ragionata dei n. interi, Pisa, 1871.
F. Tirelli, Teoria della divisibilità de' numeri, Napoli, 1875.
E. Tiberi, Teoria generale sulle condizioni di divisibilità..., Arezzo, 1890.
J. Kroupa, Casopis, Prag, 43, 1914, 117–120.
G. Schröder, Unterrichtsblätter für Math. Naturwiss., 21, 1915, 152–5.

CHAPTER XIII.

FACTOR TABLES, LISTS OF PRIMES.

Eratosthenes (third century B.C.) gave a method, called the sieve or crib of Eratosthenes, of determining all the primes under a given limit l, which serves also to construct the prime factors of numbers $<l$. From the series of odd numbers 3, 5, 7,..., strike out the square of 3 and every third number after 9, then the square of 5 and every fifth number after 25, etc. Proceed until the first remaining number, directly following that one whose multiples were last cancelled, has its square $>l$. The remaining numbers are primes.

Nicomachus and Boethius[1] began with 5 instead of with 5^2, 7 instead of with 7^2, etc., and so obtained the prime factors of the numbers $<l$.

A table containing all the divisors of each odd number ≤ 113 was printed at the end of an edition of Aratus, Oxford, 1672, and ascribed to Eratosthenes by the editor, who incorrectly considered the table to be the sieve of Eratosthenes. Samuel Horsley[2] believed that the table was copied by some monk in a barbarous age either from a Greek commentary on the Arithmetic of Nicomachus or else from a Latin translation of a Greek manuscript, published by Camerarius, in which occurs such a table to 109.

Leonardo Pisano[3] gave a table of the 21 primes from 11 to 97 and a table giving the factors of composite numbers from 12 to 100; to determine whether n is prime or not, one can restrict attention to divisors $\leq \sqrt{n}$.

Ibn Albannâ in his Talkhys[4] (end of 13th century) noted that in using the crib of Eratosthenes we may restrict ourselves to numbers $\leq \sqrt{l}$.

Cataldi[5] gave a table of all the factors of all numbers up to 750, with a separate list of primes to 750, and a supplement extending the factor table from 751 to 800.

Frans van Schooten[6] gave a table of primes to 9979.

J. H. Rahn[7] (Rhonius) gave a table of the least factors of numbers, not divisible by 2 or 5, up to 24000.

T. Brancker[8] constructed a table of the least divisors of numbers, not divisible by 2 or 5, up to 100 000. [Reprinted by Hinkley.[55]]

---

[1]Introd. in Arith. Nicomachi; Arith. Boethii, lib. 1, cap. 17 (full titles in the chapter on perfect numbers). Extracts of the parts on the crib, with numerous annotations, were given by Horsley.[2] Cf. G. Bernhardy, Eratosthenica, Berlin, 1822, 173–4.

[2]Phil. Trans. London, 62, 1772, 327–347.

[3]Il Liber Abbaci di L. Pisano (1202, revised 1228), Roma, 1852, ch. 5; Scritti, 1, 1857, 38.

[4]Transl. by A. Marre, Atti Accad. Pont. Nuovi Lincei, 17, 1863–4, 307.

[5]Trattato de' numeri perfetti, Bologna, 1603. Libri, Histoire des Sciences Math. en Italie, ed. 2, vol. 4, 1865, 91, stated erroneously that the table extended to 1000.

[6]Exercitat. Math., libri 5, cap. 5, p. 394, Leiden, 1657.

[7]Algebra, Zürich, 1659. Wallis,[10] p. 214, attributed this book to John Pell.

[8]An Introduction to Algebra, translated out of the High-Dutch [of Rahn's[7] Algebra] into English by Thomas Brancker, augmented by D. P. [=Dr. Pell], London, 1668. It is cited in Phil. Trans. London, 3, 1668, 688. The Algebra and the translation were described by G. Wertheim, Bibliotheca Math., (3), 3, 1902, 113–126.

D. Schwenter[9] gave all the factors of the odd numbers <1000.
John Wallis[10] gave a list of errata in Brancker's[8] table.
John Harris,[11] D. D., F. R. S., reprinted Brancker's[8] table.
De Traytorens[12] emphasized the utility of a factor table. To form a table showing all prime factors of numbers to 1000, begin by multiplying 2, 3,... by all other primes <1000, then multiply 2×3 by all the primes, then 2×3×5, etc.

Joh. Mich. Poetius[13] gave a table (anatomiae numerorum) of all the prime factors of numbers, not divisible by 2, 3, 5, up to 10200. It was reprinted by Christian Wolf,[14] Willigs,[19] and Lambert.[22]

Johann Gottlob Krüger[15] gave a table of primes to 100 999 (not to 1 million, as in the title), stating that the table was computed by Peter Jäger of Nürnberg.

James Dodson[16] gave the least divisors of numbers to 10000 not divisible by 2 or 5 and the primes from 10000 to 15000.

Etienne François du Tour[17] described the construction of a table of all composite odd numbers to 10000 by multiplying 3, 5,..., 3333 by 3,..., 99.

Giuseppe Pigri[18] gave all prime factors of numbers to 10000.

Michel Lorenz Willigs[19] (Willich) gave all divisors of numbers to 10000.

Henri Anjema[20] gave all divisors of numbers to 10000.

Rallier des Ourmes[21] gave as if new the sieve of Eratosthenes, placing 3 above 9 and every third odd number after it, a 7 above 49, etc. He expressed each number up to 500 as a product of powers of primes.

J. H. Lambert[22] described a method of making a factor table and gave Poetius'[13] table and expressed a desire for a table to 102 000. Lagrange called his attention to Brancker's[8] table.

Lambert[23] gave [Krüger's[15]] table showing the least factor of numbers not divisible by 2, 3, 5 up to 102000, and a table of primes to 102 000, errata in which were noted by Klügel[24].

[9]Geometria Practica, Nurnb., 1667, I, 312.
[10]Treatise of Algebra, additional treatise, Ch. III, §22, London, 1685.
[11]Lexicon Technicum, or an Universal English Dictionary of Arts and Sciences, London, vol. 2, 1710 (under Incomposite Numbers). In ed. 5, London, 2, 1736, the table was omitted, but the text describing it kept. Wallis, Opera, 2, 511, listed 30 errors.
[12]Histoire de l'Acad. Roy. Science, année 1717, Paris, 1741, Hist., 42–47.
[13]Anleitung zu der Arith. Wissenschaft vermittelst einer parallel Algebra, Frkf. u. Leipzig, 1728.
[14]Vollst. Math. Lexicon, 2, Leipzig, 1742, 530.
[15]Gedancken von der Algebra, nebst den Primzahlen von 1 bis 1 000 000,·Halle im Magd., 1746. Cf. Lambert.[23]
[16]The Calculator...Tables for Computation, London, 1747.
[17]Histoire de l'Acad. Roy. Sc., Paris, année 1754, Hist., 88–90.
[18]Nuove tavole degli elementi dei numeri dall' 1 al 10 000, Pisa, 1758.
[19]Gründliche Vorstellung der Reesischen allgemeinen Regel...Rechnungsarten, Bremen u. Göttingen, 2, 1760, 831–976.
[20]Table des diviseurs de tous les nombres naturels, depuis 1 jusqu'à 10 000, Leyden, 1767, 302 pp.
[21]Mém. de math. et de physique, Paris, 6, 1768, 485–499.
[22]Beyträge zum Gebrauche der Math. u. deren Anwendung, Berlin, 1770, II, 42.
[23]Zusätze zu den logarithmischen und trig. Tabellen, Berlin, 1770.
[24]Math. Wörterbuch, 3, 1808, 892–900.

J. Ozanam[25] gave a table of primes to 10000.

A. F. Marci[26] gave in 1772 a list of primes to 400 000.

Jean Bernoulli[26a] tabulated the primes $16n+1$ up to 21601.

L. Euler[27] discussed the construction of a factor table to one million. Given a prime $p=30a\pm t$ $(t=1, 7, 11, 13)$, he determined for each $r=1, 7, 11, 13, 17, 19, 23, 29$, the least q for which $30q+r$ is divisible by p, and arranged the results in a single table with p ranging over the primes from 7 to 1000. He showed how to use this auxiliary table to construct a factor table between given limits.

C. F. Hindenburg[28] employed in the construction of factor tables a "patrone" or strip of thick paper with holes at proper intervals to show the multiples of p, for the successive primes p.

A. Felkel[29] gave in 1776 a table of all the prime factors (designated by letters or pairs of letters) of numbers, not divisible by 2, 3, or 5, up to 408 000, requiring for entry two auxiliary tables. In manuscript[30], the table extended to 2 million; but as there were no purchasers of the part printed, the entire edition, except for a few copies, was used for cartridges in the Turkish war. The imperial treasury at Vienna, at the cost of which the table was printed, retained the further manuscript. [See Felkel.[38]]

L. Bertrand[31] discussed the construction of factor tables.

The Encyclopédie of d'Alembert, ed. 1780, end of vol. 2, contains a factor table to 100 000.

Franz Schaffgotsch[32] gave a method, equivalent to that of a stencil for each prime p, for entering the factor p in a factor table with eight headings $30m+k$, $k=1, 7, 11, 13, 17, 19, 23, 29$, and hence of numbers not divisible by 2, 3, or 5. Proofs were given by Beguelin and Tessanek, ibid., 362, 379.

The strong appeals by Lambert[23] that some one should construct a factor table to one million led L. Oberreit, von Stamford, Rosenthal, Felkel, and Hindenburg to consider methods of constructing factor tables and to prepare such tables to one million, with plans for extension to 5 or 10

[25]Recreations Math., new ed., Paris, 1723, 1724, 1735, etc., I, p. 47.

[26]Primes "in quater centenis millibus," Amstelodami, 1772.

[26a]Nouv. Mém. Ac. Berlin, année 1771, 1773, 323.

[27]Novi Comm. Acad. Petrop., 19, 1774, 132; Comm. Arith., 2, 64.

[28]Beschreibung einer ganz neuen Art nach einem bekannten Gesetze fortgehende Zahlen durch Abzählen oder Abmessen bequem u. sicher zu finden. Nebst Anwendung der Methode auf verschiedene Zahlen, besonders auf eine darnach zu fertigende Factorentafel..., Leipzig, 1776, 120 pp.

[29]Tabula omnium factorum simplicium, numerorum per 2, 3, 5 non divisibilium ab 1 usque 10 000 000 [!]. Elaborata ab Antonio Felkel. Pars I. Exhibens factores ab 1 usque 144 000, Vindobonae, 1776. Then there is a table to 408 000, given in three sections. There is a copy of this complete table in the Graves Library, University College, London. Tafel aller einfachen Factoren der durch 2, 3, 5 nicht theilbaren Zahlen von 1 bis 10 000 000. Entworfen von Anton Felkel. I. Theil. Enthaltend die Factoren von 1 bis 144 000, Wien, 1776. There is a copy of this incomplete table in the libraries of the Royal Society of London and Göttingen University.

[30]Cf. Zach's Monatliche Correspondenz, 2, 1800, 223; Allgemeine deutsche Bibliothek, 33, II, 495.

[31]Dévelop. nouveau de la partie él. math., Genève, 1774.

[32]Gesetz, welches zur Fortsetzung der bekannten Pellischen Tafeln dient, Abhand. Privatgesellschaft in Böhmen, Prag, 5, 1782, 354–382.

million. Their extended correspondence with Lambert[33] was published. Of the tables constructed by these computers, the only one published is that by Felkel.[29] The history of their connection with factor tables has been treated by J. W. L. Glaisher.[34]

Johann Neumann[35] gave all the prime factors of numbers to 100 100. Desfaviaae gave a like table in the same year.

F. Maseres[36] reprinted the table of Brancker.[8]

G. Vega[37] gave all the prime factors of numbers not divisible by 2, 3, or 5 to 102 000 and a list of primes from 102 000 to 400 031. Chernac listed errors in both tables. In Hülsse's edition, 1840, of Vega, the list of primes extends to 400 313.

A. Felkel,[38] in his Latin translation of Lambert's[23] Zusätze, gave all the prime factors except the greatest of numbers not divisible by 2, 3, 5 up to 102 000, large primes being denoted by letters. In the preface he stated that, being unable to obtain his extensive manuscript[30] in 1785, he calculated again a factor table from 408 000 to 2 856 000.

J. P. Grüson[39] gave all prime factors of numbers not divisible by 2, 3, 5 to 10500. He[39a] gave a table of primes to 10000.

F. W. D. Snell[40] gave the prime factors of numbers to 30000.

A. G. Kästner[41] gave a report on factor tables.

K. C. F. Krause[42] gave a table of 22 pages showing all products < 100 000 of two primes, a table of primes < 100 000 with letters for 01, 03, . . ., 99, and (pp. 25–28) a factor table to 10000 by use of letters for numbers < 100.

N. J. Lidonne[43] gave all prime factors of numbers to 102 000.

Jacob Struve[43a] made a factor table to 100 by de Traytorens'[12] method.

L. Chernac[44] gave all the prime factors of numbers, not divisible by 2, 3 or 5, up to 1 020 000.

J. C. Burckhardt[45] gave the least factor of numbers to 3 million. He did not compute the first million, but compared Chernac's table with a manuscript (mentioned in Briefwechsel,[33] p. 140) by Schenmarck which extended to 1 008 000. Cf. Meissel.[66]

---

[33]Joh. Heinrich Lamberts deutscher gelehrter Briefwechsel, herausgegeben von Joh. Bernoulli, Berlin, 1785, Leipzig, 1787, vol. 5.   [34]Proc. Cambridge Phil. Soc., 3, 1878, 99–138.

[35]Tabellen der Primzahlen und der Faktoren der Zahlen, welche unter 100 100, und durch 2, 3 oder 5 nicht theilbar sind, Dessau, 1785, 200 pp.

[36]The Doctrine of Permutations and Combinations. . ., London, 1795.

[37]Tabulæ logarithmico-trigonometricae, 1797, vol. 2.

[38]J. H. Lambert, Supplementa tab. log. trig., Lisbon, 1798.

[39]Pinacothéque, ou collection de Tables. . ., Berlin, 1798.

[39a]Enthüllte Zaubereyen u. Geheimnisse d. Arith., Berlin, 1796, I, 82–4.

[40]Ueber eine neue und bequeme Art, die Factorentafeln einzurichten, nebst einer Kupfertafel der einfachen Factoren von 1 bis 30000, Giessen and Darmstadt, 1800.

[41]Fortsetzung der Rechenkunst, ed. 2, Göttingen, 1801, 566–582.

[42]Factoren- und Primzahlentafel von 1 bis 100 000 neu berechnet, Jena u. Leipzig, 1804.

[43]Tables de tous les diviseurs des nombres < 102 000, Paris, 1808.

[43a]Handbuch der Math., Altona, II, 1809, 108.

[44]Cribrum Arithmeticum. . . Daventriæ, 1811, 1020 pp. Reviewed by Gauss, Göttingische gelehrte Anzeigen, 1812; Werke 2, 181–2. Errata, Cunningham.[85]

[45]Tables des diviseurs. . . 1 à 3 036 000, Paris, 1817, 1814, 1816 (for the respective three millions), and 1817 (in one volume).

P. Barlow[46] gave the prime and power of prime factors of numbers to 10000 and a list of primes to 100 103.

C. Hutton[47] gave the least factor of numbers to 10000.

Rees' Cyclopaedia, 1819, vol. 28, lists the primes to 217 219.

Peter Barlow[48] gave a two-page table for finding factors of a number $N < 100\ 000$. The primes $p = 7$ to $p = 313$ are at the head of the columns, while the 18 numbers $1000, \ldots, 9000, 10000, 20000, \ldots, 90000$ are in the left-hand column. In the body of the table is the remainder of each of the latter when divided by the primes p. To test if p is a factor of N, add its last two digits to the remainders in the line of hundreds and thousands in the column headed p and test whether the sum is divisible by p.

J. P. Kulik[49] gave a factor table to 1 million.

J. Hantschl[50] gave a factor table to 18277; J. M. Salomon,[51] to 102 011.

A. L. Crelle[52] gave the number of primes $4n + 1$ and the number of primes $4n + 3$ in each thousand up to the fiftieth.

A. Guyot[53] listed the primes to 100 000.

A. F. Möbius,[53a] using square ruled paper, inserted from right to left $0, 1, 2, \ldots$ in the top row of cells, and inserted n in each cell of the nth row below the top row whenever the corresponding number in the top row is divisible by n. We thus have a factor table. Certain numbers of the table lie in straight lines, others in parabolas, etc.

P. A. G. Colombier[53b] discussed the determination of the primes $< l'$, given those $< l$.

H. G. Köhler[54] gave a factor table to 21524.

E. Hinkley[55] gave a factor table to 100 000, listing all factors of odd numbers to 20000 and of even numbers to 12500.

F. Schallen[55a] gave the prime and prime-power factors of numbers < 10000.

F. Landry[56] gave factor and prime tables to 10000.

A. L. Crelle[57] discussed the expeditious construction of a factor table, and in particular a method of extending Chernac's[44] table to 7 million.

J. Hoüel[58] gave a factor table to 10841.

Jacob Philip Kulik (1773–1863) spent 20 years constructing a factor

[46]New Mathematical Tables, London, 1814. Errata, Cunningham.[85]

[47]Phil. and Math. Dictionary, 1815, vol. 2, 236–8.

[48]New Series of Math. Repository (ed., Th. Leybourn), London, 4, 1819, II, 30–39.

[49]Tafeln der einfachen Faktoren aller Zahlen unter 1 million, Graz, 1825.

[50]Log.-trig. Handbuch, Wien, 1827.   [51]Log. Tafeln, Wien, 1827.

[52]Jour. für Math., 10, 1833, 208.

[53]Théorie générale de la divisibilité des nombres, suivie d'applications variées et d'une table de nombres premiers compris entre 0 et 100 000, Paris, 1835.

[53a]Jour. für Math., 22, 1841, 276–284.   [53b]Nouv. Ann. Math., 2, 1843, 408–410.

[54]Log.-trig. Handbuch, Leipzig, 1848. Errata, Cunningham.[85]

[55]Tables of the prime numbers and prime factors of the composite numbers from 1 to 100 000, Baltimore, 1853. Reproduction of Brancker's[8] table.

[55a]Primzahlen-Tafel von 1 bis 10000..., Weimar, 1855. For 99 errata, see Cunningham.[85]

[56]Tables des nombres entiers non divisibles par 2, 3, 5, et 7, jusqu'à 10201, avec leurs diviseurs simples en regard, et des carrés des 1000 premiers nombres, Paris, 1855. Tables des nombres premiers, de 1 à 10000, Paris, 1855.

[57]Jour. für Math., 51, 1856, 61–99.   [58]Tables de log., Paris, 1858.

table to 100 million; the manuscript[59] has been in the library of the Vienna Royal Academy since 1867.   Lehmer[92] gave an account of the first of the eight volumes of the manuscript, listed 226 errors in the tenth million, and concluded that Kulik's manuscript is certainly not accurate enough to warrant publication, though of inestimable value in checking a newly constructed table.   Lehmer[95] gave a further account of this manuscript which he examined in Vienna.   Volume 2, running from 12 642 600 to 22 852 800 is missing.   The eight volumes contained 4,212 pages.

B. Goldberg[60] gave all factors of numbers prime to 2, 3, 5, to 251 647.

Zacharias Dase,[61] in the introduction to the table for the seventh million, printed a letter from Gauss, dated 1850, giving a brief history of previous tables and referring to the manuscript factor table for the fourth, fifth and sixth millions presented to the Berlin Academy by A. L. Crelle.   Although Gauss was confident this manuscript would be published, and hence urged Dase to undertake the seventh million, etc., the Academy found the manuscript to be so inaccurate that its publication was not advisable.   Dase died in 1861 leaving the seventh million complete and remarkably accurate, the eighth nearly complete, and a large part of the factors for the ninth and tenth millions.   The work was completed by Rosenberg, but with numerous errors.   The table for the tenth million has not been printed; the manuscript was presented to the Berlin Academy in 1878, but no trace of it was found when Lehmer[92] desired to compare it with his table of 1909.

C. F. Gauss[62] gave a table showing the number of primes in each thousand up to one million and in each ten thousand from one to three million, with a comparison with the approximate formula $\int dx/\log x$.

V. A. Lebesgue[63] discussed the formation of factor tables and gave that to 115500 constructed by Hoüel.

W. H. Oakes[64] used a complicated apparatus consisting of three tables on six sheets of various sizes and nine perforated cards (cf. Committee,[68] p. 39).

W. B. Davis[65] considered numbers in the vicinity of 10^8, and of 10^{11}.

E. Meissel[66] computed the number of primes in the successive sets of 100 000 numbers to one million and concluded that Burckhardt's[45] table gives correctly the primes to one million.

---

[59]Cited by Kulik, Abh. Böhm. Gesell. Wiss., Prag, (5), 11, 1860, 24, footnote.   A report on the manuscript was made by J. Petzval, Sitzungsberichte Ak. Wiss. Wien (Math.), 53, 1866, II, 460.   Cited by J. Perott, l'intermédiaire des math., 2, 1895, 40; 11, 1904, 103.

[60]Primzahlen- u. Faktortafeln von 1 bis 251 647, Leipzig, 1862.   Errata, Cunningham.[85]

[61]Factoren-Tafeln für alle Zahlen der siebenten Million..., Hamburg, 1862;...der achten Million, 1863;...der neunten Million (ergänzt von H. Rosenberg), 1865.

[62]Posthumous manuscript, Werke, 2, 1863, 435–447.

[63]Tables diverses pour la décomposition des nombres en leurs facteurs premiers, Mém. soc. sc. phys. et nat. de Bordeaux, 3, cah. 1, 1864, 1–37.

[64]Machine table for determining primes and the least factors of composite numbers up to 100 000, London, 1865.

[65]Jour. de Math., (2), 11, 1866, 188–190; Proc. London Math. Soc., 4, 1873, 416–7.   Math. Quest. Educ. Times, 7, 1867, 77; 8, 1868, 30–1.

[66]Math. Annalen, 2, 1870, 636–642.   Cf. 3, p. 523; 21, 1883, p. 304; 25, 1885, p. 251.

J. W. L. Glaisher[67] gave for the second and ninth millions the number of primes in each interval of 50000 and a comparison with $lix' - lix$, where $lix = \int dx/\log x$ [more precise definition at the end of Ch. XVIII].

A committee[68] consisting of Cayley, Stokes, Thompson, Smith, and Glaisher prepared the Report on Mathematical Tables, which includes (pp. 34–9) a list of factor and prime tables.

J. W. L. Glaisher[69] described in detail the method used by his father[70] and gave an account of the history of factor tables.

Glaisher[69a] enumerated the primes in the tables of Burckhardt and Dase.

Glaisher[69b] tabulated long sets of consecutive composite numbers. He[69c] enumerated the prime pairs (as 11, 13) in each successive thousand to 3 million and in the seventh, eighth, and ninth millions.

E. Lucas[69d] wrote $P(q)$ for the product of all the primes $\leqq q$, where q is the largest prime $< n$. If $xP(q) \pm 1$ are both composite, $xP(q) - n, \ldots,$ $xP(q), \ldots, xP(q) + n$ give $2n+1$ composite numbers.

Glaisher[69e] enumerated the primes $4n+1$ and the primes $4n+3$ for intervals of 10000 in the kth million for $k = 1, 2, 3, 7, 8, 9$.

James Glaisher[70] filled the gap between the tables by Burckhardt[45] and Dase[61]. The introduction to the table for the fourth million gives a history of factor tables and their construction. Lehmer[92] praised the accuracy of Glaisher's table, finding in the sixth million a single error besides two misprints.

Tuxen[71] gave a process to construct tables of primes.

Groscurth and Gudila-Godlewksi, Moscow, 1881, gave factor tables.

*V. Bouniakowsky[71a] gave an extension of the sieve of Eratosthenes.

W. W. Johnson[71b] repeated Glaisher's[70] remarks on the history of tables.

P. Seelhoff[72] gave large primes $k \cdot 2^n + 1$ ($k < 100$) and composite cases.

Simony[73] gave the digits to base 2 of primes to $2^{14} = 16384$.

L. Saint-Loup[74] gave a graphical exposition of Eratosthenes' sieve.

H. Vollprecht[75] discussed the construction of factor tables.

[67]Report British Association for 1872, 1873, trans., 19–21. Cf. W. W. Johnson, Des Moines Analyst, 2, 1875, 9–11.

[68]Report British Association for 1873, 1874, pp. 1–175. Continued in 1875, 305–336; French transl., Sphinx-Oedipe, 8, 1913, 50–60, 72–79; 9, 1914, 8–14.

[69]Proc. Cambridge Phil. Soc., 3, 1878, 99–138, 228–9.

[69a]Ibid., 17–23, 47–56; Report British Assoc., 1877, 20 (sect.). Extracts by W. W. Johnson, Des Moines Analyst, 5, 1878, 7.

[69b]Messenger Math., 7, 1877–8, 102–6, 171–6; French transl., Sphinx-Oedipe, 7, 1912, 161–8.

[69c]Ibid., 8, 1879, 28–33.

[69d]Ibid., p. 81. C. Gill, Ladies' Diary, 1825, 36–7, had noted that $xP(q)+j$ is composite for $j=2, \ldots, q-1$.

[69e]Report British Assoc., 1878, 470–1; Proc. Roy. Soc. London, 29, 1879, 192–7.

[70]Factor tables for the fourth, fifth and sixth millions, London, 1879, 1880, 1883.

[71]Tidsskrift for Mat., (4), 5, 1881, 16–25.

[71a]Memoirs Imperial Acad. Science, St. Petersburg, 41, 1882, Suppl., No. 3, 32 pp.

[71b]Annals of Math., 1, 1884–5, 15-23.

[72]Zeitschrift Math. Phys., 31, 1886, 380. Reprinted, Sphinx-Oedipe, 4, 1909, 95–6.

[73]Sitzungsber. Ak. Wiss. Wien (Math.), 96, II, 1887, 191–286.

[74]Comptes Rendus Paris, 107, 1888, 24; Ann. de l'école norm., (3), 7, 1890, 89–100.

[75]Ueber die Herstellung von Faktorentafeln, Diss. Leipzig, 1891.

C. A. Laisant[75a] would exhibit a factor table by use of shaded and un-
shaded squares on square-ruled paper without using numbers for entries.
. G. Speckmann[75b] made trivial remarks on the construction of a list of
primes.

P. Valerio[76] arranged the odd numbers prime to 5 in four columns
according to the endings 1, 3, 7, 9. From the first column cross out the first
multiple 21 of 3, then the third following number 51, etc. Similarly for the
other columns. Then use the primes 7, 11, etc., instead of 3.

J. P. Gram[77] published the computation by N. P. Bertelsen of the
number of primes to ten million in intervals of 50000 or less, which led to
the detection of numerous errors in the tables of Burckhardt[45] and Dase.[61]

G. L. Bourgerel[78] gave a table with 0, 1,..., 9 in the first row, 10,..., 19
in the second row (with 10 under 0), etc. Then all multiples of a chosen
number lie in straight lines forming a paralellogram lattice, with one branch
through 0. For example, the multiples of 3 appear in the line through 0, 12,
24, 36,..., the parallel through 3, 15, 27,..., the parallel 21, 33, 45,... ; also
in a second set of parallels 3, 12, 21, 30; 6, 15, 24, 33, 42, 51, 60; etc.

E. Suchanek[79] continued to 100 000 Simony's[73] table of primes to base 2.

D. von Sterneck[80] counted the number of primes $100\,n+1$ in each tenth of
a million up to 9 million and noted the relatively small variation from one-
fortieth of the total number of primes in the interval.

H. Vollprecht[81] discussed the determination of the number of primes $< N$
by use of the primes $< \sqrt{N}$.

A. Cunningham and H. J. Woodall[82] discussed the problem to find all
the primes in a given range and gave many successive primes > 9 million.
They[82a] listed 117 primes between $2^{24} \pm 1020$.

H. Schapira[82b] discussed algebraic operations equivalent to the sieve of
Eratosthenes.

*V. Di Girio, Alba, 1901, applied indeterminate analysis of the first
degree to define a new sieve of Eratosthenes and to factoring.

John Tennant[83] wrote numbers to the base 900 and used auxiliary tables.

A. Cunningham[83a] gave long lists of primes between $9 \cdot 10^6$ and 10^{11}.

Ph. Jolivald[84] noted that a table of all factors of the first $2n$ numbers
serves to tell readily whether a number $< 4n+2$ is prime or not.

[75a]Assoc. franç., 1891, II, 165–8.   [75b]Archiv Math. Phys., (2), 11, 1892, 439–441.
[76]La revue scientifique de France, (3), 52, 1893, 764–5.
[77]Acta Math., 17, 1893, 301–314. List of errors reproduced in Sphinx–Oedipe, 5, 1910, 49–51.
[78]La revue scientifique de France, (4), 1, 1894, 411–2.
[79]Sitzungsber. Ak. Wiss. Wien (Math.), 103, II a, 1894, 443–610.
[80]Anzeiger K. Akad. Wiss. Wien (Math.), 31, 1894, 2–4. Cf. Kronecker, p. 416 below.
[81]Zeitschrift Math. Phys., 40, 1895, 118–123.
[82]Report British Assoc., 1901, 553; 1903, 561; Messenger Math., 31, 1901–2, 165; 34, 1904–5,
    72, 184; 37, 1907–8, 65–83; 41, 1911, 1–16.   [82a]Report British Assoc., 1900, 646.
[82b]Jahresber. d. Deutschen Math. Verein., 5, 1901, I, 69–72.
[83]Quar. Jour. Math., 32, 1901, 322–342.
[83a]Ibid., 35, 1903, 10–21; Mess. Math., 36, 1907, 145–174; 38, 1908, 81–104; 38, 1909, 145–175;
    39, 1909, 33–63, 97–128; 40, 1910, 1–36; 45, 1915, 49–75; Proc. London Math. Soc., 27,
    1896, 327; 28, 1897, 377–9; 29, 1898, 381–438, 518; 34, 1902, 49.
[84]L'intermédiaire des math., 11, 1904, 97–98.

A. Cunningham [5] noted errata in various factor tables.

*J. R. Akerlund[85a] discussed the determination of primes by a machine.

Gaston Tarry[86] would use an auxiliary table (as did Barlow in 1819) to tell by the addition of two entries ($< \frac{1}{2}p$) if a given number $< N$ is divisible by a chosen prime p. For $N = 10000$, he used the base $b = 100$, and gave a table showing the numerically least residues of the numbers $r < b$ and the multiples of b for each prime $p < b$. Then $nb + r$ is divisible by p if the residues of nb and r are equal and of opposite sign. For $N = 100\ 000$, he used $b = 60060 = 2 \cdot 91 \cdot 330$ and wrote numbers in the form $mb + 330q + r$, $q < 90$, $r < 330$; or, again, $b = 20580$. Ernest Lebon[87] used such tables with the base $30030 = 2 \cdot 3 \cdot 5 \cdot 7 \cdot 11 \cdot 13$, or its product by 17.

Ernest Lebon,[88] J. Deschamps,[89] and C. A. Laisant[89a] discussed the construction of factor tables.

J. C. Morehead[90] extended the sieve of Eratosthenes to numbers $ma^k + b$ ($m = 1, 2, 3, \ldots$) in any arithmetical progression. The case $a = 2$, $b = \pm 1$, is discussed in detail, with remarks on the construction of a table to serve as a factor table for numbers $m \cdot 2^k \pm 1$.

L. L. Dines[91] treated the case $a = 6$, $b = \pm 1$, and the factorization of numbers $m \cdot 6^k \pm 1$.

D. N. Lehmer[92] gave a factor table to 10 million and listed the errata in the tables by Burckhardt, Glaisher, Dase, Dase and Rosenberg, and Kulik's tenth million, and gave references to other (shorter) lists of errata.

E. B. Escott[92a] listed 94 pairs of consecutive large numbers all of whose prime factors are small.

L. Aubry[92b] proved that a group of 30 consecutive odd numbers does not contain more than 15 primes or numbers all of whose prime factors exceed 7.

Cunningham[92c] listed the numbers of 5 digits with prime factors ≤ 11.

[85]Messenger Math., 34, 1904–5, 24–31; 35, 1905–6, 24.

[85a]Nyt Tidsskrift for Mat., Kjobenhavn, 16A, 1905, 97–103.

[86]Bull. Soc. Philomathique de Paris, (9), 8, 1906, 174–6, 194–6; 9, 1907, 56–9. Sphinx-Oedipe, Nancy, 1906–7, 39–41. Tablettes des Cotes, Gauthier-Villars, Paris, 1906. Assoc. franç. avanc. sc., 36, 1907, II, 32–42; 41, 1912, 38–43.

[87]Comptes Rendus Paris, 151, 1905, 78. Bull. Amer. Math. Soc., 13, 1906–7, 74. L'enseignement math., 9, 1907, 185. Bull. Soc. Philomathique de Paris, (9), 8, 1906, 168, 270; (9), 10, 1908, 4–9, 66–83; (10), 2, 1910, 171–7. Assoc. franç. avanc. sc., 36, 1907, II, 11–20, 49–55; 37, 1909, 33–6; 41, 1912, 44–53; 43, 1914, 29–35. Rend. Accad. Lincei, Rome, (5), 15, 1906, I, 439; 26, 1917, I, 401–5. Sphinx-Oedipe, 1908–9, 81, 97. Bull. Sc. Math. Élém., 12, 1907, 292–3. Il Pitagora, Palermo, 13, 1906–7, 81–91 (table serving to factor numbers from 30030 to 510 510). Table de caractéristiques relatives à la base 2310 des facteurs premiers d'un nombre inférieur à 30030, Paris, 1906, 32 pp. Comptes Rendus Paris, 159, 1914, 597–9; 160, 1915, 758–760; 162, 1916, 346–8; 163, 1916, 259–261; 164, 1917, 482–4.

[88]Jornal de sciencias math., phys. e nat., acad. sc. Lisbona, (2), 7, 1906, 209–218.

[89]Bull. Soc. Philomathique de Paris, (9), 9, 1907, 112–128; 10, 1908, 10–41.

[89a]Assoc. franç., 41, 1912, 32–7.

[90]Annals of Math., (2), 10, 1908–9, 88–104.   [91]Ibid., pp. 105–115.

[92]Factor table for the first ten millions, Carnegie Inst. Wash. Pub. No. 105, 1909.

[92a]Quar. Jour. Math., 41, 1910, 160–7; l'intermédiaire des math., 11, 1904, 65; Math. Quest. Educ. Times, (2), 7, 1905, 81–5.

[92b]Sphinx-Oedipe, 6, 1911, 187–8; Problem of Lionnet, Nouv. Ann. Math., (3), 2, 1883, 310.

[92c]Math. Quest. Educ. Times, (2), 21, 1912, 82–3.

E. Lebon[93] stated that he constructed in 1911 a table of residues ρ, ρ' permitting the rapid factorization of numbers to 100 million, the manuscript being in the Bibliothèque de l'Institut.

H. W. Stager[94] gave theorems on numbers which contain no factors of the form $p(kp+1)$, where $k>0$ and p is a prime, and listed all such numbers <12230.

Lehmer[95] listed the primes to ten million.

A. Gérardin[96] discussed the finding of all primes between assigned limits by use of stencils for 3, 5, 7, 11, ... He[97] described his manuscript of an auxiliary table permitting the factoring of numbers to 200 million. He[98a] gave a five-page table serving to factor numbers of the second million. Corresponding to each prime $M \leqq 14867$ is an entry P such that $N = 1\,000\,000 + P$ is divisible by M. If a value of P is not in the table, N is prime (the P's range up to 28719 and are not in their natural order). By a simple division one obtains the least odd number in any million which is divisible by the given prime $M \leqq 14867$.

C. Boulogne[98] made use of lists of residues modulis 30 and 300.

H. E. Hansen[99] gave an impracticable method of forming a table of primes based on the fact that all composite numbers prime to 6 are products of two numbers $6x \pm 1$, while such a product is $6N \pm 1$, where $N = 6xy \pm x + y$ or $6xy - x - y$. A table of values of these N's up to k serves to find the composite numbers up to $6k$. To apply this method to factor $6N \pm 1$, seek an expression for N in one of the above three forms.

N. Alliston[100] described a sieve (a modification of that by Eratosthenes) to determine the primes $4n+1$ and the primes $4n-1$.

H. W. Stager[101] expressed each number <12000 as a product of powers of primes, and for each odd prime factor gave the values >0 of k for all divisors of the form $p(kp+1)$. The table thus gives a list of numbers which include the numbers of Sylow subgroups of a group of order $\leqq 12000$.

In Ch. XVI are cited the tables of factors of a^2+1 by Euler,[4, 7] Escott,[58] Cunningham[63] and Woodall[64]; those of a^2+k^2 $(k=1, \ldots, 9)$ of Gauss[13]; those of y^n+1, $y^4 \pm 2$, $y^y \pm 1$, $x^y \pm y^x$, $2^q \pm q$, etc., of Cunningham.[68, 84-9]. Concerning the sieve of Eratosthenes, see Noviomagus[29] of Ch. I, Poretzky[66] of Ch. V, Merlin[139] and de Polignac[305-7] of Ch. XVIII. Saint-Loup[13] of Ch. XI, Reymond[151] and Kempner[152] of Ch. XIV, represented graphically the divisors of numbers, while Kulik[134] gave a graphical determination of primes.

[93]L'intermédiaire des math., 19, 1912, 237.
[94]University of California Public. in Math., 1, 1912, No. 1, 1-26.
[95]List of prime numbers from 1 to 10,006,721. Carnegie Inst. Wash. Pub. No. 165, 1914. The introduction gives data on the distribution of primes.
[96]Math. Gazette, 7, 1913-4, 192-3.
[97]Assoc. franç. avanc. sc., 42, 1913, 2-8; 43, 1914, 26-8.
[98]Ibid., 43, 1914, 17-26.
[98a]Sphinx-Oedipe, série spéciale, No. 1, Dec., 1913.
[99]L'enseignement math., 17, 1915, 93-9. Cf. pp. 244-5 for remarks by Gérardin.
[100]Math. Quest. Educat. Times, 28, 1915, 53.
[101]A Sylow factor table of the first twelve thousand numbers. Carnegie Inst. Wash. Pub. No. 151, 1916.

CHAPTER XIV.

METHODS OF FACTORING.

FACTORING BY METHOD OF DIFFERENCE OF TWO SQUARES.

Fermat[1] described his method as follows: "An odd number not a square can be expressed as the difference of two squares in as many ways as it is the product of two factors, and if the squares are relatively prime the factors are. But if the squares have a common divisor d, the given number is divisible by d and the factors by $\sqrt{d}$. Given a number n, for example 2027651281, to find if it be prime or composite and the factors in the latter case. Extract the square root of n. I get $r = 45029$, with the remainder 40440. Subtracting the latter from $2r+1$, I get 49619, which is not a square in view of the ending 19. Hence I add $90061 = 2+2r+1$ to it. Since the sum 139680 is not a square, as seen by the final digits, I again add to it the same number increased by 2, i. e., 90063, and I continue until the sum becomes a square. This does not happen until we reach 1040400, the square of 1020. For by an inspection of the sums mentioned it is easy to see that the final one is the only square (by their endings except for 499944). To find the factors of n, I subtract the first number added, 90061, from the last, 90081. To half the difference add 2. There results 12. The sum of 12 and the root r is 45041. Adding and subtracting the root 1020 of the final sum 1040400, we get 46061 and 44021, which are the two numbers nearest to r whose product is n. They are the only factors since they are primes. Instead of 11 additions, the ordinary method of factoring would require the division by all the numbers from 7 to 44021."

Under Fermat,[317] Ch. I, was cited Fermat's factorization of the number 100895598169 proposed to him by Mersenne in 1643.

C. F. Kausler[2] would add 1^2, 2^2,... to N to make the sum a square.

C. F. Kausler[3] proceeded as follows to express $4m+1$ in the form p^2-q^2. Then q is even, $q=2Q$. Set $p-q=2\beta+1$. Then $m=Q(2\beta+1)+\beta(\beta+1)$. Subtract from m in turn the pronic numbers $\beta(\beta+1)$, a table of which he gave on pp. 232–267, until we reach a difference divisible by $2\beta+1$.

Ed. Collins,[4] in factoring N by expressing it as a difference of two squares, let g^2 be the least odd or even square $>N$, according as $N \equiv 1$ or 3 (mod 4), and set $N=g^2-r$. If r is not a square, set $r=h^2-c$, where h^2 is the even or odd square just $>r$, according as r is even or odd, whence $c=4d$, $N=g^2-h^2+4d$. By trial find integers x, y such that both g^2+x and h^2+y are squares, while $x-y=4d$. Then N will be a difference of two squares.

[1]Fragment of a letter of about 1643, Bull. Bibl. Storia Sc. Mat., 12, 1879, 715; Oeuvres de Fermat, 2, 1894, 256. At the time of his letter to Mersenne, Dec. 26, 1638, Oeuvres, 2, p. 177, he had no such method.

[2]Euler's Algebra, Frankfort, 1796, III, 2. Anhang, 269–283. Cf. Kausler, De Cribro Eratosthenis, 1812.

[3]Nova Acta Acad. Petrop., 14, ad annos 1797-8 (1805), 268–289.

[4]Bull. Ac. Sc. St. Pétersbourg, 6, 1840, 84–88.

F. Landry[5] used the method of Fermat, eliminating certain squares by their endings and others by the use of moduli.

C. Henry[6] stated that Landry's method is merely a perfection of the method given in the article "nombre premier" in the Dictionnaire des Mathématiques of de Montferrier. It is improbable that the latter invented the method (based on the fact that an odd prime is a difference of two squares in a single way), since it was given by Fermat.

F. Thaarup[7] gave methods to limit the trials for x in $x^2 - y^2 = n$. We may multiply n by $f = a^2 - b^2$ and investigate $nf = X^2 - Y^2$, $X = ax - by$, $Y = bx - ay$. We may test small values of y, or apply a mechanical test based on the last digit of n.

C. J. Busk[8] gave a method essentially that by Fermat. It was put into general algebraic form by W. H. H. Hudson.[9] Let N be the given number, n^2 the next higher square. Then

$$N = n^2 - r_0 = (n+1)^2 - r_1 = \ldots,$$

where $r_1, r_2, \ldots$ are formed from r_0 by successive additions of $2n+1$, $2n+3$, $2n+5, \ldots$. Thus $r_m = r_0 + 2mn + m^2$. If r_m is a square, N is a difference of two squares. A. Cunningham (ibid., p. 559) discussed the conditions under which the method is practical, noting that the labor is prohibitive except in favorable cases such as the examples chosen by Busk.

J. D. Warner[9a] would make $N = A^2 - B^2$ by use of the final two digits.

A. Cunningham[10] gave the 22 sets of last two digits of perfect squares, as an aid to expressing a number as a difference of two squares, and described the method of Busk, which is facilitated by a table of squares.

F. W. Lawrence[11] extended the method of Busk (practical only when the given odd number N is a product of two nearly equal factors) to the case in which the ratio of the factors is approximately l/m, where l and m are small integers. If l and m are both odd, subtract from lmN in turn the squares of a, $a+1, \ldots$, where a^2 just exceeds lmN, and see if any remainder is a perfect square (b^2). If so, $lmN = (a+T)^2 - b^2$.

G. Wertheim[12] expressed in general form Fermat's method to factor an odd number m. Let a^2 be the largest square $< m$ and set $m = a^2 + r$. If $\rho \equiv 2a + 1 - r$ is a square (n^2), we eliminate r and get $m = (a+1+n) \times (a+1-n)$. If ρ is not a square, add to ρ enough terms of the arithmetic progression $2a+3$, $2a+5, \ldots$ to give a square:

$$\rho + (2a+3) + \ldots + (2a+2n-1) = s^2.$$

[5]Aux mathématiciens de toutes les parties du monde: communication sur la décomposition des nombres en leurs facteurs simples, Paris, 1867. Letter from Landry to C. Henry, Bull. Bibl. Storia Sc. Mat., 13, 1880, 469-70.

[6]Assoc. franç. av. sc., 1880, 201; Oeuvres de Fermat, 4, 1912, 208; Sphinx-Oedipe, 4, 1909, 3ᵉ Trimestre, 17-22.   [7]Tidsskrift for Mat., (4), 5, 1881, 77-85.

[8]Nature, 39, 1889, 413-5.   [9]Nature, 39, 1889, p. 510.

[9a]Proc. Amer. Assoc. Adv. Sc., 39, 1890, 54-7.

[10]Mess. Math., 20, 1890-1, 37-45. Cf. Meissner,[138] 137-8.

[11]Ibid., 24, 1894-5, 100.

[12]Zeitschrift Math. Naturw. Unterricht, 27, 1896, 256-7.

Then $2an+n^2-r=s^2$ and $m=(a+n)^2-s^2$. The method is the more rapid the smaller the difference of the two factors.

M. Neumann[13] proved that this process of adding terms leads finally to a square and hence to factors, one of which may be 1.

F. W. Lawrence[14] denoted the sum of the two factors of n by $2a$ and the difference by $2b$, whence $n=a^2-b^2$. Let q be the remainder obtained by dividing n by a chosen prime p, and write down the pairs of numbers $<p$ such that the product of two of a pair is congruent to q modulo p. If $p=7$, $q=3$, the pairs are 1 and 3, 2 and 5, 4 and 6, whence $2a\equiv4$, 0 or 3 (mod 7). Using various primes p and their powers, we get limitations on a which together determine a. The work may be done with stencils. The method was used by Lawrence[15] to show that five large numbers are primes, including 10, 11 and 12 place factors of $3^{23}-1$, $10^{29}-1$, $10^{25}-1$, respectively. The same examples were treated by other methods by D. Biddle.[16]

A. Cunningham[17] remarked that in computing by Busk's method a k for which $(s+k)^2-N$ is a square, we may use the method of Lawrence, just described, to limit greatly the number of possible forms of k.

F. J. Vaes[18] expressed N in the form a^2-b^2 by use of the square a^2 just $>N$ and then increasing a by 1, 2,..., and gave (pp. 501–8) an abbreviation of the method. He strongly recommended the method of remainders (p. 425): If p is a factor of $G=h^2-g^2$, and if $g=(G-1)/2$ has the remainder r when divided by p, then $h=(G+1)/2$ must have the remainder $r+1$, so that p is a factor of $2r+1\equiv G$. For example, let $G=80047$, whence

$$g=200^2+23=201\cdot199+24=202\cdot198+27,\dots$$

For $r=24$, 27, ·32,... we see that $2r+1$ is not a multiple of 201, 202,... until we reach $g=209\cdot191+\rho$, $\rho=104$, $2\rho+1=209$. Thus 209 divides G.

P. F. Teilhet[19] wrote $N=a^2-b$ in the form $(a+k)^2-P$, where $P=k^2+2ak+b$. Give to k successive values 1, 2,... (by additions to P), until P becomes a square v^2. To abbreviate consider the residues of P for small prime moduli.

E. Lebon[20] proceeded as had Teilhet[19] and then set $f=a+k-v$. Then

$$2kf=(a-f)^2-b,$$

and we examine primes $f<a$ to see if k is an integer.

M. Kraitchik[21] would express a given odd number A in the form y^2-x^2 by use of various moduli ρ. Let $A\equiv r$ (mod ρ) and let $a_1,\dots,a_n$ be the

[13]Zeitschrift Math. Naturw. Unterricht, 27, 1896, 493–5; 28, 1897, 248–251.

[14]Quar. Jour. Math., 28, 1896, 285–311.   French transl., Sphinx-Oedipe, 5, 1910, 98–121, with
  an addition by Lawrence on $\varrho^{2^k}+1$.

[15]Proc. Lond. Math. Soc., 28, 1897, 465–475.   French transl., Sphinx-Oedipe, 5, 1910, 130–6.

[16]Math. Quest. Educat. Times, 71, 1899, 113–4; cf. 93–99.

[17]Ibid., 69, 1898, 111.

[18]Proc. Sect. Sciences Akad. Wetenschappen Amsterdam, 4, 1902, 326–336, 425–436, 501–8
  (English); Verslagen Ak. Wet., 10, 1901–2, 374–384, 474–486, 623–631 (Dutch).

[19]L'intermédiaire des math., 12, 1905, 201–2.   Cf. Sphinx-Oedipe, 1906–7, 49–50, 55.

[20]Assoc. franç. av. sc., 40, 1911, 8–9.

[21]Sphinx-Oedipe, Nancy, Mai, 1911, numéro spécial, pp. 10–16.

quadratic residues of ρ. Then $r+x^2 \equiv a_i \pmod{\rho}$. Thus $a_i - r$ must be a quadratic residue. Reject from $a_1, \ldots, a_n$ the terms for which $a_i - r$ is not in the set. We get the possible residues of x modulo ρ. His method to factor $a^n \pm 1$ is the same as Dickson's[118] and is applied to show that the factor $(2^{73}+2^{37}+1)/(5\cdot239\cdot9929)$ of $2^{146}+1$ is a prime in case it has no factor between 10500 and 108000.

Kraitchik[22] extended the method of Lawrence.

F. J. Vaes[23] applied his[18] method to factor Mersenne's[1] number. The same was factored by various methods in L'Intermédiaire des Mathématiciens, 19, 1912, 32–5. J. Petersen, ibid., 5, 1898, 214, noted that its product by 8 equals k^2+k, where $k=898423$.

Method of Factoring by Sum of Two Squares.

Frenicle de Bessy[25] proposed to Fermat that he factor h given that

$$h = a^2+b^2 = c^2+d^2, \qquad \text{as } 221 = 100+121 = 196+25.$$

In 1647, Mersenne[61] (of Ch. I) noted that a number is composite if it be a sum of two squares in two ways.

L. Euler[26] noted that N is a prime if it is expressible as a sum of two squares in a single way, while if $N = a^2+b^2 = c^2+d^2$, N is composite:

$$N = \frac{\{(a-c)^2+(b-d)^2\}\{(a+c)^2+(b-d)^2\}}{4(b-d)^2}.$$

Euler[27] proved, that, if a number $N = 4n+1$ is expressible as the sum of two relatively prime squares in a single way, it is a prime. For, if N were composite, then $N = (a^2+b^2)(c^2+d^2)$ is the sum of the squares of $ac \pm bd$ and $ad \mp bc$, contrary to hypothesis. If $N = a^2+b^2 = c^2+d^2$, N is composite; for if we set $a = c+x$, $d = b+y$, and assume* that the common value of $2cx+x^2$ and $2by+y^2$ is of the form xyz, we get

$$2c = yz-x, \qquad 2b = xz-y, \qquad N = b^2+c^2+xyz = \tfrac{1}{4}(x^2+y^2)(1+z^2),$$

whence x^2+y^2 or $(x^2+y^2)/4$ is a factor of N. To express N as a sum of two squares in all possible ways, use is made of the final digit of N to limit the squares x^2 to be subtracted in seeking differences $N-x^2$ which are squares. Several numerical examples of factoring are treated in full.

Euler[28] gave abbreviations of the work of applying the preceding test. For example, if $4n+1 = 5m+2 = x^2+y^2$, then x and y are of the form

[22]Sphinx-Oedipe, 1912, 61–4.

[23]L'enseignement math., 15, 1913, 333–4.

[25]Oeuvres de Fermat. 2, 1894, 232, Aug. 2, 1641.

[26]Letters to Goldbach, Feb. 16, 1745, May 6, 1747; Corresp. Math. Phys. (ed., Fuss), I, 1843, 313, 416–9.

[27]Novi Comm. Ac. Petrop., 4, 1752–3, p. 3; Comm. Arith., 1, 1849, 165–173.

*Euler gave a faultless proof in the margin of his posthumous paper, Tractatus, §570, Comm. Arith., 2, 573; Opera postuma, I, 1862, 73. We have $(a+c)(a-c) = (b+d)(d-b) = pqrs$, $a+c = pq$, $a-c = rs$, $b+d = pr$, $d-b = qs$ [since, if p be the g. c. d. of $a+c$, $b+d$, then $q(a-c)$ is divisible by r, whence $a-c = rs$]. Hence $a^2+b^2 = (p^2+s^2)(q^2+r^2)/4$.

[28]Novi Comm. Ac. Petrop., 13, 1768, 67; Comm. Arith., 1, 379.

$5p \pm 1$. To express a number as $x^2 + y^2$, subtract squares in turn and seek a remainder which is a square.

N. Beguelin[29] proposed to find x such that $4p^2x^2 + 1$ is a prime by excluding the values x making the sum composite. The latter is the case if

$$4p^2x^2 + 1 = 4b^2 + (2c+1)^2, \qquad x^2 = \frac{b^2 + c^2 + c}{p^2}.$$

Set $x = q + b/p$. Then b is expressed rationally in terms of c and the known p. Taking $p = 1$, he derived a tentative process for finding a prime, of the form $4x^2 + 1$, which exceeds a given number a.

L. Euler[30] proved that $1000^2 + 3^2$ is prime since not expressible as a sum of two squares another way.

A. M. Legendre[30a] factored numbers represented as a sum of two squares in two ways.

J. P. Kulik's[30b] tables VIII and IX, relating to the ending of squares, serve to test if $4n + 1$ is a sum of two squares and hence to test if it be prime or composite.

Th. Harmuth[31] suggested testing $a^2 + b^2$ for factors, where a and b are relatively prime, by noting that it is divisible by 5 if $a \equiv \pm 1$, $b \equiv \pm 2 \pmod 5$, and similar facts for $p = 13, 17, 29, 37$, there being $p - 1$ sets of values of a, b for each prime $p = 4n + 1$.

G. Wertheim[32] explained in full Euler's[27] method of factoring.

R. W. D. Christie and A. Cunningham[33] granted $N = A^2 + B^2 = C^2 + D^2$ and showed how to find $a, \ldots, d$ so that $N = (a^2 + b^2)(c^2 + d^2)$. Similarly, if $N = x^2 + Py^2$ in two ways.

FACTORING BY USE OF BINARY QUADRATIC FORMS.

L. Euler[37] noted that a number is composite if it be expressible in two ways in the form $f = ax^2 + \beta y^2$. The product of two numbers of the form f is of the form $g = a\beta x^2 + y^2$; the product of a number of the form f by one of the form g is of the form f. If for $m > 2$ a composite number mp is expressible in a single way in the form f, there exist an infinitude of composite numbers mq expressible in a single way by f. He called (§34) a number N idoneal (numerus idoneus) if, for $a\beta = N$, every number representable by $f = ax^2 + \beta y^2$ (with ax relatively prime to βy) is a prime, the square of a prime, the double of a prime or a power of 2, so that a number representable by f in a single way is a prime. It suffices to test $N + y^2 < 4N$, y prime to N. He gave (§39, p. 208) the 65 idoneal numbers $1, 2, \ldots, 1848$ less than 10000.

[29]Nouv. Mém. Acad. Sc. Berlin, 1777, année 1775, 300.
[30]Nova Acta Petrop., 10, 1792 (1778), 63; Comm. Arith., 2, 243–8.
[30a]Théorie des nombres, ed. 3, 1830, I, 310. Simplification by Vuibert, Jour. de math. élém., 10, p. 42. Cf. l'intermédiaire des math., 1, 1894, 167, 245; 18, 1911, 256.
[30b]Tafeln der Quadrat und Kubik Zahlen … bis hundert Tausend, Leipzig, 1848.
[31]Archiv Math. Phys., 67, 1882, 215–9.
[32]Elemente der Zahlentheorie, 1887, 295–9.
[33]Math. Quest. Educat. Times, (2), 11, 1907, 52–3, 65–7, 89–90.
[37]Nova Acta Petrop., 13, 1795–6 (1778), 14; Comm. Arith., 2, 198–214.

Euler[38] used the idoneal number 232 to find all values of $a < 300$ for which $232a^2 + 1$ is a prime, by excluding the values of a for which $232a^2 + 1 = 232x^2 + y^2$, $y > 1$. .

Euler[39] noted that $N = a^2 + \lambda b^2 = x^2 + \lambda y^2$ imply

$$N = \tfrac{1}{4}(\lambda m^2 + n^2)(\lambda p^2 + q^2), \qquad a \pm x = \lambda mp, nq; \qquad y \pm b = mq, np,$$

so that $\lambda p^2 + q^2$, or its half or quarter, is a factor of N. He gave (p. 227) his[37] former table of 65 idoneal numbers. Given one representation by $ax^2 + \beta y^2$, where $\alpha\beta$ is idoneal, he sought a second representation. If $N = 4n + 2$ is idoneal, $4N$ is idoneal.

Euler[40] called $mx^2 + ny^2$ a congruent form if every number representable by it in a single way (with x, y relatively prime) is a prime, the square of a prime, the double of a prime, a power of 2, or the product of a prime by a factor of mn. Then also $mnx^2 + y^2$ is a congruent form and conversely. The product mn is called an idoneal or congruent number. His table of 65 idoneal numbers is reproduced (§18, p. 253). He stated rules for deducing idoneal numbers from given idoneal numbers. He factored numbers expressed in two ways by $ax^2 + \beta y^2$, where $\alpha\beta$ is idoneal, and noted that a composite number may be expressible in a single way in that form if $\alpha\beta$ is not idoneal.

Euler[41] proved that the first five squares are the only square idoneal numbers.

C. F. Kausler[42] proved Euler's theorem that a prime can be expressed in a single way in the form $mx^2 + ny^2$ if m, n are relatively prime. To find a prime v exceeding a given number, see whether $38x^2 + 5y^2 = v$ has a single set of positive solutions x, y; or use $1848x^2 + y^2$. As the labor is smaller the larger the idoneal number 38·5 or 1848, it is an interesting question if there be idoneal numbers not in Euler's list of 65. Cf. Cunningham.[69]

Euler[43] gave the 65 idoneal numbers n (with 44 a misprint for 45) such that a number representable in a single way by $nx^2 + y^2$ (x, y relatively prime) is a prime. By using $n = 1848$, he found primes exceeding 10 million.

N. Fuss[44] stated the principles due to Euler.[37]

E. Waring[45] stated that a number is a prime if it be expressible in a single way in the form $a^2 + mb^2$ and conversely.

A. M. Legendre[46] would express the number A to be factored, or one of its multiples kA, in the form $t^2 + au^2$, where a is as small as possible and within the limits of his Tables III–VII of the linear forms of divisors of $t^2 \pm au^2$.

---

[38]Nova Acta Petrop., 14, 1797–8 (1778), 3; Comm. Arith., 2, 215–9.

[39]Ibid., p. 11; Comm. Arith., 2, 220–242. For $\lambda = 2$, Opera postuma, I, 1862, 159.

[40]Ibid., 12, 1794 (1778), 22; Comm. Arith., 2, 249–260.

[41]Ibid., 15, ad annos 1799–1802 (1778), 29; Comm. Arith., 2, 261–2.

[42]Ibid., 156–180.

[43]Nouv. Mém. Berlin, année 1776, 1779, 337; letter to Beguelin, May, 1778; Comm. Arith., 2, 270–1.

[44]Ibid., 340–6

[45]Medit. Algebr., ed. 3, 1782, 352.

[46]Théorie des nombres, 1798, pp. 313–320; ed. 2, 1808, pp. 287–292. German transl. by Maser, 1, 329–336. Cf. Sphinx-Oedipe, 1906–7, 51.

Then the divisors of A are included among these linear forms. When $\sqrt{kA}$ is converted into a continued fraction, let $(\sqrt{kA}+I)/D$ be a complete quotient, and p/q the corresponding convergent. Then $\pm D = p^2 - kAq^2$, so that the divisors of A are divisors of $p^2 \mp D$.

C. F. Gauss[47] stated that the 65 idoneal numbers n of Euler and no other numbers have the two properties that all classes of quadratic forms of determinant $-n$ are ambiguous and that any two forms in the same genus (Geschlecht) are both properly and improperly equivalent.

Gauss[48] gave a method of factoring a number M based on the determination of various small quadratic residues of M.

Gauss[49] gave a second method of factoring M based on the finding of representations of M by forms $x^2 + D$, where D is idoneal.

F. Minding[50] gave an exposition of the method of Legendre.[46]

P. L. Tchebychef[51] gave a rapid process to find many forms $x^2 \pm ay^2$ which represent a given number A or a multiple of A. Then a table of the linear forms of the divisors of $x^2 \pm ay^2$ serves to limit the possible factors of A.

Tchebychef[52] gave theorems on the limits between which lie at least one set of integral solutions of $x^2 - Dy^2 = \pm N$. If there are two sets of solutions within the limits, N is composite. There are given various tests for primality by use of quadratic forms.

C. F. Gauss[53] left posthumous tables to facilitate factoring by use of his[49] second method.

F. Grube[54] criticized and completed certain of Euler's proofs relating to idoneal numbers, here called Euler numbers. While Gauss[47] said it is easy to prove Euler's[43] criterion for idoneal numbers, Grube could prove only the following modification: Let Ω be the set of numbers $D+n^2 \leq 4D$ in which n is prime to D. According as all or not all numbers of Ω are of the form $q, 2q, q^2, 2^\lambda$ (q a prime), D is or is not an idoneal number.

E. Lucas[55] proved that if p is a prime and k is a positive integer, and $p = x^2 + ky^2$, then $p \neq x_1^2 + ky_1^2$ for values x_1, y_1 distinct from $\pm x, \pm y$.

P. Seelhoff[56] made use of 170 determinants (including the 65 idoneal numbers of Euler and certain others of Legendre), such that every reduced form in the principal genus is of the type $ax^2 + by^2$. To factor N seek among the numbers m of which N is a quadratic residue several values

[47]Disq. Arith., 1801, Art. 303.
[48]Ibid., Arts. 329–332.
[49]Ibid., Arts. 333–4.
[50]Anfangsgründe der Höheren Arith., 1832, 185–7.
[51]Theorie der Congruenzen (in Russian), 1849; German transl. by H. Schapira, 1889, Ch. 8, pp. 281–292.
[52]Jour. de Math., 16, 1851, 257–282; Oeuvres, 1, 73.
[53]Werke, 2, 1863, 508–9.
[54]Zeitschrift Math. Phys., 19, 1874, 492–519.
[55]Nouv. Corresp. Math., 4, 1878, 36. [Euler.[37]]
[56]Archiv Math Phys., (2), 2, 1885, 329; (2), 3, 1886, 325; Zeitschrift Math. Phys., 31, 1886, 166, 174, 306; Amer. Jour. Math., 7, 1885, 264; 8, 1886, 26–44.

for which N is representable by x^2+my^2. For example, if $N-31\cdot2^{24}+1$,
$$N=x^2+7\cdot19\cdot83k^2=y^2+19\cdot83\,l^2=z^2-7r^2.$$
Eliminating $19\cdot83$ between the first two, we get $\mu N=w^2-7t^2$. This with
the third leads to factors of N. In general, when elimination of common
factors of the m's has led to representations of two multiples of N by the
same form x^2+ny^2, we may factor N unless it be prime.

H. Weber[57] computed the class invariants for the 65 determinants of
Euler and remarked that there is no known proof of the fact found by
induction by Euler and Gauss that there are only 65 determinants such that
all classes belonging to the determinant are ambiguous and hence each
genus has only one class.

T. Pepin[58] developed the theory of Gauss'[53] posthumous tables and the
means of deducing complete tables from the given abridged tables. Pepin[59]
showed how to abridge the calculations in using the auxiliary tables of Gauss
in factoring a^n-1, where a and n are primes.

D. F. Seliwanoff[60] noted that the factoring of numbers of the form
t^2-Du^2 reduces to the solution of $(D/x)=1$, all solutions of which are
easily found by use of six relations by Euler on these Jacobi symbols (D/x).

E. Lucas[61] gave a clear proof of Euler's remark that a prime can not be
expressed in two ways in the form Ax^2+By^2, if A, B are positive integers.

S. Levänen[62] showed and illustrated by examples and tables how binary
quadratic forms may be applied to factoring.

G. B. Mathews[63] gave an exposition of the subject.

T. Pepin[64] applied determinants $-8n-3$ for which each genus has three
classes of quadratic forms. The paper is devoted mainly to the solution
of $x^2+(8n+3)y^2=4A$, where A is the number to be factored.

T. Pepin[65] assumed that the given number N had been tested and found
to have no prime factor $\leqq p$. Let $\lambda x+1$, $\lambda y+1$ be the two factors of N,
each between p and N/p. The sum of the factors lies between $2\sqrt{N}$ and
$p+N/p$. Let $x-y=u$, $x+y=\rho z$. Then $(N-1)/\lambda=\lambda xy+x+y$ gives
$$\frac{4(N-1)}{\lambda^2}=\rho^2z^2+\frac{4\rho}{\lambda}z-u^2,$$
in which special values are assigned to ρ. This equation yields a quadratic
congruence for u^2 with respect to an arbitrary prime modulus, used as an
excludant. The method applies mainly to numbers $a^\lambda \pm 1$.

E. Cahen[66] used the linear divisors of x^2+Dy^2.

[57]Math. Annalen, 33, 1889, 390–410.
[58]Atti Accad. Pont. Nuovi Lincei, 48, 1889, 135–156.
[59]Ibid., 49, 1890, 163–191.
[60]Moscow Math. Soc., 15, 1891, 789; St. Petersburg Math. Soc., 12, 1899.
[61]Théorie des nombres, 1891, 356–7.
[62]Öfversigt af finska Vetenskaps-Soc. förhandlingar, 34, 1892, 334–376.
[63]Theory of Numbers, 1892, 261–271. French transl., Sphinx-Oedipe, 1907–8, 155–8, 161–70.
[64]Memorie Accad. Pontif. Nuovi Lincei, 9, I, 1893, 46–76. Cf. Pepin,[65] 332.
[65]Ibid., 17, 1900–1, 321–344; Atti, 54, 1901, 89–93. Cf. Meissner[138], 121–2.
[66]Éléments de la théorie des nombres, 1900, 324–7. Sphinx-Oedipe, 1907–8, 149–155.

A. Cunningham[67] and J. Cullen listed the 188 prime numbers $x^2+1848y^2$ between 10^7 and $101 \cdot 10^5$, with x prime to $1848y$.

A. Cunningham[68] noted that two representations of N by $\mu x^k + \nu y^k$ lead to factors of N under certain conditions.

A. Cunningham[69] recalled that an idoneal number I has the property that, if an odd number N is expressible in only one way in the form $N = mx^2 + ny^2$, where $mn = I$, and mx^2 is prime to ny^2, then N is a prime or the square of a prime. Euler's largest I is 1848. There is no larger I under 50000, a computation checked by J. Cullen. Cunningham noted on the proof-sheets of this history that this limit has been extended to 100 000.

A. Cunningham[70] noted conditions that an odd prime be expressible by $t^2 \pm qu^2$ when q or $-q$ is idoneal.

F. N. Cole[71] discussed Seelhoff's[56] method of factoring.

Al. Laparewicz[72] described and applied Gauss' [48,49] two methods.

P. Meyer[73] discussed Euler's theorem that, if n is idoneal, a number representable only once by $x^2 + ny^2$ is a prime.

R. Burgwedel[74] gave an exposition and completion of the method of Euler[37-43] and an exposition of the methods of Gauss.[48,49]

L. Valroff stated and A. Cunningham[74a] proved that $(Dx^2-a^2)(Dy^2-a^2) = Dz^2-a^2$ implies that one factor is composite unless $x^2 = y^2 = 4$ when $a = 1$, $D = 2$, and in the remaining cases if the two factors are distinct and > 1.

A. Gérardin[75] gave a method illustrated for $N = a^2 - 5 \cdot 29^2$, where $a = 6326$. We shall have a second such representation $N = (a+5x)^2 - 5y^2$ if

$$E \equiv 5x^2 + 2ax + 841 = y^2.$$

Use is made of various moduli $m = 4, 3, 7, 25, \ldots$. On square-ruled paper, mark $x = 0, 1, 2, \ldots$ at the head of the columns. On the line for modulus m, shade the square under the heading x when x makes E a quadratic non-residue of m. Then examine the column in which occurs no shaded square. Up to $x \leqq 15$, these are $x = 0$ (excluded), and $x = 4$, which gives $N = 6346^2 - 5 \cdot 227^2$ and the factor $99^2 - 5 \cdot 2^2$. The same diagram serves for all numbers $1050\,H + 671$, our N being given by $H = 38108$. To apply the method to $N = (2x)^4 + 1 = (4x^2+1)^2 - 2(2x)^2$, seek a second representation $N = (4x^2 + 2p + 1)^2 - 2(2u)^2$. The condition is $(2p+1)x^2 + \frac{1}{2}p(p+1) = u^2$, solutions of which are found for $p = 1, 8, 9, \ldots, 6^2, 35^2, \ldots$. Or we may choose x, say $x = 48$, and find $p = 8$, $u = 198$.

[67]Brit. Assoc. Reports, 1901, 552. The entry 10098201 is erroneous.
[68]Proc. London Math. Soc., 33, 1900–1, 361.
[69]Ibid., 34, 1901–2, 54.
[70]Ibid., (2), 1, 1903, 134.
[71]Bull. Amer. Math. Soc., 10, 1903–4, 134–7.
[72]Prace mat. fiz., Warsaw, 16, 1905, 45–70 (Polish).
[73]Beweis eines von Euler entdeckten Satzes, betreffend die Bestimmung von Primzahlen, Diss., Strassburg, 1906.
[74]Ueber die Eulerschen und Gausschen Methoden der Primzahlbestimmung, Diss., Strassburg, 1910, 101 pp.
[74a]Sphinx-Oedipe, 7, 1912, 60, 77–9.
[75]Wiskundig Tijdschrift, 10, 1913, 52–62.

Gérardin[76] gave a note on his machine to factor large numbers, especially those of the form $2x^4-1$.

FACTORING BY METHOD OF FINAL DIGITS.

Johann Tessanek[80] gave a tedious method of factoring N, not divisible by 2, 3, or 5, when $N/10$ is within the limits of a factor table. For example, let $N=10a+1$; its factors end in 1,1 or 3,7 or 9,9. To treat one of the four cases, consider a factor $10x+3$, the quotient being $10z+7$. Then z is the quotient of $a-2-7x$ by $10x+3$. Give to x the values 1, 2,..., and test $a-9$ for the factor 13, $a-16$ for 23, etc., by the factor table. He gave a lengthy extension[81] to divisors $100x+10f+g$. Again, to factor $N=2a+1$, given a table extending to $N/2$, note that if $2x+1$ is a divisor of N, it divides $a-x$, which falls in the table. F. J. Studnicka[82] quoted the last result.

N. Beguelin[83] would factor $N=4p+3$ by considering the final digit of $\pi = (N-11)/4$ and hence find the proper line in an auxiliary table (pp. 291-2), each line containing four fractional expressions. Proceed with each until we reach a fraction whose numerator is zero. Then its denominator is a factor of N.

Georg Simon Klügel[84] noted that a number, not divisible by 2, 3 or 5, is of the form $30x+m$ ($m=1$, 7, 11, 13, 17, 19, 23, 29). Suppose $10007 = (30x+m)(30y+n)$. Then $(m, n) = (1, 17)$, $(7, 11)$, $(13, 29)$ or $(19, 23)$. For $m=1$, $n=17$, we get

$$x=\frac{333-y}{30y+17}, \quad x<4, \ y<4.$$

But x is not integral for $y=0$, 1, 2, 3.

Johann Andreas von Segner (*ibid.*, 217-225) took two pages to prove that any number not divisible by 2 or 3 is of the form $6n\pm1$ and noted that, given a table of the least prime factor of each $6n\pm1$, he could factor any number within the limits of the table!

Sebastiano Canterzani[85] would factor $10k+1$, by noting the last digits 1, 1 or 3, 7 or 9, 9 of its factors. If one factor ends in 7, there are 10 possibilities for the digit preceding 7; if one ends in 1 or 9, there are five cases; hence 20 cases in all. A. Niegemann[85a] used the same method.

Anton Niegemann[86] gave a method of computing a table of squares arranged according to the last two digits. Thus, if $A76=(10x-6)^2$, then

[76]Assoc. franç. avanc. sc., 43, 1914, 26-8. Proc. Fifth Internat. Congress, II, 1913, 572-3; Brit. Assoc. Reports, 1912-3, 405.

[80]Abhandl. einer Privatgesellschaft in Böhmen, zur Aufnahme der Math., Geschichte,..., Prag, I, 1775, 1-64.

[81]M. Cantor, Geschichte Math., 4, 1908, 179.

[82]Casopis, 14, 1885, 120 (Fortschr. der Math., 17, 1885, 125).

[83]Nouv. Mém. Ac. Berlin, année 1777 (1779), 265-310.

[84]Leipziger Magazin für reine u. angewandte Math. (eds., J. Bernoulli und Hindenburg), 1, 1787, 199-216.

[85]Memorie dell' Istituto Nazionale Italiano, Classe di Fis. e Mat., Bologna, 2, 1810, II, 445-476.

[85a]Entwickelung... Theilbarkeit, Jahresber. Kath. Gymn. Köln., 1847-8, 23.

[86]Archiv Math. Phys., 45, 1866, 203-216.

$A0 = 10x^2 - 12x - 4$, whence $12x + 4$ is divisible by 10, so that $x = 5d - 2$. Then $A = 25d^2 - 26d + 6$. Thus if we delete the last two digits 7 , 6 of squares $A76$, we obtain numbers A whose values for $d = 1, 2, \ldots$ can be derived from the initial one 5 by successive additions of $49, 49 + 50, 49 + 2 \cdot 50, \ldots$. He gave such results for every pair of possible endings of squares.

A similar method is applied to any composite number. One case is when the last two digits are $m, 1$ and $Am1 = (10x - 1)(10y - 1)$. Then

$$A0 = 10xy - y - x - m, \qquad y + x + m = 10a, \qquad A = 10ax - x^2 - mx - a.$$

The discriminant of the last equation must be a square. A table of values of A for each a may be formed by successive additions.

G. Speckmann[87] noted that the two factors of $N = 2047$ end in 1 and 7 or 3 and 9. Treating the first case, we see that, if a and b are the digits in tens place, $b + 7a \equiv 4 \pmod{10}$, so that the factors end in 01 and 47, or 11 and 77, etc.

G. Speckmann[88] wrote the given number prime to 3 in the form $9a + b$ ($b < 9$), so that the sum of its digits is $\equiv b \pmod 9$. By use of a small auxiliary table we have the residues modulo 9 of the sums of the digits of every possible pair of factors.

R. W. D. Christie[89] and D. Biddle[90] made an extensive use of terminal digits.

E. Barbette[91] noted that $10d + u$ has a divisor $10m - 1$ if and only if $d + mu$ has that divisor. Set $d + mu = n(10m - 1)$, $d = 10d' + u'$. Then

$$mn = d' + x, \qquad 10x = mu + n + u'.$$

Eliminating n, we get a quadratic for m. Its discriminant is a quadratic function of x which is to be made a square. Similarly for $10m + 1$, $10m \pm 3$.

A. Gérardin[91a] developed Barbette's[91] method.

R. Rawson[92] found Fermat's[1] factors of a number proposed by Mersenne by writing it to the base 100 and expressing it as $(a \cdot 10^2 + 23)(b \cdot 10^2 + 3)$.

J. Deschamps[93] would use the final digits and auxiliary tables.

A. Gérardin[94] would factor N (prime to 2, 3, 5) by use of

$$N = 120n + K = (120x + a)(120y + b),$$

and a table showing, for each of the 32 values of $K < 120$, the 16 pairs a, b (each < 120) such that $ab \equiv K \pmod{120}$. He factored Mersenne's number.[1]

FACTORING BY CONTINUED FRACTIONS OR PELL EQUATIONS.

Franz von Schaffgotsch[100] would factor a by solving $az^2 + 1 = x^2$ (having

[87]Archiv Math. Phys., (2), 12, 1894, 435.   [88]Archiv. Math. Phys., 14, 1896, 441–3.
[89]Math. Quest. Educat. Times, 69, 1898, 99–104. Cf. Meissner,[138] 138–9.
[90]*Ibid.*, 87–88, 112–4; 71, 1899, 93–9; Mess. Math., 28, 1898–9, 120–149, 192 (correction). Cf. Meissner,[138] 137–8.   [91]Mathesis, (2), 9, 1899, 241.
[91a]Sphinx-Oedipe, 1906–7, [1–2, 17, 33], 49–50, 54, 65–7, 77–8, 81–4; 1907–8, 33–5; 5, 1910, 145–7; 6, 1911, 157–8.   [92]Math. Quest. Educat. Times, 71, 1899, 123–4.
[93]Bull. Soc. Philomathique de Paris, (9), 10, 1908, 10–26.
[94]Assoc. franç., 38, 1909, 145–156; Sphinx-Oedipe, Nancy, 1908–9, 129–134, 145–9; 4, 1909, 3e Trimestre, 17–25.
[100]Abh. Böhmischen Gesell. Wiss., Prag, 2, 1786, 140–7.

solutions if a is not a square) and testing $x^2 - 1$ for a factor in common with a. Further, if $ay + 1 = x^2$ does not hold for $1 < x < a - 1$, then a is a power of a prime and conversely [false if $a = 10$].

Märcker[101] noted that if there are $2n$ terms in the period of

$$\sqrt{A} = a + \frac{1}{a'} + \frac{1}{a''} + \dots$$

and $Q = 0$, $Q' = a$, $Q'' = a'P' - Q'$, ...,

$$P = 1, \qquad P' = \frac{A - Q'^2}{P}, \qquad P'' = \frac{A - Q''^2}{P'}, \dots,$$

then the nth P or its half is a factor of A. If A is a prime, then the nth P is 2.

J. G. Birch[102] derived a factor of N from a solution x of $x^2 = Ny + 1$. The continued fraction for $x/(N - x)$ is of the form

$$\frac{1}{a_0 - 1} + \frac{1}{a_1} + \frac{1}{a_2} + \dots + \frac{1}{a_2} + \frac{1}{a_1} + \frac{1}{a_0},$$

and N is the continuant defined as the determinant with $a_0, a_1, \dots, a_{n-1}$, $a_n, a_{n-1}, \dots, a_1, a_0$ in the main diagonal, elements $+1$ just above this diagonal, elements -1 just below, and zeros elsewhere. Then the continuant with the diagonal $a_0, \dots, a_{n-1}$ is a factor of N.

W. W. R. Ball[103] applied this method to a number of Mersenne.[1·]

A. Cunningham[104] noted that a set of solutions of $y^2 - Dx^2 = -1$ gives at sight factors of $y^2 + 1$.

M. V. Thielmann[105] illustrated his method by factoring $k = 36343817$. The partial denominators in the continued fraction for $\sqrt{k}$ are 1, 1, 2, 1, 1, 12056. Drop the last term and pass to the ordinary fraction $7/12$. Hence set $(12x + 7)^2 = 12^2 y + 1$. The least solution is $x = 4$, $y = 21$. Using the part of the period preceding the middle term $w = 2$, we get

$$\frac{1}{1 + 1} = \frac{P}{M}, \qquad P = 1, \qquad M = 2, \qquad Q = wM + 2P = 6, \qquad u = MQ = 12.$$

Hence $t^2 - 21u^2 = 1$ has the solution $t = 55$. For a suitably chosen n,

$$k = u^2 n^2 + 2tn + 21 = \left(2q^2 n + \frac{t \pm 1}{8}\right)\left(2M^2 n + \frac{t \mp 1}{2q^2}\right),$$

where q is the largest integer $\leq Q/2$. Here $n = 502$ and the factors of k are $2 \cdot 3^2 n + 7$ and $2 \cdot 2^2 n + 3$.

D. N. Lehmer[106] noted that if $R = pq$ is a product of two odd factors whose difference is $< 2\sqrt[4]{R}$, so that $\frac{1}{4}(p - q)^2 < \sqrt{R}$, then

$$x^2 - Ry^2 = \frac{1}{4}(p - q)^2$$

has the integral solutions $x = (p + q)/2$, $y = 1$. Hence $\frac{1}{4}(p - q)^2$ is a denominator of a complete quotient in the expansion of $\sqrt{R}$ as a continued fraction,

[101]Jour. für Math., 20, 1840, 355-9. Cf. l'intermédiaire des math., 20, 1913, 27-8.
[102]Mess. Math., 22, 1892-3, 52-5.
[103]Ibid., p. 82-3. French transl., with Birch[102], Sphinx-Oedipe, 1913, 86-9.
[104]Ibid., 35, 1905-6, 166-185; abst. in Proc. London Math. Soc., 3, 1905, xxii.
[105]Math. Annalen, 62, 1906, 401.
[106]Bull. Amer. Math. Soc., 13, 1906-7, 501-2. French transl., Sphinx-Oedipe, 6, 1911, 138-9.

in view of the theorem of Lagrange: If $x^2-Ry^2=\pm D$ has relatively prime integral solutions x, y, where $D<\sqrt{R}$, then D is a denominator of a complete quotient in the expansion of $\sqrt{R}$ as a continued fraction.

Factoring by Use of Various Moduli.

C. F. Gauss[110] gave a "method of exclusion," based on the use of various small moduli, to express a given number in a given form mx^2+ny^2.

V. Bouniakowsky[111] noted that information as to the prime factors of a number N may be obtained by comparing the solution $x=\phi(N)$ of $2^x\equiv 1$ (mod N) with the least positive solution $x=a$ found by a direct process such as the following: Since $2^a=NK+1$, multiply the given N by the unknown K, each expressed in the binary scale (base 2), add 1 and equate the result to $10\ldots0$. The digits of K are found seriatim and very simply.

H. J. Woodall[112] expressed the number N to be factored in the form $a^a+\beta^b+\ldots+r$, where $r<1000$, while a, β, $\ldots$ are small, but not necessarily distinct. Hence the residues of N with respect to various moduli are readily found by tables of residues.

F. Landry[113] employed the method of exclusion by small moduli.

D. Biddle[114] investigated factors $2\Delta p+1$ by using moduli Δ^2, $4\Delta^2$.

C. E. Bickmore, A. Cunningham and J. Cullen[115] each treated the large factor of $10^{25}+1$ by use of various moduli, and proved it is prime.

J. Cullen[115a] gave an effective graphical process to factor numbers by the use of various moduli; the numbers to be searched for in a diagram are all small.

Alfred Johnsen[116] used $R_t(p)$ to denote the numerically least residue of p modulo t. Then, for every p, t, k,

$$[R_t(k)]^2+R_t(p-k^2)\equiv R_t(p)\ (\text{mod } t).$$

If t is a factor of the given number p, the left member will be divisible by t. In practice take k^2 to be the nearest square to p, larger or smaller. For example, let $p=4699$, $k^2=4624=68^2$, $p-k^2=75$. Then

t	$[R_t(68)]^2$	$R_t(75)$	Sum
7	4	-2	2
13	9	-3	6
$\ldots$	$\ldots$	$\ldots$	$\ldots$
37	36	1	37

Thus 37 is the least factor of p.

[110]Disq. Arith., 1801, Arts. 323–6.

[111]Mém. Ac. Sc. St. Pétersbourg, Math.-Phys., (6), 2, 1841, 447–69. Extract in Bull. Ac. Sc., 6, p. 97. Cf. Nordlund.[117]

[112]Math. Quest. Educat. Times, 70, 1899, 68–71; 71, 1899, 124.

[113]Procédés nouveaux..., Paris, 1859. Cf. A. Aubry,[147] pp. 214–7.

[114]Mess. Math., 30, 1900–1, 98, 190. Math. Quest. Educat. Times, 74, 1901, 147–152.

[115]Math. Quest. Educat. Times, 72, 1900, 99–103.

[115a]Ibid., 73, 1900, 133–5; 75, 1901, 102–4. Proc. London Math. Soc., 34, 1901–2, 323–334; (2), 2, 1905, 138–141.   [116]Nyt Tidsskrift for Mat., 15 A, 1904, 109–110.

K. P. Nordlund[117] would use the exponent e to which 2 belongs modulo N [Bouniakowsky[111]]. For $N = 91$, $e = 12$ is not a divisor of $N-1$, so that N is composite, and we expect the factor 13.

L. E. Dickson[118] found the factors of $56^7 \pm 1$, $26^{13}+1$, $34^{17}+1$, $52^{13}+1$ by an expeditious method. For example, each factor of

$$b = \frac{56^7 - 1}{56 - 1} = 1 + 56N$$

is $\equiv 1 \pmod{14}$. Let $b = (1+14k)(1+14k_1)$. Then

$$k + k_1 + 14kk_1 = 4N, \qquad k + k_1 = 4 + 14h.$$

Thus k and k_1 are the roots of a quadratic whose discriminant Q is of the second degree in h. By use of various moduli which are powers of small primes, the form of h is limited step by step, until finally at most a half dozen values of h remain to be tested directly.

L. E. Dickson[119] gave further illustrations of the last method.

J. Schatunovsky[120] reduced to a minimum the number of trials in Gauss'[110] method of exclusion, taking the simplest case $m = 1$. He gave theorems on the linear forms of the factors of $a^2 + Db^2$, which lead easily to all its odd factors when D is an odd prime.

H. C. Pocklington[121] would use Fermat's theorem to tell whether N is prime or composite. Choose an integer x and find the least positive residue of x^{N-1} modulo N; if $\neq 1$, N is composite. But if it be unity, let p be a prime factor (preferably the largest) of $N-1$ and contained a times in it. Find the remainder r when x^m is divided by N, where $m = (N-1)/p$. If $r \neq 1$, let δ be the g. c. d. of $r-1$ and N. If $\delta > 1$, we have a factor of N. If $\delta = 1$, all prime factors of N are of the form $kp^a + 1$. But if $r = 1$, replace m by m/q, where q is any prime factor of m and proceed as before.

D. Biddle[121a] made use of various small moduli.

A. Gérardin[121b] used various moduli to factor 77073877.

See papers 14, 15, 21, 22, 48, 65.

Factoring Into Two Numbers $6n \pm 1$.

G. W. Kraft[122] noted that $6a + 1 = (6m+1)(6n+1)$ implies

$$n = \frac{a - m}{6m + 1}.$$

Find which $m = 1, 2, 3, \ldots$ makes n an integer.

Ed. Bartl[123] tested $6 \cdot 186 + 5$ for a prime factor less than 31, just less than its square root, by noting that 186, 185, 184, 183, 182 are not divisible by 5, 11, 17, 23, 29, respectively; while the last of 7, 13, 19 is a factor.

[117]Göteborgs Kungl. Vetenskaps. Handl., (4), 1905, VII–VIII, pp. 21–4.
[118]Amer. Math. Monthly, 15, 1908, 217–222.   [119]Quar. Jour. Math., 40, 1909, 40–43.
[120]Der Grösste Gemeinschaftliche Teiler von Algebr. Zahlen zweiter Ordnung, Diss. Strassburg, Leipzig, 1912.   [121]Proc. Cambridge Phil. Soc., 18, 1914–5, 29–30.
[121a]Math. Quest. Educat. Times, (2), 25, 1914, 43–6.
[121b]L'enseignement math., 17, 1915, 244–5.
[122]Novi Comm. Ac. Petrop., 3, ad annos 1750–1, 117–8.
[123]Zur Theorie der Primzahlen, Progr. Mies, Pilsen, 1871.

F. Landry[124] treated the possible pairs $6n \pm 1$ and $6n' \pm 1$ of factors of N. Taking for example the case of the upper signs, we have

$$6nn' + n + n' = \frac{N-1}{6} = 6q + r.$$

Set $n + n' = 6h + r$. Then $nn' = q - h$, whence

$$h = \frac{q - n'(r - n')}{6n' + 1}.$$

Give to n' values such that $6n' + 1$ is a prime $< \sqrt{N}$.

K. P. Nordlund[125] treated $6p - 1 = (6m+1)(6n-1)$ solved for m.

D. Biddle[126] applied the method to $6n \pm 1$.

Hansen,[99] of Ch. XIII, used this method.

Miscellaneous Methods of Factoring.

Matsunaga[129] wrote the number to be factored in the form $r^2 + R$. For r odd, set $r = B_1$, $B_1 - 2 = B_2$, $B_2 - 2 = B_3, \ldots$ and perform the following calculations:

$$
\begin{array}{lll}
R = Q_1 B_1 + A_1, & K_1 = 2Q_1, & K_2' = K_1 + 4, \\
A_1 + K_2' = Q_2 B_2 + A_2, & K_2 = 2Q_2 + K_2', & K_3' = K_2 + 8, \\
A_2 + K_3' = Q_3 B_3 + A_3, & K_3 = 2Q_3 + K_3', & K_4' = K_3 + 8,
\end{array}
$$

etc., until we reach $A_n = 0$; then B_n is a factor. If r is even, set $r - 1 = B_1$ and replace R by $R + 1$ in what precedes.

J. H. Lambert[130] used periodic decimals [see Lambert,[6] Ch. VII].

Jean Bernoulli[131] gave a method based on that of Lambert (Mém. de Math. Allemands, vol. 2). Let $A = a^2 + b$ have the factors $a - x$ and $a + x + y$. Then $x^2 = ay - xy - b$. Solve for x. Thus $y^2 + 4ay - 4b$ must be a square. Take $y = 1, 2, \ldots$ and use a table of squares.

J. Gough[132] gave a method to find the factors r, s of each number $f^2 - c$ between $(f-1)^2$ and f^2. For example, let $f = 3$ and make a double row for each $r = 1, \ldots, f$. In the upper row for $r = 1$, insert $2f - 1, \ldots, 1, 0$; in the lower, $(f-1)^2, \ldots, f^2$. In the upper row for $r = 2$, insert 1 (the remainder

$r=1$	$c=5$	4	3	2	1	0
	$s=4$	5	6	7	8	9
$r=2$	$c=5$		3		1	
	$s=2$		3		4	
$r=3$	$c=$					0
	$s=$					3

[124]Assoc. franç. avanc. sc., 9, 1880, 185-9.
[125]Nyt Tidsskrift for Mat., Kjobenhavn, 15 A, 1904, 36-40.
[126]Math. Quest. Educ. Times, 69, 1898, 87-8; (2), 22, 1912, 38-9, 84-6.
[129]Japanese manuscript, first half eighteenth century, Abhandl. Geschichte Math. Wiss., 30, 1912, 236-7.   [130]Nova Acta Eruditorum, 1769, 107-128.
[131]Nouv. Mém. Ac. Berlin, année 1771, 1773, 323.
[132]Jour. Nat. Phil. Chem. Arts (ed., Nicholson), 1, 1809, 1-4.

on dividing f^2 by 2), $1+2$, $1+2+2$ under 1, 3, 5 of the first row for $r=1$; in the lower row, insert 4 (the quotient), $4-1$, $4-2$. To factor f^2-c, locate the column headed by the given c; thus, for $c=3$, the factors are $s=6$, $r=1$ and $s=3$, $r=2$. Since $c=2$ occurs only in the first row, $9-2$ is prime. Joubin,[133] J. P. Kulik,[134] O. V. Kielsen,[135] and G. K. Winter[136] published papers not accessible to the author.

E. Lucas gave methods of factoring and tests for primes (Ch. XVII).

D. Biddle[137] wrote the proposed number N in the form S^2+A, where S^2 is the largest square $<N$. Write three rows of numbers, the first beginning with A, or $A-S$ if $A>S$; the second beginning with S (or $S+1$) and increasing by 1; the third beginning with S and decreasing by 1. Let A_n, B_n, C_n be the nth elements in the respective rows. Then

$$C_n = C_{n-1} - 1, \qquad B_n = B_{n-1} + 1, \qquad A_n = A_{n-1} + B_{n-1} - C_n,$$

except that, when $A_n > C_n$, we subtract C_n from A_n as often (say k times) as will leave a positive remainder, and then $B_n = B_{n-1} + 1 + k$. When we reach a value of n for which $A_n = 0$, we have $N = B_n C_n$. For example, if $N = 589 = 24^2 + 13$, the rows are

13	14	17	1	9	0	
24	25	26	28	29	31	(factors 31, 19).
24	23	22	21	20	19	

It may prove best to start with $2N$ instead of with N.

O. Meissner[138] reviewed many methods of factoring.

R. W. D. Christie[139] gave an obscure method by use of "roots." Christie[140] noted that, if $N=AB$,

$$A = (4bN + d^2 - d)/(2b), \qquad B = (4bN + d^2 + d)/2, \qquad d \equiv a - bc,$$

whence $d^2 = (B - bA)^2$.

D. Biddle[141] gave a method of finding the factors of N given those of $N+1$. Set $L=N-1$. Try to choose K and M so that $KM=N+1$ and so that $1+K$ is a factor of N. Since $2N=(1+K)M+L-M$, we will have $L-M=(1+K)m$, whence $2N=(1+K)(M+m)$. For $N=1829$, $N+1$ $=2\cdot3\cdot5\cdot61$. Take $K=30$, $M=61$. Then $m=57$, $M+m=2\cdot59$, $N=31\cdot59$. He gave (ibid., p. 43) the theoretical test that $N=S^2+A$ is composite if the sum of r terms of

$$\frac{A}{S} + \frac{N}{S(S-1)} + \frac{N}{(S-1)(S-2)} + \cdots$$

is an integer for some value of r.

[133]Sur les facteurs numériques, Havre, 1831.
[134]Abh. K. Böhm. Gesell. Wiss., 1, 1841 (2, 1842–3, 47, graphical determination of primes).
[135]Om et heel tals upplösning i factorer, Kjöbenhavn, 1841.
[136]Madras Jour. Lit. Sc., 1886–7, 13.
[137]Mess. Math., 28, 1898–9, 116–20; Math. Quest. Educat. Times, 70, 1899, 100, 122; 75, 1901, 48; extension, (2), 29, 1916, 43–6.
[138]Math. Naturw. Blätter, 3, 1906, 97, 117, 137.
[139]Math. Quest. Educat. Times, (2), 12, 1907, 90–1, 107–8.
[140]Ibid., (2), 13, 1908, 42–3, 62–3.
[141]Ibid., (2), 14, 1908, 34. The process is well adapted to factoring 2^p-1, (2), 23, 1913, 27–8.

E. Lebon[141a] would first test N for prime factors P just $< \sqrt{N}$. Let Q be the quotient and R the remainder on dividing N by P. If Q and R have a common factor, it divides N; if not, N is not divisible by any factor of Q or of R.

D. Biddle[142] considered $N = S^2 + A = (S+u)(S-v)$, wrote $uv = N_1$ and obtained like equations in letters with subscripts unity. Then treat $u_1 v_1 = N_2$ similarly, etc.

A. Cunningham[143] noted that the number of steps in Biddle's[142] process is approximately the value of k in $2^k = N$, and developed the process.

E. Lebon[144] treated the decomposition of forms

$$x^a \pm x^\beta \pm x^\gamma \pm \ldots \pm 1 \qquad (a > \beta > \gamma \ldots)$$

of degrees $\leqq 9$ into two such forms, using a table of those forms of degrees $\leqq 4$ with all coefficients positive which are not factorable. The base most used in the examples is $x = 10$. But bases 2 and 3 are considered.

E. Barbette[145] quoted from his[146] text the theorem that any integer N can be expressed in each of the four forms

$$N = \Delta_x - \Delta_y, \qquad 8N = x^2 - y^2, \qquad Nz = \Delta_x, \qquad 8Nz + 1 = y^2,$$

where $\Delta_x = x(x+1)/2$. The resulting new methods of factoring are now simplified by use of triangular and quadratic residues. The first formula implies $N = (x-y)(x+y+1)/2$. In his text, he considered the sum

$$N = (y+1) + (y+2) + \ldots + (x-1) + x = \Delta_x - \Delta_y$$

of consecutive integers. Treating four types of numbers N, he proved that this equation has 1, 2 or more than 2 sets of integral solutions x, y, according as N is a power of 2, an odd prime, or a composite number not a power of 2. He proved independently, but again by use of sums of consecutive integers, that every composite number not a power of 2 can be given the form* $N = u$ $(2v - u + 1)/2$, where u and v are integers and $v \geqq u \geqq 3$. Solving for u, and setting $x = 2v + 1$, we get $2u = x + (x^2 - 8N)^{1/2}$. Hence $x^2 - 8N = y^2$ is solvable in integers [evidently by $x = 2N + 1$, $y = 2N - 1$]. Finally, $Nz = \Delta_x$ is equivalent to $(2x+1)^2 = 8Nz + 1$. For four types of numbers N, the solutions of $y^2 = 8Nz + 1$ are found and seen to involve at least two arbitrary constants.

A. Aubry[147] reviewed various methods of factoring.

[141a]Il Pitagora, Palermo, 14, 1907–8, 96–7.
[142]Math. Quest. Educat. Times, (2), 19, 1911, 99–100; 22, 1912, 38–9; Educat. Times, 63, 1910. 500; Math. Quest. and Solutions, 2, 1916, 36–42.
[143]Ibid., (2), 20, 1911, 59–64; Educat. Times, 64, 1911, 135.
[144]Bull. soc. philomathique de Paris, (10), 2, 1910, 45–53; Sphinx-Oedipe, 1908–9, 81–3, 97–101
[145]L'enseignement math., 13, 1911, 261–277.
[146]Les sommes de p-ièmes puissances distinctes égales à une p-ième puissance, Paris, 1910, 20-76.
*This follows from the former result $N = (x-y)(x+y+1)/2$ by setting $x = v$, $y = v - u$. To give a direct proof, take u to be the least odd factor > 1 of the composite number N not a power of 2; then $q = N/u$ can be given the form $v - (u-1)/2$ by choice of v. If $v < u$, then $q < (u+1)/2 < u$, so that q has no odd factor and $q = 2^k$. But $N = 2^k U$ is of the desired form if we take $v = u/2 = N$.
[147]Sphinx-Oedipe, numéro spéc., June, 1911, 1–27. Errata and addenda, numéro spéc., Jan., 1912, 7–9, 14. L'enseignement math., 15, 1913, 202–231.

S. Bisman[148] noted that N is composite if and only if there exist two integers A, B such that $A+2B$ and $A+2BN$ divide $2(N-1)$ and $(N-1)A$, respectively. But there is no convenient maximum for the smaller integer B. To find the factor 641 of $2^{32}+1$ there are 16 cases.

A. Gérardin[149] gave a report on methods of factoring.

J. A. Gmeiner,[150] to factor a, prime to 6, determined b and ϵ so that $9a=16b+\epsilon$, $0\leqq\epsilon<16$. Let ω^2 be the largest square $<b$ and set $b=\omega^2+\rho$, $\sigma=\rho-\omega$. Hence $9a=16(\omega-x)(\omega+x+1)+\tau(x)$, where

$$\tau(x)=16\sigma+\epsilon+16x(x+1).$$

Since $\tau(x)=\tau(x-1)+32x$, we may rapidly tabulate the values of $\tau(x)$ for $x=0, 1, 2,\ldots$ If we reach the value zero, we have two factors of a. To prove that a is a prime, we need extend the table until $\omega+x+1$ is the largest square $<a$. To modify the process, use $4a=7b+\epsilon$.

A. Reymond[151] used the graphs of $y=x/n$ ($n=1, 2, 3, 5,\ldots$), marking on each the points with integral coordinates. He omitted $y=x/4$ since its integral points are on $y=x/2$. Since 17 is not the abscissa of an integral point on $y=x/n$ for $1<n<17$, 17 is a prime. [Möbius[53a] of Ch. XIII.]

A. J. Kempner[152] found, by use of a figure perspective to Reymond's[151], how to test the primality of numbers by means of the straight edge.

D. Biddle and A. Cunningham[153] factor a product N of two primes by finding $N_1<N$ and $N_2>N$ such that $N_2-N=N-N_1+2$, while each of N_1 and N_2 is a product of two even factors, the two smaller factors differing by 2 and the two larger factors differing by 2.

[148]Mathesis, (4), 2, 1912, 58–60.
[149]Assoc. franç. avanc. sc., 41, 1912, 54–7.
[150]Monatshefte Math. Phys., 24, 1913, 3–26.
[151]L'enseignement math., 18, 1916, 332–5.
[152]Amer. Math. Monthly, 24, 1917, 317–321.
[153]Math. Quest. and Solutions, 3, 1917, 21–23.

CHAPTER XV.

FERMAT NUMBERS $F_n = 2^{2^n} + 1$.

Fermat[1] expressed his belief that every F_n is a prime, but admitted that he had no proof. Elsewhere[2] he said that he regarded the theorem as certain. Later[3] he implied that it may be proved by "descent." It appears that Frenicle de Bessy confirmed this conjectured theorem of Fermat's. On several occasions Fermat[4] requested Frenicle to divulge his proof, promising important applications. In the last letter cited, Fermat raised the question if $(2k)^{2^m} + 1$ is always a prime except when divisible by an F_n. C. F. Gauss[5] stated that Fermat affirmed (incorrectly) that the theorem is true. The opposite view was expressed by P. Mansion[6] and R. Baltzer.[7]

F. M. Mersenne[8] stated that every F_n is a prime. Chr. Goldbach[9] called Euler's attention to Fermat's conjecture that F_n is always prime, and remarked that no F_n has a factor < 100; no two F_n have a common factor.

L. Euler[10] found that

$$F_5 = 2^{32} + 1 = 641 \cdot 6700417.$$

Euler[11] proved that if a and b are relatively prime, every factor of $a^{2^n} + b^{2^n}$ is 2 or of the form $2^{n+1}k + 1$ and noted that consequently any factor of F_5 has the form $64k + 1$, $k = 10$ giving the factor 641. Euler[11a] and N. Beguelin[12] used the binary scale to find the factor $641 = 1 + 2^7 + 2^9$ of F_5.

C. F. Gauss[13] proved that a regular polygon of m sides can be constructed by ruler and compasses if m is a product of a power of 2 and distinct odd primes each of the form F_n, and stated that the construction is impossible if m is not such a product. This subject will be treated under Roots of Unity.

Sebastiano Canterzani[14] treated twenty cases, each with subdivisions depending on the final digits of possible factors, to find the factor 641 of F_5,

---

[1]Oeuvres, 2, 1894, p. 206, letter to Frenicle, Aug. (?) 1640; 2, 1894, p. 309, letter to Pascal, Aug. 29, 1654 (Fermat asked Pascal to undertake a proof of the proposition, Pascal, III, 232; IV, 1819, 384); proposed to Brouncker and Wallis, June 1658, Oeuvres, 2, p. 404 (French transl., 3, p. 316). Cf. C. Henry, Bull. Bibl. Storia Sc. Mat. e Fis., 12, 1879, 500–1, 716–7; on p. 717, 42...1 should end with 7, *ibid.*, 13, 1880, 470; A. Genocchi, Atti Ac. Sc. Torino, 15, 1879–80, 803.

[2]Oeuvres, 1, 1891, p. 131 (French transl., 3, 1896, p. 120).

[3]Oeuvres, 2, 433–4, letter to Carcavi, Aug., 1659.

[4]Oeuvres, 2, 208, 212, letters from Fermat to Frenicle and Mersenne, Oct. 18 and Dec. 25, 1640.

[5]Disq. Arith., Art. 365. Cf. Werke, 2, 151, 159. Same view by Klügel, Math. Wörterbuch, 2, 1805, 211; 3, 1808, 896.

[6]Nouv. Corresp. Math., 5, 1879, 88, 122.

[7]Jour. für Math., 87, 1879, 172.

[8]Novarum Physico-Mathematicarum, Paris, 1647, 181.

[9]Corresp. Math. Phys. (ed., Fuss), I, 1843, p. 10, letter of Dec. 1729; p. 20, May 22, 1730; p. 32, July 1730.

[10]Comm. Ac. Petrop., 6, ad annos 1732–3 (1739), 103–7; Comm. Arith. Coll., 1, p. 2.

[11]Novi Comm. Petrop., 1, 1747–8, p. 20 [9, 1762, p. 99]; Comm. Arith. Coll., 1, p. 55 [p. 357].

[11a]Opera postuma, I, 1862, 169–171 (about 1770).

[12]Nouv. Mém. Ac. Berlin, année 1777, 1779, 239.

[13]Disq. Arith., 1801, Arts. 335–366; German transl. by Maser, 1889, pp. 397–448, 630–652.

[14]Mem. Ist. Naz. Italiano, Bologna, Mat., 2, II, 1810, 459–469.

and proved in the same lengthy dull manner that the quotient is a prime. An anonymous writer[15] stated that

(1)　　　　$2+1,$　　　$2^2+1,$　　　$2^{2^2}+1,$　　　$2^{2^{2^2}}+1,\ldots$

are all primes and are the only primes 2^k+1. See Malvy.[39]

Joubin[16] suggested that these numbers (1) are possibly the ones really meant by Fermat,[1] evidently without having consulted all of Fermat's statements.

G. Eisenstein[17] set the problem to prove that there is an infinitude of primes F_n.

E. Lucas[18] stated that one could test the primality of F_6 in 30 hours by means of the series $3, 17, 577, \ldots$, each term being one less than the double of the square of the preceding. Then F_n is a prime if 2^{n-1} is the rank of the first term divisible by F_n, composite if no term is divisible by F_n. Finally, if a is the rank of the first term divisible by F_n, the prime divisors of F_n are of the form $2^k q+1$, where $k=a+1$ [not $k=2^{a+1}$]. See Lucas.[22]

T. Pepin[19] stated that the method of Lucas[18] is not decisive when F_n divides a term of rank $a<2^{n-1}$; for, if it does, we can conclude only that the prime divisors of F_n are of the form $2^{a+2}q+1$, so that we can not say whether or not F_n is prime if $a+2\leqq 2^{n-2}$. We may answer the question unambiguously by use of the new theorem: For $n>1$, F_n is a prime if and only if it divides

$$k^{(F_n-1)/2}+1,$$

where k is any quadratic non-residue of F_n, as 5 or 10. To apply this test, take the minimum residues modulo F_n of

$$k^2,\quad k^4,\quad k^8,\ldots,\quad k^{2^{2^n-1}}.$$

Proof was indicated by Lucas[29] of Ch. XVII, and by Morehead.[58]

J. Pervouchine[20] (or Pervusîn) announced, November 1877, that

$$F_{12}\equiv 0 \pmod{114689=7\cdot2^{14}+1}.$$

E. Lucas[21] announced the same result two months later and proved that every prime factor of F_n is $\equiv 1 \pmod{2^{n+2}}$.

Lucas[22] employed the series $6, 34, 1154, \ldots$, each term of which is 2 less than the square of the preceeding. Then F_n is a prime if the rank of the first term divisible by F_n is between 2^{n-1} and 2^n-1, but composite if no term is divisible by F_n. Finally, if a is the rank of the first term divisible by F_n

[15]Annales de Math. (ed. Gergonne), 19, 1828–9, 256.

[16]Mémoire sur les facteurs numériques, Havre, 1831, note at end.

[17]Jour. für Math., 27, 1844, 87, Prob. 6.

[18]Comptes Rendus Paris, 85, 1877, 136–9.

[19]Comptes Rendus, 85, 1877, 329–331. Reprinted, with Lucas[18] and Landry,[29] Sphinx-Oedipe, 5, 1910, 33–42.

[20]Bull. Ac. St. Pétersbourg, (3), 24, 1878, 559 (presented by V. Bouniakowsky). Mélanges math. ast. sc. St. Pétersbourg, 5, 1874–81, 505.

[21]Atti R. Accad. Sc. Torino, 13, 1877–8, 271 (Jan. 27, 1878). Cf. Nouv. Corresp. Math., 4, 1878, 284; 5, 1879, 88. See Lucas[40] of Ch. XVII.

[22]Amer. Jour. Math., 1, 1878, 313.

and if $a < 2^{n-1}$, the prime divisors of F_n are of the form* $2^k q + 1$, where $k = a + 1$ [cf. Lucas[18]]. He noted (p. 238) that a necessary condition that F_n be a prime is that the residue modulo F_n of the term of rank $2^n - 1$ in this series is zero. He verified (p. 292) that F_5 has the factor 641 and again stated that 30 hours would suffice to test F_6.

F. Proth[23] stated that, if $k = 2^n$, $2^k + 1$ is a prime if and only if it divides $m = 3^{2^{k-1}} + 1$. He[24] indicated a proof by use of the series of Lucas defined by $u_0 = 0$, $u_1 = 1$, . . . , $u_n = 3u_{n-1} + 1$ and the facts that u_{p-1} is divisible by the prime p, while $m = u_{2^k}/u_{2^{k-1}}$. Cf. Lucas.[26]

E. Gelin[25] asked if the numbers (1) are all primes. Catalan[25] noted that the first four are.

E. Lucas[26] noted that Proth's[23] theorem is the case $k = 3$ of Pepin's.[19]

Pervouchine[27] announced, February 1878, that F_{23} has the prime factor

$$5 \cdot 2^{25} + 1 = 167772161.$$

W. Simerka[28] gave a simple verification of the last result and the fact (Pervouchine[20]) that $7 \cdot 2^{14} + 1$ divides F_{12}.

F. Landry,[29] when of age 82 and after several months' labor, found that

$$F_6 = 274177 \cdot 67280421310721,$$

the first factor being a prime. He and Le Lasseur and Gérardin[29a] each proved that the last factor is a prime (cf. Lucas[31]).

K. Broda[30] sought a prime factor p of $a^{32} + 1$ by considering

$$n = (a^{32} - 1)(a^{64} + 1)(a^{512} + a^{384} + a^{256} + a^{128} + 1).$$

Multiply by $u = (a^{32} + 1)/p$. Thus $nu = (a^{640} - 1)/p$. But $a^{640} \equiv 1 \pmod{641}$. Since each factor of n is prime to p, we take $a = 2$ and see that $2^{32} + 1$ is divisible by 641.

E. Lucas[31] stated that he had verified that F_6 is composite by his[22] test, before Landry found the factors.

P. Seelhoff[32] gave the factor $5 \cdot 2^{39} + 1$ of F_{36} and commented on Beguelin.[12]

*Lucas wrote $k = 2^{a+1}$ in error, as noted by R. D. Carmichael on the proof-sheets of this History.

[23]Comptes Rendus Paris, 87, 1878, 374.

[24]Nouv. Corresp. Math., 4, 1878, 210–1; 5, 1879, 31.

[25]Ibid., 4, 1878, 160.

[26]Ibid., 5, 1879, 137.

[27]Bull. Ac. St. Pétersbourg, (3), 25, 1879, 63 (presented by V. Bouniakowsky); Mélanges math. astr. ac. St. Pétersbourg, 5, 1874–81, 519. Cf. Nouv. Corresp. Math., 4, 1879, 284–5; 5, 1879, 22.

[28]Casopis, Prag, 8, 1879, 36, 187–8. F. J. Studnicka, ibid., 11, 1881, 137.

[29]Comptes Rendus Paris, 91, 1880, 138; Bull. Bibl. Storia Sc. Mat., 13, 1880, 470; Nouv. Corresp. Math., 6, 1880, 417; Les Mondes, (2), 52, 1880. Cf. Seelhoff, Archiv Math. Phys., (2), 2,. 1885, 329; Lucas, Amer. Jour. Math. 1, 1878, 292; Récréat. Math., 2, 1883, 235; l'intermédiaire des math., 16, 1909, 200.

[29a]Sphinx-Oedipe, 5, 1910, 37–42.

[30]Archiv Math. Phys., 68, 1882, 97.

[31]Récréations Math., 2, 1883, 233–5. Lucas,[35] 354–5.

[32]Zeitschr. Math. Phys., 31, 1886, 172–4, 380. For F_6, p. 329. French transl., Sphinx-Oedipe, 1912, 84–90.

J. Hermes[33] indicated a test for composite F_n by Fermat's theorem.

R. Lipschitz[34] separated all integers into classes, the primes of one class being Fermat numbers F_n, and placed in a new light the question of the infinitude of primes F_n.

E. Lucas[35] stated the result of Proth,[23] but with a misprint [Cipolla[46]].

H. Scheffler[36] stated that Legendre believed that every F_n is a prime(!), and obtained artificially the factor 641 of F_5. He noted (p. 167) that

$$F_n F_{n+1} \ldots F_{a-1} = 1 + 2^{2^n} + 2^{2 \cdot 2^n} + 2^{3 \cdot 2^n} + \ldots + 2^{2^a - 2^n}.$$

He repeated (pp. 173–8) the test by Pepin,[19] with $k = 3$, and (p. 178) expressed his belief that the numbers (1) are all primes, but had no proof for F_{16}.

W. W. R. Ball[37] gave references and quoted known results.

T. M. Pervouchine[38] checked his verification that F_{12} and F_{23} are composite by comparing the residues on division by $10^3 - 2$.

Malvy[39] noted that the prime $2^8 + 1$ is not in the series (1).

F. Klein[40] stated that F_7 is composite.

A. Hurwitz[41] gave a generalization of Proth's[23] theorem. Let $F_n(x)$ denote an irreducible factor of degree $\phi(n)$ of $x^n - 1$. Then if there exists an integer q such that $F_{p-1}(q)$ is divisible by p, p is a prime. When $p = 2^k + 1$, $F_{p-1}(x) = x^{2^{k-1}} + 1$.

J. Hadamard[42] gave a very simple proof of the second remark by Lucas.[21]

A. Cunningham[43] found that F_{11} has the factor 319489·974849.

A. E. Western[44] found that F_9 has the factor $2^{16} \cdot 37 + 1$, F_{18} the factor $2^{20} \cdot 13 + 1$, the quotient of F_{12} by the known factor $2^{14} \cdot 7 + 1$ has the factors $2^{16} \cdot 397 + 1$ and $2^{16} \cdot 7 \cdot 139 + 1$. He verified the primality of the factor $2^{41} \cdot 3 + 1$ of F_{38}, found by J. Cullen and A. Cunningham. He and A. Cunningham found that no more F_n have factors $< 10^6$ and similar results.

M. Cipolla[45] noted that, if q is a prime $> (9^{2^{m-2}} - 1)/2^{m+1}$ and $m > 1$, $2^m q + 1$ is a prime if and only if it divides $3^k + 1$ for $k = q \cdot 2^{m+1}$. He[46] pointed out the misprint in Lucas'[35] statement.

Nazarevsky[47] proved Proth's[23] result by using the fact that 3 is a primitive root of a prime $2^k + 1$.

[33]Archiv Math. Phys., (2), 4, 1886, 214–5, footnote.

[34]Jour. für Math., 105, 1889, 152–6; 106, 1890, 27–29.

[35]Théorie des nombres, 1891, preface, xii.

[36]Beiträge zur Zahlentheorie, 1891, 147, 151–2, 155 (bottom), 168.

[37]Math. Recreations and Problems, ed. 2, 1892, 26; ed. 4, 1905, 36–7; ed. 5, 1911, 39–40.

[38]Math. Papers Chicago Congress of 1893, I, 1896, 277.

[39]L'intermédiaire des math., 2, 1895, 41 (219).

[40]Vorträge über ausgewählte Fragen der Elementar Geometrie, 1895, 13; French transl., 1896, 26; English transl., "Famous Problems of Elementary Geometry," by Beman and Smith, 1897, 16.

[41]L'intermédiaire des math., 3, 1896, 214.

[42]Ibid., p. 114.

[43]Report British Assoc., 1899, 653–4. The misprint in the second factor has been corrected to agree with the true [44] value $2^{13} \cdot 7 \cdot 17 + 1$.

[44]Cunningham and Western, Proc. Lond. Math. Soc., (2), 1, 1903, 175; Educ. Times, 1903, 270.

[45]Periodico di Mat., 18, 1903, 331.

[46]Also in Annali di Mat., (3), 9, 1904, 141.

[47]L'intermédiaire des math., 11, 1904. 215.

A. Cunningham[47a] noted that 3, 5, 6, 7, 10, 12 are primitive roots and 13, 15, 18, 21, 30 are quadratic residues of every prime $F_n > 5$. He factored $F_4{}^4 + 8 + (F_0 F_1 F_2 F_3)^4$.

Thorold Gosset[48] gave the two complex prime factors $a \pm bi$ of the known real factors of composite F_n, $n = 5, 6, 9, 11, 12, 18, 23, 36, 38$.

J. C. Morehead[49] verified by use of the criterion of Pepin[19] with $k = 3$ that F_7 is composite, a result stated by Klein.[40]

A. E. Western[50] verified in the same way that F_7 is composite. The work was done independently and found to agree with Morehead's.

J. C. Morehead[51] found that F_{73} has the prime factor $2^{75} \cdot 5 + 1$.

A. Cunningham[52] considered hyper-even numbers

$$E_{0, n} = 2^n, \qquad E_{1, n} = 2^{E_{0, n}}, \ldots, \qquad E_{r+1, n} = 2^{E_{r, n}}.$$

For m odd, the residues modulo m of $E_{r, 0}, E_{r, 1}, \ldots$ have a non-recurrent part and then a recurring cycle.

A. Cunningham[53] gave tables of residues of $E_{1, n}, E_{2, n}, E_{r, 0}, 3^{3^n}$ and 5^{5^n} for the n's forming the first cycle for each prime modulus < 100 and for certain larger primes. A hyper-exponential number is like a hyper-even number, but with base q in place of 2. He discussed the quadratic, quartic and octic residue character of a prime modulo F_n, and of F_n modulo F_{n+x}.

Cunningham and H. J. Woodall[54] gave material on possible factors of F_n.

A. Cunningham[55] noted that, for every $F_n > 5$, $2F_n = t^2 - (F_n - 2)u^2$ algebraically, and expressed F_5 and F_6 in two ways in each of the forms $a^2 + b^2$, $c^2 \pm 2d^2$. He[56] noted that $F_n{}^3 + E_n{}^3$ is the algebraic product of $n + 2$ factors, where $E_n = 2^{2^n}$, and that $M_n = (F_n{}^3 + E_n{}^3)/(F_n + E_n)$ is divisible by M_{n-r}. If $n - m \geq 2$, $F_m{}^4 + F_n{}^2$ is composite.

A. Cunningham[57] has considered the period of $1/N$ to base 2, where N is a product $F_m F_{m-1} \ldots F_{m-r}$ of Fermat numbers.

J. C. Morehead and A. E. Western[58] verified by a very long computation that F_8 is composite. Use was made of the test by Pepin[19] with $k = 3$, which was proved to follow from the converse of Fermat's theorem.

P. Bachmann[59] proved the tests by Pepin[19] and Lucas.[22]

A. Cunningham[60] noted that every $F_n > 5$ can be represented by 4 quadratic forms of determinants $\pm G_n$, $\pm 2G_n$, where $G_n = F_0 F_1 \ldots F_{n-1}$.

Bisman[148] (of Ch. XIV) separated 16 cases in finding the factor 641 of F_5.

[47a]Math. Quest. Educ. Times, (2), 1, 1902, 108; 5, 1904, 71–2; 7, 1905, 72.

[48]Mess. Math., 34, 1905, 153–4.   [49]Bull. Amer. Math. Soc., 11, 1905, 543.

[50]Proc. Lond. Math. Soc., (2), 3, 1905, xxi.

[51]Bull. Amer. Math. Soc., 12, 1906, 449; Annals of Math., (2), 10, 1908–9, 99. French transl. in
        Sphinx-Oedipe, Nancy, 1911, 49.   [52]Report British Assoc. Adv. Sc., 1906, 485–6.

[53]Proc. London Math. Soc., (2), 5, 1907, 237–274.

[54]Messenger of Math., 37, 1907–8, 65–83.

[55]Math. Quest. Educat. Times, (2), 12, 1907, 21–22, 28–31.

[56]Ibid., (2), 14, 1908, 28; (2), 8, 1905, 35–6.

[57]Math. Gazette, 4, 1908, 263.

[58]Bull. Amer. Math. Soc., 16, 1909, 1–6. French transl., Sphinx-Oedipe, 1911, 50–55.

[59]Niedere Zahlentheorie, II, 1910, 93–95.

[60]Math. Quest. Educat. Times, (2), 20, 1911, 75, 97–98.

A. Gérardin[61] noted that $F_n = (240x + 97)(240y + 161)$ for all the F_n fully factored to date, and specified x and y more exactly in special cases.

C. Henry[62] gave references and quoted known results.

R. D. Carmichael[63] gave a test for the primality of F_n equivalent to Pepin's[19] and a further generalization (p. 65) in the direction of Hurwitz's.[41]

R. C. Archibald[64] cited many of the papers listed above and collected in a table the known factors of F_n with the exception of that given by Morehead.[51]

For a remark on F_n, see Cunningham[101] of Ch. VII.

[61]Sphinx-Oedipe, 7, 1912, 13.
[62]Oeuvres de Fermat, 4, 1912, 202–4.
[63]Annals of Math., (2), 15, 1913–4, 67.
[64]Amer. Math. Monthly, 21, 1914, 247–251.

CHAPTER XVI.

FACTORS OF $a^n \pm b^n$.

Fermat[1] stated that $(2^p+1)/3$ has no factors other than $2kp+1$ if p is an odd prime.

L. Euler[2] noted that a^4+4b^4 has the factors $a^2 \pm 2ab+2b^2$.

Euler[3] discussed the numbers a for which a^2+1 is divisible by a prime $4n+1=r^2+s^2$. Let p/q be the convergent preceding r/s in the continued fraction for r/s; then $ps-qr=\pm 1$. Thus every a is of the form $(4n+1)m \pm k$, where $k=pr+qs$.

Euler[4] gave the 161 integers $a<1500$ for which a^2+1 is a prime, and the cases $a=1, 2, 4, 6, 16, 20, 24, 34$ for which a^4+1 is a prime.

Euler[5] proved that, if m is a prime and a, b are relatively prime, a factor of a^m-b^m, not a divisor of $a-b$, is of the form $kn+1$. If $p=kn+1$ is a prime and $a=f^n \pm pa$, then a^k-1 is divisible by p. If af^n-bg^n is divisible by a prime $p=mn+1$, while f and g are not both divisible by p, then a^m-b^m is divisible by p; the converse is true if m and n are relatively prime.

Euler[6] proved the related theorems: For q an odd prime, any prime divisor of a^q-1, not a divisor of $a-1$, is of the form $2nq+1$. If a^m-1 is divisible by the prime $p=mn+1$, we can find integers x, y not divisible by p such that $A=ax^n-y^n$ is divisible by p (since the quotient of $a^m x^{mn}-y^{mn}$ by A is not divisible by p if x, y are suitably chosen).

Euler[7] treated the problem to find all integers a for which a^2+1 is divisible by a given prime $4n+1=p^2+q^2$. If a^2+b^2 is divisible by p^2+q^2, there exist integers r, s such that $a=pr+qs, b=ps-qr$. We wish $b=\pm 1$. Hence we take the convergent r/s preceding p/q in the continued fraction for p/q. Thus $ps-qr=\pm 1$, and our answer is $a=\pm(pr+qs)$. He listed all primes $P=4n+1<2000$ expressed as p^2+q^2, and listed all the a's for which a^2+1 is divisible by P. The table may be used to find all the divisors $<a$ of a given number a^2+1. He gave his[4] table and tabulated the values $a<1500$ for which $(a^2+1)/k$ is a prime, for $k=2, 5, 10$. He tabulated all the divisors of a^2+1 for $a \leqq 1500$.

N. Beguelin[8] stated that 2^n+1 has a trinary divisor $1+2^p+2^q$ only when $n=10, 24, 32$, although his examples (p. 249) contradict this statement.

Euler[9] gave a factor of $2^n \pm 1$ for various composite n's.

[1]Oeuvres, 2, 205, letter to Frenicle, Aug. (?), 1640. Bull. Bibl. St. Sc. Mat. e Fis., 12, 1879, 716.
[2]Corresp. Math. Phys. (ed., Fuss), I, 1843, p. 145; letter to Goldbach, 1742.
[3]Ibid., 242-3; letter to Goldbach, July 9, 1743.
[4]Ibid., 588-9, Oct. 28, 1752. Published, Euler.[7]
[5]Novi Comm. Petrop., 1, 1747-8, 20; Comm. Arith. Coll., 1, 57-61, and posthumous paper, ibid., 2, 530-5; Opera postuma, I, 1862, 33-35. Cf. Euler[152] of Ch. VII and the topic Quadratic Residues in Vol. III.
[6]Novi Comm. Petrop., 7, 1758-9 (1755), 49; Comm. Arith., 1, 269.
[7]Novi Comm. Petrop., 9, 1762-3, 99; Comm. Arith., 1, 358-369. French transl., Sphinx-Oedipe, 8, 1913, 1-12, 21-26, 64.
[8]Mém. Ac. Berlin, année 1777, 1779, 255. Cf. Ch. XV and Henry.[14]
[9]Posthumous paper, Comm. Arith., 2, 551; Opera postuma, I, 1862, 51.

Euler[9a] discussed the divisors of numbers of the form $fa^4 + gb^4$.

Anton Felkel[10] gave a table, incomplete as to a few entries, of the factors of $a^n - 1$, $n = 1, \ldots, 11$; $a = 2, 3, \ldots, 12$.

A. M. Legendre[11] proved that every prime divisor of $a^n + 1$ is either of the form $2nx + 1$ or divides $a^\omega + 1$ where ω is the quotient of n by an odd factor; every prime divisor of $a^n - 1$ is either of the form $nx + 1$ or divides $a^\omega - 1$ where ω is a factor of n. For n odd, the divisors must occur in $a(a^n \pm 1) = y^2 \pm a$ and are thus further limited by his tables III–XI of the linear forms of the divisors of $t^2 \pm au^2$.

C. F. Gauss[12] obtained by use of the quadratic reciprocity law the linear forms of the divisors of $x^2 - A$.

Gauss[13] gave a table of 2452 numbers of the forms $a^2 + 1$, $a^2 + 4, \ldots$, $a^2 + 81$ and their odd prime factors p, for certain a's for which the p's are all < 200.

Sophie Germain[14] noted that $p^4 + 4q^4$ has the factors $p^2 \pm 2pq + 2q^2$ [Euler[2]]. Taking $p = 1$, $q = 2^i$, we see that $2^{4i+2} + 1$ has the two factors $2^{2i+1} \pm 2^{i+1} + 1$.

F. Minding[15] gave a detailed discussion of the linear forms of the divisors of $x^2 - c$, using the reciprocity law for the case of primes. He reproduced (pp. 188–190) the discussion by Legendre.[11]

P. L. Tchebychef[16] noted that, if p is an odd prime, every odd prime factor of $a^p - 1$ is either of the form $2pz + 1$ or is a factor of $a - 1$, and moreover is a divisor of $x^2 - ay^2$. Hence, for $a = 2$, it is of the form $2pz + 1$ and also of one of the forms $8m \pm 1$. Every odd prime factor of $a^{2n+1} + 1$ is either of the form $2(2n + 1)z + 1$ or a divisor of $a + 1$ [cf. Legendre[11]].

V. A. Lebesgue[17] noted that the discussion of the linear forms of the divisors of $z^2 - D$, where D is composite, is simplified by use of Jacobi's generalization (a/b) of Legendre's symbol.

C. G. Reuschle[18] denoted $(x^{ab} - 1)/(x^a - 1)$ by $F_a(b)$. Set $a = ab + b_1$, $b = a_1b_1 + b_2$, $b_1 = a_2b_2 + b_3, \ldots$. If a, b are relatively prime,

$$\frac{1 - x^{ab}}{(x^a - 1)(x^b - 1)} = \sum_{A=0}^{b-2} x^{Aa} F_b\{a(b - 1 - A)\} + x^\beta \sum_{A=0}^{b_1 - 2} x^{Ab} F_{b_1}\{a_1(b_1 - 1 - A)\}$$

$$+ \ldots + x^{\beta + \beta_1 + \ldots + \beta_{n-2}} \sum_{A=0}^{b_{n-1} - 2} x^{Ab_{n-2}} F_{b_{n-1}}\{a_{n-1}(b_{n-1} - 1 - A)\} + x^{\beta + \ldots + \beta_{n-1}}.$$

[9a]Opera postuma, I, 1862, 161–7 (about 1773).

[10]Abhandl. d. Böhmischen Gesell. Wiss., Prag, 1, 1785, 165–170.

[11]Théorie des nombres, 1798, pp. 207–213, 313–5; ed. 2, 1808, pp. 191–7, 286–8. German transl. by Maser, p. 222.

[12]Disq. Arith., 1801, Arts. 147–150.

[13]Werke, 2, 1863, 477–495. Schering, pp. 499–502, described the table and its formation by the composition of binary forms, e. g., $(a^2 + 1)\{(a + 1)^2 + 1\} = \{a(a + 1) + 1\}^2 + 1$.

[14]Manuscript 9118 fonds français Bibl. Nat. Paris, p. 84. Cf. C. Henry, Assoc. franç. avanc. sc., 1880, 205; Oeuvres de Fermat, 4, 1912, 208.

[15]Anfangsgründe der Höheren Arith., 1832, 59–70.

[16]Theorie der Congruenzen, in Russian, 1849; in German, 1889; §49.

[17]Jour. de Math., 15, 1850, 222–7.

[18]Math. Abhandlung, Stuttgart, 1853, II, pp. 6–13.

Reuschle's[19] table A gives many factors of $a^3 \pm 1$, $a^4 \pm 1$, $a^5 \pm 1$, $a^{12} - 1$ for $a \leqq 100$, and of $a^n - 1$ for $n \leqq 42$, $a = 2, 3, 5, 6, 7, 10$.

Lebesgue[19a] proved that $x^{p-1} + \ldots + x + 1$ has no prime divisor other than the prime p and numbers of the form $kp + 1$.

Jean Plana[20] gave $3^{29} + 1 = 4 \cdot 6091q$, $3^{29} - 1 = 2 \cdot 59r$, and stated that q is a prime and that r has no factor < 52259. But Lucas[25] noted that

$$q = 523 \cdot 5385997, \qquad r = 28537 \cdot 20381027.$$

E. Kummer[21] proved that there is no prime factor, other than t and numbers $2mt \pm 1$, of the cyclotomic function

$$x^e + x^{e-1} - (e-1)x^{e-2} - (e-2)x^{e-3} + \tfrac{1}{2}(e-2)(e-3)x^{e-4} + \ldots$$

obtained from $(a^t - 1)/(a - 1)$ by setting $a + a^{-1} = x$, t being a prime $2e + 1$.

E. Catalan[22] stated that, if $n = a \mp 1$ is odd, $a^n \mp 1$ is divisible by n^2, but not by n^3. Proof by Soons, Mathesis, (3), 2, 1902, 109.

H. LeLasseur and A. Aurifeuille[23] noted that $2^{4n+2} + 1$ has the factors $2^{2n+1} \pm 2^{n+1} + 1$ [cf. Euler,[2] S. Germain[14]].

E. Lucas[24] proved that $(2^{40} + 1)/(2^8 + 1)$ is a prime and gave the factors of $30^{15} \pm 1$, $2^{41} + 1$.

Theorems by Lucas on the factors of $a^n \pm b^n$, given in various papers in 1876–8, are cited in Ch. XVII.

Lucas[25] factored $(2m)^m \pm 1$ for $m = 7, 10, 11, 12, 14, 15$, and corrected Plana.[20]

Lucas[26] gave tables due to LeLasseur and Aurifeuille of functions

$$\frac{x^n \pm y^n}{x \pm y} \quad (n \text{ odd}), \qquad \frac{x^{2m} + y^{2m}}{x^2 + y^2},$$

expressed in the form $Y^2 \pm pxyZ^2$, which is factorable if $xy = pv^2$. Factors of $x^{10} + y^{10}$ are given for various x's, y's. He gave LeLasseur's table of the proper divisors of $2^n - 1$ for all odd values of $n < 100$ except $n = 61, 67$, 71, 77, 79, 83, 85, 89, 93, 97; the proper divisors of $2^n + 1$ for n odd and < 71 (except $n = 61, 67$) and for $n = 73, 75, 81, 83, 99, 135$; the proper divisors of $2^{2k} + 1$ for $2k \leqq 74$ (except 64, 68) and for $2k = 78, 82, 84, 86, 90, 94, 102$, 126, etc. Lucas proved (pp. 790–4) that the proper divisors of $2^{4n} + 1$ are of the form $16nq + 1$, those of $a^{2abn} + b^{2abn}$ are of the form $8abnq + 1$; for n odd, those of $a^{abn} + b^{abn}$ are of the form $4abnq + 1$ if $ab = 4h + 1$, those of $a^{abn} - b^{abn}$ are of the form $4abnq + 1$ if $ab = 4h + 3$.

[19]Math. Abhandlung...Tabellen, Stuttgart, 1856. Full title in Ch. I.
[19a]Comptes Rendus Paris, 51, 1860, 11.
[20]Mem. Accad. Sc. Torino, (2), 20, 1863, 139–141.
[21]Cf. Bachmann, Kreistheilung, Leipzig, 1872.
[22]Revue de l'Instruct. publique en Belgique, 17, 1870, 137; Mélanges Math., ed. 1, p. 40.
[23]Atti R. Ac. Sc. Torino, 8, 1871; 13, 1877–8, 279. Nouv. Corresp. Math., 4, 1878, 86, 98.
      Cf. Lucas,[25] p. 238; Lucas,[26] 784.
[24]Nouv. Ann. Math., (2), 14, 1875, 523–5.
[25]Amer. Jour. Math., 1, 1878, 293.
[26]Bull. Bibl. Storia Sc. Mat. e Fis., 11, 1878, 783–798.

Lucas[27] gave the factors of 2^m+1 for $m=4n\leq60$ and for 72, 84; also for $m=4n+2\leq102$ and for 110, 114, 126, 130, 138, 150, 210.

E. Catalan[28] noted that $x^4+2(q-r)x^2+q^2$ for $x^2=(2r)^{2k+1}$ has the rational factors $(2r)^{2k+1}\pm(2r)^{k+1}+q$. The case $r=q=1$ gives LeLasseur's[23] formula. Again, $3^{6k+3}+1$ has the factors $3^{2k+1}+1$, $3^{2k+1}\pm3^{k+1}+1$.

S. Réalis[28a] deduced LeLasseur's[23] formula and $2^{4n}+2^{2n}+1=\Pi(2^{2n}\pm2^n+1)$.

J. J. Sylvester[29] considered the cyclotomic function $\psi_t(x)$ obtained by setting $a+a^{-1}=x$ in the quotient by $a^{\varphi(t)/2}$ of

(1) $$F_t(a)=\frac{(a^t-1)\Pi(a^{t/p_1p_2}-1)\cdots}{\Pi(a^{t/p_1}-1)\cdots}\qquad(t=p_1^{e_1}\cdots p_n^{e_n}),$$

where $p_1,\ldots,p_n$ are distinct primes. He stated that every divisor of $\psi_t(x)$ is of the form $kt\pm1$, with the exception that, if $t=p^j(p\mp1)/m$, p is a divisor (but not p^2). Conversely, every product of powers of primes of the form $kt\pm1$ is a divisor of $\psi_t(x)$. Proofs were given by T. Pepin, *ibid.*, 526; E. Lucas, p. 855; Dedekind, p. 1205 (by use of ideals). Lucas added that $p=2^{4h+3}-1$ and $p=2^{12h+5}-1$ are primes if and only if they divide $\psi_{p+1}(x)$ for $x=\sqrt{-1}$ and $x=3\sqrt{-1}$, respectively.

A. Lefébure[30] determined polynomials having no prime factor other than those of the form $HT+1$, where H is given. First, let $T=n^t$, where n is a prime. For A, B relatively prime integers,

$$F_n(A,\ B)=\frac{A^n-B^n}{A-B}$$

has, besides n, no prime factor except those of the form Hn^t+1, when A and B are exact n^{t-1}th powers of integers. Second, let $T=n^tm^h$, where n, m are distinct primes. The integral quotient of $F_n(u^m,\ v^m)$ by $F_n(u,\ v)$ has only prime factors of the form Hn^tm^h+1 if u, v are powers of relatively prime integers with the exponent $m^{h-1}n^{t-1}$. Similarly, if T is a product of powers of several primes.

Lefébure[31] discussed the decomposition into primes of U^R-V^R, where U, V are powers whose exponents involve factors of R.

E. Lucas[32] stated that if n and $2n+1$ are primes, then $2n+1$ is a factor of 2^n-1 or 2^n+1 according as $n\equiv3$ or $n\equiv1$ (mod 4). If n and $4n+1$ are primes, $4n+1$ is a factor of $2^{2n}+1$. If n and $8n+1=A^2+16B^2$ are primes, then $8n+1$ is a factor of $2^{2n}+1$ if B is odd, of $2^{2n}\pm1$ if B is even. Also ten theorems stating when $6n+1=4L^2+3M^2$, $12n+1=L^2+12M^2$ or $24n+1=L^2+48M^2$ are prime factors of $2^{kn}\pm1$ for certain k's.

[27]Sur la série récurrente de Fermat, Rome, 1879, 9–10. Report by Cunningham.[68]

[28]Assoc. franç. avanc. sc., 9, 1880, 228.

[28a]Nouv. Ann. Math., (2), 18, 1879, 500–9.

[29]Comptes Rendus Paris, 90, 1880, 287, 345; Coll. Math. Papers, 3, 428. Incomplete in Math. Quest. Educ. Times, 40, 1884, 21.

[30]Ann. sc. école norm. sup., (3), 1, 1884, 389–404; Comptes Rendus Paris, 98, 1884, 293, 413, 567, 613.

[31]Ann. sc. école norm. sup., (3), 2, 1885, 113.

[32]Assoc. franç. avanc. sc., 15, 1886, II, 101–2.

A. S. Bang[33] discussed $F_t(a)$ defined by (1). If p is a prime, $F_{p^k}(a)$ has only prime factors ap^k+1 if $d = a^{p^{k-1}}-1$ is prime to p, but has the factor p (and not p^2) if d is divisible by p.

Bang[34] proved that, if $a>1$, $t>2$, $F_t(a)$ has a prime factor $at+1$ except for $F_6(2)$.

L. Gianni[35] noted that if p is an odd prime dividing $a-1$ and p^r divides a^p-1, then p^{r-1} divides $a-1$.

L. Kronecker[36] noted that, if $F_n(z)$ is the function whose roots are the $\phi(n)$ primitive nth roots of unity,

$$(x-y)^{\varphi(n)}F_n\left(\frac{x+y}{x-y}\right) = G_n(x, y^2)$$

is an integral function involving only even powers of y. He investigated the prime factors q of $G_n(x, s)$ for s given. If q is prime to n and s, then q is congruent modulo n to Jacobi's symbol (s/q). The same result was stated by Bauer.[37]

J. J. Sylvester[38] called θ^m-1 the mth Fermatian function of θ.

Sylvester[39] stated that, for θ an integer $\neq 1$ or -1,

$$\theta_m \equiv \frac{\theta^m-1}{\theta-1}$$

contains at least as many distinct prime divisors as m contains divisors >1, except when $\theta = -2$, m even, and $\theta = 2$, m a multiple of 6, in which two cases the number of prime divisors may be one less than in the general case.

Sylvester[40] called the above θ_m a reduced Fermatian of index m. If $m = np^a$, n not divisible by the odd prime p, θ_m is divisible by p^a, but not by p^{a+1}, if $\theta-1$ is divisible by p. If m is odd and $\theta-1$ is divisible by each prime factor of m, then θ_m is divisible by m and the quotient is prime to m.

Sylvester[40a] stated that if $P = 1+p+\ldots+p^{r-1}$ is divisible by q, and p, r are primes, either r divides $q-1$ or $r=q$ divides $p-1$. If $P = q^j$ and p, r, j are primes, j is a divisor of $q-r$. R. W. Genese easily proved the first statement and W. S. Foster the second.

T. Pepin[41] factored various a^n-1, including $a=79$, 67, 43, $n=5$; $a=7$, $n=11$; $a=3$, $n=23$; $a=5$ or 7, $n=13$ (certain ones not in the tables by Bickmore[49]).

H. Scheffler[42] discussed the factorization of 2^r+1 by writing possible factors to the base 2, as had Beguelin.[8] He noted (p. 151) that, if $m=2^{n-1}$,

$$1+2^{(2m+1)n} = (1+2^n)^2\{1-2m+(2m-1)2^n-(2m-2)2^{2n} +\ldots-2\cdot2^{(2m-2)n}+2^{(2m-1)n}\}.$$

His formula (top p. 156), in which 2^{n-1} is a misprint for 2^{2n-1}, is equivalent to that of LeLasseur.[23]

[33]Tidsskrift for Mat., (5), 4, 1886, 70–80.   [34]Ibid., 130–137.   [35]Periodico di Mat., 2, 1887, 114.
[36]Berlin Berichte, 1888, 417; Werke, 3, I, 281–292.   [37]Jour. für Math., 131, 1906, 265–7.
[38]Nature, 37, 1888, 152.   [39]Ibid., p. 418; Coll. Papers, 4, 1912, 628.
[40]Comptes Rendus Paris, 106, 1888, 446; Coll. Papers, 4, 607.
[40a]Math. Quest. Educ. Times, 49, 1888, 54, 69.
[41]Atti Accad. Pont. Nuovi Lincei, 49, 1890, 163. Cf. Escott, Messenger Math., 33, 1903–4, 49.
[42]Beiträge zur Zahlentheorie, 1891, 147–178.

E. Lucas[43] gave algebraic factors of

$$x^6+27y^6, \qquad x^{10}-5^5y^{10}, \qquad x^{12}+6^6y^{12}.$$

K. Zsigmondy[44] proved the existence of a prime dividing $a^\gamma-b^\gamma$, but no similar binomial with a lower exponent, exceptions apart (cf. Bang,[33, 34] Birkhoff[62]).

J. W. L. Glaisher[45] gave the prime factors of $p^6-(-1)^{(p-1)/2}$ for each prime $p<100$.

T. Pepin[46] proved that $(31^7-1)/30$, $(83^5-1)/82$, $(2^{41}+1)/(3\cdot83)$ are primes.

A. A. Markoff[47] investigated the greatest prime factor of n^2+1.

W. P. Workman[48] noted the factors of $3^{6k+3}+1$ [due to Catalan[28]] and $2^{54}+1$, and stated that Lucas[43] (p. 326) gave erroneous factors of $2^{58}+1$.

C. E. Bickmore[49] gave factors of a^n-1 for $n\leqq50$, $a=2$, 3, 5, 6, 7, 10, 11, 12.

Several[49a] proved that n^n-1 is divisible by $4n+1$ if $4n+1$ is prime.

A. Cunningham[50] gave 43 primes exceeding 9 million which are factors of $(x^5\pm1)/(x\pm1)$, and factors of $3^{30}+1$, $3^{33}-1$, $3^{63}+1$, $3^{105}+1$, $5^{13}-1$, $5^{14}+1$, $5^{17}-1$, $5^{20}+1$, $5^{35}-1$.

A. Cunningham[51] considered at length the factorization of Aurifeuillians, i. e., the algebraically irreducible factors of

$$(n_1x^2)^{2n}+(2n_2y^2)^{2n}, \qquad (n_1x^2)^n+(-1)^{\frac{n+1}{2}}(n_2y^2)^n \qquad (n_1n_2=n),$$

where n_1 and x are relatively prime to n_2 and y, while n has no square factor, and is odd in the second case. Aurifeuille had found them to be expressible algebraically in the form P^2-Q^2. There are given factors of 2^n+1 for n even and $\leqq102$, and for $n=110$, 114, 126, 130, 138, 150, 210.

A. Cunningham[52] factored numbers $a^n\pm1$ by use of tables, complete to $p=101$, giving the lengths l of the periods of primes p and their powers <10000 to various bases q, so that $q^l\equiv1$ (mod p or p^k).

A. Cunningham and H. J. Woodall[53] gave factors of $N=2^x10^a\pm1$ for $x\leqq30$, $a\leqq10$, and for further sets; also, for each prime $p\leqq3001$, the least a and the least corresponding x for which p is a divisor of N. Bickmore (p. 95) gave the linear and quadratic forms of factors of N.

T. Pepin[54] factored a^7-1 for $a=37$, 41, 79; also[55] 151^5-1.

[43]Théorie des nombres, 1891, 132, exs. 2–4.
[44]Monatshefte Math. Phys., 3, 1892, 283. Details in Ch. VII, Zsigmondy.[78]
[45]Quar. Jour. Math., 26, 1893, 47.
[46]Memorie Accad. Pont. Nuovi Lincei, 9, I, 1893, 47–76.
[47]Comptes Rendus Paris, 120, 1895, 1032.   [48]Messenger Math., 24, 1895, 67.
[49]Ibid., 25, 1896, 1–44; 26, 1897, 1–38; French transl., Sphinx-Oedipe, 7, 1912, 129–44, 155–9.
[49a]Math. Quest. Educ. Times, 65, 1896, 78; (2), 8, 1905, 97.
[50]Proc. London Math. Soc., 28, 1897, 377, 379.   [51]Ibid., 29, 1898, 381–438.
[52]Messenger Math., 29, 1899–1900, 145–179. The line of $N^l=53^2$(p. 17) is incorrect.
[53]Math. Quest. Educat. Times, 73, 1900, 83–94. [Some errors.]
[54]Mem. Pont. Ac. Nuovi Lincei, 17, 1900, 321–344; errata, 18, 1901. Cf. Sphinx-Oedipe, 5, 1910, numéro spécial, 1–9. Cf. Jahrbuch Fortschritte Math., on $a=37$.
[55]Atti Accad. Pont. Nuovi Lincei, 44, 1900–1, 89.

A. Cunningham[56] factored $5^n - 1$ for $n = 75, 105$.

L. Kronecker[56a] proved that every divisor, prime to t, of (1) is $\equiv 1 \pmod{t}$.

H. S. Vandiver[56b] noted that the proof applies to the homogeneous form $F_t(a, b)$ of (1) if a, b are relatively prime.

D. Biddle[57] gave a defective proof that $3 \cdot 2^{41} + 1$ is a prime.

The Math. Quest. Educational Times contains the factorizations of:

Vol. 66 (1897), p. 97, $2^{155} - 1$ factor 31^2. Vol. 68 (1898), p. 27, p. 112, $2^{720} - 1$; p. 114, $10^{12} + 4$.

Vol. 69 (1898), p. 61, $382^4 + 1$; p. 73, $x^5 - 1$, $x = 500, 2000$; p. 117, $x^6 + y^6$; p. 118, $10^{18} + 3^3$, $3^3 . 10^{18} + 1$.

Vol. 70 (1899), p. 32, p. 69, $242^{10} + 1$; p. 47, $320^{15} - 1$; p. 64, $2^{22} + 1$, $8^{14} + 1$, $200^{18} + 1$; p. 72, $20^{14} - 1$; p. 107, $972^{15} + 1$. Vol. 71 (1899), p. 63, $x^{4n+2} - 1$; p. 72, $x^4 + y^4$.

Vol. 72 (1900), p. 61, $(3n)^{4n} - 1$ factor $24n + 1$ if prime; p. 86, $722^{10} + 1$; p. 117, $1440^{10} + 1$.

Vol. 73 (1900), p. 51, $35^{20} + 1$; p. 96, $7^{11} - 1$; p. 104, p. 114, $x^4 + y^4$.

Vol. 74 (1901), p. 27, a prime $2^q q + 1$ divides $q^k - 1$ if $k = 2^{q-1}$; p. 86, $x^{10} - 5^5 y^{10}$.

Vol. 75 (1901), p. 37, $x^4 + y^4$; p. 90, $1792^7 + 1$; p. 111, $7^{35} + 1$. [Educ. Times, (2), 54, 1901, 223, 260].

Ser. 2, Vol. 1 (1902), p. 46, $10082^6 + 1$; p. 84, $x^4 + \mu y^4$. Vol. 2 (1902), p. 33, p. 53, $N^4 + 1$; p. 118, $11^{33} + 1$.

Vol. 3 (1903), p. 49, $a^4 + b^4$ (cf. 74, 1901, 44); p. 114, $a^6 + 1$, $a = 60000$.

Vol. 6 (1904), p. 62, $96^{18} + 1$.

Vol. 7 (1905), p. 62, $208^{13} - 1$; pp. 106-7, $2^{126} + 1$.

Vol. 8 (1905), p. 50, $96^{18} + 1$; p. 64, $2^{126} + 1$.

Vol. 10 (1906), p. 36, $54^{18} + 1$, $6^{54} + 1$.

Vol. 12 (1907), p. 54, $6^{42} + 1$, $24^{30} + 1$.

Vol. 13 (1908), p. 63, 106-7, $3^{54} + 2^{54}$.

Vol. 14 (1908), p. 17, $150^{18} + 1$; p. 71, sextics; p. 96, $7^{35} + 1$.

Vol. 15 (1909), p. 57, $3^{54} + 2^{54}$; p. 33, $3^{111} + 1$, $12^{45} + 1$; p. 103, $28^{21} + 1$, $44^{11} + 1$, $6^{30} + 1$.

Vol. 16 (1909), p. 21, $19^{24} + 1$.

Vol. 18 (1910), pp. 53-5, 102-3, $x^4 + 4y^4$; pp. 69-71, $x^6 + 27y^6$; p. 93, $y^{16} - 1$.

Vol. 19 (1911), p. 103, $x^3 + y^3 = z^3 + w^3$. Vol. 23 (1913), p. 92, $(x^2 - Nx + N)^4 + N(x^2 - N)^4$.

Vol. 24 (1913), pp. 61-2, $x^{2y} \pm y^y$, $y = 5, 7, 11, 13$; pp. 71-2, $x^{12} + 2^6$, $x^{18} + 3^9$, $x^{30} + 3^{15}$.

Vol. 26 (1914), p. 23, $x^{2k} + 1$ for $k = 6n + 3 \not\equiv 3^y$; p. 39, $x^{12} + 6^6$; p. 42, $x^{10} - 5^5$, $x^{14} + 7^7$, $x^{22} + 11^{11}$, $x^{26} - 13^{13}$. Vol. 27 (1915), pp. 65-6, $45^{15} - 1$, $20^{25} - 1$, $k^{30} + 1$ for $k = 6, 8, 10$; p. 83, $x^4 + 4y^4$ (when four factors). Vol. 28 (1915), p. 72, $50^{30} + 1$. Vol. 29 (1916), p. 95, $96^{18} + 1$.

New series, vol. 1 (1916), p. 86, $x^{20} + 10^{10}$, $x^{28} + 14^{14}$; pp. 94-5, $x^{30} - 5^{15}$, $x^{30} + 15^{15}$.

Vol. 2 (1916), p. 19, $x^{30} - 5^{15}$.

Vol. 3 (1917), p. 16, $x^{15} - y^{15}$; p. 52, $x^{11} - 1$.

E. B. Escott[58] gave many cases when $1 + x^2$ is a product of two powers of primes or the double of such a product.

[56]Proc. London Math. Soc., 34, 1901, 49.
[56a]Vorlesungen über Zahlentheorie, 1, 1901, 440-1.
[56b]Amer. Math. Monthly, 10, 1903, 171.
[57]Messenger Math., 31, 1901-2, 116 (error); 33, 1903-4, 126.
[58]L'intermédiaire des math., 7, 1900, 170.

P. F. Teilhet[59] gave formulas factoring cases of $1+x^2$, as
$$(b^2+b+1)^2+1=[(b+1)^2+1](b^2+1),$$
$$4(c+1)^4+1\ \ =[(c+2)^2+(c+1)^2][(c+1)^2+c^2],$$
the last being (10, 1903, 170) a case of the known formula for the product of two sums of two squares (cf. 11, 1904, 50).

Escott[60] repeated Euler's[7] remarks on the integers x for which $1+x^2$ is divisible by a given prime. He and Teilhet (11, 1904, 10, 203) noted that any common divisor of b and $a\pm1$ divides $(a^b\pm1)/(a\pm1)$.

G. Wertheim[61] collected the theorems on the divisors of $a^m\pm1$.

G. D. Birkhoff and H. S. Vandiver[62] employed relatively prime integers $a, b\ (a>b)$ and defined a primitive divisor of $V_n=a^n-b^n$ to be one relatively prime to V_m, for all divisors m of n. They proved that, if $n\neq2$, V_n has a primitive divisor $\neq1$ except for $n=6$, $a=2$, $b=1$.

L. E. Dickson[62a] noted that $(p^4-1)(p^2-1)$ has no factor$\equiv1$ (mod p^3) if p is prime.

A. Cunningham[63] gave high primes y^2+1, $(y^2+1)/2$, y^2+y+1.

H. J. Woodall[64] gave factors of y^2+1.

J. W. L. Glaisher[65] factored $2^{2r}\pm2^r+1$ for $r\leq11$, in connection with the question of the similarity of the nth pedal triangle to a given triangle.

L. E. Dickson[66] gave a new derivation of (1), found when $F_t(a)$ is divisible by p_1 or $p_1{}^2$, where p_1 is a prime factor of t, and proved that, if a is an integer >1, $F_t(a)$ has a prime factor not dividing a^m-1 $(m<t)$ except in the cases $t=2$, $a=2^k-1$, and $t=6$, $a=2$; whence a^t-1 has a prime factor not dividing a^m-1 $(m<t)$ except in those cases [cf. Birkhoff,[62] Carmichael[75]].

Dickson[67] applied the last theorem to the theory of finite algebras and gave material on the factors of p^n-1.

A. Cunningham[68] treated at length the factorization of y^n+1 for $n=2$, 4, 8, 16, and $(y^{3n}+1)/(y^n+1)$ for $n=1, 2, 4, 8$, by means of extensive tables of solutions of the corresponding congruences modulo p. He discussed also x^n+y^n, $n=4, 6, 8, 12$.

Cunningham[68a] factored $\lambda(x^5-y^5)/(x-y)+\mu(x^6+y^6)/(x^2+y^2)$ by expressing the fractions in the form P^2-kxyQ^2, $k=5, 6$.

[59]L'intermédiaire des math., 9, 1902, 316–8.

[60]Ibid., 12, 1905, 38; cf. 11, 1904, 195–6.

[61]Anfangsgründe der Zahlenlehre, 1902, 297–303, 314.

[62]Annals of Math., 5, 1903–4, 173. Cf. Zsigmondy,[44] Dickson.[66]

[62a]Amer. Math. Monthly, 11, 1904, 197, 238; 15, 1908, 90–1.

[63]Quar. Jour. Math., 35, 1904, 10–21.

[64]Ibid., p. 95.

[65]Ibid., 36, 1905, 156.

[66]Amer. Math. Monthly, 12, 1905, 86–89.

[67]Göttingen Nachrichten, 1905, 17–23.

[68]Messenger Math., 35, 1905–6, 166–185; 36, 1907, 145–174; 38, 1908–9, 81–104, 145–175; 39, 1909, 33–63, 97–128; 40, 1910–11, 1–36. Educat. Times, 60, 1907, 544; Math. Quest. Educat. Times, (2), 13, 1908, 95–98; (2), 14, 1908, 37–8, 52–3, 73–4; (2), 15, 1909, 33–4, 103–4; (2), 17, 1910, 88, 99. Proc. London Math. Soc., 27, 1896, 98–111; (2), 9, 1910, 1–14.

[68a]Math. Quest. Educ. Times, 10, 1906, 58–9.

L. E. Dickson and E. B. Escott[69] discussed the divisibility of $p^{n/\delta}-1$ by $d(p^{n/d}-1)$, where d is a divisor of n, and δ of d.

R. D. Carmichael[70] proved that if $P^{\delta a}-R^{\delta a}$ is divisible by δa and we set $Q=(P^a-R^a)/\{a(P-R)\}$, then Q/δ is an integer if and only if a is divisible by the least integer e for which P^e-R^e is divisible by each prime factor of a not dividing $P-R$, and δ is a divisor of Q. Proof for the case $R=1$ had been given by E. B. Escott[71].

A. Cunningham[72] tabulated the factors of $y^{105} \pm 1$ for $y=2, 3, 5, 7, 12$.

K. J. Sanjana[73] considered the factors of

$$(x^{(2n+1)k} \pm 1)/(x^k \pm 1).$$

Sanjana[73a] applied his method to prove the statement of M. Kannan that
$$20^{45}-1 = 11 \cdot 19 \cdot 31 \cdot 61 \cdot 251 \cdot 421 \cdot 3001 \cdot 261451 \cdot 64008001 \cdot 3994611390415801$$
$$\cdot 4199436993616201.$$

L. E. Dickson[74] factored n^n-1 for various values of n.

R. D. Carmichael[75] employed the methods of Dickson[66] to obtain generalizations. Let $Q_n(\alpha, \beta)$ be the homogeneous form of $F_n(a)$. Let $n = \Pi p_i^{a_i}$, where the p's are distinct primes, and let c be a divisor of n and a multiple of $p_1^{a_1}$. If α, β are relatively prime, the g. c. d. of $\delta = a^{n/p_1}-\beta^{n/p_1}$ and $Q_c(\alpha, \beta)$ is 1 or p_1 and at most one $Q_c(\alpha, \beta)$ contains the factor p_1 when δ contains p_1^2; if $p_1 > 2$ divides δ, at most one $Q_c(\alpha, \beta)$ contains p_1, and no one of them contains p_1^2. If α, β are relatively prime and $c = mp_1^{a_1}$, where $m > 1$ and m is prime to p_1, then $Q_c(\alpha, \beta)$ is divisible by p_1 if and only if $\alpha^x \equiv \beta^x \pmod{p_1}$ holds for $x=m$, but not for $0<x<m$; in all other cases $Q \equiv 1 \pmod m$. If α, β are relatively prime, $Q_c(\alpha, \beta)$, and hence also $a^c-\beta^c$, has a prime factor not dividing $a^s-\beta^s(s<c)$, except in the cases (i) $c=2$, $\beta=1$, $a=2^k-1$; (ii) $Q_c(\alpha, \beta)=p=$greatest prime factor of c, and $a^{n/p} \equiv \beta^{n/p} \pmod p$; (iii) $Q_c(\alpha, \beta)=1$.

E. Miot[76] noted that LeLasseur's[23] formula is the case $m=n=1$ of

$$\left(\frac{2^{2k+1}n^2}{m}\right)^2 + m^2 = \Pi\left(m + \frac{2^{2k+1}n^2}{m} \pm 2^{k+1}n\right).$$

Welsch (p. 213) stated that the latter is no more general than the case $k=0$, which follows from the known formula for the product of two sums of two squares.

A. Cunningham[77] noted the decomposition into primes:

$$2^{77}+1 = 3 \cdot 43 \cdot 617 \cdot 683 \cdot 78233 \cdot 35532364099.$$

[69]L'intermédiaire des math., 1906, 87; 1908, 135; 18, 1911, 200. Cf. Dickson.[67]

[70]Amer. Math. Monthly, 14, 1907, 8–9.

[71]Ibid., 13, 1906, 155–6.

[72]Report British Assoc., 78, 1908, 615–6.

[73]Proc. Edinburgh Math. Soc., 26, 1908, 67–86; corrections, 28, 1909–10, viii.

[73a]Jour. Indian Math. Club, 1, 1909, 212.

[74]Messenger Math., 38, 1908, 14–32, and Dickson[118-9] of Ch. XIV.

[75]Amer. Math. Monthly, 16, 1909, 153–9.

[76]L'intermédiaire des math., 17, 1910, 102.

[77]Report British Assoc. for 1910, 529; Proc. London Math. Soc., (2), 8, 1910, xiii.

A. Cunningham[78] discussed quasi-Mersenne numbers $N_q = x^q - y^q$, with $x - y = 1$, q a prime, tabulating every prime factor < 1000 for $q < 50$, $x < 20$ if $q > 5$, $x < 50$ if $q = 5$, and treated Aurifeuillians

$$(X^q \pm Y^q)/(X \pm Y), \qquad X = \xi^2, \qquad Y = q\eta^2.$$

H. C. Pocklington[79] proved that, if n is prime, $(x^n - y^n)/(x - y)$ is divisible only by numbers of the form $mn + 1$ unless $x - y$ is divisible by n [Euler], and then is divisible only by n and numbers of the forms $mn + 1$, $n(mn + 1)$.

G. Fontené[80] stated that, if p is a prime and x, y are relatively prime, each prime factor of $(x^p - y^p)/(x - y)$ is of the form $kp + 1$, except for a factor p, occurring if $x \equiv y \pmod{p}$ and then only to the first power if $p > 2$.

G. Fontené[81] considered the homogeneous form $f_t(x, y)$ derived from (1) by setting $a = x/y$. If p^a is the highest power of a prime p dividing n,

$$f_n \equiv (f_{n/p^a})^{\varphi(p^a)}, \qquad x^n - y^n \equiv (x^{n/p^a} - y^{n/p^a})^{p^a} \pmod{p}.$$

The main theorem proved is the following: If x, y are relatively prime every prime divisor of $f_n(x, y)$ is of the form $kn + 1$, unless it is divisible by the greatest prime factor (say p) of n. It has this factor p if $p - 1$ is divisible by n/p^a and if x, y satisfy $f_{n/p^a} \equiv 0 \pmod{p}$, the latter having for each y prime to p a number of roots x equal to the degree of the congruence. In particular, if n is a power of a prime p, every prime factor of f_n is of the form $kn + 1$, with the exception of a divisor p occurring if $x \equiv y \pmod{p}$, and then to the first power if $n \neq 2$.

J. G. van der Corput[82] considered the properties of the factors of the expression derived from $a^t + b^t$ as (1) is derived from $a^t - 1$.

A. Gérardin[83] factored $a^8 + b^8$ in four numerical cases and gave

$$(a^2 + 3\beta^2)^4 + (4a\beta)^4 = \Pi\{(3a^2 \pm 2a\beta + 3\beta^2)^2 - 2(2a^2 \pm 2a\beta)^2\}.$$

A. Cunningham[84] tabulated factors of $y^4 \pm 2$, $2y^4 \pm 1$.

R. D. Carmichael[85] treated at length the numerical factors of $a^n \pm \beta^n$ and the homogeneous form $Q_n(a, \beta)$ of (1), when $a + \beta$ and $a\beta$ are relatively prime integers, while a, β may be irrational.

A. Gérardin[85a] factored $x^4 + 1$ for $x = 373, 404, 447, 508, 804, 929$; investigated $x^4 - 2$ for $x \leq 50$, $y^4 - 8$ for $y \leq 75$, $8v^4 - 1$ for $v \leq 25$, $2w^4 - 1$ for $w \leq 37$, and gave ten methods of factoring numbers $\lambda a^4 - 1$.

L. Valroff[85b] factored $2x^4 - 1$ for $101 \leq x \leq 180$, $8x^4 - 1$ for $x < 128$.

A. Gérardin[85c] expressed 622833161 (a factor of $2^{10} + 1$) as a sum of two squares in two ways to get its prime factors 2801 and 222361.

[78]Messenger Math., 41, 1911-12, 119-145.
[79]Proc. Cambr. Phil. Soc., 16, 1911, 8.
[80]Nouv. Ann. Math., (4), 9, 1909, 384; proof, (4), 10, 1910, 475; 13, 1913, 383-4.
[81]Ibid., (4), 12, 1912, 241-260.
[82]Nieuw Archief voor Wiskunde, (2), 10, 1913, 357-361.
[83]Wiskundig Tijdschrift, 10, 1913, 59.
[84]Messenger Math., 43, 1913-4, 34-57.
[85]Annals of Math., (2), 15, 1913-4, 30-70.
[85a]Sphinx-Oedipe, 1912, 188-9; 1913, 34-44; 1914, 20, 23-8, 34-7, 48.
[85b]Ibid., 1914, 5-6, 18-9, 28-30, 33, 37, 73.
[85c]Ibid., 39. Stated by E. Fauquembergue, l'intermédiaire des math., 21, 1914, 45.

A. Cunningham[86] tabulated factors of $y^y \pm 1$, $x^{zy} \pm y^{zy}$, and gave an account of printed and manuscript tables of solutions of $y^m \pm 1 \equiv 0 \pmod{p^k}$.

Cunningham[87] tabulated factors of $x^y \pm y^x$ for $x \leqq 16$ and certain y's as high as 31 when $x=2$ or 4, where x, y are relatively prime and $x > 1$, $y > 1$. Cunningham[88] noted that $x \cdot 2^x + 1$ is composite for $1 < x < 233$, $x \neq 141$. A. Cunningham and H. J. Woodall[89] tabulated factors of $2^q \pm q$ and $q \cdot 2^q \pm 1$ for $q \leqq 66$, and tabulated values of q for which one of these four functions is divisible by a given prime p or power of p. They confirmed that $x \cdot 2^x + 1$ is composite when $1 < x < 233$ except perhaps when $x = 141$. Incidentally (p. 15), the factors of $2^k \pm k - 1$ for $k \leqq 17$ are given.

For factors of $2^n - 1$ and $10^n - 1$, see Chapters I and VI. For factor tables of numbers $m \cdot 2^k \pm 1$, see Seelhoff[72] and Morehead[90] of Ch. XIII; for $m \cdot 6^k \pm 1$, Dines[91]. For factors of several numbers $a^n - 1$, see Lawrence[15], Biddle[16], and Kraitchik[21] of Ch. XIV. For the form of factors of $a^k + b^k$ when $k = 2^n$, see Euler[11] of Ch. XV. Various results in Ch. XVII relate to factors of $a^n \pm b^n$.

FACTORS OF TRINOMIALS.

Seven[95] primes p such that $(p^2 - 1)^2$ has 4 or more factors $px + 1$, $x < p$. List[96] of algebraically factorable trinomials $x^5 + xy^4 + y^5$, etc. Factors[97] of $14^8 + 14^5 + 1$, $7^8 + 2 \cdot 7^5 + 1$, etc. Conditions that $x^8 + Px^4 + c^8$ be a product of 4 rational quadratic factors.[98] Two[99] factors of $x^8 + (4m^4 + 8m^2 + 2)x^4y^4 + y^8$. Factors[100] of various trinomial expressions. For factors of $x^4 + 6bx^2 + b^2$ see Dirichlet[9] of Ch. XVII. See papers 28, 28a, 65, 89 above.

[86]Messenger Math., 45, 1915, 49–75.
[87]Ibid., 185–192.
[88]Proc. London Math. Soc., (2), 4, 1907, xviii; (2), 15, 1916–7, xxix.
[89]Messenger Math., 47, 1917, 1–38. Math. Quest. Educ. Times, (2), 10, 1906, 44.
[95]Math. Quest. Educat. Times, (2), 15, 1909, 82–3. Amer. Math. Monthly, 15, 1908, 67, 138.
  L'intermédiaire des math., 15, 1908, 121.
[96]Math. Quest. Educ. Times, (2), 16, 1909, 39–41.
[97]Ibid., 65–6.
[98]Ibid., (2), 18, 1910, 64–5; (2), 22, 1912, 20–1.
[99]Sphinx-Oedipe, 6, 1911, 8–9.
[100]Math. Quest. Educ. Times, 72, 1900, 26–8; 74, 1901, 130–1; (2), 6, 1904, 97; 19, 1911, 85;
  20, 1911, 25–6, 76–8; 22, 1912, 54–61. Math. Quest. and Solutions, 3, 1917, 66; 4, 1917,
  13, 39; 5, 1918, 38, 50–1.

CHAPTER XVII.

RECURRING SERIES; LUCAS' u_n, v_n.

Leonardo Pisano[1], or Fibonacci, employed, in 1202 (revised manuscript, 1228), the recurring series 1, 2, 3, 5, 8, 13, ... in a problem on the number of offspring of a pair of rabbits. We shall write U_n for the nth term, and u_n for the $(n+1)th$ term of 0, 1, 1, 2, 3, 5, ... derived by prefixing 0, 1 to the former series.

Albert Girard[2] noted the law $u_{n+2} = u_{n+1} + u_n$ for these series.

Robert Simson[3] noted that this series is given by the successive convergents to the continued fraction for $(\sqrt{5}+1)/2$. The square of any term is proved to differ from the product of the two adjacent terms by ± 1.

L. Euler[4] noted that $(a+\sqrt{b})^k = A_k + B_k\sqrt{b}$ implies

$$A_k = \tfrac{1}{2}\{(a+\sqrt{b})^k + (a-\sqrt{b})^k\}, \qquad B_k = \frac{1}{2\sqrt{b}}\{(a+\sqrt{b})^k - (a-\sqrt{b})^k\}.$$

J. L. Lagrange[5] noted that the residues of A_k and B_k with respect to any modulus are periodic.

Lagrange[6] proved that if the prime p divides no number of the form $t^2 - au^2$, then p divides a number of the form

$$\{(t+u\sqrt{a})^{p+1} - (t-u\sqrt{a})^{p+1}\}/\sqrt{a}.$$

A. M. Legendre[7] proved that, if $\phi^2 - A\psi^2 = 1$, then $(\phi+\psi\sqrt{A})^q - 1$ is of the form $r+s\sqrt{A}$, where r and s are divisible by a prime ω, not dividing $A\psi$, for

$$q = \omega - 1 \text{ if } \left(\frac{A}{\omega}\right) = +1, \qquad q = \omega + 1 \text{ if } \left(\frac{A}{\omega}\right) = -1.$$

C. F. Gauss[8] proved [Lagrange's[6] result] that, if b is a quadratic non-residue of the prime p, then B_{p+1} is divisible by p for every integral value of a. If e is a divisor of $p+1$, then B_e is divisible by p for $e-1$ values of a, being a factor of B_{p+1}.

G. L. Dirichlet[9] proved that, if b is an integer not a square and x is any integer prime to b, and if U, V are polynomials in x, b such that

$$(x+\sqrt{b})^n = U + V\sqrt{b},$$

then U and V have no common odd divisors. If n is an odd prime, no prime of which b is a quadratic residue is a factor of V unless it be of the form $2mn+1$. No prime of which b is a quadratic non-residue is a factor of V unless it be of the form $2mn-1$. Lagrange[6] had proved conversely that a

---

[1]Scritti, I, 1857 (Liber Abbaci), 283–4.

[2]L'Arithmétique de Simon Stevin de Bruges, par Albert Girard, Leyde, 1634, p. 677. Les Oeuvres Math. de Simon Stevin, 1634, p. 169.

[3]Phil. Trans. Roy. Soc. London, 48, I, 1753, 368–376; abridged edition, 10, 1809, 430–4.

[4]Novi Comm. Acad. Petrop., 18, 1773, 185; Comm. Arith., 1, 554.

[5]Additions to Euler's Algebra, 2, 1774, §§ 78–9, pp. 599–607. Euler, Opera Omnia, (1), 1, 619.

[6]Nouv. Mém. Ac. Berlin, année 1775 (1777), 343; Oeuvres, 3, 782–3.

[7]Théorie des nombres, 1798, p. 457; ed. 2, 1808, p. 429; ed. 3, 1830, vol. 2, Art. 443, pp. 111–2.

[8]Disq. Arith., 1801, Art. 123. [9]De formis linearibus, Breslau, 1827; Werke, 1, 51. Cf. Kronecker.[54]

prime of which b is a non-residue, and having the form $2mn-1$, will divide V. If $b=-n$, where n is a prime $4m+3$, no prime divides V unless it is of the form $kn \neq 1$, and conversely. The divisors of U are discussed for the case n a power of 2; in particular, of $U = x^4 + 6bx^2 + b^2$ when $n=4$.

J. P. M. Binet[10] noted that the number of terms of a solution v_n, expressed as a function of $r_1, r_2, \ldots$, of the equation $v_{n+2} = v_{n+1} + r_n v_n$ in finite differences is

$$\frac{1}{\sqrt{5}} \left\{ \left(\frac{1+\sqrt{5}}{2} \right)^{n+1} - \left(\frac{1-\sqrt{5}}{2} \right)^{n+1} \right\}.$$

This equals U_n as shown by taking each r_n to be unity.

G. Lamé[11] used the series of Pisano[1] to prove that the number of divisions necessary to find the g. c. d. of two integers by the usual process of division does not exceed five times the number of digits in the smaller integer. Lionnet[12] added that the number of divisions does not exceed three times it when no remainder exceeds half the corresponding divisor. See also Serret, Traité d'Arithmétique; C. J. D. Hill, Acta Univ. Lundensis, 2, 1865, No. 1; E. Lucas, Nouv. Corresp. Math., 2, 1876, 202, 214; 4, 1878, 65, and Théorie des Nombres, 1891, 335, Ex. 3; P. Bachmann, Niedere Zahlentheorie, 1902, 116–8; L. Grosschmid, Math.-Naturwiss. Blätter, 8, 1911, 125–7, for an elementary proof by induction; Math. és Phys. Lapok, 23, 1914, 5–9; R. D. Carmichael, Theory of Numbers, p. 24, Ex. 2.

H. Siebeck[13] considered the recurring series defined by

$$N_r = aN_{r-1} + cN_{r-2}, \qquad N_0 = 0, \ N_1 = 1,$$

for a, c relatively prime. By induction,

$$N_r = a^{r-1} + \binom{r-2}{1} a^{r-3}c + \binom{r-3}{2} a^{r-5}c^2 + \ldots + \frac{r}{2} a^\beta c^\gamma,$$

where $\beta = 0$ or 1, $\gamma = (r-1)/2$ or $(r-2)/2$, according as r is odd or even;

$$N_{rm} = rc^{r-1} N_{m-1}^{r-1} N_m N_1 + \binom{r}{2} c^{r-2} N_{m-1}^{r-2} N_m^2 N_2 + \ldots + N_m^r N_r,$$

whence N_{rm} is divisible by N_m. If p and q are relatively prime, N_p and N_q are relatively prime and conversely. If p is a prime, $b = a^2 + 4c$, and $s = (b/p)$ is Legendre's symbol, then

$$N_p \equiv s, \qquad N_{p-s} \equiv 0 \pmod{p},$$

so that either N_{p+1} or N_{p-1} is divisible by p.

J. Dienger[14] considered the question of the number of terms of the series of Pisano with the same number of digits and the problem to find the rank of a given term.

A. Genocchi[15] took a and b to be relatively prime integers and proved that B_{mn} is divisible by B_m and that the quotient Q has no odd divisor in

[10]Comptes Rendus Paris, 17, 1843, 563.
[11]Ibid., 19, 1844, 867–9. Cf. Binet, pp. 937–9.
[12]Complément des éléments d'arithmétique, 1857, 39–42.
[13]Jour. für Math., 33, 1846, 71–6.   [14]Archiv Math. Phys., 16, 1851, 120–4.
[15]Annali di Mat., (2), 2, 1868–9, 256–267. Cf. Genocchi[22, 51].

common with B_m other than a divisor of n. If p is an odd divisor of B_m and if h is the least k for which B_k is divisible by p, then h is a divisor of m. If p is an odd prime, B_{p-1} or B_{p+1} is divisible by p according as b is a quadratic residue or non-residue of p, whatever be the value of a. This is used to prove the existence of primes of the two forms $n^i z \pm 1$ (n a prime >2) and the existence of an infinitude of primes of each of the forms $mz \pm 1$ [Ch. XVIII].

E. Lucas[16] stated without proof theorems on the series of Pisano.[1] The sum of the first n terms equals $U_{n+2} - 2$; the sum of those terms taken with alternate signs equals $(-1)^n U_{n-1}$. Also

$$U^2_{n-1} + U^2_n = U_{2n}, \quad U_n U_{n+1} - U_{n-1} U_{n-2} = U_{2n}, \quad U^3_n + U^3_{n+1} - U^3_{n-1} = U_{3n+2}.$$

We have the symbolic formulas

$$U^{n+p} = U^{n-p}(U+1)^p, \qquad U^{n-p} = U^n (U-1)^p,$$

where, after expansion, exponents are replaced by subscripts. From E. Catalan's Manuel des Candidats à l'École Polytechnique, I, 1857, 86, he quoted

$$U_n = \frac{n+1}{2^n}\left\{1 + \frac{5}{3}\binom{n}{2} + \frac{5^2}{5}\binom{n}{4} + \ldots\right\}.$$

Lucas[17] employed the roots a, b of $x^2 = x+1$ and set

$$u_n = \frac{a^n - b^n}{a-b}, \qquad v_n = a^n + b^n = \frac{u_{2n}}{u_n} = u_{n-1} + u_{n+1}.$$

The u's form the series of Pisano with the terms 0, 1 prefixed, so that $u_0 = 0$, $u_1 = u_2 = 1$, $u_3 = 2$. Since $5u_n^2 - v_n^2 = \pm 4$, u_n and v_n have no common factor other than 2. If p is a prime $\neq 2$, 5, we have $u_p \equiv \pm 1$, $v_p \equiv 1$ (mod p). We have the symbolic formulas

$$u_{n+2kp} = u^n\{v_k u^k - (-1)^k\}^p, \qquad (-1)^{kp} u_{n-kp} = u^n\{v_k - u^k\}^p.$$

Given a law $U_{n+k} = A_0 U_{n+p} + \ldots + A_p U_n$ of recurrence, we can replace the symbol U^k by $\phi(U)$, where

$$\phi(u) = A_0 u^p + A_1 u^{p-1} + \ldots + A_{p-1} u + A_p,$$

since $U_{n+kp} = U^n\{\phi(U)\}^p$, symbolically.

E. Lucas[18] stated theorems on the series of Pisano. We have

$$2^n\sqrt{5}\,u_n = (1+\sqrt{5})^n - (1-\sqrt{5})^n, \qquad u_{n+1} = 1 + \binom{n}{1} + \binom{n-1}{2} + \ldots,$$

and his[16] symbolic formulas with u's in place of U's. u_{pq} is divisible by u_p and u_q, and by their product if p, q are relatively prime. Set $v_n = u_{2n}/u_n$. Then

$$v_{n+2} = v_{n+1} + v_n, \qquad v_{4n} = v^2_{2n} - 2, \qquad v_{4n+2} = v^2_{2n+1} + 2.$$

[16]Nouv. Corresp. Math., 2, 1876, 74-5.

[17]Ibid., 201-6.

[18]Comptes Rendus Paris, 82, 1876, 165-7.

If the term of rank $A+1$ in Pisano's series is divisible by the odd number A of the form $10p\pm3$ and if no term whose rank is a divisor of $A+1$ is divisible by A, then A is a prime. If the term of rank $A-1$ is divisible by $A \doteq 10p\pm1$ and if no term of rank a divisor of $A-1$ is divisible by A, then A is a prime. It is stated that $A=2^{127}-1$ is a prime since $A=10p-3$ and u_k is never divisible by A for $k=2^n$, except for $n=127$.

Lucas[19] employed the roots a, b of a quadratic equation $x^2-Px+Q=0$, where P, Q are relatively prime integers. Set

$$u_n=\frac{a^n-b^n}{a-b}, \qquad v_n=a^n+b^n, \qquad \delta=a-b.$$

The quotients of $\delta u_n\sqrt{-1}$ and v_n by $2Q^{n/2}$ are functions analogous to the sine and cosine. It is stated that

(1) $$u_{2n}=u_n v_n, \qquad v_n{}^2-\delta^2 u_n{}^2=4Q^n,$$

(2) $$2u_{m+n}=u_m v_n+u_n v_m, \qquad u_n{}^2-u_{n-1}u_{n+1}=Q^{n-1}.$$

Not counting divisors of Q or δ^2, we have the theorems:

(I) u_{pq} is divisible by u_p, u_q, and by their product if p, q are relatively prime.

(II) u_n, v_n are relatively prime.

(III) If d is the g. c. d. of m, n, then u_d is the g. c. d of u_m, u_n.

(IV) For n odd, u_n is a divisor of x^2-Qy^2.

By developing u_{np} and v_{np} in powers of u_n and v_n, we get formulas analogous to those for $\sin nx$ and $\cos nx$ in terms of $\sin n$ and $\cos n$, and thus get the law of apparition of primes in the recurring series of the u_n [stated explicitly in Lucas[20]], given by Fermat when δ is rational and by Lagrange when δ is irrational. The developments of $u_n{}^p$ and $v_n{}^p$ as linear functions of u_n, $u_{2n}, \ldots$ are like the formulas of de Moivre and Bernoulli for $\sin^p x$ and $\cos^p x$ in terms of $\sin kx$, $\cos kx$. Thus—

(V) If n is the rank of the first term u_n containing the prime factor p to the power λ, then u_{pn} is the first term divisible by $p^{\lambda+1}$ and not by $p^{\lambda+2}$; this is called the law of repetition of primes in the recurring series of u_n.

(VI) If p is a prime $4q+1$ or $4q+3$, the divisors of u_{pn}/u_n are divisors of x^2-py^2 or $\delta^2 x^2+py^2$, respectively.

(VII) If $u_{p\pm1}$ is divisible by p, but no term of rank a divisor of $p\pm1$ is divisible by p, then p is a prime.

Lucas[20] proved the theorems stated in the preceding paper. Theorems II and IV follow from (1_2) and (2_2), while (2_1) shows that every factor common to u_{m+n} and u_m divides u_n and conversely.

(VIII) If a and b are irrational, but real, u_{p+1} or u_{p-1} is divisible by the prime p, according as δ^2 is a quadratic non-residue or residue of p (law of apparition of primes in the u's). If a and b are integers, u_{p-1} is divisible by p. Hence the proper divisors of u_n are of the form $kn+1$ if δ is rational, $kn\pm1$ if δ is irrational.

---

[19]Comptes Rendus Paris, 82, 1876, pp. 1303–5.

[20]Sur la théorie des nombres premiers, Atti R. Accad. Sc. Torino (Math.), 11, 1875–6, 928–937.

The law V of repetition of primes follows from

$$\delta^{p-1}u_n{}^p = u_{pn} - Q^n \binom{p}{1} u_{(p-2)n} + Q^{2n} \binom{p}{2} u_{(p-4)n} - \ldots \pm Q^{tn} \binom{p}{t} u_n,$$

where $t = (p-1)/2$. Special cases of the law are due to Arndt,[33] p. 260, and Sancery,[61] each quoted in Ch. VII. Theorem VII, which follows from VIII, gives a test for the primality of $2^n \pm 1$ which rests on the success of the operation, whereas Euler's test for $2^{31} - 1$ was based on the failure of the operation. The work to prove that $2^{31} - 1$ is prime is given, and it is stated that $2^{67} - 1$ was tested and found composite,[21] contrary to Mersenne. Finally, $x^2 + Qy^2$ is shown to have an infinitude of prime divisors.

A. Genocchi[22] noted that Lucas' u_n, v_n are analogous to his[15] B_n, A_n. [If we set $a = a + \sqrt{b}$, $\beta = a - \sqrt{b}$, we have

$$u_k = \frac{a^k - \beta^k}{a - \beta} = B_k, \qquad v_k = a^k + \beta^k = 2A_k.]$$

Lucas[23] stated that, if $4m+3$ is prime, $p = 2^{4m+3} - 1$ is prime if the first term of the series 3, 7, 47,..., defined by $r_{n+1} = r_n{}^2 - 2$, which is divisible by p is of rank $4m+2$; but p is composite if no one of the first $4m+2$ terms is divisible by p. Finally, if a is the rank of the first term divisible by p, the divisors of p are of the form $2^a k \pm 1$, together with the divisors of $x^2 - 2y^2$. There are analogous tests by recurring series for the primality of

$$3 \cdot 2^{4m+3} - 1, \qquad 2 \cdot 3^{4m+2} \pm 1, \qquad 2 \cdot 3^{4m+3} - 1, \qquad 2 \cdot 5^{2m+1} + 1.$$

Lucas[24] proposed as an exercise the determination of the last digit in the general term of the series of Pisano and for the series defined by $u_{n+2} = au_{n+1} + bu_n$; also the proof of VIII: If p is a prime,

$$u_{p-1} = \frac{(a + \sqrt{b})^{p-1} - (a - \sqrt{b})^{p-1}}{\sqrt{b}}$$

is divisible by p if b is a quadratic residue of p, excepting values of a for which $a^2 - b$ is divisible by p; and the corresponding result [of Lagrange[6] and Gauss[8]] for u_{p+1}. Moret-Blanc[25] gave a proof by use of the binomial theorem and omission of multiples of p.

Lucas[26] wrote s_n for the sum of the nth powers of the roots of an equation whose coefficients are integers, the leading one being unity. Then $s_{np} - s_n{}^p$ is an integral multiple of p. Take $n = 1$. Then $s_1 = 0$ implies $s_p \equiv 0 \pmod{p}$. It is stated that if $s_1 = 0$ and if s_k is divisible by p for $k = p$, but not for $k < p$, then p is a prime.

[21] A. Cunningham, Proc. Lond. Math. Soc., 27, 1895–6, 54, remarked that, while primality is proved by Lucas' process by the success of the procedure, his verification that a number is composite is indirect and proved by the failure of the process and hence is liable to error.

[22] Atti. R. Accad. Sc. Torino, 11, 1875–6, 924.

[23] Comptes Rendus Paris, 83, 1876, 1286–8.

[24] Nouv. Ann. Math., (2), 15, 1876, 82.

[25] Ibid., (2), 20, 1881, 258 [p. 263, for primality of $2^{31} - 1$].

[26] Assoc. franç. avanc. sc., 5, 1876, 61–67. Cf. Lucas[18].

By use of (1) and (2), theorems I–IV are proved. Theorem VIII is stated, and VII is proved. Employing two diagrams and working to base 2, he showed that $2^{31}-1$ is a prime.

Lucas[27] considered a product $m=p^{\omega}r^{\rho}\ldots$ of powers of primes, no one dividing Q. Set $\Delta=(a-b)^2$, $(\Delta/p)=0$, $\pm 1\equiv\Delta^{(p-1)/2}$ (mod p),

$$\psi(m)=p^{\omega-1}r^{\rho-1}\ldots\left[p-\left(\frac{\Delta}{p}\right)\right]\left[r-\left(\frac{\Delta}{r}\right)\right]\ldots$$

Then $u_t\equiv 0$ (mod m) for $t=\psi(m)$. The ranks n of terms u_n divisible by m are multiples of a certain divisor μ of $\psi(m)$. This μ is the exponent to which a or b belongs modulo m. The case $b=1$ gives Euler's generalization of Fermat's theorem. The primality test[23] is reproduced and applied to show that $2^{19}-1$ is a prime.

Lucas[28] considered the series of Pisano. Taking a, $b=(1\pm\sqrt{5})/2$, we have $u_1=u_2=1$, $u_3=2$, etc. According as n is odd or even the divisors of u_{3n}/u_n are divisors of $5x^2-3y^2$ or $5x^2+3y^2$; those of u_{4n}/u_{2n} are divisors of $5x^2-2y^2$ or $5x^2+2y^2$; those of v_{3n}/v_n are divisors of x^2+3y^2 or x^2-3y^2; those of v_{2n} are divisors of x^2+2y^2 or x^2-2y^2; those of u_{2n}/u_n are divisors of x^2+5y^2 or x^2-5y^2. The law V of repetition of primes and theorem III are stated. The law VIII of apparition of primes now takes the following form: If p is a prime $10q\pm 1$, u_{p-1} is divisible by p; if p is a prime $10q\pm 3$, u_{p+1} is divisible by p. The test[18] for the primality of A is given and applied to show that $2^{127}-1$ and $2^{31}-1$ are primes. There is a table of prime factors of u_n for $n\leqq 60$. Finally, $4u_{pn}/u_n$ is expressible in the form x^2-py^2 or $5x^2+py^2$ according as the prime p is of the form $4q+1$ or $4q+3$.

Lucas[29] considered the series defined by $r_{n+1}=r_n^2-2$,

$$\sqrt{5}\,r_1=\left(\frac{1+\sqrt{5}}{2}\right)^A-\left(\frac{1-\sqrt{5}}{2}\right)^A,\qquad r_2=\left(\frac{1+\sqrt{5}}{2}\right)^A+\left(\frac{1-\sqrt{5}}{2}\right)^A.$$

Let $A\equiv 3$ or 9 (mod 10), $q\equiv 0$ (mod 4); or $A\equiv 7$, 9 (mod 10), $q\equiv 1$(mod 4); or $A\equiv 1$, 7 (mod 10), $q\equiv 2$ (mod 4); or $A\equiv 1$, 3 (mod 10), $q\equiv 3$ (mod 4). Then $p=2^qA-1$ is a prime if the rank of the first term divisible by p is q; if a ($a<q$) is the rank of the first term divisible by p, the divisors of p are either of the form* $2aAk+1$, or of the forms of the divisors of x^2-2y^2 and x^2-2Ay^2. Corresponding tests are given for 2^qA+1 and 3^qA-1. The first part of the theorem of Pepin[30] for testing the primality of $a_n=2^{2^n}+1$ follows from theorem VII with $a=5$, $b=1$, $p=a_n$; the second part follows from the reciprocity theorem and the form of a_n-1.

For $A=p$, let the above r_1 become r. When $p\equiv 7$ or 9 (mod 10) and p is a prime, then $2p-1$ is a prime if and only if $r\equiv 0$ (mod $2p-1$). When $p=4q+3$ is a prime, $2p+1$ is a prime if and only if $2^p\equiv 1$ (mod $2p+1$). When $p=4q+3$ is a prime, $2p-1$ is a prime if and only if

[27]Comptes Rendus Paris, 84, 1877, 439–442. Corrected by Carmichael.[89]
[28]Bull. Bibl. Storia Sc. Mat. e Fis., 10, 1877, 129–170. Reprinted as "Recherches sur plusieurs ouvrages de Léonard de Pise." Cf. von Sterneck[21] of Ch. XIX.
[29]Assoc. franç. avanc. sc., 6, 1877, 159–166. *Corrected to $2^qAK\pm 1$ in Lucas[39]; see Lucas.[42]
[30]Comptes Rendus Paris, 85, 1877, 329–331. See Ch. XV, Pepin[19], Lucas[18, 22] Proth.[23]

$$\frac{1}{\sqrt{2}}\{(1+\sqrt{2})^p - (1-\sqrt{2})^p\} \equiv 0 \pmod{2p-1}.$$

To test the primality of $p = 2^{4q+1} - 1$, use $x^2 - 4x + 1 = 0$ with the roots $2 \pm \sqrt{3}$. Then if p is a prime, u_{p+1} is divisible by p. We use the residues of the series 2, 7, 97,... defined by $r_{n+1} = 2r_n^2 - 1$.

Lucas[31] stated that $p = 2^{4m+3} - 1$ is a prime if the rank of the first term[23] of 3, 7, 47,... divisible by p is between $2m$ and $4m+2$. To test $P = 2^{4q+1} - 1$, form the series

$$r_1 = 1, \qquad r_2 = -1, \qquad r_3 = -7, \qquad r_4 = 17, \ldots, \qquad r_{n+1} = 2r_n^2 - 3^{2^{n-1}};$$

if l is the least integer for which r_l is divisible by P, then P is a prime when l is comprised between $2q$ and $4q+1$, composite when $l > 4q+1$.

Lucas[32] expressed u_n, v_n as polynomials in P and $\Delta = P^2 - 4Q = \delta^2$, obtained various relations between them corresponding to relations between sine and cosine; in particular,

$$u_{n+2} = Pu_{n+1} - Qu_n, \qquad u_{n+2r} = v_r u_{n+r} - Q^r u_n,$$

and formulas derived from them by replacing u by v; also symbolic formulas generalizing those[16] for the series of Pisano.

In the second paper, u_{n+1}, v_n are expressed as determinants of order n whose elements are Q, $-P$, 2, 1, 0. There is given a continued fraction for $u_{(n+1)r}/u_{nr}$, from which is derived (1_2) and generalizations. The same fraction is developed into a series of fractions.

Lucas[33] noted that u_{nr} is divisible by u_r since

$$\frac{u_{nr}}{u_r} = v_{(n-1)r} + Q^r v_{(n-3)r} + Q^{2r} v_{(n-5)r} + \ldots + Q^{tr} v_r,$$

where $t = \frac{1}{2}n - 1$ if n is even, $t = \frac{1}{2}(n-1)$ if n is odd, the final factor being then absent. Proof is given for (2_1) and $2v_{m+n} = v_m v_n + \Delta u_n u_m$. From these are derived new formulas by changing the sign of n and applying

$$u_{-n} = -u_n/Q^n, \qquad v_{-n} = v_n/Q^n.$$

To show that

$$[m, n] = \frac{u_{m+n} u_{m+n-1} \ldots u_{m+1}}{u_n u_{n-1} \ldots u_1}$$

is integral, apply (2_1) repeatedly to get

$$2[m, n] = [m-1, n]v_n + [m, n-1]v_m.$$

Finally, sums of squares of functions u_n, v_n are found.

Lucas[34] gave a table of the linear forms $4\Delta + r$ of the odd divisors of $x^2 + \Delta y^2$ and $x^2 - \Delta y^2$ for $\Delta = 1, \ldots, 30$. By use of (1_2), it is shown that the terms of odd rank in the series u_n are divisors of $x^2 - Qy^2$; the terms of even or odd rank in the series v_n are divisors of $x^2 + \Delta y^2$ or $x^2 + Q\Delta y^2$, respectively.

[31]Messenger Math., 7, 1877–8, 186.
[32]Sur la théorie des fonctions numériques simplement périodiques, Nouv. Corresp. Math., 3, 1877, 369–376, 401–7. These and the following five papers were reproduced by Lucas.[38]
[33]Ibid., 4, 1878, 1–8, continuation of preceding.
[34]Ibid., pp. 33–40.

Lucas[35] proved III by use of (2_1) and gave

$$u_{n+1} = \phi_n u_1 - Q\phi_{n-1}u_0,$$

$$\phi_n = P^n - \binom{n-1}{1}P^{n-2}Q + \binom{n-2}{2}P^{n-4}Q^2 - \binom{n-3}{3}P^{n-6}Q^3 + \dots;$$

$$u_{np} = \delta^{p-1}u_n{}^p + pQ^n\delta^{p-3}u_n{}^{p-2} + \frac{p(p-3)}{2}Q^{2n}\delta^{p-5}u_n{}^{p-4} + \dots$$

Lucas[36] determined the quadratic forms of divisors of v_{2n} from

$$v_{2n} = \Delta u_n{}^2 + 2Q^n, \qquad v_{2n} = v_n{}^2 - 2Q^n.$$

In the last, take $Q = 2q^2$, $n = 2\mu + 1$; thus $v_{4\mu+2}$ factors if Q is the double of a square. As a special case we have the result by H. LeLasseur (p. 86):

$$2^{2(2q+1)} + 1 = (2^{2q+1} + 2^{q+1} + 1)(2^{2q+1} - 2^{q+1} + 1).$$

In the first expression for v_{2n}, take $n = \mu + 1$, $\Delta = \pm 2h^2$, $Q = \mp g^2$; thus $v_{4\mu+2}$ factors when $Q\Delta$ is of the form $-2t^2$. Similarly, $v_{4\mu}$ factors if $\Delta = -2t^2$. Lucas[37] gave the formulas

$$\frac{u_{3n}}{u_n} = \Delta u_n{}^2 + 3Q^n, \qquad \frac{v_{3n}}{v_n} = v_n{}^2 - 3Q^n,$$

developments of $u_n{}^p$, $v_n{}^p$ as linear functions of v_{kn}, $k = p$, $p-2$, $p-4, \dots$, and complicated developments of u_{nr}, v_{nr}.

Lucas[38] reproduced the preceding series of seven papers, added (p. 228) a theorem on the expression of $4u_{pr}/u_r$ as a quadratic form, a proof (p. 231) of his[26] test for primality by use of the s_k, and results on primes and perfect numbers cited elsewhere.

Lucas[39] considered series u_n of the first kind (in which the roots a, b are relatively prime integers) and deduced Fermat's theorem and the analogue $u_t \equiv 0 \pmod{m}$, $t = \phi(m)$, of Euler's generalization. Proof is given of the earlier theorems VII, VIII and (p. 300) of his[27] generalization of the Euler-Fermat theorem. The primality test[23] is stated (p. 305) and applied to show that $2^{31} - 1$ and $2^{19} - 1$ are primes. It is stated (page 309) that $p = 2^{4q+3} - 1$ is prime if and only if

$$3 \equiv 2 \cos \pi/2^{2q+1} \pmod{p},$$

after rationalizing with respect to the radicals in the value of the cosine. The primality tests[29] are given (page 310), with similar ones for $3^q A + 1$, $2 \cdot 5^q A + 1$. The tests[29] for the primality of $2p+1$ are given (p. 314). The primality test[29] for $2^{4q+1} - 1$ is proved (pp. 315-6).

Lucas[40] reproduced his[36] earlier results, and for $p = 3, 5, 7, 11, 13, 17$, expressed v_{pr}/v_{2r} in the form $x^2 - 2pQ^r y^2$, and, for p a prime $\leqq 31$, expressed

[35]Nouv. Corresp. Math., 4, 1878, 65-71.
[36]Ibid., pp. 97-102.
[37]Ibid., pp. 129-134, 225-8.
[38]Amer. Jour. Math., 1, 1878, 184-220. Errors noted by Carmichael.[39]
[39]Ibid., pp. 289-321.
[40]Atti R. Accad. Sc. Torino, 13, 1877-8, 271-284.

u_{pr}/u_r in the form $\Delta x^2 \pm pQ^r y^2$. The prime factors of $3^{29} \pm 1$ are given on p. 280. The proper divisors of $2^{4n}+1$ are known to be of the form $8nq+1$; it is shown that q is even. Thus for $2^{32}+1$ the first divisor to be tried is 641, for $2^{2^{12}}+1$ the first one is 114689; in each case the division is exact (cf. Ch. XV). The following is a generalization: If the product of two relatively prime integers a and b is of the form $4h+1$, the proper divisors of $a^{2abn}+b^{2abn}$ are of the form $8abnq+1$. A primality test for $2^{4q+3}-1$ is given. Finally, $p = 2^{4nq+2n+1}-1$ is a prime if and only if

$$\left(2^n+\sqrt{2^{2n}+1}\right)^{\frac{p+1}{2}} + \left(2^n-\sqrt{2^{2n}+1}\right)^{\frac{p+1}{2}} \equiv 0 \pmod{p}.$$

T. Pepin[41] gave a test for the primality of $q = 2^n-1$. Let

$$u_1 \equiv \frac{2(a^2-b^2)}{a^2+b^2} \pmod{q}$$

and form the series $u_1, u_2, \ldots, u_{n-1}$ by use of

$$u_{a+1} \equiv u_a^2 - 2 \pmod{q}.$$

Then q is a prime if and only if u_{n-1} is divisible by q. This test differs from that by Lucas[23] in the choice of u_1.

E. Lucas[42] reproduced his[29] test for the primality of $2^q A - 1$, etc., and the test at the end of another paper,[40] with similar tests for $2^{4q+3}-1$ and $2^{12q+5}-1$.

G. de Longchamps[43] noted that, if $d_k = u_k - au_{k-1}$,

$$d_p = b^{p-1}, \qquad d_p d_q = b^{p+q-2},$$

with the generalization

$$\prod_{j=1}^{x} d_{p_j} = d_s, \qquad s = p_1 + \ldots + p_x - x + 1.$$

Take $p_1 = \ldots = p_x = p$. Hence

$$(u_p - au_{p-1})^x = u_{px-x+1} - au_{px-x}.$$

There is a corresponding theorem for the v's.

J. J. Sylvester[44] considered the g. c. d. of u_x, u_{x+1} if

$$u_x = (2x-1)u_{x-1} - (x-1)u_{x-2}.$$

E. Gelin[45] stated and E. Cesàro[46] proved by use of $U_{n+p} = U_p U_n + U_{p-1}U_{n-1}$ that, in the series of Pisano, the product of the means of four consecutive terms differs from the product of the extremes by ± 1; the fourth power of the middle term of five consecutive terms differs from the product of the other four terms by unity.

[41]Comptes Rendus Paris, 86, 1878, 307–310.
[42]Bull. Bibl. Storia Sc. Mat. e Fis., 11, 1878, 783–798. The further results are cited in Ch. XVI.
    Comptes Rendus, 90, 1880, 855–6, reprinted in Sphinx-Oedipe, 5, 1910, 60–1.
[43]Nouv. Corresp. Math., 4, 1878, 85; errata, p. 128.
[44]Comptes Rendus Paris, 88, 1879, 1297; Coll. Papers, 3, 252.
[45]Nouv. Corresp. Math., 6, 1880, 384.
[46]*Ibid.*, 423–4.

Magnon,[47] in reply to Lucas, proved that

$$\frac{1}{a_1}-\frac{1}{a_2}+\frac{1}{a_3}-\ldots=\frac{2}{1-\sqrt{5}},$$

if a_n-1 is the sum of the squares of the first $n-1$ terms of Pisano's series. H. Brocard[48] studied the arithmetical properties of the U's defined by $U_{n+1}=U_n+2U_{n-1}$, $U_0=1$, $U_1=3$, in connection with the nth pedal triangle. E. Cesàro[49] noted that if U_n is the nth term of Pisano's series, then $(2U+1)^n-U^{3n}=0$, symbolically. E. Lucas[50] gave his[23] test for the primality of $2^{4q+3}-1$. A. Genocchi[51] reproduced his[15] results. M. d'Ocagne[52] proved for Pisano's series that [Lucas[16]]

$$\sum_{i=0}^{p}u_i=u_{p-2}-1,\qquad \Sigma(-1)^iu_i=(-1)^pu_{p-1}-1,\qquad \lim_{p=\infty}\frac{u_p}{u_{p-i}}=\left(\frac{1+\sqrt{5}}{2}\right)^i,$$

$$u_pu_i-u_{p+1}u_{i-1}=(-1)^{i+1}u_{p-i+1},\qquad u_pu_{p-1}=u^2_p-u^2_{p-1}+(-1)^p.$$

The main problem treated is that to insert p terms $a_1,\ldots,a_p$ between two given numbers $a_0=a$, $a_{p+1}=b$, such that $a_i=a_{i-1}+a_{i-2}$. The solution is

$$a_i=\frac{bu_i+(-1)^iau_{p+1-i}}{u_{p+1}}.$$

Most of the paper is devoted to the question of the maximum number of negative terms in the series of a's.

E. Catalan[52a] proved that $U_n{}^2-U_{n-p}U_{n+p}=(-1)^{n-p+1}U^2_{p-1}$ for Pisano's series.

E. Lucas[53] stated, apropos of sums of squares, that

$$u_{2n+1}=u^2_{n+1}+u_n{}^2,\qquad v_{2n}=u^2_{n-1}+2u^2_n+u^2_{n+1},$$

$$v_{4n+2}=v^2_{2n+1}+2,\qquad v_{4n}=(2u_n)^2+u^2_n+2.$$

L. Kronecker[54] obtained Dirichlet's[9] theorems by use of modular systems. Lucas[54a] proved that, if $u_n=(a^n-b^n)/(a-b)$,

$$u^n_{p-1}-u_{(p-1)n}/u_n$$

is divisible by u_p when p is a prime and n is odd and not divisible by p, and by $u_p{}^2$ when $n=2p+1$.

L. Liebetruth[55] considered the series $P_1=1$, $P_2=x,\ldots,P_n=xP_{n-1}-P_{n-2}$, and proved any two consecutive terms are relatively prime, and

$$P_n=P_\lambda P_{n-\lambda+1}-P_{\lambda-1}P_{n-\lambda}\qquad (\lambda<n).$$

Taking $n=2\lambda$, $3\lambda,\ldots$, we see that P_λ is a common factor of $P_{2\lambda}$, $P_{3\lambda},\ldots$. The g. c. d. of P_m, P_n is P_d, where d is the g. c. d. of m, n. Next,

[47]Nouv. Corresp. Math., 6, 1880, 418–420.   [48]Nouv. Corresp. Math., 6, 1880, 145–151.
[49]Ibid., 528; Nouv. Ann. Math., (3), 2, 1883, 192; (3), 3, 1884, 533. Jornal de Sc. Math.
     Astr., 6, 1885, 17.
[50]Récréations mathématiques, 2, 1883, 230.   [51]Comptes Rendus Paris, 98, 1884, 411–3.
[52]Bull. Soc. Math. France, 14, 1885–6, 20–41.
[52a]Mém. soc. roy. sc. Liège, (2), 13, 1886, 319–21 (=Mélanges Math., II).
[53]Mathesis, 7, 1887, 207; proofs, 9, 1889, 234–5.
[54]Berlin Berichte, 1888, 417–423; Werke, 3, I, 281–292. Cf. Kronecker[36] of Ch. XVI.
[54a]Assoc. franç. avanc. sc., 1888, II, 30.   [55]Beitrag zur Zahlentheorie, Progr., Zerbst, 1888.

$$P_1+P_3+\ldots+P_{2n-1}=P_n{}^2, \qquad P_2+P_4+\ldots+P_{2n}=P_nP_{n+1}.$$

If $P_n \equiv P_m \pmod{P_\lambda}$ then $n \equiv m \pmod{2\lambda}$. Also,

$$P_n = x^{n-1} + \sum_{k=1}^{} (-1)^k \frac{(n-k-1)\ldots(n-2k)}{1\cdot 2\ldots k} x^{n-2k-1}.$$

If a_n/b_n is the nth convergent to $\frac{1}{x}-\frac{1}{x}+\frac{1}{x}-\ldots$, then $a_{n+2}=xa_{n+1}-a_n$, $b_n = a_{n+1}$. Hence $a_n = P_n$ if $a_1 = 1$, $a_2 = x$.

Sylvester stated and W. S. Foster[55a] proved that if $f(\theta)$ is a polynomial with integral coefficients and $u_{x+1}=f(u_x)$, $u_1=f(0)$, and δ is the g. c. d. of r, s, then u_δ is the g. c. d. of u_r, u_s.

A. Schönflies[56] considered the numbers $n_0=1$, $n_1,\ldots, n_q$ defined by

$$n_\lambda = n^\lambda - n^{\lambda-1} + n^{\lambda-2} - \ldots + (-1)^\lambda \qquad (\lambda = 0, 1, \ldots)$$

and proved geometrically that if n_{r-1} is the least of these numbers which has a common factor with n_q, then r is a divisor of $q+1$, while a relation

$$mn_i \equiv mn_{r+i} \pmod{n_q}$$

holds for every index i.

L. Gegenbauer[57] gave a purely arithmetical proof of this theorem.

E. Lucas[58] gave an exposition of his theory, with an introduction to recurring series.

M. Frolov[59] used a table of quadratic residues of composite numbers to factor Lucas' numbers v_n.

D. F. Seliwanov[60] proved Lucas' results on the factors of u_n, v_n.

E. Catalan[61] gave the first 43 terms of the series of Pisano, noted that U_n divides $U_{2 +1}$, that U_{2n} is a sum of two squares, and treated the series

$$u_n = au_{n-1}+u_{n-2}, \qquad u_1 = a, \qquad u_2 = a^2+1.$$

Fontès[61a] proved theorems stated by Lucas[58] (p. 127), and found in an elementary way the general term of Pisano's series, as given by Binet[11].

E. Maillet[61b] proved that a necessary condition that every positive integer, exceeding a certain limit, shall equal (up to a limited number of units) the sum of the absolute values of a finite number of terms of a recurring series, satisfying an irreducible law of recurrence with integral coefficients, is that all the roots of the corresponding generating equation be roots of unity.

W. Mantel[62] noted that, if the denominator $F(x)$ of the generating fraction of a recurring series is irreducible modulo p, a prime, the residues modulo p of the terms of the recurring series repeat periodically, and the length of a period is at most $p^n - 1$; the proof is by use of Galois' generalization of Fermat's theorem. The case of a reducible $F(x)$ is also treated.

[55a]Math. Quest. Educ. Times, 50, 1889, 54–5.   [56]Math. Annalen, 35, 1890, 537.

[57]Denkschriften Ak. Wiss. Wien (Math.), 57, 1890, 528.

[58]Théorie des nombres, 1891, 299–336; 30; 127, ex. 1. A pamphlet, published privately by Lucas in 1891, is cited in l'intermédiaire des math., 5, 1898, 58.

[59]Assoc. franç. avanc. sc., 21, 1892, 149.

[60]Math. Soc. Moscow, 16, 1892, 469–482 (in Russian).

[61]Mém. Acad. R. Belgique, 45, 1883; 52, 1893–4, 11–14.

[61a]Assoc. franç. avanc. sc., 1894, II, 217–221.   [61b]Assoc. franç. avanc. sc., 1896, II, 78–89

[62]Nieuw Archief voor Wiskunde, Amsterdam, 1, 1895, 172–184.

R. W. D. Christie[63] stated that, for the recurring series defined by $a_{n+1} = 3a_n - a_{n-1}$, $2m-1$ is a prime if and only if $a_m - 1$ is divisible by $2m-1$. The error of this test was pointed out by E. B. Escott.[64]

S. Réalis[64a] noted that two of N consecutive terms of 7, 13, 25, ..., $3(n^2+n)+7$, ... are divisible by N if N is a prime $6m+1$.

C. E. Bickmore[64b] discussed factors of u_n in the final series of Catalan[61]. He[64c] and others gave known formulas and properties of Pisano's series.

R. Perrin[65] employed $v_n = v_{n-2} + v_{n-3}$, $v_0 = 3$, $v_1 = 0$, $v_2 = 2$. Then v_n is divisible by n if n is a prime. This was verified to be not true when n is composite for a wide range of values of n. The same subject was considered by E. Malo[66] and E. B. Escott[67] who noted that Perrin's test is incomplete. Several[67a] discussed the computation of Pisano's u_n for large n's.

E. B. Escott[67b] computed $\Sigma 1/u_n$. E. Landau[67c] had evaluated $\Sigma 1/u_{2h}$ in terms of the sum of Lambert's[7] series of Ch. X, and $\Sigma 1/u_{2h+1}$ in relation to theta series.

A. Tagiuri[68] employed the series $u_1 = 1$, $u_2 = 1$, $u_3 = 2$, ... of Leonardo and the generalization U_1, U_2, ..., where $U_n = U_{n-1} + U_{n-2}$, with $U_1 = a$, $U_2 = b$ both arbitrary. Writing e for $a^2 + ab - b^2$, it is proved that

$$U_{n+s} = u_{s+1}U_n + u_s U_{n-1}, \qquad U_n^2 - U_{n-k}U_{n+k} = (-1)^{n-k}u_k^2 e,$$
$$U_n U_s - U_{n-k}U_{s+k} = (-1)^{n-k}u_k u_{k+s-n}e.$$

$\{U_{n+s} + (-1)^s U_{n-s}\}/U_n$ is an integer independent of a, b, n; it equals $u_{s+1} + u_{s-1}$. It is shown that u_r is a multiple of u_s if and only if r is a multiple of s.

Tagiuri[69] obtained analogous results for the series defined by $V_n = hV_{n-1} + lV_{n-2}$, and the particular series v_n obtained by taking $v_1 = 1$, $v_2 = h$. If h and l are relatively prime, v_r is a multiple of v_s if and only if r is a multiple of s. Let $\Phi(v_i)$ be the number of terms of the series of v's which are $\leq v_i$ and prime to it; if $h > 1$, $\Phi(v_i)$ is Euler's $\phi(i)$; but, if $h = 1$, $\Phi(v_i) = \phi(i) + \phi(i/2)$, the last term being zero if i is odd. If i and j are relatively prime, $\Phi(v_{ij}) = \Phi(v_i)\Phi(v_j)$.

Tagiuri[70] proved that, for his series of v's, the terms between v_{kp} and $v_{k(p+1)}$ are incongruent modulo v_k if $h > 1$, and for $h = 1$ except for $v_{kp+1} \equiv v_{kp+2}$. If μ is not divisible by k and ϵ is the least solution of $l^{2k\epsilon} \equiv 1 \pmod{v_k}$, then

$$v_x \equiv v_\mu \pmod{v_k} \qquad \text{if } x \equiv \mu \pmod{4k\epsilon}.$$

If μ is not divisible by k, and k is odd, and ϵ_1 is the least positive solution of $l^{k\epsilon} \equiv 1 \pmod{v_k}$, then $v_x \equiv v_\mu \pmod{v_k}$ if $x \equiv \mu \pmod{2k\epsilon_1}$.

A. Emmerich[71] proved that, in the series of Pisano,

[63]Nature, 56, 1897, 10.   [64]Math. Quest. Educat. Times, 3, 1903, 46; 4, 1903, 52
[64a]Math. Quest. Educat. Times, 66, 1897, 82–3; cf. 72, 1900, 40, 71.
[64b]Ibid., 71, 1899, 49–50.   [64c]Ibid., 111; 4, 1903, 107–8; 9, 1906, 55–7.
[65]L'intermédiaire des math., 6, 1899, 76–7.
[66]Ibid., 7, 1900, 281, 312.   [67]L'intermédiaire des math., 8, 1901, 63–64.
[67a]Ibid., 7, 1900, 172–7.   [67b]Ibid., 9, 1902, 43–4.
[67c]Bull. Soc. Math. France, 27, 1899, 198–300.   [68]Periodico di Mat., 16, 1901, 1–12.
[69]Peridico di Mat., 97–114.   [70]Ibid., 17, 1902, 77–88, 119–127.
[71]Mathesis, (3), 1, 1901, 98–9.

$u_{n+5} \equiv u_n \pmod{2}$, $u_{n+5} \equiv 3u_n \pmod{5}$, $u_{n+60} \equiv u_n \pmod{10}$,

so that u_0, u_3, u_6, u_9, ... alone are even, u_0, u_5, u_{10}, ... are multiples of 5.

J. Wasteels[72] proved that two positive integers x, y, for which $y^2 - xy - x^2$ equals $+1$ or -1, are consecutive terms of the series of Pisano. If $5x^2 \pm 4$ is a square, x is a term of the series of Pisano. These are converses of theorems by Lucas.[17]

G. Candido[73] treated u_n, v_n, by algebra and function-theory.

E. B. Escott[74] proved the last result in Lucas' paper.[40]

A. Arista[75] expressed $\Sigma_{n=1}^{n=\infty} u_n^{-1}$ in finite form.

M. Cipolla[76] gave extensive references and a collection of known formulas and theorems on u_n, v_n. His application to binomial congruences is given under that topic.

G. Candido[77] gave the necessary and sufficient conditions, involving the u_k, that a polynomial x has the factor $x^2 - Px + Q$, whose roots are a, b.

A. Laparewicz[78] treated the factoring of $2^m \pm 1$ by Lucas' method.[39]

E. B. Escott[78a] showed the connection between Pisano's series and the puzzle to convert a square into a rectangle with one more (or fewer) units of area than the square.

E. B. Escott[79] applied Lucas' theory to the case $u_n = 2u_{n-1} + u_{n-2}$.

L. E. Dickson[79a] proved that if z_k is the sum of the kth powers of the roots of $a^m + p_1 a^{m-1} + \ldots + p_m = 0$, where the p's are integers and $p_1 = 0$, then, in the series defined by $z_{x+m} + p_1 z_{x+m-1} + \ldots + p_m z_x = 0$, z_t is divisible by t if t is a prime.

E. Landau[80] proved theorems on the divisors of U_m, V_m, where
$$(x+i)^m = U_m(x) + iV_m(x), \quad i = \sqrt{-1}.$$

P. Bachmann[81] treated at length recurring series.

C. Ruggieri[82] used Pisano's series for u_{-n} to solve for ξ and η
$$a\xi^2 + b\xi\eta + c\eta^2 = k, \qquad b^2 - 4ac = 5m^2.$$

E. Zeuthen[83] proposed a problem on the series of Pisano.

H. Mathieu[84] noted that in $1, 3, 8, \ldots, x_{n+1} = 3x_n - x_{n-1}$, the expressions $x_n x_{n+1} + 1$, $x_{n-1}x_{n+1} + 1$ are squares.

Valroff[85] stated in imperfect form theorems of Lucas.

A. Aubry[86] gave a summary of results by Genocchi[15] and Lucas.

[72]Mathesis, (3), 2, 1902, 60–62.
[73]Periodico di Mat., 17, 1902, 320–5; l'intermédiaire des math., 23, 1916, 175–6.
[74]L'intermédiaire des math., 10, 1903, 288.   [76]Giornale di Mat., 42, 1904, 186–196.
[76]Rendiconto Ac. Sc. Fis. e Mat. Napoli, (3), 10, 1904, 135–150.
[77]Periodico di Mat., 20, 1905, 281–285.
[78]Wiadomosci Matematyczne, Warsaw, 11, 1907, 247–256 (Polish).
[78a]The Open Court, August, 1907. Reproduced by W. F. White, A Scrap-Book of Elementary
    Mathematics, Notes, Recreations, Essays, The Open Court Co., Chicago, 1908, 109–113.
[79]L'intermédiaire des math., 15, 1908, 248–9.   [79a]Amer. Math. Monthly, 15, 1908, 209.
[80]Handbuch...Verteilung der Primzahlen, I, 1909, 442–5.
[81]Niedere Zahlentheorie, II, 1910, 55–96, 124.   [82]Periodico di Mat., 25, 1910, 266–276.
[83]Nyt Tidsskr. for Math., Kjobenhavn, A 22, 1911, 1–9. Solution by Fransen and Damm.
[84]L'intermédiaire des math., 18, 1911, 222; 19, 1912, 87–90; 23, 1916, 14 (generalizations).
[85]Ibid., 19, 1912, 145, 212, 285.   [86]L'enseignement math., 15, 1913, 217–224

R. Niewiadomski[87] noted that, for a series of Pisano,

$$U^Z_{N\pm a} \equiv U^Z_{\pm a+1} \text{ or } -U^Z_{\pm a-1} \pmod{N},$$

according as the prime $N = 10m \pm 1$ or $10m \pm 3$. He showed how to compute rapidly distant terms of the series of Pisano and similar series, and factored numerous terms.

L. Bastien[88] employed a prime p and integer $a_1 < p$ and determined $a_2, a_3, \ldots$, each $< p$, by means of $a_1a_2 \equiv Q$, $a_2 + a_3 \equiv P$, $a_3a_4 \equiv Q$, $a_4 + a_5 \equiv P$, ... (mod p). Then

$$a_{2h+1} \equiv \frac{K_{h+1}a_1 - QK_h}{K_h a_1 - QK_{h-1}} \pmod{p}, \quad K_{h+1} = PK_h - QK_{h-1}.$$

The types of series are found and enumerated. Every divisor of K_p is of the form $\lambda p \pm 1$. Some of Lucas' results are given.

R. D. Carmichael[89] generalized many of Lucas'[38, 39] theorems and corrected several. The following is a generalization (p. 46) of Fermat's theorem: If $a + \beta$ and $a\beta$ are integers and $a\beta$ is prime to $n = p_1^{a_1} \ldots p_k^{a_k}$, where $p_1, \ldots, p_k$ are distinct primes, $u_\lambda = (a^\lambda - \beta^\lambda)/(a - \beta)$ is divisible by n when λ is the l. c. m. of

$$(3) \qquad p_i^{a_i-1}\{p_i - (a, \beta)_{p_i}\} \qquad (i = 1, \ldots, k).$$

Here, if p is an odd prime, the symbol $(a, \beta)_p$ denotes 0, $+1$ or -1, according as $(a - \beta)^2$ is divisible by p, is a quadratic residue of p, or is a quadratic non-residue of p; while $(a, \beta)_2$ denotes $+1$ if $a\beta$ is even, 0 if $a\beta$ is odd and $a + \beta$ is even, and -1 if $a\beta(a + \beta)$ is odd. In particular, if ϕ is the product of the numbers (3), $u_\phi \equiv 0 \pmod{n}$, which is the corrected form of the theorem of Lucas'[27].

Relations have been noted[90] between terms of recurring series defined by one of the equations

$$u_n + u_{n+1} = u_{n+2}, \quad u_n + u_{n+2} = u_{n+3}, \quad v_{n+1} + v_{n-1} = 4v_n, \quad v_1 = 1, v_2 = 3.$$

E. Malo[91] and Prompt[92] considered the residues with respect to a prime modulus $10m \pm 1$ of the series $u_0, u_1, u_2 = u_0 + u_1, \ldots, u_n = u_{n-1} + u_{n-2}$.

A. Boutin[93] noted relations between terms of Pisano's series.

A. Agronomof[94] treated $u_n = u_{n-1} + u_{n-2} + u_{n-3}$.

Boutin[95] and Malo[95] treated sums of terms of Pisano's series.

A. Pellet[96] generalized Lucas'[28] law of apparition of primes.

A. Gérardin[97] proved theorems on the divisors of terms of Pisano's series.

[87]L'intermédiaire des math., 20, 1913, 51, 53–6.
[88]Sphinx-Oedipe, 7, 1912, 33–38, 145–155.
[89]Annals of Math., (2), 15, 1913, 30–70.
[90]Math. Quest. Educat. Times, 23, 1913, 55; 25, 1914, 89–91.
[91]L'intermédiaire des math., 21, 1914, 86–8.
[92]Ibid., 22, 1915, 31–6.   [93]Mathesis, (4), 4, 1914, 125.
[94]Mathesis, (4), 4, 1914, 126.   [95]L'intermédiaire des math., 23, 1916, 42–3.
[96]L'intermediaire des math., 23, 1916, 64–7   [97]Nouv. Ann. Math., (4), 16, 1916, 361–7.

E. Piccioli[98] noted that in Pisano's series $1, 1, 2, 3, \ldots,$

$$u_k = \binom{k}{0} + \binom{k-1}{1} + \binom{k-2}{2} + \ldots + t, \qquad t = \binom{\frac{1}{2}(k+1)}{\frac{1}{2}(k-1)} \text{ or } \binom{\frac{1}{2}k}{\frac{1}{2}k},$$

according as k is odd or even.

T. A. Pierce[99] proved for the two functions $\Pi_{i=1}^{i=n}(1 \pm a_i{}^m)$ of the roots a_i of an equation with integral coefficients properties analogous to those of Lucas' u_n, v_n.

ALGEBRAIC THEORY OF RECURRING SERIES.

J. D. Cassini[100] and A. de Moivre[101] treated series whose general term is a sum of a given number of preceding terms each multiplied by a constant. D. Bernoulli[102] used such recurring series to solve algebraic equations. J. Stirling[103] permitted variable multipliers.

L. Euler[104] studied ordinary recurring series and their application to solving equations.

J. L. Lagrange[105] made the subject depend on the integration of linear equations in finite differences, treating also recurring series with an additive term. The general term of such a series was found by V. Riccati.[106]

P. S. Laplace[107] made systematic use of generating functions and applied recurring series to questions on probability.

J. L. Lagrange[108] noted that if $Ay_k + By_{k+1} + \ldots + Ny_{k+n} = 0$ is the recurring relation and if $A + Bt + \ldots + Nt^n = 0$ has distinct roots $a, \beta, \ldots$, the general term of the series is $y_x = aa^x + b\beta^x + \ldots$. For the case of multiple roots he stated a formula which G. F. Malfatti[109] proved to be erroneous; the latter gave a new process explained for 2, 3 or 4 equal roots.

Lagrange[110] had noticed independently his error and now gave the general term of a recurring series in the case of multiple roots by a more direct process than that of Malfatti.

Pietro Paoli[111] investigated the sum of a recurring series.

[98]Periodico di Mat., 31, 1916, 284–7.
[99]Annals of Math., (2), 18, 1916, 53–64.
[100]Histoire acad. roy. sc. Paris, année 1680, 309.
[101]Phil. Trans. London, 32, 1722, 176; Miscellanea analytica, 1730, 27, 107–8; Doctrine of chances, ed. 2, 1738, 220–9.
[102]Comm. Acad. Petrop., 3, ad annum 1728, 85–100.
[103]Methodus differentialis, London, 1730, 1764.
[104]Introductio in analysin infinitorum, 1748, I, Chs. 4, 13, 17.   Cf. C. F. Degen, Det K. Danske Vidensk. Selskabs Afhand., 1, 1824, 135; Oversigt...Forhand., 1818–9, 4.
[105]Miscellanea Taurinensia, 1, 1759, Math., 33–42; Oeuvres, I, 23–36.
[106]Mém. présentés div. sav. Paris, 5, 1768, 153–174; Comm. Bonon., 5, 1767.   Cf. M. Cantor, Geschichte Math., IV, 1908, 261.
[107]Mém. sav. étr. ac. sc. Paris, 6, année 1771, 1774, p. 353; 7, année 1773, 1776; Oeuvres, VIII, 5–24, 69–197.   Mém. ac. roy. sc. Paris, année, 1779, 1782, 207; Oeuvres, X, 1–89 (année 1777, 99).
[108]Nouv. Mém. Ac. Berlin, année 1775, 1777, 183–272; Oeuvres, IV, 151.
[109]Mém. mat. fis. soc. Ital., 3, 1786–7, 571.
[110]Nouv. Mém. Ac. Sc. Berlin, années 1792–3, 247; Oeuvres, V, 625–641 (p. 639 on the error).
[111]Mem. Acad. Mantova, 1, 1795, 121.   See Partitions in Vol. III of this History.

J. B. Fourier's[111a] error in applying recurring series to the solution of numerical equations was pointed out by R. Murphy.[111b]

P. Frisiani[111c] applied recurring series to the solution of equations.

E. Betti[111d] employed doubly recurring series to solve equations in two unknowns, by extending the method of Bernoulli.[102]

W. Scheibner[112] considered a series with a three-term recursion formula, deduced the linear relation between any three terms, not necessarily consecutive, and applied his results to continued fractions and Gauss' hypergeometric series.

D. André[113] deduced the generating equation of a recurring series V_i from that of a recurring series U_i, given a linear homogeneous relation between the terms V_i multiplied by constants and the terms U_n, $U_{n-1}, \ldots$, multiplied by polynomials in n.

D. André[114] considered a series $U_1, U_2, \ldots$, with

$$U_n = u_n + \sum_{k=1}^{\lambda_n} A_k^{(n)} U_{n-k},$$

where u_n, λ_n are given functions of n, λ_n being an integer $\leq n-1$, while $A_k^{(n)}$ is a given function of k, n. It is proved that

$$U_n = \sum_{p=1}^{n} \Psi(n, p) u_p, \qquad \Psi(n, p) = \Sigma A_{k_1}^{(n_1)} A_{k_2}^{(n_2)} \ldots,$$

where the second summation extends over all sets of integral solutions of

$$k_1 + k_2 + \ldots = n - p, \qquad n_1 = k_1 + p, \qquad n_t = k_t + n_{t-1} \qquad (0 < k_t < \lambda_{nt}).$$

Application is made to eight special types of series.

D. André[115] discussed the sums of the series whose general terms are

$$\frac{u_n x^n}{n(n+1)\ldots(n+p-1)}, \qquad \frac{u_n x^{an+\beta}}{(an+\beta)!},$$

where u_n is the general term of any recurring series.

G. de Longchamps[115a] proved the first result by Lagrange[108] and expressed y_x as a symmetric function of the distinct roots $a, \beta, \ldots$. He[115b] reduced $U_n = A_1 U_{n-1} + \ldots + A_g U_{n-g} + f(n)$, where f is a polynomial of degree p, to the case $f(n) \equiv 0$ by making a substitution $U_n = V_n + \lambda_0 n^p + \ldots + \lambda_p$.

C. A. Laisant[115c] studied the ratios of consecutive terms of recurring series, in particular for Pisano's series.

[111a]Analyse des équations, Paris, 1831.
[111b]Phil. Mag., (3), 11, 1837, 38–40.
[111c]Effemeridi Astronomiche di Milano, 1850, 3.
[111d]Annali di Sc. Mat. Fis., 8, 1857, 48–61.
[112]Berichte Gesell. Wiss. Leipzig (Math.), 16, 1864, 44–68.
[113]Bull. Soc. Math. France, 6, 1877–8, 166–170.
[114]Ann. sc. l'école norm. sup., (2), 7, 1878, 375–408; 9, 1880, 209–226.   Summary in Bull. des Sc. Math., (2), 1, I, 1877, 350–5.
[115]Comptes Rendus Paris, 86, 1878, 1017–9; 87, 1878, 973–5.
[115a]Assoc. franç., 9, 1880, 91–6.
[115b]Ibid., 1885, II, 94–100.
[115c]Bull. des Sc. Math., (2), 5, I, 1881, 218–249.

M. d'Ocagne[116] considered the recurring series U_i with

$$U_n = a_1 U_{n-1} + a_2 U_{n-2} + \ldots + a_p U_{n-p},$$

and with $U_0, \ldots, U_{p-1}$ arbitrary; and the series u with the same law, but with $u_i = 0$ $(i = 0, \ldots, p-2)$, $u_{p-1} = 1$. Then

$$U_n = U_0 u_{n+p-1} + (U_1 - a_1 U_0) u_{n+p-2} + \ldots + (U_{p-1} - a_1 U_{p-2} - \ldots - a_{p-1} U_0) u_n.$$

For each series he found the sum of any fixed number of consecutive terms and the limit of that sum.

M. d'Ocagne[117] treated $u_{p+n} = u_{p+n-1} + \ldots + u_n$. He[118] discussed the convergents to a periodic continued fraction by use of $u_n = a_1 u_{n-1} + (-1)^k u_{n-2}$, $u_0 = 0$, $u_1 = 1$.

·L. Gegenbauer[118a] found the solution P_m of $g_n P_n = 2^\lambda u_n P_{n-1} + \psi_n P_{n-2}$, where

$$P_0 = 1, \quad P_1 = 2^\lambda, \quad g_n = 2^{\kappa n a} u_{1n}, \quad \psi_n = 2^{2\lambda + \sigma + a\sigma n} u_{2n}.$$

S. Pincherle[118b] applied $p_{n+1}(x) = (x - a_n)(x - \beta_n) p_n(x)$ to developments in series.

E. Study[118c] showed how to express the general term of a recurring series as a sum of the general terms of simpler recurring series, exhibited explicitly the general term when $n = 3$, and applied the theory to bilinear forms.

M. d'Ocagne[119] considered a recurring series with the law of recurrence

$$(A_1, \ldots, A_p): \qquad Y_n + A_1 Y_{n-1} + \ldots + A_p Y_{n-p} = 0$$

of order p and generating equation

$$\Phi(x) = x^p + A_1 x^{p-1} + \ldots + A_p = 0.$$

Set

$$Q_i(x) = x^i + A_1 x^{i-1} + \ldots + A_i, \qquad \Psi(x) = Y_{p-1} + Q_1(x) Y_{p-2} + \ldots + Q_{p-1}(x) Y_0.$$

The existence of a common root a of $\Phi(x) = 0$, $\Psi(x) = 0$ is a necessary and sufficient condition that the Y's satisfy also a law of recurrence of order $p-1$, viz., $(Q_1(a), \ldots, Q_{p-1}(a))$, and then the initial law of recurrence is said to be reducible to one of order $p-1$.

M. d'Ocagne[120] considered the series with the law of recurrence

$$u^i_n = a^i_0 u^i_{n-1} + a^i_1 u^i_{n-2} + \ldots + a^i_{p_i-1} u^i_{n-p_i}$$

and generating equation

$$\phi_i(x) = x^{p_i} - a^i_0 x^{p_i-1} - \ldots - a^i_{p_i-1},$$

[116]Nouv. Ann. Math., (3), 2, 1883, 220–6; 3, 1884, 65–90; 9, 1890, 93–7; 11, 1892, 526–532 (5, 1886, 257–272). Bull. Soc. Math. France, 12, 1883–4, 78–90 (case $p=2$); 15, 1886–7, 143–4; 19, 1890–1, 37–9 (minor applications). Nieuw Archief voor Wiskunde, 17, 1890, 229–232 (applications to sin ma as function of sin a and cos a).
[117]Comptes Rendus Paris, 104, 1887, 419–420; errata, 534.
[118]Ibid., 108, 1889, 499–501.
[118a]Sitzungsber Ak. Wiss. Wien (Math.), 97, IIa, 1888, 82–89.
[118b]Atti R. Accad. Lincei, Rendiconti, 5, 1889, I, 8–12, 323–7.
[118c]Monatshefte Math. Phys., 2, 1891, 22–54.
[119]Bull. Soc. Math. France, 20, 1892, 121–2.
[120]Comptes Rendus Paris, 115, 1892, 790–2; errata, 904.

such that, for $i=0$, $u_0=\ldots=u_{p-2}=0$, $u_{p-1}=1$. If $\phi_0(x)=\phi_1(x)\cdots\phi_m(x)$,

$$u^0_{n+p-1}=\Sigma u^1_{n_1+p_1-1}\cdots u^m_{n_m+p_m-1},$$

summed for all combinations of n's for which $n_1+\ldots+n_m=n$. Application is made to the sum of a recurring series with a variable law of recurrence.

M. d'Ocagne[121] reproduced the last result, and gave a connected exposition of his earlier results and new ones.

R. Perrin[122] considered a recurring series U of order p with the terms $u_0, u_1, \ldots$. The general term of the kth derived series of U is defined to be

$$u_n^{(k)} = \begin{vmatrix} u_n & u_{n+1} & \cdots & u_{n+k} \\ u_{n+1} & u_{n+2} & \cdots & u_{n+k+1} \\ \cdots & \cdots & \cdots & \cdots \\ u_{n+k} & u_{n+k+1} & \cdots & u_{n+2k} \end{vmatrix}.$$

If any term of the $(p-1)$th derived series is zero, the law of recurrence of the given series U is reducible (to one of lower order). If also any term of the $(p-2)$th derived series is zero, continue until we get a non-vanishing determinant; then its order is the minimum order of U. This criterion is only a more convenient form of that of d'Ocagne.[119, 121]

E. Maillet[123] noted that a necessary condition that a law of recurrence of order p be reducible to one of order $p-q$ is that $\Phi(x)$ and $\Psi(x)$ of d'Ocagne[119] have q roots in common, the condition being also sufficient if $\Phi(x)=0$ has only distinct roots. He found independently a criterion analogous to that of Perrin[122] and studied series with two laws of recurrence.

J. Neuberg[124] considered $u_n=au_{n-1}+bu_{n-2}$ and found the general term of the series of Pisano.

C. A. Laisant[125] treated the case F a constant of d'Ocagne's[121] $u_k\{f(u)\}=F(k)$.

S. Lattès[126] treated $u_{n+p}=f(u_{n+p-1},\ldots,u_n)$, where f is an analytic function.

M. Amsler[127] discussed recurring series by partial fractions.

E. Netto,[127a] L. E. Dickson,[127b] A. Ranum,[128] and T. Hayashi[129] gave the general term of a recurring series. N. Traverso[130] gave the general term for $Q_n=(n-1)(Q_{n-1}+Q_{n-2})$ and $u_n=au_{n-1}+bu_{n-2}$.

Traverso[131] applied the theory of combinations with repetitions to express, as a function of p, the solution of $Q_m=p(Q_{m-1}+Q_{m-2}+\ldots+Q_{m-n})$.

[121]Jour. de l'école polyt., 64, 1894, 151–224.

[122]Comptes Rendus Paris, 119, 1894, 990–3.

[123]Mém. Acad. Sc. Toulouse, (9), 7, 1895, 179–180, 182–190; Assoc. franç., 1895, III, 233 [report with miscellaneous Dioph. equations of order n, Vol. 11]; Nouv. Ann. Math., (3), 14, 1895, 152–7, 197–206.

[124]Mathesis, (2), 6, 1896, 88–92; Archivo de mat., 1, 1896, 230.

[125]Bull. Soc. Math. France, 29, 1901, 145–9.   [126]Comptes Rendus Paris, 150, 1910, 1106–9.

[127]Nouv. Ann. Math., (4), 10, 1910, 90–5.   [127a]Monatshefte Math. Phys., 6, 1895, 285–290.

[127b]Amer. Math. Monthly, 10, 1903, 223–6.

[128]Bull. Amer. Math. Soc., 17, 1911, 457–461.

[129]Ibid., 18, 1912, 191–2.

[130]Periodico di Mat., 29, 1913–4, 101–4; 145–160.

[131]Ibid., 31, 1915–6, 1–23, 49–70, 97–120, 145–163, 193–207.

F. Nicita[132] found many relations like $2a_n{}^2 - b_n{}^2 = -(-1)^n$ between the two series $a_1 = 1$, $a_2 = 2$, .., $a_n = \frac{1}{2}(a_{n+1} - a_{n-1})$, ...; $b_1 = 1$, $b_2 = 3$, ..., $b_n = \frac{1}{2}(b_{n+1} - b_{n-1})$, ...
Reference may be made to the text by A. Vogt[133] and to texts and papers on difference equations cited in Encyklopädie der Math. Wiss., I, 2, pp. 918, 935; Encyclopédie des Sc. Math., I, 4, 47–85.

A. Weiss[134] expressed the general term t_k of a recurring series of order r linearly in terms of t_q, t_{q-1}, ..., t_{q-r+1}, where q is an integer.

W. A. Whitworth[135] proved that, if $c_0 + c_1 x + c_2 x^2 + \ldots$ is a convergent recurring series of order r whose first $2r$ terms are given, its scale of relation and sum to infinity are the quotients of certain determinants.

H. F. Scherk[136] started with any triangle ABC and on its sides constructed outwards squares $BCED$, $ACFG$, $ABJH$. Join the end points to form the hexagon $DEFGHJ$. Then construct squares on the three joining lines EF, GH, JD and again join the end points to form a new hexagon, etc. If a_i, b_i, c_i are the lengths of the joining lines in the ith set, $a_{n+1} = 5a_{n-1} - a_{n-3}$. The nth term is found as usual.

Sylvester[137] solved $u_x = u_{x-1} + (x-1)(x-2)u_{x-2}$. A. Tarn[138] treated recurring series connected with the approximations to $\sqrt{2}$, $\sqrt{3}$, $\sqrt{5}$.

V. Schlegel[139] called the development of $(1 - x - x^2 - \ldots - x^n)^{-1}$ the $(n-1)$th series of Lamé; each coefficient is the sum of the n preceding. For $n = 2$, the series is that of Pisano.

References on the connection between Pisano's series and leaf arrangement and golden section (Kepler, Braun, etc.) have been collected by R. C. Archibald.[140]

Papers by C. F. Degen,[141] A. F. Svanberg,[142] and J. A. Vész[143] were not available for report.

---

[132]Periodico di Mat., 32, 1917, 200–210, 226–36.
[133]Theorie der Zahlenreihen u. der Reihengleichung, Leipzig, 1911, 133 pp.
[134]Jour. für Math., 38, 1849, 148–157.
[135]Oxford, Cambridge and Dublin Mess. Math., 3, 1866, 117–121; Math. Quest. Educ. Times, 3, 1865, 100–1.
[136]Abh. Naturw. Vereine zu Bremen, 1, 1868, 225–236.
[137]Math. Quest. Educ. Times, 13, 1870, 50.
[138]Math. Quest. and Solutions, 1, 1916, 8–12.
[139]El Progreso Mat., 4, 1894, 171–4.
[140]Amer. Math. Monthly, 25, 1918, 232–8.
[141]Mém. Acad. Sc. St. Pétersbourg, 1821–2, 71. .
[142]Nova Acta R. Soc. Sc. Upsaliensis, 11, 1839, 1.
[143]Értekez. a Math., Magyar Tudom. Ak. (Math. Memoirs Hungarian Ac. Sc.), 3, 1875, No. 1.

CHAPTER XVIII.

THEORY OF PRIME NUMBERS.

Existence of an Infinitude of Primes.

Euclid[1] noted that, if p were the greatest prime, and $M = 2 \cdot 3 \cdot 5 \ldots p$ is the product of all the primes $\leqq p$, then $M + 1$ is not divisible by one of those primes and hence has a prime factor $> p$, thus involving a contradiction.

L. Euler[2] deduced the theorem from the [invalid] equation

$$\sum_{n=1}^{\infty} \frac{1}{n} = \Pi \left(1 - \frac{1}{p} \right)^{-1},$$

the left member being infinite and the right finite if there be only a finite number of primes. Euler[3] concluded from the same equation that "the number of primes exceeds the number of squares."

Euler[4] modified Euclid's[1] argument slightly. The number of integers $< M$ and prime to M is $\phi(M) = 2 \cdot 4 \ldots (p - 1)$, so that they include integers which are either primes $> p$ or have prime factors $> p$.

The theorem follows from Tchebychef's[261] proof of Bertrand's postulate.

L. Kronecker[5] noted that we may rectify Euler's[2] proof by using

$$\sum_{n=1}^{\infty} \frac{1}{n^s} = \Pi \left(1 - \frac{1}{p^s} \right)^{-1} \qquad (s > 1),$$

where p ranges over all primes > 1. If there were only a finite number of p's, the product would remain finite when s approaches unity, while the sum increases indefinitely. He also gave the proof a form leading to an interval from m to n within which there exists a new prime however great m is taken.

R. Jaensch[6] repeated Euler's[2] argument, also ignoring convergency.

E. Kummer[7] gave essentially Euler's[4] argument.

J. Perott[8] noted that, if $p_1, \ldots, p_n$ are the primes $\leqq N$, there are 2^n integers $\leqq N$ which are not divisible by a square, and

$$2^n > N - \sum_{k=1}^{n} \left[\frac{N}{p_k^2} \right] > N \left(1 - \sum \frac{1}{p_k^2} \right) > N \left(2 - \frac{\pi^2}{6} \right) > \frac{N}{3}.$$

Hence there exist infinitely many primes.

L. Gegenbauer[8a] proved the theorem by means of $\sum_{n=1}^{n=\infty} n^{-s}$.

---

[1]Elementa, IX, 20; Opera (ed., Heiberg), 2, 1884, 388–91.
[2]Introductio in analysin infinitorum, 1, Ch. 15, Lausanne, 1748, p. 235; French transl. by
    J. B. Labey, 1, 218.
[3]Comm. Acad. Petrop., 9, 1737, 172–4.
[4]Posthumous paper, Comm. Arith. Coll., 2, 518, Nos. 134–6; Opera Postuma, I, 1862, 18.
[5]Vorlesungen über Zahlentheorie, I, 1901, 269–273, Lectures of 1875–6.
[6]Die Schwierigeren Probl. Zahlentheorie, Progr. Rastenburg, 1876, 2.
[7]Monatsber. Ak. Wiss. Berlin für 1878, 1879, 777–8.
[8]Bull. sc. math. et astr., (2), 5, 1881, I, 183–4.
[8a]Sitzungsber. Ak. Wiss. Wien (Math.), 95, II, 1887, 94–6; 97, IIa, 1888, 374–7.

J. Perott[9] applied the theory of commutative groups to show that, if $q_1, \ldots, q_n$ are primes, there exist at least $n-1$ primes between q_n and $M = q_1 \cdots q_n$.

T. J. Stieltjes[10] expressed the product P of the primes $2, 3, \ldots, p$ as a product AB of two factors in any way. Since $A + B$ is not divisible by $2, \ldots, p$, there exists a prime $> p$.

J. Hacks[11] proved the existence of an infinitude of primes by use of his formula (Ch. XI, Hacks[14]) for the number of integers $\leq m$ not divisible by a square.

C. O. Boije af Gennäs[12] showed how to find a prime exceeding the nth prime $p_n > 2$. Take $P = 2^{\nu_1} 3^{\nu_2} \ldots p_n^{\nu_n}$, each $\nu_i \geq 1$. Express P as a product of relatively prime factors δ, P/δ, where $Q = P/\delta - \delta > 1$. Since Q is divisible by no prime $\leq p_n$, it is a product of powers of primes $q_i \geq p_n + 2$. Take δ so that $Q < (p_n + 2)^2$. Then Q is a prime.

Axel Thue[13] proved that, if $(1 + n)^k < 2^n$, there exist at least $k + 1$ primes $< 2^n$.

J. Braun[13a] noted that the sum of the inverses of the primes $\leq p$ is, for $p \geq 5$, an irreducible fraction > 1; hence the numerator contains at least one prime $> p$. He attributed to Hacks a proof by means of $\Pi(1 - 1/p^2)^{-1} = \Sigma s^{-2} = \pi^2/6$; the product would be rational if there were only a finite number of primes, whereas π is irrational.

E. Cahen[14] proved the "identity of Euler" used by Kronecker.[5] Störmer[288] gave a proof.

A. Lévy[15] took a product P of k of the first n primes $p_1, \ldots, p_n$ and the product Q of the remaining $n - k$. Then $P + Q$ is either prime or has a prime factor $> p_n$; likewise for $P - Q$. If p_n is a prime such that $p_n + 2$ is composite, there exist at least n primes $> p_n$, but $\leq 1 + p_1 p_2 \ldots p_n$. When

$$\pm \frac{1}{p_1} \pm \ldots \pm \frac{1}{p_n}$$

is reduced to a simple fraction, the numerator has no factor in common with $p_1 \ldots p_n$; hence there is a prime $> p_n$. He considered (pp. 242-4) the primes defined by $x(x-1) - 1$ for consecutive integers x.

A. Auric[16] assumed that $p_1, \ldots, p_k$ give all the primes. Then the number of integers $< n = \Pi p_i^{a_i}$ is

$$< \Pi(a_i + 1) < \left(\frac{\log n p_k}{\log p_1} \right)^k,$$

which is small in comparison with n, whence k increases indefinitely with n.

[9]Amer. Jour. Math., 11, 1888, 99-138; 13, 1891, 235-308, especially 303-5.
[10]Annales fac. sc. de Toulouse, 4, 1890, 14, final paper.
[11]Acta Math., 14, 1890-1, 335.
[12]Öfversigt K. Sv. Vetenskaps-Akad. Förhand., Stockholm, 50, 1893, 469-471.
[13]Archiv for Math. og Natur., Kristiania, 19, 1897, No. 4, 1-5.
[13a]Das Fortschreitungsgesetz der Primzahlen durch eine transcendente Gleichung exakt dargestellt, Wiss. Beilage Jahresbericht, Gymn., Trier, 1899, 96 pp.
[14]Éléments de la théorie des nombres, 1900, 319-322.
[15]Bull. de Math. Élémentaires, 15, 1909-10, 33-34, 80-82.
[16]L'intermédiaire des math., 22, 1915, 252.

G. Métrod[17] noted that the sum of the products $n-1$ at a time of the first n primes >1 is either a prime or is divisible by a prime greater than the nth. He also repeated Euler's[4] proof.

INFINITUDE OF PRIMES IN A GENERAL ARITHMETICAL PROGRESSION.

L. Euler[20] stated that an arithmetical progression with the first term unity contains an infinitude of primes.

A. M. Legendre[21] claimed a proof that there is an infinitude of primes $2mx+\mu$ if $2m$ and μ are relatively prime.

Legendre[22] noted that the theorem would follow from the following lemma: Given any two relatively prime integers A, C, and any set of k odd primes $\theta, \lambda, \ldots, \omega$ [not divisors of A], and denoting the zth odd prime by $\pi^{(z)}$, then among $\pi^{(k-1)}$ consecutive terms of the progression $A-C$, $2A-C$, $3A-C, \ldots$ there occurs at least one divisible by no one of the primes $\theta, \ldots, \omega$. Although Legendre supposed he had proved this lemma, it is false [Dupré[28]].

G. L. Dirichlet[23] gave the first proof that $mz+n$ represents infinitely many primes if m and n are relatively prime. The difficult point in the proof is the fact that

$$\sum_{n=1}^{\infty} \frac{\chi(n)}{n} \neq 0,$$

where $\chi(n)=0$ if n, k have a common factor >1, while, in the contrary case, $\chi(n)$ is a real character different from the chief character of the group of the classes of residues prime to k modulo k. This point Dirichlet proved by use of the classes of binary quadratic forms.

Dirichlet[24] extended the theorem to complex integers.

E. Heine[25] proved "without higher calculus" Dirichlet's result

$$\lim_{\rho=0} \rho\left\{\frac{1}{(b+a)^{1+\rho}}+\frac{1}{(b+2a)^{1+\rho}}+\ldots\right\}=\frac{1}{a}.$$

A. Desboves[26] discussed the error in Legendre's[22] proof.

L. Durand[27] gave a false proof.

A. Dupré[28] showed that the lemma of Legendre[22] is false and gave (p. 61) the following theorem to replace it: The mean number of terms,

---

[17]L'intermédiaire des math., 24, 1917, 39–40.
[20]Opusc. analytica, 2, 1785 (1775), 241; Comm. Arith., 2, 116–126.
[21]Mém. ac. sc. Paris, année 1785, 1788, 552.
[22]Théorie des nombres, ed. 2, 1808, p. 404; ed. 3, 1830, II, p. 76; Maser, 2, p. 77.
[23]Bericht Ak. Wiss. Berlin, 1837, 108–110; Abhand. Ak. Wiss. Berlin, Jahrgang 1837, 1839, Math., 45–71; Werke, 1, 1889, 307–12, 313–42. French transl., Jour. de Math., 4, 1839, 393–422. Jour. für Math., 19, 1839, 368–9; Werke, 1, 460–1. Zahlentheorie, §132, 1863; ed. 2, 1871; 3, 1879; 4, 1894 (p. 625, for a simplification by Dedekind).
[24]Abhand. Ak. Wiss. Berlin, Jahrgang 1841, 1843, Math., 141–161; Werke, 1, 509–532. French transl., Jour. de Math., 9, 1844, 245–269.
[25]Jour. für Math., 31, 1846, 133–5.
[26]Nouv. Ann. Math., 14, 1855, 281.
[27]Ibid., 1856, 296.
[28]Examen d'une proposition de Legendre, Paris, 1859. Comptes Rendus Paris, 48, 1859, 487.

prime to $\theta, \lambda, \ldots, \omega$, contained in $\pi^{(k-1)}$ consecutive terms of the progression is $\geq P^{-1}Q\pi^{(k-1)}-2$, where $P=3\cdot5\cdot7\cdot11\ldots$, $Q=(3-1)(5-1)\ldots$

J. J. Sylvester[29] gave a proof.

V. I. Berton[29a] found h such that between x and xh occur at least $2g$ primes each of one of the $2g$ linear forms $2py+r_i$, where $r_1, \ldots, r_{2g}$ are the integers $<2p$ and prime to $2p$.

C. Moreau[30] noted the error in Legendre's[22] proof.

L. Kronecker[5] (pp. 442–92) gave in lectures, 1886–7, the following extension* of Dirichlet's theorem (in lectures, 1875–6, for the case m a prime): If μ is any given integer, we can find a greater integer ν such that, if m, r are any two relatively prime integers, there exists at least one prime of the form $hm+r$ in the interval from μ to ν (p. 11, pp. 465–6). Moreover (pp. 478–9), there is the same mean density of primes in each of the $\phi(m)$ progressions $mh+r_i$, where the r_i are the integers $<m$ and prime to m.

I. Zignago[31] gave an elementary proof.

H. Scheffler[32] devoted 31 pages to a revision of Legendre's insufficient proof and gave a process to determine all primes under a given limit.

G. Speckmann[33] failed in an attempt to prove the theorem.

P. Bachmann[34] gave an exposition of Dirichlet's[23] proof.

Ch. de la Vallée-Poussin[35] obtained without computations, by use of the theory of functions of a complex variable, a proof of the difficult point in Dirichlet's[23] proof. He[36] proved that the sum of the logarithms of the primes $hk+l\leq x$ equals $x/\phi(k)$ asymptotically and concluded readily that the number of primes $hk+l\leq x$ equals, asymptotically,

$$\frac{1}{\phi(x)}\cdot\frac{x}{\log x}.$$

F. Mertens[37] proved the existence of an infinitude of primes in an arithmetical progression by elementary methods not using the quadratic reciprocity theorem or the number of classes of primitive binary quadratic forms. He supplemented the theorem by showing how to find a constant c such that between x and cx there lies at least one prime of the progression for every $x\geq1$ [cf. Kronecker,[5] pp. 480–96].

---

[29]Proc. London Math. Soc., 4, 1871, 7; Messenger Math., (2), 1, 1872, 143–4; Coll. Math. Papers, 2, 1908, 712–3.

[29a]Comptes Rendus Paris, 74, 1872, 1390.

[30]Nouv. Ann. Math., (2), 12, 1873, 323–4. Also, A. Piltz, Diss., Jena, 1884.

*Improvements in the exposition were made by the editor, Hensel (cf. p. 508).

[31]Annali di Mat., (2), 21, 1893, 47–55.

[32]Beleuchtung u. Beweis eines Satzes aus Legendre's Zahlentheorie [II, 1830, 76], Leipzig, 1893.

[33]Archiv Math. Phys., (2), 12, 1894, 439–441. Cf. (2), 15, 1897, 326–8.

[34]Die analytische Zahlentheorie, 1894, 51, 74–88.

[35]Mém. couronnés...acad. roy. sc. Belgique, 53, 1895–6, No. 6, 24–9.

[36]Annales de la soc. sc. de Bruxelles, 20, 1896, II, 281–361. Cf. 183–256, 361–397; 21, 1897, I, 1–13, 60–72; II, 251–368.

[37]Sitzungsber. Ak. Wiss. Wien (Math.), 106, 1897, II a, 254–286. Parts published earlier, ibid., 104, 1895, II a, 1093–1121, 1158–1166; Jour. für Math., 78, 1874, 46–62; 117, 1897, 169–184.

F. Mertens[38] gave a proof, still simpler than his[37] earlier one, of the difficult point in Dirichlet's[23] proof. The proof is very elementary, involving computations of finite sums.

F. Mertens[39] gave a simplification of Dirichlet's[24] proof of his generalization to complex primes.

H. Teege[40] proved the difficult point in Dirichlet's[23] proof.

E. Landau[41] proved that the number of prime ideals of norm $\leq x$ of an algebraic field equals the integral-logarithm $Li(x)$ asymptotically. By specialization to the fields defined by $\sqrt{-1}$ or $\sqrt{-3}$, we derive theorems[42] on the number of primes $4k \pm 1$ or $6k \pm 1 \leq x$.

L. E. Dickson[43] asked if $a_i n + b_i$ $(i = 1, \ldots, m)$ represent an infinitude of sets of m primes, noting necessary conditions.

H. Weber[44] proved Dirichlet's[24] theorem on complex primes.

E. Landau[45] simplified the proofs by de la Vallée-Poussin[35] and Mertens.[38]

E. Landau[46,47] simplified Dirichlet's[23] proof. Landau[48] proved that, if k, l are relatively prime, the number of primes $ky + l \leq x$ is

$$\frac{1}{\phi(k)} \int_2^x \frac{du}{\log u} + O(xe^t), \qquad t \equiv -\sqrt[\gamma]{\log x},$$

where γ is a constant depending on k. For O see Pfeiffer[90] of Ch. X.

A. Cunningham[49] noted that, of the N primes $\leq R$, approximately $N/\phi(n)$ occur in the progressions $nx + a$, $a < n$ and prime to n, and gave a table showing the degree of approximation when $R = 10^5$ or $5 \cdot 10^5$, with n even and < 1928. Within these limits there are fewer primes $nx + 1$ than primes $nx + a$, $a > 1$.

INFINITUDE OF PRIMES REPRESENTED BY A QUADRATIC FORM.

G. L. Dirichlet[55] gave in sketch a proof that every properly primitive quadratic form (a, b, c), $a, 2b, c$ with no common factor, represents an infinitude of primes.

Dirichlet[56] announced the extension that among the primes represented by (a, b, c), an infinitude are representable by any given linear form $Mx + N$, with M, N relatively prime, provided a, b, c, M, N are such that the linear and quadratic forms can represent the same number.

[38]Sitzungsber. Ak. Wiss. Wien (Math.), 108, 1899, II a, 32–37.

[39]Ibid., 517–556. Polish transl. in Prace mat. fiz., 11, 1900, 194–222.

[40]Mitt. Math. Gesell. Hamburg, 4, 1901, 1–11.

[41]Math. Annalen, 56, 1903, 665–670.

[42]Sitzungsber. Ak. Wiss. Wien (Math.), 112, 1903, II a, 502–6.

[43]Messenger Math., 33, 1904, 155.

[44]Jour. für Math., 129, 1905, 35–62. Cf. p. 48.

[45]Sitzungsber. Akad. Berlin, 1906, 314–320.

[46]Rend. Circ. Mat. Palermo, 26, 1908, 297.

[47]Handbuch . . . Verteilung der Primzahlen, I, 1909, 422–35.

[48]Sitzungsber. Ak. Wiss. Wien (Math.), 117, 1908, IIa, 1095–1107.

[49]Proc. London Math. Soc., (2), 10, 1911, 249–253.

[55]Bericht Ak. Wiss. Berlin, 1840, 49–52; Werke, 1, 497–502. Extract in Jour. für Math., 21, 1840, 98–100.

[56]Comptes Rendus Paris, 10, 1840, 285–8; Jour. de Math., 5, 1840, 72–4; Werke, 1, 619–623.

H. Weber[57] and E. Schcring[58] completed Dirichlet's[55] proof of his first theorem. A. Meyer[59] completed Dirichlet's[56] proof of his extended theorem. F. Mertens[60] gave an elementary proof of Dirichlet's[56] extended theorem.

Ch. de la Vallée-Poussin[36] proved that the number of primes $\leq x$ representable by a properly primitive definite positive or indefinite[61] irreducible binary quadratic form is asymptotic to $gx/\log x$, where g is a constant; and the same for primes belonging also to a linear form compatible with the character of the quadratic form.

L. Kronecker[5] (pp. 494-5) stated a theorem on factorable forms in several variables which represent an infinitude of primes.

ELEMENTARY PROOFS OF THE EXISTENCE OF AN INFINITUDE OF PRIMES $mz+1$, FOR ANY GIVEN m.

V. A. Lebesgue[65] gave a proof for the case m a prime, using the fact that $x^{m-1} - x^{m-2}y + \ldots + y^{m-1}$ has besides the possible factor m only prime factors $2km+1$. A like method applies[65a] to $2mz-1$.

J. A. Serret[66] gave an incomplete proof for any m.

F. Landry[67] gave a proof like Lebesgue's.[65] If θ is the largest prime $2km+1$ and if x is the product of all of them, x^m+1 is divisible by no one of them. Since $(x^m+1)/(x+1)$ has no prime divisor not of the form $2km+1$, there exists at least one $>\theta$.

A. Genocchi[68] proved the existence of an infinitude of primes $mz \pm 1$ and $n^iz \pm 1$ for n a prime by use of the rational and irrational parts of $(a+\sqrt{b})^k$.

L. Kronecker[5] (pp. 440-2) gave in lectures, 1875-6, a proof for the case m a prime; the simple extension in the text to any m was added by Hensel.

E. Lucas gave a proof by use of his u_n (Lucas,[39] p. 291, of Ch. XVII).

A. Lefébure[30] of Ch. XVI stated that the theorem follows from his results.

L. Kraus[69] gave a proof.

A. S. Bang[70] and Sylvester[33] proved it by use of cyclotomic functions.

K. Zsigmondy[78] of Ch. VII gave a proof. Also, E. Wendt,[71] and Birkhoff and Vandiver[62] of Ch. XVI.

---

[57]Math. Annalen, 20, 1882, 301–329. Elliptische Functionen (=Algebra, III), ed. 2, 1908, 613–6.
[58]Werke, 2, 1909, 357–365, 431–2.
[59]Jour. für Math., 103, 1888, 98–117. Exposition by Bachmann,[34] pp. 272–307.
[60]Sitzungsber. Ak. Wiss. Wien (Math.), 104, 1895, IIa, 1093–1153, 1158. Simplification, ibid., 109, 1900, IIa, 415–480.
[61]Cf. E. Landau, Jahresber. D. Math. Verein., 24, 1915, 250–278.
[65]Jour. de Math., 8, 1843, 51, note. Exercices d'analyse numérique, 1859, 91.
[65a]Jour. de Math., (2), 7, 1862, 417.
[66]Jour. de Math., 17, 1852, 186–9.
[67]Deuxième mémoire sur la théorie des nombres, Paris, 1853, 3.
[68]Annali di mat., (2), 2, 1868–9, 256–7. Cf..Genocchi[22, 51] of Ch. XVII.
[69]Casopis Math. a Fys., 15, 1886, 61–2. Cf. Fortschritte, 1886, 134–5.
[70]Tidsskrift for Math., (5), 4, 1886, 70–80, 130–7. See Bang[33, 34], Ch. XVI.
[71]Jour. für Math., 115, 1895, 85.

N. V. Bervi[72] proved that the ratio of the number of integers $cm+1$ not $>n$ and not a product of two integers of that form to the number of all primes not $>n$ has the limit unity for $n=\infty$.

H. C. Pocklington[73] proved that, if n is any integer, there is an infinitude of primes $mn+1$, an infinitude not of this form if $n>2$, and an infinitude not of the forms $mn\pm1$ if $n=5$ or $n>6$.

E. Cahen[74] proved the theorem for m an odd prime.

J. G. van der Corput[75] proved the theorem.

ELEMENTARY PROOFS OF THE EXISTENCE OF AN INFINITUDE OF PRIMES IN
SPECIAL ARITHMETICAL PROGRESSIONS.

J. A. Serret[66] for the common difference 8 or 12, and for $10x+9$.

V. A. Lebesgue[80] for $4n\pm1$, $8n+k$ $(k=1, 3, 5, 7)$. Lebesgue[81] for the same and $2^m n+1$, $6n-1$. Also, by use of infinite series, for the common difference 8 or 12.

E. Lucas[82] for $5n+2$, $8n+7$.

J. J. Sylvester[83] for the difference 8 or 12 and[84] for $p^k x-1$, p a prime.

A. S. Bang[85] for the differences 4, 6, 8, 10, 12, 14, 18, 20, 24, 30, 42, 60.

E. Lucas[86] for $4n\pm1$, $6n-1$, $8n+5$.

R. D. von Sterneck[87] for $an-1$.

K. Th. Vahlen[88] for $mz+1$ by use of Gauss' periods of roots of unity. Also, if m is any integer and p a prime such that $p-1$ is divisible by a higher power of 2 than $\phi(m)$ is, while k is a root of $km+1\equiv -1 \pmod{p}$, the linear form $mpx+km+1$ represents an infinitude of primes; known special cases are $mx+1$ and $2px-1$.

J. J. Iwanow[89] for the difference 8 or 12.

E. Cahen[14] (pp. 318–9) for $4x\pm1$, $6x\pm1$, $8x+5$. K. Hensel[5] (pp. 438–9, 508) for the same forms. M. Bauer[90] for $an-1$.

E. Landau[47] (pp. 436–46) for $kn\pm1$.

I. Schur[91] proved that if $l^2\equiv 1 \pmod{k}$ and if one knows a prime $>\phi(k)/2$ of the form $kz+l$, there exists an infinitude of primes $kz+l$; for example,

$$2^n z+2^{n-1}\pm1, \quad 8mz+2m+1, \quad 8mz+4m+1, \quad 8mz+6m+1,$$

where m is any odd number not divisible by a square.

K. Hensel[92] for $4n\pm1$, $6n\pm1$, $8n-1$, $8n\pm3$, $12n-1$, $10n-1$.

[72]Mat. Sbornik (Math. Soc. Moscow), 18, 1896, 519.
[73]Proc. Cambr. Phil. Soc., 16, 1911, 9–10.   [74]Nouv. Ann. Math., (4), 11, 1911, 70–2.
[75]Nieuw Archief voor Wiskunde, (2), 10, 1913, 357–361 (Dutch).
[80]Nouv. Ann. Math., 15, 1856, 130, 236.
[81]Exercices d'analyse numérique, 1859, 91–5, 103–4, 145–6.
[82]Amer. Jour. Math., 1, 1878, 309.   [83]Comptes Rendus Paris, 106, 1888, 1278–81, 1385–6.
[84]Assoc. franç. av. sc., 17, 1888, II, 118–120.
[85]Nyt Tidsskrift for Math., Kjobenhavn, 1891, 2B, 73–82.
[86]Théorie des nombres, 1891, 353–4.   [87]Monatshefte Math. Phys., 7, 1896, 46.
[88]Schriften phys.-ökon. Gesell. Königsberg, 38, 1897, 47.
[89]Math. Soc. St. Petersburg, 1899, 53–8 (Russian).
[90]Jour. für Math., 131, 1906, 265–7; transl. of Math. és Phys. Lapok, 14, 1905, 313.
[91]Sitzungsber. Berlin Math. Gesell., 11, 1912, 40–50, with Archiv M. P.
[92]Zahlentheorie, 1913, 304–5.

R. D. Carmichael[93] for $p^k n - 1$ (p an odd prime) and $2^k \cdot 3n - 1$.

M. Bauer's[94] paper was not available for report.

POLYNOMIALS REPRESENTING NUMEROUS PRIMES.

Chr. Goldbach[100] noted that a polynomial $f(x)$ cannot represent primes exclusively, since the constant term would be unity, whereas it is $f(p)$ in $f(x+p)$.

L. Euler[101] proved this by noting that, if $f(a) = A$, $f(nA + a)$ is divisible by A.

Euler[102] noted that $x^2 - x + 41$ is a prime for $x = 1, \ldots, 40$.

Euler[103] noted that $x^2 + x + 17$ is a prime for $x = 0, 1, \ldots, 15$ and [error] 16; $x^2 + x + 41$ is a prime for $x = 0, 1, \ldots, 15$.

A. M. Legendre[104] noted that $x^2 + x + 41$ is a prime for $x = 0, 1, \ldots, 39$, that $2x^2 + 29$ is a prime for $x = 0, 1, \ldots, 28$, and gave a method of finding such functions. [Replacing x by $x + 1$ in Euler's[102] function, we get $x^2 + x + 41$.] If $\beta^2 + 2(a + \beta)x - 13x^2$ is a square only when $x = 0$, and a and β are relatively prime, then $a^2 + 2a\beta + 14\beta^2$ is a prime or double a prime. He gave many such results.

Chabert[104a] stated that $3n^2 + 3n + 1$ represents many primes for n small.

G. Oltramare[105] noted that $x^2 + ax + b$ has no prime divisor $\leq \mu$ and hence is a prime when $< \mu^2$, if $a^2 - 4b$ is a quadratic non-residue of each of the primes $2, 3, \ldots, \mu$. The function $x^2 + ax + (a^2 + 163)/4$ is suitable to represent a series of primes. Taking $x = 0$, $a = u/v$, he stated that $u^2 + 163v^2$ or its quotient by 4 gives more than 100 primes between 40 and 1763.

H. LeLasseur[106] verified that, for a prime A between 41 and 54000, $x^2 + x + A$ does not represent primes exclusively for $x = 0, 1, \ldots, A - 2$.

E. B. Escott[107] noted that $x^2 + x + 41$ gives primes not only for $x = 0, 1, \ldots, 39$, but also[108] for $x = -1, -2, \ldots, -40$. Hence, replacing x by $x - 40$, we get $x^2 - 79x + 1601$, a prime for $x = 0, 1, \ldots, 79$. Several such functions are given.

Escott[109] examined values of A much exceeding 54000 in $x^2 + x + A$ without finding a suitable $A > 41$. Legendre's[104] first seven formulas for primes give composite numbers for $a = 2$, the eighth for $a = 3$, etc. Escott found that $x^3 + x^2 + 17$ is a prime for $x = -14, -13, \ldots, +10$. In $x^3 - x^2 - 17$ replace x by $x - 10$; we get a cubic which is a prime for $x = 0, 1, \ldots, 24$.

[93]Annals of Math., (2), 15, 1913, 63–5.   [94]Archiv Math. Phys., (3), 25, 1916, 131–4.
[100]Corresp. Math. Phys. (ed., Fuss), I, 1843, 595, letter to Euler, Nov. 18, 1752.
[101]Novi Comm. Acad. Petrop., 9, 1762–3, 99; Comm. Arith., 1, 357.
[102]Mém. de Berlin, année 1772, 36; Comm. Arith., 1, 584.
[103]Opera postuma, I, 1862, 185. In Pascal's Repertorium Höheren Math., German transl. by
Schepp, 1900, I, 518, it is stated incorrectly to be a prime for the first 17 values of x; likewise by Legendre, Théorie des nombres, 1798, 10; 1808, 11.
[104]Théorie des nombres, 1798, 10, 304–312; ed. 2, 1808, 11, 279–285; ed. 3, 1830, I, 248–255;
German transl. by Maser, I, 322–9.   [104a]Nouv. Ann. Math., 3, 1844, 250.
[105]Mém. l'Inst. Nat. Genevois, 5, 1857, No. 2, 7 pp.
[106]Nouv. Corresp. Math., 5, 1879, 371; quoted in l'intermédiaire des math., 5, 1898, 114–5.
[107]L'intermédiaire des math., 6, 1899, 10–11.
[108]The same 40 primes as for $x = 0, \ldots, 39$, as noted by G. Lemaire, ibid., 16, 1909, p. 197.
[109]Ibid., 17, 1910, 271.

E. Miot[110] stated that $x^2-2999x+2248541$ is a prime for $1460 \leqq x \leqq 1539$.

G. Frobenius[111] proved that the value of $x^2+xy+py^2$ is a prime if $<p^2$, that of $2x^2+py^2$ (y odd) if $<p(2p+1)$, that of x^2+2py^2 (x odd) if $<p(p+2)$, and noted cases in which an indefinite form $x^2+xy-qy^2$ is a prime.

Lévy[15] examined x^2-x-1. He[112] considered $f(x)=ax^2+abx+c$, where a, b, c are integers, $0 \leqq a < 4$. Giving to x the values 0, 1, 2, ..., we get a set of integers such that, for every n exceeding a certain value, $f(n)$ is either prime or admits a prime factor which divides a number $f(p)$, where $p < n$. For example, if for $f(x)=x^2-x+41$ we grant that $f(0)$, $f(1)$, $f(2)$, $f(3)$ and $f(4)$ are primes, we can conclude that $f(x)$ is prime for $\dot{x} \leqq 40$. Likewise when 41 is replaced by 11 or 17. Again, $2x^2-2x+19$ and $3x^2-3x+23$ give successions of 18 and 22 primes respectively. Bouniakowsky[35] of Ch. XI considered polynomials which represent an infinitude of primes.

Braun[13a] proved that there exists no quotient of two polynomials such that the greatest integer contained in its numerical value is a prime for all integral values $>k$ of the variable.

GOLDBACH'S EMPIRICAL THEOREM: EVERY EVEN INTEGER IS A SUM OF
TWO PRIMES.

Chr. Goldbach[120] conjectured that every number N which is a sum of two primes is a sum of as many primes including unity as one wishes (up to N), and that every number >2 is a sum of three primes.

L. Euler[121] remarked that the first conjecture can be confirmed from an observation previously communicated to him by Goldbach that every even number is a sum of two primes. Euler expressed his belief in the last statement, though he could not prove it. From it would follow that, if n is even, n, $n-2$, $n-4$, ... are the sums of two primes and hence n a sum of 3, 4, 5, ... primes.

R. Descartes[122] stated that every even number is a sum of 1, 2 or 3 primes.

E. Waring[123] stated Goldbach's theorem and added that every odd number is either a prime or is a sum of three primes.

L. Euler[124] stated without proof that every number of the form $4n+2$ is a sum of two primes each of the form $4k+1$, and verified this for $4n+2 \leqq 110$.

[110]L'intermédiaire des math., 19, 1912, 36. [From X^2+X+41 by setting $X=x-1500$.]

[111]Sitz. Ak. Wiss. Berlin, 1912, 966–980.

[112]Bull. Soc. Math. France, 1911, Comptes Rendus des Séances. Extract in Sphinx-Oedipe, 9, 1914, 6–7.

[120]Corresp. Math. Phys. (ed., P. H. Fuss), 1, 1843, p. 127 and footnote; letter to Euler, June 7, 1742.

[121]Ibid., p. 135; letter to Goldbach, June 30, 1742. Cited by G. Eneström, Bull. Bibl. Storia Sc. Mat. e Fis., 18, 1885, 468.

[122]Posthumous manuscript, Oeuvres, 10, 298.

[123]Meditationes Algebraicae, 1770, 217; ed. 3, 1782, 379. The theorem was ascribed to Waring by O. Terquem, Nouv. Ann. Math., 18, 1859, Bull. Bibl. Hist., p. 2; by E. Catalan, Bull. Bibl. Storia Sc. Mat. e Fis., 18, 1885, 467; and by Lucas, Théorie des Nombres, 1891, 353.

[124]Acta Acad. Petrop., 4, II, 1780 (1775). 38; Comm. Arith. Coll., 2, 1849, 135.

A. Desboves[125] verified that every even number between 2 and 10000 is a sum of two primes in at least two ways; while, if the even number is the double of an odd number, it is simultaneously a sum of two primes of the form $4n+1$ and also a sum of two primes of the form $4n-1$.

J. J. Sylvester[126] stated that the number of ways of expressing a very large even number n as a sum of two primes is approximately the ratio of the square of the number of primes $<n$ to n, and hence bears a finite ratio to the quotient of n by the square of the natural logarithm of n. [Cf. Stäckel[132]].

F. J. E. Lionnet[127] designated by x the number of ways $2a$ can be expressed as a sum of two odd primes, by y the number of ways $2a$ can be expressed as a sum of two distinct odd composite numbers, by z the number of odd primes $<2a$, and by q the largest integer $\leq a/2$. He proved that $q+x=y+z$ and argued that it is very probable that there are values of n for which $q=y+z$, whence $x=0$.

N. V. Bougaief[127a] noted that, if $M(n)$ denotes the number of ways n can be expressed as a sum of two primes, and if θ_i denotes the ith prime >1,

$$\Sigma_i (n-3\theta_i)M(n-\theta_i)=0.$$

G. Cantor[128] verified Goldbach's theorem up to 1000. His table gives the number of decompositions of each even number <1000 as a sum of two primes and lists the smaller prime.

V. Aubry[129] verified the theorem from 1002 to 2000.

R. Haussner[130] verified the law up to 10000 and announced results observed by a study of his[131] tables up to 5000. His table I (pp. 25–178) gives the number ν of decompositions of every even n up to 3000 as a sum $x+y$ of two primes and the values of x ($x\leq y$), as in the table by Cantor. His table II (pp. 181–191) gives ν for $2<n<5000$; this table and further computations enable him to state that Goldbach's theorem is true for $n<10000$. Let $P(2\rho+1)$ be the number of all odd primes 1, 3, 5, . . . which are $\leq 2\rho+1$, and set

$$\xi(2\rho+1)=P(2\rho+1)-2P(2\rho-1)+P(2\rho-3), \qquad P(-1)=P(-3)=0.$$

Then the number of decompositions of $2n$ into a sum of two primes x, y ($x\leq y$) is

$$\sum_{\rho=0}^{n-1} P(2n-2\rho-1)\xi(2\rho+1).$$

If $\epsilon=1$ or -1 according as n is a prime or not,

$$\nu=\tfrac{1}{2}\sum_{\rho=1}^{n-1} P(2n-2\rho-1)\xi(2\rho+1)+\frac{\epsilon}{2}.$$

[125]Nouv. Ann. Math., 14, 1855, 293.
[126]Proc. London Math. Soc., 4, 1871–3, 4–6; Coll. M. Papers, 2, 709–711.
[127]Nouv. Ann. Math., (2), 18, 1879, 356. Cf. Assoc. franç. av. sc., 1894, I, p. 96.
[127a]Comptes Rendus Paris, 100, 1885, 1124.
[128]Assoc. franç. av. sc., 1894, 117–134; l'intermédiaire des math., 2, 1895, 179.
[129]L'intermédiaire des math., 3, 1896, 75; 4, 1897, 60; 10, 1903, 61 (errata, p. 166, p. 283).
[130]Jahresbericht Deutschen Math.-Verein., 5, 1896, 62–66. Verhandlungen Gesell. Deutscher Naturforscher u. Aerzte, 1896, II, 8.
[131]Nova Acta Acad. Caes. Leop.-Carolinae, 72, 1899, 1–214.

Table III gives the values of P and ξ for each odd number $2\rho+1 < 5000$. P. Stäckel[132] noted that Lionnet's[127] argument is not conclusive, and designated by G_{2n} the number of all decompositions of $2n$ as a sum of two primes (counting $p+q$ and $q+p$ as two different decompositions). If P_k is the number of all odd primes from 1 to k,

$$\sum_{n=1}^{\infty} G_{2n}x^{2n} = (\Sigma x^p)^2 = (1-x^2)^2 \left(\sum_{\nu=0}^{\infty} P_{2\nu+1}x^{2\nu+1} \right)^2,$$

where p ranges over all the odd primes. Approximations to G_{2n} for n large in terms of Euler's ϕ-function are

$$\frac{P^2_{2n}}{\phi(2n)}, \qquad \frac{[P(2n-\sqrt{2n})-P(\sqrt{2n})]^2}{n-\sqrt{2n}} \cdot \frac{n}{\phi(2n)},$$

where $P(k)$ is written for P_k for convenience in printing. Lack of agreement with Sylvester[126] is noted; cf. Landau.[135] It is stated that the truth of Goldbach's theorem is made very probable [but not proved[133]].

Sylvester[133a] stated that any even integer $2n$ is a sum of two primes, one $> n/2$ and the other $< 3n/2$, whence it is possible to find two primes whose difference is less than any given number and whose sum is twice that number. F. J. Studnicka[134] discussed Sylvester's statement.

Sylvester[134a] stated that, if N is even and $\lambda, \ldots, \omega$ are the θ primes $> \frac{1}{4}N$ and $< \frac{3}{4}N$ (excluding $\frac{1}{2}N$ if it be prime), the number of ways of composing N [by addition] with two of these primes is the coefficient of x^N in

$$\left(\frac{1}{1-x^\lambda} + \ldots + \frac{1}{1-x^\omega} \right)^r / r(r-1)\theta^{r-2} \qquad (r \geq 2).$$

E. Landau[135] noted that Stäckel's approximation to G_n is

$$\mathfrak{G}_n = \frac{n^2}{\log^2 n\phi(n)},$$

and showed that $\Sigma_{n=1}^x G_n$ has the true approximation $\frac{1}{2}x^2/\log^2 x$. By a longer analysis, he proved that if we use Stäckel's $\mathfrak{G}_n$ to form the sum, we do not obtain a result of the correct order of magnitude.

L. Ripert[136] examined certain large even numbers.

E. Maillet[137] proved that every even number ≤ 350000 (or 10^6 or $9 \cdot 10^6$) is, in default by at most 6 (or 8 or 14), the sum of two primes.

A. Cunningham[138] verified Goldbach's theorem for all numbers up to 200 million which are of the forms

$$(4 \cdot 3)^n, \quad (4 \cdot 5)^n, \quad 2 \cdot 10^n, \quad 2^n(2^n \mp 1), \quad a \cdot 2^n, \quad 2a^n, \quad (2a)^n, \quad 2(2^n \mp a),$$

for $a = 1, 3, 5, 7, 9, 11$. He reduced the formula of Haussner for ν to a form more convenient for computation.

[132]Göttingen Nachrichten, 1896, 292–9.　　[133]Encyclopédie des sc. math., I, 17, p. 339, top.
[133a]Nature, 55, 1896–7, 196, 269.　　[134]Casopis, Prag, 26, 1897, 207–8.
[134a]Educ. Times, Jan. 1897. Proof by J. Hammond, Math. Quest. Educ. Times, 26, 1914, 100.
[135]Göttingen Nachrichten, 1900, 177–186.
[136]L'intermédiaire des math., 10, 1903, 67, 74, 166 (errors, p. 168).
[137]Ibid., 12, 1905, 107–9.　　[138]Messenger Math., 36, 1906, 17–30.

J. Merlin[139] considered the operation $A(b, a)$ of effacing from the natural series of integers all the numbers $ax+b$. The effect of carrying out one of the two sets of operations $A(r_1, p_1)$, $A(r_i, p_i)$, $A(r'_i, p_i)$, $i=2,\ldots, n$, where p_n is the nth prime >1, is equivalent to constructing a crib of Eratosthenes up to p_n. It is stated that in every interval of length $\nu p_n \log p_n$ there is at least one number not effaced, if ν is independent of n. It is said to follow that, for a sufficiently large, there exist two primes having the sum $2a$. Under specified assumptions, there exist an infinitude of n's for which $p_{n+1}-p_n=2$.

M. Vecchi[140] wrote p_n for the nth odd prime and called p_h and p_{h+a} of the same order if $p^2{}_h>p_{h+a}$. Then $2n>132$ is a sum of two primes of the same order in $[\frac{1}{2}(\phi+1)]$ ways if and only if there exist ϕ numbers not $>n-p_{m+1}+1$ and not representable in any of the forms

$$a_i+3x, \qquad b_i+5x,\ldots, \qquad l_i+p_m x \qquad (i=1,\ 2),$$

where p_{m+1} is the least prime p for which $p^2+p>2n$, and the known terms a_i, $\ldots$ are the residues with respect to the odd prime occurring as coefficient of x.

*G. Giovannelli, Sul teorema di Goldbach, Atri, 1913.

THEOREMS ANALOGOUS TO GOLDBACH'S.

Chr. Goldbach[145] stated empirically that every odd number is of the form $p+2a^2$, where p is a prime and a is an integer $\geqq 0$. L. Euler[146] verified this up to 2500. Euler[124] verified for $m=8N+3\leqq 187$ that m is the sum of an odd square and the double of a prime $4n+1$.

J. L. Lagrange[147] announced the empirical theorem that every prime $4n-1$ is a sum of a prime $4m+1$ and the double of a prime $4h+1$.

A. de Polignac[148] conjectured that every even number is the difference of two consecutive primes in an infinitude of ways. His verification up to 3 million that every odd number is the sum of a prime and a power of 2 was later[148a] admitted to be in error for 959.

M. A. Stern[149] and his students found that $53\cdot109=5777$ and $13\cdot641=5993$ are neither of the form $p+2a^2$ and verified that up to 9000 there are no further exceptions to Goldbach's[145] assertion. Also, 17, 137, 227, 977, 1187 and 1493 are the only primes <9000 not of the form $p+2b^2$, $b>0$. Thus all odd numbers <9000, which are not of the form $6n+5$, are of the form $p+2b^2$.

E. Lemoine[150] stated empirically that every odd number >3 is a sum of a prime p and the double of a prime π, and is also of the forms $p-2\pi$ and $2\pi'-p'$.

---

[139]Comptes Rendus Paris, 153, 1911, 516–8.　Bull. des. sc. math., (2), 39, I, 1915, 121–136.　In a prefatory note, J. Hadamard noted that, while the proof has a lacuna, it is suggestive.

[140]Atti Reale Accad. Lincei, Rendiconti, (5), 22, II, 1913, 654–9.

[145]Corresp. Math. Phys. (ed., Fuss), 1, 1843, 595; letter to Euler, Nov. 18, 1752.

[146]Ibid., p. 596, 606; Dec. 16, 1752.

[147]Nouv. Mém. Ac. Berlin, année 1775, 1777, 356; Oeuvres, 3, 795.

[148]Nouv. Ann. Math., 8, 1849, 428 (14, 1855, 118).

[148a]Comptes Rendus Paris, 29, 1849, 400, 738–9.

[149]Nouv. Ann. Math., 15, 1856, 23.　　　[150]L'intermédiaire des math., 1, 1894, 179; 3, 1896, 151

H. Brocard[151] gave an incorrect argument by use of Bertrand's postulate that there exists a prime between any two consecutive triangular numbers. G. de Rocquigny[152] remarked that it seems true that every multiple of 6 is the difference of two primes of the form $6n+1$. Brocard[153] verified this property for a wide range of values. L. Kronecker[154] remarked that an unnamed writer[148] had stated empirically that every even number can be expressed in an infinitude of ways as the difference of two primes. Taking 2 as the number, we conclude that there exist an infinitude of pairs of primes differing by 2. L. Ripert[155] verified that every even number <10000 is a sum of a prime and a power, every odd one except 1549 is such a sum. E. Maillet[156] commented on de Polignac's conjecture that every even number is the difference of two primes. E. Maillet[157] proved that every odd number <60000 (or $9 \cdot 10^6$) is, in default by at most 8 (or 14), the sum of a prime and the double of a prime.

PRIMES IN ARITHMETICAL PROGRESSION.

E. Waring[165] stated that if three primes (the first of which is not 3) are in arithmetical progression, the common difference d is divisible by 6, except for the series 1, 2, 3 and 1, 3, 5. For 5 primes, the first of which is not 5, d is divisible by 30; for 7 primes, the first not 7, d is divisible by $2 \cdot 3 \cdot 5 \cdot 7$; for 11 primes, the first not 11, d is divisible by $2 \cdot 3 \cdot 5 \cdot 7 \cdot 11$; and similarly for any prime number of primes in arithmetical progression, a property easily proved. Hence by continually adding d to a prime, we reach a number divisible by 3, 5,..., unless d is divisible by 3, 5,....

J. L. Lagrange[166] proved that if 3 primes, no one being 3, are in arithmetical progression, the difference d is divisible by 6; for 5 primes, no one being 5, d is divisible by 30. He stated that for 7 primes, d is divisible by $2 \cdot 3 \cdot 5 \cdot 7$, unless the first one is 7, and then there are not more than 7 consecutive prime terms in a progression whose difference is not divisible by $2 \cdot 3 \cdot 5 \cdot 7$.

E. Mathieu[167] proved Waring's statement.

M. Cantor[168] proved that if $P = 2 \cdot 3 \ldots p$ is the product of all the primes up to the prime p, there is no arithmetical progression of p primes, no one of which is p, unless the common difference is divisible by P. He conjectured that three successive primes are not in arithmetical progression unless one of them is 3.

A. Guibert[169] gave a short proof of the theorem stated thus: Let $p_1, \ldots, p_n$ be primes $\geqq 1$ in arithmetical progression, where n is odd and >3. Then no prime >1 and $\leqq n$ is a p_i. If n is a prime and is a p_i, then $i=1$.

[151]L'intermediaire des math., 4, 1897, 159.   Criticism by E. Landau, 20, 1913, 153.
[152]Ibid., 5, 1898, 268.   [153]L'intermédiaire des math., 6, 1899, 144.
[154]Vorlesungen über Zahlentheorie, 1, 1901, 68.
[155]L'intermédiaire des math., 10, 1903, 217–8.   [156]Ibid., 12, 1905, 108.
[157]Ibid., 13, 1906, 9.   [165]Meditationes Algebraicae, 1770; ed. 3, 1782, 379.
[166]Nouv. Mém. Ac. Berlin, année 1771, 1773, 134–7.   [167]Nouv. Ann. Math., 19, 1860, 384–5.
[168]Zeitschrift Math. Phys., 6, 1861, 340–3.
[169]Jour. de Math., (2), 7, 1862, 414–6.

The common difference is divisible by each prime $\leq n$, and by n itself if n is a prime not in the series.

H. Brocard[169a] gave several sets of five consecutive odd integers, four of which are primes. Lionnet[169b] had asked if the number of such sets is unlimited.

G. Lemaire[170] noted that $7+30n$ and $107+30n$ $(n=0, 1, \ldots, 5)$ are all primes; also $7+150n$ and $47+210n$ $(n=0, \ldots, 6)$.

E. B. Escott[171] found conditions that $a+210n$ $(n=0, 1, \ldots, 9)$ be all primes and noted that the conditions are satisfied if $a=199$.

Devignot[172] noted the primes $47+210n$, $71+2310n$ $(n=0, 1, \ldots, 6)$.

A. Martin[173] gave numerous sets of primes in arithmetical progression.

Tests for Primality.

The fact that n is a prime if and only if it divides $1+(n-1)!$ was noted by Leibniz,[7] Lagrange,[18] Genty,[24] Lebesgue,[85] and Catalan,[106] cited in Chapter III, where was discussed the converse of Fermat's theorem in furnishing a primality test. Tests by Lucas, etc., were noted in Ch. XVII. Further tests have been noted under Cipolla[172] and[176] Cole[173] of Ch. I, Sardi[273] of Ch. III, Lambert[6] of Ch. VI, Zsigmondy[79] of Ch. VII, Gegenbauer[80, 92] of Ch. X, Jolivald[84] of Ch. XIII, Euler,[26-43] Tchebychef,[52] Schaffgotsch[100] and Biddle[141] of Ch. XIV, Hurwitz[41] and Cipolla[45] of Ch. XV. See also the papers by von Koch,[235] Hayashi,[239, 240] Andreoli,[244] and Petrovitch[245] of the next section.

L. Euler[177] gave a test for the primality of a number $N=4m+1$ which ends with 3 or 7. Let R be the remainder on subtracting from $2N$ the next smaller square $(5n)^2$ which ends with 5. To R add $100(n-1)$, $100(n-3)$, $100(n-5)$, ... If among R and these sums there occurs a single square, N is a prime or is divisible by this square. But if no square occurs or if two or more squares occur, N is composite. For example, if $N=637$, $(5n)^2=1225$, $R=49$; among 49, 649, 1049, 1249 occurs only the square 49; hence N is a prime or is divisible by 49 $[N=49\cdot13]$.

W. L. Kraft[178] noted that $6m+1$ is a prime if m is of neither of the forms $6xy\pm(x+y)$; $6m-1$ is a prime if $m\neq6xy+x-y$.

A. S. de Montferrier[179] noted that an odd number A is a prime if and only if $A+k^2$ is not a square for $k=1, 2, \ldots, (A-3)/2$.

M. A. Stern[180] noted that n is a prime if and only if it occurs $n-1$ times in the $(n-1)$th set, where the first set is 1, 2, 1; the second set, formed by inserting between any two terms of the first set their sum, is 1, 3, 2, 3, 1; etc.

[169a]Nouv. Ann. Math., (3), 15, 1896, 389–90.   [169b]Nouv. Ann. Math., (3), 1, 1882, 336.
[170]L'intermédiaire des math., 16, 1909, 194–5.
[171]Ibid., 17, 1910, 285–6.
[172]Ibid., 45–6.   [173]School Science and Mathematics, 13, 1913, 793–7,
[176]Doubt as to the sufficiency of Cole's test has been expressed, Proc. London Math. Soc., (2).
    16, 1917–8.   [177]Opera postuma, I, 188–9 (about 1778).
[178]Nova Acta Acad. Petrop., 12, 1801, hist., p. 76, mem., p. 217.
[179]Corresp. Math. Phys. (ed., Quetelet), 5, 1829, 94–6.
[180]Jour. für Math., 55, 1858, 202.

L. Gegenbauer[181] noted that $4n+1$ is a prime if

$$\left[\frac{4n+1-y^2}{4y}\right]=\left[\frac{4n-3-y^2}{4y}\right]$$

for every odd y, $1<y\leq\sqrt{4n+1}$, and gave two similar tests for $4n+3$.
D. Gambioli[182] and O. Meissner[183] discussed the impracticability of the
test by the converse of Wilson's theorem.
J. Hacks[184] gave the characteristic relations for primes p:

$$\sum_{y=1}^{p-1}\sum_{s=1}^{p-1}\left[\frac{ys}{p}\right]=\left(\frac{p-1}{2}\right)^2(p-2),\qquad \sum_{y=1}^{p-1}\left\{\sum_{s=1}^{(p-1)/2}\left[\frac{ys}{p}\right]+\sum_{s=1}^{[y/2]}\left[\frac{ps}{y}\right]\right\}=\left(\frac{p-1}{2}\right)^3.$$

K. Zsigmondy[185] noted that a number is a prime if and only if not
expressible in the form $a_1a_2+\beta_1\beta_2$, where the a's and β's are positive integers
such that $a_1+a_2=\beta_1-\beta_2$. An odd number C is a prime if and only if
$C+k^2$ is not a square for $k=0,1,\ldots,[(C-9)/6]$.
R. D. von Sterneck[185a] gave several criteria for the $(s+1)$th prime by use
of partitions into elements formed from the first s primes.
H. Laurent[185b] noted that

$$(e^{2\pi i\Gamma(z)/z}-1)/(e^{-2\pi i/z}-1)$$

equals 0 or 1 according as z is composite or prime.
Fontebasso[186] noted that N is a prime if not divisible by one of the
primes $2, 3,\ldots, p$, where $N/p<p+4$.
H. Laurent[187] proved that if we divide

$$F_n(x)\doteq\prod_{j=1}^{n-1}(1-x^j)(1-x^{2j})\ldots(1-x^{(n-1)j})$$

by $(x^n-1)/(x-1)$, the remainder is 0 or n^{n-1} according as n is composite or
prime. If we take x to be an imaginary root of $x^n=1$, $F_n(x)$ becomes 0 or
n^{n-1} in the respective cases.
Helge von Koch[188] used infinite series to test whether or not a number is a
power of a prime.
Ph. Jolivald[189] noted that, since every odd composite number is the
difference of two triangular numbers, an odd number N is a prime if and
only if there is no odd square, with a root $\leq(2N-9)/3$, which increased by
$8N$ gives a square.
S. Minetola[190] noted that, if $k-n$ is divisible by $2n+1$, then $2k+1$ is
composite. We may terminate the examination when we reach a prime
$2n+1$ for which $(k-n)/(2n+1)\leq n$.
A. Bindoni[191] added that we may stop with a prime giving $(k-n)$

[181]Sitzungsber. Ak. Wiss. Wien (Math.), 99, IIa, 1890, 389.
[182]Periodico di Mat., 13, 1898, 208–212.   [183]Math. Naturw. Blätter, 3, 1906, 100
[184]Acta Mathematica, 17, 1893, 205.   [185]Monatsh. Math. Phys., 5, 1894, 123–8.
[185a]Sitzungsber. Ak. Wiss. Wien (Math.), 105, IIa, 1896, 877–882.
[185b]Comptes Rendus Paris, 126, 1898, 809–810.   [186]Suppl. Periodico di Mat., 1899, 53.
[187]Nouv. Ann. Math., (3), 18, 1899, 234–241.
[188]Öfversigt Veten.-Akad. Förhand., 57, 1900, 789–794 (French).
[189]L'intermédiaire des math., 9, 1902, 96; 10, 1903, 20.
[190]Il Boll. Matematica Giorn. Sc.-Didat., Bologna, 6, 1907, 100–4.   [191]Ibid., 165–6.

$\div (2n+1) \leqq n+2a-1$, where a is the difference between $2n+1$ and the next greater prime.

F. Stasi[192] noted that N is a prime if not divisible by one of the primes $2, 3, \ldots, p$, where $N/p < p+2a$ and a is the difference between p and the prime just $> p$.

E. Zondadari[193] noted that

$$\frac{\sin^2 \pi x}{(\pi x)^2 (1-x^2)^2} \prod_{n=2}^{\infty} \frac{\pi x}{n \sin \pi x/n}$$

is zero when $x = \pm p$ (p a prime) and not otherwise.

A. Chiari[194] cited known tests for primes, as the converse of Wilson's theorem.

H. C. Pocklington[195] employed single valued functions $\phi(x)$, $\psi(x)$, vanishing for all positive integers x (as $\phi = \psi = \sin \pi x$), and real, finite and not zero for all other positive values of x. Then, for the gamma function Γ,

$$\phi^2(x) + \psi^2 \left(\frac{1+\Gamma(x)}{x} \right)$$

is zero if and only if x is a prime [Wigert[236a]].

E. B. Escott[196] stated that if we choose $a_1, \ldots, a_n, b$ so that the coefficients of $x^{2n}, x^{2n-2}, \ldots, x^2$ in the expansion of

$$(x^n + a_1 x^{n-1} + \ldots + a_n)^2 (x+b)$$

are all zero, then all the remaining coefficients, other than the first and last, are divisible by $2n+1$ if and only if $2n+1$ is a prime.

J. de Barinaga[197] concluded from Wilson's theorem that if $(P-1)!$ is divided by $1+2+\ldots+(P-1) = P(P-1)/2$, the remainder is $P-1$ when P is a prime, but is zero when P is composite (not excluding $P=4$ as in the converse of Wilson's theorem). Hence on increasing by unity the least positive residues $\neq 0$ obtained on dividing $1 \cdot 2 \ldots x$ by $1+2+\ldots+x$, for $x = 1, 2, 3, \ldots$, we obtain the successive odd primes $3, 5, \ldots$.

M. Vecchi[140] noted that, if $x \geqq 1$, $N > 2$ is a prime if and only if it be of the form $2^x \pi' - \pi$, where π is the product of all odd primes $\leqq p$, p being the largest odd prime $\leqq [\sqrt{N}]$, and where π' is a product of powers of primes $> p$ with exponents $\geqq 0$. Again, $N > 121$ is a prime if and only if of the form $\pi - 2^y \pi'$ where $y \geqq 1$.

Vecchi[198] gave the simpler test: $N > 5$ is a prime if and only if $a - \beta = N$, $a+\beta = \pi$, for a, β relatively prime, where π is the product of all the odd primes $\leqq [\sqrt{N}]$.

G. Rados[199] noted that p is a prime if and only if $\{2!3! \ldots (p-2)! (p-1)!\}^4 \equiv 1 \pmod{p}$.

Carmichael[93] gave several tests analogous to those by Lucas.

---

[192]Il Boll. Matimatica Giorn. Sc.-Didat., Bologna, 6, 1907, 120–1.
[193]Rend. Accad. Lincei, (5), 19, 1910, I, 319–324.   [194]Il Pitagora, Palermo, 17, 1910–11, 31–33.
[195]Proc. Cambr. Phil. Soc., 16, 1911, 12.   [196]L'intermédiaire des math., 19, 1912, 241–2.
[197]Revista de la Sociedad Mat. Española, 2, 1912, 17–21.
[198]Periodico di Mat., 29, 1913, 126–8.   [199]Math. és Termés Ertesitö, 34, 1916, 62–70·

NUMBER OF PRIMES BETWEEN ASSIGNED LIMITS.

Formula (5) of Legendre in Ch. V implies that if θ, λ, ... are the primes $\leqq \sqrt{n}$, the number of primes $\leqq n$ and $> \sqrt{n}$ is one less (if unity be counted a prime) than

$$n - \Sigma \left[\frac{n}{\theta} \right] + \Sigma \left[\frac{n}{\theta \lambda} \right] - \dots$$

Statements or proofs of this result have been given by C. J. Hargreave,[205] E. de Jonquières,[206] R. Lipschitz,[207] J. J. Sylvester,[208] E. Catalan,[209] F. Rogel,[210] J. Hammond[211] with a modification, H. W. Curjel,[211a] S. Johnsen,[212] and L. Kronecker.[213]

E. Meissel[214] proved that if $\theta(m)$ is the number of primes (including unity) $\leqq m$ and if

$$\Phi(p_1{}^{n_1} \dots p_m{}^{n_m}) = (-1)^{n_1 + \dots + n_m} \frac{(n_1 + n_2 + \dots + n_m)!}{n_1! \dots n_m!},$$

$$1 = \Phi(1)\theta \left[\frac{m}{1} \right] + \Phi(2)\theta \left[\frac{m}{2} \right] + \dots + \Phi(m)\theta \left[\frac{m}{m} \right].$$

E. Meissel[215] wrote $\Phi(m, n)$ for Legendre's formula for the number of integers $\leqq m$ which are divisible by no one of the first n primes $p_1 = 2, \dots, p_n$. Then

$$\Phi(m, n) = \Phi(m, n-1) - \Phi\left(\left[\frac{m}{p_n} \right], n-1 \right).$$

Let $\theta(m)$ be the number of primes $\leqq m$. Set $n + \mu = \theta(\sqrt{m})$, $n = \theta(\sqrt[3]{m})$. Then

$$\theta(m) = \Phi(m, n) + n(\mu+1) + \frac{\mu(\mu-1)}{2} - 1 - \sum_{s=1}^{\mu} \theta \left(\frac{m}{p_{n+s}} \right),$$

which is used to compute $\theta(m)$ for $m = k \cdot 10^6$, $k = 1/2$, 1, 10.

Meissel[216] applied his last formula to find $\theta(10^8)$.

Lionnet[216a] stated that the number of primes between A and $2A$ is $< \theta(A)$.

N. V. Bougaief[217] obtained from $\theta(n) + \theta(n/2) + \theta(n/3) + \dots = \Sigma[n/p]$, by inversion (Ch. XIX),

$$\theta(x) = \Sigma \left[\frac{n}{a} \right] - 2\Sigma \left[\frac{n}{ab} \right] + 3\Sigma \left[\frac{n}{abc} \right] - \dots - \Sigma \left[\frac{n}{a^2} \right] + \Sigma \left[\frac{n}{a^2 b} \right] - \Sigma \left[\frac{n}{a^2 bc} \right] + \dots,$$

where a, b, ... range over all primes.

[205]Lond. Ed. Dub. Phil. Mag., (4), 8, 1854, 118–122.
[206]Comptes Rendus Paris, 95, 1882, 1144, 1343; 96, 1883, 231.
[207]Ibid., 95, 1882, 1344–6; 96, 1883, 58–61, 114–5, 327–9.
[208]Ibid., 96, 1883, 463–5; Coll. Math. Papers, 4, p. 88.
[209]Mém. Soc. Roy. Sc. de Liège, (2), 12, 1885, 119; Mélanges Math., 1868, 133–5.
[210]Archiv Math. Phys., (2); 7, 1889, 381–8.　　　　[211]Messenger Math., 20, 1890–1, 182.
[211a]Math. Quest. Educ. Times, 67, 1897, 27.
[212]Nyt Tidsskrift for Mat., Kjobenhavn, 15 A, 1904, 41–4.
[213]Vorlesungen über Zahlentheorie, I, 1901, 301–4.　　　　[214]Jour. für Math., 48, 1854, 310–4.
[215]Math. Ann., 2, 1870, 636–642.　Outline in Mathews' Theory of Numbers, 273–8, and in G. Wertheim's Elemente der Zahlentheorie, 1887, 20–25.
[216]Ibid., 3, 1871, 523–5.　Corrections, 21, 1883, 304.
[216a]Nouv. Ann. Math., 1872, 190.　Cf. Landau, (4), 1, 1901, 281–2.
[217]Bull. sc. math. astr., 10, I, 1876, 16.　Mat. Sbornik (Math. Soc. Moscow), 6, 1872–3, I, 180.

P. de Mondésir[218] wrote N_p for the number of multiples of the prime p which are $<2N$ and divisible by no prime $<p$. Then the number of primes $<2N$ is $N-\Sigma N_p+n+1$, where n is the number of primes $<\sqrt{2N}$. Also,

$$N_p=\left[\frac{N}{p}\right]-\Sigma\left[\frac{N}{ap}\right]+\Sigma\left[\frac{N}{abp}\right]-\ldots,$$

where $a, b, \ldots$ are the primes $<p$. By this modification of Legendre's formula, he computed the number 78490 of primes under one million.

*L. Lorenz[219] discussed the number of primes under a given limit.

Paolo Paci[220] proved that the number of integers $\leqq n$ divisible by a prime$<\sqrt{n}$ is

$$N=\Sigma\left[\frac{n}{r}\right]-\Sigma\left[\frac{n}{rs}\right]+\ldots\pm\left[\frac{n}{2\cdot3\cdot5\ldots p}\right],$$

where $r, s, \ldots$ range over all the H primes $2, 3, \ldots, p$ less than $\sqrt{n}$. Thus there are $n-1-N+H$ primes from 1 to n. The approximate value of N is

$$n\left\{\Sigma\frac{1}{r}-\Sigma\frac{1}{rs}+\ldots\right\}=n\left\{1-\frac{\phi(2\cdot3\ldots p)}{2\cdot3\ldots p}\right\}.$$

K. E. Hoffmann[221] denoted by N the number of primes $<m$, by λ the number of distinct prime factors of m, by μ the number of composite integers $<m$ and prime to m. Evidently $N=\phi(M)-\mu+\lambda$. To find N it suffices to determine μ. To that end he would count the products $<m$ by twos, by threes, etc. (with repetitions) of the primes not dividing m.

J. P. Gram[222] proved that the number of powers of primes $\leqq n$ is

$$P(n)=\Sigma\left[\frac{n}{a}\right]-2\Sigma\left[\frac{n}{ab}\right]+3\Sigma\left[\frac{n}{abc}\right]-\ldots.$$

[Cf. Bougaief.[217]] Of the two proofs, one is by inversion from

$$P(n)+P\left(\frac{n}{2}\right)+P\left(\frac{n}{3}\right)+\ldots=\Sigma\left[\frac{n}{p}\right]+\Sigma\left[\frac{n}{p^2}\right]+\Sigma\left[\frac{n}{p^3}\right]+\ldots.$$

E. Cesàro[223] considered the number x of primes $\leqq qn$ and $>n$, where q is a fixed prime. Let $\omega_1, \ldots, \omega_\nu$ be the primes $\leqq n$ other than 1 and q. Let $q^k\leqq n<q^{k+1}$. Then

$$k+2+x=qn-\Sigma\left[\frac{qn}{\omega_1}\right]+\Sigma\left[\frac{qn}{\omega_1\omega_2}\right]-\ldots.$$

Let $l_{r,s}$ be the number of the $[qn/(\omega_1\ldots\omega_s)]$ which give the remainder r when divided by q. Set $t_s=\Sigma jl_{j,s}$. Then

$$x=(k+1)q-(k+2)-t_1+t_2-t_3+\ldots.$$

[218]Assoc. franç. av. sc., 6, 1877, 77–92. Nouv. Corresp. Math., 6, 1880, 256.
[219]Tidsskr. for Math., Kjobenhavn, (4), 2, 1878, 1–3.
[220]Sul numero de numeri primi inferiori ad un dato numero, Parma, 1879, 10 pp.
[221]Archiv Math. Phys., 64, 1879, 333–6.
[222]K. Danske Vidensk. Selskabs. Skrifter, (6), 2, 1881–6, 183–288; résumé in French, 289–308.
      See pp. 220–8, 296–8.
[223]Mém. Soc. Sc. Liège, (2), 10, 1883, 287–8.

E. Catalan[224] obtained the preceding results for the case $q=2$; then t_1 is the number of odd quotients $[2n/\beta]$, t_2 the number of odd quotients $[2n/(\beta\gamma)], \ldots$, where $\beta, \gamma, \ldots$ are the primes >2 and $\leq n$.

L. Gegenbauer[225] gave eight formulas, (29)–(36), of the type of Legendre's, a special case of one being

$$\underset{x}{\Sigma} S_k\left(\left[\frac{n}{x}\right]\right)\mu(x) = 1 + L_k(n), \qquad S_k(n) \equiv \underset{t=1}{\overset{n}{\Sigma}} t^k,$$

where x ranges over the integers divisible by no prime $>\sqrt{n}$, while $\mu(x)$ is Merten's function (Ch. XIX) and $L_k(n)$ is the sum of the kth powers of all primes $>\sqrt{n}$ but $\leq n$. The case $k=0$ is Legendre's formula. The case $k=1$ is Sylvester's[208]

E. Meissel[226] computed the number of primes $<10^9$.

Gegenbauer[226a] gave complicated expressions for $\theta(n)$, one a generalization of Bougaief's.[217]

A. Lugli[227] wrote $\phi(n, i)$ for the number of integers $\leq n$ which are divisible by no one of the first i primes $p_1=2$, $p_2=3, \ldots$. If i is the number of primes $\leq \sqrt{n}$ and if s is the least integer such that

$$s-1 = \psi[\sqrt{n/p_s}],$$

the number $\psi(n)$ of primes $\leq n$, excluding 1, is proved to satisfy

$$\psi(n) = \left[\frac{n}{2}\right] - \overset{s-1}{\underset{j=2}{\Sigma}} \phi\left(\left[\frac{n}{p_j}\right], j-1\right) - \overset{i}{\underset{j=s}{\Sigma}} \psi\left[\frac{n}{p_j}\right] + \tfrac{1}{2}(i^2 - s^2 - i + 5s + 6).$$

This method of computing $\psi(n)$ is claimed to be simpler than that by Legendre or Meissel.

J. J. van Laar[227a] found the number of primes <30030 by use of the primes <1760.

C. Hossfeld[228] gave a direct proof of

$$\Phi(gp_1 \ldots p_n \pm r, n) = g(p_1-1) \ldots (p_n-1) \pm \Phi(r, n),$$

the case of the upper signs being due to Meissel.[215]

F. Rogel[229] gave a modification and extension of Meissel's[215] formula.

H. Scheffler[230] discussed the number of primes between p and q.

J. J. Sylvester[231] stated that the number of primes $>n$ and $<2n$ is

$$n - \Sigma H\frac{n}{a} + \Sigma H\frac{n}{ab} - \Sigma H\frac{n}{abc} + \ldots,$$

if $a, b, \ldots$ are the primes $\leq \sqrt{2n}$ and Hx denotes x when its fractional part

[224]Mém. Soc. Sc. Liège, Mém. No. 1.
[225]Sitzungsber. Ak. Wiss. Wien (Math.), 89, II, 1884, 841–850; 95, II, 1887, 291–6.
[226]Math. Annalen, 25, 1885, 251–7.
[216a]Sitzungsber. Ak. Wiss. Wien (Math.), 94, II, 1886, 903–10.
[227]Giornale di Mat., 26, 1888, 86–95.
[227a]Nieuw Archief voor Wisk., 16, 1889, 209–214.
[228]Zeitschrift Math. Phys., 35, 1890, 382–4.
[229]Math. Annalen, 36, 1890, 304–315.   [230]Beiträge zur Zahlentheorie, 1891, 187.
[231]Lucas, Théorie des nombres, 1891, 411–2.   Proof by H. W. Curjel, Math. Quest. Educ. Times, 57, 1892, 113.

is 1/2, but the nearest integer to x in the contrary case. L. Gegenbauer[231a] gave a proof and generalization.

Sylvester[231b] noted that, if $\theta(u)$ is the number of primes $\leqq u$, and if $p_1, \ldots, p_i$ be the primes $\leqq \sqrt{x}$, and $q_1, \ldots, q_j$ those between $\sqrt{x}$ and x, then $\Sigma\theta(x/p) - \Sigma\theta(x/q) = \{\theta(\sqrt{x})\}^2$.

H. W. Curjel[231c] noted that the number of primes $>p$ and $<p^2$ is $\geqq p$ if p is a prime $\geqq 5$. We have only to delete from $1, 2, \ldots, p^2$ multiples of $2, 3, 5, \ldots$, or p.

L. Gegenbauer[232] considered the integers x divisible by no square and formed of the odd primes $\leqq m$, when $n \geqq m \geqq \sqrt{2n}$. Of the numbers $[2n/x]$ which are of one of the forms $4s+1$ and $4s+2$, count those in which x is formed of an even number of primes and those in which x is formed of an odd number; denote the difference of the counts by a. He stated that the interval from $m+1$ to n (limits included) contains $a-1$ more primes than the interval from $n+1$ to $2n$.

He gave (pp. 89–93) an expression for the sum of the values taken by an arbitrary function $g(x)$ when x ranges over the primes among the first n terms of an arithmetical progression; in particular, he enumerated the primes $\leqq n$ of the form $4s+1$ or $4s-1$.

F. Graefe[233] would find the number of primes $<m = 10000$ by use of tables showing for each prime p, $5 \leqq p \leqq \sqrt{m}$, the values of n for which $6n+1$ or $6n+5$ is divisible by p.

P. Bachmann[234] quoted de Jonquières,[206] Lipschitz,[207] Sylvester,[208] and Cesàro.[223]

H. von Koch[235] wrote $f(x) = (x-1)(x-2)\ldots(x-n)$,

$$\theta(x) = \prod_{\lambda=2}^{n}\left\{1 - \frac{\rho(x)}{\lambda}\right\}, \qquad \rho(x) = \sum_{\mu, \nu=2} \frac{f(x)}{(x-\mu\nu)f'(\mu\nu)} \qquad (\mu\nu \leqq n),$$

and proved that, for positive integers $x \leqq n$, $\theta(x) = 1$ or 0 according as x is prime or composite. The number of primes $\leqq m \leqq n$ is $\theta(1) + \ldots + \theta(m)$.

A. Baranowski[236] noted the formula, simpler than Meissel's,[215]

$$\psi(n) = \phi[n, \psi(\sqrt{n})] + \psi(\sqrt{n}) - 1$$

for computing the number $\psi(n)$ of primes $\leqq n$.

S. Wigert[236a] noted that the number of primes $<n$ is

$$\frac{1}{2\pi i}\int \frac{f'(x)dx}{f(x)}, \quad \text{where } f(x) = \sin^2\pi x + \sin^2\pi\left(\frac{1+\Gamma(x)}{x}\right),$$

[231a]Denkschr. Akad. Wiss. Wien (Math.), 60, 1893, 47.
[231b]Math. Quest. Educ. Times, 56, 1892, 67–8.
[231c]Ibid., 58, 1893, 127.
[232]Monatshefte Math. Phys., 4, 1893, 98.
[233]Zeitschrift Math. Phys., 39, 1894, 38–50.
[234]Die Analytische Zahlentheorie, 1894, 322–5.
[235]Comptes Rendus Paris, 118, 1894, 850–3.
[236]Bull. Int. Ac. Sc. Cracovie, 1894, 280–1 (German). Cf. *Rozprawy Akad. Umiej., Cracovie, (2), 8, 1895, 192–219.
[236a]Öfversigt K. Vetensk. Ak. Förhand., Stockholm, 52, 1895, 341–7.

since the only real zeros of $f(x)$ are the primes. The integration extends over a closed contour enclosing the segment of the x-axis from 1 to n and narrow enough to contain no complex zero of $f(x)$.

T. Levi-Civita[236b] gave an analytic formula, involving definite integrals and infinite series, for the number of primes between a and β.

L. Gegenbauer[237] gave formulas, similar to that by von Koch,[225] for the number of primes $4s \pm 1$ or $6s \pm 1$ which are $\leqq n$.

A. P. Minin[237a] wrote $\psi(y) = 0$ or 1 according as y is composite or prime; then
$$\theta(n-1) = [n-2] + [n-5] + [n-7] + \ldots - \Sigma\psi(x-1)[n-x],$$
summed for all composite integers x.

Gegenbauer[237b] proved that Sylvester's[231] expression for the number of primes $> n$ and $< 2n$ equals $\Sigma\mu(x)[m/x+1/2]$, where x takes those integral values $\leqq 2n$ which are products of primes $\leqq \sqrt{2n}$.

F. Rogel[238] gave a recursion formula for the number of primes $\leqq m$.

T. Hayashi[239] wrote Rf/q for the remainder obtained on dividing f by q. By Laurent's[187] result, $-RF_n(x)/(x^n-1)n^{n-2} = 0$ or 1 according as n is composite or prime. Hence the sum of the jth powers of the primes between s and t is
$$-R \sum_{n=s}^{t} \frac{F_n(x)}{(x^n-1)n^{n-j-2}},$$
which becomes the number of primes for $j = 0$. If a is a primitive nth root of unity, Wilson's theorem shows that
$$\sum_{j=0}^{n-1} a^{jm} = n \text{ or } 0 \qquad (m = (n-1)!+1),$$
according as n is prime or composite. Hence $Rx^{(n-1)!}/(x^n-1) = 1$ or 0 according as n is prime or composite. Thus
$$R \sum_{n=s}^{t} x^{(n-1)!}/(x^n-1)$$
is the number of primes between s and t.

Hayashi[240] reproduced the second of his two preceding results and gave it the form
$$\int_0^{2\pi} r^{n-m} \frac{\{\cos (m-n)\theta - r^n\cos m\theta\} d\theta}{1 - 2r^n \cos n\theta + r^{2n}} = 2\pi \text{ or } 0,$$
according as n is prime or not, and gave a direct proof.

J. V. Pexider[241] investigated the number $\psi(x)$ of primes $\leqq x$. Write
$$\Delta\left[\frac{n}{\mu}\right] = \left[\frac{n}{\mu}\right] - \left[\frac{n-1}{\mu}\right], \qquad \delta_\mu = \Delta\left[\frac{ak}{\mu}\right].$$

[236b]Atti R. Accad. Lincei, Rendiconti, (5), 4, 1895, I, 303–9.
[237]Monatshefte Math. Phys., 7, 1896, 73.
[237a]Bull. Math. Soc. Moscow, 9, 1898, No. 2; Fortschritte, 1898, 165.
[237b]Monatshefte Math. Phys., 10, 1899, 370–3.
[238]Archiv Math. Phys., (2), 17, 1900, 225–237.
[239]Jour. of the Phys. School in Tokio, 9, 1900; reprinted in Abhand. Gesch. Math. Wiss., 28, 1900, 72–5.
[240]Archiv Math. Phys., (3), 1, 1901, 246–251.
[241]Mitt. Naturforsch. Gesell. Bern, 1906, 82–91.

Hence the number of integers $\leq x$ which are divisible by a, but not by $a-1$, $a-2$, ..., 2, is

$$\sigma_a = \sum_{k=1}^{[x/a]} \prod_{\mu=2}^{a-1} (1-\delta_\mu).$$

The number $\Psi(x)$, of primes $\leq x$ and $>\nu=[\sqrt{x}]$ is $[x]-1-\Sigma_{a=2}^{a=\nu}\sigma_a$. Let $p_1, \ldots, p_a$ be the primes $\leq \sqrt{a}$. Let p_ω be the greatest prime $\leq \nu$. Then

$$\Psi(x)+1=[x]-\left[\frac{x}{2}\right]-\sum_{a=3}^{\omega}\sum_{k=1}^{[x/pa]}\prod_{\mu=2}^{a-1}\left\{1-\Delta\left[\frac{p_a k}{p_\mu}\right]\right\},$$

from which follows Legendre's formula.

S. Minetola[242] obtained a formula to compute the number of primes $\leq K=2k+1$, not presupposing a knowledge of any primes >2, by considering the sets of positive integers n, n', ... for which

$$(2n+1)(2n'+1)\leq K, \qquad (2n+1)(2n'+1)(2n''+1)\leq K, \ldots$$

F. Rogel[243] started with Legendre's formula for the number $A(z)$ of primes $\leq z$, introduced the remainders $t-|t|$, and wrote $R_n(z)$ for the sum of these partial remainders. He obtained relations between values of the A's and R's for various arguments z, and treated sums of such values. For arbitrary x's (p. 1815),

$$\sum_{\nu=1}^{p_{n+1}-1} (x_{\nu+1}-x_\nu)A(\nu) = -\Sigma x_p + x_{p_{n+1}}A(p_n),$$

summed for the primes p between 1 and the nth prime p_n. By special choice of the x's, we get formulas involving Euler's ϕ-function (p. 1818), and the number or sum of the divisors of an integer. See Rogel[22] of Ch. XI.

G. Andreoli[244] noted that, if x is real, and Γ is the gamma function,

$$\Phi(x) = \sin^2\frac{(\Gamma(x)+1)\pi}{x}+\sin^2\pi x$$

is zero if and only if x is a prime. Hence the number of primes $<n$ is

$$\frac{1}{2\pi i}\int_1^n \frac{\Phi'(x)dx}{\Phi(x)}.$$

The sum of the kth powers of the primes $<n$ is given asymptotically.

M. Petrovitch[245] used a real function $\theta(x, u)$, like

$$a \cos 2\pi x + b \cos 2\pi u - a - b,$$

which is zero for every pair of integers x, u, and not if x or u is fractional. Let $\Phi(x)$ be the function obtained from $\theta(x, u)$ by taking

$$u = \{1+\Gamma(x)\}/x.$$

Thus $y=\Phi(x)$ cuts the x-axis in points whose abscissas are the primes.

[242]Giornale di Mat., 47, 1909, 305-320.
[243]Sitzungsber. Ak. Wiss. Wien (Math.), 121, 1912, II a, 1785-1824; 122, 1913, II a, 669-700.
[244]Rendiconti Accad. Lincei, (5), 21, II, 1912, 404-7. Wigert.[236a]
[245]Nouv. Ann. Math., (4), 13, 1913, 406-10.

E. Landau[246] indicated errors in l'intermédiaire des mathématiciens on the approximate number of primes $ax+b<N$.

*M. Kössler[247] discussed the relation between Wilson's theorem and the number of primes between two limits.

See Cesàro[64] of Ch. V, Gegenbauer[12] of Ch. XI, and papers 62–81 of Ch. XIII.

BERTRAND'S POSTULATE.

J. Bertrand[260] verified for numbers $<6\ 000\ 000$ that for any integer $n>6$ there exists at least one prime between $n-2$ and $n/2$.

P. L. Tchebychef[261] obtained limits for the sum $\theta(z)$ of the natural logarithms of all primes $\leqq z$ and deduced Bertrand's postulate that, for $x>3$, there exists a prime between x and $2x-2$. His investigation shows that for every $\epsilon>1/5$ there exists a number ξ such that for every $x\geqq\xi$ there exists at least one prime between x and $(1+\epsilon)x$.

A. Desboves,[262] assuming an unproved theorem of Legendre's,[22] concluded the existence of at least two primes between any number >6 and its double, also between the squares of two consecutive primes; also at least p primes between $2n$ and $2n-k$ for p and k given and n sufficiently large, and hence between a sufficiently large number and its square.

F. Proth[263] claimed to prove Bertrand's postulate.

J. J. Sylvester[264] reduced Tchebychef's ϵ to 0.16688.

L. Oppermann[265] stated the unproved theorem that if $n>1$ there exists at least one prime between $n(n-1)$ and n^2, and also between n^2 and $n(n+1)$, giving a report on the distribution of primes.

E. C. Catalan[266] proved that Bertrand's postulate is equivalent to

$$\frac{(2n)!}{n!\,n!}>a^a\beta^b\ldots\pi^p,$$

where $a,\ldots,\pi$ denote the primes $\leqq n$, while a is the number of odd integers among $[2n/a]$, $[2n/a^2],\ldots$, b the number among $[2n/\beta]$, $[2n/\beta^2],\ldots$. He noted (p. 31) that if the postulate is applied to $b-1$ and $b+1$, we see the existence between $2b$ and $4b$ of at least one even number equal to the sum of two primes.

J. J. Sylvester[267] reduced Tchebychef's ϵ to 0.092; D. von Sterneck[268] to 0.142.

[246]L'intermédiaire des math., 20, 1913, 179; 15, 1908, 148; 16, 1909, 20–1.
[247]Časopis, Prag, 44, 1915, 38–42.
[260]Jour. de l'école roy. polyt., cah. 30, tome 17, 1845, 129.
[261]Mém. Ac. Sc. St. Pétersbourg, 7, 1854 (1850), 17–33, 27; Oeuvres, 1, 49–70, 63. Jour. de Math., 17, 1852, 366–390, 381. Cf. Serret, Cours d'algèbre supérieure, ed. 2, 2, 1854, 587; ed. 6, 2, 1910, 226.
[262]Nouv. Ann. Math., 14, 1855, 281–295.
[263]Nouv. Corresp. Math., 4, 1878, 236–240.
[264]Amer. Jour. Math., 4, 1881, 230.
[265]Oversigt Videnskabs Selsk. Forh., 1882, 169.
[266]Mém. Soc. R. Sc. Liège, (2), 15, 1888 (=Mélanges Math., III), 108–110.
[267]Messenger Math., (2), 21, 1891–2, 120.
[268]Sitzungsb. Akad. Wiss. Wien, 109, 1900, IIa, 1137–58.

T. J. Stieltjes stated and E. Cahen[269] proved that we may take ϵ to be any positive number however small, since $\theta(z)$ is asymptotic[315-6] to z.

H. Brocard[270] stated that at least four primes lie between the squares of two consecutive primes, the first being >3. He remarked that this and the similar theorem by Desboves[262] can apparently be deduced from Bertrand's postulate; but this was denied by E. Landau.[271]

E. Maillet[272] proved there is at least one prime between two consecutive squares $<9 \cdot 10^6$ or two consecutive triangular numbers $\leq 9 \cdot 10^6$.

E. Landau[47] (pp. 89–92) proved Bertrand's postulate and hence the existence of a prime between x (excl.) and $2x$ (incl.) for every $x \geq 1$.

A. Bonolis[273] proved that, if $x>13$ is a number of p digits and a is the least integer $>x/\{10(p+1)\}$, there exist at least a primes between x and $[\frac{3}{2}x-2]$, which implies Bertrand's postulate. If $x>13$ is a number of p digits and β is the greatest integer $<x/(3p-3)$, there are fewer than β primes from x to $[\frac{3}{2}x-2]$.

MISCELLANEOUS RESULTS ON PRIMES.

H. F. Scherk[280] stated the empirical theorems: Every prime of odd rank (the nth prime 1, 2, 3, 5, . . . being of rank n) can be composed by addition and subtraction of all the smaller primes, each taken once; thus

$$13 = 1+2-3-5+7+11 = -1+2+3+5-7+11.$$

Every prime of even rank can be composed similarly, except that the next earlier prime is doubled; thus

$$17 = 1+2-3-5+7-11+2 \cdot 13 = -1-2+3-5+7-11+2 \cdot 13.$$

Märcker[281] noted that, if $a, b, \ldots, m$ are the primes between 1 and A and if p is their product, all the primes from A to A^2 are given by

$$p\left(\frac{a}{a}+\frac{\beta}{b}+\ldots+\frac{\mu}{m}+n\right),$$

and each but once if each numerator is positive and less than its denominator.

O. Terquem[282] noted that the primes $<n^2$ are the odd numbers not included in the arithmetical progressions q^2, q^2+2q, $q^2+4q, \ldots$ up to n^2, for $q=3, 5, \ldots, n-1$.

H. J. S. Smith[283] gave a theoretical method of finding the primes between the xth prime P_x and P^2_{x+1}, given the first x primes.

C. de Polignac[283a] considered the primes $\leq x$ in a progression $Km+h$.

[269]Comptes Rendus Paris, 116, 1893, 490; Thèse, 1894, 45; Ann. École Normale, (3), 11, 1894.
[270]L'intermédiaire des math., 11, 1904, 149.
[271]Ibid., 20, 1913, 177.
[272]Ibid., 12, 1905, 110-3.
[273]Atti Ac. Sc. Torino, 47, 1911–12, 576-585.
[280]Jour. für Math., 10, 1833, 201.
[281]Ibid., 20, 1840, 350.
[282]Nouv. Ann. Math., 5, 1846, 609.
[283]Proc. Ashmolean Soc., 3, 1857, 128-131; Coll. Math. Papers, 1, 37.
[283a]Comptes Rendus Paris, 54, 1862, 158-9.

E. Dormoy[284] noted that, if 2, 3, ..., r, s, t, u are the primes in natural order, all primes (and no others) $< u^2$ are given by

$$2 \cdot 3 \ldots stm + D_t a_t + tC_t D_s a_s + tsC_t C_s D_r a_r + \ldots$$
$$+ tsr \ldots 7 \cdot 5C_t C_s C_r \ldots C_5 D_3 a_3 + ts \ldots 5 \cdot 3C_t C_s \ldots C_3,$$

where C_t is found from the quotients obtained in finding the g. c. d. of t and $2 \cdot 3 \ldots rs$ by a rule which if applied to four quotients a, b, c, d consists in forming in turn 1, $p = dc + 1$, $pb + d$, $(pb + d)a + p = C_t$. Further, $D_t = tC_t \pm 1$, the sign being $+$ or $-$ according as there is an odd or an even number of operations in the g. c. d. process.

C. de Polignac[284a] wrote p_n for the nth prime and discussed the expressibility of all numbers, under a specified limit and divisible by no one of $p_1, \ldots, p_{n-1}$, in the form

$$(p_2, p_3, \ldots, p_{n-1}, p_n) + (p_3, p_4, \ldots, p_n p_1) + \ldots + (p_1, \ldots, p_{n-1}),$$

where $(a, b, \ldots)$ denotes $\pm a^\alpha b^\beta \ldots$. For example, every number < 53 and divisible by neither 2 nor 3 is of the form $\pm 3^\alpha \pm 2^\beta$.

J. J. Sylvester[285] proved that if m is prime to i and not less than n, the product $(m+i)(m+2i) \ldots (m+ni)$ is divisible by some prime $> n$.

A. A. Markow[286] found a fragment in a manuscript by Tchebychef aiding him to prove the latter's result that if μ is the greatest prime divisor of $(1+2^2)(1+4^2) \ldots (1+4N^2)$, then μ/N increases without limit with N (cf. Hermite, Cours, ed. 4, 1891, 197).

J. Iwanow[287] generalized the preceding theorem as follows: If μ is the greatest prime divisor of $(A+1^2) \ldots (A+L^2)$, then μ/L increases without limit with L.

C. Störmer[288] concluded the existence of an infinitude of primes from Tchebychef's[286] result and used the latter to prove that $i(i-1)(i-2) \ldots (i-n)$ is neither real nor purely imaginary if n is any integer $\neq 3$, and $i = \sqrt{-1}$.

E. Landau[47] (pp. 559–564) discussed the topics in the last three papers.

Braun[13a] proved that the $(n+1)$th prime is the only root $x \neq 1$ of

$$x \cdot \prod_{i=1}^{n} a_i^{[x/a_i] + [x/a_i^2] + \cdots} = x!,$$

where $a_1 = 2$, a_2, ..., a_n are the first n primes.

C. Isenkrahe[289] expressed a prime in terms of the preceding primes.

R. Le Vavasseur[290] noted that all primes between p_n and p^2_{n+1}, where p_n is the nth prime, are given by $\Sigma_{i=1}^{i=n} q_i w_i P_n / p_i$ (mod P_n), where $P_n = p_1 p_2 \cdots p_n$ and $w_i P_n / p_i \equiv 1$ (mod p_i).

[284]Comptes Rendus Paris, 63, 1866, 178–181.
[284a]Comptes Rendus Paris, 104, 1887, 1688–90.
[285]Messenger Math., 21, 1891–2, 1–19, 192. Math. Quest. Educ. Times, 56, 1892, 25.
[286]Bull. Acad. Sc. St. Pétersbourg, 3, 1895, 55–8.
[287]Ibid., 361–6.
[288]Archiv Math. og Natur., Kristiania, 24, 1901–2, No. 5.
[289]Math. Annalen, 53, 1900, 42.
[290]Mém. Ac. Sc. Toulouse, (10), 3, 1903, 36–8.

O. Meissner[291] stated that, if $n+1$ successive integers $m, \ldots, m+n$ are given, we can not in general find another set $m_1, \ldots, m_1+n$ containing a prime $m_1+\nu$ corresponding to every prime $m+\nu$ of the first set. But for $n=2$, it is supposed true that there exist an infinitude of prime pairs.

G. H. Hardy[292] noted that the largest prime dividing a positive integer x is

$$\lim_{r=\infty} \lim_{m=\infty} \lim_{n=\infty} \sum_{\nu=0}^{m} [1 - (\cos\{(\nu!)^r \pi/x\})^{2n}].$$

C. F. Gauss,[293] in a manuscript of 1796, stated empirically that the number $\pi_2(x)$ of integers $\leq x$ which are products of two distinct primes, is approximately $x \log \log x/\log x$.

E. Landau[294] proved this result and the generalization

$$\pi_\nu(x) = \frac{1}{(\nu-1)!} \cdot \frac{x(\log \log x)^{\nu-1}}{\log x} + O\left\{\frac{x(\log \log x)^{\nu-2}}{\log x}\right\},$$

where $\pi_\nu(x)$ is the number of integers $\leq x$ which are products of ν distinct primes; also related formulas for $\pi_\nu(x)$.

Several writers[295] gave numerous examples of a sum of consecutive primes equal to an exact power.

E. Landau[296] proved that the probability that a number of n digits be a prime, when n increases indefinitely, is asymptotically equal to $1/(n \log 10)$.

J. Barinaga[297] expressed the sum of the first n primes as a product of distinct primes for $n=3, 7, 9, 11, 12, 16, 22, 27, 28$, and asked if there is a general law.

Coblyn[298] noted as to prime pairs that, when $4(6p-2)!$ is divided by $36p^2-1$, the remainder is $-6p-3$ if $6p-1$ and $6p+1$ are both primes, zero if both are composite, $-2(6p+1)$ if only $6p-1$ is prime, and $6p-1$ if only $6p+1$ is prime.

J. Hammond[299] gave formulas connecting the number of odd primes $<2n$, and the number of partitions of $2n$ into two distinct primes or into two relatively prime composite numbers.

V. Brun[300] proved that, however great a is, there exist a successive composite numbers of the form $1+u^2$. There exist a successive primes no two of which differ by 2. He determined a superior limit for the number of primes $<x$ of a given class.

---

[291]Archiv Math. Phys., 9, 1905, 97.

[292]Messenger Math., 35, 1906, 145.

[293]Cf. F. Klein, Nachrichten Gesell. Wiss. Göttingen, 1911, 26–32.

[294]*Ibid.*, 361–381; Handbuch...der Primzahlen, I, 1909, 205–211; Bull. Soc. Math. France, 28, 1900, 25–38.

[295]L'intermédiaire des math., 18, 1911, 85–6.

[296]*Ibid.*, 20, 1913, 180.

[297]L'intermédiaire des math., 20, 1913, 218.

[298]Soc. Math. de France, C. R. des Séances, 1913, 55.

[299]Proc. London Math. Soc., (2), 15, 1916–7, Records of Meetings, Feb. 1916, xxvii.

[300]Nyt Tidsskrift for Matematik, B, 27, 1916, 45–58.

DIATOMIC SERIES.

A. de Polignac[305] crossed out the multiples of 2 and 3 from the series of natural numbers and obtained the "table a_2":

$$(0) \ 1 \ (2) \ (3) \ (4) \ 5 \ (6) \ 7 \ (8) \ (9) \ (10) \ 11 \ldots.$$

The numbers of terms in the successive sets of consecutive deleted numbers are $1, 3, 1, 3, 1, \ldots$, which form the "diatomic series of 3." Similarly, after deleting the multiples of the first n primes, we get a table a_n and the diatomic series of the nth prime P_n. That series is periodic and the terms after 1 of the period are symmetrically distributed (two terms equidistant from the ends are equal), while the middle term is 3. Let π_n denote the product of the primes $2, 3, \ldots, P_n$. Then the number of terms in the period is $\phi(\pi_n)$. The sum of the terms in the period is $\pi_n - \phi(\pi_n)$ and hence is the number of integers $< \pi_n$ which are divisible by one or more primes $\leqq P_n$. As applications he stated that there exists a prime between P_n and P_n^2, also between a^n and a^{n+1}. He[306] stated that the middle terms other than 3 of a diatomic series tend as n increases to become $1, 3, 7, 15, \ldots, 2^m - 1, \ldots$.

J. Deschamps[307] noted that, after suppressing from the series of natural numbers the multiples of the successive primes $2, 3, \ldots, p$, the numbers left form a periodic series of period $2 \cdot 3 \ldots p$; and similar theorems. Like remarks had been made previously by H. J. S. Smith.[308]

ASYMPTOTIC DISTRIBUTION OF PRIMES.

P. L. Tchebychef's[261] investigation shows that for x sufficiently large the number $\pi(x)$ of primes $\leqq x$ is between $0 \cdot 921Q$ and $1 \cdot 106Q$, where $Q = x/\log x$. He[314] proved that the limit, if existent, of $\pi(x)/Q$ for $x = \infty$ is unity. J. J. Sylvester[267] obtained by the same methods the limits $0 \cdot 95Q$ and $1 \cdot 05Q$.

By use of the function $\zeta(s) = \Sigma_{n=1}^{n=\infty} n^{-s}$ of Riemann, J. Hadamard[315] and Ch. de la Vallée-Poussin[316] independently proved that the sum of the natural logarithms of all primes $\leqq x$ equals x asymptotically. Hence follows the fundamental theorem that $\pi(x)$ is asymptotic to Q, i. e.,

$$\lim_{x=\infty} \pi(x) \cdot \frac{\log x}{x} = 1.$$

[305]Recherches nouvelles sur les nombres premiers, Paris, 1851, 28 pp. Abstract in Comptes Rendus Paris, 29, 1849, 397–401, 738–9; same in Nouv. Ann. Math., 8, 1849, 423–9. Jour. de Math., 19, 1854, 305–333.

[306]Nouv. Ann. Math., 10, 1851, 308–12.

[307]Bull. Soc. Philomathique de Paris, (9), 9, 1907, 102–112

[308]Proc. Ashmolean Soc., 3, 1857, 128–131; Coll. Math. Papers, 1, 36.

[314]Mém. Ac. Sc. St. Pétersbourg, 6, 1851, 146; Jour. de Math., 17, 1852, 348; Oeuvres, 1, 34.

[315]Bull. Soc. Math. de France, 24, 1896, 199–220.

[316]Annales de la Soc. Sc. de Bruxelles, 20, II, 1896, 183–256.

Now Q is asymptotic to the "integral logarithm of x":

$$Lix = \lim_{\delta=0} \left(\int_0^{1-} \frac{du}{\log u} + \int_{1+\delta}^x \frac{du}{\log u} \right),$$

so that the latter is asymptotic to $\pi(x)$. De la Vallée-Poussin[317] proved that Lix represents $\pi(x)$ more exactly than $x/\log x$ and its remaining approximations

$$\frac{x}{\log x} + \frac{x}{\log^2 x} + \ldots + \frac{(m-1)!\,x}{\log^m x}.$$

The history of this extensive subject is adequately presented in the luminous and exhaustive text by E. Landau,[47] in which is given (pp. 908–961) a complete list of references. The reader may consult the article by J. Hadamard,[318] the extensive report by G. Torelli,[319] the summaries by Landau,[320] also G. H. Hardy and J. E. Littlewood,[321] and the recent papers 42–44 of Chapter XIX.

---

[317]Mém. Couronnés Acad. Roy. Belgique, 59, 1889, 1–74.

[318]Encyclopédie des sc. math., tome I, vol. 3, pp. 310–345.

[319]Atti R. Accad. Sc. Fis. Mat., Napoli, (2), 11, 1902, No. 1, 222 pp.

[320]Proc. Fifth Internat. Congress, Cambridge, I, 1913, 93–108.   Math. Zeitschrift, 1, 1918, 1–24, 213–9.

[321]Acta Math., 41, 1917, 119–196.

CHAPTER XIX.

INVERSION OF FUNCTIONS; MÖBIUS' FUNCTION $\mu(n)$; NUMERICAL INTEGRALS AND DERIVATIVES.

INVERSION; FUNCTION $\mu(n)$.

A. F. Möbius[1] defined the function $\mu(n)$ to be zero if n is divisible by a square >1, but to be $(-1)^k$ if n is a product of k distinct primes >1, while $\mu(1) = 1$. He employed the function in the reversion of series:

$$F(x) = \sum_{s=1}^{\infty} \frac{f(sx)}{s^n} \text{ implies } f(x) = \sum_{s=1}^{\infty} \mu(s) \frac{F(sx)}{s^n}.$$

His results were expressed in more general form by Glaisher[69] and cited in Chapter X. See also E. Meissel,[2] who[3] noted that

$$(1) \qquad \sum_{j=1}^{m} \mu(j) \left[\frac{m}{j} \right] = 1.$$

R. Dedekind[4] proved that, if $F(m) = \Sigma f(d)$, where d ranges over all the divisors of m, then

$$(2) \qquad f(m) = F(m) - \Sigma F\left(\frac{m}{a}\right) + \Sigma F\left(\frac{m}{ab}\right) - \Sigma F\left(\frac{m}{abc}\right) + \ldots,$$

where the summations extend over all the combinations $1, 2, 3, \ldots$ at a time of the distinct prime factors $a, b, \ldots, k$ of m. The proof follows from a distribution of all the factors of m into two sets S and T. Put all the divisors of m into set S; all divisors of m/a into set T, all of m/b into T, etc.; all divisors of $m/(ab)$ into S, all of $m/(ac)$ into S, etc.; all of $m/(abc)$ into T, etc. Then, with the exception of m itself, every divisor of m occurs as often in the set S as in the set T. In particular, for Euler's $\phi(m)$, $m = \Sigma\phi(d)$, whence

$$\phi(m) = m - \Sigma\frac{m}{a} + \Sigma\frac{m}{ab} - \ldots = m\left(1 - \frac{1}{a}\right)\left(1 - \frac{1}{b}\right)\ldots.$$

For another example, see Dedekind[71] of Ch. VIII. Similarly, $F(m) = \Pi f(d)$ implies

$$f(m) = \frac{F(m) \; \Pi F\left(\dfrac{m}{ab}\right) \ldots}{\Pi F\left(\dfrac{m}{a}\right) \Pi F\left(\dfrac{m}{abc}\right) \ldots}.$$

J. Liouville[5] stated simultaneously with Dedekind the inversion theorem for sums and made the same application to $\phi(m)$.

Liouville[6] stated the theorem for sums as a problem.

[1] Jour. für Math., 9, 1832, 105; Werke, 4, 591. He wrote a_n for $\mu(n)$.
[2] Ibid., 48, 1854, 301–316.
[3] Observationes quaedam in theoria numerorum, Berlin, 1850, pp. 3–6.
[4] Jour. für Math., 54, 1857, pp. 21, 25.
[5] Jour. de Mathématiques, (2), 2, 1857, 110–2.
[6] Nouv. Ann. Math., 16, 1857, 181–2.

B. Merry[7] gave a proof by noting that, if d is any divisor of m, and if q of the prime factors of m occur to the same power in d as in m, then $f(d)$ occurs once in $F(m)$, q times in $\Sigma F(m/a)$, $q(q-1)/2$ times in $\Sigma F(m/ab)$, etc. Thus the coefficient of $f(d)$ in (2) is

$$1-q+\frac{q(q-1)}{1\cdot 2}-\ldots=(1-1)^q=0$$

if $q>0$, but is unity if $q=0$, i. e., if $d=m$. This proof is only another way of stating Dedekind's proof.

R. Dedekind[8] gave another form and proof of his theorems. Let

$$m\left(1-\frac{1}{a}\right)\left(1-\frac{1}{b}\right)\ldots=\Sigma\nu_1-\Sigma\nu_2,$$

where ν_1 ranges over the positive terms of the expanded product and $-\nu_2$ over the negative terms. A simple proof shows that, if ν is any divisor $<m$ of m, there are as many terms ν_1 divisible by ν as terms ν_2 divisible by ν. Thus

$$\Sigma f(\nu)=F(m),\qquad \Pi f(\nu)=F(m)$$

imply, respectively,

$$f(m)=\Sigma F(\nu_1)-\Sigma F(\nu_2),\qquad f(m)=\frac{\Pi F(\nu_1)}{\Pi F(\nu_2)}.$$

Liouville[8a] wrote $F(n)=\Sigma f(n/D^\mu)$, where D ranges over those divisors of $n=a^\alpha b^\beta\ldots$ for which D^μ divides n. Then

$$f(n)=F(n)-\Sigma F(n/a^\mu)+\Sigma F(n/a^\mu b^\mu)-\ldots.$$

E. Laguerre[9] expressed (2) in the form

(3)
$$f(m)=\Sigma\mu\left(\frac{m}{d}\right)F(d),$$

where d ranges over the divisors of m. Let

$$\sum_{n=1}^{\infty}f(n)\frac{x^n}{1-x^n}=\sum_{n=1}^{\infty}F(n)x^n,$$

whence $F(m)=\Sigma f(d)$. For $m=\Pi p^a$, where the p's are distinct primes, let $f(m)=\Pi f(p^a)$, and $f(p^n)=p^{n-1}(p-1)$. Then

$$F(m)=\Pi\{1+f(p)+\ldots+f(p^{a-1})\}=\Pi p^a=m.$$

The hypotheses are satisfied if f is Euler's function ϕ. This discussion deduces $\Sigma\phi(d)=m$ from the usual expression of type (3) for $\phi(m)$, rather than the reverse as claimed.

N. V. Bougaief[10] proved (1).

F. Mertens[11] defined $\mu(n)$ and noted that $\Sigma\mu(d)=0$ if $n>1$, where d ranges over the divisors of n.

[7]Nouv. Ann. Math. 16, 1857, 434.
[8]Dirichlet's Zahlentheorie, mit Zusätzen von Dedekind, 1863, §138; ed. 2, 1871, p. 356; ed. 4, 1894, p. 360.
[8a]Jour. de Math., (2), 8, 1863, 349.
[9]Bull. Soc. Math. France, 1, 1872-3, 77-81.
[10]Mat. Sbornik (Math. Soc. Moscow), 6, 1872-3, 179. Cf. Sterneck.[19]
[11]Jour. für Math., 77, 1874, 289; 78, 1874, 53.

E. Cesàro[12] proved formulas, quoted in Ch. X, which include (3) as a special case. His erroneous evaluation of the mean of $\mu(n)$ is cited there. Cesàro[13] reproduced the general formula just cited and extended it to three pairs of functions:

$$\Sigma f_1(d)F_1\left(\frac{n}{d}\right) = \Sigma f_2(d)F_2\left(\frac{n}{d}\right) = \Sigma f_3(d)F_3\left(\frac{n}{d}\right),$$

$$F_1(n) = \Sigma f_2(d)f_3\left(\frac{n}{d}\right), \qquad F_2 = \Sigma f_3 f_1, \qquad F_3 = \Sigma f_1 f_2.$$

where, in each, d ranges over the divisors of n.

Cesàro[14] noted that, if $h(n)+k(n)=1$ and

$$H(n) = h(p)h(q)\ldots, \qquad K(n) = k(p)k(q)\ldots,$$

where $p, q, \ldots$ are the prime factors of n, then

$$H(n) = \Sigma \mu(d)K(d), \qquad K(n) = \Sigma \mu(d)H(d).$$

For $h(n) = k(n) = 1/2$, then $H(n) = K(n)$ is the reciprocal of the number of divisors, without square factors, of n.

Cesàro[15] treated the inversion of series. Let $\Omega(x) = 1$ or 0, according as x is or is not in a given set Ω of integers. Let $\Omega(x)\Omega(y) = \Omega(xy)$. Let $\epsilon_i(x)$ be functions such that $\epsilon_a\{\epsilon_\beta(x)\} = \epsilon_{a\beta}(x)$ for every pair of indices a, β. Then

$$F(x) = \sum_\omega h(\omega)f\{\epsilon_\omega(x)\},$$

where ω ranges over all the numbers of Ω, implies that

$$f(x) = \sum_\omega H(\omega)F\{\epsilon_\omega(x)\},$$

if the sum $\Sigma h(d)H(n/d)$, for d ranging over the divisors of n, equals 1 or 0 according as $n=1$ or $n>1$. Cf. Möbius[1].

N. V. Bougaief[16] considered the function $\nu(x)$ with the value log p if x is a power of a prime p, the value 0 in all other cases. Then, if d ranges over the divisors of n, $\Sigma\nu(d) = \log n$ implies $\Sigma\mu(d) \log d = -\nu(n)$.

H. F. Baker[17] gave a generalization of the inversion formula, the statement of which will be clearer after the consideration of one of his applications of it. Let $a_1,\ldots, a_n$ be distinct primes and S any set of positive integers. For $k \leqq n$, let $F(a_1,\ldots, a_k)$ denote the set of all the numbers in S which are divisible by each of the primes $a_{k+1}, a_{k+2},\ldots, a_n$, so that $F(a_1,\ldots, a_n) = S$. For $k=0$, write $F(0)$ for F, so that $F(0)$ consists of the numbers of S which are divisible by $a_1,\ldots, a_n$. Returning to the general $F(a_1,\ldots, a_k)$, we divide it into sub-sets. Those of its numbers which are divisible by no one of $a_1,\ldots, a_k$ form the sub-set $f(a_1,\ldots, a_k)$. Those divisible by a_1, but by no one of $a_2,\ldots, a_k$, form the sub-set $f(a_2, a_3,\ldots, a_k)$.

[12]Mém. soc. roy. sc. de Liège, (2), 10, 1883, No. 6, pp. 26, 47, 56–8.
[13]Giornale di Mat., 23, 1885, 168 (175).
[14]Ibid., 25, 1887, 14–19. Cf. 1–13 for a type of inversion formulas.
[15]Annali di Mat., (2), 13, 1885, 339; 14, 1886–7, 141–158.
[16]Comptes Rendus Paris, 106, 1888, 652–3. Cf. Cesàro, ibid., 1340–3; Cesàro,[12] pp. 315–320; Bougaief, Mat. Sbornik (Math. Soc. Moscow), 13, 1886–8, 757–77; 14, 1888–90, 1–44, 169–201; 18, 1896, 1–54; Kronecker[30] (p. 276); Berger[17a] (pp. 106–115); Gegenbauer[12] of Ch. XI—all on $\Sigma\mu(d) \log d$.
[17]Proc. London Math. Soc., 21, 1889–90, 30–32.

Those divisible by a_1 and a_2, but by no one of $a_3, \ldots, a_k$, form the sub-set $f(a_3, \ldots, a_k)$. Finally, those divisible by $a_1 \ldots, a_k$ form the sub-set designated $f(0)$. Thus

$$F(a_1, a_2, \ldots, a_k) = f(a_1, a_2, \ldots, a_k) + \sum_1 f(a_2, a_3, \ldots, a_k) + \sum_2 f(a_3, a_4, \ldots, a_k)$$
$$+ \ldots + \sum_{n-1} f(a_1) + f(0),$$

where $\sum$ indicates that the summation extends over all combinations of $a_1, \ldots, a_r$ taken* $k - r$ at a time.

When we have any such set or function $f(a_1, \ldots, a_k)$, uniquely determined by $a_1, \ldots, a_k$, independently of their order, and we define F by the foregoing formula, then we have the inverse formula

$$f(a_1, a_2, \ldots, a_n) = F(a_1, a_2, \ldots, a_n) - \sum_1 F(a_2, a_3, \ldots, a_n) + \sum_2 F(a_3, a_4, \ldots, a_n)$$
$$- \ldots + (-1)^{n-1} \sum_{n-1} F(a_1) + (-1)^n F(0),$$

where $\sum_r$ now indicates that the summation extends over all the combinations of $a_1, \ldots, a_n$ taken $n - r$ at a time. The proof is just like that by B. Merry for Dedekind's formula.

To give an example, let $n = 2$, $a_1 = 2$, $a_2 = 3$, $S = 3, 4, 6, 8$. Then $F(a_1) = 3, 6$; $f(a_1) = 3$, $f(0) = 6$; $F(a_2) = 4, 6, 8$; $f(a_2) = 4, 8$. Thus

$$F(a_1, a_2) - F(a_1) - F(a_2) + F(0) = S - (3, 6) - (4, 6, 8) + 6 \equiv 0 = f(a_1, a_2).$$

A. Berger[17a] called f_1 conjugate to f_2 if $\sum f_1(d) f_2(d) = 1$ for $k = 1$, 0 for $k > 1$, when d ranges over the divisors of k. Let $g(mn) = g(m)g(n)$, $g(1) = 1$. Write $h(k) = \sum f(d) f_1(\delta) g(\delta)$, where $d\delta = k$. Then $f(k) = \sum f_2(n) g(d) h(\delta)$. Dedekind's inversion formula is a special case. For, if $f_1(n) \equiv 1$, then $f_2(n) \equiv \mu(n)$.

K. Zsigmondy[18] stated that if, for every positive integer r,

$$\sum_r f(r_c) = F(r),$$

where c ranges over all combinations of powers $\leq r$ of the relatively prime positive integers $n_1, \ldots, n_\rho$, while r_c denotes the greatest integer $\leq r/c$, then

$$f(r) = F(r) - \sum_n F(r_n) + \sum_{n,n'} F(r_{nn'}) - \ldots,$$

where the summation indices $n, n', \ldots$ range over the combinations of $n_1, \ldots, n_\rho$ taken $1, 2, \ldots$ at a time.

R. D. von Sterneck[19] noted that, if d ranges over the divisors of n $\sum \theta(d) = \psi(n)$ implies that

$$F(m) \equiv \psi(1) + \ldots + \psi(m) = \sum_{=1}^{m} \theta(j) \left[\frac{m}{j} \right].$$

Taking $m = 1, \ldots, n$ and solving, we get $\theta(n)$ expressed as a determinant of order n, whence

$$\theta(n) = \psi(n) - \sum \psi(d_1) + \sum \psi(d_2) - \ldots + (-1)^\nu \psi(d_\nu),$$

if $n = p_1^{a_1} \ldots p_\nu^{a_\nu}$ and d_ρ is derived from n by reducing ρ exponents by unity.

*Here and in the statement of the theorem occur confusing misprints for k and n.
[17a]Nova Acta Regiae Soc. Sc. Upsaliensis, (3), 14, 1891, No. 2, 46, 104.
[18]Jour. für Math., 111, 1893, 346. Applied in Ch. V, Zsigmondy.[77]
[19]Monatshefte Math. Phys., 4, 1893, 53–6.

P. Bachmann[20] proved that $f(n) = \sum_{k=1}^{k=\infty} F(kn)$ implies that

$$F(1) = \sum_{n=1}^{\infty} \mu(n)f(n).$$

Write $X = [x/n]$. Taking $F(n) = X$, nX, $\Phi(X)$, whence $f(n) = T(X)$, $n\sigma(X)$, $D(X)$, respectively, we obtain Lipschitz's[50] (Ch. X) formulas:

$$[x] = \sum_{n=1}^{[x]} \mu(n)T\left[\frac{x}{n}\right] = \Sigma\mu(n)n\sigma\left[\frac{x}{n}\right], \qquad \Phi[x] = \Sigma\mu(n)D\left[\frac{x}{n}\right].$$

Let $F(n)$ be zero if n is not a divisor of P and write $\psi(P/n)$ for $F(n)$. Hence if d divides P, $f(d) = \Sigma\psi(P/kd)$ implies $\psi(P) = \Sigma\mu(d)f(d)$, where k ranges over the divisors of P/d, and d over those of P.

D. von Sterneck[21] considered a function $f(n)$ with the properties: (i) $f(1) = 1$; (ii) the g. c. d. of $f(m)$ and $f(n)$ is $f(d)$ if d is the g. c. d. of m and n; (iii) for primes p, other than specified ones, one of the numbers $f(p\pm 1)$ is divisible by p; (iv) the g. c. d. of $f(\rho n)/f(n)$ and $f(n)$ divides ρ. Then if $L(n)$ is the l. c. m. of the values of f for all the divisors $< n$ of n, $F(n) = f(n) \div L(n)$ is an integer which can be given the form

$$F(n) = \frac{f(n)\Pi f\left(\dfrac{n}{p_i p_j}\right)\Pi f\left(\dfrac{n}{p_i p_j p_k p_l}\right)\cdots}{\Pi f\left(\dfrac{n}{p_i}\right)\Pi f\left(\dfrac{n}{p_i p_j p_k}\right)\cdots}, \qquad n = \Pi p_i^{l_i}.$$

The four properties hold for the function defined by the recursion formula $f(n) = af(n-1) + \beta f(n-2)$, where a and β are relatively prime, with the initial conditions $f(1) = 1, f(2) = a$. For $a = 2x, \beta = b - x^2$, we have[22]

$$f(n) = \frac{(x+\sqrt{b})^n - (x-\sqrt{b})^n}{2\sqrt{b}}.$$

The case $a = \beta = 1$ was discussed by Lucas[28] of Ch. XVII, and his test for primality holds for the present generalization.

The four properties hold also for

$$f(n) = \frac{a^n - b^n}{a - b},$$

if a, b are relatively prime;[23] then $f(p-1)$ is divisible by p if p is a prime not dividing a, b or $a-b$.

K. Zsigmondy[24] gave a generalized inversion formula. Let N be any multiple of the relatively prime integers $n_1, \ldots, n_s$. Set

$$F(m, N, \epsilon) = \Sigma f\left(\left[\frac{m}{d}\right], \frac{N}{d}, \epsilon d\right),$$

where d ranges over those divisors $\leq m$ of N which are products of powers

[20]Die Analytische Zahlentheorie, 1894, 310.
[21]Monatshefte Math. Phys., 7, 1896, 37, 342.
[22]Dirichlet, Werke, 1, 47–62. See Dirichlet,[9] Ch. XVII.
[23]Zsigmondy, Monatshefte Math. Phys., 3, 1892, 265.
[24]Ibid., 7, 1896, 190–3.

of $n_1, \ldots, n_s$. Then, if a ranges over those divisors d which are divisible by no one of $\nu_1, \ldots, \nu_{s'}$, chosen from $n_1, \ldots, n_s$,

$$\sum_a f\left(\left[\frac{m}{a}\right], \frac{N}{a}, \epsilon a\right) = F(m, N, \epsilon) - \sum_{\nu} F\left(\left[\frac{m}{\nu}\right], \frac{N}{\nu}, \epsilon\nu\right)$$
$$+ \sum_{\nu, \nu'} F\left(\left[\frac{m}{\nu\nu'}\right], \frac{N}{\nu\nu'}, \epsilon\nu\nu'\right) - \ldots$$

The left member equals $F(m, N, \epsilon)$ constructed for the numbers other than $\nu_1, \ldots, \nu_{s'}$ of the set $n_1, \ldots, n_s$. For $s' = s$, we have

$$f(m, N, \epsilon) = F(m, N, \epsilon) - \sum_n F\left(\left[\frac{m}{n}\right], \frac{N}{n}, \epsilon n\right) + \ldots$$

The latter becomes the series in Bachmann[20] when $m = \infty$, $N = 0$, $\epsilon = 1$, while $n_1, n_2, \ldots$ are primes.

F. Mertens[25] considered $\sigma(n) = \mu(1) + \mu(2) + \ldots + \mu(n)$ and proved that

$$\sum_{j=g+1}^{n} \mu(j)\left[\frac{n}{j}\right] = \sigma\left(\frac{n}{1}\right) + \sigma\left(\frac{n}{2}\right) + \ldots + \sigma\left(\frac{n}{g}\right) - g\sigma(g),$$
$$\sigma(n) = 2\sigma(g) - \sum_{r, s=1}^{g} \mu(r)\mu(s)\left[\frac{n}{rs}\right], \qquad g = [\sqrt{n}].$$

By means of a table (pp. 781–830) of the values of $\sigma(n)$ and $\mu(n)$ for $n < 10000$, it is verified that $|\sigma(n)| < \sqrt{n}$ for $1 < n < 10000$.

D. von Sterneck[26] verified the last result up to 500 000, and for 16 larger values under 5 million.

A. Berger[27] noted that, if $g(m)g(n) = g(mn)$, $g(1) = 1$,
$$\sum \mu(d)g(d) = \Pi\{1 - g(p)\} \qquad (n > 1),$$
where d ranges over all divisors of n, p over the prime divisors of n. If $\sum g(m)$ is absolutely convergent,
$$\sum_{k=1}^{\infty} \mu(k)g(k) = \Pi\{1 - g(p)\},$$
where p ranges over all primes.

D. von Sterneck[28] noted that, if $\theta(x) \leqq 1$ for every x and if
$$\sum_{x=1}^{n} \theta(x)\left[\frac{n}{x}\right] = f(n),$$
then
$$\left|\sum_{k=1}^{n} \theta(k)\right| < \frac{n}{9} + 6 + \left|f(n) - f\left[\frac{n}{2}\right] - f\left[\frac{n}{3}\right] - f\left[\frac{n}{6}\right]\right|.$$

In particular, $|\sum \mu(k)| < 8 + n/9$.

D. F. Seliwanov[29] gave Dedekind's formula with application to $\phi(n)$.

H. von Koch[29a] defined $\mu(k)$ by use of infinite determinants.

[25]Sitzungsber. Ak. Wiss. Wien (Math.), 106, II a, 1897, 761–830.
[26]Ibid., 835–1024; 110, II a, 1901, 1053–1102; 121, II a, 1912, 1083–96; Proc. Fifth Intern. Congress Math., 1912, I, 341–3.
[27]Öfversigt Vetenskaps-Akad. Förhand., Stockholm, 55, 1898, 579–618.
[28]Monatshefte Math. Phys., 9, 1898, 43–5.
[29]Math. Soc. St. Pétersbourg, 1899, 120.
[29a]Öfversigt K. Vetensk.-Akad. Förhand., Stockholm, 57, 1900, 659–68.

E. B. Elliott[96] of Ch. V gave a generalization of $\mu(n)$.

L. Kronecker[30] defined the function $\rho(n, k)$ of the g. c. d. (n, k) of n, k to be 1 if $(n, k)=1$, 0 if $(n, k)>1$, and proved for any function $f(n, k)$ of (n, k) the identity

$$\sum_{k=1}^{n} \rho(n, k)f(n, k) = \sum_{d}\sum_{k=1}^{n/d} \mu(d)f(n, kd),$$

where d ranges over the divisors of n. The left member is thus the sum of the values of $f(n, k)$ for $k<n$ and prime to n. Set

$$F(n, d) = \sum_{k=1}^{n/d} f(n, kd), \qquad \Phi(n, d) = \sum_{k=1}^{n/d} \rho\left(\frac{n}{d}, k\right)f(n, kd).$$

Thus when d ranges over the divisors of n,

$$F(n, 1) = \sum_{d}\Phi(n, d), \qquad \Phi(n, 1) = \Sigma\mu(d)F(n, d)$$

are consequences of each other. The same is true (p. 274) for

$$h(n) = \Sigma f(d)g\left(\frac{n}{d}\right), \qquad f(n) = \Sigma\mu(d)g(d)h\left(\frac{n}{d}\right),$$

if $g(rs) = g(r)g(s)$. Application is made (p. 335) to mean values.

E. M. Lémeray[31] gave a generalized inversion theorem. Let $\psi_2(a, b)$ be symmetrical in a, b and such that the function ψ_3 defined by

$$\psi_3(a, b, c) = \psi_2\{a, \psi_2(b, c)\}$$

is symmetrical in a, b, c. Then the function

$$\psi_4(a, b, c, d) = \psi_3\{a, b, \psi_2(c, d)\}$$

will be symmetrical in a, b, c, d and similarly for $\psi_k(a_1,\ldots, a_k)$. For example,

$$\psi_2(a, b) = a\sqrt{1+b^2}+b\sqrt{1+a^2}, \qquad \psi_3 = abc+\Sigma a\sqrt{1+b^2}\sqrt{1+c^2}.$$

Let $v=\Omega(y, u)$ be the solution of $y=\psi_2(u, v)$ for v. The theorem states that, if $d_1,\ldots, d_k$ are the divisors of $m = p^\alpha q^\beta r^\gamma\ldots$ and if $F(m)$ be defined by

$$F(m) = \psi_k\{f(d_1),\ldots, f(d_k)\},$$

we have inversely $f(m) = \Omega(G, H)$, where

$$G = \psi_\mu\left\{F(m), F\left(\frac{m}{pq}\right), F\left(\frac{m}{pr}\right),\ldots, F\left(\frac{m}{pqrs}\right),\ldots\right\},$$

$$H = \psi_\nu\left\{F\left(\frac{m}{p}\right), F\left(\frac{m}{q}\right),\ldots, F\left(\frac{m}{pqr}\right),\ldots\right\},$$

where μ is the number of combinations of the distinct prime factors p, $q,\ldots$ of m taken $0, 2, 4,\ldots$ at a time, and ν the number taken $1, 3, 5,\ldots$ at a time.

L. Gegenbauer[32] defined $\mu(x)$ to be $+1$ if x is a unit of the field $R(i)$ of complex integers or a product of an even number of distinct primes of

[30]Vorlesungen über Zahlentheorie, I, 1901, 246–257. His ϵ_n is $\mu(n)$.

[31]Nouv. Ann. Math., (4), 1, 1901, 163–7.

[32]Verslag. Wiss. Ak. Wetenschappen, Amsterdam, 10, 1901–2, 195–207 (German.) English transl. in Proc. Sect. Sc. Ak. Wet., 4, 1902, 169–181

$R(i)$, -1 if a product of an odd number, 0 if x is divisible by the square of a prime of $R(i)$. Let $\{m\}$ denote a complete set of residues $\neq 0$ of complex integers modulo m. Then the sum of the values of $f(x)$ for all complex integers x relatively prime to a given one n, which are in $\{m\}$, equals $\Sigma\mu(d)\Sigma f(dx)$, where d ranges over all divisors of n in $\{m\}$, and x ranges over $\{m/d\}$. This is due, for the case of real numbers, to Nazimov[167] of Chapter V. Again, $\Sigma\mu(d) = 1$ or 0 according as norm n is 1 or >1. Also $\Sigma f(d) = F(n)$ implies $\Sigma\mu(d)F(n/d) = f(n)$.

J. C. Kluyver[33] employed Kronecker's[30] identity for special functions f and obtained known results like

$$\Sigma \cos \frac{2\pi\nu}{n} = \mu(n), \qquad \Pi \, 2\sin\frac{\pi\nu}{n} = e^{\gamma(n)},$$

where ν ranges over the integers $<n$ and prime to n, while $\gamma(n)$ is Bougaief's[16] function $\nu(n)$.

P. Fatou[34] noted that Merten's $\sigma(n)$ does not oscillate between finite limits. E. Landau[35] proved that it is at most of the order of ne^t, where $t = -a\sqrt{\log n}$. Landau[36] noted that Furlan[37] made a false use of analysis and ideal theory to obtain a result of Landau's on Merten's[25] $\sigma(n)$.

O. Meissner[37a] employed primes p_i, q_i. For $n = \Pi p_i{}^{e_i}$ set $Z(n) = \Pi e_i{}^{p_i}$ and $Z_2(n) = Z\{Z(n)\}$. Then $Z(n) = n$ only if n is $\Pi p_i{}^{p_i}$ or 16 or $\Pi p_i{}^{q_i} q_i{}^{p_i}$. Next, $Z_2(n) = n$ in these three cases and when the exponents e_i in n are distinct primes; otherwise, $Z_2(n) < n$. We have $[1/Z(n)] = \mu^2(n)$.

R. Hackel[38] extended the method of von Sterneck[28] and obtained various closer approximations, one[39] being

$$\left| \Sigma\theta(k) \right| < \frac{n}{26} + 152 + \left| \Sigma f\left[\frac{n}{a}\right] - \Sigma f\left[\frac{n}{b}\right] \right|,$$

where $a = 1, 6, 10, 14, 105;\ b = 2, 3, 5, 7, 11, 13, 385, 1001$.

W. Kusnetzov[40] gave an analytic expression for $\mu(n)$.

K. Knopp[160] of Ch. X gave many formulas involving $\mu(n)$.

A. Fleck[40a] generalized $\mu(m) \equiv \mu_1(m)$ by setting

$$\mu_k(m) = \prod_{i=1}^{\lambda} (-1)^{a_i}\binom{k}{a_i}, \qquad m = \prod_{i=1}^{\lambda} p_i{}^{a_i}.$$

Using the zeta function (12) of Ch. X, and ϕ_k of Fleck[125] of Ch. V, we have

$$\sum_{d|m} \mu_k(d) = \mu_{k-1}(m), \qquad \sum_{m=1}^{\infty} \frac{\mu_{k-1}(m)}{m^s} = \zeta(s) \sum_{m=1}^{\infty} \frac{\mu_k(m)}{m^s}, \qquad \phi_k(m) = \sum_{d|m} d\mu_{k+1}\left(\frac{m}{d}\right).$$

[33]Verslag. Wiss. Ak. Wetenschappen, Amsterdam, 15, 1906, 423–9.   Proc. Sect. Sc. Ak. Wet., 9, 1906, 408–14.   [34]Acta Math., 30, 1906, 392.
[35]Rend. Circ. Mat. Palermo, 26, 1908, 250.   [36]Rend. Circ. Mat. Palermo, 23, 1907, 367–373.
[37]Monatshefte Math. Phys., 18, 1907, 235–240.
[37a]Math. Naturw. Blätter, 4, 1907, 85–6.
[38]Sitzungsber. Ak. Wiss. Wien (Math.), 118, 1909, II a, 1019–34.
[39]Sylvester, Messenger Math., (2), 21, 1891–2, 113–120.
[40]Mat. Sbornik (Math. Soc. Moscow), 27, 1910, 335–9.
[40a]Sitzungsber. Berlin Math. Gesell., 15, 1915, 3–8.

The theorem $\Sigma_{n=1}^{\infty}\mu(n)/n = 0$ and other results on sums involving $\mu(n)$ play an important rôle in the theory of the asymptotic distribution of primes. In accord with the plan of not entering into details on that topic (Ch. XVIII), the reader is referred for the former topic to the history and exposition by E. Landau,[41] and to the subsequent papers by A. Axer,[42] E. Landau,[43] and J. F. Steffensen.[44]

Proofs of (2) or (3) are given in the following texts:

P. Bachmann, Die Lehre von der Kreistheilung, 1872, 8–11; Die Elemente der Zahlentheorie, 1892, 40–4; Grundlehren der Neueren Zahlentheorie, 1907, 26–9.

T. J. Stieltjes, Théorie des nombres, Ann. fac. Toulouse, 4, 1890, 21.

Borel and Drach, Introd. théorie des nombres, 1895, 24–6.

E. Cahen, Éléments de la théorie des nombres, 1900, 346–350.

E. Landau,[41] 577–9.

NUMERICAL INTEGRALS AND DERIVATIVES.

N. V. Bougaief[55] (Bugaiev) called $F(n)$ the numerical integral of $f(n)$ if $F(m) = \Sigma f(\delta)$, summed for all the divisors δ of m, and called $f(n)$ the numerical derivative function of $F(n)$, denoted by $DF(n)$ symbolically.

Granting that there is, for every n, the development

$$F(n) = a_1[n] + a_2\left[\frac{n}{2}\right] + a_3\left[\frac{n}{3}\right] + \cdots$$

where $[x]$ is the largest integer $\leq x$, then a_k is the numerical derivative of $F(k) - F(k-1)$. He developed $[n^{1/2}]$, $[n^{1/3}]$, etc.

N. V. Bougaief,[56] after amplifying the preceding remarks, proved that

$$\underset{d\delta=n}{\Sigma}\ \theta(\delta)\chi(d) = \psi(n), \qquad \theta(n)\theta(m) = \theta(nm)$$

imply

$$\chi(n) = \theta(n)D\left\{\frac{\psi(n)}{\theta(n)}\right\}.$$

Writing $D^{-1}\theta(d)$ for $\Sigma\theta(d)$, summed for the divisors d of n, we have

$$D^{\mu}\Sigma\chi(\delta)\theta(d) = \Sigma\chi(\delta)D^{\mu}\theta(d),$$

for any integer μ, positive or negative. There are formulas like

[41]Handbuch...Verteilung der Primzahlen, II, 1909, 567–637, 676–96, 901–2.

[42]Prace mat. fiz., Warsaw, 21, 1910, 65–95; Sitzungsber. Ak. Wiss. Wien (Math.), 120, 1911, II a, 1253–98.

[43]Sitzungsber. Ak. Wiss. Wien. (Math.), 120, 1911, IIa, 973–88; Rend. Circ. Mat. Palermo, 34, 1912, 121–31.

[44]Analytiske Studier..., Diss., Kjobenhavn, 1912, 148 pp. Fortschritte, 43, 262–3. Extract in Acta Math., 37, 1914, 75–112.

[55]Journal de la Soc. Philomatique de Moscou, 5, 1871.

[56]Theory of numerical derivatives, Moscow, 1870–3, 222 pp. Extracts from Mat. Sbornik (Math. Soc. Moscow), 5, I, 1870–2, 1–63; 6, 1872–3, I, 133–180, 199–254, 309–360 (reviewed in Bull. Sc. Math. Astr., 3, 1872, 200–2; 5, 1873, 296–8; 6, 1874, 314–6). Résumé by Bougaief, Bull. Sc. Math. Astr., 10, I, 1876, 13–32.

$$n = \sum_{u=1}^{n} \mu^2(u)[\sqrt{n/u}], \qquad \sum_{u=1}^{n} \phi(u) = \tfrac{1}{2} + \tfrac{1}{2}\Sigma\mu(u)\left[\frac{n}{u}\right]^2,$$

$$\Sigma\theta\left(\frac{n}{a}\right) \equiv \theta(\sqrt{n}) \ (\text{mod } 2), \qquad \Sigma\theta\left(\frac{n}{a^2}\right) - \Sigma\theta\left(\frac{n}{ab}\right) \equiv \theta(\sqrt[3]{n}) \ (\text{mod } 3),$$

where $\theta(n)$ is the number of primes $a \leq n$. Other special results were cited under 155, Ch. V; 6, Ch. XI; 217, Ch. XVIII.

E. Cesàro[56a] treated $\Sigma f(\delta)$ in connection with median and asymptotic formulas.

Bougaief[57] treated numerical integrals, noting formulas like

$$\sum_{d|n} \xi\left(\frac{n}{d}\right)\psi(d) = \sum_{d|\frac{n}{a}} \psi(d) + \sum_{d|\frac{n}{b}} \psi(d) + \ldots,$$

where $\xi(n)$ is the number of prime factors $a, b, \ldots$ of $n = a^\alpha b^\beta \ldots$,

$$\sum_{d|n} \psi(d) = \sum_{d|\frac{n}{a}} \psi(d) + \sum_{d|n} \psi(a^\alpha d) = \sum_{d|\frac{n}{a}} \psi(ad) + \sum_{d|\frac{n}{a^\alpha}} \psi(d).$$

Bougaief[58] gave a large number of formulas of the type

$$\Sigma\psi(d)\left[\frac{n}{d}\right] = \Sigma^n\psi(d) + \Sigma^{[n/2]}\psi(d) + \Sigma^{[n/3]}\psi(d) + \ldots,$$

where, on the left, d ranges over all the divisors of m; while, on the right, d ranges over those divisors of m which do not exceed $n, [n/2], \ldots$, respectively.

Bougaief[59] gave the relation

$$\sum_{d|n} \theta(\sqrt{d}) = \Sigma_p \xi_0\left(\frac{n}{p^2}, n\right),$$

where p ranges over all primes $\leq \sqrt{n}$, and $\xi_k(m, n)$ is the sum of the kth powers of all divisors $\leq m$ of n, so that ξ_0 is their number, and $\theta(t)$ is the number of primes $\leq t$.

L. Gegenbauer[60] noted that the preceding result is a case of

$$\sum_{d|n} g(d)h\left(\frac{n}{d}\right)\sum_{\lambda=1}^{d} f(\lambda) = \sum_{\lambda=1}^{n} f(\lambda)\left\{\sum_{d_\lambda} g(d_\lambda)h\left(\frac{n}{d_\lambda}\right)\right\},$$

where d_λ ranges over the divisors $\geq \lambda$ of n. Special cases are

$$\Sigma_d d^\rho\theta_k\left(d^{\frac{1}{\mu}}\right) = \Sigma_a a^k\bar{\xi}_\rho(a^\mu, n), \qquad \Sigma_d d^\rho\theta_k\left\{\left(\frac{n}{d}\right)^{\frac{1}{\mu}}\right\} = \Sigma_a a^k\xi_\rho\left(\frac{n}{a^\mu}, n\right),$$

where $\bar{\xi}_\rho(m, n)$ is the sum of the ρth powers of the divisors $\geq m$ of n.

[56a]Giornale di Mat., 25, 1887, 1–13.
[57]Mat. Sbornik (Math. Soc. Moscow), 14, 1888–90, 169–197; 16, 1891, 169–197 (Russian).
[58]Ibid., 17, 1893–5, 720–59.
[59]Comptes Rendus Paris, 119, 1894, 1259.
[60]Monatshefte Math. Phys., 6, 1895, 208.

Bougaief[60a] noted that, for an arbitrary function ψ,

$$\Sigma\psi(d)\left[\frac{n^2+a}{d}\right] = \sum_{u=1}^{n}\sum_{n=1}^{[(n^2+a)/u]}\psi(d), \quad \sum_{u=1}^{n}\sum_{n=1}^{[n^2/u]}\psi(d) = n\sum_{u=1}^{n}\sum_{n=1}^{[n/u]}\psi(d).$$

N. V. Bervi[61] treated numerical integrals extended over solutions of indeterminate equations, in particular for $n = a + b(x+y) + cxy$, $b^2 = b + ac$. Bougaief[62] considered definite numerical integrals, viz., sums over all divisors, between a and b, of n. He expressed sums of $[x]$, the greatest integer $\leq x$, as sums of values of $\zeta(n, m)$, viz., the number of divisors $\leq n$ of m. Also sums of ζ's expressed as $\zeta_i(1) + \zeta_i(2) + \ldots + \zeta_i(n)$, where $\zeta_i(n)$ is the number of the divisors of n which are ith powers.

I. I. Cistiakov[62a] (Tschistiakow) treated the second numerical derivative. Bougaief[62b] gave 13 general formulas on numerical integrals.

Bougaief[63] gave a method of transforming a sum taken over $1, 2, \ldots, n$ into a sum taken over all the divisors of n. He obtains various identities between functions.

D. J. M. Shelly,[64] using distinct primes $a, b, \ldots$, called

$$N' = N\left(\frac{a}{a} + \frac{\beta}{b} + \ldots\right)$$

the derivative of $N = a^a b^\beta \ldots$. Similar definitions are given for derivatives of fractions and for the case of fractional exponents $a, \beta, \ldots$. The primes are the only integers whose derivatives are unity.

[60a]Comptes Rendus Paris, 120, 1895, 432–4.
[61]Mat. Sbornik (Math. Soc. Moscow), 18, 1896, 519; 19, 1897, 182.
[62]Ibid., 18, 1896, 1–54 (Russian); see Jahrb. Fortschritte Math., 27, 1896, p. 158.
[62a]Ibid., 20, 1899, 595; see Fortschritte, 1899, 194.
[62b]Ibid., 549–595. Two of the formulas are given in Fortschritte, 1899, 194.
[63]Ibid., 21, 1900, 335, 499; see Fortschritte, 31, 1900, 197.
[64]Asociación española, Granada, 1911, 1–12.

CHAPTER XX.

PROPERTIES OF THE DIGITS OF NUMBERS.

John Hill[1] noted that $139854276 = 11826^2$ is formed of the nine digits permuted and believed erroneously that it is the only such square.

N. Brownell[1a] found 169 and 961 as the squares whose three digits are in reverse order and whose roots are composed of the same digits in reverse order. The least digit in the roots is also the least in the squares, while the greatest digit in the roots is one-third of the greatest in the squares and one-half of the digit in the tens place.

W. Saint[1b] proved that every odd number N not divisible by 5 is a divisor of a number $11...1$ of $D \leq N$ digits [by a proof holding only for N prime also to 3]. For, let $1...1$ (to D digits) have the quotient q and remainder r when divided by D. This remainder r must recur if the number of digits 1 be increased sufficiently. Hence let $1...1$ (to $D+d$ digits) give the remainder r and quotient Q when divided by D. By subtraction, $D(Q-q) = 1...10...0$ (with d units followed by D zeros). Hence if $1...1$ (to d digits) were not divisible by every odd number $\leq D$ and prime to 5 [and to 3], there would be a remainder R; then $R0...0$ (with D zeros) would be divisible by an odd number prime to 5 [and to 3], which is impossible.

P. Barlow[1c] stated, and several gave inadequate proofs, that no square has all its digits alike. He[1d] stated and proved that $111111111^2 = 123456789987654321$ is the largest square such that if unity be subtracted from each of its digits and again from each digit of the remainder, etc., all zeros being suppressed, each remainder is a square. Denote $(10^k - 1)/(10-1)$ by $\{k\}$. Then $\{\frac{1}{2}(x+1)\}^2$ has x digits and exceeds $\{x\}$ by $10\{\frac{1}{2}(x-1)\}^2$. Since zeros are suppressed we have a square as remainder, and the process can be repeated. It is stated that therefore the property holds only for 1^2, 11^2, 111^2, ...

Several[1e] found that 135 is the only number N composed of three digits in arithmetical progression such that the digits will be reversed if 132 times the middle digit be added to N.

W. Saint[1f] found the least integral square ending with the greatest number of equal digits. The possible final digits are 1, 4, 5, 6, 9. Any square is of the form $4n$ or $4n+1$. Hence the final digit is 4. If the square terminated with more than three 4's, its quotient by 4 would be a square ending with two 1's, just proved to be impossible. Of the numbers ending with

---

[1] Arithmetic, both in Theory and Practice, ed. 4, London, 1727, 322.
[1a] The Gentleman's Diary, or Math. Repository, London, 1767; Davis' ed., 2, 1814, 123.
[1b] Jour. Nat. Phil. Chem. Arts (ed., Nicholson), London, 24, 1809, 124–6.
[1c] The Gentleman's Diary, or Math. Repository, London, 1810, 38–9, Quest. 952.
[1d] Ibid., 1810, 39–40, Quset. 953.
[1e] Ibid., 1811, 33–4, Quest. 960.
[1f] Ladies' Diary, 1810–11, Quest., 1218; Leybourn's M. Quest. L. D., 4, 1817, 139–41.

three 4's, the least is 1444. J. Davey discussed only numbers of 3 or 4 digits of which the last 2 or 3 are equal, respectively.

Several[1ʼ] found that the squares 169 and 961 are composed of the same digits in reverse order, have roots of two digits in reverse order, while the sum of the digits in each square equals the square of the sum of the digits in each root; finally, the sum of the digits in each root equals the square of their difference.

An anonymous writer[2] proposed the problem to find a number n given the product of n by the number obtained from n by writing its digits in reverse order [Laisant[18]].

P. Tédenat[3] considered the problem to find a number of n digits whose square ends with the same n digits in the same order. If a is such a number of $n-1$ digits, so that $a^2 = 10^{n-1}b+a$, we can find a digit A to annex at the left of a to obtain a desired number $10^{n-1}A+a$ of n digits. Squaring the latter, we obtain the condition $(2a-1)A \equiv -b \pmod{10}$.

J. F. Français[4] noted the solutions

$$x=2^n p=5^n q+1, \qquad x^2=10^n pq+x,$$
$$y=5^n r=2^n s+1, \qquad y^2=10^n rs+y,$$

in which the resulting condition $2^n p - 5^n q = 1$ or $5^n r - 2^n s = 1$ is to be satisfied. Special solutions are given by $n=1$, $p=3$; $n=2$, $p=19$; $n=3$, $p=47$; $n=4$, $p=586$; etc., to $n=7$.

J. D. Gergonne[5] generalized the problem to base B. Then

$$x(x-1) = B^n y.$$

Let p, q be relatively prime and set $B^n = pq$. Then $x=pt$, $x-1=qu$, or vice versa. The condition $pt-qu=1$ is solved for t, u. When $B=10$, $n=20$, the least u is 81199.

Anonymous writers[6] stated and proved by use of the decimal fraction for $1/n$ that every number divides a number of the form $9\ldots90\ldots0$.

A. L. Crelle[7] proved the generalization: Every number divides a number obtained by repeating any given set of digits and affixing a certain number of zeros, as $23\ldots230\ldots0$.

Several[7a] found a square whose root has two digits, their quotient being equal to their difference. By $x/y=x-y$, $x=y+1+1/(y-1)$, an integer, whence $y=2$, $x=4$. Thus the squares are 24^2 or 42^2.

The[7b] three digits of a number are in geometrical progression; the product of the sum of their cubes by the cube of their sum is 1663129; if the number obtained by reversing the digit be divided by the middle digit, the

---

[1ʼ]Ladies' Diary, 1811–12, Quest. 1231; Leybourn, *l. c.*, 153–4.
[2]Annales de Math. (ed., Gergonne), 3, 1812–3, 384.
[3]*Ibid.*, 5, 1814–5, 309–321. Problem proposed on p. 220.
[4]*Ibid.*, 321–2.
[5]*Ibid.*, 322–7.
[6]*Ibid.*, 19, 1828–9, 256; 20, 1829–30, 304–5.
[7]*Ibid.*, 20, 1829–30, 349–352; Jour. für Math., 5, 1830, 296.
[7a]Ladies' Diary, 1820, 36, Quest. 1347.
[7b]*Ibid.*, 1822, 33, Quest. 1374.

quotient is $46\frac{1}{3}$. By the last condition, the middle digit must be 3, since not a higher multiple of 3. Hence the number is 931.

To find a symmetrical number $abcba$ of five digits whose square exhibits all ten digits, W. Rutherford[7c] noted that the square is divisible by 9 since the sum of the digits is divisible by 9. Hence the sum of the digits of the number is divisible by 3. Also $a \geq 3$. Taking $c = a + b$, $c = 8$, he got 35853. J. Sampson noted also the answers 84648, 97779.

J. A. Grunert[8] proved by use of Euler's generalization of Fermat's theorem that[6] every number divides $9 \ldots 90 \ldots 0$.

Drot[8a] asked for the values of x for which N^x has the same final k digits as N, when $k = 1$, 2 or 3.

J. Bertrand[8b] discussed the numbers of digits of certain numbers.

A. G. Emsmann[9] treated a number b of n digits to base 10 equal to the product of the sum of its digits by a, and such that if another number of n digits be subtracted from b the remainder shall equal the number obtained by writing the digits of b in reverse order.

J. Booth[10] noted that a number of six digits formed by repeating any set of three digits is divisible by 7, 11, 13 [since by 1001].

G. Bianchi[10a] noted various numerical relations like $10^9 = 11111111 + 8.1111111 + 8.9.111111 + \ldots + 8.9^6.1 + 9^8 = 2222222 + \ldots + 7.8^6.2 + 8^8$, $98 = (12 - 1 - 0)9 - 1$, $987 = (123 - 12 - 1)9 - 3$, $9876 = (1234 - 123 - 13)9 - 6$.

C. M. Ingleby[11] added the digits of a number N written to base r, then added the digits of this sum, etc., finally obtaining a number, designated SN, of a single digit; and proved that $S(MN) = S(SM \cdot SN)$.

P. W. Flood[11a] proved that 64 is the only square the sum of whose digits less unity and product plus unity are squares.

G. Cantor[12] employed any distinct positive integers a, b, $\ldots$, considered the system of integers in which a occurs $\bar{a}$ times, b occurs $\bar{b}$ times, etc., and called a system simple if every number can be expressed in a single way in the form $\alpha a + \beta b + \ldots$, where $\alpha = 0$, 1, $\ldots$, $\bar{a}$; $\beta = 0$, 1, $\ldots$, $\bar{b}$; $\ldots$. A system is simple if and only if each basal number k divides the next one l and if k occurs $\bar{k} = (l/k) - 1$ times.

G. Barillari[13] noted that, if 10 belongs to the exponent m modulo b, the number $P = \alpha\beta \ldots \lambda \alpha\beta \ldots \lambda \ldots$, obtained by repeating h times ($h > 1$) any set of n digits, is divisible by b if b is prime to $10^n - 1$ and if nh is a multiple

[7c]Ladies' Diary, 1835, 38, Quest. 1576.
[8]Jour. für Math., 5, 1830, 185–6.
[8a]Nouv. Ann. Math., 4 1845, 637–44; 5, 1846, 25. For references to tables of powers, 13, 1854, 424–5.
[8b]Ibid., 8, 1849, 354.
[9]Abhandlung über eine Aufgabe aus der Zahlentheorie, Progr. Frankfurt, 1850, 36 pp.
[10]Proc. Roy. Soc. London, 7, 1854–5, 42–3.
[10a]Proprieta e rapporti de' numeri interi e composti colle cifre semplici . . . , Modena, 1856. Same in Mem. di Mat. e di Fis. Soc. Ital. Sc., Modena, (2), 1, 1862, 1–36, 207.
[11]Oxford, Cambr. and Dublin Messenger Math., 3, 1866, 30–31.
[11a]Math. Quest. Educ. Times, 7, 1867, 30.
[12]Zeitschrift Math. Phys., 14, 1869, 121–8.
[13]Giornale di Mat., 9, 1871, 125–135.

of m, but P is not divisible by b if nh is not a multiple of m. If b divides $10^n - 1$, P is divisible by b when $h = b$, but not divisible by b when h is not a multiple of b.

A. Morel[14] proved that the numbers ending with 12, 38, 62 or 88 are the only ones whose squares end with two equal digits.

H. Hoskins[14a] found the sum of the 117852 numbers of 7 digits which can be formed with the digits 1, 1, 2, 2, 2, 2, 2, 3, 3, 4, 5, 6, 7.

J. Plateau[15] noted that every odd number not ending with 5 has a multiple of the form 11...1 [Saint[1b]].

P. Mansion[16] proved the theorem of Plateau.

J. W. L. Glaisher[17] deduced Crelle's[7] theorem from Plateau's.[15]

C. A. Laisant[18] treated a problem[2] on reversing digits.

G. R. Perkins[18a] and A. Martin[19] stated that all powers of numbers ending with 12890625 end with the same digits.

E. Catalan[20] noted that the g. c. d. of two numbers of the form 1...1 of n and n' digits is of like form and has Δ digits, where Δ is the g. c. d. of n and n'.

Lloyd Tanner,[20a] generalizing Martin's[19] question, found how many numbers N of n digits to the base r end with the same digits as their squares, i. e., $N^2 - N = Kr^n$. If r^n is the product of q powers of primes, there are $2^q - 2$ values of N. He[20b] found numbers M and N with n digits to the base r such that the numbers formed by prefixing M to N and N to M have a given ratio.

J. Plateau[21] proposed the problem to find two numbers whose product has all its digits alike. Angenot noted that

$$\frac{b^{pq}-1}{b^p-1}, \qquad \frac{b^p-1}{b-1}$$

give a solution for base b. Catalan[21] noted that Euler's theorem

$$\frac{b^{\varphi(n)}-1}{b-1} = nm$$

for n prime to b, furnishes a solution n, m.

Lloyd Tanner[22] stated and Laisant proved that 87109376 and 12890625 are the only numbers of 8 digits whose squares end with the same 8 digits.

[14]Nouv. Ann. Math., (2), 10, 1871, 44–6, 187–8.

[14a]Math. Quest. Educ. Times, 15, 1871, 89–91.

[15]Bull. Acad. Roy. de Belgique, (2), 16, 1863, 62; 28, 1874, 468–476.

[16]Nouv. Corresp. Math., 1, 1874–5, 8–12; Mathesis, 3, 1883, 196–7. Bull. Bibl. Storia Sc. Mat., 10, 1877, 476–7.

[17]Messenger Math., 5, 1875–6, 3–5.

[18]Mém. soc. sc. phys. et nat. de Bordeaux, (2), 1, 1876, 403–11.

[18a]Math. Miscellany, Flushing, N. Y., 2, 1839, 92.

[19]Math. Quest. Educat. Times, 26, 1876, 28.

[20]Mém. Societé Sc. Liège, (2), 6, 1877, No. 4.

[20a]Messenger Math., 7, 1877–8, 63–4. Cases $r \leq 12$, Math. Quest. Educ. Times, 28, 1878, 32–4.

[20b]Math. Quest. Educ. Times, 29, 1878, 94–5.

[21]Nouv. Corresp. Math., 4, 1878, 61–63.

[22]Ibid., 5, 1879, 217; 6, 1880, 43.

Moret-Blanc[23] proved that 1, 8, 17, 18, 26, 27 are the only numbers equal to the sum of the digits of their cubes.

C. Berdellé[23a] considered the last n digits of numbers, in particular of 5^k.

E. Cesàro[24] noted that the sum of the pth powers of ten consecutive integers ends with 5 unless p is a multiple of 4, when it ends with 3.

F. de Rocquigny[25] noted that if a number of n digits equals the sum of the 2^n-1 products of its digits taken 1, 2,..., n at a time, its final $n-1$ digits are all 9.

E. Cesàro[26] considered the period of the digits of rank n in powers of 5.

Lists[26a] have been given of squares formed by the nine digits >0, or the ten digits, not repeated.

O. Kessler[27] gave a table of divisors of numbers formed by repeating a given set of digits a small number of times.

T. C. Simmons[27a] noted that, if the sum of the digits of n is 10, that of $2n$ is 11 unless each digit of n is <5 or two are 5. For 4 digits the numbers of each type are counted.

J. S. Mackay[28] treated the last subject.

E. Lemoine[29] considered numbers like $A = 8607004053$ such that, if a is the number derived by reversing the digits of A, the sum $A + a = 12111011121$ reverses into itself.

M. d'Ocagne[30] considered the sum $\sigma(N)$ of the digits of the first N integers. If $N_p = a_p \cdot 10^p + \ldots + a_1 \cdot 10 + a_0$ and $d = a_p \cdot 10^p - 1$, then

$$\sigma(d) = 10^{p-1} \cdot 5 a_p (a_p - 1 + 9p), \qquad \sigma(N_p) = \sigma(d) + (N_{p-1} + 1) a_p + \sigma(N_{p-1}).$$

Hence

$$\sigma(N_p) = \tfrac{1}{2} a_0 (a_0 + 1) + \sum_{i=1}^{p} a_i \{ 10^{i-1} \cdot 5 (a_i - 1 + 9i) + N_{i-1} + 1 \}.$$

The number of digits in $1, \ldots, N$ is $(p+1)(N+1) - (10^{p+1} - 1)/9$. See the next paper.

M. d'Ocagne[31] noted that, in writing down the natural numbers $1, \ldots, N$, where N is composed of n digits, the total number of digits written is $n(N+1) - 1_n$, where $1_n = 1 \ldots 1$ (to n digits).

E. Barbier[31a] asked what is the 10^{1000}th digit written if the series of natural numbers be written down.

[23]Nouv. Ann. Math., (2), 18, 1879, 329; proposed by Laisant, 17, 1878, 480.

[23a]Assoc. franç., 8, 1879, 176–9.

[24]Nouv. Corresp. Math., 6, 1880, 519; Mathesis, 1888, 103.

[25]Les Mondes, 53, 1880, 410–2.

[26]Nouv. Corresp. Math., 4, 1878, 387; Nouv. Ann. Math., (3), 2, 1883, 144, 287; 1884, 160.

[26a]Math. Magazine, 1, 1882–4; 69–70; l'intermédiaire des math., 4, 1897, 168; 14, 1907, 135; Sphinx-Oedipe, 1908–9, 35; 5, 1910, 64; Educ. Times, March, 1905.   Math. Quest. Educ. Times, 52, 1890, 61; (2), 8, 1905, 83–6 (with history).

[27]Zeitschrift Math. Phys., 28, 1883, 60–64.

[27a]Math. Quest. Educ. Times, 41, 1884, 28–9, 64–5.

[28]Proc. Edinburgh Math. Soc., 4, 1885–6, 55–56.

[29]Nouv. Ann. Math., (3), 4, 1885, 150–1.

[30]Jornal de sc. math. e ast., 7, 1886, 117–128.

[31]Ibid., 8, 1887, 101–3; Comptes Rendus Paris, 106, 1888, 190.

[31a]Comptes Rendus Paris, 105, 1887, 795, 1238.

L. Gegenbauer[31b] proved generalizations of Cantor's[12] theorems, allowing negative coefficients. Given the distinct positive integers $a_1, a_2, \ldots$, every positive integer is representable in a single way as a linear homogeneous function of $a_1, a_2, \ldots$ with integral coefficients if each a_λ is divisible by $a_{\lambda-1}$ and the quotient equals the number of permissible values of the coefficients of the smaller of the two.

R. S. Aiyar and G. G. Storr[31c] found the number p_n of integers the sum of whose digits (each >0) is n, by use of $p_n = p_{n-1} + \ldots + p_{n-9}$.

E. Strauss[32] proved that, if $a_1, a_2, \ldots$ are any integers >1, every positive rational or irrational number <1 can be written in the form

$$\frac{a_1}{a_1} + \frac{a_2}{a_1 a_2} + \frac{a_3}{a_1 a_2 a_3} + \ldots \qquad (a_1 < a_1, \; a_2 < a_2, \ldots),$$

the a's being integers, and in a single way except in the case in which all the a_i, beginning with a certain one, have their maximum values, when also a finite representation exists.

E. Lucas[33] noted that the only numbers having the same final ten digits as their squares are those ending with ten zeros, nine zeros followed by 1, 8212890625 and 1787109376. He gave (ex. 4) the possible final nine digits* of numbers whose squares end with 224406889. He gave (p. 45, exs. 2, 3) all the numbers of ten digits to base 6 or 12 whose squares end with the same ten digits. Similar special problems were proposed by Escott and Palmstrom in l'Intermédiaire des Mathématiciens, 1896, 1897.

J. Kraus[34] discussed the relations between the digits of a number expressed to two different bases.

A. Cunningham[34a] called N an agreeable number of the mth order and nth degree in the r-ary scale if the m digits at the right of N are the same as the m digits at the right of N^n when each is expressed to base r; and tabulated all agreeable numbers to the fifth order and in some cases to the tenth. A number N of m digits is completely agreeable if the agreement of N with its nth power extends throughout its m digits, the condition being $N^n \equiv N \pmod{r^m}$.

E. H. Johnson[34b] noted that, if a and $r-1$ are relatively prime and $aa\ldots a$ (to $r-1$ digits to base r) is divided by $r-1$, there appear in the quotient all the digits $1, 2, \ldots, r-1$ except one, which can be found by dividing the sum of its digits by $r-1$.

C. A. Laisant[34c] stated that, if $N = 123\ldots n$, written to base $n+1$, be multiplied by any integer $<n$ and prime to n, the product has the digits of N permuted.

[31b]Sitzungsber. Ak. Wiss. Wien (Math.), 95, 1887, II, 618–27.

[31c]Math. Quest. Educ. Times, 47, 1887, 64.   [32]Acta Math., 11, 1887–8, 13–18.

[33]Théorie des nombres, 1891, p. 38.   Cf. Math. Quest. Educ. Times, (2), 6, 1904, 71–2.

*Same by Kraitchik, Sphinx-Oedipe, 6, 1911, 141.

[34]Zeitschr. Math. Phys., 37, 1892, 321–339; 39, 1894, 11–37.

[34a]British Assoc. Report, 1893, 699.   [34b]Annals of Math., 8, 1893–4, 160–2.

[34c]L'intermédiaire des math., 1894, 236; 1895, 262.   Proof by "Nauticus," Mathesis, (2), 5, 1895, 37–42.

Tables of primes to the base 2 are cited under Suchanek[80] of Ch. XIII. There is a collection[34d] of eleven problems relating to digits. To find[34e] the number < 90 which a person has in mind, ask him to annex a declared digit and to tell the remainder on division by 3, etc.

T. Hayashi[35] gave relations between numbers to the base r:

$$123\ldots\{r-1\}\cdot(r-1)+r=1\ldots1 \text{ (to } r \text{ digits)},$$
$$\{r-1\}\{r-2\}\ldots321\cdot(r-1)-1=\{r-2\}\{r-2\}\ldots \text{ (to } r \text{ digits)}.$$

Several writers[36] proved that

$$123\ldots\{r-1\}\cdot(r-2)+r-1=\{r-1\}\ldots321.$$

T. Hayashi[37] noted that if $A=10+r(10)^2+r^2(10)^3+\ldots$ be multiplied or divided by any number, the digits of each period of A are permuted cyclically.

A. L. Andreini[37a] found pairs of numbers N and p (as 37 and 3) such that the products of N by all multiples $\leq(B-1)p$ of p are composed of p equal digits to the base $B\leq12$, whose sum equals the multiplier.

P. de Sanctis[38] gave theorems on the product of the significant digits of, or the sum of, all numbers of n digits to a general base, or the numbers beginning with given digits or with certain digits fixed, or those of other types.

A. Palmstrom[39] treated the problem to find all numbers with the same final n digits as their squares. Two such numbers ending in 5 and 6, respectively, have the sum 10^n+1. If the problem is solved for n digits, the $(n+1)$th digit can be found by recursion formulæ. There is a unique solution if the final digit (0, 1, 5 or 6) is given.

A. Hauke[40] discussed obscurely $x^m\equiv x$ (mod s^r) for x with r digits to base s. If $m=2$, while r and s are arbitrary, there are 2^ν solutions, ν being the number of distinct prime factors of s.

G. Valentin and A. Palmstrom[41] discussed $x^k\equiv x$ (mod 10^n), for $k=2, 3, 4, 5$.

G. Wertheim[42] determined the numbers with seven or fewer digits whose squares end with the same digits as the numbers, and treated simple problems about numbers of three digits with prescribed endings when written to two bases.

[34d]Sammlung der Aufgaben...Zeitschr. Math. Naturw. Unterricht, 1898, 35–6.

[34e]Math. Quest. Educ. Times, 63, 1895, 92–3.

[35]Jour. of the Physics School in Tokio, 5, 1896, 153–6, 266–7; Abhand. Geschichte der Math. Wiss., 28, 1910, 18–20.

[36]Jour. of the Physics School in Tokio, 5, 1896, 82, 99–103; Abhand., 16–18.

[37]Ibid., 6, 1897, 148–9; Abhand., 21.

[37a]Periodico di Mat., 14, 1898–9, 243–8.

[38]Atti Accad. Pont. Nuovi Lincei, 52, 1899, 58–62; 53, 1900, 57–66; 54, 1901, 18–28; Memorie Accad. Pont. Nuovi Lincei, 19, 1902, 283–300; 26, 1908, 97–107; 27, 1909, 9–23; 28, 1910, 17–31.

[39]Skrifter udgivne af Videnskabs, Kristiania, 1900, No. 3, 16 pp.

[40]Archiv Math. Phys., (2), 17, 1900, 156–9.

[41]Forhandlinger Videnskabs, Kristiania, 1900–1, 3–9, 9–13.

[42]Anfangsgründe der Zahlenlehre, 1902, 151–3.

C. L. Bouton[43] discussed the game *nim* by means of congruences between sums of digits of numbers to base 2.

H. Piccioli[43a] employed $N = a_1 \ldots a_n$ of $n \geqq 3$ digits and numbers $a_{i_1} \ldots a_{i_n}$ and $a_{j_1} \ldots a_{j_n}$ obtained from N by an even and odd number of transpositions of digits. Then $\Sigma a_{i_1} \ldots a_{i_n} = \Sigma a_{j_1} \ldots a_{j_n}$.

If[43b] a number of n digits to base R has r fixed digits, including the first, and the sum of these r is $\equiv -a \pmod{R-1}$, the number of ways of choosing the remaining digits so that the resulting number shall be divisible by $R-1$ is the number of integers of $n-r$ or fewer digits whose sum is $\equiv a \pmod{R-1}$ and hence is $N+1$ or N, according as $a = 0$ or $a > 0$, where $N = (R^{n-r} - 1)/(R-1)$.

G. Metcalfe[43c] noted that 19 and 28 are the only integers which exceed by unity 9 times the integral parts of their cube roots.

A. Tagiuri[44] proved that every number prime to the base g divides a number $1 \ldots 1$ to base g (generalization of Plateau's[15] theorem).

If[44a] A, B, C have 2, 3, 4 digits respectively and A becomes A' on reversing its digits, and $2A - 1 = A'$, $3B - 2A + 10 = B'$, $4C - B + 1 + [B/10] = C'$, then $A = 37$, $B = 329$, $C = 2118$.

P. F. Teilhet[45] proved that we can form any assigned number of sets, each including any assigned number of consecutive integers, such that with the digits of the qth power of any one of these integers we can form an infinitude of different qth powers, provided $q < m$, where m is any given integer.

L. E. Dickson[45a] determined all pairs of numbers of five digits such that their ten digits form a permutation of $0, 1, \ldots, 9$ and such that the sum of the two numbers is 93951.

A. Cunningham[45b] found cases of a number expressible to two bases by a single digit repeated three or more times. He[45c] noted that all 10 digits or all > 0 occur in the square of 10101010101010101 or of $1 \ldots 1$ (to 9 digits), each square being unaltered on reversing its digits.

He[45d] and T. Wiggins expressed each integer $\leqq 140$ by use of four nines, as $13 = 9 + \sqrt{9} + 9/9$, allowing also $.9 = 1$, $(\sqrt{9})!$, and the exponent $\sqrt{9}$, and cited a like table using four fours.

If[45e] $r \equiv 1 \pmod{q}$, $1 \ldots 1$ (with q^n digits to base r) is divisible by q^n.

If[45f] the square of a number n of r digits ends with those r digits, then $10^r + 1 - n$ has the same property. Also, $(n-1)^3$ ends with the same r digits

[43]Annals of Math., (2), 3, 1901–2, 35–9. Generalized by E. H. Moore, 11, 1910, 93–4.
[43a]Nouv. Ann. Math., (4), 2, 1902, 46–7.
[43b]Math. Quest. Educ. Times, (2), 1, 1902, 119–120.
[43c]Ibid, 63–4.
[44]Periodico di Mat., 18, 1903, 45.
[44a]Math. Quest. Educ. Times, (2), 5, 1904, 82–3.
[45]L'intermédiaire des math., 11, 1904, 14–6.
[45a]Amer. Math. Monthly, 12, 1905, 94–5.
[45b]Math. Quest. Educ. Times, (2), 8, 1905, 78.
[45c]Ibid, 10, 1906, 20.   [45d]Math. Quest. Educ. Times, 7, 1905, 43–46.
[45e]Ibid., 7, 1905. 49–50.   [45f]Ibid., 7, 1905, 60–61.

as $n-1$. If the cube of a number n of r digits ends with those r digits, $10^r - n$ has the same property.

P. Zühlke[46] proved the three theorems of Palmstrom[39] and gave all solutions of $x^p \equiv x \pmod{10^3}$ for $p = 3, \ldots, 12$.

M. Koppe[47] noted that by prefixing a digit to a solution 0, 1, 5 or 6 of $x^2 \equiv x \pmod{10}$ we get solutions of $x^2 \equiv x \pmod{10^2}$, then for 10^3, etc. We can pass from a solution with n digits for 10^n to solutions with $2n$ digits for 10^{2n}. He treated also $x^5 \equiv x \pmod{10^n}$.

G. Calvitti[48] treated the problem: Given a number A, a set C of γ digits, and a number p prime to the base g, to find the least number x of times the set C must be repeated at the right of A to give a number $N_x \equiv A \pmod{p}$. The condition is $G(N_1 - N_0) \equiv 0 \pmod{p}$, where

$$G = \frac{g^{x\gamma} - 1}{g^\gamma - 1}.$$

If $N_1 - N_0 \equiv 0$, any x is a solution. If not, the least value λ of x makes $G \equiv 0 \pmod{p/\rho}$, where ρ is the g. c. d. of $N_1 - N_0$ and p. Then λ is the l. c. m. of $\lambda_1, \ldots, \lambda_k$, where λ_i is the least root of $G \equiv 0 \pmod{p_i}$, if p/ρ is the product of $p_1, \ldots, p_k$, relatively prime in pairs. Hence the problem reduces to the case of a power of a prime p. Write $(a)_x$ for $(a^x - 1)/(a-1)$. It is shown that the least root of $(a)_x \equiv 0 \pmod{p^k}$ is mp^{k-t}, where m is the least root of $(a)_x \equiv 0 \pmod{p}$, and p^t is the highest power of p dividing $(a)_m$. Given any set C of digits and any number p prime to the base g, there exist an infinitude of numbers $C \ldots C$ divisible by p.

A. Gérardin[48a] added 220 to the sum of its digits, repeated the operation 18 times and obtained 418; 9 such operations on 284 gave 418. A. Boutin stated that if a and b lead finally to the same number, neither a nor b is divisible by 3, or both are divisible by 3 and not by 9, or both are divisible by 9.

E. Malo[49] considered periodicity properties of A and a in

$$5^k = 10^m A_{n,p} + a_p \qquad (a_p < 10^m, \quad k = n \cdot 2^{m-2} + p, \quad 0 \leq p \leq 2^{m-2} - 1),$$

and solved Cesàro's[26] three problems on the digits of powers of 5.

A. L. Andreini[50] noted that the squares of A and B end with the same p digits if and only if the smaller of $r+s$ and $u+v$ equals p, where

$$A + B = a \cdot 2^r \cdot 5^u, \qquad A - B = \beta \cdot 2^s \cdot 5^v.$$

[46]Sitz. Berlin Math. Gesell., 4, 1905, 10–11 (Suppl., Archiv Math. Phys., (3), 8, 1905).

[47]Ibid., 5, 1906, 74–8. (Suppl., Archiv, (3), 11, 1907.)

[48]Periodico di Mat., 21, 1906, 130–142.

[48a]Sphinx-Oedipe, 1, 1906, 19, 47–8. Cf. l'interméd. math., 22, 1915, 134, 215.

[49]Sur certaines propriétés arith. du tableau des puissances de 5, Sphinx-Oedipe, 1906–7, 97–107; reprinted, Nancy, 1907, 13 pp., and in Nouv. Ann. Math., (4), 7, 1907, 419–431.

[50]Il Pitagora, Palermo, 14, 1907–8, 39–47.

W. Jänichen[50b] stated that, if $q_p(x)$ denotes the sum of the digits of x to the base p and if p is a prime divisor of n, then, for μ as in Ch. XIX,

$$\sum_{d|n} \mu(d) q_p\left(\frac{n}{d}\right) = 0.$$

E. N. Barisien[50c] noted that the sum of all numbers of n digits formed with p distinct digits $\neq 0$, of sum s, is

$$s(p+1)^{n-2}\{p(10^{n-1}-1)/9 + (p+1)10^{n-1}\}.$$

A. Gérardin[50d] listed all the 124 squares formed of 7 distinct digits.

Several writers[51] treated the problem to find four consecutive numbers $a, b = a+1, c = a+2, d = a+3$, such that $(a)_1 = 11 \ldots 1$ (to a digits) is divisible by $a+1$, $(b)_1$ by $2b+1$, $(c)_1$ by $3c+1$, $(d)_1$ by $4d+1$.

A. Cunningham and E. B. Escott[52] treated the problems to find integers whose squares end with the same n digits or all with n given digits; to find numbers having common factors with the numbers obtained by permuting the digits cyclically, as

$$259 = 7 \cdot 37, \qquad 592 = 16 \cdot 37, \qquad 925 = 25 \cdot 37.$$

E. N. Barisien[53] noted that the squares of 625, 9376, 8212890625 end with the same digits, respectively. R. Vercellin[54] treated the same topic.

E. Nannei[55] discussed a problem by E. N. Barisien: Take a number of six digits, reverse the digits and subtract; to the difference add the number with its digits reversed; we obtain one of 13 numbers $0, 9900, \ldots, 1099989$. The problem is to find which numbers of six digits leads to a particular one of these 13, and to generalize to n digits.

Several writers[56] examined numbers of 6 digits which become divisible by 7 after a suitable permutation of the digits; also[57] couples of numbers, as 18 and 36, 36 and 54, whose g. c. d. 18 is the sum of their digits.

E. N. Barisien[58] gave ten squares not changed by reversing the digits, as $676 = 26^2$.

A. Witting[59] noted that, besides the evident ones 11 and 22, the only numbers of two digits whose squares are derived from the squares of the numbers with the digits interchanged by reversing the digits are 12 and 13. Similarly for the squares of 102 and 201, etc. Also,

$$102 \cdot 402 = 201 \cdot 204, \qquad 213 \cdot 936 = 312 \cdot 639, \qquad 213 \cdot 624 = 312 \cdot 426.$$

A. Cunningham[60] treated three numbers L, M, N of l, m, n digits, respectively, such that $N = LM$, and N has all the digits of L and M and no others.

[50b]Archiv Math. Phys., (3), 13, 1908, 361. Proof by G. Szegö, 24, 1916, 85–6.
[50c]Sphinx-Oedipe, 1907–8, 84–86. For $p=n$, Math. Quest. Educ. Times, 72, 1900, 126–8.
[50d]Ibid., 1908–9, 84–5.
[51]L'intermédiaire des math., 16, 1909, 219; 17, 1910, 71, 203, 228, 286 [136].
[52]Math. Quest. Educat. Times, (2), 15, 1909, 27–8, 93–4.
[53]Suppl. al Periodico di Mat., 13, 1909, 20–21.   [54]Suppl. al Periodico di Mat., 14, 1910–11, 17–20.
[55]Ibid., 13, 1909, 84–88.   [56]L'intermédiaire des math., 17, 1910, 122, 214–6, 233–5.
[57]Ibid., 170, 261–4; 18, 1911, 207.   [58]Mathesis, (3), 10, 1910, 65.
[59]Zeitschrift Math.-Naturw. Unterricht, 41, 1910, 45–50.
[60]Math. Quest. Educat. Times, (2), 18, 1910, 23–24.

D. Biddle[61] applied congruences to find numbers like 15 and 93 whose product 1395 has the same digits as the factors.

P. Cattaneo[62] considered numbers Q (and C) whose square (cube) ends with the same digits as the number itself. No $Q > 1$ ends with 1. No two Q's with the same number of digits end with 5 or with 6. All Q's $< 10^{14}$ are found. A single C of n digits ends with 4 or 6. Any Q is a C. Any $Q-1$ is a C. If N is a Q with n digits and if $2N-1$ has n digits, it is a C.

M. Thié,[62a] using all nine digits > 0, found numbers of 2, 3 or 4 digits with properties like $12 \cdot 483 = 5796$.

Pairs[62b] of cubes 3^3, 6^6 and 375^3, 387^3 whose sums of digits are squares, 3^2 and 6^2.

T. C. Lewis[63] discussed changes in the digits of a number to base r not affecting its divisibility by p.

Numbers[64] B and B^n having the same sum of digits.

Pairs[65] of primes like $23 \cdot 89 = 29 \cdot 83$.

Cases[66] like $7 \cdot 9403 = 65821$ and $3 \cdot 1458 = 6 \cdot 0729$, where the digits 0, 1,..., 9 occur without repetition.

N^{pn+1} ending[67] with the same digits as N.

Numbers[68] like $512 = (5+1+2)^3$, $47045881000000 = (47+4+58+81)^6$.

All[69] numbers like $2 \cdot 5 \cdot 27 = 1 \cdot 18 \cdot 15$, $2+5+27 = 1+15+18$.

Number[70] divisible by the same number reversed.

Number[71] an exact power of the sum of its digits; two numbers each an exact power of the sum of the digits of the other.

Solve[72] $KN + P = N'$, N' derived from N by reversing the digits.

Symmetrical numbers (ibid., p. 195).

F. Stasi[73] proved that, if a, b are given integers and a has m digits, we can find a multiple of b of the form

$$10^\rho(a \cdot 10^{mi} + a \cdot 10^{m(i-1)} + \ldots + a), \qquad \rho \geqq 0.$$

Taking b prime to a and to 10, we see that b divides $10^{mi} + \ldots + 1$. The case $m = 1$ gives the result of Plateau.[15]

Cunningham[73a] and others wrote N_1 for the sum of N and its digits to base r, N_2 for the sum of N_1 and its digits, etc., and found when N_m is divisible by $r-1$.

[61]Math. Quest. Educ. Times, (2), 19, 1911, 60–2. Cf. (2), 17, 1900, 44.
[62]Periodico di Mat., 26, 1911, 203–7.
[62a]Nouv. Ann. Math., (4), 11, 1911, 46.
[62b]Sphinx-Oedipe, 6, 1911, 62.
[63]Messenger Math., 41, 1911–12, 185–192.
[64]L'intermédiaire des math., 18, 1911, 90–91; 19, 1912, 267–8.
[65]Ibid., 1911, 121, 239.   [66]Ibid., 19, 1912, 26–7, 187.
[67]Ibid., 50–1, 274–9.
[68]Ibid., 77–8, 97.
[69]Ibid., 125, 211.
[70]Ibid., 128.
[71]Ibid., 137–9, 202; 20, 1913, 80–81.
[72]Ibid., 221.
[73]Il Boll. Matematica Gior. Sc.-Didat., 11, 1912, 233–5.
[73a]Math. Quest. Educ. Times, (2), 21, 1912, 52–3.

A. Cunningham[74] listed 63 symmetrical numbers $a_0a_1a_2a_1a_0$ each a product of two symmetrical numbers of 3 digits, and all numbers n^3, $n < 10000$, and all n^5, n^7, n^9, n^{11}, $n < 1000$, ending with 2, 7, 8, symmetrical with respect to 2 or 3 digits, as $618^3 = 236029032$.

Pairs[75] of numbers whose l. c. m. equals the product of the digits.

Pairs[76] of biquadrates, cubes and squares having the same digits.

*P. de Sanctis[77] noted a property of numbers to the base $h^2 + 1$.

L. von Schrutka[78] noted that 15, 18, 45 in $7 \cdot 15 = 105$, $6 \cdot 18 = 108$ and $9 \cdot 45 = 405$ are the only numbers of two digits which by the insertion of zero become multiples.

G. Andreoli[79] considered numbers N of n digits to the base k whose rth powers end with the same n digits as N. Each decomposition of k into two relatively prime factors gives at most two such N's. If the base is a power of a prime, there is no number > 1 whose square ends with the same digits.

Welsch[80] discussed the final digits of pth powers.

H. Brocard[81] discussed various powers of a number with the same sum of digits.

A. Agronomof[82] wrote $\overline{N}$ for the number obtained by reversing the digits of N to base 10 and gave several long formulas for $\Sigma_{j=1}^{j=N} \overline{j}$.

The[82a] only case in which $N^2 - \overline{N}^2$ is a square for two digits is $65^2 - 56^2 = 33^2$. There is no case for three digits.

R. Burg[83] found the numbers N to base 10 such that the number obtained by reversing its digits is a multiple kN of N, in particular for $k = 9, 4$.

E. Lemoine[84] asked a question on symmetrical numbers to base b.

H. Sebban[85] noted that 2025 is the only square of four digits which yields a square 3136 when each digit is increased by unity. Similarly, 25 is the only one of two digits.

R. Goormaghtigh[86] noted that this property of the squares of 5, 6 and 45, 56 is a special case of $A^2 - B^2 = 1 \ldots 1$ (to $2p$ digits), where $A = 5 \ldots 56$, $B = 4 \ldots 45$ (to p digits). Again, the factorizations $11111 = 41 \cdot 271$, $1111111 = 239 \cdot 4649$ yield the answers 115^2, 156^2 and 2205^2, 2444^2.

[74]L'intermédiaire des math., 20, 1913, 42–44.
[75]Ibid., 80.
[76]Ibid., 124, 262, 283–4.
[77]Atti Accad. Romana Nuovi Lincei, 66, 1912–3, 43–5.
[78]Archiv Math. Phys., (3), 22, 1914, 365–6.
[79]Giornale di Mat., 52, 1914, 53–7.
[80]L'intermédiaire des math., 21, 1914, 23–4, 58.
[81]Ibid., 22, 1915, 110–1. Objections by Maillet, 23, 1916, 10–12.
[82]Suppl. al Periodico di Mat., 19, 1915, 17–23.
[82a]Sphinx-Oedipe, 9, 1914, 42.
[83]Sitzungsber. Berlin Math. Gesell., 15, 1915, 8–18.
[84]Nouv. Ann. Math., (4), 17, 1917, 234.
[85]L'intermédiaire des math., 24, 1917, 31–2.
[86]Ibid., 96. Cf. H. Brocard, 25, 1918, 35–8, 112–3.

Several[86a] gave $9 \cdot n! + n + 1 = 1 \ldots 1$ for $n \leq 9$, with generalization to any base.

E. J. Moulton[87] found the number of positive integers with $r+1$ digits fewer than p of which are unity (or zero). L. O'Shaughnessy[88] found the number of positive integers $< 10^t$ which contain the digit 9 exactly r times. Books[89] on mathematical recreations may be consulted.

F. A. Halliday[90] considered numbers N formed by annexing the digits of B to the right of A, such that $N = (A+B)^2$, as for $81 = (8+1)^2$. Set $N = A \cdot 10^n + B$. Then $A(10^n - 1) = (A+B)(A+B-1)$, so that it is a question of the factors of $10^n - 1$.

*J. J. Osana[91] discussed numbers of two and three digits.

E. Gelin[92] listed 450 problems, many being on digits.

[86a]L'intermédiaire des math., 25, 1918, 44–5.
[87]Amer. Math. Monthly, 24, 1917, 340–1.
[88]Ibid., 25, 1918, 27.
[89]E. Lucas, Arithmétique amusante, 1895.   E. Fourrey, Récréations Arithmétiques, 1899.
    W. F. White, Scrap-Book of Elem. Math., etc.
[90]Math. Quest. and Solutions, 3, 1917, 70–3.
[91]Revista Soc. Mat. Española, 5, 1916, 156–160.
[92]Mathesis, (2), 6, 1896, Suppl. of 34 pp.

AUTHOR INDEX.

The numbers refer to pages. Those in parenthesis relate to cross-references. Those in brackets refer to editors or translators. The other numbers refer to actual reports.

467

CH. II. FORMULAS FOR THE NUMBER AND SUM OF DIVISORS, PROBLEMS OF FERMAT AND WALLIS.

Hain, 52
Halcke, 58
Henry, [53, 54, 56]
Hévélius, 54
Hoppe, 51

Kersey, 51
Kraft, 53
Kronecker, 54

Landau, 57
Lionnet, 52, 58
Lucas, 55–58

Mersenne, 51, 53
Minin, 52
Moreau, 57
Moret-Blanc, 57

Newton, 51 (53)

Ozanam, 56

Peano, 53
Plato, 51
Prestet, 52
Pujo, 52

Rudio, 54

Stifel, 51

Vacca, 57
Van Schooten, 51, 55

Wallis, 51, 53–56
Waring, 52, 54, 55
Wertheim, 54
Winsheim, 51
Wolff, 54

CH. III. FERMAT'S AND WILSON'S THEOREMS, GENERALIZATIONS AND CONVERSES; SYMMETRIC FUNCTIONS OF 1, 2,..., $p-1$ MODULO p.

Allardice, 96
Anton, 75
Anonymous, 67, 70, 92
Arévalo, 83
Arndt, 71, 72 (77, 81, 82)
Arnoux, 81, 94
Aubry, 81, 101 (71, 103)
Axer, 86

Bachmann, 72, 81, 82, 95 (78)
Banachiewicz, 94
Bauer, 88, 89
Beaufort, 62
Beaujeux, 75
Binet, 67
Birkenmajer, 96
Blissard, 74
Borel and Drach, 85, 87 (89)
Bossut, [63]
Bottari, 83
Bouniakowsky, 67, 73, 92, 95 (89, 93)
Brennecke, 69, 73 (80)
Bricard, 81 (75, 82)

Cahen, 80, 103
Candido, 80
Cantor, 62
Capelli, 80, 89
Caraffa, 69
Carmichael, 82, 94 (78, 86, 95)
Carré, 60
Catalan, 71, 77, 98
Cauchy, 67, 69, 70, 95 (86)
Cayley, 76 (75, 83)
Cesàro, 98
Chinese, 59, 91
Cipolla, 93, 94
Concina, 101
Cordone, 85
Crelle, 68, 70, 71, (72)

Cunningham, 94

D'Alembert, 63
Daniëls, 78
Dedekind, [74]
De la Hire, 60
Del Beccaro, 79
De Paoli, 67 (79, 82)
D'Escamard, 81
Desmarest, 73
Dickson, 80, 85, 89
Dirichlet, 66, 74 (65, 68, 71, 73, 77, 80, 81, 84)
D'Ocagne, 79, 98
Donaldson, 82
Durège, 73

Earnshaw, 70
Eisenstein, 95
Epstein, (86)
Escott, 93, 94, 103
Euler, 60–65, 92 (66, 67, 69, 72–74, 76, 77, 81–83)

Fergola, 96 (97, 98)
Fermat, 59 (60, 94)
Ferrers, 79, 99
Fontebasso, 74
Franel, 93
Frattini, 84
Frost, 96 (100)

Garibaldi, 77
Gauss, 60, 65, 84 (67, 69, 71, 72, 74, 75, 82, 83, 101)
Gegenbauer, 86, 93, 99
Genese, 78
Genty, 64
Gerhardt, [59]
Glaisher, 99, 100 (102, 103)
Goldbach, 92
Gorini, 70
Grandi, 85

Graves, 73
Gruber, 87
Grunert, 66–68, 71, 72 (65, 73, 81)

Harris, 81
Hayashi, 93
Heal, 77
Heather, 72
Hensel, 91, 101
Horner, 66, 69 (68, 71, 73, 74, 82)

Illgner, 83
Irwin, 103
Ivory, 65 (66–69, 71, 74)

Jacobi, 90 (91)
Janssen van Raay, 101
Jeans, 93 (59, 91)
Jolivald, 93
Jorcke, 76

Kantor, 84
Klügel, [67]
Koenigs, 85
Korselt, 93
Kossett, 93
Kraft, 60
Kronecker, 80, 88

Lacroix, 65, [63]
Lagrange, 62 (59, 64, 73, 74, 97, 99)
Laisant, 75 (76, 78)
Lambert, 61, 92
Laplace, 63 (67, 73, 82)
Lebesgue, 74, 96
Legendre, 64 (80)
Leibniz, 59, 60, 91 (65, 70, 94)
Leudesdorf, 96
LeVavasseur, 88

CH. IV. RESIDUE OF $(u^{p-1}-1)/p$ MODULO p.

CH. V. EULER'S Φ-FUNCTION, GENERALIZATIONS, FAREY SERIES.

CH. VI. PERIODIC DECIMAL FRACTIONS; FACTORS OF $10^n \pm 1$

Sornin, 164
Stammer, 166
Stasi, 178
Sturm, 166
Suffield, 166

Tagiuri, 176

Telosius, 179
Thibault, 164 (168)

Van den Broeck, 172
Van Henekeler, 165

Wallis, 159 (160)
Weixer, 179

Welsch, 179
Wertheim, 161
Westerberg, 163
Wiley, 179
Workman, 176 (168)
Wucherer, 161
Young, 165

CH. VII. PRIMITIVE ROOTS, BINOMIAL CONGRUENCES.

Alagna, 217 (218)
Alasia, 190 (210)
Allegret, 190
Amici, 197, 216, 217
Anonymous, 204
Arndt, 187, 188, 208, 209 (193)
Arnoux, 199, 218

Bachmann, 194, 199, 218
Barillari, 192 (193)
Barinaga, 203
Bellavitis, 193
Bennett, 195
Berger, 214 (215)
Besant, 217
Bháscara, 204
Bindoni, 199
Bougaief, 213
Bouniakowsky, 191, 192, 212 (204)
Brennecke, (208)
Bukaty, 212 (210)
Burckhardt, (185, 201)
Buttel, 190

Cahen, 198
Calvitti, (204)
Carmichael, 200, 202
Cauchy, 184, 186, 187, 209 (188, 190, 194, 195, 198, 200, 212, 213)
Cayley, 191
Chabanel, 202
Christie, 199
Cipolla, 200, 218-221
Colebrooke, [204]
Concina, 222
Contejean, 194
Creak, 222
Crelle, 185, 209 (186, 188, 190, 208)
Cunningham, 198-204, 217-222 (185, 189, 190, 213)

Daniëls, 194
Da Silva, 190, 210
De Jonquières, 197
Demeczky, 201

Desmarest, 188, 189, 210 (214)
Dickstein, 210, 215
Dirichlet, 185, 191, 211 (198, 214)
Dittmar, 212
Dupain, 192

Epstein, 200
Erlerus, 186, 208 (196)
Euler, 181, 204, 205 (222)

Foglini, 199
Fontené, 201
Forsyth, 193
Frattini, 193
Frégier, 183
Friedmann, 219
Frolov, 196

Gauss, 182, 194, 195, 207 (183-185, 187, 188, 193, 194, 197, 198, 209, 210, 213, 214)
Gazzaniga, 213
Gegenbauer, 194, 196, 215
Gérardin, 222
Goldberg, 192
Gorgas, 211
Grave, 201
Grigoriev, 199
Grosschmid, 221

Hacken, 194
Hanegraeff, 210
Heime, 191
Hill, 192
Hofmann, 193
Hoüel, 191
Hurwitz, (203)

Ivory, 184 (190)

Jacobi, 185 (188, 190-192, 198, 201, 203, 211)
Japanese, 204

Keferstein, 194
Korkine, 201 (203, 221)

Kraitchik, 202
Krediet, 203
Kronecker, 192, 198
Kulik, 189
Kunerth, 213

Lacroix, 183 [208]
Ladrasch, 212
Lagrange, 181, 205 (182, 206, 207, 214, 216)
Laisant, 193
Lambert, 181
Landau, 201
Landry, 190
Laplace, 208
Lazzarini, (222)
Lebesgue, 184, 188-192, 208, 211 (196, 204)
Legendre, 182, 205-207 (185, 187, 188, 194, 208, 213-215, 219, 222)
Leibniz, 215
Libri, 208
Lucas, 194, 213 (198, 200, 202, 203, 218)

Maillet, 202
Mann, 215
Marcolongo, 214
Maser, [182, 206, 207]
Massarini, 188
Mathews, 195, 215
Matsunaga, 204
Maximoff, 222
Mayer, 216
Meissner, 219
Mertens, 198 (192)
Meyer, 211
Miller, 198, 201, 203
Minding, 185
Moreau, 198
Murphy, 186

Nordlund, 200

Oltramare, 189, 190 (191)
Ostrogradsky, 185 (186, 188)

Pepin, 197, 213

CH. VIII. HIGHER CONGRUENCES.

CH. IX. DIVISIBILITY OF FACTORIALS, MULTINOMIAL COEFFICIENTS.

CH. X. SUM AND NUMBER OF DIVISORS.

CH. XI. MISCELLANEOUS THEOREMS ON DIVISIBILITY, GREATEST COMMON DIVISOR, LEAST COMMON MULTIPLE.

AUTHOR INDEX 477

Stouff, 345
Stuyvaert, 344
Sylvester, 342
Szenic, 345

Tagiuri, 343
Tarry, (346)
Terquem, 344
Tiberi, 346
Tirelli, 346

Transon, 339
Tucker, 341

Unferdinger, 345

Valerio, 342
Van Langeraad, 344
Vincenot, 344
Volterrani, 346

Walenn, 345
Wertheim, 345
Widmann, 337
Wilbraham, 339
Wronski, 344

Young, 344

Zbikowski, 339
Zeipel, 339
Zuccagni, 345

CH. XIII. FACTOR TABLES, LISTS OF PRIMES.

Akerlund, 355
Alliston, 356
Anjema, 348
Aratus, 347
Aubry, 355

Barlow, 351 (355)
Beguelin, 349
Bernhardy, 347
Bernoulli, 349 [350]
Bertelsen, 354
Bertrand, 349
Boethius, 347
Boulogne, 356
Bouniakowsky, 353
Bourgerel, 354
Brancker, 347 (348, 350)
Burckhardt, 350 (352–5)

Camerarius, 347
Cataldi, 347
Cayley, 353
Chernac, 350 (351)
Colombier, 351
Crelle, 351–2
Cunningham, 354–5 (350–2, 356)

D'Alembert, 349
Dase, 352 (353–5)
Davis, 352
De Polignac, (356)
Deschamps, 355
Desfaviaae, 350
De Traytorens, 348 (350)
Di Girio, 354
Dines, 355
Dodson, 348
Du Tour, 348

Eratosthenes, 347 (348, 353–6)
Escott, 355 (356)
Euler, 349 (356)

Felkel, 349, 350

Gauss, 350, 352 (356)
Gérardin, 356
Gill, 353
Glaisher, 350, 353 (355)
Goldberg, 352
Gram, 354
Groscurth, 353
Grüson, 350
Gudila-Godlewski, 353
Guyot, 351

Hansen, 356
Hantschl, 351
Harris, 348
Hindenburg, 349
Hinkley, 351 (347)
Horsley, 347
Hoüel, 351–2
Hülsse, [350]
Hutton, 351

Ibn Albannâ, 347

Jäger, 348
Johnson, 353
Jolivald, 354

Kästner, 350
Kempner, (356)
Klügel, 348
Köhler, 351
Krause, 350
Kronecker, (354)
Krüger, 348
Kulik, 351 (355–6)

Laisant, 354–5
Lambert, 348, 350 (349)
Landry, 351
Lebesgue, 352
Lebon, 355–6
Lehmer, 352–3, 355–6
Leonardo Pisano, 347
Libri, 347
Lidonne, 350

Lionnet, 355
Lucas, 353

Marci, 349
Marre, [347]
Maseres, 350
Meissel, 352 (350)
Merlin, (356)
Möbius, 351
Morehead, 355

Neumann, 350
Nicomachus, 347
Noviomagus, (356)

Oakes, 352
Oberreit, 349
Ozanam, 349

Pell, [347]
Perott, 352
Petzval, 352
Pigri, 348
Poetius, 348
Poretzky, (356)

Rahn (Rhonius), 347
Rallier des Ourmes, 348
Rees, 351
Reymond, (356)
Rosenberg, 352 (355)
Rosenthal, 349

Saint-Loup, 353 (356)
Salomon, 351
Schaffgotsch, 349
Schallen, 351
Schapira, 354
Schenmarck, 350
Schwenter, 348
Seelhoff, 353
Simony, 353 (354)
Snell, 350
Speckmann, 354
Stager, 356
Struve, 350
Suchanek, 354

Tarry, 355
Tennant, 354
Tessanek, 349
Tuxen, 353

Valerio, 354

Van Schooten, 347
Vega, 350
Vollprecht, 353–4
Von Stamford, 349
Von Sterneck, 354

Wallis, 348 (347)
Wertheim, [347]
Willigs (Willich), 348
Wolf, 348
Woodall, 354 (356)

Ch. XIV. Methods of Factoring.

Aubry, 373 (369)

Ball, 368
Barbette, 367, 373
Bartl, 370
Beguelin, 361, 366
Bernoulli, 371
Bickmore, 369
Biddle, 359, 367, 369–374
Birch, 368
Bisman, 374
Bouniakowsky, 369 (370)
Burgwedel, 365
Busk, 358 (359)

Cahen, 364
Canterzani, 366
Cantor, [366]
Christie, 361, 367, 372
Cole, 365
Collins, 357
Cullen, 365, 369
Cunningham, 358–9, 361, 365, 368–9, 373–4 (362)

De Bessy (see Frenicle)
De Montferrier, 358
Deschamps, 367
Dickson, 370 (360)

Euler, 360–2 (363–5)

Fermat, 357 (358, 367)
Frenicle, 360
Fuss, 362

Gauss, 363, 369 (364–5, 370)
Gérardin, 365–7, 370, 374

Gmeiner, 374
Gough, 371
Grube, 363

Hansen, (371)
Harmuth, 361
Henry, 358
Hudson, 358

Johnsen, 369
Joubin, 372

Kausler, 357, 362
Kempner, 374
Kielsen, 372
Klügel, 366
Kraft, 370
Kraitchik, 359, 360
Kulik, 361, 372

Lagrange, 369
Lambert, 371
Landry, 358, 369, 371
Laparewicz, 365
Lawrence, 358–360
Lebon, 359, 373
Legendre, 361–2 (363)
Lehmer, 368
Levänen, 364
Lucas, 363–4 (372)

Märcker, 368
Mathews, 364
Matsunaga, 371
Meissner, 372 (358, 364, 367)
Mersenne, 357, 360 (367–8)
Meyer, 365
Minding, 363

Möbius, (374)

Neumann, 359
Niegemann, 366
Nordlund, 370–1 (369)

Pepin, 364
Petersen, 360
Pocklington, 370

Rawson, 367
Reymond, 374

Schaffgotsch, 367
Schatunovsky, 370
Seelhoff, 363 (365)
Seliwanoff, 364
Speckmann, 367
Studnička, 366

Tchebychef, 363
Teilhet, 359
Tessanek, 366
Thaarup, 358
Thielmann, 368

Vaes, 359, 360
Valroff, 365
Von Segner, 366
Vuibert, 361

Waring, 362
Warner, 358
Weber, 364
Wertheim, 358, 361
Winter, 372
Woodall, 369

Ch. XV. Fermat Numbers $F_n = 2^{2^n} + 1$.

Anonymous, 376
Archibald, 380

Bachmann, 379
Ball, 378
Baltzer, 375
Beguelin, 375 (377)
Bisman, 379
Broda, 377

Canterzani, 375

Carmichael, 377, 380
Catalan, 377
Cipolla, 378
Cullen, 378
Cunningham, 378–380

Eisentein, 376
Euler, 375

Fermat, 375 (376)
Frenicle, 375

Gauss, 375
Gelin, 377
Genocchi, 375
Gérardin, 377, 380
Goldbach, 375
Gosset, 379

Hadamard, 378
Henry, 375, 380
Hermes, 378
Hurwitz, 378 (380)

CH. XVI. FACTORS OF $A^n \pm B^n$.

CH. XVII. RECURRING SERIES; LUCAS' u_n, v_n.

CH. XVIII. THEORY OF PRIME NUMBERS.

Ch. XIX. Inversion of Functions; Möbius' Function $\mu(n)$; Numerical Integrals and Derivatives.

CH. XX. PROPERTIES OF THE DIGITS OF NUMBERS.

SUBJECT INDEX.

A CATALOG OF SELECTED
DOVER BOOKS
IN SCIENCE AND MATHEMATICS

Math–Decision Theory, Statistics, Probability

ELEMENTARY DECISION THEORY, Herman Chernoff and Lincoln E. Moses. Clear introduction to statistics and statistical theory covers data processing, probability and random variables, testing hypotheses, much more. Exercises. 364pp. 5⅜ x 8½. 65218-1

STATISTICS MANUAL, Edwin L. Crow et al. Comprehensive, practical collection of classical and modern methods prepared by U.S. Naval Ordnance Test Station. Stress on use. Basics of statistics assumed. 288pp. 5⅜ x 8½. 60599-X

SOME THEORY OF SAMPLING, William Edwards Deming. Analysis of the problems, theory, and design of sampling techniques for social scientists, industrial managers, and others who find statistics important at work. 61 tables. 90 figures. xvii +602pp. 5⅜ x 8½. 64684-X

LINEAR PROGRAMMING AND ECONOMIC ANALYSIS, Robert Dorfman, Paul A. Samuelson and Robert M. Solow. First comprehensive treatment of linear programming in standard economic analysis. Game theory, modern welfare economics, Leontief input-output, more. 525pp. 5⅜ x 8½. 65491-5

PROBABILITY: An Introduction, Samuel Goldberg. Excellent basic text covers set theory, probability theory for finite sample spaces, binomial theorem, much more. 360 problems. Bibliographies. 322pp. 5⅜ x 8½. 65252-1

GAMES AND DECISIONS: Introduction and Critical Survey, R. Duncan Luce and Howard Raiffa. Superb nontechnical introduction to game theory, primarily applied to social sciences. Utility theory, zero-sum games, n-person games, decision-making, much more. Bibliography. 509pp. 5⅜ x 8½. 65943-7

INTRODUCTION TO THE THEORY OF GAMES, J. C. C. McKinsey. This comprehensive overview of the mathematical theory of games illustrates applications to situations involving conflicts of interest, including economic, social, political, and military contexts. Appropriate for advanced undergraduate and graduate courses; advanced calculus a prerequisite. 1952 ed. x+372pp. 5⅜ x 8½. 42811-7

FIFTY CHALLENGING PROBLEMS IN PROBABILITY WITH SOLUTIONS, Frederick Mosteller. Remarkable puzzlers, graded in difficulty, illustrate elementary and advanced aspects of probability. Detailed solutions. 88pp. 5⅜ x 8½. 65355-2

PROBABILITY THEORY: A Concise Course, Y. A. Rozanov. Highly readable, self-contained introduction covers combination of events, dependent events, Bernoulli trials, etc. 148pp. 5⅜ x 8¼. 63544-9

STATISTICAL METHOD FROM THE VIEWPOINT OF QUALITY CONTROL, Walter A. Shewhart. Important text explains regulation of variables, uses of statistical control to achieve quality control in industry, agriculture, other areas. 192pp. 5⅜ x 8½. 65232-7

Math–Geometry and Topology

ELEMENTARY CONCEPTS OF TOPOLOGY, Paul Alexandroff. Elegant, intuitive approach to topology from set-theoretic topology to Betti groups; how concepts of topology are useful in math and physics. 25 figures. 57pp. 5⅜ x 8½. 60747-X

COMBINATORIAL TOPOLOGY, P. S. Alexandrov. Clearly written, well-organized, three-part text begins by dealing with certain classic problems without using the formal techniques of homology theory and advances to the central concept, the Betti groups. Numerous detailed examples. 654pp. 5⅜ x 8½. 40179-0

EXPERIMENTS IN TOPOLOGY, Stephen Barr. Classic, lively explanation of one of the byways of mathematics. Klein bottles, Moebius strips, projective planes, map coloring, problem of the Koenigsberg bridges, much more, described with clarity and wit. 43 figures. 210pp. 5⅜ x 8½. 25933-1

CONFORMAL MAPPING ON RIEMANN SURFACES, Harvey Cohn. Lucid, insightful book presents ideal coverage of subject. 334 exercises make book perfect for self-study. 55 figures. 352pp. 5⅝ x 8¼. 64025-6

THE GEOMETRY OF RENÉ DESCARTES, René Descartes. The great work founded analytical geometry. Original French text, Descartes's own diagrams, together with definitive Smith-Latham translation. 244pp. 5⅜ x 8½. 60068-8

PRACTICAL CONIC SECTIONS: The Geometric Properties of Ellipses, Parabolas and Hyperbolas, J. W. Downs. This text shows how to create ellipses, parabolas, and hyperbolas. It also presents historical background on their ancient origins and describes the reflective properties and roles of curves in design applications. 1993 ed. 98 figures. xii+100pp. 6½ x 9¼. 42876-1

THE THIRTEEN BOOKS OF EUCLID'S ELEMENTS, translated with introduction and commentary by Thomas L. Heath. Definitive edition. Textual and linguistic notes, mathematical analysis. 2,500 years of critical commentary. Unabridged. 1,414pp. 5⅜ x 8½.. Three-vol. set. Vol. I: 60088-2 Vol. II: 60089-0 Vol. III: 60090-4

GEOMETRY OF COMPLEX NUMBERS, Hans Schwerdtfeger. Illuminating, widely praised book on analytic geometry of circles, the Moebius transformation, and two-dimensional non-Euclidean geometries. 200pp. 5⅝ x 8¼. 63830-8

DIFFERENTIAL GEOMETRY, Heinrich W. Guggenheimer. Local differential geometry as an application of advanced calculus and linear algebra. Curvature, transformation groups, surfaces, more. Exercises. 62 figures. 378pp. 5⅜ x 8½. 63433-7

CURVATURE AND HOMOLOGY: Enlarged Edition, Samuel I. Goldberg. Revised edition examines topology of differentiable manifolds; curvature, homology of Riemannian manifolds; compact Lie groups; complex manifolds; curvature, homology of Kaehler manifolds. New Preface. Four new appendixes. 416pp. 5⅜ x 8½. 40207-X

History of Math

THE WORKS OF ARCHIMEDES, Archimedes (T. L. Heath, ed.). Topics include the famous problems of the ratio of the areas of a cylinder and an inscribed sphere; the measurement of a circle; the properties of conoids, spheroids, and spirals; and the quadrature of the parabola. Informative introduction. clxxxvi+326pp; supplement, 52pp. 5⅜ x 8½. 42084-1

A SHORT ACCOUNT OF THE HISTORY OF MATHEMATICS, W. W. Rouse Ball. One of clearest, most authoritative surveys from the Egyptians and Phoenicians through 19th-century figures such as Grassman, Galois, Riemann. Fourth edition. 522pp. 5⅜ x 8½. 20630-0

THE HISTORY OF THE CALCULUS AND ITS CONCEPTUAL DEVELOP-MENT, Carl B. Boyer. Origins in antiquity, medieval contributions, work of Newton, Leibniz, rigorous formulation. Treatment is verbal. 346pp. 5⅜ x 8½. 60509-4

THE HISTORICAL ROOTS OF ELEMENTARY MATHEMATICS, Lucas N. H. Bunt, Phillip S. Jones, and Jack D. Bedient. Fundamental underpinnings of modern arithmetic, algebra, geometry, and number systems derived from ancient civilizations. 320pp. 5⅜ x 8½. 25563-8

A HISTORY OF MATHEMATICAL NOTATIONS, Florian Cajori. This classic study notes the first appearance of a mathematical symbol and its origin, the competition it encountered, its spread among writers in different countries, its rise to popularity, its eventual decline or ultimate survival. Original 1929 two-volume edition presented here in one volume. xxviii+820pp. 5⅜ x 8½. 67766-4

GAMES, GODS & GAMBLING: A History of Probability and Statistical Ideas, F. N. David. Episodes from the lives of Galileo, Fermat, Pascal, and others illustrate this fascinating account of the roots of mathematics. Features thought-provoking references to classics, archaeology, biography, poetry. 1962 edition. 304pp. 5⅜ x 8½. (Available in U.S. only.) 40023-9

OF MEN AND NUMBERS: The Story of the Great Mathematicians, Jane Muir. Fascinating accounts of the lives and accomplishments of history's greatest mathematical minds–Pythagoras, Descartes, Euler, Pascal, Cantor, many more. Anecdotal, illuminating. 30 diagrams. Bibliography. 256pp. 5⅜ x 8½. 28973-7

HISTORY OF MATHEMATICS, David E. Smith. Nontechnical survey from ancient Greece and Orient to late 19th century; evolution of arithmetic, geometry, trigonometry, calculating devices, algebra, the calculus. 362 illustrations. 1,355pp. 5⅜ x 8½. Two-vol. set. Vol. I: 20429-4   Vol. II: 20430-8

A CONCISE HISTORY OF MATHEMATICS, Dirk J. Struik. The best brief history of mathematics. Stresses origins and covers every major figure from ancient Near East to 19th century. 41 illustrations. 195pp. 5⅜ x 8½. 60255-9

Mathematics

FUNCTIONAL ANALYSIS (Second Corrected Edition), George Bachman and Lawrence Narici. Excellent treatment of subject geared toward students with background in linear algebra, advanced calculus, physics, and engineering. Text covers introduction to inner-product spaces, normed, metric spaces, and topological spaces; complete orthonormal sets, the Hahn-Banach Theorem and its consequences, and many other related subjects. 1966 ed. 544pp. 6⅛ x 9¼. 40251-7

ASYMPTOTIC EXPANSIONS OF INTEGRALS, Norman Bleistein & Richard A. Handelsman. Best introduction to important field with applications in a variety of scientific disciplines. New preface. Problems. Diagrams. Tables. Bibliography. Index. 448pp. 5⅜ x 8½. 65082-0

VECTOR AND TENSOR ANALYSIS WITH APPLICATIONS, A. I. Borisenko and I. E. Tarapov. Concise introduction. Worked-out problems, solutions, exercises. 257pp. 5⅝ x 8¼. 63833-2

THE ABSOLUTE DIFFERENTIAL CALCULUS (CALCULUS OF TENSORS), Tullio Levi-Civita. Great 20th-century mathematician's classic work on material necessary for mathematical grasp of theory of relativity. 452pp. 5⅝ x 8¼. 63401-9

AN INTRODUCTION TO ORDINARY DIFFERENTIAL EQUATIONS, Earl A. Coddington. A thorough and systematic first course in elementary differential equations for undergraduates in mathematics and science, with many exercises and problems (with answers). Index. 304pp. 5⅜ x 8½. 65942-9

FOURIER SERIES AND ORTHOGONAL FUNCTIONS, Harry F. Davis. An incisive text combining theory and practical example to introduce Fourier series, orthogonal functions and applications of the Fourier method to boundary-value problems. 570 exercises. Answers and notes. 416pp. 5⅜ x 8½. 65973-9

COMPUTABILITY AND UNSOLVABILITY, Martin Davis. Classic graduate-level introduction to theory of computability, usually referred to as theory of recurrent functions. New preface and appendix. 288pp. 5⅜ x 8½. 61471-9

ASYMPTOTIC METHODS IN ANALYSIS, N. G. de Bruijn. An inexpensive, comprehensive guide to asymptotic methods–the pioneering work that teaches by explaining worked examples in detail. Index. 224pp. 5⅜ x 8½ 64221-6

APPLIED COMPLEX VARIABLES, John W. Dettman. Step-by-step coverage of fundamentals of analytic function theory–plus lucid exposition of five important applications: Potential Theory; Ordinary Differential Equations; Fourier Transforms; Laplace Transforms; Asymptotic Expansions. 66 figures. Exercises at chapter ends. 512pp. 5⅜ x 8½. 64670-X

INTRODUCTION TO LINEAR ALGEBRA AND DIFFERENTIAL EQUATIONS, John W. Dettman. Excellent text covers complex numbers, determinants, orthonormal bases, Laplace transforms, much more. Exercises with solutions. Undergraduate level. 416pp. 5⅜ x 8½. 65191-6

CALCULUS OF VARIATIONS WITH APPLICATIONS, George M. Ewing. Applications-oriented introduction to variational theory develops insight and promotes understanding of specialized books, research papers. Suitable for advanced undergraduate/graduate students as primary, supplementary text. 352pp. 5⅜ x 8½.
64856-7

COMPLEX VARIABLES, Francis J. Flanigan. Unusual approach, delaying complex algebra till harmonic functions have been analyzed from real variable viewpoint. Includes problems with answers. 364pp. 5⅜ x 8½.
61388-7

AN INTRODUCTION TO THE CALCULUS OF VARIATIONS, Charles Fox. Graduate-level text covers variations of an integral, isoperimetrical problems, least action, special relativity, approximations, more. References. 279pp. 5⅜ x 8½.
65499-0

COUNTEREXAMPLES IN ANALYSIS, Bernard R. Gelbaum and John M. H. Olmsted. These counterexamples deal mostly with the part of analysis known as "real variables." The first half covers the real number system, and the second half encompasses higher dimensions. 1962 edition. xxiv+198pp. 5⅜ x 8½.
42875-3

CATASTROPHE THEORY FOR SCIENTISTS AND ENGINEERS, Robert Gilmore. Advanced-level treatment describes mathematics of theory grounded in the work of Poincaré, R. Thom, other mathematicians. Also important applications to problems in mathematics, physics, chemistry, and engineering. 1981 edition. References. 28 tables. 397 black-and-white illustrations. xvii+666pp. 6⅛ x 9¼.
67539-4

INTRODUCTION TO DIFFERENCE EQUATIONS, Samuel Goldberg. Exceptionally clear exposition of important discipline with applications to sociology, psychology, economics. Many illustrative examples; over 250 problems. 260pp. 5⅜ x 8½.
65084-7

NUMERICAL METHODS FOR SCIENTISTS AND ENGINEERS, Richard Hamming. Classic text stresses frequency approach in coverage of algorithms, polynomial approximation, Fourier approximation, exponential approximation, other topics. Revised and enlarged 2nd edition. 721pp. 5⅜ x 8½.
65241-6

INTRODUCTION TO NUMERICAL ANALYSIS (2nd Edition), F. B. Hildebrand. Classic, fundamental treatment covers computation, approximation, interpolation, numerical differentiation and integration, other topics. 150 new problems. 669pp. 5⅜ x 8½.
65363-3

THREE PEARLS OF NUMBER THEORY, A. Y. Khinchin. Three compelling puzzles require proof of a basic law governing the world of numbers. Challenges concern van der Waerden's theorem, the Landau-Schnirelmann hypothesis and Mann's theorem, and a solution to Waring's problem. Solutions included. 64pp. 5⅜ x 8½.
40026-3

THE PHILOSOPHY OF MATHEMATICS: An Introductory Essay, Stephan Körner. Surveys the views of Plato, Aristotle, Leibniz & Kant concerning propositions and theories of applied and pure mathematics. Introduction. Two appendices. Index. 198pp. 5⅜ x 8½.
25048-2

CATALOG OF DOVER BOOKS

TENSOR CALCULUS, J.L. Synge and A. Schild. Widely used introductory text covers spaces and tensors, basic operations in Riemannian space, non-Riemannian spaces, etc. 324pp. 5⅜ x 8¼. 63612-7

ORDINARY DIFFERENTIAL EQUATIONS, Morris Tenenbaum and Harry Pollard. Exhaustive survey of ordinary differential equations for undergraduates in mathematics, engineering, science. Thorough analysis of theorems. Diagrams. Bibliography. Index. 818pp. 5⅜ x 8½. 64940-7

INTEGRAL EQUATIONS, F. G. Tricomi. Authoritative, well-written treatment of extremely useful mathematical tool with wide applications. Volterra Equations, Fredholm Equations, much more. Advanced undergraduate to graduate level. Exercises. Bibliography. 238pp. 5⅜ x 8½. 64828-1

FOURIER SERIES, Georgi P. Tolstov. Translated by Richard A. Silverman. A valuable addition to the literature on the subject, moving clearly from subject to subject and theorem to theorem. 107 problems, answers. 336pp. 5⅜ x 8½. 63317-9

INTRODUCTION TO MATHEMATICAL THINKING, Friedrich Waismann. Examinations of arithmetic, geometry, and theory of integers; rational and natural numbers; complete induction; limit and point of accumulation; remarkable curves; complex and hypercomplex numbers, more. 1959 ed. 27 figures. xii+260pp. 5⅜ x 8½. 42804-4

POPULAR LECTURES ON MATHEMATICAL LOGIC, Hao Wang. Noted logician's lucid treatment of historical developments, set theory, model theory, recursion theory and constructivism, proof theory, more. 3 appendixes. Bibliography. 1981 ed. ix+283pp. 5⅜ x 8½. 67632-3

CALCULUS OF VARIATIONS, Robert Weinstock. Basic introduction covering isoperimetric problems, theory of elasticity, quantum mechanics, electrostatics, etc. Exercises throughout. 326pp. 5⅜ x 8½. 63069-2

THE CONTINUUM: A Critical Examination of the Foundation of Analysis, Hermann Weyl. Classic of 20th-century foundational research deals with the conceptual problem posed by the continuum. 156pp. 5⅜ x 8½. 67982-9

CHALLENGING MATHEMATICAL PROBLEMS WITH ELEMENTARY SOLUTIONS, A. M. Yaglom and I. M. Yaglom. Over 170 challenging problems on probability theory, combinatorial analysis, points and lines, topology, convex polygons, many other topics. Solutions. Total of 445pp. 5⅜ x 8½. Two-vol. set.
Vol. I: 65536-9 Vol. II: 65537-7